Probability Based High Temperature Engineering

Leo Razdolsky

Probability Based High Temperature Engineering

Creep and Structural Fire Resistance

Springer

Leo Razdolsky
LR Structural Engineering, Inc.
Lincolnshire, IL
USA

ISBN 978-3-319-41907-7 ISBN 978-3-319-41909-1 (eBook)
DOI 10.1007/978-3-319-41909-1

Library of Congress Control Number: 2016946310

Printed on acid-free paper

This Springer imprint is published by Springer Nature
The registered company is Springer International Publishing AG Switzerland

Preface

The proposed manuscript, *Probability Based High Temperature Engineering: Creep and Structural Fire Resistance*, bridges the disciplines of aerospace engineering (AE), structural engineering (SE), materials science engineering (MSE), and fire protection engineering (FPE) by offering a screening tool that can be used by engineers to perform the assessments of high-temperature creep as a structural fire resistance factor and its impact on structures. In many cases this type of analysis requires establishing creep models based on the design fires and the fuel packages and running these models to estimate the degradation of strength of structural components. A proper understanding of the uncertainties involved in the modeling of creep process is quintessential for safe and sustainable construction. The gap between the available laboratory standard fire test data and the typical real fire scenario further aggravates the situation. The broad scope of composition, mechanical, and environmental parameters characterizing each creep test in the database calls for the use of stochastic models. As the first step, stochastic models for all the required input parameters as well as the correlation fields are needed to properly calibrate safety factors for creep, as well as to provide the basis for realistic probabilistic assessment of reliability. The internationally available literature contains many, more or less involved, theories and models that have been developed to predict engineering creep.

All structures contain various uncertain properties. Modern standards customarily represent uncertainties in terms of semi-probabilistic checking concepts. By contrast, a full probabilistic reliability analysis can be performed, which determines the failure probability of the system based on the applied probability methods for specified limit states. This limit states are described by a given set of deterministic functions and analyzed independently from the random parameters used in the correspondent probabilistic reliability analysis. Standard methods in a reliability analysis compute results of this limit state function in order to estimate the failure probability, independent of the type of problem definition. This independency permits a separate implementation of the reliability analysis. Therefore, the correspondent probabilistic reliability analysis would be applicable for different fields of interest, such as dynamical reliability problems or design in foundation engineering.

Safety, reliability, and risk are key issues in a world of increasingly complex engineering facilities and infrastructure. The existence of natural and man-made hazards calls for the consideration of safety and risk in both the design and the preliminary stages. This book is focusing on providing solutions for practical applications and dealing with the practical challenges involved in incorporating structural safety and reliability in engineering practice. While there is an underlying theoretical framework in the areas of structural safety and reliability, translation from theory to practice in engineering is still urgently needed.

Despite the great development of research on creep of various materials and designs now, almost no books are devoted to the phenomenological theory of creep that is uniting disparate theories developed in relation to the calculation of various types of structures and facilities under the influence of thermal load from fire or any other high-temperature loads.

While there is an underlying theoretical framework in the areas of structural safety and reliability, translation from theory to practice in engineering is still urgently needed.

This book is focusing on providing solutions for practical applications and dealing with the practical challenges involved in incorporating structural safety and reliability in engineering practice. The main attention is paid to approximate methods (exact analytical solution in this case is not required, since the ultimate goal is to solve the creep constitutive equations in a probabilistic formulation) of solution of integral and differential equations considering creep effect of the high temperature from fire. Keeping in mind mainly engineering applications and practical calculations, the author did not seek large mathematical rigor and cared more about the visibility and accessibility of presentation, without, however, being overly simplistic. To achieve this goal the author has decided to include in each chapter a large number of examples to illustrate the theoretical basis of the material presented. The worked examples have all been programmed using simple POLYMATH software to insure their accuracy. Hence the numerical outputs quoted in this book are taken directly from POLYMATH solutions and the numbers have been rounded to the approximate number of significant figures at each output. The final solutions quoted in this book can therefore be compared directly with those obtained using other computer software (for instance MATHCAD or MATLAB). An in-depth explanation is presented through examples (that are presented in a simple step-by-step computational form) of computing the ultimate strength capacity of standard structural systems.

The main objective of this book is to provide the intelligent structural engineer practitioner with the approximate analysis of creep deformations and its effect on structural analysis and design. The results show that the developed approach is capable to reproduce basic features of high-temperature creep processes in engineering structures. The book contains a significant amount of original material, as well as substantially revised from previously published by author.

This book is similar to the previous book of the author ("Probability Based Structural Fire Load" and published by the Cambridge University Press in 2014) by the method of determining the statistical data for solving creep problems through the use of methods of applied probability theory. Namely, similar to the previous book integral dimensionless constitutive creep equation, first solved by numerical methods in a deterministic setting, and then declaring one of the dimensionless parameters included in the creep equation as a random variable, we obtain the discrete solutions of this equation, which, in turn, are realization of a random creep process.

Well-developed methods for approximating the failure probability are FORM and SORM (First-Order and Second-Order Reliability Methods). These are analytical solutions converting the integration into an optimization problem. In order to simplify the calculation the distribution functions of the random variables and the limit state function are transformed into a standardized Gaussian space. This transformation is defined via the cumulative distribution function $\Phi^*(z)$, where z is the transformed and standardized Gaussian variables. This leads to a simplified formulation of the failure probability $P_f \approx \Phi^*(-\beta)$, where β is the so-called reliability index.

This book will be a useful tool to aerospace engineers, structural engineers, materials science engineers, and fire protection engineers. It also has practical application in academia for AE, SE, MSE, and FPE students. The book will be used as guidance tool to determine the effect of high-temperature creep on structural fire resistance and which model variables require a greater degree of analysis. This can help in determining the most economical structural design for a given fire scenario.

This book will serve a wide range of readers, in particular, graduate students, professors, scientists, and engineers. Thus, the book should be considered not only as a graduate textbook, but also as a reference handbook to those working or interested in areas of applied probability in continuum mechanics, stress analysis, materials science, and fire protection design. In addition, the book provides an extensive coverage of great many practical problems and numerous references to the literature.

Lincolnshire, IL, USA Leo Razdolsky

The original version of the book was revised: Belated correction from author to include the list of references has been incorporated. The erratum to the book is available at https://doi.org/10.1007/978-3-319-41909-1_10

Contents

Chapter 1
Introduction

Notation

k	The thermal conductivity that has the dimensions W/m K or J/m s K
T	Temperature
d	Thickness in the direction if heat flow
ρ	Air density and
c	Specific heat capacity
K	Number of collisions which result in a reaction per second
A	Total number of collisions
E	Activation energy
R	Ideal gas constant
P	Losses of heat due to thermal radiation
e	Emissivity factor
σ	Stefan–Boltzmann constant (σ = 5.6703(10^{-8}) W/m^2 K^4)
T_o	Ambient temperature
A_v	Area of openings
c_{p}	Average specific heat at constant pressure
t	Time
$\vec{v}(u; v; w)$	Velocity vector
D	Diffusion coefficient (m^2/s)
p	Pressure
v	Kinematic viscosity; $v = \mu/\rho$
θ	Dimensionless temperature
τ	Dimensionless time
h	Height of the compartment (m)
a	Thermal diffusivity (m^2/s)
Time	$t = \frac{h^2}{a}\tau$ (s)
Temperature	$T = \frac{RT_*^2}{E}\theta + T_*$ (K), where T_* = 600 K is the base line temperature

L. Razdolsky, *Probability Based High Temperature Engineering*,
DOI 10.1007/978-3-319-41909-1_1

Coordinates	$\bar{x} = x/h$ and $\bar{z} = z/h$ "x" and "z" dimensionless coordinates
Velocities	$\bar{u} = \frac{v}{h}u\,(\mathrm{m/s})$ and $\bar{w} = \frac{v}{h}w\,(\mathrm{m/s})$ Horizontal and vertical components velocity accordingly
v	Kinematic viscosity (m^2/s)
u and w	Dimensionless velocities
$Pr = v/a$	Prandtl number
$Fr = \dfrac{gh^3}{va}$	Froude number
g	Gravitational acceleration
$Le = a/D = Sc/Pr$	The Lewis number
$Sc = v/D$	The Schmidt number
$\overline{\beta} = \dfrac{RT_*}{E}$	Dimensionless parameter
$\overline{\gamma} = \dfrac{c_p RT_*^2}{QE}$	Dimensionless parameter
$P = \dfrac{e\sigma K_v (\beta T_*)^3 h}{\lambda}$	Thermal radiation dimensionless coefficient
$\sigma = 5.67(10^{-8})$ ($W/m^2\ K^4$)	Stefan–Boltzmann constant
$K_v = A_o h/V$	Dimensionless opening factor
A_o	Total area of vertical and horizontal openings
$\delta = \left(\frac{E}{RT_*^2}\right) Qz \left(\exp\left(-\frac{E}{RT_*}\right)\right)$	Frank-Kamenetskii's parameter
C	[1 − $P(t)/P_o$]—Concentration of the burned fuel product in the fire compartment
$\overline{W} = \dfrac{v}{h}W$	Vertical component of gas velocity
$\bar{U} = \dfrac{v}{h}U$	Horizontal component of gas velocity
$b = L/h$	"L" and "h"—Length (width) and height of fire compartment accordingly
W; U	Dimensionless velocities
R_n	Nominal strength
S_i	Nominal load
φ	Resistance factor
γ	Load factor
R_c	Characteristic value for the resistance
A	Design variable (e.g., the cross-sectional area of the steel rod considered previously)
G_c	Characteristic value for the permanent load
S	Characteristic value for the variable load
ψ_1	Partial safety factor for the permanent load
ψ_2	Partial safety factor for the variable load
β	Reliability index

S	Probability space
A	Set of outcomes (events) to which a probability is assigned
$P(E_2 \vert E_1)$	Conditional probability
$\Phi^*(.)$	Denotes the cumulative distribution function of standard normal distribution
μ_A, μ_B, σ_A, σ_B	Mean and variance of A and B, respectively
$J(t, t')$	Compliance function (often also called the creep function)
T_M	Melting point of the metal
$\varepsilon(t)$	Strain
$\sigma(t)$	Stress
$\overline{\sigma}(t) = E(t)\varepsilon(t)$	Instantaneous (elastic) stress
ε_e	Instantaneous (elastic) strain
ε_c	Creep strain
ε_T	Thermal expansion due to high temperature effect
$K(t, t') = \partial J(t, t')/\partial t'$	Retardation function (memory function)
$R(t, t')$	Relaxation function (also called the relaxation modulus

1.1 Creep Phenomena in Structural Engineering

Apart from nineteenth-century steam boilers, machines and equipment for high temperature operation have been developed principally in the twentieth century. Energy conversion systems based on steam turbines, gas turbines, high-performance automobile engines, and jet engines provide the technological foundation for modern society. As a problem in metal use it has grown steadily in importance because engineers have persistently raised their operating temperatures. In designing missiles, data are needed at higher temperatures and stresses and shorter time (5–60 min) than are determined for creep test. In many fields, until now it is one of the half-dozen most important of these problems. Consequences of the growing success in producing alloys that deform very slowly even at high stress and temperature are that the problem of creep failure has loomed larger. In studies of the engineering (structural) creep (degradation of strength of materials) of metals, it has become apparent that the creep behavior at elevated temperatures is different from that at low temperatures. In order to provide the answers for these converging requirements, a comparatively new field of high temperature engineering creep deformations has emerged. It has been imperative to develop limited basic knowledge, theories, and experimental results into usable data in an expedient fashion. A part of this larger study is the investigation of the probability-based engineering creep phenomena.

Structural engineers bear the ultimate responsibility to ensure that all "structures" (be they buildings, bridges, aerospace structural systems, shopping malls, sports stadia, factories, transport hubs, nuclear power stations etc.) must comply with the required standards of "safety" under "loading," both in normal use and in foreseeable exceptional conditions (e.g., natural and man-made hazards such as high winds, earthquake, fire, blast, etc.). This is an extraordinary responsibility and is taken extremely seriously by the profession and building codes and standards are devised and updated based on new research and experience. This makes the profession of structural engineering extremely conservative. Another reason for conservatism is the large variety and complexity of loading that a structure could be subjected to during its life and furthermore the complexity of the response of the materials the structure is made of and the structure as a whole to the loading. The severity of the loading cannot be precisely determined and neither can the response. Fire is among the most unpredictable of hazards and really should be considered in a probabilistic framework.

Currently dealing with all uncertainties in case of fire (in codes and standards) is realized by ascribing mathematically determined factors to overestimate loads and underestimate the "strength" of the materials. This broad brush approach to uncertainty has worked well in the twentieth century, but methodologies are now available for a more customized approach, where the design process could explicitly account for uncertainties in loads and structural response for each specific structure (if its importance so warrants). This in turn allow engineers to demonstrate "safety" using "alternative means" based on "first principles" and advanced "calculation procedures," while maintaining the more "prescriptive" traditional approaches. The former approach has come to be known rather loosely as "performance based structural engineering" (PBSE). Routine use of PBSE in industry is still a rarity, however the movement is irreversible and sooner or later most developed nations will move to PBSE type approaches for all high-value projects. Because of the high degree of uncertainty associated with natural or man-made hazards, PBSE approaches are the most sensible way of ensuring structural safety against them.

At the root of the structural safety problem is the uncertain nature of the man-made and environmental forces that act on structures, of material properties that are changing quite rapidly under high temperature conditions, and of structural analysis procedures that are no more than models of reality. The natural consequence of uncertainty is risk. Structural engineering relies heavily on analysis and computation rather than on testing because of the scale and uniqueness of typical projects both public and private sectors. Structural codes are linked to computational methods of safety assessment, and their primary purpose is to manage risk and maintain safety of structural systems at socially acceptable levels.

The basic notions underlying the probability-based load requirements and resistance criteria are relatively simple. Structural failure occurs if the resistance, R, is less than the structural action, S, due to the applied loads. If R and S are modeled as random variables, the limit state (or failure) probability can be computed as the probability that R is less than S [1, 2].

Much of the early history of structural reliability revolved around difficulties in performing this computation. If a desired or target limit state probability for design can be established (by assessing historically acceptable designs, professional consensus, or legislative or regulatory fiat), then structural design should strive to achieve solutions yielding limit state probabilities close to that target value. Design solutions with higher limit state probabilities are unacceptable from a safety point of view; designs with lower probabilities are needlessly expensive.

In probability-based limit states design, the structural reliability formulation is presented in such a way as to make it practical for design by engineers who may not be familiar with reliability concepts or have access to the necessary statistical data. Structural safety requires that

$$\text{Required strength} < \text{Design strength} \tag{1.1}$$

where the required strength is determined from structural analysis utilizing the specified design loads, and the design strength is calculated from principles of structural mechanics with specified material strengths and structural element dimensions. With the performance requirement that the member reliability should exceed target reliability, Eq. (1.1) can be restated for practical design purposes as

$$\sum_i \gamma_i S_i < \varphi R_n \tag{1.2}$$

In this equation, R_n is the nominal strength corresponding to the limit state of interest and S_i is the nominal load. These strengths and loads traditionally have been provided in codes and standards, and most engineers are familiar with them. The factors φ and γ are resistance and load factors that reflect (1) uncertainty in strength and load, and (2) consequence of failure, reflected in the target reliability measure. The right-hand side of Eq. (1.2) is the purview of each material specification (steel, concrete, engineered wood, etc.).

The left-hand side is defined for all construction materials by the national load standard referenced by the Model Codes and other regulatory documents in the United States.

The probabilistic approach to structural safety continues to resonate in the structural engineering community. The aftermath of natural and man-made disasters during the past two decades, rapid evolution of design and introduction of new materials and technologies, and heightened expectations on the part of the public, all have made judgmental approaches to ensuring safety of the built environment increasingly difficult to defend. The traditional practice of setting safety factors and revising codes based solely on experience does not work in this environment, where such trial and error approaches to managing uncertainty and safety may have unacceptable consequences. In an era in which standards for public safety are set in an increasingly public forum, more systematic and quantitative approaches to engineering for public safety are essential. The probabilistic approach addresses this need, and in the past two decades has been widely accepted worldwide as a new paradigm, for design of new structures and evaluation of existing facilities.

History shows that fire is a frequent and deadly event that strikes structural systems. During the late 1960s and 1970s, a number of natural disasters occurred worldwide that caused extensive loss of life and property damage and focused the attention of the structural engineering community and the public on the need to advance building practices for disaster mitigation.

Among the more notable of these were the structural failure investigations that followed the San Fernando, California, Earthquake of 1971, the Managua, Nicaragua, Earthquake of 1972, and the Miyagi-ken-oki Earthquake of 1978; the investigation of snow and rain load conditions prior to the collapse of the Hartford Civic Arena roof in 1978; and the evaluations of wind loads, wind load effects, and building performance following Hurricane Camille on the Gulf Coast (1969) and Cyclone Tracy in Darwin, Australia (1974); the First Interstate Bank Building in Los Angeles, One Meridian Plaza in Philadelphia, and Buildings 5 and 7 at the World Trade Center after 9/11 shows that burnouts can occur in buildings. When a burnout occurs, there is a potential for partial or even complete collapse of the structure.

These and other investigations of building performance revealed a number of deficiencies in the provisions for structural safety appearing in the codes of practice of the time, and emphasized the need for improvements in design for natural hazards.

The late 1960s also witnessed the beginnings of the move toward a new philosophy of structural design in the United States, Canada, and Western Europe. The shortcomings of allowable stress design were recognized in many quarters, and a search was underway for more rational approaches to distinguish between various conditions (termed limit states) that affect building performance, to ensure safety under rare but high-hazard conditions, and to maintain function under day-to-day conditions. Concurrently, the new field of structural reliability was developing around the notion that many of the uncertainties in loads and strengths could be modeled probabilistically. Advances were being made in first-order reliability analysis, stochastic load modeling and supporting statistical databases. Several probabilistic code formats were suggested [3], including an early version of Load and Resistance Factor Design (LRFD) for steel buildings.

Here is an example how the uncertainty associated with natural or man-made fire is characterized by Howard Baum, NIST Fellow Emeritus [4]: "The idea that the dynamics of a fire might be studied numerically dates back to the beginning of the computer age. Indeed, the fundamental conservation equations governing fluid dynamics, heat transfer, and combustion were first written down over a century ago. Despite this, practical mathematical models of fire (as distinct from controlled combustion) are relatively recent due to the inherent complexity of the problem.

The difficulties revolve about three issues: First, there are an enormous number of possible fire scenarios to consider due to their accidental nature. Second, the physical insight and computing power required to perform all the necessary calculations for most fire scenarios are limited. Any fundamentally based study of fires must consider at least some aspects of bluff body aerodynamics, multi-phase flow, turbulent mixing and combustion, irradiative transport, and conjugate heat transfer; all of which are active research areas in their own right. Finally, the "fuel" in most fires was never intended as such. Thus, the mathematical models and the data needed to characterize the degradation of the condensed phase materials that supply the fuel may not be available. Indeed, the mathematical modeling of the physical and chemical transformations of real materials as they burn is still in its infancy.

In order to make progress, the questions that are asked have to be greatly simplified. To begin with, instead of seeking a methodology that can be applied to all fire problems, we begin by looking at a few scenarios that seem to be most amenable to analysis. Hopefully, the methods developed to study these "simple" problems can be generalized over time so that more complex scenarios can be analyzed. Second, we must learn to live with idealized descriptions of fires and approximate solutions to our idealized equations.

Finally, the methods should be capable of systematic improvement. As our physical insight and computing power grow more powerful, the methods of analysis can grow with them. Now it could be, of course, that the mathematical modeling of fire dynamics is just incomplete, that it gives a coarse description of a reality that is actually much finer. If that were the case we should join the large number of people in their search for a finer mathematical model of physical reality. However, it has become clear that the search for such underlying 'hidden variable' models runs into certain difficulties: they must at least allow us to see the "chemical transformations of real materials as they burn", which is very close to impossible. And even if that would not disturb us, (which it does), they have not been very successful in the prediction of new phenomena. It seems that we must accept the inherent complicity of fire dynamics theory.

The beauty of the probabilistic approach is that the probability-based structural fire protection engineering does not predict the result of physical experiments with certainty, but yields probabilities for their possible outcomes; therefore we do not have to search for a 100 % guaranty answer, which does not exist anyway. However, even if it does exist (with very good approximation of a real fire scenario in any particular case), still it would not have any practical value in general population of such fires, because it is limited to this event which will not be repeatable again. Therefore the mathematical FDS modeling is applicable to the structural fire investigation processes, but not to the structural fire design stage, when the precise value of heat release rate (HRR) is not known in advance. This statement is supported again in the same reference [4]: "Because the model was originally designed to analyze industrial-scale fires, it the numerical and physical parameters. Current research is aimed at improving this situation, but it is safe to say that modeling fire growth and spread will always require a higher level of user skill and judgment than that required for modeling the transport of smoke and heat from specified fires can be used reliably when the HRR of the fire is specified and the transport of heat and exhaust products is the principal aim of the simulation. In these cases, the model predicts flow velocities and temperatures to accuracy within 10–20 % of experimental measurements, depending on the resolution of the numerical grid 2. However, for fire scenarios where the HRR is predicted rather than specified, the uncertainty of the model is higher. There are several reasons for this: (1) properties of real materials and real fuels are often unknown or difficult to obtain, (2) the physical processes of combustion, radiation and solid phase heat transfer are more complicated than their mathematical representations in FDS, (3) The results of calculations are sensitive to both."

On the other hand, the experimental data is available from very limited number of real fire test results and it is fare to say that one cannot expect to have a large number of reliable statistical information regarding structural fire load in tall and super tall buildings or other structural systems (it is not practical and cost prohibited proposition). That is true also with aerospace engineering systems, nuclear power plants, etc. The main goal of the applied theory of probability (as we know) is to use its "means and methods" that requires *minimum* statistical data in order to obtain necessary results (mean value, median, variance, autocorrelation functions, etc.). At the end, of course, one should check these results against the limited test data (for example standard uniaxial tension test data at elevated temperature or "real" fire test results).

It is always preferable to have first a deterministic solution (as it is done by the author in case of probability-based structural fire load) of a specific creep deformation phenomenological problem and then build the probabilistic model based on randomness nature of some parameters and variables. We are not going to discuss physical modeling of a specific deterioration (creep) phenomenon; rather we treat deterioration as a stochastic process and review the inherent probabilistic structures in different models. For a general overview of a specific deterioration phenomenon; and mechanical mechanisms, readers may refer to the special tutorial series [5]. In general, the concept of creep and degradation under the action of high temperatures in this book will be considered interchangeable as creep deformation consists of a number of components: the impact of random temperature, reduction of the material modulus of elasticity, as well as modification and redistribution of structural rigidity as a whole.

The literature review of stochastic deterioration models in this book is to be proceeded in the order of model complexity. Starting from the simplest random variable models, the discussion is followed by marginal distribution models, second-order process models, and finally full distribution models. The full distribution models are further divided into cumulative damage models, pure jump process models, and stochastic differential models.

Reliability, to simply put, is the ability of a physical object (e.g., an electronic device, a bridge, a product line, etc.) to perform its required function under stated conditions for a specified period of time. Opposite to reliability is failure, referring to the event of failing to perform the required function or failing to conform to performance standards.

Probabilistic reliability theory defines reliability as the probability that the object performs its required function throughout its service life under specified conditions. Engineering reliability analysis is concerned with finding the reliability R or the probability of failure $Pr(f)$ of a feature, structure, or system. As a system is considered reliable unless it fails, the reliability and probability of failure sum to unity: $R + Pr(f) = 1$. In the engineering reliability literature, the term failure is used to refer to any occurrence of an adverse event under consideration, including simple events such as maintenance items. To distinguish adverse but non-catastrophic events (which may require repairs and associated expenditures) from events of catastrophic failure (as used in the dam safety context), the term probability of unsatisfactory performance *Pr*(*U*) is sometimes used. The probabilities of failure are calculated using FORM/SORM methods [6]. Using FORM, the probability of failure corresponding to the stress failure mode, P_{f1}, usually is 9.3×10^{-5}, while the probability of failure corresponding to the displacement failure mode, P_{f2}, is 2.65×10^{-3}. Note that the system probability of failure, P_f is approximated as the sum of the probabilities of failure corresponding to the two different failure modes P_{f1} and P_{f2}. Particular attention is paid to the introduction and generalized use of probabilistic concepts. Although most of the specific considerations refer to engineering creep of steel, the fundamental concepts apply to other materials used in structural engineering: aluminum alloys, composites, timber, etc.

Since any structural element in a fire compartment deteriorates during fire event and the environment in which this element works always changes, reliability is also a time-related concept. The time at which the object fails to perform the required function is called the failure time, or lifetime. The probability distribution of lifetime characterizes the structural system reliability over time and can be expressed by probability density function (pdf), cumulative distribution function (CDF), survival function (SF), or failure rate function (also known as hazard rate function). The relationship among these functions can be found in many textbooks of reliability theory, for example, in [7]. The SF denotes the reliability at any given time and is thus also called reliability function.

Reliability theory evolved apart from the mainstream of probability and statistics. It was originally a tool to help nineteenth-century maritime insurance and life insurance companies compute profitable rates to charge their customers. The reliability theory did not join engineering until the end of the Second World War. But once engineers found out the utility of reliability theory, they advanced the theory in two different approaches at an almost isolated manner. Safety being their major concern in design, civil and structural engineers defined the reliability as the probability of the structural strength being greater than the stress applied from loads on the structure [8]. They expressed the reliability as the following mathematical form: $p_r = \int_{R>S} f(r,s)\mathrm{d}r\mathrm{d}s$, in which p_r denotes the reliability, R and S denote the random resistance (strength) and stress, respectively; $f(r, s)$ denotes their joint probability distribution. The lower case of R and S represent a realization of the corresponding random variable. To calculate the reliability, one first establishes probabilistic models for the strength and the stress separately.

Depending on the nature of randomness, the stress may be modeled by either a random variable or a stochastic process. For time-variant variables such as high temperature–time fire load and degradation strength and stiffness, extreme value analysis is usually employed to find the statistical distributions of their maximum or minimum values during the nominal design life, assuming that the stochastic processes are stationary. The strength and stress may be further modeled if necessary as functions of some basic random variables (dimensionless temperature and time). The reliability is then calculated using first-order or second-order reliability methods, or simulation techniques. This is called the *Stress–Strength Interference* (SSI) *approach*. Although the SSI approach is traditionally used in structural engineering, strength and stress should be better understood as the capacity and demands accordingly.

The advantage of the SSI approach is that it provides the sensitivity information during the reliability analysis. This is important because from the sensitivity information, engineers would know the direction of optimizing their designs in order to achieve the reliability and cost target.

1.1.1 Stochastic Modeling of Creep (Deterioration)

High Temperature Engineering Creep modeling is closely related to failure modeling in the context of risk and reliability analysis. Creep-related failures can be classified into shock failures and first passage failures. A shock failure, also called a traumatic or "hard" failure in literature, occurs when a traumatic event (e.g., severe fire, earthquakes, tornadoes, tsunami, etc.) happens; the creep process reaches to some threshold level, but then decays and reaches again after curtain period of time, etc. Each time during this traumatic event, when the creep process reaches the given threshold level (stress or deformation), the structural system collects some damage (residual stress or inelastic deformation). In contrast, a first passage failure relates directly to the continuous deterioration process and is thus also classified as a "soft" failure. It occurs when the deterioration process reaches to some threshold over which that the system no longer works. This literature review places emphasis on the first passage type of failures except explicitly mentioned otherwise. For a review of general stochastic failure models, see, for example [9]. The review of stochastic deterioration models in the book is to be proceeded in the order of model complexity. Starting from the simplest random variable models, the discussion is followed by marginal distribution models, second-order process models, and finally full distribution models. The full distribution models are further divided into cumulative damage models, pure jump process models, and stochastic differential models.

1.1.2 Random Variable Model

A random variable model describes the randomness of the deterioration process by a finite-dimensional random vector θ as $X(t) = g\ (t;\ \theta)$, where g is a deterministic function. Once the probability distribution of θ is determined, the distribution of $X(t)$ is also known using, for example, transformation techniques for functions of random variables [10]. The distribution of the first passage time, defined as $Pr\{T \le t\} = Pr\{X(t) \ge \varsigma, X(s) < \varsigma,\ \text{for}\ 0 \le s < t\}$ where ς is the predetermined failure threshold (for example, dimensionless stress or deformation), that can be computed in a straightforward manner.

Usually the functional form of g is suggested from empirical and experimental studies. For example, the well-known unity checks of combined axial and bending stresses [11]. A simple stochastic "Power Law" (creep compliance function) replaces the property parameters materials (PPT) with random variables in order to capture the scatter in stress intensity, material properties, and environmental (temperature) factors.

Two special forms of random variable models were extensively used in time-dependent structural reliability analysis. The first one relates to the deterioration modeling of structural resistance $R(t)$ that assumes a random initial resistance R_0 and a deterministic and possibly nonlinear deterioration curve $g(t)$ [12–15].

The other special random variable model is the so-called random rate model, in which the deterioration (creep) is assumed a linear function of temperature (time) with a random deterioration rate, i.e., $X(t) = B\theta$ in which B denotes the deterioration rate. If there is another constant A added in the linear model above and both the intercept A and the rate B are normally distributed, then the distribution of the first passage time is called Bernstein's distribution, a three-parameter normal distribution with variance a function of time [16].

$$F_T(t) = Pr\{A + Bt \geq \varsigma\} = 1 - \Phi^*\left[\frac{\varsigma - \mu_A - \mu_B t}{\sqrt{\sigma_A^2 + \sigma_B^2 t^2}}\right] \tag{1.3}$$

where $\Phi^*(.)$ denotes the CDF of standard normal distribution, μ_A, μ_B, σ_A, σ_B are mean and variance of A and B, respectively.

Motivations of using random variable models are twofold. First, the random variable models are simple. Second, they are directly related to statistical analysis of deterioration data. Given a set of deterioration data, an analyst's first response may be using regression techniques that fit the data with some kind of curves! That idea is exactly what appears in (1.3) if some or all of the model parameters are grouped to model the random effects across samples. In this context, the random variable models are also called *general degradation path* models [17].

1.1.3 Second-Order Process Model

A second-order process model, as indicated by the name, specifies the first two moments of deterioration, i.e., the mean and autocorrelation function. Since the deterioration must be nonnegative and nonstationary, direct specification of its autovariance functions is inconvenient. To bypass this difficulty, an auxiliary nonnegative stationary process $Y(t)$ is multiplied, for example, with the mean deterioration curve $g(t)$ as $X(t) = Y(t)g(t)$. Note, however, it is not easy in the real world to collect sufficient data for an accurate estimation of the correlation or covariance functions. Therefore, strong assumptions (e.g., a constant coefficient of correlation along temperature or time) may have to be made [18, 19].

1.1.4 Cumulative Damage/Shock Model

In a cumulative damage (CD) model (see Fig. 1.1), deterioration is deemed to be caused by shocks and accumulates additively. CD models are also called shock models in the literature. Let us denote by D_i the damage size caused by the ith shock and by $N(t)$—the number of shocks up to time t.

Then the deterioration, $X(t)$, is expressed by $X(t) = \sum_{i=1}^{N(t)} D_i$.

Suppose the times between two successive shocks are modeled by T_1, T_2.... We have

$$N(t) = \min\{n = 0, 1, 2, \ldots | \sum_{i=1}^{n+1} T_i > t\}$$

That is, $N(t)$ is a counting process that counts the number of shocks up to time t. Therefore, the CD model specifies two probability laws: one for the counting process $N(t)$, or equivalently for the inter-occurrence time W_i, and the other for the damage size D_i each shock causes. The simplest example of the CD model is the compound Poisson process in which $N(t)$ follows a Poisson process, or W_i follows an exponential distribution, whereas D_i is a nonnegative random variable. When the

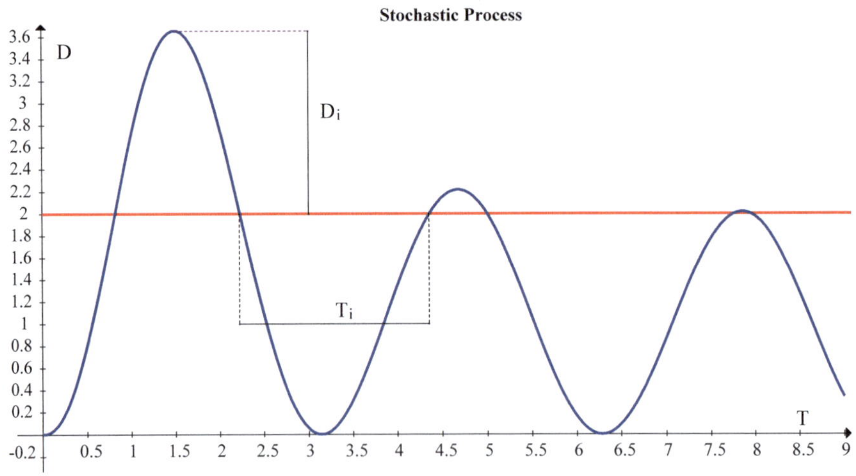

Fig. 1.1 Stochastic process

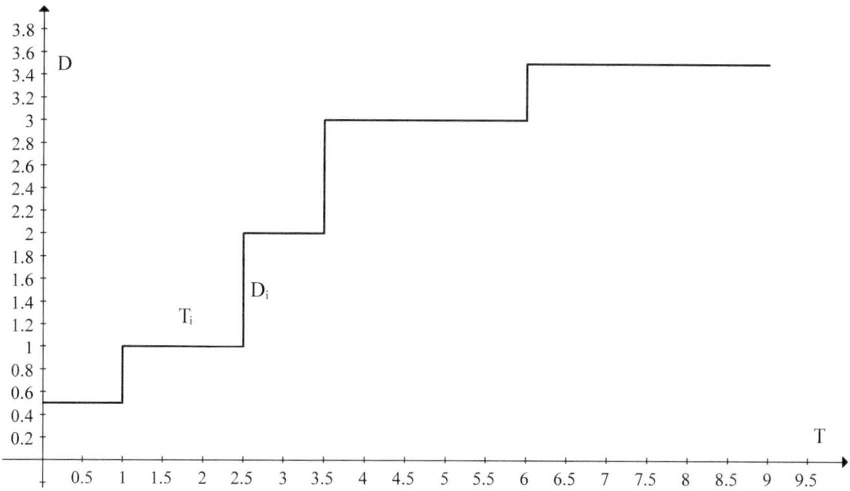

Fig. 1.2 Typical cumulative damage models

damage size D_i is fixed or follows a Dirac distribution, the compound Poisson process reduces to a simple Poisson process scaled by the value of D_i.

Figure 1.2 shows several typical cumulative damage models according to different assumptions on T_i and D_i. If T_i is fixed, say $T_i = 1$, and D_i is discretely distributed, $X(t)$ is a discrete-time Markov chain model, as the discrete distribution of D_i stipulates a transition probability: $Pr\{X_{t+1} = j|X_t = i\} = Pr\{D_i = j - i\}$. If T_i is not fixed but follows an exponential distribution, then the deterioration becomes a continuous-time Markov chain model. Yet if T_i follows a general distribution, a semi-Markov chain characterizes the deterioration. If both T_i and D_i follow some general distributions, the deterioration is said, in the terminology of Morey [20] to follow a compound renewal model. The SF, $S(t)$, or probability that a component will survive t units of time in a CD model has the following general mathematical form: $S(t) = \sum_{k=0}^{\infty} Pr\{M > k\}Pr\{N(t) = k\}$ in which M denotes the random number of shocks survived. For the first passage type of failures, $Pr\{M > k\}$ equals the kth convolution of distribution function of D_i, i.e., $Pr\{M > k\} = F_1(\zeta) * F_2(\zeta) * \ldots F_k(\zeta)$ in which F_i is the CDF of D_i and ς is the failure threshold. For a shock-type failure, i.e., shocks either make the component fail if $D_i > \varsigma$, or have no influence otherwise $Pr\{M > k\} = [1 - F_1(\zeta)][1 - F_2(\zeta)] \ldots [1 - F_k(\zeta)] = \prod_{i=1}^{k} [1 - F_i(\zeta)]$.

This model is also called an *extreme shock model* [21].

1.2 High Temperature Engineering Creep: Deterministic Approach

It is always preferable to have first a deterministic solution (as it is done by the author in case of probability-based structural fire load) of a specific creep deformation phenomenological (rheological) problem and then build the probabilistic model based on randomness nature of some parameters and variables. We are not going to discuss here physical modeling of a specific deterioration (creep) phenomenon; rather we treat deterioration as a cumulative action of an infinitively large number of viscoelastic elements.

The fact that due to high temperature the creep function cannot be expressed as a function of a single variable, $t - t'$, causes a major complication in structural analysis and makes the Laplace transform methods ineffective. Therefore, numerical methods must be used. As will be shown below, for efficient numerical solutions, the integral-type creep equations will be substituted by the system of linear (or nonlinear) algebraic equations or other simplified methods (successive approximation, Galerkin method, etc.) will be used. The kernel of integral equation will be also approximated to be an exponential type degenerate kernel.

Under real fire conditions, creep becomes a dominant factor and influences fire resistance of structural members. Significant forces develop in restrained elements and these forces induce high stresses. The extent of creep deformations is affected by magnitude and rate of corresponded temperature–time load application. The detailed analysis of temperature–time curves for the various categories of fire severity (deterministic approach) is described by the author [22]. A probabilistic approach to the same problem is presented by the author [23]. Below are the basic differential equations (in a very short form) obtained by the author in order to find these temperature–time functions as well as basic notations, dimensionless parameters, dependent and independent dimensionless variables that are also used in the current book. The differential equations for heat and mass transfer in a fire compartment can be written as follows [22]:

$$c_p\rho\frac{\partial T}{\partial t} = \operatorname{div}(\lambda \operatorname{grad}T - c_p\rho\vec{v}\nabla T) + Qze^{-E/RT} - \frac{e\sigma A_v(T^4 - T_o^4)}{V} \tag{1.4}$$

$$\frac{\partial C_i}{\partial t} = D_i\Delta C_i - \operatorname{div}\vec{v}C_i - \frac{v_i}{v_1}Qze^{-E/RT} \tag{1.5}$$

The mass fractions are defined as follows:

$$C_{mi} = \frac{M_iC_i}{\sum_k M_kC_k} = \frac{M_iC_i}{\rho} \tag{1.6}$$

where "i" and "k"—are gas components numbers; M_k—molecular weights.

For the binary mixture of gas species

$$C_{m1} + C_{m2} = 1 \tag{1.7}$$

The Fick's Law for the multi-mass fractions mixtures diffusion process can be written as follows:

$$g = -D\rho \text{grad} C_{mi} + \vec{v} C_{mi} \tag{1.8}$$

The most general form of the Navies–Stokes equation is

$$\rho\left(\frac{\partial \vec{v}}{\partial t} + \vec{v} \cdot \nabla \vec{v}\right) = -\nabla p + \nabla S_{ij} + \vec{f} \tag{1.9}$$

This is a statement of the conservation of momentum in a fluid and it is an application of Newton's second law to a continuum. A very significant feature of the Navies–Stokes equations is the presence of convective acceleration: the effect of time independent acceleration of a fluid with respect to space, represented by the quantity: $\vec{v}\,\nabla\vec{v}$.

Regardless of the flow assumptions, a statement of the conservation of mass is generally necessary. This is achieved through the mass continuity equation, given in its most general form as follows:

$$\frac{\partial \rho}{\partial t} + \nabla \cdot (\rho \vec{v}) = 0 \tag{1.10}$$

The Navies–Stokes equations are strictly a statement of the conservation of momentum. In order to fully describe fluid flow, more information is needed: this may include boundary conditions, the conservation of mass, the conservation of energy, and an equation of state. Let us rewrite now Eqs. (1.4), (1.5), (1.9), and (1.10) in a dimensionless form (assuming for Newtonian fluids: pressure p = const. and air density ρ = const.):

$$\frac{\partial \theta}{\partial \tau} + Pr\left(u\frac{\partial \theta}{\partial x} + w\frac{\partial \theta}{\partial z}\right) = \nabla^2 \theta + \delta(1 - C)^k e^{\frac{\theta}{1+\beta\theta}} - P\theta^4 \tag{1.11}$$

$$\frac{\partial C}{\partial \tau} + Pr\left(u\frac{\partial C}{\partial x} + w\frac{\partial C}{\partial z}\right) = Le\nabla^2 C + \gamma\delta(1 - C)^k e^{\frac{\theta}{1+\beta\theta}} \tag{1.12}$$

$$\frac{\partial u}{\partial \tau} + Pr\left(u\frac{\partial u}{\partial x} + w\frac{\partial u}{\partial z}\right) = \frac{4}{3} Pr \nabla^2 u \tag{1.13}$$

$$\frac{\partial w}{\partial \tau} + Pr\left(u\frac{\partial w}{\partial x} + w\frac{\partial w}{\partial z}\right) = \frac{4}{3} Pr \nabla^2 w + Fr \tag{1.14}$$

$$\frac{\partial u}{\partial x} + \frac{\partial w}{\partial z} = 0 \tag{1.15}$$

Initial conditions are as follows:

$$\tau = 0;\; C(0, x, z) = 0;\; u(0, x, z) = \text{const};\; w(0, x, z) = 0;\; \theta = \theta_o \tag{1.16}$$

Boundary conditions are as follows:

$$\begin{aligned} &x = 0;\; x = 1;\; z = 0;\; z = 1 \\ &\theta = 0;\quad \frac{\partial C}{\partial n} = 0 \end{aligned} \tag{1.17}$$

$\overline{\gamma} = \frac{c_p R T_*^2}{QE}$—Dimensionless parameter that characterizes the amount of fuel burned in the compartment before the temperature had reached the baseline point of $T_* = 300$ °C ($0 < \gamma < 1$). If this parameter is small, then the fire will have a flashover point, and if it is large—the fire will proceed in a steady-state motion until the decay stage.

Parameter "δ" is calculated based on [24]:

$$\delta = 12.1(\ln\theta_*)^{0.6} \tag{1.18}$$

$C = [1 - P(t)/P_o]$—Concentration of the burned fuel product in the compartment.

The Prandtl number, *Pr*, in Eqs. (1.11) and (1.12) is the ratio of the kinematic viscosity and the thermal diffusivity. *Pr* is most insensitive to temperature in gases made up of the simplest molecules. However if the molecular structure is very complex (such as in case of long-chain hydrocarbons, for example), *Pr* might reach values on the order of 10^5. Obviously all solutions of differential equations have to be obtained in dimensionless forms (temperature "θ" and time "τ") and then should be transferred in real temperature and time variables (see "Notation").

Based on SFPE guide [2011] and "Swedish" fire curves [25, 26] for the post-flashover realistic fire exposure we can standardize fires as follows:

The direct solution of Eqs. (1.11) and (1.12) is the "normal" way of solving the problem (obtaining the temperature–time function in a fire compartment). However, in case of developed fire in a large building volume the mathematical modeling of the physical and chemical transformations of real materials are known only with a small degree of confidence. At the same time based on many full fire test results data one can expect that curtain parameters such as the maximum temperature, type of temperature–time function, etc., are well known. On the other hand some other parameters [for example, parameter "γ" from Eq. (5.4)] are known with some degree of approximation. From a physical point of view this parameter characterizes the ratio of heat loses (for example, considerable quantities of soot) during the

Table 1.1 Fire severity

Category	Fuel load L (M J/m^2)	Max. temperature T_{max} (K)	Max. dimensionless temperature θ_{max}	Parameter "γ" from Eq. (05.04)
Ultra fast	$500 < L < 700$	$1020 < T_{max} < 1300$	$7.0 < \theta_{max} < 11.67$	$0 < \gamma < 0.05$
Fast	$300 < L < 500$	$880 < T_{max} < 1020$	$4.67 < \theta_{max} < 7.0$	$0.05 < \gamma < 0.175$
Medium	$100 < L < 300$	$820 < T_{max} < 880$	$3.67 < \theta_{max} < 4.67$	$0.175 < \gamma < 0.275$
Slow	$50 < L < 100$	$715 < T_{max} < 820$	$1.92 < \theta_{max} < 3.67$	$0.275 < \gamma < 1.0$

Note If fuel load $L > 700$, select $\gamma = 0$

development stages of a fire (incipient and free-burning) divided by total energy released (heat rate) [22]:

$$\overline{\gamma} = \frac{c_p R T_*^2}{QE}$$

If, for example, the heat rate of a chemical reaction is large l, then the parameter γ is small. Therefore parameter γ has a bounded variation between 0 and 1.

It is also important to underline here that for any given value of parameter "γ" from the interval [0;1] the only one solution of Eqs. (1.11) and (1.12) exists and the temperature–time function in this case has the only one maximum value. It can be seen by observation (see below), that this maximum temperature value increases when the parameter "γ" decreases from 1 to 0. On the other hand the maximum gas temperature in a real fire compartment and the fuel load are defining the category of the fire severity (see Table 1.1), therefore there is a correlation between the fire severity category and the value of parameter "γ." In order to establish this correlation the Mathematical Optimum Control Theory will be used here. For mathematical background of this theory—see Refs. [27, 28].

It should be noted that the strain values differ from dimensionless temperature only by a constant (the coefficient of linear expansion of the material, which is assumed to be independent of temperature changes). Therefore, all conclusions concerning the temperature–time dependency at high temperatures can be repeated for strain (deformation). Thus, the main task of the engineering theory of creep and its application to the structural calculation and design is finding the stress–temperature (strain) functions (in the deterministic and probabilistic formulation).

Creep is the progressive time-dependent inelastic deformation under constant load and temperature. Relaxation is the time-dependent decrease of stress under the condition of constant deformation and temperature. For many structural materials, for example steel, both the creep and the relaxation can be observed above a certain critical temperature. The creep process is accompanied by many different slow microstructural rearrangements including dislocation movement, aging of microstructure, and grain boundary cavitations.

The above definitions of creep and relaxation are related to the case of a homogeneous stress states realized in standard material testing. Under "creep in structures" we understand time-dependent changes of strain and stress states taking place in structural components as a consequence of external loading and temperature.

Examples of these changes include progressive deformations, relaxation and redistribution of stresses, local reduction of material strength, etc. Furthermore, the strain and stress states are inhomogeneous and multiaxial in most cases. The scope of "creep modeling for structural analysis" is to develop a tool which allows simulating the time-dependent behavior in engineering structures up to the critical state of creep failure.

The creep deformation of metals first came clearly into focus about half a century ago. As a problem in metal use it has grown steadily in importance because engineers have persistently raised their operating temperatures in many fields, until now it is one of the half-dozen most important of these problems. Besides the abundance of technical data that has necessarily been accumulated, there have been many studies in the last fifteen years of the physics of creep. These studies have shown that there are several different creep régimes depending mainly on the temperature. If T_M is the melting point of the metal in question, the different régimes roughly cover the temperature ranges 0–0.3 T_M, 0.3–0.5 T_M, 0.5–0.9 T_M and 0.9–1.0 T_M. The bottom range includes the so-called logarithmic creep and the top range creep by diffusion. Both are quite well understood but neither is particularly important and they are dealt with briefly. It is the middle two temperature ranges in which creep worries engineers; they have therefore received much more attention and the creep behavior in them is described more fully in this book.

Consequences of the growing success in producing alloys that deform very slowly even at high stress and temperature is that the problem of creep failure has loomed larger.

Despite the great development of research on creep of various materials and designs are now almost no books devoted to the general theory of creep and uniting disjointed theories developed in relation to the calculation of various types of structures and facilities under the influence of thermal load from fire or any other high temperature loads. Author set himself the task to present an "abstract" (phenomenological) creep theory, suitable in its various versions for all materials (including but not limited to metal, anisotropic and composite materials).

Therefore in this book almost no experimental data relating to certain specific materials are provided. The reader can find this information in the extensive literature, partly given in the bibliography. The main attention is paid to the approximate methods of the probability-based creep theory and its application to structural design, which is very different from other, sections of solid mechanics, in particular the classical theory of elasticity. Keeping in mind mainly engineering applications and practical calculations, the author did not seek large mathematical rigor and care more about visibility and accessibility of presentation, without going, does however, to oversimplify.

The objective of this manuscript is to present an extensive overview about the deterministic and corresponding probabilistic theoretical modeling and numerical analysis of high temperature creep and strength of structures. The study deals with three principal topics including integral constitutive equations for creep in structural materials under uniaxial and multiaxial stress states, structural mechanics standard models: beams, plates, frames and three-dimensional solids, and approximate numerical procedures for the solution of creep mechanics. Within the framework of the deterministic constitutive modeling (that is presented in dimensionless form) various extensions of the von Mises-Odqvist type creep theory are discussed to take into account stress state effects, anisotropy, composite materials as well as hardening and limit state processes. Transient creep effect is described by the introduction of dimensionless hardening state variables and the additional optimum control functional.

Governing probability-based equations for creep are introduced to formulate the structural reliability assessment and approximate numerical algorithms. The statistical data necessary for probabilistic analysis of the strain-time function is similar to the dimensionless temperature–time function in the event of a real fire, since it is both a source of structural loads as well as the source of statistical data for the dimensionless strain (multiplied by a constant—coefficient of thermal conductivity, which is assumed to be independent of temperature). Probabilistic characteristics of the dimensionless stress (within the correlation theory of random processes) are then calculated from the general non-invariant in time nonlinear integral equation of creep. To verify the subroutine, several benchmark problems are developed and solved by special approximate numerical methods.

The analysis of the material behavior can be based on different experimental observations, for example, macroscopic and microscopic. The engineering approach is related to the stress–strain analysis of structures and mostly based on the standard mechanical tests. Although the mechanisms by which metals, plastics, and ceramics deform in creep are different, the phenomena which are observed in these materials are similar. The basic mathematical laws of creep often seem to apply across the material classes, even though the physic-chemical phenomena are different. Studies of creep behavior of rod-shaped samples in tension show that the creep response, characterized in terms of the time-dependent strain at the time-dependent stress and temperature, may be represented by a general law that is described by the integral Volterra equation of the second kind. Constitutive equations deal with continuum concepts of stress and strain, not with material microstructure. Knowledge of the microstructure allows one to explore the causal mechanisms responsible for material behavior.

Approximate methods of structural creep analyses are appropriate. The solution of deterministic and probabilistic classical phenomenological creep equations using different approximate methods of integration is very important, and it is presented in this book. The extended theory satisfies the basic requirements for a theory of transient creep at elevated temperatures: that the transient creep is closely connected with the subsequent steady creep, and that the apparent exponent of the time in the transient region is permitted wide variations between 0 and 1. From this theory, it is possible to construct nondimensional creep curves which extend continuously from the transient region into the steady-state region. The corresponding family of creep curves for any metal may be obtained from the nondimensional family by use of appropriate constants [29]. The constants required are those obtained from steady creep measurements, together with two additional constants which represent the difference between the phases. This transient component, of the creep could be very important in some applications. An understanding **of** the nature of transient creep would be highly desirable not only from the academic point, of view but also because of the possibility of predicting the magnitude of the transient creep.

The existing theories of *transient creep* are inadequate to account for the actual behavior of polycrystalline metals at elevated temperatures. The so-called "exhaustion theory" of Mott and a more recent theory of Mott [30], which also involves the movement of dislocations but is intended to apply to higher temperatures, likewise requires that the transient creep strain be proportional to the *cube* root of the time. Although some experimental data do indicate an apparent cube root relation, this result, usually is true for only a part of the time range. Actually, when the logarithm of the creep strain is plotted against **the** logarithm of the time, the apparent exponent or the time may vary anywhere from a value in the neighborhood of one-sixth to unity.

One feature which seems to be characteristic of all the existing theories is that *tile* mechanism which is supposed to give rise to the *transient creep* is distinctly different from that which gives rise to the subsequent steady creep. On the other hand, the experimental work of Dorn and his associates [31] indicates a strong interconnection between the transient and the steady *creep* at elevated temperatures. When the creep strain is plotted against an appropriate parameter, a single curve is obtained regardless of whether the creep is in the transient or the steady region. If the mechanism for transient creep were different from that for steady creep, one would expect two different temperature dependencies and consequently two curves instead of one would be required. Thus, any proposed theory of transient creep which is intended to apply at elevated temperatures must satisfy at least two elementary requirements: It must show a close connection with the steady creep which follows, and it must permit the apparent exponent of the time in the transient region to vary over a wide range between 0 and 1. In the treatment of the elevated temperature behavior of metals in reference [29], the physical properties necessary to describe the behavior at constant temperature wore elasticity and viscosity.

The metal was thus conceived to consist of a single phase specified by the elasticity and viscosity constants. Suppose, however, that the metal could be more accurately described as consisting of two phases, each with its own set of elasticity and viscosity constants. Then, with such a metal, it is conceivable that, after application of a stress, the two phases would come to action at different rates and would thus result, in an overall creep rate which would vary with the time; or, the metal would exhibit transient creep.

The total strain of a uniaxial loaded specimen at time t may be subdivided as $\varepsilon(t) = \varepsilon_e(t) + \varepsilon_c(t) + \varepsilon_T(t)$ in which ε_e is the instantaneous (elastic) strain, which is reversible if the stress is small, ε_c is the creep strain, ε_T is the thermal expansion due to high temperature effect. As a consequence of creep, the stress in redundant structures usually varies with time even if the temperature load is constant. The calculation of creep caused by variable thermal stress is greatly facilitated by the principle of superposition. This principle is equivalent to the hypothesis of linearity of the constitutive equation that relates the stress and strain histories, states that the response to a sum of two stress (or strain) histories is the sum of the responses to each of them taken separately.

According to this principle, the strain caused by stress history may be obtained by decomposing the history into small increments $d\sigma(t')$ applied at times t', and summing the corresponding strains which equal $d\sigma(t')\ J(t, t')$, where $J(t, t')$ is the compliance function (often also called the creep function).

$$\varepsilon(t) = \int_0^t J(t,t')d\sigma(t') + \varepsilon_e(t) \tag{1.19}$$

This equation is a general uniaxial constitutive relation defining viscoelastic material. The integral in this equation should be understood as the Stieltjes integral, which is preferable to the usual Riemann integral since it applies not only for continuous but also discontinuous stress histories (see Fig. 1.3). When $d\sigma(t')$ is continuous, we may substitute $d\sigma(t') = [d\sigma(t')/dt']$ which yields the usual (Riemann) integral. We will assume here that the temperature–time curve is presented by a continuous function in case of any fire event, therefore after substituting $d\sigma(t') = [d\sigma(t')/dt']$ and integrating by parts, one may transform Eq. (1.19) to the following equivalent form:

$$\varepsilon(t)E(t) = \sigma(t) + \int_0^t K(t,t')\sigma(t')dt' \tag{1.20}$$

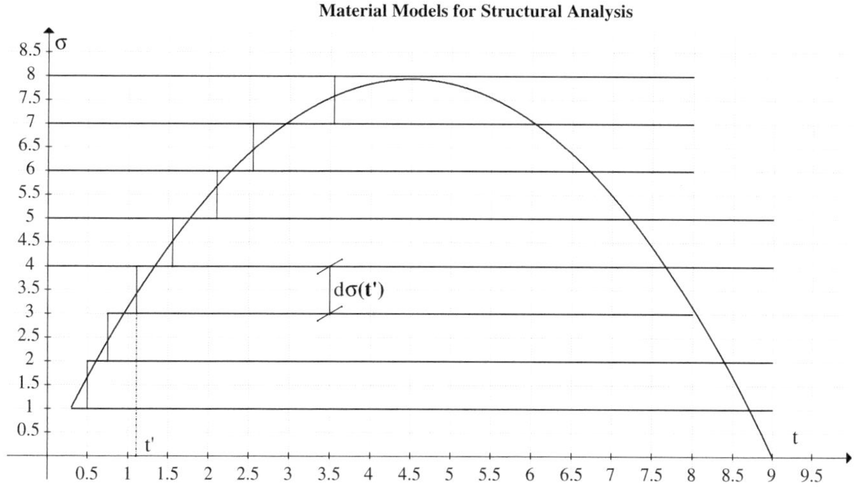

Fig. 1.3 Decomposition of stress history

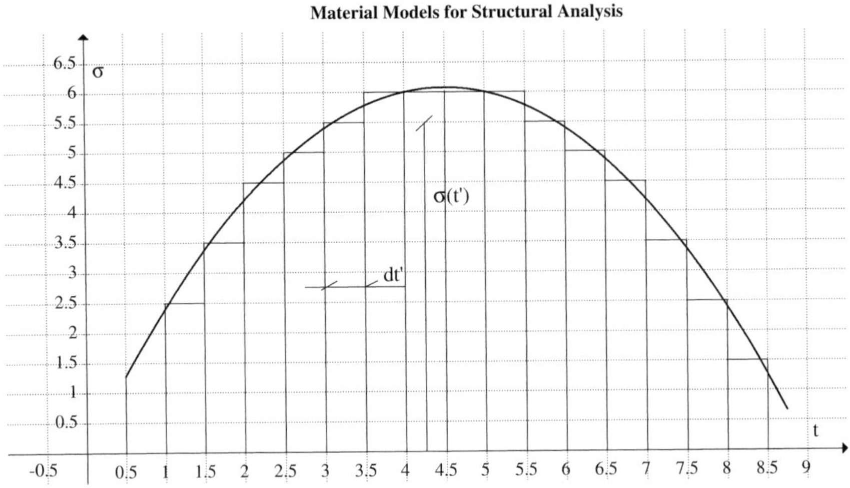

Fig. 1.4 Decomposition of stress impulses

where $K(t, t') = -\ \partial J(t, t')/\ \partial t'$. Geometrically, this equation means that the stress history is decomposed into vertical strips each of which is considered as an impulse function of stress (see Fig. 1.4). Thus, $K(t, t')$ represents the strain at time t caused by a unit stress impulse at time t' and is called the stress impulse memory function.

The principle of superposition may be equivalently expressed in terms of the relaxation function, $R(t, t')$ (also called the relaxation modulus), which represents the uniaxial stress σ at time t caused by a unit constant axial strain imposed at time t' and held constant afterwards. Equation (1.21) represents the integral equation for the stress history.

$$\sigma(t) = E(t)\varepsilon(t) + \int_0^t R(t, t')\varepsilon(t')\mathrm{d}t' \qquad (1.21)$$

When the strain history is given, Eq. (1.21) represents a Volterra integral equation for the strain history $\varepsilon(t)$. By solving this equation for the strain history specified as a step function (a constant unit strain imposed at age t'), one may calculate the stress histories for various t' (relaxation curves), and thus obtain the relaxation function. For realistic forms of $K(t, t')$, this solution must be carried out numerically. Conversely, Eq. (1.20) represents a Volterra integral equation for $\sigma(t)$.

By solving this equation for the stress history in the form of a step function, i.e., a constant unit stress applied at age t', one may calculate the individual creep curves, which together define the compliance function. Equation (1.20) is said to be the resolvent of Eq. (1.21) and vice versa. Functions $K(t, t')$ and $R(t, t')$, called the kernels of the integral equations, are complementary to each other, and if one of them is specified the other one follows.

Multiaxial generalization of all the preceding relations is obtained easily, by virtue of the fact that the material is essentially isotropic, Based on the hypothesis of linearity (principle of superposition), Eqs. (1.20) and (1.21) are generalized as

$$\varepsilon(t)E(t) = B\sigma(t) + \int_0^t BK(t,t')\sigma(t')\mathrm{d}t' \tag{1.22}$$

where $\sigma = (\sigma_{11}, \sigma_{22}, \sigma_{33}, \sigma_{12}, \sigma_{13}, \sigma_{23})^{\mathrm{tr}}$; $\varepsilon = (\varepsilon_{11}, \varepsilon_{22}, \varepsilon_{33}, \varepsilon_{12}, \varepsilon_{13}, \varepsilon_{23})^{\mathrm{tr}}$ and

$$B = \begin{bmatrix} 1-v-v & 0 & 0 & 0 \\ 1-v & 0 & 0 & 0 \\ 1 & 0 & 0 & 0 \\ & 1+v & 0 & 0 \\ & & 1+v & 0 \\ & & & 1+v \end{bmatrix} \tag{1.23}$$

The numerical subscription of σ and ε denote the components of the stress and strain tensors in Cartesian coordinates x_i (i = 1, 2, 3), superscript tr denotes the transpose of a matrix, and v is the Poisson ratio generalized for viscoelastic behavior, with $v(t, t)$ representing the elastic Poisson ratio at time t. The multiaxial stress–strain relations may also be written without matrix symbolism, as separate relations for the volumetric components and for the deviatory components of the stress and strain tensors (see Chap. 6). These equations are similar to Eqs. (1.20) and (1.21), the uniaxial compliance function $J(t, t')$ being replaced by the volumetric compliance function. Numerical creep analysis of large structural systems may be greatly facilitated, and analytical solutions of some problems may be rendered possible, if the integral-type constitutive equations are converted to a differential-type form consisting of a system of first-order ordinary differential equations in time. Such a conversion is possible if the kernel $K(t, t')$ or $R(t, t')$ has the degenerate form, i.e., consists of a sum of products of functions of single variables t and t'. The most general forms of the degenerate kernels may be written as

$$K(t,t') = \sum_{i=1}^{n} f_i(t)\varphi_i(t') \tag{1.24}$$

where $f_i(t)$ and $\varphi_i(t')$ are linearly independent functions. It is easy to prove that any kernel $K(t, t')$ can be substituted by the expansion (1.24), if the solution of Eq. (1.20) exists and converges (see Chap. 2). If n = 1, then $K(t, t') = f(t)\varphi(t')$. Substituting this in Eq. (1.20) and denoting $\overline{\sigma}(t) = \varepsilon(t)E(t)$ we have now:

$$\overline{\sigma}(t) = \sigma(t) + f(t)\int_0^t \varphi(t')\sigma(t')\mathrm{d}t' \tag{1.25}$$

Differentiating Eq. (1.25) once with respect to time and using the Leibniz rule for differentiating, we have

$$\frac{d\overline{\sigma}(t)}{dt} = \frac{d\sigma(t)}{dt} + f(t)\varphi(t)\sigma(t) + \frac{df(t)}{dt}\int_0^t \varphi(t')\sigma(t')dt' \tag{1.26}$$

Denoting the integral in the right-hand side of Eqs. (1.25) and (1.26) by Z, and excluding it from this system of equations, we obtain the linear differential equation of the first order.

$$\begin{aligned} &\frac{d\sigma(t)}{dt}f(t) + \left[f^2(t)\varphi(t) - \frac{df}{dt}\right]\sigma(t) = f(t)\frac{d\overline{\sigma}(t)}{dt} - \frac{df}{dt}\overline{\sigma}(t) \\ &\sigma(t=0) = \overline{\sigma}(t=0) = 0 \;-\; \text{Initial condition} \\ &\overline{\sigma}(\mathrm{t}) = E(t)\varepsilon(t) \;-\; \text{instantaneous (elastic) stress} \end{aligned} \tag{1.27}$$

The solution of Eq. (1.27) is as follows:

$$\sigma(t) = \overline{\sigma}(t) - f(t)\int_0^t \overline{\sigma}(\tau)\varphi(\tau)[\exp(-\int_\tau^t f(s)\varphi(s)ds)]d\tau \tag{1.28}$$

The expression $R = R(t,\tau) = f(t)\varphi(\tau)\left[\exp\left(-\int_\tau^t f(s)\varphi(s)ds\right)\right]$ is the resolvent of the kernel $K(t, \tau)$.

In the particular case where the kernel $K(t, t')$ may be represented as $K(t, t') = \varphi(t')$, the integral Eq. (1.25) becomes:

$$\overline{\sigma}(t) = \sigma(t) + \int_0^t \varphi(t')\sigma(t')dt' \tag{1.29}$$

Differentiating Eq. (1.29) once with respect to time, we have

$$\frac{d\overline{\sigma}(t)}{dt} = \frac{d\sigma(t)}{dt} + \varphi(t)\sigma(t) \text{ or: } \dot{E}\varepsilon + E\dot{\varepsilon} = \dot{\sigma} + \varphi(t)\sigma(t) \tag{1.30}$$

Equation (1.30) is similar to the differential equation describing the creep behavior for standard linear viscoelastic model.

The solution of Eqs. (1.25) and (1.26) can be also obtained as

$$\begin{cases} \dfrac{\mathrm{d}\sigma(t)}{\mathrm{d}t} = \dfrac{\mathrm{d}\overline{\sigma}(t)}{\mathrm{d}t} - f(t)\varphi(t)\sigma(t) - \dfrac{\mathrm{d}f(t)}{\mathrm{d}t} Z \\ \dfrac{\mathrm{d}Z}{\mathrm{d}t} = \varphi(t)\sigma(t) \end{cases} \tag{1.31}$$

$$\sigma(t=0) = Z(t=0) = 0 - \text{Initial condition}$$

The most general forms of the degenerate kernels may be written as

$$K(t,t') = \sum_{i=1}^{n} f(1/\alpha_i)[1 - \mathrm{e}^{-\alpha_i t}]\mathrm{e}^{-\alpha_i t'}; \quad R(t,t') = \sum_{i=1}^{n} \varphi(1/\beta_i)\mathrm{e}^{-\beta_i t'} \tag{1.32}$$

$$t_{ci} = 1/\alpha_i; \quad t_{ri} = 1/\beta_i$$

Here $f(.)$ and $\varphi(.)$ are functions of one variable, called the reduced times. The times t_{ci} and t_{ri} are constants and called the retardation and relaxation times respectfully. The expansion in Eq. (1.32) represents a series of real exponentials, called the Dirichlet series. In general, functions $f(.)$ or $\varphi(.)$ can be calculated by the method of least squares. As for t_{ci} and t_{ri}, however, they cannot be calculated from measured creep data but must be suitably chosen in advance. The choice of times t_{ci} and t_{ri} cannot be arbitrary but must satisfy certain conditions. The values must not be spaced too sparsely, and they must cover the entire time range of interest.

The t_{ci} and t_{ri}—values that give a close fit of given $K(t, t')$ or $R(t, t')$ experimental data are not unique. Equally good fits of the given compliance function data can be obtained for many possible choices of t_{ci} and t_{ri}—values which are spaced in time scale differently and cover the entire time range of interest. The plot of $f(.)$ or $\varphi(.)$ versus t_{ci} or t_{ri} is called the retardation spectrum or the relaxation spectrum.

The Dirichlet series expansion should be regarded only as an approximation to the compliance function motivated by computational convenience, rather than as a fundamental law. The expansion (1.32) contains many material parameters defining all the functions $f(.)$ or $\varphi(.)$. The Integral-type constitutive equations in this case are as follows:

$$\overline{\sigma}(t) = \varepsilon(t)E(t) = \sigma(t) + \int_0^t \int_0^1 f(\alpha)[1 - \mathrm{e}^{-\alpha t}]K(t,t')\mathrm{e}^{-t'}\sigma(t')\mathrm{d}\alpha\mathrm{d}t' \tag{1.33}$$

$$0 < \alpha < 1$$

$$\sigma(t) = E(t)\varepsilon(t) + \int_0^t \int_0^1 \varphi(\varsigma)R(t,t')\mathrm{e}^{-\beta t'}\varepsilon(t')\mathrm{d}t' \tag{1.34}$$

$$0 < \varsigma < 1$$

The functions $f(\alpha)$ and $\varphi(\zeta)$ are of importance for two reasons. First, they provide a first impression of the response of a material under stress or strain as a function of time; they are, therefore, of interest when the material is used in load-bearing constructions.

On the other hand knowledge of these functions and of the spectra of relaxation (or retardation) times derived from them is very helpful for obtaining insight into the *mechanisms* by which they are originated. Analysis of the time dependency of mechanical properties thus provides a powerful tool to investigate the relations between structure and material properties.

Creep and relaxation experiments are carried out on time scales ranging from several seconds up to several hours. The integral Eq. (1.20) and methods for its solution are greatly simplified if in this equation to replace the independent variable, namely $t = \theta$, where θ is the dimensionless temperature [22]. Since creep is thermally activated process, the creep law (1.20) may naturally be assumed to include the well-known Arrhenius equation.

Equation (1.20) for high temperature creep, proposed below, requires major modifications and additions.

1. The kernel of the integral equation should be
   - multiplied by an exponential function, which is the Arrhenius law;
   - it is presented by an analytic (infinitely differentiable or holomorphic) function;
   - it is defined on the square $a < x < b$; $a < s < b$, and vanishing in the triangle $a < x < s < b$;
   - it is assumed here for simplicity that $x = \theta$ [change of variables in Eq. (1.20)]. The real functions $x = f(\theta)$ are based on fire severity scenario and used in corresponding chapters;
   - can be substituted by a degenerate kernel $K(\theta, s, \alpha) = a(\theta, \alpha)b(s)$
2. Due to the fact that the entire interval temperature change θ in case of flashover fire usually is big ($0 < \theta < 10$ or 300 °C < T < 900 °C), it is proposed here to divide it into smaller subintervals, and solving the integral Eq. (1.20) in each of these subinterval. These solutions then must satisfy the continuity condition on the borders of two neighboring subintervals.
3. The numerical solution of Eq. (1.20) may be obtained by successive approximation method, i.e., as the limit of a mean-square-convergent sequence.
4. The contracting mapping principle implies that if $|K(x,y)| \leq M$ and $|\lambda| < \frac{1}{M(b-a)}$ or $|\lambda| < \left(\int_a^b \int_a^b K^2(x,s)\varphi_n(s))\mathrm{d}x\mathrm{d}s\right)^{-1/2}$, then the considered integral equation has a unique solution in the space $L_2\,[a, b]$.
5. Control function $f(\alpha)$ in each subinterval is a monotone nondecreasing continuous function.

6. Assuming that the control function $f(\alpha)$ and its derivatives are continuous, the condition of the Pontryagin maximum is no different from the well-known Euler–Lagrange conditions for finding extreme values of the functional.

The mixed Volterra–Fredholm integral equation with the degenerate kernel is now as follows:

$$\bar{\sigma}(\theta) = \varepsilon(\theta)E(\theta) = \sigma(\theta) + \delta \int_0^\theta \int_0^1 f(\alpha)[1 - \mathrm{e}^{-\alpha\theta}]\mathrm{e}^{\frac{s}{(1+\beta s)}}\mathrm{e}^{-s}\sigma(s)\mathrm{d}\alpha \mathrm{d}s$$
$$0<\alpha<1; E(\theta) \approx E_0\mathrm{e}^{-0.15\theta};\quad \varepsilon(\theta) = 7.02(10^{-4})\theta;\quad \delta = 1.0;\quad 0<\theta<10 \tag{1.35}$$

Equation (1.22) is only approximate because, strictly speaking, creep process consists of several chemical reactions, each of which probably has different activation energy. Nevertheless, it has been proven [22] that the rates at which the creep process is evolving at various temperatures conform to the Frank-Kamenetskii approximation of the Arrhenius equation at temperature $T_* = 600$ °K [23]. The dependence of elastic modulus E upon temperature T is based on experimental data presented in [32].

The functions $f(\alpha)$ and $\varphi(\beta)$ can be obtained based on application of Mathematical Theory of Optimum Processes [27].

The idea of design optimization suggests that for a given set of possible designs, there exists a design that is the best or optimal [33, 34]. Design parameters (often referred to as design variables or design functions) define the creep process or structure of interest and thus provide a means for changing it. The idea of optimal control problems in creep at high temperatures is to find a control function, which reflects the effect of mechanical properties of the material (in integral form), but also contributed to obtaining the most "favorable" solution of the original task of finding the functional dependence of stresses and strains in the temperature creep.

According to the constitutive relations (stress–strain relations), structural optimization is based on time-dependent constitutive relations and high temperature dependent effects.

The set of all possible designs that can be generated by adjusting the design variables between their upper and lower limits is called the design space.

Function $\sigma(\theta, \alpha)$ and $f(\alpha)$ are the basic state variables of the creep process, corresponding to the dimensionless temperature and to the material property parameter (MPP), respectively.

Then, this first equation ruling the evolution of $\sigma(\theta, \alpha)$ is coupled with the kinematic functional for the phase variables. This functional can be selected, for example, so that it represents the stress rate with increasing temperature, which in turn may indicate a decrease in the duration of the transient creep. Differentiating Eq. (1.22) with respect to α we have

$$\sigma'_{\alpha}(\theta,\alpha) = -\int_0^{\theta} e^{\frac{s}{(1+\beta s)}} e^{-s}\sigma(s)\mathrm{d}s \int_0^1 \left[f'(\alpha)[1-e^{-\theta\alpha}] + f(\alpha)[\theta e^{-\theta\alpha}]\right]\mathrm{d}\alpha \tag{1.36}$$

We need to specify a payoff (or reward) criterion. Let us define the *payoff functional*

$$P = \int_0^1 \left[f'(\alpha)[1-e^{-\theta\alpha}] + f(\alpha)[\theta e^{-\theta\alpha}]\right]\mathrm{d}\alpha \tag{1.37}$$

Our aim is to find a control function $f(\alpha)$, which *maximizes* the payoff. In other words, we want $P[f^{*}(\alpha)] \geq P[f(\alpha)]$ for all controls $f(\alpha) \in A$. Such a control $f^{*}(\alpha)$ is called *optimal*.

The Euler–Lagrange equation for the functional (1.37) is

$$\begin{aligned} &f''(\alpha)\left[1-e^{-\theta\alpha}\right] + \left[\theta e^{-\theta\alpha}\right] = 0; \\ &f'(\alpha) = -\ln\left[1-e^{-\theta\alpha}\right]; \quad f(0) = 1 \end{aligned} \tag{1.38}$$

We can find now the solution of Eq. (1.38) for any given discrete dimensionless temperatures θ_1, θ_2, …, θ_n (using POLYMATH software). Let us say that the portioning points are θ_i (i = 1, 2, …, 10).

The solutions of Eq. (1.38), optimal control functions $f(\alpha)$ for each value of θ (in other words for the subinterval $[i-1, \imath]$) are shown in Fig. 1.5. This figure shows that the impact of optimal control (spectrum of parameters α) on the final solution of

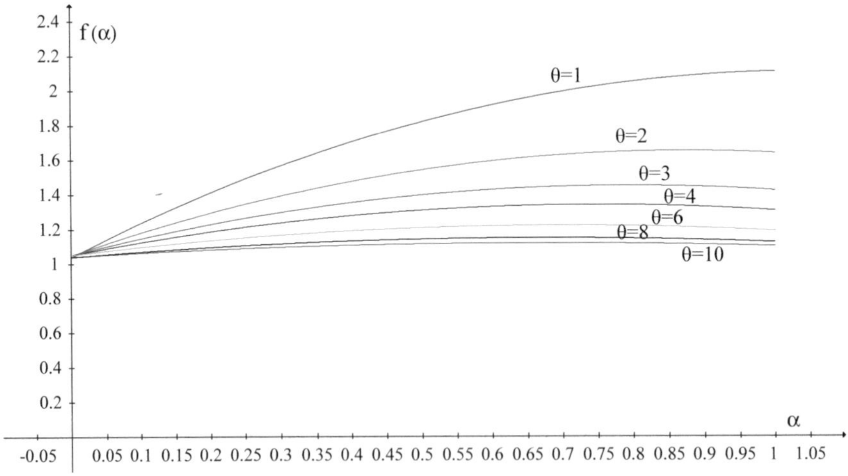

Fig. 1.5 Optimal control functions $f(\alpha)$

Eq. (1.35) is significantly reduced when the temperature θ increases. Therefore, special attention should be paid to the choice of parameters α_i and the convergence of the approximate solutions at $(0 < \theta < 6)$. Substituting the function $f(\alpha)$ in Eq. (1.35) and integrating with respect to the independent variable α, we obtain a system of Volterra integral equations of the second type, which have only two independent variables: θ and s [for each subinterval $(i - 1, i)$].

Approximate methods of numerical integration of Volterra equations by strips (or piecewise) method are presented below.

1.2.1 Incremental Quasi-elastic Stress–Strain Relations (Piecewise Collocation Method)

The most effective approach to numerical step-by-step structural creep analysis is to approximate the stress–strain relation for the time step as an incremental quasi-elastic relation, and then solve the structural creep problem as a sequence of elasticity problems. This can be done both for the integral-type and the differential-type formulations.

Let the total temperature interval $[0 < \theta < 10]$ be subdivided by discrete temperatures θ_i $(i = 0, 1, 2, \ldots)$ into temperature–time steps $\Delta\theta_i = \theta_i - \theta_{i-1}$. Temperature $\theta_0 = 0$ *coincides* with the gas temperature of 300 °C in a fire compartment.

Under constant temperature loads, the strains and stresses vary at a rate which increases (decreases) roughly as the inverse of retardation (relaxation) time, and for this reason it is advantageous to use temperature steps $\Delta\theta_i$ constant. When, for example, Gauss–Legendre algorithm described below (see Chaps. 3 and 4) is used, normally two points per subinterval suffice.

Identification of the material parameters from given test data at various temperatures may be carried out combining the Dirichlet series expansions at reference temperature with the determination of energy activation E. Unfortunately, it is found that the data presently available are insufficient in scope for unique determination of energy activation. (The average value for $E = 12{,}000$ cal/mol has been used in [22]). Thus the purpose of the activation energy concept, as introduced here, should be seen in reduction of number of unknown material parameters.

The main difficulty in formulating a creep law at variable temperature is due to the fact that temperature rise accelerates not only creep effect but also deteriorates the instantaneous modulus of elasticity. The modulus of elasticity deteriorates at rate that is dependable on temperature θ. This effect may be expressed in terms of a change of the time scale, considering that all material parameters (e.g., E and η), rather than being functions of actual time t, are functions of dimensionless temperature θ in a fire compartment.

For many practical purposes, the structural creep analysis need not be very accurate. As a matter of fact, it makes no sense to do it accurately if the stochastic nature of creep is ignored and no measures to reduce the statistical uncertainty are taken. Approximate methods of structural creep analysis are then appropriate. The simplest approach is to obtain the instantaneous modulus of elasticity variation as function of temperature. The dependence of elastic modulus E upon temperature θ is based on experimental data presented in [32]. It has been also assumed here that the strain ε is linearly proportional to the temperature load. Equation (1.35) is a nonlinear integral equation of the second kind (stress–strain relationship is not linear!). However, a change of variables in the integral equation reduces the equation to a linear equation with respect to the unknown stress function $\sigma(\theta)$. When considering numerical methods for integral equations, particular attention should be paid to the character of the kernel, which is usually the main factor governing the choice of an appropriate quadrature formula or system of approximating functions. The principal result of the theory of Volterra equations of the second kind may be described as follows. For each complex $\delta \neq \infty$ there exists a square-integrable solution of the Volterra equation of the second kind which is, moreover, unique. This solution may be obtained by successive approximation, i.e., as the limit of a mean-square-convergent sequence. The contracting mapping principle implies that if

$$\left|e^{\frac{\theta}{1+\beta\theta}}K(\boldsymbol{\theta},s)\right| \leq M \text{ and } |\boldsymbol{\lambda}| < \frac{1}{10M} \text{ or } |\delta| < \left(\int_0^{10}\int_0^{10} e^{\frac{2\theta}{1+\beta\theta}}K^2(x,s)\mathrm{d}x\mathrm{d}s\right)^{-1/2} \tag{1.39}$$

Then the considered integral equation has a unique solution in the space L_2 [0, 10] which can be constructed by the method of successive approximation. In the case of a continuous kernel.

$K(\theta, s)$ and $f = \boldsymbol{\varepsilon}(\boldsymbol{\theta})E(\boldsymbol{\theta}) \in C\,[0, 10]$, this sequence converges uniformly on [0, 10] to a unique continuous solution. Equations of type (1.35) and their nonlinear counterparts arise in the theory of parabolic boundary value problems. Actually few numerical methods for (1.35) are known. In Chap. 4, we give a numerical solution obtained by continuous dimensionless temperature collocation and temperature discretization method. Nonlinear Volterra equations is the name sometimes given to Volterra equations in which the product $K(\theta, s)\ \sigma(s)$ has been replaced by some function $K(\theta, s, \sigma(s))$ which is nonlinear with respect to $\sigma(s)$. Equations of this type are frequently encountered in theoretical and in applied studies. In the case of nonlinear Volterra equations it may be shown, if certain assumptions are made with respect to $K(\theta, s, \sigma(s))$, that successive approximations converge on an interval $[\theta, \theta + \Delta\theta]$, where $\Delta\theta$ is sufficiently small. Approximate solutions of nonlinear Volterra equations are found by using the Taylor expansion formula; it is sufficient to replace $K(\theta, s, \sigma(s))$ by $K(\theta, s)\ \sigma(s)$. If $K(\theta, s, \sigma(s))$ is independent of θ, for example, $K(\theta, s, \sigma(s)) = a(\theta)\ b(s)\ f[\sigma(s)]$.

1.2.2 Strip Method (Volterra Integral Equations)

If the interval [*a*, *b*] is large ([0, 10] in our case), then sometimes the inequality (1.39) is not satisfied and therefore the contracting mapping principle is not applicable. The strip method is based on replacing the kernel in a special way by a degenerate kernel, evaluating the resolvent of the degenerate integral equation and then improving the approximate solution through the use of a rapidly convergent iterative algorithm. To construct the degenerate kernel, divide the square $[a \leq \theta \leq b, a \leq s \leq b]$ into N strips $\left\{\frac{b-a}{N} i \leq \boldsymbol{\theta} \leq \frac{b-a}{N}(i+1), a \leq s \leq b\right\} \quad i = 0, 1, \ldots N-1$. In each strip, say the *i*th, the function $K(\theta, s)$ is approximated in the mean square, or uniformly, by functions $K_i(\theta, s) = P_i(\theta)Q_i(s)$. In this case, $K_i(\boldsymbol{\theta}, s) = K(\xi_i, s) \quad \xi_i \in \left\{\frac{b-a}{N} i \leq \boldsymbol{\theta} \leq \frac{b-a}{N}(i+1)\right\}, a \leq s \leq b$.

The function $K_i(\theta, s)$ is now used to construct a degenerate kernel

$$K_N(\boldsymbol{\theta}, s) = \sum_{\mathrm{i}=0}^{N-1} P_i(\boldsymbol{\theta}) Q_i(s)$$

$$P_i(\boldsymbol{\theta}) = \begin{cases} P_i(\boldsymbol{\theta}), & \boldsymbol{\theta} \in \left[\frac{b-a}{N} i \leq \boldsymbol{\theta} \leq \frac{b-a}{N}(i+1)\right] \\ 0, & \boldsymbol{\theta} \notin \left[\frac{b-a}{N} i \leq \boldsymbol{\theta} \leq \frac{b-a}{N}(i+1)\right] \end{cases} \tag{1.40}$$

$$Q_i(s) = \begin{cases} Q_i(s), & s \in \left[\frac{b-a}{N} i \leq s \leq \frac{b-a}{N}(i+1)\right] \\ 0, & \mathrm{s} \notin \left[\frac{b-a}{N} i \leq s \leq \frac{b-a}{N}(i+1)\right] \end{cases}$$

The approximate solution of the Eq. (1.35) with the degenerate kernel (1.40), generally, is better the larger the number N of strips and the function $K_N(\theta, s)$ is the better approximation of $K(\theta, s)$ in each strip. The approximate solution $\sigma_N(\theta)$ can be further improved by using the iterative algorithm that insures the continuation of the final solution $\sigma(\theta)$ in the whole interval [*a*, *b*] at the same time and

$$\tilde{\sigma}_N(\theta) = \sum_{i=0}^{N} \tilde{\sigma}_i(\theta); \quad \sigma(\theta) = \lim_{N \to \infty} \tilde{\sigma}_N(\theta) \tag{1.41}$$

For example, let us say that the temperature interval [0, 10] has been partitioned by integer numbers 0, 1, 2, …, 10. In each subinterval $[i, i + 1]$ of approximate stress function $\tilde{\sigma}_i(\theta)$ is shifted to the right by 1, therefore the additional stress from creep deformation is zero at partitioned point *i*. Thus the stress–temperature (strain) diagram is composed of the stress function form previous interval $[i - 1, i]$ and additional stress function from the next interval $[i, i + 1]$. That in turn ensures the continuity of the whole stress–strain diagram. This type of contingency of stresses and strains at partitioned points guarantees the convergence of approximate solution (1.41) in L_2 space. The approximate solution (1.26) in each subinterval $[i, i + 1]$ can be obtained by for example using the Galerkin method or Gauss–Legendre method. The approximate stress function must contain material properties parameters (MPP)

α_1, α_2 .... α_n. Another way of solving the problem is to present the relationship between the degenerate kernel type and MPP by having the approximate function "$f_i(\theta, \alpha)$" as a product of a given functions $f(\alpha)$ and $A_i[1 - \exp(-\alpha\theta)]$, where A_i is unknown parameter. However, it should be noted that in both cases the inequality (1.39) must be satisfied for each subinterval. This in turn means that subintervals may be of different lengths, and their minimum number may also be dictated by the inequality (1.39). Strip method often used for solving engineering creep problems, when the structure is exposed to excessive heat from fire, because in this case the maximum temperature change over a very wide range ($0 < \theta < 10$ or 300 °C < T < 900 °C), and therefore the inequality (1.39) is not satisfied on the whole interval [a, b].

If the dimensionless temperature θ in the integral Eq. (1.35) is a deterministic variable, the dimensionless stress and $\sigma(\theta)$ is also a deterministic function of temperature. Otherwise (dimensionless temperature θ—random function of time), the dimensionless stress $\sigma(\theta)$ is a random function of temperature, and therefore the time.

1.3 High Temperature Engineering Creep: Probabilistic Approach

General principles of the theory of reliability of structural systems outlined above now require implementation. First of all you need to create statistical data characterizing the creep process, based on the solutions of Volterra integral equations (in a deterministic setting) obtained above (the same way as it was done by the author in the case of probability-based structural fire load).

As will be confirmed later, no loss in the generality of material representation is incurred if one restricts attention to the special case of a series of real exponentials called Dirichlet series (see Eq. 1.20 or 1.21). The parameters t_{ci} and t_{ri} in Eq. (1.19) are constants and called retardation and relaxation times respectfully. The determination of these parameters from the test data should not be attempted, because they are not unique and substantially different values might give equally close data fits. One may intuitively anticipate the fact that the spectrum of retardation or relaxation times is actually continuous and may be characterized by the rate of temperature increase in a fire compartment. A suitable choice of retardation time distribution should follow the (HRR) in dimensionless time scale.

Many studies have been undertaken to deduce the spectrum of retardation or relaxation times from the known compliance function of the material, but experimental data generally exhibit random scatter which does not allow taking higher derivatives except perhaps the second derivative. Therefore, certain methods have been developed to modify the creep data to allow taking higher order derivatives. Instead of numerical differentiation of the test data, one must differentiate a smooth continuous compliance function, which matches the experimental data well enough.

The main difficulty in formulating a creep law at variable temperature is due to the fact that temperature rise accelerates not only creep but also deteriorating effect (reduction of instantaneous modulus of elasticity). In order to consider both of these effects, the time t in Eq. (1.23) is replaced with reduced time θ (dimensionless temperature).

The function $E[\theta(\tau)]$ is of importance for two reasons. First, it provides a first impression of the response of a material under temperature stress or strain as a function of time. On the other hand, knowledge of this function and of the spectra of relaxation (or retardation) times derived from them is very helpful for obtaining insight into the *molecular mechanisms* by which they are originated. Analysis of the temperature dependency of mechanical properties thus provides a powerful tool to investigate the relations between structure and properties.

Creep-sensitive structures are still being designed for the mean creep properties. However, the only meaningful approach is to design them for a certain suitable probability (such as 95 %) that a specified strength limit or deflection limit will not be exceeded. The main reason is that a moderate coefficient of variation of material and environmental characteristics can translate in a very large (or very small) coefficient of variation of deflection, stress, etc.

Despite numerous contributions to the literature over the last 15 years, introduction of probabilistic models is requiring further research, especially on the practical side, as well as education. An important probabilistic problem in the prediction of creep effects in case of real fire event is the verification of results based on standard creep tests and measurements. Various statistical regression approaches as well as Bayesian approaches are introduced in this book, and they are presented in a simple, step-by-step form, understandable for a practicing Structural Engineer that is not necessarily familiar with applied probability theory.

The experimental data is available from very limited number of real fire test results and it is fare to say that one cannot expect to have a large number of reliable statistical information regarding structural fire load in tall and super tall buildings or other structural systems (it is not practical and cost prohibited proposition). That is true also with aerospace engineering systems, nuclear power plants, etc. The main goal of the mathematical theory of probability (as we know) is to use its "means and methods" that requires minimum statistical data in order to obtain necessary results (mean value, median, variance, autocorrelation functions, etc.). At the end, of course, one should check these results against the limited test data (for example standard fire test data or "real" fire test results).

1.3.1 Case Studies

Mandarin Oriental Hotel Fire

Fire in buildings can have a severe impact in terms of both human safety and potential economic loss. This is especially true in the case of fires of such severity that the building structure is damaged. Concrete buildings are traditionally regarded as safe in a fire situation as concrete is nonflammable and exhibits highly insulating material properties. The majority of current research relating to the impact of fire on structures examines other forms of construction. The current research seeks to redress the balance by using a systematic approach to examine effects of fire on a holistic steel structure in simplified but realistic temperature exposures. Computer modeling is used extensively. Computational Fluid Dynamics (CFD) analyses are used to explore likely fire temperature and duration in localized areas. Structural Finite Element Modeling (FEM) is used to develop a hierarchy of models, beginning with simple structural forms and progressing logically to more detailed structures. This produces a systematic and comprehensive analysis of the reaction of the structure to fire for comparison to the real, observable damage to the building and assessment of generic failure behaviors. The structural model produced is used with a number of variations in support condition, fire spread rate and extent, and fire protection.

Skyscraper-Turned-Torch Remains Standing

On February 9, 2009, in the middle of the Lunar New Year, the distinctive 40-story Mandarin Oriental hotel in Beijing's Television Cultural Centre (TVCC) erupted in sparks and flames that consumed the building from top to bottom in an intense fire lasing for several hours. Blamed on a ground-based fireworks display gone afoul or illegal fireworks operations inside of the building, the fire started around the tower's top and proceeded downward around the tower's sides while fireworks continued to burst dramatically above the blaze. The spectacular torching of the Mandarin tower further underscores the anomalous nature of the official explanation of the total collapse of the Twin Towers and WTC 7—an explanation that primarily blames fires for causing the "global collapse" of those structures. It is yet another example of a severe fire that failed to induce even the partial collapse of a skyscraper. Although less similar to the WTC towers than other skyscrapers ravaged by fires, the Mandarin tower is notable for the magnitude of the fire it withstood—a fire that dwarfed the fires that preceded the "collapses" of each of the WTC skyscrapers. Coming exactly 2711 days after 9/11/2001, the burning of the Hotel Mandarin Oriental, which was unoccupied pending its completion, killed one firefighter. Since 9-11 there has been a greater interest in the safety of tall buildings and how increased safety can be achieved without compromising on aesthetics or unnecessary costs. The events of 9-11 changed the perceptions of structural designers, contractors, and owners with respect to safety and security issues. Everybody had a reaction. Tall building design moved out of the technical domain and now also forms part of the realm of public interest, due to the heightened awareness of building performance since 9-11. Codes and standards have historically evolved as a result of reactions to major events. One example of this is the call for much longer periods of fire resistance on tall buildings immediately after 9-11. This is an understandable emotion-driven response but we would propose instead that designing a structure with fire as a design load provides a more robust design solution. Simply increasing fire proofing thickness without understanding the actual structural response to heat provides no guarantees of increased safety.

There are two issues to be of major concern and critical importance for building design.

1. It must be stressed that it is not possible to design for every conceivable or inconceivable event. Therefore a threat and risk assessment can be used to quantify real risks, in order to develop suitable mitigation measures, on a project by project basis.
2. Understanding the role of structure and its real response to fire along with the performance of fire proofing materials in real events is also key—even more so as events such as the Madrid fire enhance our understanding of real structural performance.

1.3.2 An Integrated Design Approach

Fire is another loading condition that must be accommodated in the design of a building. Owing to the prescriptive way in which fire-resistant design has been handled historically, designers tend not to think of fire as a loading condition. Rather, they commonly handle fire design by prescribing fireproofing systems that have been calibrated by standard testing of full-size building components. These prescriptive rules for producing a fire-resistant structure tend to mask the effects of a major fire on real structural systems.

Although performance-based fire-resistant design has been available for many years, it tends to be used only by a few very knowledgeable designers. Historically, structural codes have not considered fire as a load. Rather, it was considered as a hazard that could be mitigated by prescriptive rules. High-rise buildings have an excellent record of protecting life when fires occur. Loss of life in high-rise office and apartment buildings has been extremely low.

Despite the excellent record, experience with the First Interstate Bank Building in Los Angeles, One Meridian Plaza in Philadelphia, and Buildings 5 and 7 at the World Trade Center after 9/11 shows that burnouts can occur in buildings. When a burnout occurs, there is a potential for partial or even complete collapse of the structure. Performance-based procedures can be used to help mitigate the risk of collapse and, at the same time, produce a cost-effective design.

However, many engineers are skeptical of the accuracy of failure probability predictions for the following reasons. Data on statistical variability in material properties, geometry and thermal loading distributions are not always available in full. Uncertainties like variability in failure stress can be characterized well from results of coupon tests. Therefore, it is assumed here that the nominal strength (resistance) of the structural system (or elements) has a deterministic value.

As a first step to overcome the problem of insufficient statistical data, it is proposed in this book a probabilistic design optimization procedure confined to failure stresses, for which the AISC, ICC and ASCE 7 already require probabilistic characterization. We assume that estimates based on the historical record are available for the probability of failure of structural components designed deterministically according to the AISC code. With such estimates, the proposed approach can obtain probabilistic designs without the need for the detailed statistical distributions for uncertainties detailed above.

The design process can be summarized here in an integrated design approach: establishing the real risks, analyzing their impact on the structure performance using the tools available to us, and developing designs to accommodate this.

The basic probabilistic approach can be summarized as the statistical definition of dimensionless resistance values required for structural analysis methods, statistical definition of the resulting internal forces and strength of the structure associated with predicted failure modes, and evaluation of the resulting probability of structural failure. This chapter presents the fundamental theory and concepts behind probabilistic methods and list the general steps involved in performing a probabilistic structural analysis.

The failure mode can change from one location to another on the structural system, with some areas being subject to more than one failure mode. This is where engineering know-how is essential. Creep failure under static temperature loading application (steady or steadily increasing) can be caused by exceeding the resistance capacity of the whole structure or from a single component.

1.3.3 Definition of Acceptable Probability of Failure (Target Probability)

The acceptable probability of failure is the criterion to which the results of the probabilistic analysis will be compared to determine if the design is acceptable. Specification of this acceptable, or target, probability of failure for the total structure is a complex issue that generally will not be decided upon by the engineer performing the probabilistic analysis. Legal, technical, and socioeconomic considerations are involved. The agency certifying the structure should be responsible for setting this overall specification for the structure. Proposed failure probability values seen most often in literature [6] in case of fire range from $1(10^{-5})$ to $1 \times (10^{-7})$ per year, but this issue remains unresolved at the present time.

If a structural engineer is performing a probabilistic analysis on only a portion of the structure, there exists the challenge to set a target probability of failure for that structural element (beam; girder, connection etc.), given the whole structural system target level. Depending on the complexity (and dependency) of the structural components, this task could range from being straightforward (e.g., all secondary beams are equally critical and independent) to requiring use of fault tree analysis methods (the girder might fail by itself or only after the secondary beams have failed, etc.) to account for redundant load paths. Modeling system probability is discussed further in this book.

It is important to note that traditional structural analysis and finite element theory are not being supplanted, but rather are an integral part of the probabilistic design process. The probabilistic structure must be built around the existing structural analysis process. Optimally, probabilistic analysis codes should be interfaced to these structural analysis programs and procedures so that the structural analysis output can be directly fed to the probabilistic program and vice versa.

1.3.4 Structural Reliability Assessment

This section begins with the basic mathematical formulation of the problem, along with simple examples for assessing structural reliability. Then more complex (real world) problems are (from random variables representing maximum dimensionless temperature to random functions representing fire as a stochastic process) formulated and four major definitions of reliability index are presented in this book.

The basic problem for probabilistic analysis remains to formulate expressions defining the temperature load on the structure and the resistance (or strength) of the structure. For a typical design condition, both load and strength can be plotted in the same horizontal axis as shown in Fig. 1.6. The mean strength, obviously, is greater than the mean applied load. However, the overlap of PDFs suggests that it is possible for strength to be less than applied load, which is the condition for failure. This illustration conveys the essence of probabilistic structural analysis.

A technically accurate description of the load-resistance curve overlap is shown in Fig. 1.4, showing the load and strength along the horizontal and vertical axes, respectively. The line drawn represents the scenarios where load = strength, or

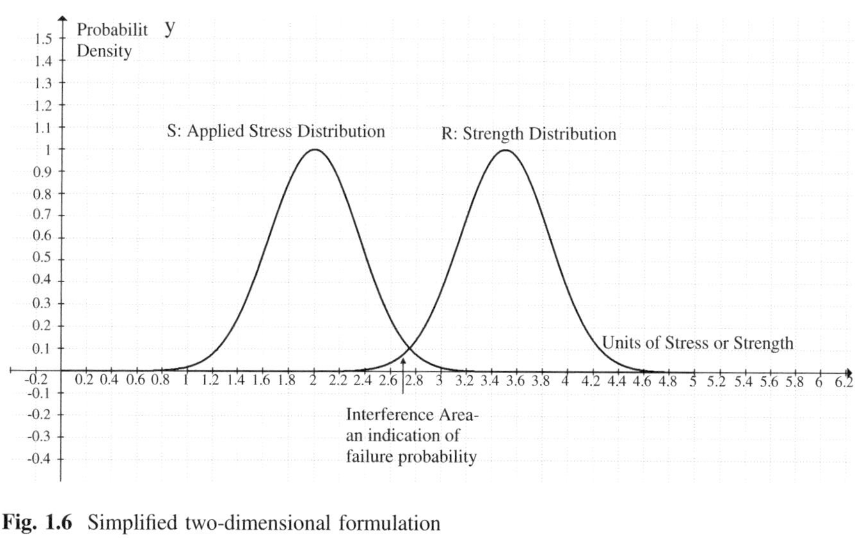

Fig. 1.6 Simplified two-dimensional formulation

$g(R, S) = R - S = 0$. This is often referred to as the "limit state" that separates the failure region ($g < 0$) from the safe region ($g > 0$). The function $g(R, S)$ is commonly referred to as the *performance function*. The probability of failure is defined as the volume under the surface (failure region where $g < 0$).

The probability of failure is defined as $P_f = P[g(R, S) \leq 0]$. The statistical variation of R and S are described by the probability density functions $f_R(r)$ and $f_S(s)$, respectively. The overlap region is quantitatively obtained from the following expression:

$$P_f = \iint_{\Omega} f_{R,S}(r, s)\mathrm{d}r\mathrm{d}s \tag{1.42}$$

where $f_{R,S}(r, s)$ is the joint density function and Ω is the failure set, i.e., the set of all values of R and S such that $g(R, S) \leq 0$. If the variables R and S are statistically independent (changing one has no effect on the other), then the joint density function is expressed as the product of individual density functions as follows:

$f_{R,S}(r, s) = f_R(r)f_S(s)$ and thus

$$P_f = \iint_{\Omega} f_R(r)f_S(s)\mathrm{d}r\mathrm{d}s \tag{1.43}$$

1.3.5 Limit State Design

Limit state design (LSD) refers to a design method used in structural engineering. A limit state is a condition of a structure beyond which it no longer fulfills the relevant design criteria [13]. The condition may refer to a degree of temperature loading or other actions on the structure, while the criteria refer to structural integrity, fitness for use, durability, or other design requirements. A structure designed by LSD is proportioned to sustain all actions likely to occur during its design life, and to remain fit for use, with an appropriate level of reliability for each limit state. Building codes based on LSD implicitly define the appropriate levels of reliability by their prescriptions. A clear distinction is made between the Ultimate State (US) and the Ultimate Limit State (ULS). The US is a physical situation that involves either excessive deformations leading and approaching collapse of the component under consideration or the structure as a whole, as relevant, or maximum temperatures exceeding pre agreed values (for example, columns and beams fire ratings). While the ULS is not a physical situation but rather an agreed computational condition that must be fulfilled, among other additional criteria, in order to comply with the engineering demands for strength and stability under design loads. A structure is deemed to satisfy the ultimate limit state criterion if all factored internal forces: bending and torsional moments, shear and axial tension or compression forces are below the factored resistances calculated for the section under consideration. Complying with the design criteria of the ULS is considered as the minimum requirement (among other additional demands) to provide a proper structural fire safety. In addition to the ULS check mentioned above, an Ultimate State (US) computational check must be performed in case of structural fire design. The aim is to prove that under the action of characteristic temperature design loads (unfactored), and/or while applying certain (unfactored) magnitudes of imposed deformations, or stochastic vibrations from temperature gradients, etc. the structural behavior complies with, and does not exceed, the US design criteria values, specified in the relevant standard in power [14]. These criteria involve various temperature limits, deformations limits, flexibility (or rigidity) limits due to degradation of strength of materials subjected to high temperatures, dynamic behavior limits, as well as crack control requirements (crack width) and other arrangements concerned with the durability of the structure. In view of nonstructural issues it might involve also limits applied to heat transmission that might also affect the structural design. This calculation check is performed at the elastic zone, where characteristic (unfactored) actions are applied and the structural behavior is purely elastic. These US design criteria (temperature exceeding a given level) are also commonly referred to as the performance function (criteria). To satisfy the ultimate limit state, the structure must not collapse when subjected to the peak design load for which it was designed. A structure is deemed to satisfy the ultimate limit state criteria if all factored bending, shear and tensile or compressive stresses are below the factored resistances calculated for the section under consideration. The factored stresses referred to are found by applying magnification factors to the loads on the section. Reduction factors are applied to determine the various factored resistances of the section. The limit state criteria can also be set in terms of load rather than stress: using this approach the structural element being analyzed (e.g., a beam or a column or other load-bearing element, such as walls) is shown to be safe when the "Magnified" loads are less than the relevant "Reduced" resistances. The load and resistance factors are determined using statistics and a preselected probability of failure. While arguably not philosophically superior to permissible or allowable stress design, it does have the potential to produce a more consistently designed structure as each element is intended to have the same probability of failure. However, it should be underlined here that the structural engineer has the ability to design some portions (most important ones) of the whole structural system with the preselected smaller probability of failure, while designing the rest of the structure with the uniformly higher probability of failure. In practical terms this normally results in a more efficient structure, and as such, it can be argued that LSD is superior from a practical engineering viewpoint. Probabilistic analysis is concerned with assessing the probability that a design will satisfy one or more performance criteria. It is up to the analyst to formulate what constitutes acceptable performance, or conversely failure, for the design under consideration. For example, one may want to determine the probability that the maximum internal force developed in a component under high temperature loading exceeds the component's resistance. Most formulations of what constitutes failure have two components: a capacity (also referred to as a resistance), commonly denoted R, and a load, denoted S. Failure is then defined as occurring when the load exceeds the capacity: $S > R$. Note that both the load and the capacity may be random

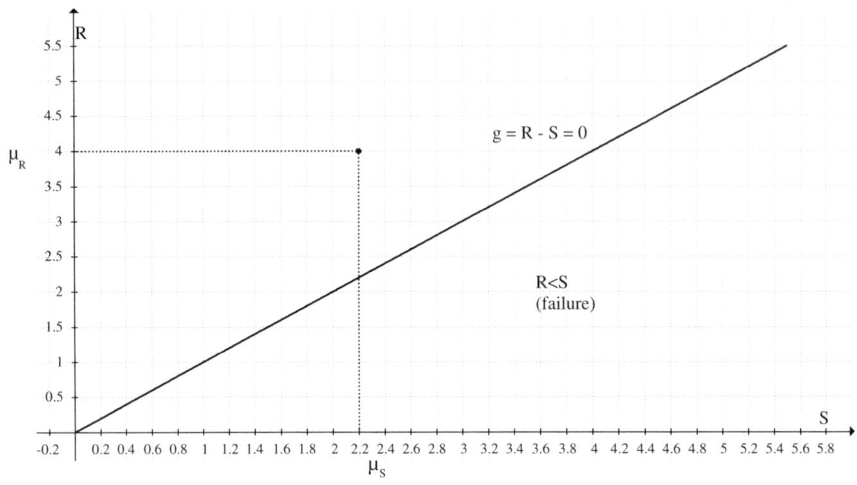

Fig. 1.7 Limit state

quantities. Further, the probabilistic methods of obtaining stochastic structural resistance are the main objective of this book and therefore the structural design load of a structural system is the "companion" objective. The purpose of the *limit state* concept is to provide a unified framework for expressing probability of failure definitions. In terms of the load and capacity, the limit state is given by $R = S$. However, the limit state is traditionally formulated by grouping all random quantities together, so this limit state is more commonly written as $R - S = 0$. This is an important concept, and the reason this is done is so that all the information about the performance of the system needed to determine whether or not it is safe or unsafe is captured by one quantity. This quantity is referred to as the *performance function* and is commonly denoted by Z. Most generally, the performance function may be expressed in terms of all of the basic random variables in the problem: $Z = g(X_1, X_2, \ldots, X_n)$. One of the simplest limit states exists when the basic random variables themselves are the capacity and load of the design. In this case the limit state is given by $Z = g(R, S) = R - S = 0$. Note that the limit state is formulated such that the unsafe region corresponds to $Z < 0$. This limit state is depicted in Fig. 1.7, with the failure region shaded. Note that the mean values of the random variables are located in the safe region.

$R - S$ is not the only possible limit state formulation for this problem. For more complicated problems, often the stochastic resistance is not a basic random variable, but it is a function of other basic variables (parameters property materials—PPT). The limit state concept allows any failure definition to be expressed in a unified manner:

- Analyst must define what constitutes a safe or unsafe design, based on some measure of system performance
- Limit state expresses boundary between safe and unsafe region of design space

- Performance function is expressed in terms of all random quantities (basic variables)
- By convention, failure occurs when performance function is less than a critical value (which is often 0)
- Creep deformations and stresses due to high temperature effect may be functions of other basic variables and be presented as a stochastic process.

1.3.6 Partial Safety Factor ψ and Reliability Index β

The aim of this section is to introduce the most common techniques of structural reliability analysis, namely, First-Order Reliability Methods (FORM). Different cases of limit state functions and probabilistic characteristics of basic random variables are then introduced with increasing generality. Furthermore, FORM results are related to partial safety factors used in common design code. The introduced method of structural reliability theory provides strong tools for the calculation of failure probabilities for individual failure modes or components.

For illustrative purposes, we will first consider the case where the limit state function $g(x)$ is a linear function of the basic random variables X. Then we may write the limit state function as

$$g(x) = a_0 + \sum_{i=1}^{n} b_i x_i \tag{1.44}$$

If the basic random variables are normally distributed we furthermore have that the linear safety margin Z defined through

$$Z = a_0 + \sum_{i=1}^{n} b_i X_i \tag{1.45}$$

is also normally distributed with mean value and variance

$$\begin{aligned} \mu_Z &= a_0 + \sum_{i=1}^{n} b_i \mu_{Xi} \\ \sigma_Z^2 &= \sum_{i=1}^{n} b_i^2 \sigma_X^2 + \sum_{i=1}^{n} \sum_{j=1;j\neq i}^{n} \rho_{ij} b_i b_j \sigma_i \sigma_j \end{aligned} \tag{1.46}$$

where ρ_{ij} are the correlation coefficients between the variables X_i and X_j.

Defining the failure event by Eq. (1.42) we can write the probability of failure as $P_{\mathrm{f}} = P(g(X) \leq 0) = P(Z \leq 0)$, which in this simple case reduces to the evaluation of the standard normal distribution function: $P_{\mathrm{f}} = \Phi^*(-\boldsymbol{\beta})$, where β the so-called reliability index (due to Cornell [35] and Basler [36]) is given as: $\boldsymbol{\beta} = \frac{\boldsymbol{\mu}_Z}{\boldsymbol{\sigma}_Z}$.

Then the reliability index β has the simple geometrical interpretation as the smallest distance from the line (or generally the hyperplane) forming the boundary between the safe domain and the failure domain, i.e., the domain defined by the failure event. It should be noted that this definition of the reliability index [37] does not depend on the limit state function but rather the boundary between the safe domain and the failure domain.

The point on the failure surface with the smallest distance to origin is commonly denoted the design point or most likely failure point. It is seen that the evaluation of the probability of failure in this simple case reduces to some simple evaluations in terms of mean values and standard deviations of the basic random variables, i.e., the first- and second-order information. The limit state equation in this case is a straight line equation, therefore the distance between the strait line $L\ Ax + By + C = 0$ and the origin is given by $d(0, L) = |C|/\sqrt{A^2 + B^2}$. Applying this equation to the limit state equation yields the following value for the distance:

$$d(0, L) = \beta = |\mu_R - \mu_S|/\sqrt{\sigma_R^2 + \sigma_S^2} \tag{1.47}$$

The reliability index β is usually used in obtaining the ultimate design load in LRFD method [11].

Let us introduce now new parameter (the *partial safety factor*—load factor) $\psi = \mu_R/\mu_S$. Then the reliability index β yields

$$\beta = \frac{|\mu_R - \mu_S|}{\sqrt{\sigma_R^2 + \sigma_S^2}} = \frac{|\psi - 1|}{\sqrt{V_R^2\psi^2 + V_S^2}}; \tag{1.48}$$

where $V_R = \sigma_R/\mu_R$ and $V_S = \sigma_S/\mu_S$.

If the material properties are deterministic, then $V_R = 0$ and (1.48) is reduced

$$\Psi = 1 + \beta V_S. \tag{1.49}$$

Parameter V_S is called a load variation parameter (*coefficient of variation*), and V_R is called a material variation parameter.

It can be seen from (1.48) and (1.49) that the reliability index β and the *partial safety factor* $\psi = \mu_R/\mu_S$ are functions of two parameters of normal PDF (mean value and standard deviation) and *coefficient of variation* V_S.

Probability-based limit states design is based on the notion that the reliability index implied by a given structural design should be the same (or almost the same) for all design loads and their combinations. This requirement can be met by using average "safety factor" $\Psi = 1.67(0.9) = 1.5$ [39], therefore the reliability index β from (4.10) is: $\beta = 0.5/0.1 = 5.0$ and vise versa: $\Psi = 1 + 2.433(0.1) = 1.24$.

When the limit state function is nonlinear in the basic random variables R and S, the situation is not as simple as outlined in the previous. An obvious approach is, however, considering statistical linearization method explained in detail in Chap. 2, to represent the failure domain in terms of the boundary between the safe domain

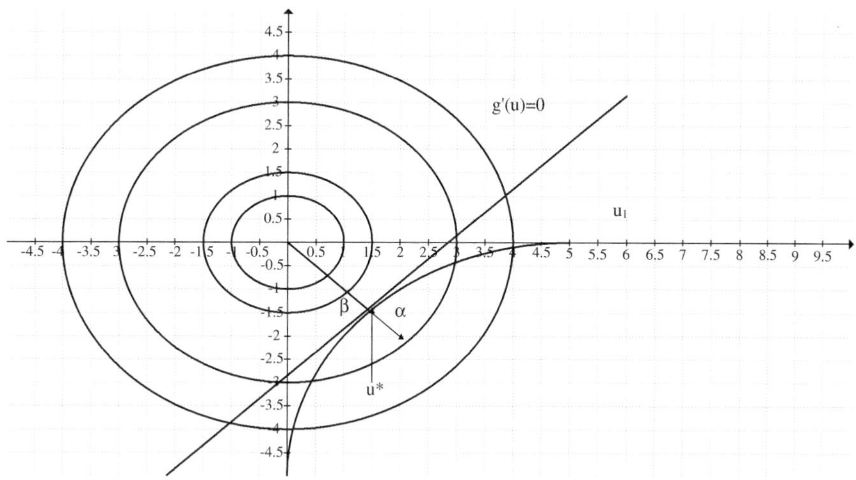

Fig. 1.8 Illustration of the linearization of the failure surface in standard normal space

and the failure domain, i.e., the failure surface, but the question remain how to do this appropriately. Hasofer and Lind [38] suggested performing this linearization in the design point of the failure surface represented in normalized space. The situation is illustrated in the two-dimensional space in Fig. 1.8.

In Fig. 1.8, a principal sketch is given illustrating that the failure surface is linearized in the design point **u*** by the strait line. The α-vector is the outward directed normal vector to the failure surface in the design point **u***, i.e., the point on the linearized failure surface with the shortest distance—β—to the origin. As the limit state function is in general nonlinear one does not know the design point in advance and this has to be found iteratively, e.g., by solving the following optimization problem: This problem may be solved in a number of different ways. Provided that the limit state function is differentiable the following simple iteration scheme may be followed. The results given in Equation ($g(x)$) have been applied to study the statistical characteristics in accordance with some differentiable function $f(x)$, i.e.: $g = f(X)$, where $x = (x_1, x_2, x_n)$ is a vector of realizations of the random variables **X** representing measurement uncertainties with mean values $\mu = (\mu(x_1), \mu(x_2), \mu(x_n))$ and Cov $[X_i;X_j] = \rho_{ij}\ \sigma_{Xi}\ \sigma_{Xj}$ where σ_{Xi} are the standard deviations and ρ_{ij} the correlation coefficients. The idea is to approximate the function $f(x)$ by its Taylor expansion including only the linear terms, i.e.:

$$g(\vec{x}) = g(\vec{x}_0) + \sum_{i=1}^{n} (x_i - x_{i,0}) \frac{\partial g(\vec{x})}{\partial x_i} |_{x=x_0} \tag{1.50}$$

where $\vec{x}_0 = (x_{1,0}, x_{2,0}, \ldots, x_{n,0})$.

The point $\vec{x}_0$ is the point in which the linearization is performed normally chosen as the mean value point and the partial gradient, $\frac{\partial g(\vec{x})}{\partial x_i}|_{x=x_0}$, $i = 1, 2, n$ are the first-order partial derivatives of $g(x)$ taken in $x = x_0$. From Eq. (1.50) it is seen that the mean value can be assessed by: $\mu_g = g(\mu_x)$ and its variance $\mathrm{Var}(g) = D_g$ can be determined by

$$\begin{aligned} D_g(\vec{x}) &= \sum_{i=1}^{n}\left[\frac{\partial g(\vec{x})}{\partial x_i}|_{x=x_0}\right]^2 \sigma_{X_i}^2 + 2\sum_{i<j}^{n}\left[\frac{\partial g(\vec{x})}{\partial x_i}|_{x=x_0}\right]\left[\frac{\partial g(\vec{x})}{\partial x_j}|_{x=x_0}\right]\rho_{ij}\sigma_{X_i}\sigma_{X_j} \\ \mu_g(\vec{x}) &= g(\mu_X) \end{aligned} \tag{1.51}$$

The formula (1.51) is reduced, if all random variables are not correlated with each other ($\rho_{ij} = 0$).

$$D_g(\vec{x}) = \sum_{i=1}^{n}\left[\frac{\partial g(\vec{x})}{\partial x_i}|_{x=x_0}\right]^2 \sigma_{X_i}^2 \tag{1.52}$$

The linearization method can be improved by approximating function $g(x)$ by its Taylor expansion including not only the linear terms. Since we have only one independent variable (dimensionless temperature θ) in our case, the modification of statistical linearization is given below for some differentiable function $g(\theta)$, i.e.:

$$g(\theta) = g(\mu_\theta) + \frac{\mathrm{d}g(\mu_\theta)}{\mathrm{d}\theta}(\theta - \mu_\theta) + \frac{1}{2}\frac{\mathrm{d}^2 g(\mu_\theta)}{\mathrm{d}\theta^2}(\theta - \mu_\theta)^2 \tag{1.53}$$

The mean and variance of random variable g can be obtained now as follows:

$$\begin{aligned} \mu_g &= g(\mu_\theta) + \frac{1}{2}\frac{\mathrm{d}^2 g(\mu_\theta)}{d\theta^2} D_\theta \\ D_g &= \left[\frac{\mathrm{d}g(\mu_\theta)}{\mathrm{d}\theta}\right]^2 D_\theta + \frac{1}{4}\left[\frac{\mathrm{d}^2 g(\mu_\theta)}{\mathrm{d}\theta^2}\right]^2 D[\hat{\theta}^2] + \left[\frac{\mathrm{d}g(\mu_\theta)}{\mathrm{d}\theta}\right]\left[\frac{\mathrm{d}^2 g(\mu_\theta)}{\mathrm{d}\theta^2}\right] K[\hat{\theta}\hat{\theta}^2] \\ &= \left[\frac{\mathrm{d}g(\mu_\theta)}{\mathrm{d}\theta}\right]^2 D_\theta + \frac{1}{4}\left[\frac{\mathrm{d}^2 g(\mu_\theta)}{\mathrm{d}\theta^2}\right]^2 [\mu_4(\theta) - D_\theta^2] + \left[\frac{\mathrm{d}g(\mu_\theta)}{\mathrm{d}\theta}\right]\left[\frac{\mathrm{d}^2 g(\mu_\theta)}{\mathrm{d}\theta^2}\right] \mu_3(\theta) \end{aligned} \tag{1.54}$$

where $K\left[\hat{\theta}\hat{\theta}^2\right]$—cross-correlation function of centered random variables $\hat{\theta}$ and $\hat{\theta}^2 \mu_3(\theta)$ and $\mu_4(\theta)$—third and fourth moments of random variables θ

It has been assumed here that the random variable θ (dimensionless temperature) has normal probability distribution; therefore formula (1.54) can be reduced as

$$D_g = \left[\frac{\mathrm{d}g(\mu_\theta)}{\mathrm{d}\theta}\right]^2 D_\theta + \frac{1}{2}\left[\frac{\mathrm{d}^2 g(\mu_\theta)}{\mathrm{d}\theta^2}\right]^2 D_\theta^2 \tag{1.55}$$

Formula (1.55) can be used for the error approximation of the statistical linearization method, if the random variable θ has close normal pdf.

The main steps to performing this probability-based high temperature creep analyses are as follows:

Case 1 Ultimate design stress value is a random variable.

1. For each selected PPM α_i find the solution $\sigma_i(\theta)$ of integral Eq. (1.25) and present the maximum stress results in a tabulated form. Let us say that the dimensionless temperature interval [0, 10] has been partitioned by integer numbers 0, 1, 2… 10. In each subinterval $[i, i + 1]$ the approximate stress function $\tilde{\sigma}_i(\theta)$ is computed, and the stress–temperature (strain) diagram is composed of the stress function form previous interval $[i - 1, i]$ and additional stress function from the next interval $[i, i + 1]$ (see Strip method above) (Table 1.2).
2. Compose the maximum stress data from Table 1.3 and present the results in a tabulated form.
3. Compute the mean and standard deviation values from statistical data above (see Table 1.3).
4. Transform the design variables into standard normal distribution.
5. Identify the most probable point (MPP), or design point. For a given limit state function, g, the main contribution to failure probability is from the region where g is closest to the origin in the transformed design variable space. The MPP is defined as the closest point to the origin in the transformed space.
6. Develop an approximation to the performance function (g-function) around the MPP. Thus the g-function is approximated by a simply defined surface through that point (MPP). Usually the g-function is a deterministic function prescribed by a Structural Code.
7. Compute probability of failure using the defined g-function and compare it with the target probability.
8. Compute the reliability index β and the dimensionless design temperature $\theta_{\max} = \mu_\theta + \beta(\sigma_\theta)$.
9. Provide structural design calculations (check) for the new design temperature $\theta_{\max}$.

Case 2 Ultimate design stress is a random process.

One can consider now the statistical data (see Table 1.2) as realizations of a stochastic process. Let us find now the estimates of elements of autocorrelation matrix: the variances and correlation moments. In order to obtain the variances the following steps are required:

1. Compute the sum of squares of matrix column elements;
2. Divide this sum by number of rows ($n = 10$ in our case);
3. In order to get the unbiased estimate multiply the result from p. 2 by the ratio $[n/n - 1]$;
4. Repeat computations (pp. 1, 2, 3) for each matrix column. These are the diagonal members in correlation matrix;

Table 1.2 Dimensionless stresses (statistical data)

θ / $\sigma(\theta)$	Θ_1	Θ_2	…	Θ_κ	……..	Θ_{10}	"α" value
$\sigma_1(\theta)$	$\sigma_1(\theta_1)$	$\sigma_1(\theta_2)$		$\sigma_1(\theta_\kappa)$		$\sigma_1(\theta_{10})$	α_1
$\sigma_2(\theta)$	$\sigma_2(\theta_1)$	$\sigma_2(\theta_2)$		$\sigma_2(\theta_\kappa)$		$\sigma_2(\theta_{10})$	α_2
…	…	…	…	…	…	…	…
$\sigma_{10}(\theta)$	$\sigma_{10}(\theta_1)$	$\sigma_{10}(\theta_2)$	…	$\sigma_{10}(\theta_\kappa)$		$\sigma_{10}(\theta_{10})$	α_{10}
Mean value $\sigma_\theta(\theta_k)$	Σ_{av}		Total:	μ_σ	σ_σ		
$D_1 = \frac{1}{10}\sum_{i=1}^{n=10}[\sigma_i(\theta_k)]^2$							
$D = \frac{10}{9}\left[D_1 - (m_\sigma(\theta_k))^2\right]$ Variance "$D_\sigma(\theta_k)$"							
Standard deviation $\sigma_\sigma(\theta_k) = \sqrt{D}$							

Table 1.3 Maximum dimensionless stresses (statistical data)

Value	Max. stress value	Deviation	Variance	Mean value	Standard deviation	"α" value
1	σ_1					α_1
2	σ_2					α_2
…	…	…	…	…	…	…
10	σ_{10}					α_{10}
Aver.	σ_{av}		Total	μ_σ	σ_σ	

5. Compute the sum of products of two different matrix column elements (the second moments are centered, therefore $m_\sigma(\theta_k) \equiv 0$);
6. Divide this sum by number of rows (n = 10 in our case);
7. In order to get the unbiased estimate multiply the result from p. 6 by the ratio [n/ $n - 1$];
8. Repeat computations (pp. 5, 6, 7) for all combinations of matrix column. These are the diagonal members in correlation matrix.
9. If the random function has an ergodic character, then only *one* random realization function is sufficient enough in order to obtain the correlation function. The mean value can be calculated as follows:

$$\boldsymbol{\mu}_{\boldsymbol{\sigma}} = \frac{1}{\boldsymbol{\theta}_{\max}} \int_0^{\boldsymbol{\theta}_{\max}} \boldsymbol{\sigma}(\boldsymbol{\tau})\mathrm{d}\boldsymbol{\tau} \tag{1.56}$$

In order to calculate the autocorrelation function in this case the chosen function has to be centered, and then the autocorrelation function can be computed as follows [39]:

$$K_{\boldsymbol{\sigma}}(\boldsymbol{\theta}) = \frac{1}{\theta_{\max} - \boldsymbol{\theta}} \int_0^{\boldsymbol{\theta}_{\max}-\boldsymbol{\theta}} [\widehat{\sigma}(\boldsymbol{\tau})\widehat{\boldsymbol{\sigma}}(\boldsymbol{\theta}+\boldsymbol{\tau})]\mathrm{d}\boldsymbol{\tau} \tag{1.57}$$

where $\boldsymbol{\theta}$ = 0, 1, 2, … 10 and $\boldsymbol{\theta}_{\max}$ = **10**.

The resistance of structural system where the stress-temperature-time function presents a random process may fail when the maximum stress in a critical member reaches a sufficiently high level. This type of failure is generally by overstress or by excessive permanent deformation rendering the structural element or system inoperative. If the structural fire resistance has a finite probability of exceeding the high level, then failure is possible, and an important problem is to find the probability that the system can operate without failure for some given time (duration of fire event) or range of dimensionless temperatures corresponding to some given fire severity category. More precisely, the following problem is considered. Given a continuous and differentiable random function $\sigma(\theta)$, one wishes to find the probability that the value $\sigma = a$ will not be exceeded in the temperature interval $(0, \theta)$. This problem is called the first occurrence time problem and the probability density $P(a, \theta)$ is the first occurrence density [40]. The probability of failure in $(0, \theta)$ is unity if $X(0) > a$, and the probability of failure in $(0, \theta)$ is (if $\sigma < a$):

$$P[\boldsymbol{\sigma}(\boldsymbol{\theta}_j) > a] = \int_{\mathbf{a}}^{\infty} f(\boldsymbol{\sigma}|\boldsymbol{\theta}_j)\mathrm{d}\boldsymbol{\sigma} \tag{1.58}$$

The average first occurrence of temperature "θ" where the stress value σ is above a given level "a" for stationary processes is defined as follows [40]:

$$\overline{\boldsymbol{\theta}}_a = \boldsymbol{\theta}_{\max} \int_{\mathbf{a}}^{\infty} f(x)\mathrm{d}x \tag{1.59}$$

where $f(x)$—probability density of the maximum stress ordinates.

The average number of the occurrences above a given level "a" for stationary processes during the same temperature interval is defined as follows [40]:

$$\overline{n}_a = \boldsymbol{\theta}_{\max} \int_{0}^{\infty} vf(a, v)\mathrm{d}v \tag{1.60}$$

The average number of temperature intervals for all occurrences above a given level "a" for stationary processes is defined as follows:

$$\overline{\boldsymbol{\tau}} = \frac{\int_a^{\infty} f(x)\mathrm{d}x}{\int_0^{\infty} vf(a, v)\mathrm{d}v} \tag{1.61}$$

The average area between stationary random curve and the horizontal line $y = a$ for the first occurrences above a given level "a" is [41]

$$\overline{\boldsymbol{s}\boldsymbol{v}}_a = \overline{x} - \int_{-\infty}^{a} xf(x)\mathrm{d}x - a\overline{\boldsymbol{\tau}\boldsymbol{v}}_a \tag{1.62}$$

where $\overline{\mathbf{v}}_a = \frac{\overline{n}_a}{T}$.

Many other creep deformation problems can be formulated (based on autocorrelation theory of random functions) depending on different structural limit state conditions.

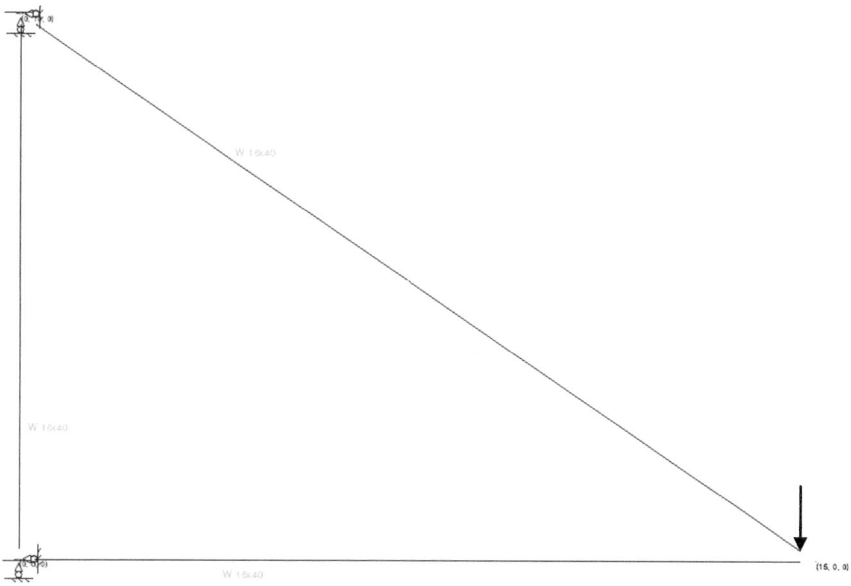

Fig. 1.9 Structural steel frame model

Example 1.1 The example consists of a steel frame depicted in the Fig. 1.9, with all structural elements W 16x 40 (cross-sectional area A = 11.8 in^2; Z_x = 73 in^3; F_y = 50 ksi), subjected to dead load (denoted P = 50k.) and static maximum temperature load has normal distribution with parameters μ_T = 1000 °F and σ_T = 100 °F. Define the performance function and compute probability of structural failure. The computer output in this case is as follows:

Analysis Results

Translations						
Translations [ft]				Rotations [Rad]		
Node	TX	TY	TZ	RX	RY	RZ
Condition dl = Dead load						
1	0.00000	0.00000	0.00000	0.00000	0.00000	−0.00029
2	0.00000	0.00000	0.00000	0.00000	0.00000	−0.00011
3	−0.00319	−0.01312	0.00000	0.00000	0.00000	−0.00108
Condition tl = temp load						
1	0.00000	0.00000	0.00000	0.00000	0.00000	0.00403
2	0.00000	0.00000	0.00000	0.00000	0.00000	0.00330
3	*0.14819*	*0.21829*	0.00000	0.00000	0.00000	0.01986

Member Forces

Condition dl = Dead load						
	M33 [Kip * ft]	V2 [Kip]	M22 [Kip * ft]	V3 [Kip]	Axial [Kip]	Torsion [Kip * ft]
Member 1						
0 %	13.07	2.24	0.00	0.00	0.00	0.00
25 %	7.48	2.24	0.00	0.00	0.00	0.00
50 %	1.89	2.24	0.00	0.00	0.00	0.00
75 %	−3.71	2.24	0.00	0.00	0.00	0.00
100 %	−9.30	2.24	0.00	0.00	0.00	0.00
Member 2						
0 %	−9.30	−0.40	0.00	0.00	87.72	0.00
25 %	−7.48	−0.40	0.00	0.00	87.72	0.00
50 %	−5.65	−0.40	0.00	0.00	87.72	0.00
75 %	−3.83	−0.40	0.00	0.00	87.72	0.00
100 %	−2.00	−0.40	0.00	0.00	87.72	0.00
Member 3						
0 %	*−13.07*	−1.01	0.00	0.00	*−72.76*	0.00
25 %	−9.30	−1.01	0.00	0.00	−72.76	0.00
50 %	−5.53	−1.01	0.00	0.00	−72.76	0.00
75 %	−1.77	−1.01	0.00	0.00	−72.76	0.00
100 %	2.00	−1.01	0.00	0.00	−72.76	0.00
Condition tl = temp load						
	M33 [Kip * ft]	V2 [Kip]	M22 [Kip * ft]	V3 [Kip]	Axial [Kip]	Torsion [Kip * ft]
Member 1						
0 %	−214.17	−41.30	0.00	0.00	0.00	0.00
25 %	−110.92	−41.30	0.00	0.00	0.00	0.00
50 %	−7.67	−41.30	0.00	0.00	0.00	0.00
75 %	95.59	−41.30	0.00	0.00	0.00	0.00
100 %	198.84	−41.30	0.00	0.00	0.00	0.00
Member 2						
0 %	198.84	11.40	0.00	0.00	42.04	0.00
25 %	147.47	11.40	0.00	0.00	42.04	0.00
50 %	96.11	11.40	0.00	0.00	42.04	0.00
75 %	44.74	11.40	0.00	0.00	42.04	0.00
100 %	−6.62	11.40	0.00	0.00	42.04	0.00
Member 3						
0 %	*214.17*	13.84	0.00	0.00	*−41.30*	0.00
25 %	162.28	13.84	0.00	0.00	−41.30	0.00
50 %	110.40	13.84	0.00	0.00	−41.30	0.00
75 %	58.51	13.84	0.00	0.00	−41.30	0.00
100 %	6.62	13.84	0.00	0.00	−41.30	0.00

$P_R = 0.9(50)11.8 = 531$ k.; $M_R = 0.9(50)73/12 = 274$ ft-k

The deterministic performance function in this case is defined by Eq. H2-1 [14]

$$\left|\frac{P_S}{P_R}+\frac{M_S}{M_R}\right| \leq 1.0; \tag{1.63}$$

where $P_S = P_{D.L.} + P_T; M_S = M_{D.L.} + M_T; P_R = \phi_c P_n; M_R = \phi_b M_n.$

After substituting computer output data into (1.14) we have

$$\begin{aligned} g &= \left|\frac{72.76 + 0.041(T_{\max})}{531} + \frac{13.07 + 0.274(T_{\max})}{274}\right| - 1.0 \\ g &= -0.815 + 0.001077(T_{\max}) = 0 \quad \Rightarrow T_{\max} = 756.7\,^{\circ}\mathrm{F} \end{aligned} \tag{1.64}$$

where $a = -0.815;\ b = 0.001077.$

The probability of structural failure in this case is

$$P_{\mathrm{f}} = 1 - \Phi^*(z) = \frac{1}{\sqrt{2\pi}} \int_{-\infty}^{z} \mathrm{e}^{-\frac{z^2}{2}} \mathrm{d}z = 1 - 0.007 = 0.993 \tag{1.65}$$

where $z = \frac{756.7 - 1000}{100} = -2.433$

$$\beta = 2.43 \quad \text{and} \quad T_{\max} = 1000 + (2.43)100 = 1243\,^{\circ}\mathrm{F}.$$

In any real post-flashover fire scenario, the ultimate resistance of a structural element is affected due to reduction of steel yielding stresses [even though the ultimate strength is presented as a deterministic parameter in Eq. (1.63)]. The latest test results of steel material at elevated temperatures are taken from [14], and the reduction coefficient for yielding stress F_y as function of temperature is presented below (see Fig. 1.10).

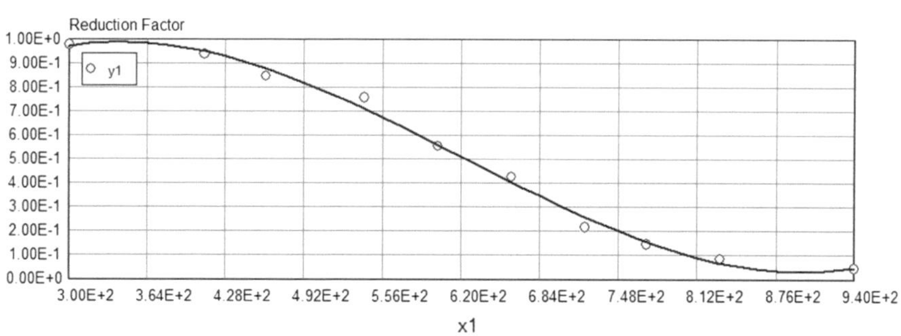

Fig. 1.10 Reduction factor of F_y at elevated temperatures (nonlinear approximation)

Model: $y1 = a0 + a1 * x1 + a2 * x1^2 + a3 * x1^3$

Variable	Value
$a0$	−0.5802925
$a1$	0.0104841
$a2$	−2.111E − 05
$a3$	1.135E − 08

$$y1 = -0.58 + 0.0105T - 2.11(10^\wedge - 5)T^\wedge 2 + 1.135(10^\wedge - 8)T^\wedge 3 \quad 300 < T < 940\,^\circ\mathrm{C} \tag{1.66}$$

The deterministic performance function g in this example is

$$g = 0.185(50) + 0.001077(T_{\max})(50) - (50)y1 = 0$$

where $y1$ = 0.814 from Eq. (1.18) @ T = 500 °C ⇒ $T_{\max}$ = 584 °F.

The probability of structural failure (without creep) in this case is

$$P_{\mathrm{f}} = 1 - \Phi^*(z) = \frac{1}{\sqrt{2\pi}} \int_{-\infty}^{z} \mathrm{e}^{-\frac{z^2}{2}} \mathrm{d}z = 1 - 0.00 = 1.0$$

where $z = \frac{584-1000}{100} = -4.16$.

The reliability index β is increasing with the degradation of steel strength due to high temperature effect (compare with β = 2.43 from [22]—without degradation of steel strength).

The dimensionless stress–temperature random function has an ergodic character (see Chap. 4 for discussion on this subject), therefore only *one* random realization function is sufficient enough in order to obtain the mean value and correlation function of resistance R in this case. Let us choose the random realization function that corresponds to α = 0.33 (the point from the interval of 0 < α < 1). The solution of the Volterra integral equation (1.23) with the degenerate kernel is $\sigma(\theta) = \theta \mathrm{e}^{-0.211\theta}$. The mean value can be calculated now as follows:

$$\begin{aligned}\boldsymbol{\mu}_{\boldsymbol{\sigma}} &= \frac{1}{\boldsymbol{\theta}} \int_0^{\boldsymbol{\theta}} \boldsymbol{\sigma}(\boldsymbol{\tau}) \mathrm{d}\boldsymbol{\tau} = \frac{1}{10} \int_0^{10} \left(\boldsymbol{\tau} \mathrm{e}^{-0.211\boldsymbol{\tau}}\right) \mathrm{d}\boldsymbol{\tau} = \frac{1}{10} \mathrm{e}^{-0.211\boldsymbol{\tau}} \left[-\frac{\boldsymbol{\tau}}{0.211} - \frac{1}{0.211^2} \right]_0^{10} \\ &= \frac{1}{10} [22.46 - 5.75] = 1.67\end{aligned}$$

where $\boldsymbol{\theta} = \boldsymbol{\theta}_{\max}$ = 10– maximum dimensionless; $\boldsymbol{\mu}_{\boldsymbol{\sigma}=1.67}$ corresponds to T = 460 °F.

Therefore the probability of structural failure (with creep) in this case is

$$P_{\mathrm{f}} = 1 - \Phi^{*}(z) = \frac{1}{\sqrt{2\pi}} \int_{-\infty}^{z} \mathrm{e}^{-\frac{z^2}{2}} \mathrm{d}z = 1 - 0.00 = 1.0$$

where $z = \frac{460-1000}{100} = -5.4 > 4.16$.

If the fire severity and statistical data are given (for example, Medium severity; $\mu_\theta = 4.16$; $\sigma_\theta = 0.33$—see Chap. 5), then the reliability index $\beta = 5.4$ and maximum design dimensionless temperature $\theta_{\max}$ value is $\theta_{\max} = \mu_\theta + \beta\sigma_\theta = 4.16 + 5.4(0.33) = 5.942 = 656.5\,^{\circ}\mathrm{C}$.

References

1. Ellingwood B, Galambos TV, MacGregor JG, Allin Cornell C (1980) Development of a probability-based load criterion for american national standard A58, NBS Special Publication 577, National Bureau of Standards, Washington, DC
2. Allin Cornell C (1969) A probability-based structural code. J Am Concr Inst 66:974–985
3. Lind NC (1973) The design of structural design norms. J Struct Mech 1(3)
4. NIST Special Publication 1018-5Fire Dynamics Simulator (Version 5) Technical reference guide, 2008 vol 1: Mathematical Model
5. Dasgupta A, Pecht M (1991) Material failure mechanism and damage models. IEEE Trans Reliab 40:531–536
6. Hasofer AM, Lind NC (1974) An exact and invariant first order reliability format. J Eng Mech Div Proc ASCE 100(EMl), 111–121
7. Gertsbakh I (2000) Reliability theory with application to preventive maintenance. Springer, Berlin
8. Freudenthal AM (1947) The safety of structures. Trans ASCE 112:125–180
9. Singpurwalla ND (1995) Survival in dynamic environment. Stat Sci 10:86–103
10. Razdolsky L (2014) Probability based structural fire load. Cambridge University Press, London, UK
11. AISC, American Institute of Steel Construction, 13th edn. AISC, Chicago, IL. ISBN 1-56424-055-X
12. Kameda H, Koike T (1975) Reliability theory of deteriorating structures. J Struct Div ASCE 101(1):295–310
13. Oswald GF, Schueller GI (1984) Reliability of deteriorating structures. Eng Fracture Mech 20 (3):479–488
14. Enright MP, Frangopol DM (1998). Failure time prediction of deteriorating fail-safe structures. J Struct Eng 124(12):1448–1457
15. Melchers RE (1999) Structural reliability analysis and prediction, 2nd edn. Wiley, NY
16. Gertsbakh IB, Kordonskiy KB (1969) Models of failures. Springer, Berlin
17. Lu CJ, Meeker WQ (1993) Using degradation measure to estimate a time-to-failure distribution. Technometrics 35(2):161–174 (Quality and Reliability Engineering International Vol 30, Issue 8, Article first published online: 11 Jun 2013)
18. Li CQ, Melchers RE (2005a) Time-dependent reliability analysis of corrosion-induced concrete cracking. ACI Struct J 102(4):543–549
19. Li CQ, Melchers RE (2005b) Time-dependent risk assessment of structural deterioration caused by reinforcement corrosion. ACI Struct J 102(5):754–762

20. Morey RC (1966) Some stochastic properties of a compound renewal damage model. Oper Res 14(5):902–908
21. Gut A, Äusler H (1999) Extreme shock models. Extremes 2(3):295–307
22. Razdolsky L (2012) Structural fire load, theory and principles. McGraw Hill, New York, USA
23. Razdolsky L (2012) Probability based structural fire load. Cambridge University Press, London, UK
24. Frank-Kamenestkii DA (1969) Diffusion and heat transfer in chemical kinetics. Plenum Press, New York
25. SFPE Engineering Guide Standard on Calculating Fire Exposure to Structures (2011) Society of fire protection engineers, Bethesda, MD, 20814
26. Magnusson SE, Thelandersson S (1970) Temperature-time curves of complete process of fire development in enclosed spaces. Acts Polytechnica Scandinavia
27. Pontryagin LS (1987) The mathematical theory of optimal processes hardcover. Gordon and Breach Science Publishers
28. Athans MA, Falb PL (1966) Optimal control. McGraw–Hill, New York
29. Stowell, EZ (1958) A phenomenological relation between stress, strain rate, and temperature for metals at elevated temperatures. NACA Rep 1343
30. Mott NP, Nabarro FRN (1948) Dislocation theory and transient creep. Rep. of a conference on the strength of solids (HH Wills Phys. Lab., Univ Bristol, July 7–9, 1947), The Physical Soc, 1948, pp 1–19
31. Sherby OD, Trozera TA, Darn JE (1956) The effects of creep stress history at high temperatures on the creep of aluminum alloys. In: Proceedings of ASTM, vol 56, pp 789–806
32. AISC, American Institute of Steel Construction, 13th edn, AISC, Chicago, IL. ISBN 1-56424-055-X
33. Smith DE (1997) Design optimization. In: Dieter GE (ed) ASM handbook: materials selection and design, vol 20. ASM International, Materials Park
34. Rozvany GIN (1989) Structural design via optimality criteria. Kluwer Academic Publishers, London
35. Cornell CA (1967) Bounds on the reliability of structural systems. J Struct Div ASCE 93
36. Basler E (1961) Untersuchungen über den Sicherheitsbegriff von Bauwerken. Schweizer Archiv für angewandte Wissenschaft und Technik
37. Hasofer AM, Lind NC (1974) An exact and invariant first order reliability format. J Eng Mech Div Proc ASCE 100(EMl), 111–121
38. McCormac JC (2008) Structural steel design, 4th edn. Pearson Prentice Hall, Upper Saddle River. ISBN 978-0-13-221816-0
39. Wentzel E, Ovcharov L (1986) Applied problems in probability theory. Mir Publishers, Moscow
40. Sveshnokov AA (1978) Problems in probability theory, mathematical statistics and theory of random functions. Dover Publications, New York

Chapter 2
Integral Volterra Equations

2.1 Overview

Any functional equation in which the unknown function appears under the sign of integration is called an integral equation. Integral equations arise in a great many branches of science; for example, in potential theory, acoustics, elasticity, fluid mechanics, irradiative transfer, theory of population, etc. In many instances the integral equation originates from the conversion of a boundary value problem or an initial value problem associated with a partial or an ordinary differential equation, but many problems lead directly to integral equations and cannot be formulated in terms of differential equations. Integral equations are of many types; here, we attempt to indicate some of the main distinguishing features with particular regard to the use and construction of algorithms.

In the classical theory of integral equations one distinguishes between Fredholm equations and Volterra equations. In a Fredholm equation the region of integration is fixed, whereas in a Volterra equation the region is variable. Thus, the equation

$$\varphi(x) - \lambda \int_a^b K(x,s)\varphi(s)\mathrm{d}s = \mathrm{cf}(x); \quad a \leq x \leq b \tag{2.1}$$

is an example of Fredholm equation, and the equation

$$\varphi(x) - \lambda \int_a^x K(x,s)\varphi(s)\mathrm{d}s = \mathrm{cf}(x); \quad a \leq x \tag{2.2}$$

L. Razdolsky, *Probability Based High Temperature Engineering*,
DOI 10.1007/978-3-319-41909-1_2

is an example of a Volterra equation. Here, the forcing function f(x) and the kernel function.

$K(x, s, \varphi(s))$ are prescribed, while $\varphi(x)$ is the unknown function to be determined. (More generally the integration and the domain of definition of the functions may extend to more than one dimension.) The parameter λ is often omitted; it is, however, of importance in certain theoretical investigations (e.g., stability) and in the eigenvalue problem discussed below. If in (1.1) or (1.2), $c = 0$, the integral equation is said to be of the first kind. If $c = 1$, the equation is said to be of the second kind. Equations (1.1) and (1.2) are linear if the kernel $K(x, s, \varphi(s)) = k(x, s)\ \varphi(s)$, otherwise they are nonlinear. Note: in a linear integral equation, $k(x, s)$ is usually referred to as the kernel. We adopt this convention throughout. These two types of equations are broadly analogous to problems of initial- and boundary value type for an ordinary differential equation (ODE); thus the Volterra equation, characterized by a variable upper limit of integration, is amenable to solution by methods of marching type whilst most methods for treating Fredholm equations lead ultimately to the solution of an approximating system of simultaneous algebraic equations. For comprehensive discussion of numerical methods, see [1, 2]. In what follows, the term "integral equation" is used in its general sense, and the type is distinguished when appropriate.

2.1.1 Structure of Kernel

When considering numerical methods for integral equations, particular attention should be paid to the character of the kernel, which is usually the main factor governing the choice of an appropriate quadrature formula or system of approximating functions. Various commonly occurring types of singularity call for individual treatment. Likewise, provision can be made for cases of symmetry, periodicity or other special structure, where the solution may have special properties and/or economies may be affected in the solution process. We note in particular the following cases to which we shall often have occasion to refer in the description of individual algorithms.

(a) A linear integral equation with a kernel $k(x, s) = k(s, x)$ is said to be symmetric. This property plays a key role in the theory of Fredholm integral equations.
(b) If $k(x, s) = k(a + b - x, a + b - s)$ in a linear integral equation, the kernel is called Centro-symmetric.
(c) If in Eqs. (1.1) or (1.2) the kernel has the form $K(x, s), \varphi(s) = k(x - s)\ g(s, \varphi(s))$, the equation is called a convolution integral equation; in the linear case $g(s, \varphi(s)) = \varphi(s)$.
(d) If the kernel in (1.1) has the form

$$K(x,\,s,\,\varphi(s)) = K1(x,\,s,\,\varphi(s)),\quad a \le s \le x,$$

$$K(x,\,s,\,\varphi(s)) = K2(x,\,s,\,\varphi(s)),\quad x \le s \le b,$$

where the functions $K1$ and $K2$ are well behaved, whilst K or its s-derivative is possibly discontinuous, may be described as discontinuous or of '*split*' type; in the linear case $K(x, s, \varphi(s)) = k(x, s)\varphi(s)$ and consequently $K1 = k1(\varphi(s))$ and $K2 = k2(\varphi(s))$. Examples are the commonly occurring kernels of the type $k(x - s)$ and the Green's functions (influence functions) which arise in the conversion of ODE boundary value problems to integral equations. It is also of interest to note that the Volterra Eq. (1.2) may be conceived as a Fredholm equation with kernel of split type, with $K2(x, s, \varphi(s)) \equiv 0$; consequently methods designed for the solution of Fredholm equations with split kernels are also applicable to Volterra equations.

2.1.2 Singular and Weakly Singular Equations

An integral equation may be called singular if either

(a) its kernel contains a singularity, or
(b) the range of integration is infinite,

and it is said to be weakly singular if the kernel becomes infinite at $s = x$. Sometimes a solution can be affected by a simple adaptation of a method applicable to a nonsingular equation: for example, an infinite range may be truncated at a suitably chosen point. In other cases, however, theoretical considerations will dictate the need for special methods and algorithms. Examples are:

(i) Integral equations with singular kernels of Cauchy type;
(ii) Equations of Wiener–Hopf type;
(iii) Various dual integral equations arising in the solution of boundary value problems of mathematical physics;
(iv) The well-known Abel integral equation, an equation of Volterra type, whose kernel contains an inverse square-root singularity at $s = x$.

2.1.3 Eigenvalue Problem

Closely connected with the linear Fredholm integral equation of the second kind is the eigenvalue problem represented by the homogeneous equation

$$\varphi(x) - \lambda \int_a^b K(x,s)\varphi(s)\mathrm{d}s = 0;\quad a \le x \le b \tag{2.3}$$

If λ is chosen arbitrarily this equation in general possesses only the trivial solution $\varphi(x) = 0$. However, for a certain critical set of values of λ, the characteristic values or eigenvalues (the latter term is sometimes reserved for the reciprocals $\mu = 1/\lambda$), there exist nontrivial solutions $\varphi(\mathrm{x})$, termed characteristic functions or eigenfunctions, which are of fundamental importance in many investigations. The analogy with the Eigen problem of linear algebra is readily apparent, and indeed most methods of solution of Eq. (1.3) entail reduction to an approximately equivalent algebraic problem.

2.1.4 Nonlinear Fredholm Equation of the Second Kind

$$\varphi(x) - \lambda \int_a^b K(x, s, \varphi(s))\mathrm{d}s = f(x); \quad a \le x \le b \tag{2.4}$$

The numerical solution of Eq. (1.4) is usually accomplished either by simple iteration or by a more sophisticated iterative scheme based on Newton's method; in the latter case it is necessary to solve a sequence of linear integral equations. Convergence may be demonstrated subject to suitable conditions of Lipchitz continuity of the function K with respect to the argument φ. Examples of Fredholm type (for which the provision of algorithms is contemplated) are:

(a) the Uryson equation

$$\varphi(x) - \int_0^1 K(x, s, \varphi(s))\mathrm{d}s = 0; \quad 0 \le x \le 1 \tag{2.5}$$

(b) the Hammerstein equation a nonlinear integral equation of the type

$$\varphi(x) - \int_a^b F(x, s), g(s, \varphi(s))\mathrm{d}s = 0; \quad a \le x \le b \tag{2.6}$$

Where $K(x, s)$ and $g(x, \varphi(s))$ are given functions, while $\varphi(s)$ is the unknown function. Named after Hammerstein [3], who considered the case where $K(x, s)$ is a symmetric and positive Fredholm kernel, i.e., all its eigenvalues are positive. If, in addition, the function $g(x,\ \varphi(s))$ is continuous and satisfies the condition $|g(x, \varphi(s))| \leq C_1|s| + C_2$, where C_1 and C_2 are positive constants and C_1 is smaller than the first eigenvalue of the kernel $K(x, s)$, the Hammerstein equation has at least one continuous solution. If, on the other hand, $g(x,\ \varphi(s))$ happens to be a non-decreasing function of s for any fixed x from the interval $[a, b]$, Hammerstein's equation cannot have more than one solution. This property holds also if $g(x, \varphi(s))$ satisfies the condition.

$|G(x,\ \varphi\ (s_1)) - g(x,\ \varphi\ (s_2))| \leq C|s_1 - s_2|$, where the positive constant C is smaller than the first eigenvalue of the kernel $K(x, s)$. A solution of the Hammerstein equation may be constructed by the method of successive approximation.

2.1.5 Integral Equations Volterra

Volterra equations may be regarded as a special case of Fredholm equations, with the kernel $K(x, s)$ defined on the square $a < x < b$; $a < s < b$, and vanishing in the triangle $a < x < s < b$. A Volterra equation of the second kind without free term is called a homogeneous Volterra equation. The expression $\int_a^x K(x,s)\varphi(s)\mathrm{d}s$ defines an integral operator acting in L_2; it is known as the Volterra operator. The principal result of the theory of Volterra equations of the second kind may be described as follows. For each complex $\lambda \neq \infty$ there exists a square-integrable solution of the Volterra equation of the second kind which is, moreover, unique. This solution may be obtained by successive approximation, i.e., as the limit of a mean-square-convergent sequence.

Consider the Volterra integral equation of the first kind

$$\int_a^x K(x,s)\varphi(s)\mathrm{d}s = f(x); \quad a \leq x \tag{2.7}$$

Clearly it is necessary that $f(a) = 0$; otherwise no solution to (1.7) can exist. The following types of Volterra integral equations of the first kind occur in real life problems: equations with unbounded kernel at $s = x$, equations with sufficiently smooth kernel.

In general, a nonsingular Volterra equation of the first kind presents less computational difficulty than the Fredholm Eq. (1.4) with a smooth kernel. A Volterra equation of the first kind may, under suitable conditions, be converted by differentiation to one of the second kind or by integration by parts to an equation of the second kind for the integral of the wanted function.

A very general Volterra equation of the second kind is given by

$$\varphi(x) - \int_a^x K(x, s, \varphi(s))\mathrm{d}s = f(x); \quad a \leq x \tag{2.8}$$

The resemblance of Volterra equations to ODEs suggests that the underlying methods for ODE problems can be applied to Volterra equations. Indeed this turns out to be the case. The main advantages of implementing these methods are their well-developed theoretical background, i.e., convergence and stability. Many Volterra integral equations arising in real-life problems have a convolution kernel [4].

However, a subclass of these equations which have kernels of the form $K(x - s) = \sum_{j=0}^{N} \boldsymbol{\lambda}_j (x - s)^j$ where λ_j are real numbers can be converted into a system of linear or nonlinear ODEs. Equations of type (1.2) were first systematically studied by Volterra [5, 6]. The principal result of the theory of Volterra equations of the second kind may be described as follows. For each complex $\lambda \neq \infty$ there is a square-integrable solution of the Volterra equation of the second kind which is, moreover, unique. This solution may be obtained by successive approximation (cf. Sequential approximation, method below), i.e., as the limit of a mean-square-convergent sequence:

The contracting-mapping principle implies that if $|K(x, y)| \leq M$ and $|\boldsymbol{\lambda}| < \frac{1}{M(b-a)}$ or $|\boldsymbol{\lambda}| < \left(\int_a^b \int_a^b K^2(x, s)\varphi_n(s))\mathrm{d}x\mathrm{d}s \right)^{-1/2}$, then the considered integral equation has a unique solution in the space L_2[a, b] which can be constructed by the method of successive approximation (1.3). Function φ_0 is an arbitrary square-integrable function. In the case of a continuous kernel *K*(*x*, *s*) and $f \in C[a, b]$, this sequence converges uniformly on [*a*, *b*] to a unique continuous solution. In practical applications of Volterra equations of the second kind, it is very important that its solution be found at least approximately, e.g., by the method of successive approximation. However, other methods are usually more convenient, and some of them such will now be described. Let *f* and *K* be continuous functions. The interval [*a*, *b*] is subdivided into *N* equal parts with the aid of partitioning points x_i, and $x_0 = a$, $x_N = b$. To find the approximate value of $\varphi(x_i)$, the integral over the interval is replaced by a quadrature sum:

$$\int_a^b F(x)\mathrm{d}x = \sum_{k=1}^{N} A_k F(x_k) + \boldsymbol{\rho} \tag{2.9}$$

Where: A_k and x_k are constants for a given interval [a, b] and the quadrature sum formula. For example:

1. *Using the rectangle formula with nodes* $x_0, \ldots, x_{i-1}$:

$$\int_{x_0}^{x_i} K(x_i, s)\varphi(s)\mathrm{d}s \approx \frac{b-a}{N}\sum_{j=0}^{i-1} K(x_i, x_j)\varphi(x_j) \tag{2.10}$$

$$x_1 = a;\ x_2 = a + (b-a)/n; \ldots x_N = a + (N-1)(b-a)/n;\ A_k = (b-a)/N$$

The approximate value of $\varphi(x_i)$ is then obtained using collocation:

$$\tilde{\varphi}(x_i) = \lambda\frac{b-a}{N}\sum_{j=0}^{i-1} K(x_i, x_j)\tilde{\varphi}(x_j) + f(x_i); \quad \tilde{\varphi}(x_0) = f(a)$$

The values of the approximate solution at the points on [a, b] situated between the partitioning points may be found, for example, from the relation

$$\tilde{\varphi}(x) \simeq \lambda\frac{b-a}{N}\sum_{j=0}^{i-1} K(x, x_j)\tilde{\varphi}(x_j) + f(x); \quad x_{j-1} < x \le x_j$$

For $N \to \infty$ this approximate solution converges uniformly to the exact solution of the Volterra equation of the second kind.

2. *Using the trapezoidal formula*:

$$x_1 = a;\ x_2 = a + (b-a)/(N-1); x_N = b;\ A_1 = A_N = (b-a)/2(N-1);$$
$$A_2 = \ldots = A_{N-1} = (b-a)/(N-1);$$

3. *Using tangent formula*:

$$x_1 = a + (b-a)/2N;\ x_2 = a + 3(b-a)/2N; \ldots x_N = a + (2N-1)(b-a)/2N;\ A_k = (b-a)/N;$$

4. *Using Chebyshev formula*:

$$x_k = (b-a)/2 + x_k^{(N)}(b-a)/2;\ A_k = (b-a)/N;$$

Chebyshev nodes in the interval (−1, 1) are:

$$x_k^{(N)} = \cos[(2k-1)\pi/2\mathrm{N}];\ k = 1, 2, \ldots, N$$

5. *Using Gauss–Legendre quadrature formula* [7]:

$$x_k = a + (b-a)x_k^{(N)};\ A_k = (b-a)A_k^{(N)};$$

$x_k^{(N)}$—root of P_N—Legendre polynomials,

For the simplest integration problem stated above, the associated polynomials are Legendre polynomials, $P_N(x)$, and the method is usually known as Gauss–Legendre quadrature. With the N-th polynomial normalized to give $P_N(1) = 1$, the k-th Gauss node, x_k, is the k-th root of P_N are listed in Table 2.1.

Change of interval

An integral over [a, b] must be changed into an integral over [−1, 1] before applying the Gaussian quadrature rule. This change of interval can be done in the following way:

$$\int_a^b f(x)\mathrm{d}x = \frac{b-a}{2}\int_{-1}^{1} f\left(\frac{b-a}{2}z + \frac{a+b}{2}\right)\mathrm{d}z$$

Table 2.1 Roots and weights of Legendre polynomials

Number of points (N)	Points x_k	Weights $A_k^{(N)}$
1	0	2
2	$\pm\sqrt{\frac{1}{3}}$	1
3	0	8/9
	$\pm\sqrt{\frac{3}{5}}$	5/9
4	$\pm\sqrt{\frac{3}{7}-\frac{2}{7}\sqrt{\frac{6}{5}}}$	$\frac{18+\sqrt{30}}{36}$
	$\pm\sqrt{\frac{3}{7}+\frac{2}{7}\sqrt{\frac{6}{5}}}$	$\frac{18-\sqrt{30}}{36}$
5	0	128/225
	$\pm\frac{1}{3}\sqrt{5-2\sqrt{\frac{10}{7}}}$	$\frac{322+13\sqrt{70}}{900}$
	$\pm\frac{1}{3}\sqrt{5+2\sqrt{\frac{10}{7}}}$	$\frac{322-13\sqrt{70}}{900}$

Applying the Gaussian quadrature rule then results in the following approximation:

$$\int_a^b f(x)\mathrm{d}x \approx \frac{b-a}{2}\sum_{k=1}^{N} A_k f\left(\frac{b-a}{2}z_k + \frac{a+b}{2}\right) \tag{2.11}$$

After applying the formula (1.11) to the integral in the left-hand side of Eq. (1.2) we obtain the equality

$$\tilde{\varphi}(x) - \boldsymbol{\lambda}\sum_{k=1}^{N} A_k K(x, x_k)\tilde{\varphi}(x_k) = f(x) + \lambda\rho(x) \tag{2.12}$$

The approximate value of $\varphi(x_i)$ is then obtained using collocation

$$\tilde{\varphi}(x_i) - \boldsymbol{\lambda}\sum_{k=1}^{N} A_k K(x_i, x_k)\tilde{\varphi}(x_k) = f(x_i) + \lambda\rho_i; \quad i = 1, 2, \ldots, N \tag{2.13}$$

Discarding the right-hand side of Eq. (1.12) small value $\lambda\rho_i$, we obtain the following system of equations with N unknowns, which can be written as follows in the expanded form:

$$\begin{cases} \tilde{\varphi}(x_1)[1 - \boldsymbol{\lambda} A_1 K(x_1, x_1)] - \lambda\tilde{\varphi}(x_2)A_2K(x_1, x_2) - \ldots - \lambda\tilde{\varphi}(x_N)A_NK(x_1, x_N) = f(x_1) \\ -\lambda\tilde{\varphi}(x_1)A_1K(x_2, x_1) + \tilde{\varphi}(x_2)[1 - \boldsymbol{\lambda} A_2K(x_2, x_2)] - \ldots - \lambda\tilde{\varphi}(x_N)A_NK(x_2, x_N) = f(x_2) \\ \ldots \\ -\lambda\tilde{\varphi}(x_1)A_1K(x_n, x_1) - \lambda\tilde{\varphi}(x_2)A_2K(x_n, x_2) - \ldots + \tilde{\varphi}(x_N)[1 - \lambda A_NK(x_N, x_N)] = f(x_N) \end{cases} \tag{2.14}$$

Solving this system of Eq. (2.14), we find approximations for the values of $\tilde{\varphi}(x_1); \tilde{\varphi}(x_2); \ldots \tilde{\varphi}(x_N)$ unknown function $\varphi(x_1); \varphi(x_2); \ldots \varphi(x_N)$. Obviously, the accuracy of the result obtained by replacing the integral Eq. (1.2) with the systems of linear Eq. (1.14), will be higher than a smaller error we make by replacing the integral by the sum. Accurate error estimation is presented in [8, 9].

Finally, it should be noted that the method of replacing the integral equation to a system of algebraic also suitable for the approximate solution of the problem of finding the eigenvalues (or characteristic values) and eigenfunctions (or characteristic functions). To solve this problem, instead of the system of Eq. (2.13) is necessary to consider the corresponding homogeneous system.

$$\tilde{\varphi}(x_i) - \boldsymbol{\lambda}\sum_{k=1}^{N} A_k K(x_i, x_k)\tilde{\varphi}(x_k) = 0; \quad i = 1, 2, \ldots, N \tag{2.15}$$

Determinant of this system will be

$$\Delta(\lambda) = \begin{vmatrix} [1-\lambda A_1 K(x_1,x_1)] - \lambda A_2 K(x_1,x_2) - \ldots - \lambda A_N K(x_1,x_N) \\ -\lambda A_1 K(x_2,x_1) + [1-\lambda A_2 K(x_2,x_2)] - \ldots - \lambda A_N K(x_2,x_N) \\ \ldots \\ -\lambda A_1 K(x_n,x_1) - \lambda A_2 K(x_n,x_2) - \ldots + [1-\lambda A_N K(x_N,x_N)] \end{vmatrix} \tag{2.16}$$

Equating to zero the determinant (2.16) and solving the corresponding equation $\Delta(\lambda) = 0$, we find the values of λ at which the system of Eq. (2.16) has a solution $\{\tilde{\varphi}(x_i)\}$ that is not identically zero. These values are in fact the approximate eigenvalues values of λ. Further, taking these values λ equal to the roots of the Eq. (2.16), and obtaining for these values the corresponding independent solution of (2.15), we can obtain approximate expressions for the eigenfunctions of the equation.

Everything said so far also applies to Volterra equations whose kernel $K(x, s)$ is a matrix of dimension r $\times$ r, and where φ and f are r-dimensional vector-functions. The name Volterra equation or generalized Volterra equation is also given to a more general integral equation (GIE), of the form:

$$\varphi(P) - \lambda \int\limits_{D(P)} K(P,Q)\varphi(Q)\mathrm{d}Q = f(P) \tag{2.17}$$

Assuming that the successive approximations such are in some sense convergent (e.g., uniformly or on the average) on the domain of definition of the functions φ and f for all $\lambda \neq \infty$. Here, P and Q are points of the n-dimensional Euclidean space; $D(P)$ is the domain of integration, which usually depends on the point P, and $D(P) \subseteq D$ for any P. The following equation may serve as an example:

$$\varphi(x,y) - \lambda \int\limits_a^x \int\limits_a^b K(x,y,\xi,\eta)\varphi(\xi,\eta)\mathrm{d}\xi\mathrm{d}\eta = f(x,y) \tag{2.18}$$

If the function $K(x, y, \xi, \eta)$ is square-integrable for $a \leq x \leq b$, $a \leq y \leq b$, $a \leq \xi \leq b$, $a \leq \eta \leq b$, while $f(x, y)$ is square-integrable for $a \leq x \leq b$, $a \leq y \leq b$, the sequence (1.3) is mean-square-convergent for $\lambda \neq \infty$. Generalized Volterra equations of the first kind usually cannot be reduced to Volterra equations of the second kind, though this may be possible in special cases. GIE (or mixed Volterra–Fredholm integral equations and Fredholm integral equations on the unbounded interval [0, ∞) occur in mechanics and many related fields of engineering and mathematical physics [10]. Recently, many papers have been devoted to the existence of solutions of nonlinear functional integral equations [11].

Equations of type (2.18) and their nonlinear counterparts arise in the theory of parabolic boundary value problems. Actually few numerical methods for (2.18) are known. In [11] a method for numerical treatment of (2.18) is given and also shown existence and unicity of the solution of (2.18) independently of the value chosen for λ. In Chap. 4, we give a numerical solution obtained by continuous time collocation and time discretization method.

Nonlinear Volterra equations is the name sometimes given to Volterra equations in which the product $K(x, s)\ \varphi(s)$ has been replaced by some function $K(x, s, \varphi(s))$ which is nonlinear with respect to $\varphi(s)$. Equations of this type are frequently encountered in theoretical and in applied studies. Thus, the Cauchy problem for an ordinary differential equation may be readily reduced to the problem of solving a nonlinear Volterra equation. The application of potential theory to boundary value problems for equations of parabolic type reduces such problems to a generalized Volterra equation. In the case of nonlinear Volterra equations it may be shown, if certain assumptions are made with respect to $K(x, s, \varphi(s))$, that successive approximations of type (2.3) converge on an interval $[a, a + \Delta a]$, where Δa is sufficiently small. Approximate solutions of nonlinear Volterra equations are found by using the recurrence relation (2.5); it is sufficient to replace $K(x, s, \varphi(s))$ by $K(x, s)\ \varphi(s)$. If $K(x, s, \varphi(s))$ is independent of x, for example, $K(x, s, \varphi(s)) = a(x)\ b(s)\ f[\varphi(s)]$.

2.2 Reduction of ODEs to the Volterra Integral Equation (IE)

In mathematics, an *integral equation* is an equation in which an unknown function appears under an integral sign. There is a close connection between differential and integral equations, and some problems may be formulated either way. Consider an ordinary differential equation (ODE) of the first order $\frac{\mathrm{d}y}{\mathrm{d}x} = f(x;y)$, where $f(x; y)$ is defined and continuous in a two-dimensional domain G which contains the point $(x0; y0)$. Integrating the ODE subject to the initial condition $y(x0) = y0$, we obtain

$$\varphi(x) = y_0 + \int_{x_0}^{x} f(t, \varphi(t))\mathrm{d}t \tag{2.19}$$

This is the *Volterra integral equation of the second kind* with respect to the unknown function $\varphi(t)$. This equation is equivalent to the initial value problem of ODE. Note that this is generally a *nonlinear* integral equation with respect to $\varphi(t)$.

Consider now a linear ODE of the second order with variable coefficients

$$y'' + A(x)y' + B(x)y = g(x) \tag{2.20}$$

The initial condition

$$y(x_0) = y_0; y'(x_0) = y_1 \tag{2.21}$$

$A(x)$ and $B(x)$ are given functions continuous in an interval G which contains the point x_0. Integrating y'' in (2.20) we obtain

$$y'(x) = -\int_{x_0}^{x} A(t)y'(x)\mathrm{d}x - \int_{x_0}^{x} B(x)y(x)\mathrm{d}x + \int_{x_0}^{x} g(x)\mathrm{d}x + y_1 \tag{2.22}$$

Integrating the first integral on the right-hand side in (1.22) by parts yields

$$y'(x) = -A(x)y(x) - \int_{x_0}^{x} [B(x) - A'(x)]y(x)\mathrm{d}x + \int_{x_0}^{x} g(x)\mathrm{d}x + A(x_0)y_0 + y_1$$

Integrating a second time, we obtain

$$\begin{aligned} y(x) = &-\int_{x_0}^{x} A(x)y(x)\mathrm{d}x - \int_{x_0}^{x}\int_{x_0}^{x} [B(t) - A'(t)]y(t)\mathrm{d}x\mathrm{d}t + \int_{x_0}^{x}\int_{x_0}^{x} g(t)\mathrm{d}x\mathrm{d}t \\ &+ [A(x_0)y_0 + y_1](x - x_0) + y_0 \end{aligned} \tag{2.23}$$

Using the relationship $\int_{x_0}^{x}\int_{x_0}^{x} g(t)\mathrm{d}x\mathrm{d}t = \int_{x_0}^{x} (x-t)g(t)\mathrm{d}t$ we transform (2.23) to obtain

$$\begin{aligned} y(x) = &-\int_{x_0}^{x} \{A(t) + (x-t)[B(t) - A'(t)]\}y(t)\mathrm{d}t \\ &+ \int_{x_0}^{x} (x-t)g(t)\mathrm{d}t + [A(x_0)y_0 + y_1](x - x_0) + y_0 \end{aligned} \tag{2.24}$$

Separate the known functions in (2.24) and introduce the notation for the *kernel function*

$$\begin{aligned} K(x,t) &= -A(t) + (t-x)[B(t) - A'(t)], \\ f(x) &= \int_{x_0}^{x} (x-t)g(t)\mathrm{d}t + [A(x_0)y_0 + y_1](x - x_0) + y_0 \end{aligned} \tag{2.25}$$

Then (2.25) becomes

$$y(x) = f(x) + \int_{x_0}^{x} K(x;t)y(t))\mathrm{d}t \tag{2.26}$$

This is the Volterra IE of the second kind with respect to the unknown function $y(t)$. This equation is equivalent to the initial value problem (2.22) and (2.23). Note that here we obtain a *linear* integral equation with respect to $y(x)$.

Example 2.1 Consider a homogeneous linear ODE of the second order with constant coefficients

$$y'' + \omega^2 y = 0 \tag{2.27}$$

Initial conditions (at $x_0 = 0$)

$$Y(0) = 0;\ y'(0) = \omega \tag{2.28}$$

We see that here: $A(x) \equiv 0; B(x) \equiv \boldsymbol{\omega}^2; g(x) \equiv 0; y_0 = 0; y_0' = \boldsymbol{\omega}$. We use the same notation as in (2.25). Substituting into (2.24) and calculating $K(x;\ t) = \omega^2 (t - x)$; $f(x) = \omega x$; we find that the IE (2.26), equivalent to the initial value problem (2.27) and (2.28), takes the form (2.29).

$$y(x) = \boldsymbol{\omega} x - \boldsymbol{\omega}^2 \int_0^x (x - t)y(t)\mathrm{d}t \tag{2.29}$$

The solution of the differential Eqs. (2.27) and (2.28) is: $y(x) = \sin\ (\omega x)$. Therefore, the solution of IE (2.11) is the same.

The following theorems apply to Volterra equations of the first kind. If $f(s)$ and $K(x,\ s)$ are differentiable, $K(x,x) \neq 0\, x \in [a,b]$, and if $K(x,\ x)$ and $K'_x(x,\ x)$ are square-summable on $[a,\ b]$ and on $a < x < b$; $a < s < b$, respectively, a Volterra equation of the first kind is equivalent to the Volterra equation of the second kind obtained by differentiation of the Volterra equation of the first kind and having the form:

$$\varphi(x) + \int_a^x \frac{K_x'(x,s)}{K(x,x)} \varphi(s)\mathrm{d}s = \frac{f'(x)}{K(x,x)} f(x)$$

If $K(x,\ x) = 0$ at least at one point, the solution of the Volterra equation of the first kind must be more thoroughly investigated. If, on the other hand, $K(x,x) \equiv 0$, then the differentiation operation may be repeated under certain conditions. If the differentiation is impossible or does not result in a Volterra equation of the second kind, this equation of the first kind may be solved, for example, using a regularization algorithm.

In practical applications of Volterra equations of the second kind it is very important that its solution be found at least approximately, e.g., by the method of successive approximation. However, other methods are usually more convenient, and one such method will now be described. Let f and K be continuous functions. The interval $[a, b]$ is subdivided into N equal parts with the aid of partitioning points x_i, and $x_0 = a$, $x_N = b$. To find the approximate value of $\varphi(x_i)$, the integral over the interval is replaced by a quadrature sum, for example, using the rectangle formula with nodes $x_0 \ldots x_{i-1}$:

2.3 Sequential Approximation Method (*Method of Successive Approximation*)

This is a method to construct an approximating equation for approximate (and numerical) solutions of certain kinds of linear and nonlinear integral equations. However, the main type of integral equations suitable for solving by this method is linear one-dimensional integral Fredholm equations of the second kind. The method as applied to such equations consists of an approximation which replaces the kernel $K(x, s)$ of the integral Eq. (1.2) by a degenerate kernel of the type $K_N(x,s) = \sum_{n=1}^{N} a_n(x)b_{n(s)}$ followed by the solution of the Fredholm degenerate integral equation

$$\tilde{\varphi}(x) - \lambda \int_a^b K_N(x,s)\tilde{\varphi}(s)\mathrm{d}s = f(x); \quad a \le x \le b; \quad a \le s \le b \tag{2.30}$$

Solving (2.30) is reduced to solving a system of linear algebraic equations. The degenerate kernel $K_N(x,s)$ may be found from the kernel $K(x,s)$ in several ways, e.g., by expanding the kernel into a Taylor series or a Fourier series (for other methods see Strip method (integral equations)—see below). The method of degenerate kernels may be applied to systems of integral equations of the type (2.2), to multi-dimensional equations with relatively simple domains of integration and to certain nonlinear equations of Hammerstein type (cf. Hammerstein equation—see above). In many cases, the good convergence properties of the approximations constructed by this method allow one to apply it to practical computations.

We seek a solution of the integral Eq. (2.2) in the form of a series arranged in powers of parameter λ.

$$\varphi(x) = \tilde{\varphi}_0(x) + \lambda\tilde{\varphi}_1(x) + \lambda^2\tilde{\varphi}_2(x) + \ldots \tag{2.31}$$

Substituting this series in Eq. (2.2), we find

$$\varphi(x) = \varphi_0(x) + \lambda\varphi_1(x) + \lambda^2\varphi_2(x) + \ldots = f(x) + \lambda\int_a^b K(x,y)[\varphi_0(y) + \lambda\varphi_1(y) + \lambda^2\varphi_2(y) + \ldots]\mathrm{d}y$$

Equating the coefficients of equal powers of λ, we obtain

$$\begin{aligned} \varphi_0(x) &= f(x) \\ \varphi_1(x) &= \int_a^b K(x,y)[\varphi_0(y)]\mathrm{d}y \\ \varphi_2(x) &= \int_a^b K(x,y)[\varphi_1(y)]\mathrm{d}y \\ &\ldots \end{aligned} \tag{2.32}$$

From these equations, we can determine sequentially all the functions $\varphi_1(x)$, $\varphi_2(x)$, i.e., if we introduce the so-called *iterated kernel*, we can write the following expression for the functions $\varphi_n(x)$

$$\begin{aligned} \varphi_n(x) &= f(x) + \sum_{m=1}^{n}\lambda^m\int_a^b K_m(x,y)f(y)\mathrm{d}y \\ K_1(x,y) &= K(x,y); \quad K_m(x,y) = \int_a^b K_{m-1}(y,t)K(x,t)\mathrm{d}t \end{aligned} \tag{2.33}$$

$K_m(x; y)$ is called *the m-th iterated kernel*. One can easily prove that the iterated kernels satisfy a more general relationship

$$K_m(x,y) = \int_a^b K_r(x,t)K_{m-r}(y,t)\mathrm{d}t; \quad r = 1,2,\ldots,m-1 \quad (m = 2,3.\ldots) \tag{2.34}$$

Assume that there exists a constant C_1 such that

$$\sqrt{\int_a^b |K(x,y)|^2\mathrm{d}y} \leq C_1 \quad \forall x \in [a,b] \quad \lambda \leq \frac{1}{C_1} \tag{2.35}$$

Then the sequence φ_n of successive approximations (2.31) converges uniformly for all λ, satisfying (2.35). The limit of the sequence φ_n is the unique solution to IE (2.2).

Let us introduce now the notation for *the resolvent*

$$\Gamma(x, y, \lambda) = \sum_{m=1}^{n} \lambda^{m-1} \int_a^b K_m(x, y) \mathrm{d}y \tag{2.36}$$

Changing the order of summation and integration in (2.36), we obtain the solution in the compact form

$$\varphi(x) = f(x) + \lambda \int_a^b \Gamma(x, y, \lambda) f(y) \mathrm{d}y \tag{2.37}$$

One can show that the resolvent satisfies the IE

$$\Gamma(x, y, \lambda) = K(x, y) + \lambda \int_a^b \Gamma(t, y, \lambda) K(x, t) \mathrm{d}t \tag{2.38}$$

To study the convergence properties of the sequence (2.36) and to prove the existence of a solution to (1.2), the contracting-mapping principle formulated below is widely used. This principle implies that if inequality (2.35) holds and the kernel $K(x, y)$ is limited ($|K(x, y)| \leq M$), then the sequences (2.36) converge and the considered integral equation has a unique solution in the space $L_2[a, b]$ which can be constructed by the method of successive approximation.

If λ satisfies (2.35), then the series (2.36) can be successfully used to approximate solution of integral Eq. (2.2). Often required quadrature (2.32) can be performed accurately, then the resulting series (2.31) converges at least as a geometric progression with ratio $|\lambda|M(b - a)$. The error, which will occur, if we restrict ourselves only to "n" members of the series (2.31) can be estimated easily. That is, assuming $f(x) \leq N$ is easy to get sequentially that $|\varphi_n(x)| < NM^n(b - a)^n$, and so the remainder of the series (1.31) after "n" members.

$$|\lambda^n \varphi_n(x) + \lambda^{n+1} \varphi_{n+1}(x) + \ldots| < \frac{NM^n(b-a)^n |\lambda|^n}{1 - M(b-a)^{|\lambda|}} \tag{2.39}$$

Practically, estimating of the error of the solution can be done differently. First of all, with the help of some interpolation formula (for example the method of least squares), one can, using the numbers found $\varphi(x_i)$, obtain an approximate expression for $\varphi(x)$ in the whole interval $[a, b]$. To test how well the function $\varphi(x)$ satisfies the integral Eq. (2.8), it can be substituted into the integral equation, and then this integral should be replaced by a sum using the other quadrature formula, and check what the error for several values of x_i.

Power series (2.39) is an expansion in the λ near the point $\lambda = 0$ and will therefore converge to the first eigenvalue of the kernel $K(x, y)$, i.e., $|\lambda| < |\lambda_1|$. Consequently, the power series (1) cannot be used if $|\lambda| > |\lambda_1|$, since it diverges, and difficult to use, if λ is close to λ_1, since then the series converges slowly. In these cases, it is often convenient (when we know the approximate location of the eigenvalues of the integral equation), apply the so-called method of analytic continuation for the approximate solution of Eq. (1.8). We shall assume for convenience that $\lambda_1 = -1$, since this can be achieved by replacing λ with $\bar{\lambda} = -\frac{\lambda}{\lambda_1}$. Solution with $|\lambda| < 1$ is given by a convergent series.

$$\varphi = \varphi_0 + \lambda\varphi_1 + \lambda^2\varphi_2 + \ldots$$

With this series, we can calculate the value of the function φ (x, λ) and its derivatives with respect to λ (for example, when $\lambda = 0.5$). In turn, this will enable the expansion of the solution of Eq. (1) in powers of $(\lambda - 1/2)$. This expansion can be obtained directly by substituting in the power series λ on $(\lambda' + 1/2)$ (referring to $\lambda' = \lambda - 1/2$), and expanding power series in powers of λ'. The distance between point $\lambda = 1/2$ and eigenvalue $\lambda = -1$ is equal to 1.5 and, therefore, the resulting series in powers of λ' will have a radius of convergence of 1.5 and will allow us to, for example, to compute the solution for $\lambda = 1$, which was not possible with the original series. Rebuilt series will converge at $\lambda = 1$, i.e. $\lambda' = 0.5$, unless this value itself is not an eigenvalue. This method is called the method of *analytic continuation*. Rebuilt power series is as follows.

$$\varphi = \varphi_0 + (\lambda' + \frac{1}{2})\varphi_1 + (\lambda' + \frac{1}{2})^2\varphi_2 + \ldots = [\varphi_0 + (\frac{1}{2})\varphi_1 + (\frac{1}{2})^2\varphi_2 + (\frac{1}{2})^3\varphi_3 + \ldots] \\ + [\varphi_1 + \varphi_2 + \frac{3}{4}\varphi_3 + \ldots]\lambda' + [\varphi_2 + \frac{3}{2}\varphi_3 + \ldots](\lambda')^2 + \ldots \tag{2.40}$$

2.4 Linear Volterra IE of the Second Kind

2.4.1 *Basic Steps of the Method of Moments (MoM) Galerkin Method*

We restrict our attention here to the MoM applications to solving integral equations, where the unknowns are stress functions. The basic idea of the MoM is as follows. The unknown quantity (φ) is expanded in terms of a set of linearly independent known functions, φ_n (referred to as basis or expansion functions), i.e., it is approximated by the following finite series $\varphi \approx \sum_{n=1}^{N} a_n\varphi_n$, where a_n are unknown coefficients yet to be determined. The expansion functions should be chosen, usually based on experience, so that reasonable approximation of φ is obtained with a small number of terms, N.

When function φ is substituted into (2.18), one obtains the approximate equation

$$L\left(\sum_{n=1}^{N} a_n \varphi_n\right) \approx f \tag{2.41}$$

Note that Eq. (2.41) cannot be exactly satisfied at all points, as we have a finite number of terms in the series. Exceptions are rare examples that do have analytical solutions, but which are not of our interest here. The unknown coefficients (a_n) should now be determined such that Eq. (2.41) is satisfied in a sense. Hence, a measure is needed describing the degree of accuracy to which the left side and the right side of Eq. (2.41) match.

In the MoM, this measure is obtained in the following way. Both sides of Eq. (2.41) are multiplied by a known, properly selected function, referred to as the weighting function, w_n, and the results integrated over a spatial region. This integration is a special, but very frequent case of an inner product of two functions, f and φ_n, which is denoted by $<f\ \varphi>$. Generally, the inner product of elements f and φ of a given space is a scalar, which satisfies the following conditions:

$$\begin{gathered}\langle \varphi, f\rangle = \langle f, \varphi\rangle; \langle \alpha\varphi + \beta f, h\rangle = \alpha\langle \varphi, h\rangle + \beta\langle f, h\rangle; \langle \varphi, \varphi^*\rangle > 0 \\ \text{if } \varphi \neq 0, \text{and} \langle \varphi, \varphi^*\rangle = 0 \quad \text{if } \varphi = 0,\end{gathered}$$

where α and β are arbitrary scalars, and h is another element of the same space.

Equation (1.41) represents a system of N ordinary linear equations in N unknowns, and it can be solved using various techniques. To prepare a computer code that uses the MoM to solve a complex high temperature creep problem, usually requires a lot of work and experience. Often, codes are specialized for certain classes of problems. In most cases, a useful measure of accuracy of the solution obtained does not exist. In spite of this deficiency, the MoM is the most powerful tool available nowadays for analysis of fairly general creep deformation field problems that involve linear integral equations Volterra of second type.

The expansion and testing functions can be arbitrary. However, to provide an efficient solution, the expansion functions should be selected such that the solution can be well approximated by a relatively small number of functions. Similarly, the weighting functions should provide a reliable measure of discrepancy between the two sides of equation. On the other hand, all these functions should be selected bearing in mind complexity and speed of computations, and flexibility to accommodate to a wide range of problems [12].

Expansion and testing functions may coincide, i.e., we can take $\varphi_n = w_n$; $n = 1, 2, \ldots, N$. In this case, we have a Galerkin solution, which is equivalent to the Rayleigh-Ritz variational method, often used in the finite-element approach.

Both expansion and testing functions can be divided into two categories. The first category is sub-domain functions. The domain, where the unknown function (φ) is defined, is divided into a number of small sub-domains.

Each basis function is defined only on one sub-domain (i.e., it is assumed zero elsewhere), and it is a very simple function. Such a choice simplifies evaluation of integral equation, and it can relatively easily accommodate an arbitrary memory function. However, it may result in instabilities as the approximation of the unknown function is discontinuous or has discontinuous derivatives, and it may require a large number of basis functions for an accurate solution.

The piecewise-constant approximation is discontinuous. A better approximation is the piecewise approximation, which is continuous, but has a discontinuous first derivative. Analytically, this approximation can be constructed in two ways. For simplicity, we consider a one-dimensional expansion. The first way is assuming a linear function on a sub-domain, and then matching the approximations on adjacent sub-domains to obtain continuity. Alternatively, a sub-domain memory function (the kernel) can be approximated by the piecewise function, which is continuous, but has a discontinuous first derivative.

More sophisticated functions can be designed using more complicated sub-domain functions and introducing additional constraints. An expansion function that closely resembles creep function may expedite the numerical solution. Approximation by expansion functions involved in the MoM means not only an approximation of the unknown function, but also of the material property parameters (MPP). The approximation of the MPP means a modification of their spectra of the sub-domains where the unknown function is defined.

In practice, however, the entire domain is divided into a small number of relatively large sub-domains. For example, in case of high temperature creep analysis the dimensionless temperature θ ranges from 0 to 10 (for metallic materials) and it is divided into ten segments.

The expansion and testing functions are then defined on these large sub-domains. This procedure is referred to as the almost-entire-domain approximation.

This kind of functions may well accommodate complex temperature creep analysis problems and yield good results with a smaller number of unknowns.

The more complicated the basis functions are, the more analytical preparation is usually required before starting to write the computer code. A set of basis functions is usually suitable for a certain class of problems, but not for a general structure. Hence, a code customized for a class of problems is usually more efficient than a general code.

2.4.2 Linear Integral Equations (IEs) with Degenerate Kernels

The importance of degenerate integral equations in the general theory of Fredholm equations is based on the fact that the solution of any Fredholm equation of the second kind can be approximated by solutions of degenerate integral equations in the mean-square (and certain other) metrics to any degree of accuracy. Their degenerate kernels approximate the kernel of the initial equation in one sense or another. An IE (1.8) with *a degenerate kernel* $K(x,y) = \sum_{i=1}^{n} a_i(x)b_i(x)$ can be represented, by changing the order of summation and integration, in the form:

$$L(\varphi) = \varphi(x) - \lambda \sum_{i=1}^{n} a_i(x) \int_a^x b_i(y)\varphi(y)\mathrm{d}y = f(x) \tag{2.42}$$

Here, one may assume that functions $a_i(x)$ (and $b_i(y)$) are linearly independent (otherwise, the number of terms in (2.42) can be reduced). It is easy to solve such IEs with degenerate and separable kernels. Denote

$$\tilde{\varphi}(x) = f(x) + \lambda \sum_{i=1}^{n} A_i a_i(x) \tag{2.43}$$

A_i is unknown constants. Equation (1.42) becomes:

The problem reduces to the determination of unknowns A_i. To this end, substitute (2.43) into Eq. (2.42) to obtain, after simple algebra,

$$f(x) + \lambda \sum_{i=1}^{n} A_i a_i(x) - \lambda \sum_{i=1}^{n} a_i(x) \int_a^x \left[b_j(y)\{f(y) + \lambda \sum_{j=1}^{n} A_j a_j(y)\} \right] \mathrm{d}y = f(x) \tag{2.44}$$

Denoting $\psi_j(x) = \int_0^x b_j(y)a_j(y)\mathrm{d}y$; $\vartheta_i(x) = \int_a^x b_i(y)f(y)\mathrm{d}y$ and substituting into (1.44) we have

$$\sum_{i=1}^{n} a_i(x) \left[A_i - \sum_{i=1}^{n} \{\vartheta_i(x) + \lambda \sum_{j=1}^{n} A_j \psi_j(x)\} \right] = 0 \tag{2.45}$$

Since functions $a_i(x)$ are linearly independent, equality (2.45) yields

$$A_i - \sum_{i=1}^{n} \{\boldsymbol{\vartheta}_i(x) + \boldsymbol{\lambda} \sum_{j=1}^{n} A_j \boldsymbol{\psi}_j(x)\} = 0 \tag{2.46}$$

For a function $\tilde{\varphi}(x)$ to represent the exact solution of the Eq. (2.42), it is necessary that the operator $L(\varphi)$ is equals identically zero, and this is equivalent to the fact that it (the integral equation) must be orthogonal to all the functions a_i that is:

$$\int_a^b \left[\sum_{i=1}^{n} \{A_i - \vartheta_i(x) - \lambda \sum_{j=1}^{n} A_j \psi_j(x)\} \right] a_i(x)\mathrm{d}x = 0 \tag{2.47}$$

Denoting $a_{ij} = \int_a^b \psi_j(x) a_i(x)\mathrm{d}x; f_i = \int_a^b \vartheta_i(x) a_i(x)\mathrm{d}x$ and changing the order of summation and integration in (2.47) we obtain the solution in the compact form

$$A_i - \lambda \sum_{j=1}^{n} a_{ij} A_j = f_i \quad i = 1, 2 \ldots n \tag{2.48}$$

Assume now that functions $\varphi(x)$; $f(x)$; $a(x)$ and $b(x)$ in Eq. (2.41) are continuously differentiable functions, and λ is a parameter. Differentiating the Eq. (2.42) we obtain

$$L'(\varphi) = \varphi'(x) - \lambda \sum_{i=1}^{n} a_i(x) b_i(y)\varphi(y) - \lambda \sum_{i=1}^{n} a_i'(x) \int_a^x b_i(y)\varphi(y)\mathrm{d}y = f'(x) \tag{2.49}$$

Denoting $Z_i = \int_a^x b_j(y)\varphi(y)\mathrm{d}y$ or : $Z_i' = b_j(y)\varphi(y)$ and substituting in (1.47) we have now (n + 1) ordinary differential equations (ODE) of first order with zero initial conditions.

$$\begin{aligned} \frac{\mathrm{d}\varphi}{\mathrm{d}x} &= f'(x) + \lambda \sum_{i=1}^{n} a_i(x) b_i(x)\varphi(x) + \lambda \sum_{i=1}^{n} a_i'(x) Z_i(x) \\ \frac{\mathrm{d}Z_i}{\mathrm{d}x} &= b_i(x)\varphi(x);\ \varphi(\mathrm{x} = \mathrm{a}) = 0;\ Z(x = a) = 0; \quad i = 1, 2, \ldots, n \end{aligned} \tag{2.50}$$

It should be noted that the above inequality (2.35) must be satisfied. Otherwise, the interval [a, b] should be divided into a finite number of smaller intervals, and the solution of the integral Eq. (1.8) is found by applying the so-called strip method.

2.5 Special Types of Integral Equations

2.5.1 Strip Method (Integral Equations)

If the interval [a, b] is large, then in many cases the inequality (1.35) is not satisfied and therefore the contracting-mapping principle is not applicable. This method is based on replacing the kernel in a special way by a degenerate kernel, evaluating the resolvent of the degenerate equation and then improving the approximate solution through the use of a rapidly convergent iterative algorithm. To construct the degenerate kernel, divide the square [$a \le x \le b$, $a \le s \le b$] into N strips

$\{\frac{b-a}{N}i \le x \le \frac{b-a}{N}(i+1), a \le s \le b\} i = 0, 1, \ldots N-1$. In each strip, say the i-th, the function $K(x, s)$ is approximated in the mean square, or uniformly, by functions $K_i(x, s) = P_i(x)Q_i(s)$. In this case, $K_i(x,s) = K(\xi_i, s)\xi_i \in \{\frac{b-a}{N}i \le x \le \frac{b-a}{N}(i+1)\}$, $a \le s \le b$.

The function $K_i(x, s)$ is now used to construct a degenerate kernel:

$$K_N(x,s) = \sum_{i=0}^{N-1} P_i(x)Q_i(s)$$
$$P_i(x) = \begin{cases} P_i(x), & x \in \left[\frac{b-a}{N}i \le x \le \frac{b-a}{N}(i+1)\right] \\ 0, & x \notin \left[\frac{b-a}{N}i \le x \le \frac{b-a}{N}(i+1)\right] \end{cases} \tag{2.51}$$
$$Q_i(s) = \begin{cases} Q_i(s), & s \in \left[\frac{b-a}{N}i \le s \le \frac{b-a}{N}(i+1)\right] \\ 0, & s \notin \left[\frac{b-a}{N}i \le s \le \frac{b-a}{N}(i+1)\right] \end{cases}$$

The solution of the equation with the degenerate kernel (2.51) approximates the solution of Eq. (2.2), generally, the larger the number N of strips and the function $K_N(x, s)$ the better the approximation of $K(x, s)$ in each strip [13]. The approximate solution $\varphi_N(x)$ can be further improved by using the iterative algorithm that insures the continuation of the final solution $\varphi(x)$ in the whole interval $[a, b]$ at the same time. The contracting-mapping principle is of special interest for nonlinear equations; see e.g., [14].

2.5.2 Power Series Solution for Integral Equations

In many cases if the Kernel of the integral equation is of the form $K(xs)$ and the Mellin transform of $K(t)$ exists we can find the solution of the integral equation

$$g(s) = s\int_0^\infty K(st)f(t)\mathrm{d}t \tag{2.52}$$

In a form of a power series

$$f(t) = \sum_0^\infty \frac{a_n}{M(n+1)}t^n$$
$$g(s) = \sum_{n=0}^\infty a_n s^{-n}; \quad M(n+1) = \int_0^\infty t^n K(t)\mathrm{d}t \tag{2.53}$$

Where $g(s)$ is the Z-transform of the function $g(s)$ and $M(n+1)$ is the Mellin transform of the Kernel [15].

2.5.3 *Hammerstein Integral Equation*

A nonlinear integral equation of the type

$$\varphi(x) + \int_a^b K(x,s)f[s,\varphi(s)]\mathrm{d}s = 0; \quad a \le x \le b \tag{2.54}$$

Where $K(x, s)$ and $f(x, s)$ are given functions, while $\varphi(x)$ is the unknown function. Named after Hammerstein [16], who considered the case where $K(x, s)$ is a symmetric and positive Fredholm kernel, i.e., all its eigenvalues are positive. If, in addition, the function $f(x, s)$ is continuous and satisfies the condition $|f(x,s)| \le C_1|s| + C_2$, where C1 and C2 are positive constants and C_1 is smaller than the first eigenvalue of the kernel $K(x, s)$, the Hammerstein equation has at least one continuous solution. If, on the other hand, $f(x, s)$ happens to be a non-decreasing function of s for any fixed x from the interval $[a, b]$, Hammerstein's equation cannot have more than one solution. This property holds also if $f(x, s)$ satisfies the condition $|f(x,s_1) - f(x,s_2)| \le C|s_1 - s_2|$, where the positive constant C is smaller than the first eigenvalue of the kernel $K(x, s)$. A solution of the Hammerstein equation may be constructed by the method of successive approximation.

Below are examples of solutions of Volterra and Fredholm integral equations of the second kind (and the mixed type Fredholm–Volterra equation), which are intended to acquaint the reader with the approximate numerical methods for solving these equations used in the subsequent chapters of this book, as well as give an engineering assessment of convergence of these solutions by comparison with known exact solutions. Of the large number of examples available in mathematical physics are selected only those that is close to the phenomenological theory of creep of materials and its practical application to the structural analysis and design. Some examples are solved by applying different numerical methods (more than one) in order to compare them and to obtain upper and lower bounds with respect to the exact solution of the given integral equation.

2.6 Examples

Example 2.2 (Method #1) Data: Consider a Volterra IE of the 2nd kind

$$f(x) = \varphi(x) - \lambda \int_0^x \mathrm{e}^{x-y}\varphi(y)\mathrm{d}y \tag{2.55}$$

The kernel $K(x, y) = e^{x-y}$, $f(x)$ is a given continuously differentiable function, and $\lambda\lambda\psi$ is a parameter.

Solution: method #1.

Differentiating (1.54) we obtain

$$f(x) = \varphi(x) - \lambda \int_0^x e^{x-y}\varphi(y)dy$$

$$f'(x) = \varphi'(x) - \lambda[\varphi(x) + \int_0^x e^{x-y}\varphi(y)dy]$$

Subtracting term wise we obtain an ordinary differential equation (ODE) of the first order

$$\varphi'(x) - (\lambda + 1)\varphi(x) = f'(x) - f(x) = F(x) \tag{2.56}$$

with respect to the (same) unknown function $\varphi(x)$. Setting $x = 0$ in (2.54) we obtain the initial condition for the unknown function $\varphi(x)$: $\varphi(0) = f(0)$. Thus, we have obtained the initial value problem for $\varphi(x)$. Integrating (2.56) subject to the initial condition we obtain the Volterra IE (1.54), which is therefore equivalent to the initial value problem (2.56).

Later (see Chap. 4) equation similar to (2.55) will represent the simplest case of creep deformations.

$$f(x) = \varphi(x) - \lambda \int_0^x e^{-\alpha(x-y)}\varphi(y)dy \tag{2.57}$$

where $\alpha-$material property parameter (MPP) and $\lambda = -1$

The corresponding differential Eq. (1.57) is:

$$\begin{aligned} &\varphi'(x) - (\lambda - \alpha)\varphi(x) = f'(x) + \alpha f(x) = F(x, \alpha) \\ &\text{where} f(x) = xe^{-0.15x}; f'(x) = e^{-0.15x}(1 - 0.15x); \alpha = 0.33 \\ &\text{initial condition: } \varphi(0) = 0 \end{aligned} \tag{2.58}$$

Solution of Eq. (2.58) in this case is as follows (using POLYMATH software):

Calculated values of DEQ variables

	Variable	Initial value	Minimal value	Maximal value	Final value
1	t	0	0	1.0	1.0
2	y	0	0	0.5595	0.5595

Differential equation

$$1 \quad d(y)/d(t) = -1.33 * y + (\exp(-0.15 * t)) * 1.0 * (1 - 0.15 * t) + t * \exp(-0.15 * t)) * 0.33$$

See Fig. 2.1.

The Stress–Temperature function can be approximated by quadratic parabola: **Model**: $y = a0 + a1 * t + a2 * t\wedge 2 = 0.0083 + 0.892t - 0.347t^2$

Variable	Value
$a0$	0.00883
$a1$	0.892
$a2$	−0.347

See Fig. 2.2.

Method #2 Equation (1.55) is as follows:

$$\varphi'(x) = f'(x) - \lambda\varphi(x) + \lambda\alpha \int_0^x e^{-\alpha(x-y)}\varphi(y)dy$$
$$\text{or}: \begin{cases} \varphi'(x) = f'(x) - \lambda\varphi(x) + \lambda\alpha e^{-\alpha x} Z \\ Z' = e^{\alpha y}\varphi(y) \end{cases} \quad \varphi(0) = 0;\ Z(0) = 0 \tag{2.59}$$

Consider now the larger interval $[a, b] = [0, 10]$. If inequality (1.35) does not hold then the strip method will be applied: the whole interval is divided in 10 equal subintervals.

Partitioning points are as follows: $x_0 = 0$; $x_1 = 1$; $x_{10} = 10$ Let us find the solution of Eq. (1.59) using POLYMATH software for (initial condition for subinterval "i" is the end point of the solution of subinterval "$i - 1$"):

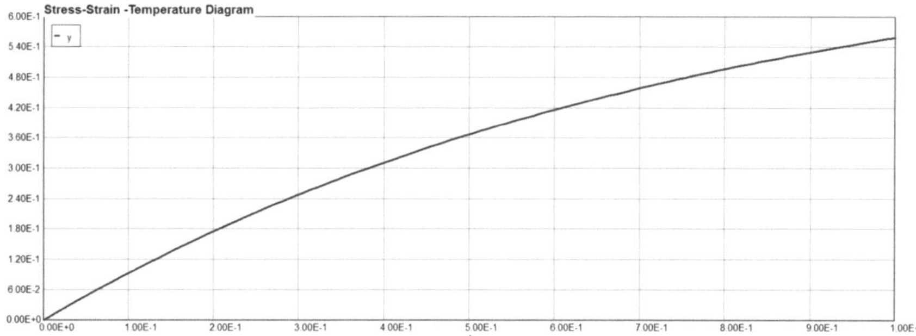

Fig. 2.1 Stress–Temperature diagram

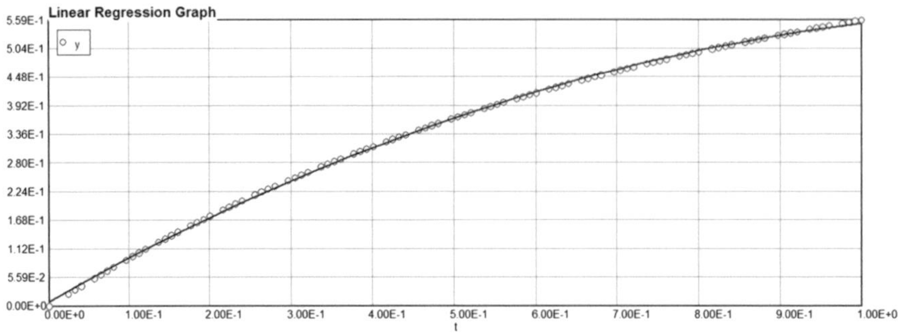

Fig. 2.2 Stress–Temperature (Strain) function

Subinterval [0, 1]

Calculated values of DEQ variables

	Variable	Initial value	Minimal value	Maximal value	Final value
1	t	0	0	1.0	1.0
3	$y1$	0	0	0.5594981	0.5594981
4	z	0	0	0.4189734	0.4189734

Differential equations

$$2 \quad d(y1)/d(t) = -1.0 * y1 + (\exp(-0.15 * t)) * 1.0 * (1 - 0.15 * t) + z * (\exp(-0.33 * t)) * 0.33 * 1 + 0$$

$$3 \quad d(z)/d(t) = y1 * (\exp(0.33 * t)) + 0$$

Note: One can see that methods 1 and 2 obviously provide identical results for function φ.

Subinterval [1, 2]

Calculated values of DEQ variables

	Variable	Initial value	Minimal value	Maximal value	Final value
1	t	1.0	1.0	2.0	2.0
3	$y1$	0.56	0.56	0.7081798	0.7081798
4	z	0.42	0.42	1.498877	1.498877

Differential equations

2 $d(y1)/d(t) = -1.0 * y1 + (\exp(-0.15 * t)) * 1.0 * (1 - 0.15 * t) + z * (\exp(-0.33 * t)) * 0.33 * 1$

3 $d(z)/d(t) = y1 * (\exp(0.33 * t))$

Subinterval [2, 3]

Calculated values of DEQ variables

	Variable	Initial value	Minimal value	Maximal value	Final value
1	t	2.0	2.0	3.0	3.0
3	$y1$	0.71	0.71	0.7376851	0.7376851
4	z	1.5	1.5	3.172535	3.172535

Differential equations

2 $d(y1)/d(t) = -1.0 * y1 + (\exp(-0.15 * t)) * 1.0 * (1 - 0.15 * t) + z * (\exp(-0.33 * t)) * 0.33 * 1$

3 $d(z)/d(t) = y1 * (\exp(0.33 * t))$

Subinterval [3, 4]

Calculated values of DEQ variables

	Variable	Initial value	Minimal value	Maximal value	Final value
1	t	3.0	3.0	4.0	4.0
3	$y1$	0.74	0.7293919	0.7401333	0.7293919
4	z	3.2	3.2	5.547809	5.547809

Differential equations

2 $d(y1)/d(t) = -1.0 * y1 + (\exp(-0.15 * t)) * 1.0 * (1 - 0.15 * t) + z * (\exp(-0.33 * t)) * 0.33$

3 $d(z)/d(t) = y1 * (\exp(0.33 * t))$

Subinterval [4, 5]

Calculated values of DEQ variables

	Variable	Initial value	Minimal value	Maximal value	Final value
1	t	4.0	4.0	5.0	5.0
3	$y1$	0.74	0.7049854	0.74	0.7049854
4	z	5.5	5.5	8.700056	8.700056

Differential equations

2 $d(y1)/d(t) = -1.0 * y1 + (\exp(-0.15 * t)) * 1.0 * (1 - 0.15 * t)$
$+ z * (\exp(-0.33 * t)) * 0.33 * 1$
3 $d(z)/d(t) = y1 * (\exp(0.33 * t))$

Subinterval [5, 6]

Calculated values of DEQ variables

	Variable	Initial value	Minimal value	Maximal value	Final value
1	t	5.0	5.0	6.0	6.0
3	$y1$	0.71	0.6703445	0.71	0.6703445
4	z	8.7	8.7	12.95057	12.95057

Differential equations

2 $d(y1)/d(t) = -1.0 * y1 + (\exp(-0.15 * t)) * 1.0 * (1 - 0.15 * t)$
$+ z * (\exp(-0.33 * t)) * 0.33 * 1$
3 $d(z)/d(t) = y1 * (\exp(0.33 * t))$

Subinterval [6, 7]

Calculated values of DEQ variables

	Variable	Initial value	Minimal value	Maximal value	Final value
1	t	6.0	6.0	7.0	7.0
3	$y1$	0.67	0.6311913	0.67	0.6311913
4	z	13.0	13.0	18.57578	18.57578

Differential equations

2 $d(y1)/d(t) = -1.0 * y1 + (\exp(-0.15 * t)) * 1.0 * (1 - 0.15 * t)$
$+ z * (\exp(-0.33 * t)) * 0.33 * 1$
3 $d(z)/d(t) = y1 * (\exp(0.33 * t))$

Subinterval [7, 8]

Calculated values of DEQ variables

	Variable	Initial value	Minimal value	Maximal value	Final value
1	t	7.0	7.0	8.0	8.0
3	$y1$	0.63	0.5900709	0.63	0.5900709
4	z	18.6	18.6	25.87087	25.87087

Differential equations

2 $d(y1)/d(t) = -1.0 * y1 + (\exp(-0.15 * t)) * 1.0 * (1 - 0.15 * t) + z * (\exp(-0.33 * t)) * 0.33 * 1$

3 $d(z)/d(t) = y1 * (\exp(0.33 * t))$

Subinterval [8, 9]

Calculated values of DEQ variables

	Variable	Initial value	Minimal value	Maximal value	Final value
1	*t*	8.0	8.0	9.0	9.0
3	*y*1	0.6	0.5534172	0.6	0.5534172
4	*z*	25.9	25.9	35.44505	35.44505

Differential equations

2 $d(y1)/d(t) = -1.0 * y1 + (\exp(-0.15 * t)) * 1.0 * (1 - 0.15 * t) + z * (\exp(-0.33 * t)) * 0.33 * 1$

3 $d(z)/d(t) = y1 * (\exp(0.33 * t))$

Subinterval [9, 10]

Calculated values of DEQ variables

	Variable	Initial value	Minimal value	Maximal value	Final value
1	*t*	9.0	9.0	10.0	10.0
3	*y*1	0.55	0.5087269	0.55	0.5087269
4	*z*	35.4	35.4	47.59758	47.59758

Differential equations

2 $d(y1)/d(t) = -1.0 * y1 + (\exp(-0.15 * t)) * 1.0 * (1 - 0.15 * t) + z * (\exp(-0.33 * t)) * 0.33 * 1$

3 $d(z)/d(t) = y1 * (\exp(0.33 * t))$

Let us check if inequality (1.35) holds for the whole interval [0, 10]:

$$|\lambda| = 1 < \frac{1}{|e^{-0.33x}|\sqrt{\int\limits_0^{10} [0.33(e^{-(10-x)})]^2 dx}} = \frac{1}{\sqrt{00.1648}} = 2.46$$

Table 2.2 Stress–Temperature data

θ	0	1	2	3	4	5	6	7	8	9	10
φ	0	0.56	0.71	0.74	0.73	0.71	0.67	0.63	0.59	0.55	0.51
y	0	0.56	0.71	0.74	0.73	0.70	0.66	0.63	0.58	0.54	0.50

Therefore, the solution can be found in this case ($\alpha = 0.33$) without subdividing the whole interval in ten subintervals. However, on the other hand it proves that this form of strip method application works properly. The comparison results are provided in Table 2.2.

Calculated values of DEQ variables

	Variable	Initial value	Minimal value	Maximal value	Final value
1	t	0	0	10.0	10.0
2	y	0	0	0.7364601	0.5006154

Differential equations

1 $\mathrm{d}(y)/\mathrm{d}(t) = -1.33 * y + (\exp(-0.15 * t)) * 1.0 * (1 - 0.15 * t) + t * (\exp(-0.15 * t)) * 0.33$

2 $\mathrm{d}(z)/\mathrm{d}(t) = y1 * (\exp(0.33 * t))$

See Fig. 2.3.

If Eq. (1.55) has a form: $\varphi'(x) = f'(x) - \lambda n\varphi(x) + \lambda \sum_{i=1}^{n} \alpha_i \int_0^x \mathrm{e}^{-\alpha_i(x-y)}\varphi(y)\mathrm{d}y$ then

$$\begin{cases} \varphi'(x) = f'(x) - \lambda n\varphi(x) + \lambda \sum\limits_{i=1}^{n} \alpha_i Z_i \\ Z_i' = \mathrm{e}^{\alpha_i x}\varphi(x) \end{cases} \tag{2.60}$$
$$\varphi(0) = 0; Z_i(0) = 0;$$

Example 2.3 Data: Consider a Volterra IE of the second kind

$$l\varphi(x) = x - \int_0^x (x - t)\varphi(t)\mathrm{d}t 0 \le x \le 1$$

$$\text{Exact solution} \to \varphi(x) = \sin x$$

The exact solution is easy to obtain by differentiating the integral equation above twice. We have now: $\varphi''(x) + \varphi(x) = 0$. Initial conditions: $\varphi(0) = 0; \varphi'(0) = 1$. Solution $\varphi(x) = \sin x$.

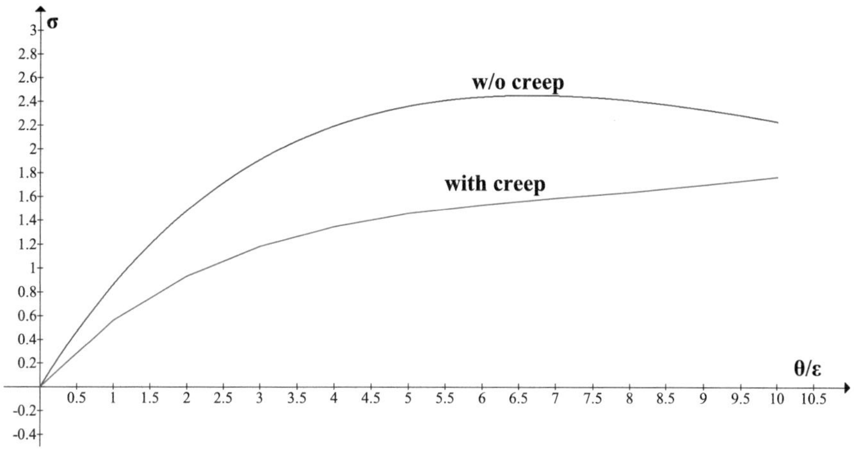

Fig. 2.3 Stress–Temperature (Strain) diagram

Method #1 *Gauss quadrature sum: n = 2*

K(x, t) = x-t f(x) = x; λ = -1; x1 = 0.2113; x_2 = 0.7887; $K_{1,1}$ = 1; $K_{1,2}$ = 0; $K_{2,1}$ = 0.2887; $K_{2,2}$ = 1; f_1 = 0.2113; f_2 = 0.7887;

$$\varphi(x_1) = \frac{\begin{vmatrix} 00.2113 & 0 \\ 0.7887 & 1 \end{vmatrix}}{\begin{vmatrix} 1 & 0 \\ 0.2887 & 1 \end{vmatrix}} = 0.2113; \quad \varphi(x_2) = \frac{\begin{vmatrix} 1 & 0.2113 \\ 0.2887 & 0.7887 \end{vmatrix}}{\begin{vmatrix} 1 & 0 \\ 0.2887 & 1 \end{vmatrix}} = 0.7277$$

$$\varphi(x) = x - \frac{1}{2}[(x - 0.2113)0.2113 + (x - 0.7887)0.7277] = 0.5x + 0.309$$

$$\varphi(x = 1) = 0.809; \quad \sin(1) = 0.841 \quad (\text{error: } 30.96\%)$$

See Fig. 2.4.

Method #2 *Galerkin Method*

$\varphi(x) = Ax + B$

$$\varphi_1 = x; \varphi_2 = 1$$

$$\begin{cases} \int_0^1 x(Ax + B)\mathrm{d}x + \int_0^1 \int_0^x t(At + B)(x - t)\mathrm{d}t\mathrm{d}x = \int_0^1 x^2 \mathrm{d}x \\ \int_0^1 (Ax + B)\mathrm{d}x + \int_0^1 \int_0^x (At + B)(x - t)\mathrm{d}t\mathrm{d}x = \int_0^1 x\mathrm{d}x \end{cases}$$

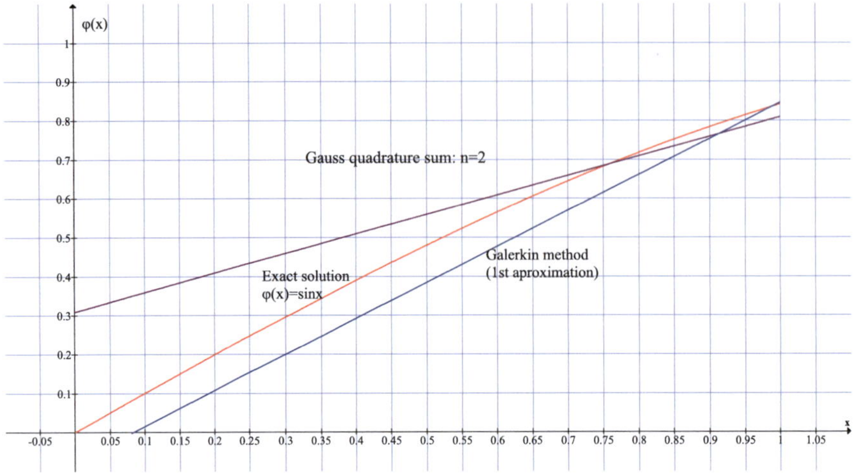

Fig. 2.4 Numerical methods comparison (Gauss and Galerkin)

$5A + 8B = 4$
$4A + 9B = 3$
$A = 12/13;\ B = -1/13$
$\varphi(x = 1) = 0.846\ \sin(1) = 0.841$ (error: 0.6 %)

Method #3 *Successive approximation (Sequential approximation method)*

1. $\varphi_0 = x;\ \lambda = -1$

$$\varphi_1 = \int_0^x (x-t)t\mathrm{d}t = \frac{x^3}{3!}$$

$$\varphi_2 = \int_0^x (x-t)\frac{t^3}{3!}\mathrm{d}t = \frac{x^5}{5!}$$

$$\ldots$$

$$\varphi(x) = \varphi_0 + \lambda\varphi_1 + \lambda^2\varphi_2 + \ldots = x - \frac{x^3}{3!} + \frac{x^5}{5!} - \ldots = \sin x$$

Example 2.4 Data: Consider a Volterra IE of the second kind

$$\varphi(x) = \mathrm{e}^x + \int_0^x \varphi(t)\mathrm{d}t \quad 0 \le x \le 1$$

$$\text{Exact solution} \rightarrow \varphi(x) = \mathrm{e}^x(1+x)$$

Successive approximation (Sequential approximation method)

1. $\varphi_0 = e^x$; $\lambda = 1$

$$\varphi_1 = \int_0^x e^t dt = e^x - 1$$

$$\varphi_2 = \int_0^x (e^t - 1)dt = e^x - 1 - x$$

$$\varphi_3 = \int_0^x (e^t - 1 - x)dt = e^x - 1 - x - \frac{x^2}{2}$$

...

$$\varphi_n(x) = \varphi_0 + \lambda\varphi_1 + \lambda^2\varphi_2 + \ldots = e^x + e^x - n - (n-1)x - \sum_{k=3}^{n} \frac{x^{k-1}}{k-1}$$

$$\varphi_3(x = 1) = 4e - 3 - 2 - \frac{1}{2} = 5.373$$

$$\varphi(x = 1) = e^x(1 + x) = 5.436 \approx 5.373(1.2\%)$$

Example 2.5 Consider the following nonlinear Volterra integral equation (with the exact solution $x(t) = \sin t$.)

$$x(t) - \frac{1}{2}\int_0^t x^2(s)ds = \sin t + \frac{1}{8}\sin 2t - \frac{1}{4}t; \quad 0 < t < 10$$

Inequality (1.35) is $|-1/2| < 1$. Therefore: $0 < t < \infty$.
Solution:
Differentiating the integral equation above, we obtain

$$x'(t) = (\cos t) + \frac{1}{4}(\cos 2t) - \frac{1}{4} + \frac{1}{2}x^2(t); \quad x(0) = 0$$

Differential equations

$d(u)/d(t) = \cos(t) + 0.25 * \cos(2 * t) - 0.25 + 0.5 * u^2$

Calculated values of DEQ variables

	Variable	Initial value	Minimal value	Maximal value	Final value
1	t	0	0	10.0	10.0
2	u	0	−0.9977534	0.9989641	−0.5440211

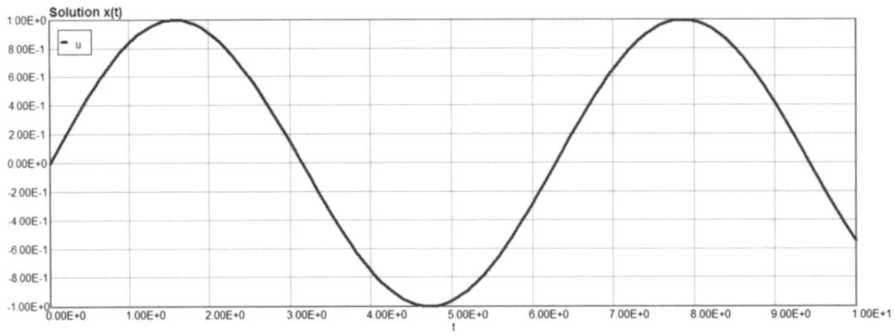

Fig. 2.5 Solution of nonlinear Volterra integral equation

See Fig. 2.5.

Example 2.6 Data: Consider the linear Volterra integral equation. The analytical solution of the above problem is given by $u(x) = \exp(-x)$

$$u(x) = \cos x - \sin x + 2\int_0^x [\sin(x-t)]u(t)\mathrm{d}t; \quad 0<x<1$$

Solution:

Differentiating the integral equation above we obtain

$$u'(x) = -\sin x - \cos x + 2(\cos x)Z_1 + 2(\sin x)Z_2$$

$$\frac{\mathrm{d}Z_1}{\mathrm{d}x} = (\cos x)u; \ \frac{\mathrm{d}Z_2}{\mathrm{d}x} = (\sin x)u; \quad u(0) = 1; \ Z_1(0) = Z_2(0) = 0$$

Using POLYMATH software for ODE system above, we have

Calculated values of DEQ variables

	Variable	Initial value	Minimal value	Maximal value	Final value
1	t	0	0	10.0	10.0
2	u	1.0	6.737E-05	1.0	6.737E-05

Differential equations

1 $\mathrm{d}(u)/\mathrm{d}(t) = -\sin(t) - \cos(t) + 2 * (\cos(t)) * z3 + 2 * (\sin(t)) * z4$

2 $\mathrm{d}(z1)/\mathrm{d}(t) = (\cos(t)) * u$

3 $\mathrm{d}(z2)/\mathrm{d}(t) = (\sin(t)) * u$

Calculated values of DEQ variables

	Variable	Initial value	Minimal value	Maximal value	Final value
1	*t*	0	0	1.0	1.0
2	*u*	1.0	0.3678794	1.0	0.3678794

The analytical solution $u(x = 1) = e^{-1} = 0.36788$

Example 2.7 Data: Consider the nonlinear Fredholm integral equation.

$$u(x) = \frac{7}{8}x + \frac{1}{2}\int_0^1 \mathrm{xtu}^2(t)\mathrm{d}t$$

The analytical solution of the above integral equation is given by $u(x) = x$.

$$\tilde{u}(x) = \frac{7}{8}x + Ax; \quad a(x) = x; \quad b(t) = t;$$

$$Ax = \frac{x}{2}\int_0^1 t\left[\frac{7}{8}t + At\right]^2 \mathrm{d}t; \quad A = \frac{1}{2}\int_0^1 t^3\left[\frac{7}{8} + A\right]^2 \mathrm{d}t = \frac{\left[\frac{7}{8} + A\right]^2}{8}$$

$$A^2 - 6.25A + 0.765 = 0; \quad A = 0.125; \quad \tilde{u} = 1.0x = 1.0x$$

Example 2.8 Data: Consider the nonlinear Fredholm integral equation.

$$\varphi(x) - \frac{1}{2}\int_0^1 e^{xs}\varphi(s)\mathrm{d}s = 1 - \frac{1}{2x}(e^x - 1)$$

The analytical solution of the problem is given by $\varphi(x) = 1$. Let us apply the Gauss quadrature formula for the approximate solution of integral equation above. Replacing the integral equation to a system of algebraic equations for $n = 2$, and taking into account that there are $\lambda = 0.5$, $A_1 = A_2 = 0.5$, we obtain the system (2.14) in this case in the form of

$$\left(1 - \frac{1}{4}K_{1,1}\right)\tilde{\varphi}(x_1) - \frac{1}{4}K_{1,2}\tilde{\varphi}(x_2) = f_1$$

$$-\frac{1}{4}K_{2,1}\tilde{\varphi}(x_1) + \left(1 - \frac{1}{4}K_{2,2}\right)\tilde{\varphi}(x_2) = f_2 \ldots$$

According to the above, for x_1 and x_2 are taken Gauss abscissas for the interval [0, 1],

$x_1 = 0.2113$, $x_2 = 0.7887$. Calculating the value of $K_{\iota,\kappa} = K(x_i, x_k)$ and $f_i = f(x_i)$ and substituting them into the system of Eq. (2.14), we obtain.

$$\begin{cases} 0.7386\tilde{\varphi}(x_1) - 0.2954\tilde{\varphi}(x_2) = 0.4434 \\ -0.2954\tilde{\varphi}(x_1) + 0.5343\tilde{\varphi}(x_2) = 0.2384 \end{cases}$$
$$\tilde{\varphi}(x_1) = 0.9997; \quad \tilde{\varphi}(x_2) = 0.9990$$

Therefore, an approximate solution of Eq. (2.14) in any other points in the interval (0.1) is given by the following equation

$$\tilde{\varphi}(x_1) = \frac{1}{4}\left(e^{0.2113x}0.9997 + 0.999e^{0.7887x}\right) + 1 - \frac{1}{2x}(e^x - 1)$$

We can, for example, calculate the values: $\tilde{\varphi}(x = 0) = 0.997$; $\tilde{\varphi}(x = 1) = 0.9991$.

In other words, an approximate solution of almost coincides with the exact solution.

The following Examples 2.9, 2.10, and 2.11 have a twofold purpose. First, they are directly related to the problems of material creep at high temperatures, due to the fact that the creep function contains the Arrhenius law (exponentially increasing function of dimensionless temperature). Second, in dealing with these examples we use the numerical methods for solving Volterra equations of the second kind, which will be used in subsequent chapters of this book in solving practical engineering creep problems. At the same time, this chapter provides a qualitative assessment of solution of integral equations. For example, it appears that the Galerkin method provides a very conservative estimate of the solution (the lower bound).

Example 2.9 Data: The IE $\varphi(x) - \lambda\int_a^x K(x,s)\varphi(s)\mathrm{d}s = f(x)$ with the degenerate kernel $K(x,y) = \sum_{i=1}^n a_i(x)b_i(y)$ is:

$$K_N(\theta, s) = \sum_{n=1}^{N} a_n(\theta)b_n(s) \quad \alpha_1 = 0.33; \quad \alpha_2 = 0.99$$
$$a_n(\theta) = A_1(1 - e^{-\alpha_1\theta}) + A_2(1 - e^{-\alpha_2\theta})$$
$$b_n(s) = \left[\exp\left(\frac{s}{1 + 0.1s}\right)\right][e^{-s} + e^{-2s}]$$

Then the contracting-mapping principle implies that if $|\lambda| < \left(\int_a^b \int_a^b K^2(x,s)\right.$ $\left.\varphi(s)\mathrm{d}s\mathrm{d}x\right)^{-1/2}$. The solution of the degenerate integral Eq. (1.14) is uniformly convergent in this case, since

$$|\lambda = -1| < \left(\int_a^b \int_a^b \mathrm{K}^2(\mathrm{x},\mathrm{s})\varphi(\mathrm{s})\mathrm{d}s\mathrm{d}x\right)^{-1/2}$$

$$= \left(\int_0^1 \int_0^1 \sum_{i=1}^{2} \{(1-\mathrm{e}^{-\alpha_1 x}) + (1-\mathrm{e}^{-\alpha_2 x})\}\left\{\left[\exp\left(\frac{s}{1+0.1s}\right)\right][\mathrm{e}^{-s}+\mathrm{e}^{-2s}]\right\}\mathrm{d}s\mathrm{d}x\right)^{-1/2} = 3.34$$

The approximate solution of integral equation is as follows:

$$\tilde{\varphi}(x) + \int_0^x K_N(x,s)\tilde{\varphi}(s)\mathrm{d}s = f(x)$$

The corresponding dummy differential equations are as follows:

$$\frac{\mathrm{d}\psi_{11}(x)}{\mathrm{d}x} = a_1(x)b_1(x) = (1-\mathrm{e}^{-\alpha_1\theta})\left[\exp\left(\frac{x}{1+0.1x}\right)\right]\mathrm{e}^{-x}$$

$$\frac{\mathrm{d}\psi_{12}(x)}{\mathrm{d}x} = a_1(x)b_2(x) = (1-\mathrm{e}^{-\alpha_1\theta})\left[\exp\left(\frac{x}{1+0.1x}\right)\right]\mathrm{e}^{-2x}$$

$$\frac{\mathrm{d}\psi_{21}(x)}{\mathrm{d}x} = a_2(x)b_1(x) = (1-\mathrm{e}^{-\alpha_2\theta})\left[\exp\left(\frac{x}{1+0.1x}\right)\right]\mathrm{e}^{-x}$$

$$\frac{\mathrm{d}\psi_{22}(x)}{\mathrm{d}x} = a_2(x)b_2(x) = (1-\mathrm{e}^{-\alpha_2\theta})\left[\exp\left(\frac{x}{1+0.1x}\right)\right]\mathrm{e}^{-2x}$$

$$\frac{\mathrm{d}\vartheta_1(x)}{\mathrm{d}x} = f(x)b_1(x) = x\mathrm{e}^{-0.15x}\left[\exp\left(\frac{x}{1+0.1x}\right)\right]\mathrm{e}^{-x}$$

$$\frac{\mathrm{d}\vartheta_2(x)}{\mathrm{d}x} = f(x)b_2(x) = x\mathrm{e}^{-0.15x}\left[\exp\left(\frac{x}{1+0.1x}\right)\right]\mathrm{e}^{-2x}$$

Solutions are (using POLYMATH software):

Calculated values of DEQ variables

	Variable	Initial value	Minimal value	Maximal value	Final value
1	t	0	0	1.0	1.0
2	$y1 = \psi_{11}$	0	0	0.1417315	0.1417315
3	$y2 = \psi_{12}$	0	0	0.0760957	0.0760957
4	$y3 = \psi_{21}$	0	0	0.0760957	0.0760957

(continued)

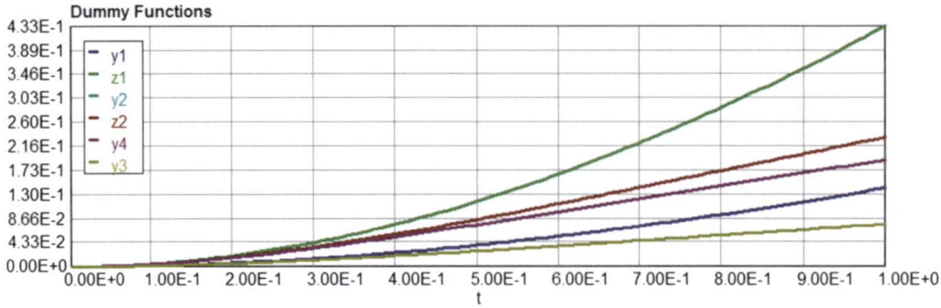

Fig. 2.6 Dummy equation

(continued)

	Variable	Initial value	Minimal value	Maximal value	Final value
5	$y4 = \psi_{22}$	0	0	0.1911314	0.1911314
6	$z1 = \vartheta_1(x)$	0	0	0.4327624	0.4327624
7	$z2 = \vartheta_2(x)$	0	0	0.232214	0.232214

Differential equations

1 $\mathrm{d}(y1)/\mathrm{d}(t) = (\exp((t)/(1.0+0.1*t)))*(\exp(-(1*t)))\hat{}1*(1-\exp(-0.33*(t-0)))*(t\hat{}0)$
2 $\mathrm{d}(z1)/\mathrm{d}(t) = (t\hat{}1)*(\exp(-1*(t)))\hat{}1*(\exp(-0.15*t))\hat{}1*(\exp((t)/(1.0+0.1*t)))$
3 $\mathrm{d}(y2)/\mathrm{d}(t) = (\exp((t)/(1.0+0.1*t)))*(\exp(-(2*t)))\hat{}1*(1-\exp(-0.33*(t-0)))*(t\hat{}0)$
4 $\mathrm{d}(z2)/\mathrm{d}(t) = (\exp((t)/(1.0+0.1*t)))*(t)*(\exp(-0.15*t))*(\exp(-2*t))$
5 $\mathrm{d}(y4)/\mathrm{d}(t) = (\exp((t)/(1.0+0.1*t)))*(\exp(-(2*t)))\hat{}1*(1-\exp(-0.99*(t-0)))$
6 $\mathrm{d}(y3)/\mathrm{d}(t) = (\exp((t)/(1.0+0.1*t)))*(\exp(-(2*t)))\hat{}1*(1-\exp(-0.33*(t-0)))*(t\hat{}0)$

See Fig. 2.6.

Model: $y1 = a * (1 - \exp(-1 * t))$

Variable	Initial guess	Value
a	1.0	0.153

Model: $y2 = a * (1 - \exp(-1 * t))$

Variable	Initial guess	Value
a	1.0	0.0934

Model: $y3 = a * (1 - \exp(-1 * \mathrm{t}))$

Variable	Initial guess	Value
a	1.0	0.0934

Model: $y4 = a * (1 - \exp(-1 * t))$

Variable	Initial guess	Value
a	1.0	0.243

Model: $z1 = a * (1 - \exp(-1 * t))$

Variable	Initial guess	Value
a	1.0	0.466

Model: $z2 = a * (1 - \exp(-1 * t))$

Variable	Initial guess	Value
a	1.0	0.285

$$f_1 = 0.466 \int_0^1 [1 - \mathrm{e}^{-0.33s}][1 - \mathrm{e}^{-s}]\mathrm{d}s = 0.0322 \quad \alpha_1 = 0.33; \quad \alpha_2 = 0.99$$

$$f_2 = 0.285 \int_0^1 [1 - \mathrm{e}^{-s}][1 - \mathrm{e}^{-0.99s}]\mathrm{d}s = 0.0476$$

$$b_{11} = 1 + 0.153 \int_0^1 A_1(1 - \mathrm{e}^{-0.33s})[1 - \mathrm{e}^{-s}]\mathrm{d}s = 1.01$$

$$b_{12} = 0.0934 \int_0^1 A_1(1 - \mathrm{e}^{-0.33s})[1 - \mathrm{e}^{-s}]\mathrm{d}s = 0.00646$$

$$b_{21} = 0.0934 \int_0^{10} A_2(1 - e^{-0.99s})[1 - \mathrm{e}^{-s}]\mathrm{d}s = 0.0156$$

$$b_{22} = 1 + 0.243 \int_0^{10} A_2(1 - \mathrm{e}^{-0.99s})[1 - \mathrm{e}^{-s}]\mathrm{d}s = 1.0406$$

1.01	0.00646	0.0322
0.0156	1.0406	0.378

Linear Equations Solution

	Variable	Value
1	$x1 = A_1$	0.0295606
2	$x2 = A_2$	0.3628088

The equations

[1] $1.01 \cdot x1 + 0.00646 \cdot x2 = 0.0322$

[2] $0.0156 \cdot x1 + 1.0406 \cdot x2 = 0.378$

$$\sigma_1(\theta) = f(\theta) - \left[0.0296(1 - \mathrm{e}^{-0.33\theta}) + 0.363(1 - \mathrm{e}^{-0.99\theta})\right]$$
$$\sigma_1(\theta = 1) = 0.86 - \left[0.0296(1 - \mathrm{e}^{-0.33}) + 0.363(1 - \mathrm{e}^{-0.99})\right] = 0.624$$
$$\sigma_{10}(\theta = 10) = 2.23 - \left[0.0296(1 - \mathrm{e}^{-3.3}) + 0.363(1 - \mathrm{e}^{-9.9})\right] = 1.838$$

Example 2.10 Data: Successive approximation ($b(s)$—Dirichlet series)

The IE $\varphi(x) - \lambda \int_a^x K(x,s)\varphi(s)\mathrm{d}s = f(x)$ with the degenerate kernel $K(x,y) = \sum_{i=1}^n a_i(x)b_i(y)$ is:

$$a(x) = (1 - \mathrm{e}^{-\alpha_1 x});\ b(s) = \mathrm{e}^{-\alpha_1 s} + \mathrm{e}^{-2\alpha_1 s} + \mathrm{e}^{-3\alpha_1 s} + \ldots \alpha_1 = 0.33$$
$$K_N(x,s) = (1 - \mathrm{e}^{-\alpha_1 x})(e^{-\alpha_1 s} + \mathrm{e}^{-2\alpha_1 s} + \mathrm{e}^{-3\alpha_1 s} + \ldots) = (1 - \mathrm{e}^{-\alpha_1 x})[(\mathrm{e}^{-\alpha_1 s})/(1 - \mathrm{e}^{-\alpha_1 s})]$$
$$\tilde{\varphi}(x) + \int_0^x K_N(x,s)\tilde{\varphi}(s)\mathrm{d}s = f(x) x \in [0,1] \alpha_1 = 0.33 \quad 0 < x < 1$$
$$\tilde{\varphi}_0(x) = f(x) = x\mathrm{e}^{-0.15x}$$
$$\tilde{\varphi}_1(x) = \int_0^x K_N(x,s)\tilde{\varphi}_0(s)\mathrm{d}s$$
$$= \int_0^x (1 - \mathrm{e}^{-\alpha_1 x})[\mathrm{e}^{-\alpha_1 s} + \mathrm{e}^{-\alpha_1 2s} + \mathrm{e}^{-\alpha_1 3s} + \ldots]\left[\exp\left(\frac{s}{1 + 0.1s}\right)\right] s\mathrm{e}^{-0.15s}\mathrm{d}s$$
$$\text{Dummy equation:} \frac{\mathrm{d}y}{\mathrm{d}s} = \left[\frac{\mathrm{e}^{-(0.15+\alpha_1)s}}{1 - \mathrm{e}^{-\alpha_1 s}}\right]\left[\exp\left(\frac{s}{1 + 0.1s}\right)\right][s]$$
$$\text{Solution} : \tilde{\varphi}_1 = (1 - \mathrm{e}^{-\alpha_1 x})(-0.211 + 40.2x)$$
$$\text{1st approximation } \varphi_1(x) = x\mathrm{e}^{-0.15x} - (1 - \mathrm{e}^{-\alpha_1 x})(-0.211 + 40.2x)$$
$$\max|\varphi_0(x) - \varphi_1(x)| = |\varphi_0(x = 1) - \varphi_1(x = 1)| = 0.26$$

Differential equations

$$8 \quad \mathrm{d}(y6)/\mathrm{d}(t) = (\exp((t)/(1.0 + 0.1 * t)))\hat{}1 * t * \exp(-(0.48 * t))\hat{}1/(1 - \exp(-(0.33 * t))\hat{}1)$$

Calculated values of DEQ variables

	Variable	Initial value	Minimal value	Maximal value	Final value
1	t	0.001	0.001	1.0	1.0
7	$y6$	0	0	4.188095	4.188095

Model: $y6 = a0 + a1 * \mathrm{t}$ $\varphi_{11} = -0.211 + 4.2x$

Variable	Value
$a0$	−0.2109213
$a1$	4.200333

2nd approximation:

$$\tilde{\varphi}_2(x) = \int_0^x K_N(x,s)\tilde{\varphi}_1(s)\mathrm{d}s$$

$$= (1 - \mathrm{e}^{-\alpha_1 x}) \int_0^x \left[\exp\left(\frac{s}{1+0.1s}\right)\right]\left[\frac{\mathrm{e}^{-(\alpha_1)s}}{1-\mathrm{e}^{-\alpha_1 s}}\right](1-\mathrm{e}^{-\alpha_1 s})(-0.211+4.2\mathrm{s})\mathrm{d}s$$

$$\text{Dummy equation:} \frac{\mathrm{d}y}{\mathrm{d}s} = \left[\frac{\mathrm{e}^{-(\alpha_1)s}}{1-\mathrm{e}^{-\alpha_1 s}}\right]\left[\exp\left(\frac{s}{1+0.1s}\right)\right](1-\mathrm{e}^{-\alpha_1 s})(-0.211+40.2\mathrm{s})$$

$$\text{Solution : } \tilde{\varphi}_2 = (1-\mathrm{e}^{-\alpha_1 x})(0.04\text{–}0.676x + 3.183x^2)$$

$$\varphi_{\mathrm{tot}}(x) = x\mathrm{e}^{-0.15x} - (1-\mathrm{e}^{-\alpha_1 x})(-0.211+40.2x) + (1-\mathrm{e}^{-\alpha_1 x})(0.04\text{–}0.676x+3.183x^2)$$

$$= x\mathrm{e}^{-0.15x} - (1-\mathrm{e}^{-\alpha_1 x})(-0.171+30.524x-3.183x^2)$$

$$max|\varphi_1(x) - \varphi_2(x)| = |\varphi_1(x=1) - \varphi_2(x=1)| = 0.0478 \rightarrow \text{small}$$

$$\varphi_{\mathrm{tot}}^{(2)}(x=1) = 0.812$$

See Fig. 2.7.

Model: $y5 = a0 + a1 * t + a2 * t^2$ $\varphi_{22} = 0.04 - 0.676x + 3.183x^2$

Variable	Value
$a0$	0.0405221
$a1$	−0.6765522
$a2$	3.182798

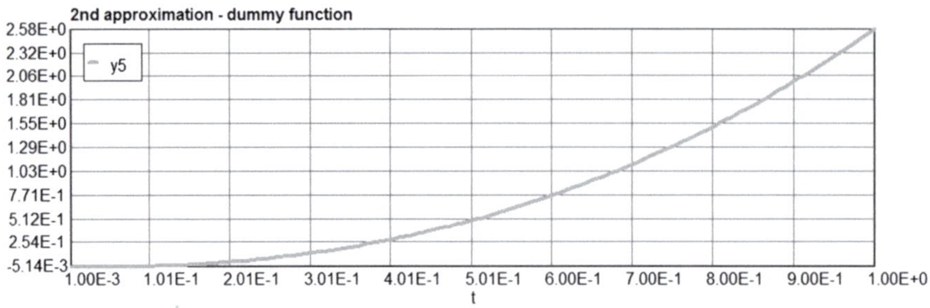

Fig. 2.7 Dummy equation (second approximation)

See Fig. 2.8.

Example 2.11 Data: Successive approximation (Spectra Method)

The IE $\varphi(x) - \lambda \int_a^x K(x,s)\varphi(s)\mathrm{d}s = f(x)$ with the degenerate kernel $K(x,y) = \sum_{i=1}^n a_i(x)b_i(y)$ is:

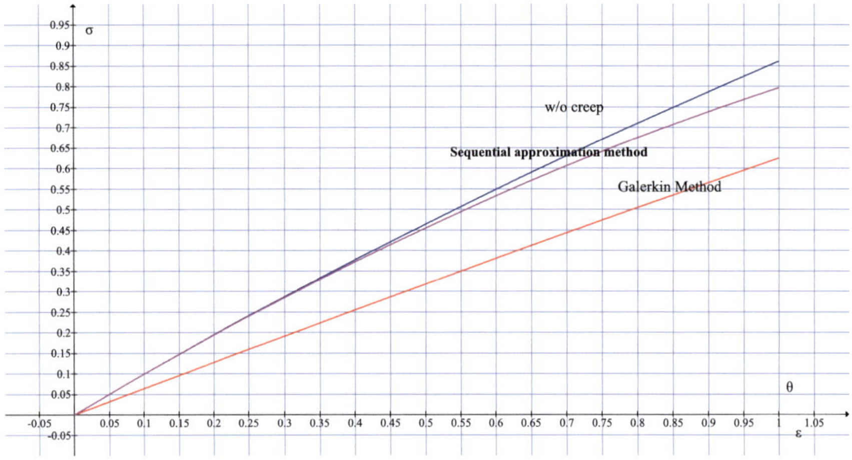

Fig. 2.8 Stress–Temperature/Strain diagram

$$\tilde{\varphi}(x) + \int_0^x K_N(x,s,\alpha)\tilde{\varphi}(s)\mathrm{d}s = f(x)x \in [0,1] 0 < \alpha < \infty$$

$$\tilde{\varphi}_0(x) = f(x) = x\mathrm{e}^{-0.15x}$$

$$\tilde{\varphi}_1(x) = \int_0^x K_N(x,s)\tilde{\varphi}_0(s)\mathrm{d}s = \int_0^x\int_0^\infty F(\alpha)K_N(x,s)\tilde{\varphi}_0(s)\mathrm{d}s\mathrm{d}\alpha$$

$$= \int_0^\infty \mathrm{e}^{-\alpha}(1-\mathrm{e}^{-\alpha x})\mathrm{d}\alpha \int_0^x [\mathrm{e}^{-s}+\mathrm{e}^{-2s}+\mathrm{e}^{-3s}+\ldots]\left[\exp\left(\frac{s}{1+0.1s}\right)\right]s\mathrm{e}^{-0.15s}\mathrm{d}s$$

$$\text{Dummy equation:} \frac{\mathrm{d}y}{\mathrm{d}s} = \left[\frac{\mathrm{e}^{-(1.15)s}}{1-\mathrm{e}^{-s}}\right]\left[\exp\left(\frac{s}{1+0.1s}\right)\right][s]$$

$$\text{Solution}: \tilde{\varphi}_1 = \int_0^\infty \mathrm{e}^{-\alpha}(1-\mathrm{e}^{-\alpha x})(-0.0244+10.147x)\mathrm{d}\alpha$$

$$\text{1st approximation } \varphi_1(x) = x\mathrm{e}^{-0.15x} - \frac{x}{x+1}(-0.0244+10.147x)$$

$$\varphi_{\mathrm{tot}}^{(1)}(x) = 0.86 - 0.56 = 0.3$$

2nd approximation:

$$\tilde{\varphi}_2(x) = \int_0^x K_N(x,s)\tilde{\varphi}_1(s)\mathrm{d}s$$

$$= \int_0^\infty \mathrm{e}^{-\alpha}(1-\mathrm{e}^{-\alpha x})\mathrm{d}\alpha \int_0^x \left[\exp\left(\frac{s}{1+0.1s}\right)\right]\left[\frac{\mathrm{e}^{-s}}{1-\mathrm{e}^{-s}}\right](-0.0244+1.147s)\frac{s}{s+1}\mathrm{d}s$$

$$\text{Dummy equation:} \frac{\mathrm{d}y}{\mathrm{d}s} = \left[\frac{\mathrm{e}^{-s}}{1-\mathrm{e}^{-s}}\right]\left[\exp\left(\frac{s}{1+0.1s}\right)\right]\left(\frac{s}{s+1}\right)(-0.0244+10.147\mathrm{s})$$

$$\text{Solution}: \tilde{\varphi}_2 = \left(\frac{x}{x+1}\right)(-0.00466+0.0338x+0.409x^2)$$

$$\varphi_{\mathrm{tot}}(x) = x\mathrm{e}^{-0.15x} - \left(\frac{x}{x+1}\right)[(-0.0244+10.147x)-(-0.00466+0.0338x+0.409x^2)]$$

$$= x\mathrm{e}^{-0.15x} - \left(\frac{x}{x+1}\right)(-0.02+10.113x-0.409x^2)$$

$$\varphi_{\mathrm{tot}}^{(2)}(x=1) = 0.518$$

Calculated values of DEQ variables

	Variable	Initial value	Minimal value	Maximal value	Final value
1	*t*	0.001	0.001	1.	1.
4	*y*3	0	0	1.140278	1.140278

Differential equations

6 $\mathrm{d}(y3)/\mathrm{d}(t) = t * (\exp((t)/(1.0+0.1*t))) * (\exp(-(1.15*t)))^\wedge 1/(1-\exp(-1*(t-0)))$

Model: $y3 = a0 + a1 * t$

Variable	Value
$a0$	−0.0244
$a1$	1.147

2nd approximation:

Calculated values of DEQ variables

	Variable	Initial value	Minimal value	Maximal value	Final value
1	t	0.001	0.001	1.0	1.0
	$y5$	0	−0.0002337	0.4346849	0.4346849

Differential equations

7 $d(y5)/d(t) = ((\exp(t/(1.0 + 0.1 * t)))) * (t/(1 + t)) * (\exp(-(1 * t))) * (-0.0244 + 1.147 * t) * (1/(1 - (\exp(-(1 * t)))))$

See Fig. 2.9.

Model: $y5 = a0 + a1 * t + a2 * t\hat{}2$

Variable	Value
$a0$	−0.00466
$a1$	0.0338
$a2$	0.409

$$\tilde{\varphi}(x) + \int_0^x K_N(x, s, \alpha)\tilde{\varphi}(s)ds = f(x)x \in [0, 1] \quad 0 < \alpha < \infty$$

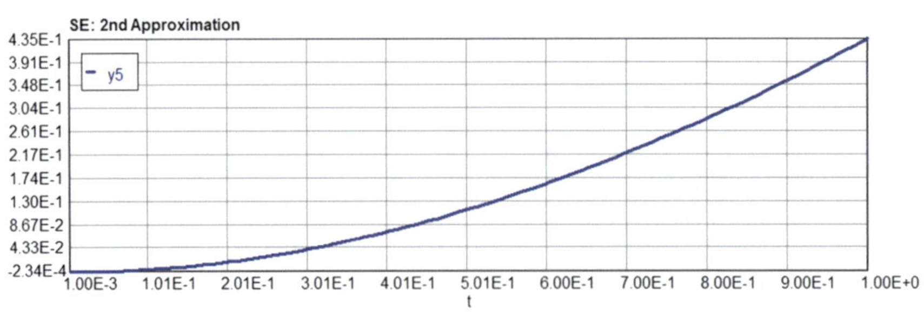

Fig. 2.9 Dummy Equation (second approximation—successive approximation)

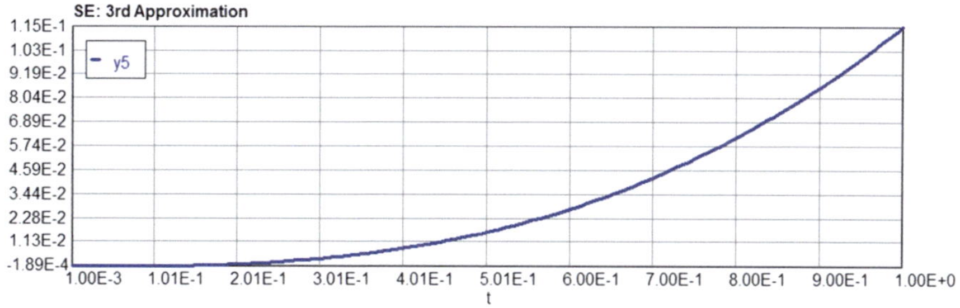

Fig. 2.10 Figure 1.9 Dummy Equation (3nd approximation—successive approximation)

3rd approximation:

$$\tilde{\varphi}_3(x) = \int_0^x K_N(x,s)\tilde{\varphi}_2(s)\mathrm{d}s =$$

$$\text{Dummy equation:} \frac{\mathrm{d}y}{\mathrm{d}s} = [\frac{\mathrm{e}^{-s}}{1-\mathrm{e}^{-s}}][\exp(\frac{s}{1+0.1s})](\frac{s}{s+1})(-0.00466+0.0338x+0.409x^2)$$

$$\text{Solution}: \tilde{\varphi}_3 = (\frac{x}{x+1})(0.00466-0.0604x+0.167x^2)$$

$$\varphi_{\text{tot}}(\mathrm{x}) = x\mathrm{e}^{-0.15x} - (\frac{x}{x+1})[(-0.02+1.113x-0.409x^2)+(0.00466-0.0604x+0.167x^2)] =$$

$$= x\mathrm{e}^{-0.15x} - (\frac{x}{x+1})(-0.0153+1.05x-0.242x^2)$$

$$\varphi_{\text{tot}}^{(3)}(x=1) = 0.464 \approx 0.518(110.6\%)$$

Calculated values of DEQ variables

	Variable	Initial value	Minimal value	Maximal value	Final value
1	t	0.001	0.001	1.0	1.0
	$y5$	0	−0.0001892	0.1149818	0.1149818

Differential equations

7 $\mathrm{d}(y5)/\mathrm{d}(t) = ((\exp(t/(1.0+0.1*t)))) * (t/(1+t)) * (\exp(-(1*t))) * (-0.00466+0.0338*t+0.409*t^\wedge 2) * (1/(1-(\exp(-(1*t)))))$

See Fig. 2.10.

Model: $y5 = a0 + a1 * t + a2 * t$^2

Variable	Value
$a0$	0.00466
$a1$	−0.0604
$a2$	0.167

References

1. Atkinson KE (1976) A survey of numerical methods for the solution of Fredholm integral equations of the second kind, SIAM
2. Baker CTH (1977) The numerical treatment of integral equations, Oxford University Press, Chap. 4
3. Hammerstein A (1930) Nichtlineare Integralgleichungen nebst Anwendungen. Acta Math 54:117–176
4. Brunner H, van der Houwen PJ (1986) The numerical solution of Volterra equations CWI monographs. North-Holland, Amsterdam
5. Volterra V (1896) "Sulla inversione degli integrali definiti" *Rend Accad Lincei* 5, 177–185, 289–300
6. Volterra V (1897) Sopra alcune questioni di inversione di integrali definiti. Ann di Math 25 (2):139–187
7. Abramowitz M, Stegun IA (eds) (1965) Chapter 25.4, Integration, Handbook of mathematical functions with formulas, graphs, and mathematical tables, Dover, New York, ISBN: 978-0486612720, MR: 0167642
8. Petrovskii IG (1957) Lectures on the theory of integral equations, Graylock
9. Kantorovich LV, Krylov VI (1958) Approximate methods of higher analysis, Noordhoff (Translated from Russian)
10. Bana's J, Regan DO (2008) On existence and local attractivity of solutions of a quadratic Volterra integral equation of fractional order. J Math Anal Appl 34:573–582
11. Miller RK, Nohel JA, Wong JSW (1969) A stability theorem for nonlinear mixed integral equations. J Math Anal Appl 25:446–449
12. Sarkar K, Djordjevic AR, Arvas E (1985) On the choice of expansion and weighting functions in the numerical solution of operator equations. IEEE Trans Antennas Propagat AP-33:988–996
13. V.I. Smirnov, "A course of higher mathematics", **4**, Addison-Wesley (1964) (Translated from Russian)
14. Cronin J (1964) Fixed points and topological degree in nonlinear analysis, Amer Math Soc
15. Kendall E, Atkinson KE (1997) The numerical solution of integral equations of the second kind. Cambridge monographs on applied and computational mathematics
16. Hammerstein A (1930) Nichtlineare Integralgleichungen nebst Anwendungen. Acta Math 54:117–176

Chapter 3
Phenomenological Time Invariant Creep Models

Notations

k	The thermal conductivity that has the dimensions W/m K or J/m s K
T	Temperature in K
Q	Activation energy
R	Ideal gas constant
T_o	Ambient temperature
t	Time
$t^* = t_r$	Creep rupture (failure) time
$\vec{v}(u; v; w)$	Velocity vector
D	Diffusion coefficient (m^2/sec)
p	Pressure
ν	Kinematic viscosity $\nu = \mu/\rho$
θ	Dimensionless temperature
τ	Dimensionless time
h	Height of the compartment (m)
a	Thermal diffusivity (m^2/sec)
Time	$t = \frac{h^2}{a}\tau$ (sec)
Temperature	$T = \frac{RT_*^2}{E}\theta + T_*$ (K), where $T_* = 600$°K is the base line temperature
Coordinates	$\bar{x} = x/h$ and $\bar{z} = z/h$—"x" and "z"—dimensionless coordinates
Velocities	$\bar{u} = \frac{\nu}{h}u$ (m/sec) and $\bar{w} = \frac{\nu}{h}w$ (m/sec)—horizontal and vertical components velocity accordingly; ν—kinematic viscosity (m^2/sec); "u" and "w"—dimensionless velocities
$J(t, t')$	Compliance function (often also called the creep function)
T_M	Melting point of the metal
$\varepsilon(t)$	Strain
$\sigma(t)$	Stress
$\overline{\sigma}(t) = E(t)\varepsilon(t)$	Instantaneous (elastic) stress
σ_{eq}	Equivalent stress

L. Razdolsky, *Probability Based High Temperature Engineering*,
DOI 10.1007/978-3-319-41909-1_3

σ^*	Ultimate stress
ε_e	Instantaneous (elastic) strain
ε_c	Creep strain
$\dot{\varepsilon}$	Strain rate
ϕ	Resistance factor
ε_T	Thermal expansion due to high temperature effect
$K(t, t') = -\partial J(t, t')/\partial t'$	Retardation function (memory function)
$R(t, t')$	Relaxation function (also called the relaxation modulus)
E	Modulus of Elasticity
E_{rel}	Relaxation Modulus
H	Long-term Modulus of Elasticity
α	Material Property Parameter (MPP)
t_r	Time to rupture
η	Viscosity parameter
$n = \eta/E$	Relaxation time
$f(\alpha)$	Retardation spectrum of the model
$\varphi(\alpha)$	Relaxation spectrum of the model

3.1 Introduction

All materials exhibit time-dependent behavior. The stress and strain induced when a load is applied are a function of time. In the most general form can be thought of as a 3-dimensional surface. The stress–strain–time relationship, or constitutive law, can be determined by loading a specimen with constant stress (creep) or constant strain (stress relaxation or isometric). We can also construct isochronous curve by cross section of the surface with constant time levels as shown in Fig. 3.1.

When a material is subjected to a constant load, it deforms continuously (see Fig. 3.2). The initial strain is roughly predicted by its stress–strain modulus. The material will continue to deform with time indefinitely or until rupture or yielding

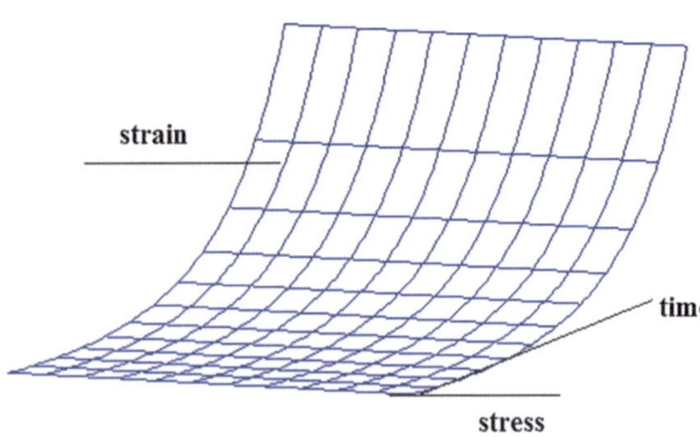

Fig. 3.1 Constant stress–strain–time coordinates

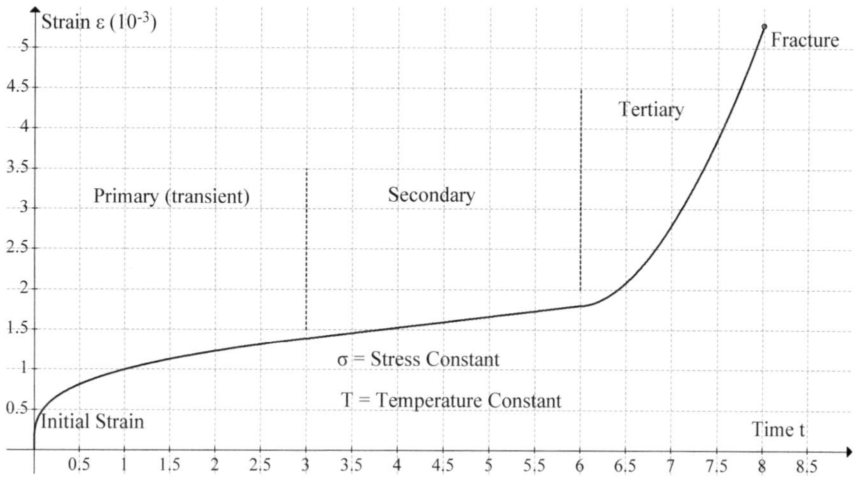

Fig. 3.2 Creep curve: constant load and temperature is applied

causes failure. The primary region is the early stage of loading when the creep rate decreases rapidly with time. Then it reaches a steady state, which is called the secondary creep stage followed by a rapid increase (tertiary stage) and fracture. This phenomenon of deformation under load with time is called creep. Of course, this is an idealized curve.

Some materials do not have secondary stage, while tertiary creep only occurs at high stresses and for ductile materials [1, 2]. All plastics creep to a certain extent. The degree of creep depends on several factors, such as type of material, magnitude of load, temperature, and time. The standard test method for creep characterization is ASTM D2990. In this test procedure, the dimensional changes that occur during time under a constant static load are measured.

If the applied load is released before the creep rupture occurs, an immediate elastic recovery equal to the elastic deformation, followed by a period of secondary recovery is observed (Fig. 3.3). The material in most cases does not recover to the original shape and a permanent deformation remains. The magnitude of the permanent deformation depends on length of time, amount of stress applied, and temperature.

The creep rupture is basically similar to a creep test with the exception that it is continued until the material fails. Since higher loads are used, creep rates are higher and the material fails in a shorter time. This test is useful in establishing a safe envelope inside which a creep test can be conducted. The basic information obtained from the stress rupture test is the time to failure at a given stress.

Based on this data, a safe stress can be determined below which it is safe to operate, given the time requirement of the end use application. The construction of the creep rupture envelope is shown in Fig. 3.4. Test is conducted under constant stresses and the points of the onset of tertiary stage are connected to form the creep rupture envelope [3].

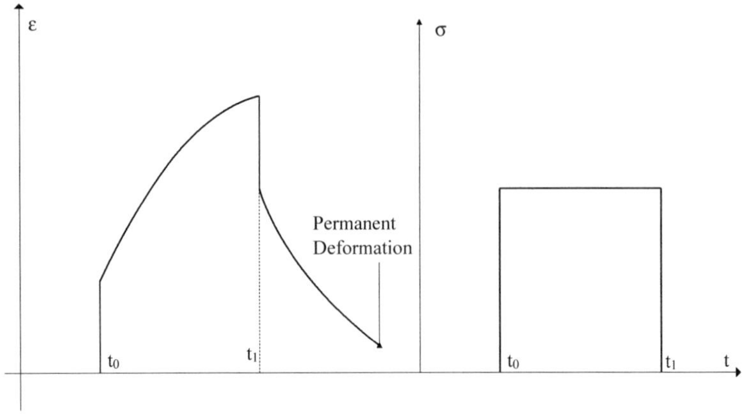

Fig. 3.3 Creep curve with recovery. A constant load is applied at t_0 and removed at t_1

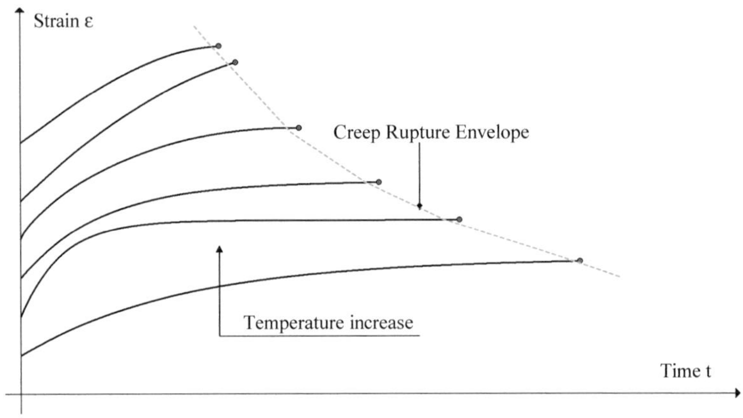

Fig. 3.4 Creep rupture envelope

3.1.1 Stress Relaxation, Constant Strain

Stress relaxation is defined as a gradual decrease in stress with time under a constant deformation or strain. This behavior is studied by applying a constant deformation to the specimen and measuring the stress required to maintain that strain as a function of time.

Stress relaxation test can be used for some practical applications. For example, low stress relaxation is desired for friction bolted connections. The stress data obtained from stress relaxation test can be used to calculate transient modulus by simply dividing the stress at a particular time by the applied strain. The 3-dimensional stress–strain–time relationships can also be constructed by stress–strain relaxation test as shown in Fig. 3.1. However, stress relaxation test is more difficult to perform than creep test and has limited practical applications. As a result, stress relaxation test D2991 is dropped by ASTM in 1992.

3.1.2 Limit State Design

Structural Design can be divided into two categories, design for strength and design for deformations. The strength of a component is limited by the yield strength and rupture resistance of the material from which it is made. As shown in Fig. 3.4, a creep rupture envelope can be obtained from creep test. For an expected life time, the maximum allowable stress can be decided from the creep rupture envelope line. Design for deflections with creep curves proceeds by establishing the maximum strain acceptable ε_{max}, thereby establishing a horizontal line on the creep diagram correspondingly.

Design basis selection depends on the specific application [4]. Usually, strain or dimension requirement is more critical, and design for deformations is favored in this case. If the dimension precision of the structural component under discussion is not so important compared as strength, design for strength is then used accordingly. For complicated structures, both can be used for design criteria to ensure successful material performance during service time.

Since the dead and live loads of the structural system are nearly constant over time, creep behavior, i.e., strain under transient high temperature load, is the main consideration for our structure design. Our goal is to get stress–temperature–strain–time relationship (constitutive law) and the creep rupture data for our structural design. Since the service time is not very long (for instance 4–6 h in case of fire event) property prediction methods should be used for obtaining constitutive law. There are several test methods for evaluating long-term properties of materials, including creep test ASTM-D2990, hydrostatic test D1598, D2837, and DMA (Dynamic Mechanical Analysis) [5–9]. In the following sections, constitutive law for creep behavior is described; methods for superposition of temperature–time data are presented.

3.1.3 Linear and Nonlinear Viscoelasticity

The time-dependent material behavior is often referred to as *viscoelasticity*. If a constant load σ_1 is applied to a viscoelastic specimen, the time-dependent strain is recorded as ε_1 as shown in Fig. 3.3. After some period of time, the load is removed. Suppose the specimen is allowed to recover and a larger stress σ_2 is applied and the strain ε_2 is recorded.

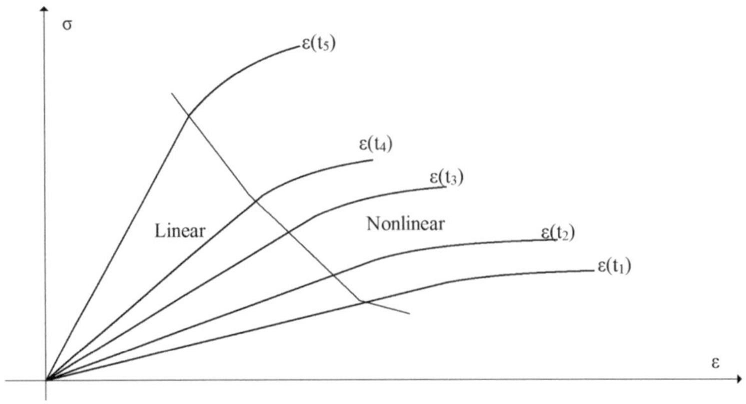

Fig. 3.5 Linear-nonlinear transition of stress–strain relationship (different time/temperature levels)

In general, for stress σ, the creep compliance $C(t)$ can be given as the ratio of strain to stress at a certain time.

$$C(t) = \varepsilon(t)/\sigma \tag{3.1}$$

This property is often characterized as *linear viscoelasticity*. In the linear range, the compliance is independent of stress, which means that the compliance is the same whether stresses used in the creep test are σ_1, σ_2 or other stress levels. A transition from linear to nonlinear viscoelasticity is shown in Fig. 3.5.

However, metal matrix materials (MMM) generally exhibit linear viscoelastic property at low stresses such that the corresponding strain is below $\sim 0.5\ (10^{-2})$. At higher stress levels, the material will assume *nonlinear viscoelastic* behaviors which will not obey the linear relation between stress and strain described by Eq. (3.1).

Since nonlinear behavior is very important to determine material behavior at higher temperature stress levels, some models have been suggested for different kind of materials [10–16].

3.2 Linear Viscoelastic Material: Constitutive Equations

Viscoelastic materials are those for which the relationship between stress and strain depends on time. Linear viscoelastic materials are those for which there is a linear relationship between stress and strain (at any given time). Linear viscoelasticity is a theory describing the behavior of such ideal materials. Linear viscoelasticity is a reasonable approximation to the time-dependent behavior of polymers, metals, and ceramics at relatively low temperatures and under relatively low stress.

3.2.1 Definitions

The deformation resulting from a given load will depend on the properties of the material. It may be reversible (elastic or recoverable deformation), or irreversible (viscous, plastic or permanent deformation, or flow), or it may comprise both a recoverable and a permanent part. We wish to express the behavior of a material in form of a *constitutive equation*, i.e., an equation which specifies the properties of the material in a manner which is independent of the size or shape (i.e., the geometry) of the body and depends only on its material nature. Constitutive equations are also referred to as (rheological) *equations of state*. There are five different stages of transformation of a material subjected to the transient temperature–time loading condition (see Fig. 3.6).

Five Regions of Viscoelastic Behavior:

A. *Glass*. The material is rigid, yet brittle if not reinforced by chemical cross-links or crystallites.
B. *Glass Transition Zone*. The material starts to become compliant over time (at constant temperature), or in a narrow temperature range. The glass transition temperature, T_g identifies this zone. The modulus of elasticity drops from 10^9 to 10^6 Pa. Onset of large scale, coordinated molecular motion during heating of a glass. Onset of lag between perturbation and materials response during cooling (departure from equilibrium state during cooling) is a single most important materials parameter for materials design. Different measurement techniques, different cooling, or other perturbation rates result in measurements of slightly different T_g's.
C. *Rubber*. The material is very flexible, capable of being stretched to several times its original dimensions without breaking. Commercial rubbers are chemically cross-linked to keep them from softening at elevated temperatures.

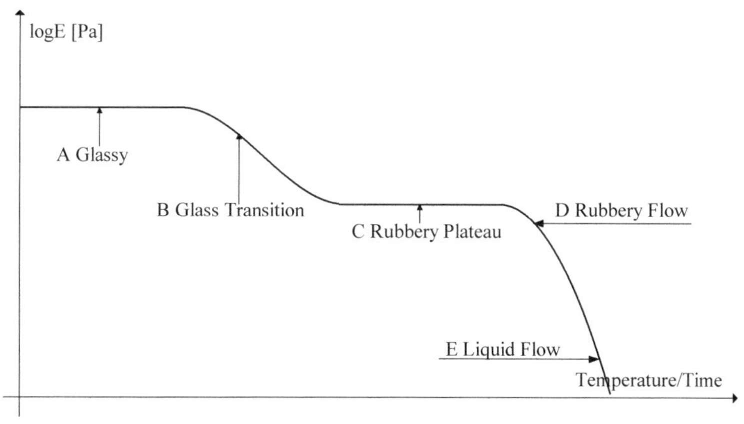

Fig. 3.6 Five regions of viscoelastic behavior

D. *Rubbery Flow Zone*. The material becomes tacky, and will spread like a liquid if stress is applied.
E. *Melting Zone*. As the material is heated beyond the rubbery flow zone its viscosity steadily decreases. As the viscosity (resistance to flow) decreases, so does the materials modulus. This is the processing region of the viscoelastic curve.

If the material is not cross-linked, the stiffness exhibits a short plateau due to the ability of molecular entanglements to act as network junctions; at still higher temperatures the entanglements slip and the material becomes a viscous liquid. Neither the glassy nor the rubbery modulus depends strongly on time, but in the vicinity of the transition near T_g time effects can be very important. Clearly, a plot of modulus versus temperature, such as is shown in Fig. 3.7, is a vital tool in materials science and engineering. It provides a map of a vital engineering property, and is also a fingerprint of the molecular motions available to the material.

As mentioned at the outset, temperature has a dramatic influence on rates of viscoelastic response, and in practical work it is often necessary to adjust a viscoelastic analysis for varying temperature. This strong dependence of temperature can also be useful in experimental characterization: if for instance a viscoelastic transition occurs too quickly at room temperature for easy measurement, the experimenter can lower the temperature to slow things down.

The ratio of strain to stress is called the "compliance" J, and in the case of time-varying strain arising from a constant stress the ratio is the "creep compliance":

$$J_{cr}(t) = \varepsilon(t)/\sigma_0. \tag{3.2}$$

A typical form of this function is shown in Fig. 3.8, plotted against the logarithm of time.

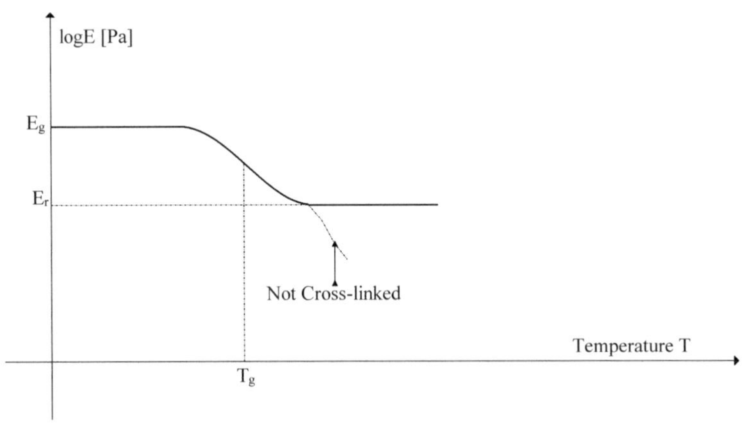

Fig. 3.7 Modulus of elasticity and temperature relationship

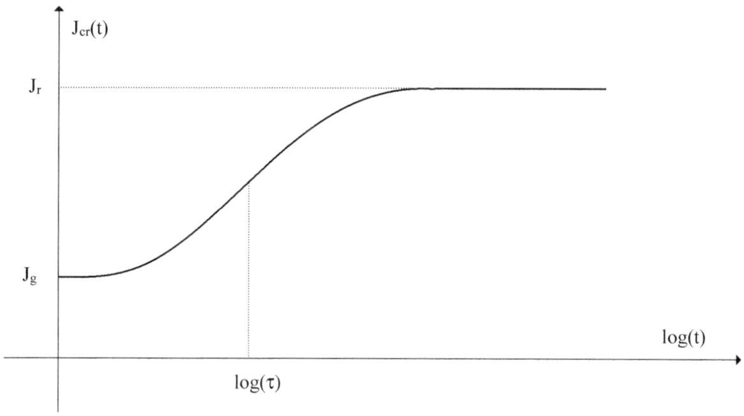

Fig. 3.8 Typical creep compliance function $J_{crp}(t)$

The logarithmic form of the plot changes the shape of the curve drastically, stretching out the short-time portion of the response and compressing the long-time region. Upon loading, the material strains initially to the "glassy" compliance J_g; this is the elastic deformation corresponding to bond distortion. In time, the compliance rises to equilibrium or "rubbery" value J_r, corresponding to the rubbery extension of the material. The value along the abscissa labeled "log τ" marks the inflection from rising to falling slope, and τ is called the "relaxation time" of the creep process.

Stress relaxation

Analogously with creep compliance, one may superimpose the relaxation curves by means of the "relaxation modulus," defined as

$$E_{\mathrm{rel}}(t) = \sigma(t)/\varepsilon_0, \tag{3.3}$$

This function is plotted against log time in Fig. 3.9. At short times, the stress is at a high plateau corresponding to a "glassy" modulus E_g, and then falls exponentially to a lower equilibrium "rubbery" modulus E_r as the molecules gradually accommodate the strain by conformational extension rather than bond distortion.

Creep and relaxation are both manifestations of the same molecular mechanisms, and one should expect that $E_{rel}(t)$ and $J_{cr}(t)$ are related. However, even though $E_g = 1/C_g$ and $E_r = 1/C_r$, in general $E_{rel}(t) \neq 1/J_{cr}(t)$. In particular, the relaxation response moves toward its equilibrium value more quickly than does the creep response.

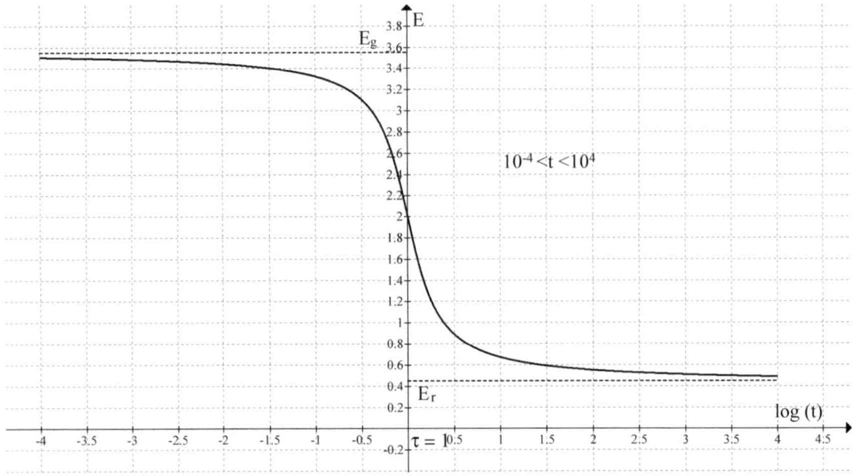

Fig. 3.9 The stress relaxation modulus $E_{rel}(t)$. Here $\tau = 1$

3.2.2 Closed-Cycle Condition

This section examines the mathematical consequences of the physical expectation, that the response of a linear viscoelastic material to harmonic loading ought to be harmonic, of the same frequency as the excitation, but out of phase with it. This so-called closed-cycle condition, that: "the steady-state response to harmonic loading also be harmonic," is satisfied by materials that do not age. That is, by materials whose property functions depend only on one time scale: the time measured from when the load was first applied, irrespective of the time elapsed since their manufacturing.

As will be shown in what follows, the closed-cycle condition requires that the kernels of the constitutive integrals, R and K, depend only on the difference of their arguments, and also, that all transients die out. In other words, the closed-cycle condition requires that $K(t, s) = K(t - s)$ and $R(t - s)$; as has been assumed without proof in our derivations.

According to the fading memory principle, closed-cycle condition is met for arbitrary excitations only in the limit as $t \to \infty$, if the kernel of the integral is bounded, as indicated above.

Summarizing: for the response of a viscoelastic material to a cyclic excitation to also be periodic, its material property functions must depend on the difference between current time and loading time. That is, the closed-cycle condition (that the response to periodic excitation be periodic) can only be satisfied by *non-aging* materials.

The terms "storage" and "loss" can be understood more readily by considering the mechanical work done per loading cycle.

3.2.3 Relationship Between Relaxation Modulus and Compliance

Expressions (3.2) and (3.3) relate stresses to strains, through the corresponding relaxation modulus and creep compliance. This suggests that the two expressions may be combined in some form to obtain the relationship between the two property functions.

It is reasonable to expect that the values of J and E_{rel} at $t = 0$, as well as at sufficiently long times, might be reciprocals of each other. An important application of this is in the derivation of constitutive equations for materials that are termed hyper-viscoelastic. *Hyperelastic* materials are truly elastic in the sense that if a load is applied to such a material and then removed, the material returns to its original shape without any dissipation of energy in the process. In other word, hyperelastic material stores energy during loading and releases exactly the same amount of energy during unloading. There is no path dependence. Equations for hyper-viscoelastic materials are derived from those of hyperelastic materials. A material is termed hyperelastic, if there exists a potential function of the strains, say, W, such that each individual stress component in such a material may be computed as the derivative of W with respect to the corresponding strain [17]. Since both the glassy or short-term and the equilibrium or long-term responses of a viscoelastic solid are elastic, either can be used to define the potential function.

3.2.4 Principle of Fading Memory

Loosely speaking, we say that a material has fading memory if the influence of an action on its response becomes less important as time goes by. Accordingly, the mathematical implications of the fading memory hypothesis—often called principle—can be established by loading and unloading a viscoelastic system, and monitoring its response after the load is removed. Before establishing the consequences of the principle of fading memory on a rigorous basis, we develop them by examining the response of a viscoelastic material to the relaxation and creep experiments; with which we are already familiar. The results of these experiments are the relaxation modulus and the creep compliance. Geometrically, the fading memory hypothesis simply requires that the relaxation modulus and creep compliance be monotonically decreasing and increasing functions of their arguments, respectively, and also that the absolute values of their slopes decrease monotonically. In addition the experimental observations indicate that:

- The relaxation modulus decreases with observation time and is bounded by the glassy modulus for fast processes and by the equilibrium modulus for very slow processes.

- The creep compliance increases with observation time and is bounded by the glassy and equilibrium compliances for very fast and slow processes, respectively.

3.2.5 The Maxwell Model [18]

The word viscoelastic is derived from the words "viscous" + "elastic" that are, a viscoelastic material exhibits both elastic and viscous behavior. We can build up a theory of linear viscoelasticity by considering simple linear elements such as the (elastic) linear spring and the (viscous) linear dashpot. We will start with simple models and increase the complexity until we have an infinite number of elements. In general, the more elements we have, the more accurate will our model be in describing the real response of real materials. That said, the more complex the model, the more material parameters there are which need to be evaluated by experiment—the determination of a large number of material parameters might be a difficult, if not an impossible, task.

We will first examine these models with an emphasis on the physical responses of the spring and dashpot elements. Later on, we will reexamine them from a more mathematical perspective, using the substitution method. We will only look at 1-dimensional models. We will examine ways of extending our analysis to 3-dimensions later on.

We can gauge the viability of our models by seeing whether they predict the observed responses of real models to some simple loads/strains. In particular, do they predict the creep, recovery and relaxation stress/strain curves discussed above?

The simplest way to create a model of a material is to suppose that it consists of nothing but a linear spring of stiffness E. The response of this material to a creep-recovery test is to undergo an instantaneous elastic strain upon loading, to maintain that strain so long as the load is applied, and then to undergo an instantaneous de-straining upon removal of the load. The *constitutive equation* for the linear elastic solid is then simply $\varepsilon = \frac{1}{E}\sigma$.

An important point to make, even though it is obvious and self-evident, is that the spring reacts instantly to the load, and when the load is removed it again reacts instantly. The response may also be written in the form $\varepsilon = J\sigma_0$ and J is known as the *compliance function* (the inverse of the stiffness).

The response of our ideal material is obviously not very representative of the response to a real material. There is no creep stage, inelastic recovery or permanent creep strain.

The Linear Viscous Fluid (the "linear dash pot")

Imagine next a material which responds like a viscous dashpot; the dashpot is a piston cylinder, filled with a viscous fluid. The strain is achieved by dragging the piston through the fluid. By definition, the dashpot responds with a strain rate proportional to stress:

$\dot{\varepsilon} = (1/\eta)\sigma$, where η is the *viscosity* parameter of the material. This is the typical response of many *fluids*; the larger the stress, the faster the straining (as can be seen by pushing your hand through water at different speeds).

The strain due to a suddenly applied load σ may be obtained by integrating the constitutive equation above. Assuming zero initial strain, one has $\varepsilon = (1/\eta)\ \sigma t$. The strain is seen to increase linearly and without bound so long as the stress is applied. Note that there is no movement of the dashpot at the onset of load; it takes time for the strain to build up. When the load is removed, there is no stress to move the piston back through the fluid, so that any strain built up is permanent. The slope of the creep line is σ/η.

The linear relationship between the stress and strain during the creep test may be expressed in the form $\varepsilon = \sigma\ J(t)$, where $J(t) = 1/\eta$. J here is called the *creep (compliance) function* ($J = 1/E$ for the elastic spring).

First, consider an ideal incompressible viscous fluid (an incompressible "Newtonian" fluid) bounded by a movable upper plate and a fixed lower plate. Flow of the fluid may be induced by the application of a shear stress τ to the upper plate. The fluid in contact with the upper plate will "stick" to it and move quickly, but the fluid at the lower fixed plate will not move. A velocity gradient is thus established and it can be shown (and verified experimentally) that the velocity gradient is related to the applied shear by a constant η, the viscosity of the fluid. Thus: $\frac{\mathrm{dv}}{\mathrm{dy}} = \frac{1}{\eta}\tau$. Now: $v = \mathrm{d}(u_x)/\mathrm{d}t$, and shear strain is related to the displacements through $\gamma = \mathrm{d}(u_x)/\mathrm{d}y$. Thus $\frac{\mathrm{d}v}{\mathrm{d}y} = \frac{\mathrm{d}}{\mathrm{d}y}\left(\frac{\mathrm{d}u_x}{\mathrm{d}t}\right) = \frac{\mathrm{d}}{\mathrm{d}t}\left(\frac{\mathrm{d}u_x}{\mathrm{d}y}\right) = \frac{\mathrm{d}\gamma}{\mathrm{d}t}$, and it follows that the shear strain rate is proportional to the shear stress, $\frac{\mathrm{d}\gamma}{\mathrm{d}t} = \dot{\gamma} = \frac{1}{\eta}\tau$.

This idea of viscous fluid flow is used in the analysis of viscoelastic materials. We can imagine a bar of material acting like a fluid when under tension. In that case we have normal stress and strain rate $\dot{\varepsilon} = \frac{1}{\eta}\sigma$. That is, the larger the stress, the faster the material strains. The strain due to a suddenly applied load σ_0 may be obtained by integrating this expression; we have (assuming zero initial strain) $\varepsilon = \frac{1}{\eta}\sigma_0 t$. The strain is seen to increase linearly and without bound so long as the stress is applied. Note that there is *no movement* of the dashpot at the onset of load. It takes time for the strain to build up. When the load is removed, there is no stress to move the piston back through the fluid, so that any strain built up is permanent. Note that the strain is proportional to stress, as it must be for a linear material. Again, this linear relationship between stress and strain may be emphasized by rewriting the response in the form $\varepsilon_0 = \sigma_0 J(t)$, $J(t) = t/\eta$. The compliance J is now a function of time, and is called the *creep compliance function* (or simply "creep function"). The creep and recovery response of the material is as shown on Fig. 3.10. The slope of the creep line is σ_0/η. There is no instantaneous strain, creep of ever decreasing strain rate, elastic recovery or inelastic recovery, but there is a permanent strain (all the creep strain is permanent).

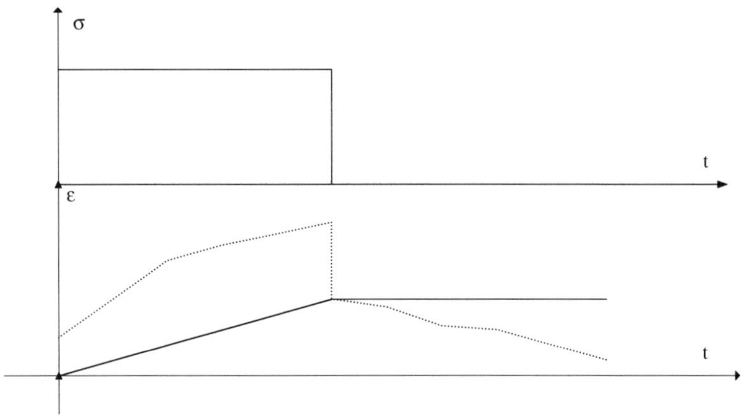

Fig. 3.10 Creep and recovery response of the viscous materials

The creep data are most accurately presented as the plot of strain versus time for various stresses and temperatures. A simple model is a Maxwell unit [18], which consists of a Hookean spring and a Newtonian dashpot as show. The spring models the elastic response while the dashpot models the viscous (or time dependent) response to load. The time dependence of viscoelastic response is analogous to the time dependence of reactive electrical circuits, and both can be described by identical ordinary differential equations in time (a convenient way of developing these relations while also helping visualize molecular motions employs "spring-dashpot" models). These mechanical analogs use elastic springs that described by $\sigma = k\varepsilon$, where σ and ε are analogous to the spring force and displacement, and the spring constant k is analogous to the Young's modulus E; k therefore has units of N/m^2. The spring models the instantaneous bond deformation of the material, and its magnitude will be related to the fraction of mechanical energy stored reversibly as strain energy. The entropic uncoiling process is fluid like in nature, and can be modeled by a "Newtonian dashpot," in which the stress produces not a strain but a strain rate: $\sigma = \eta\,\dot{\varepsilon}$. Here the overdot denotes time differentiation and η is a viscosity with units of N-s/m^2. In many of the relations to follow, it will be convenient to employ the ratio of viscosity to stiffness: $n = \eta/k$. The unit of n is time, and it will be seen that this ratio is a useful measure of the response time of the material's viscoelastic response. The "Maxwell" solid is a mechanical model in which a Hookean spring and a Newtonian dashpot are connected in series. The spring should be visualized as representing the elastic or energetic component of the response, while the dashpot represents the conformational or entropic component. In a series connection such as the Maxwell model, the stress on each element is the same and equal to the imposed stress, while the total strain is the sum of the strain in each element. We can divide the total strain into two separate strains, one for the spring (ε_1) and one for the dashpot (ε_2). The stress σ will be the same in both elements.

We thus have the three equations:

$$\varepsilon_E = \frac{1}{E}\sigma;\ \dot{\varepsilon}_\eta = \frac{1}{\eta}\sigma \text{ and } \varepsilon_E + \varepsilon_\eta = \varepsilon \tag{3.4}$$

The sub-indexes η and E indicate dashpot and spring respectively.

We can in principal eliminate ε_1 and ε_2, and be left with one constitutive equation relating ε to σ. Differentiating the first and third equations, and putting the first and second into the third, gives the constitutive relation for the Maxwell model:

$$\begin{aligned} &\dot{\varepsilon} = \frac{1}{E}\dot{\sigma} + \frac{1}{\eta}\sigma \quad \text{or:} \quad \sigma + \frac{\eta}{E}\dot{\sigma} = \eta\dot{\varepsilon} \\ &\mathrm{n} = \frac{\eta}{E} - \text{Relaxation time;} \quad \varepsilon(t) = \sigma(t) = 0 \quad \text{for} \quad t < \tau_0 \end{aligned} \tag{3.5}$$

Solutions of Eq. (3.5) may be determined considering either stress or strain as the controlled variable. In the first case we have directly

$$\varepsilon(t) = \frac{\sigma(t)}{E} + \frac{1}{\eta}\int\limits_{\tau_0}^{t} \sigma(\tau)\mathrm{d}\tau; \quad \text{or} \quad E\varepsilon(t) = \sigma(t) + \frac{1}{n}\int\limits_{\tau_0}^{t} \sigma(\tau)\mathrm{d}\tau; \quad n = \frac{\eta}{E} \tag{3.6}$$

Since the strain response is unbounded for $t \to \infty$, one says that the Maxwell model exhibits unbounded creep and sometimes refers to it as Maxwell fluid.

Considering now the strain history as given, we obtain from (3.5), using the general solution for first-order differential equations.

$$\sigma(t) = E\int\limits_{\tau_0}^{t} \mathrm{e}^{-\frac{E}{\eta}(t-\tau)}\dot{\varepsilon}(\tau)\mathrm{d}\tau; \quad n = \frac{\eta}{E} \tag{3.7}$$

At a constant strain in time (ε = const.) the right hand side of (3.5) vanishes and the stress will vary according to the law:

$$\sigma = \sigma_0(\mathrm{e}^{-t/n}) \tag{3.8}$$

where—the stress σ_0 is the stress at the initial time $t = 0$ (see Fig. 3.11).

Here the significance of $n \equiv \eta/k$ as a characteristic "relaxation time" is evident; it is physically the time needed for the stress to fall to 1/e of its initial value. The smaller the relaxation time, the faster the relaxation process, even though total relaxation takes theoretically an infinite time. The relaxation modulus E_{rel} may be obtained from this relation directly, noting that initially only the spring will deform and the initial stress and strain are related by $\sigma_0 = k\ \ \varepsilon_0$. So, $E_{\mathrm{rel}}(t) = (\sigma_0/\varepsilon_0)\exp(-t/n) = k\exp(-t/n)$. This important function is plotted schematically in Fig. 3.11. The two adjustable parameters in the model, k and τ, can

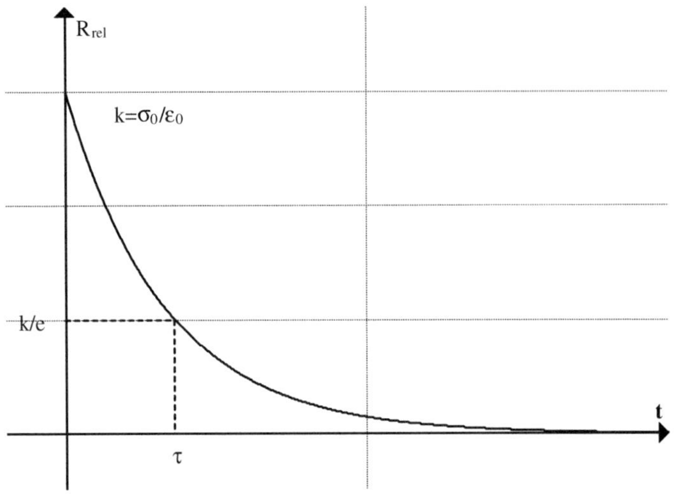

Fig. 3.11 Stress relaxation

be used to force the model to match an experimental plot of the relaxation modulus at two points. The spring stiffness k would be set to the initial modulus E_0, and τ would be chosen to force the value k/e to match the experimental data at $t = \tau$.

The relaxation time τ is strongly dependent on temperature and other factors that affect the mobility of the material, and is roughly inverses to the rate of molecular motion. Above $0.5T_{\text{melt}}$, τ is very short; below $0.5T_{\text{melt}}$ it is very long.

As mentioned above, temperature has a dramatic influence on rates of viscoelastic response, and in practical work it is often necessary to adjust a viscoelastic analysis for varying temperature. This strong dependence of temperature can also be useful in experimental characterization: if for instance a viscoelastic transition occurs too quickly at room temperature for easy measurement, the experimenter can lower the temperature to slow things down.

In some materials, especially "simple" ones, the relation between time and temperature can be described by correspondingly simple models. Such materials are termed "thermorheologically simple."

For such simple materials, the effect of lowering the temperature is simply to shift the viscoelastic response (plotted against log time) to the right without change in shape. This is equivalent to increasing the relaxation time τ without changing the creep compliances. A "time-temperature shift factor" $a_T(T)$ can be defined as the horizontal shift that must be applied to a response curve, say $J_{cr}(t)$—creep function, measured at an arbitrary temperature T in order to move it to the curve measured at some reference temperature T_* (the onset of temperature effect on creep deformations).

$$\log[a_T(T)] = \log[n(T)] - \log[n(T_*)]$$

A series of creep or relaxation data taken over a range of temperatures can be converted to a single "master curve" via this horizontal shifting. A particular curve is chosen as reference, and then the other curves shifted horizontally to obtain a single curve spanning a wide range of log time.

Each curve produces its own value of $a_T(T)$, so that a_T becomes a tabulated function of temperature. The master curve is valid only at the reference temperature, but it can be used at other temperatures by shifting it by the appropriate value of log (a_T).

In the above we assume a single relaxation time. If the model contains multiple relaxation times, thermorheologically simplicity demands that all have the same shift factor, since otherwise the response curve would change shape as well as position as the temperature is varied. This is a very serious limitation, because the Generalized Maxwell model (also known as the Maxwell–Wiechert model [19]) considers several Maxwell elements that are assembled in parallel. It takes into account that the relaxation which does not occur at a single time, but in a set of times. Due to molecular segments of different lengths with shorter ones contributing less than longer ones, there is a varying time distribution. The Wiechert model shows this by having as many spring–dashpot Maxwell elements as are necessary to accurately represent the distribution. This leads to a distribution of relaxation times, which in turn produces a relaxation spread over a much longer time than can be modeled accurately with a single relaxation time. When the engineer considers it necessary to incorporate this effect, the Wiechert model can have as many spring-dashpot Maxwell elements as are needed to approximate the distribution satisfactorily.

A new spectrum-based model for describing the behavior of time-dependent materials is presented. In this chapter, unlike most prior modeling techniques, the time-dependent response of viscoelastic materials is not expressed through the use of series. Instead, certain criteria have been imposed to select a spectrum function that has the potential of describing a wide range of material behavior [20]. Another consequence of choosing the spectrum function of the type used in this chapter is to have a few closed form analytic solutions in the theory of linear viscoelasticity. The Laplace transform technique is used to obtain the necessary formulae for viscoelastic Lame' functions, relaxation and bulk modules, creep bulk and shear compliance, as well as Poisson's ratio. Using the Elastic–Viscoelastic Correspondence Principle (EVCP), material constants appearing in the proposed model are obtained by comparing the experimental data with the solution of the integral equation for a simple tensile test. The resulting viscoelastic functions describe the material properties which can then be used to express the behavior of a material in other loading configurations.

The total stress σ transmitted by the model is the stress in the isolated spring (of stiffness k_e) plus that in each of the Maxwell spring-dashpot arms. The material constants (k_e and the various k_j and τ_j) can be selected by forcing the predicted values of relaxation modulus $E_{rel}(t)$ to match those determined experimentally.

Change of Variable in Integrals

In this section, we present the substitution technique used with definite integrals. For functions of two or more variables, there is a similar process we can use. It is a little bit more involved though. In addition, in higher dimensions, a change of variable can also be used to simplify the region of integration.

You will recall in one-dimensional calculus, when given an integral of the form $\int g'(\theta)f[g(\theta)]\mathrm{d}\theta$, we performed the change of variable $\tau = g(\theta)$ which gave us $\mathrm{d}\tau = g'(\theta)\mathrm{d}\theta$ and thus $\int g'(\theta)f[g(\theta)]\mathrm{d}\theta = \int f(\tau)\mathrm{d}\tau$. We can write this slightly differently as follows. Since $\tau = g(\theta)$, $\mathrm{d}\tau/\mathrm{d}\theta = g'(\theta)$, hence, we have $\int f(\tau)\mathrm{d}\tau = \int f[g(\theta)]\frac{\mathrm{d}\tau}{\mathrm{d}\theta}\mathrm{d}\theta$. Obviously, we can extend the change of variable theorem for double integrals or multidimensional integrals. We then look at several well-known creep models to see how one can benefit from a change of variable. These benefits include using a change of variable to simplify an integrand.

As an application, we will look at Maxwell model; Kelvin–Voigt model and Standard Linear Model. Note that most of the explanations given below will be without rigorous mathematical prove and for regions in the xy-plane or for functions of two variables. The change of variable method can also be applied in other situations; however, in such cases there usually is not a clean litmus test that will tell you whether or not a given substitution will help solve the differential equation. Rather one has to resort to trial and error in order to find an appropriate substitution. There is at least a guiding principle though: you want to look for a substitution that will end up simplifying the differential (or integral) equation.

It is important to understand that the creep function K or relaxation function R is given and it is a function of time. At the same time strain, modulus of elasticity and Arrhenius law are functions of dimensionless temperature in Eq. (3.6). We must find an appropriate change of variable in which the integral on the right is simpler. As stated, the theorem is deceiving. It makes it look like the original integral is easier than the integral on the right which appears to contain much more. But this is not the case in practice. We know that the ultimate goal in case of high temperature engineering creep is to obtain the stress–strain function. In order to compute this functional relationship in the case of Maxwell model, we will substitute the independent variable time t (or dimensionless time τ) in the differential Eq. (3.5) with function $\tau = \tau\ (\theta)$. Given the fact that in case of temperature creep strain is linearly proportional to the dimensionless temperature θ and the coefficient of linear expansion assumed to be unchanged with temperature rise, it can be stated that in the Eq. (3.5), the function $\varepsilon(t)$ is a given function, and the stress function $\sigma(t)$—unknown. As an example, we can take the temperature–time function for a fast fire from [21] (Fig. 3.12):

$$\theta = -1.29 + 380\tau - 6047\tau^2 + 35830\tau^4 \tag{3.9}$$

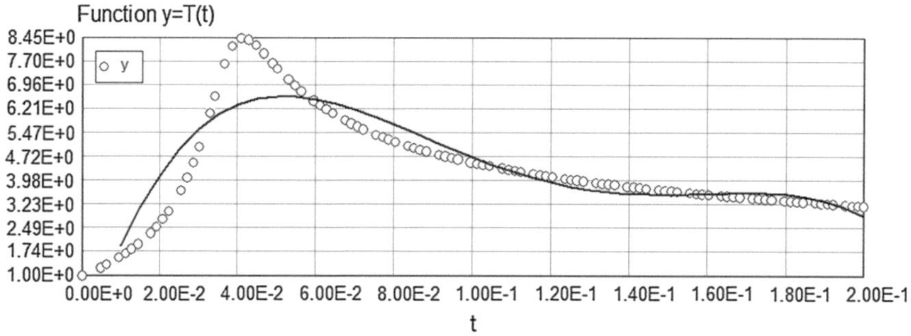

Fig. 3.12 Typical temperature–time curve

Model: $t = a0 + a1 * y + a2 * y\hat{}2$

Variable	Value
$a0$	0.0096088
$a1$	0.0202256
$a2$	−0.0016026

The inverse function θ^{-1} is $\tau = 0.01 + 0.02\theta - 0.0016\theta^2$. The first derivative of the inverse function is $d\tau/d\theta = 0.02 - 0.0032\theta$.

Using these expressions and performing the change of variables in the differential Eq. (3.2), we obtain.

$$\frac{d\sigma}{d\theta} = -\frac{1}{n}\sigma(0.02 - 0.0032\theta) + E_0 A \tag{3.10}$$

The numerical solution of the Eq. (3.10) is given below (using POLYMATH software).

$$\frac{d\sigma}{d\theta} = -100\sigma(0.02 - 0.0032\theta) + 2.9(7.02)$$

$$n = 0.01; \quad A = 7.02(10^{-4}); \quad E_0 = 2.9(10^4)$$

Calculated values of DEQ variables

	Variable	Initial value	Minimal value	Maximal value	Final value
1	x	0	0	8	8
2	y	0	0	123.4979	123.4979

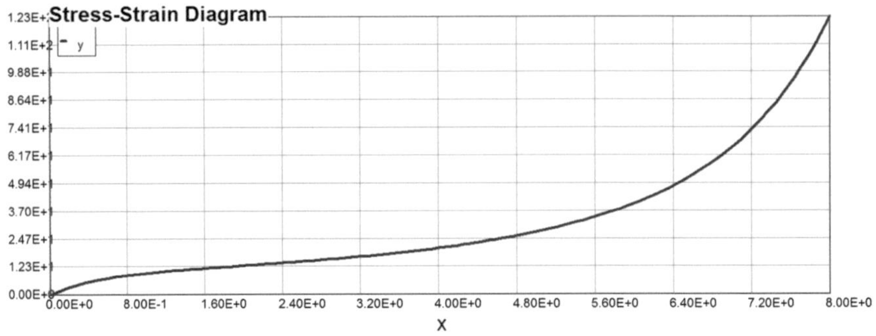

Fig. 3.13 Thermal creep (Maxwell model)

Differential equations

$$1 \quad \mathrm{d}(y)/\mathrm{d}(x) = -100 * y * (0.02 \; - \; 0.0032 * x) \; + \; 2.9 * 7.02$$

From Fig. 3.13 one can see now that the thermal creep (Maxwell model) has three characteristic zones: primary, secondary, and tertiary). It should also be noted that the stress–temperature function differs from the stress–strain diagram only by the scale factor on the horizontal axis. Now consider the case when the relaxation time is constant, but the modulus of elasticity E is a function of temperature θ, i.e. $E = E_0 \exp(-0.15\theta)$ [22] (Fig 3.14). The numerical solution of the Eq. (3.6) is:

$$\begin{aligned} \frac{\mathrm{d}\sigma}{\mathrm{d}\theta} &= -\frac{1}{n}\sigma(0.02 \; - \; 0.0032\theta) + EA \\ \frac{\mathrm{d}\sigma}{\mathrm{d}\theta} &= -100\sigma(0.02 \; - \; 0.0032\theta) + 2.9(7.02)\mathrm{e}^{-0.15\theta} \end{aligned} \tag{3.11}$$

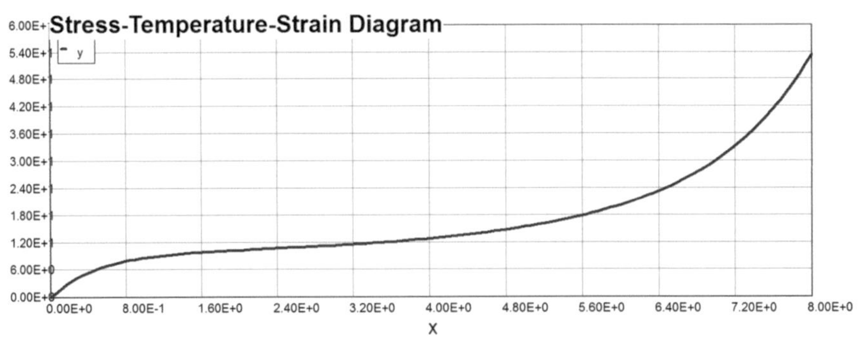

Fig. 3.14 Stress–temperature diagram

Calculated values of DEQ variables

	Variable	Initial value	Minimal value	Maximal value	Final value
1	x	0	0	8	8
2	y	0	0	53.44049	53.44049

Differential equations

1 d(y)/d(x) = −100 ∗ y ∗ (0.02 − 0.0032 ∗ (x − 0)) + 2.9 ∗ 7.02 ∗ 1 ∗ (exp(−0.15 ∗ (x − 0)))^1

Numerical solution can be approximated now as follows:
Model: y = 3.16 + 5.5 * x − 0.818 * x^2 − 0.0866 * x^3 + 0.0246 * x^4

Variable	Value
$a0$	3.161792
$a1$	5.509431
$a2$	−0.818518
$a3$	−0.0866128
$a4$	0.0246396

$$\begin{aligned}\sigma &= 3.16 + 5.5\theta - 0.82\theta^2 - 0.086\theta^3 + 0.025\theta^4 \\ \theta &= -1.29 + 380\tau - 6047\tau^2 + 35830\tau^4\end{aligned} \tag{3.12}$$

Consider now the temperature decay process (cool off of a material). In this case in order to get a relaxation curve due to the fact that modulus of elasticity E is increasing while the temperature reduces to zero, we need to change the sign of θ (symmetry with respect to “y” axis) from plus to minus and the interval θ [0, 8] to θ [8, 16]. The result is presented now on Fig. 3.15.

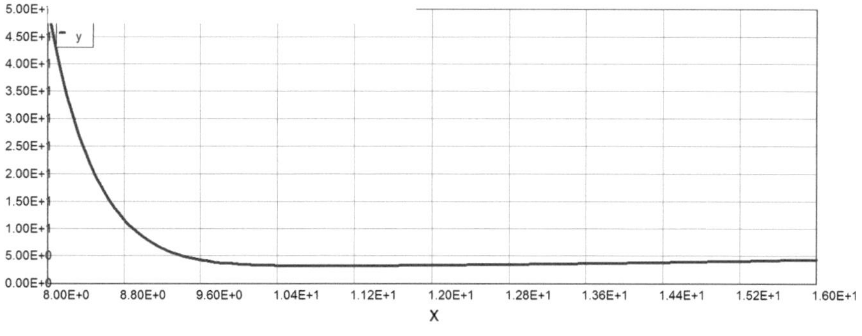

Fig. 3.15 Maxwell model relaxation curve

Calculated values of DEQ variables

	Variable	Initial value	Minimal value	Maximal value	Final value
1	x	8	8	16	16
2	y	50	3.218826	50	4.37016

Differential equations

1 d(y)/d(x) = −100 * y * (0.02 + 0.0032 * (x − 8)) + 2.9 * 7.02 * 0.3 * (exp(0.15 * (x − 8)))^1

Model: y = 3754 − 1223.5 * x + 147.92 * x^2 − 7.86 * x^3 + 0.155 * x^4

Variable	Value
$a0$	3754.625
$a1$	−1223.521
$a2$	147.9179
$a3$	−7.859728
$a4$	0.1549715

Finally, what is the response of this model to a sudden load σ_0? Physically, we know that the spring will stretch immediately and that the dashpot will take time to react. Thus we will have an initial strain $\varepsilon_0(0) = \sigma_0(0)/E$. We then expect the dashpot to take up the stress, and so for the strain to increase linearly with slope $\sigma_0(0)/\eta$. The stress-rate is zero so our constitutive law becomes as follows:

$$\dot{\varepsilon} = \frac{\sigma_0}{\eta} \Rightarrow \varepsilon(t) = \frac{\sigma_0}{\eta} t + C \Rightarrow \varepsilon(t) = \sigma_0\left(\frac{1}{\eta} t + \frac{1}{E}\right) \tag{3.13}$$

We again emphasize the linearity of the response by writing the above in terms of a creep compliance function $\varepsilon(t) = \sigma_0 J(t)$ where $J(t) = t/\eta + 1/E$. Now when we remove the load, the spring will again react immediately, but the dashpot will have no tendency to recover. Thus there is an immediate elastic recovery $\sigma_0(0)/E$, with the "creep" strain due to the dashpot remaining. The full creep and recovery response is as shown below (see Fig. 3.16).

There is creep, but not of the ever decreasing strain rate type, there is no delayed elasticity, but there is the elastic response, the permanent strain, and we seem to be getting nearer the real thing.

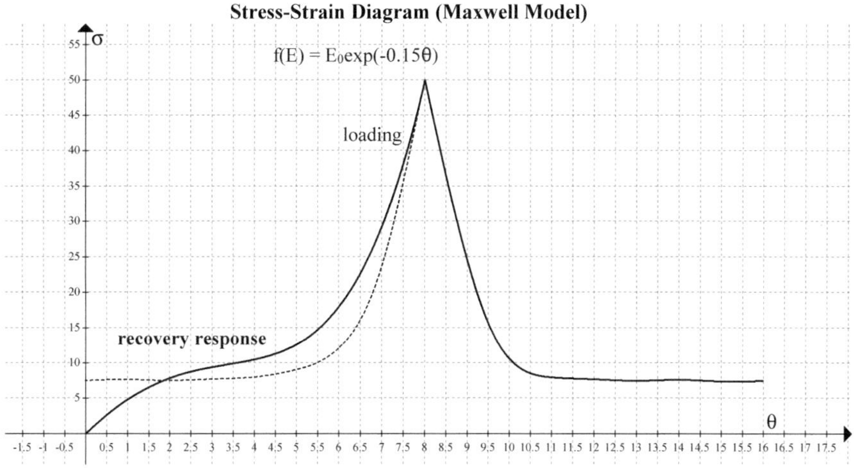

Fig. 3.16 Full creep and recovery response (Maxwell model)

3.2.6 *The Kelvin–Voigt Model*

Consider next the other two-element model, the Kelvin (or Voigt) model [23], which consists of a spring and dashpot in parallel. It is assumed there is no bending in this type of parallel arrangement, so that the strain experienced by the spring is the same as that experienced by the dashpot. This time, however, here there is no reason why the stresses in both portions should be the same. We thus have the following three equations:

$$\varepsilon = \frac{1}{E}\sigma_1;\ \dot{\varepsilon} = \frac{1}{\eta}\sigma_2 \text{ and } \sigma_1 + \sigma_2 = \sigma \tag{3.14}$$

Eliminating σ_1 and σ_2, then leaves us with the constitutive law which is in standard form.

$$\sigma = E\varepsilon + \eta\dot{\varepsilon} \tag{3.15}$$

If we apply a sudden load σ_0 to the Kelvin–Voigt model, the spring will want to stretch immediately, but is held back by the dashpot, which cannot react immediately. All the stress is thus initially taken up by the dashpot (there is no stress in the spring because if there was there would have to be at least some strain). Our creep curve thus starts with an initial slope σ_0/η. Some strain then starts to take place, and so the stress starts to decrease in the dashpot and increase in the spring—the stress is being transferred from the dashpot to the spring. The slope of the creep curve is now σ_2/η, where σ_2 is the stress in the dashpot—thus the slope of the creep curve is ever decreasing as σ_2 decreases. In the limit when $\sigma_2 = 0$ (which will happen after an infinite amount of time!), the spring will carry all the stress and thus the

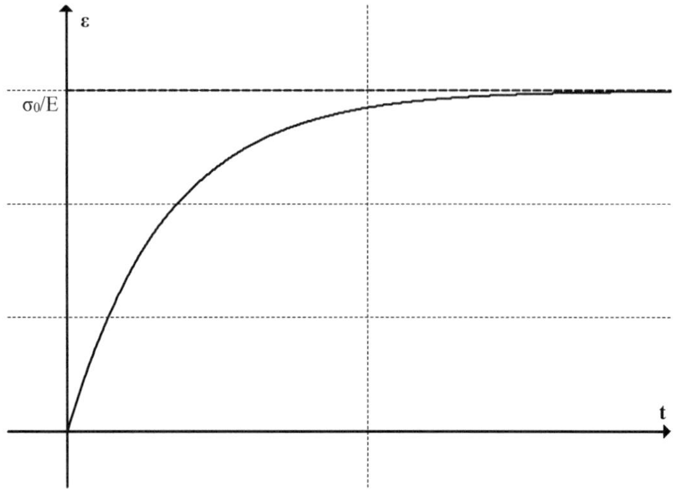

Fig. 3.17 Kelvin–Voigt model (The response function)

maximum strain is σ_0/E. We have a first order non-homogeneous ODE Eq. (3.15) with the initial condition ε (0) = 0, which we can solve to get the response which agrees with the above physical reasoning.

$$\varepsilon(t) = \frac{\sigma_0}{E}\left(1 - e^{-\frac{E}{\eta}t}\right) \tag{3.16}$$

The response (3.16) is exponential, and ε asymptotically approaches the value σ_0/E (see Fig. 3.17).

The value E can be considered as modulus of elasticity of infinitely prolonged action of a constant load (long-term modulus of elasticity). The value η/E in formula (3.16) is often called retardation time and is analogous in meaning to the relaxation time: an estimate of the time required for the creep process to approach completion. If the movement does not start from zero but from some initial deformation ε_0, the Eq. (3.16) will be presented as:

$$\varepsilon(t) = \frac{\sigma_0}{E}\left(1 - e^{-\frac{E}{\eta}t}\right) + \varepsilon_0 e^{-\frac{E}{\eta}t} \tag{3.17}$$

Along with the increase of deformation under the law (3.15) there has been a gradual decrease in the initial deformation of the same attenuation coefficient of the exponential function. Formula (3.17) can be represented as:

$$\varepsilon = \frac{\sigma_0}{E} - \left(\frac{\sigma_0}{E} - \varepsilon_0\right)e^{-\frac{E}{\eta}t} \tag{3.18}$$

When we unload, the spring will want to contract but again the dashpot will hold it back. But the spring will eventually pull the dashpot back to its original zero position

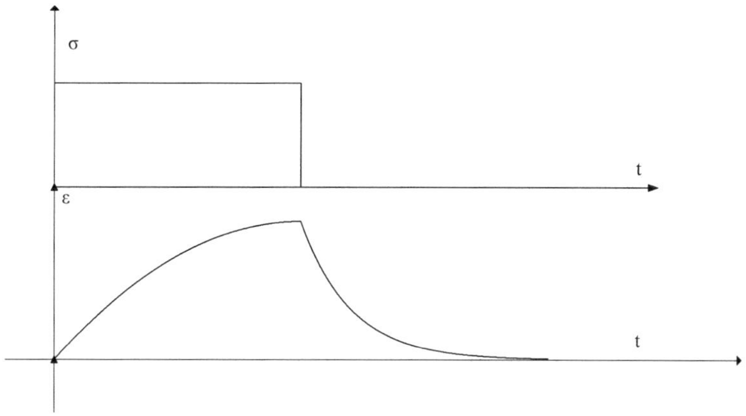

Fig. 3.18 Full creep and recovery response (Kelvin–Voigt model)

given time (we expect full recovery). Thus suppose we unload at time $t = \tau$ .The strain at this time is $\varepsilon(\tau) = (\sigma_0/E)(1 - \mathrm{e}^{-(E/\eta)\tau})$. Our constitutive law, with zero stress, reduces to $0 = E\varepsilon + \eta\dot{\varepsilon}$. Solving this equation, we get $\varepsilon(\tau) = C(\mathrm{e}^{-(E/\eta)\tau})$, but this t is measured from the point where "zero load" begins. If we want to measure time from the onset of load, we must replace this t with $t - \tau$. Using the aforementioned "initial" condition then leads to

$$\varepsilon(\tau) = (\sigma_0/E)(\mathrm{e}^{-(E/\eta)(t-\tau)})\,(1 - \mathrm{e}^{-(E/\eta)\tau}) \quad t > \tau \tag{3.19}$$

The creep and recovery response is as shown below (see Fig. 3.18). We have a transient-type creep and inelastic recovery, but no permanent strain [24].

At constant strain ε = const., as follows from the Eq. (3.19) the stress σ remains constant over time. This result is contrary to the behavior of real materials. Usually, in order to keep a predetermined strain for a long time requires a lot of stress at first, and then less and less. This phenomenon is called stress relaxation. Material deforms under the law (3.19) is non-relaxing material. In Nature and Technology such material apparently does not occur.

Let us analyze now the Kelvin–Voigt model response to a high temperature loading condition in a matter similar to the Maxwell model response presented above. After replacing the independent dimensionless time τ in the differential Eq. (3.15) with the function $\theta = \theta\ (\tau)$ we have:

$$\sigma = E\varepsilon + \eta\dot{\varepsilon} = E_0 A\left[\theta + n\frac{1}{0.02 - 0.0032\theta}\right]$$

The stress–temperature diagram in this case (n = 0.01; E_0 = const.) is presented below (see Fig. 3.19).

If n = 0.01 = const. but $E = E_0\ (\exp\ (-0.15\theta))$, then we have (Fig. 3.20):

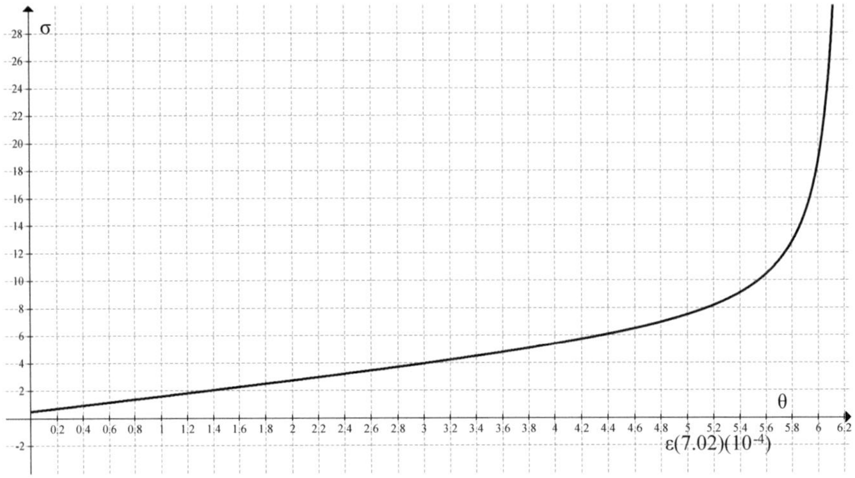

Fig. 3.19 Stress–temperature diagram (n = 0.01; E_0 = const.)

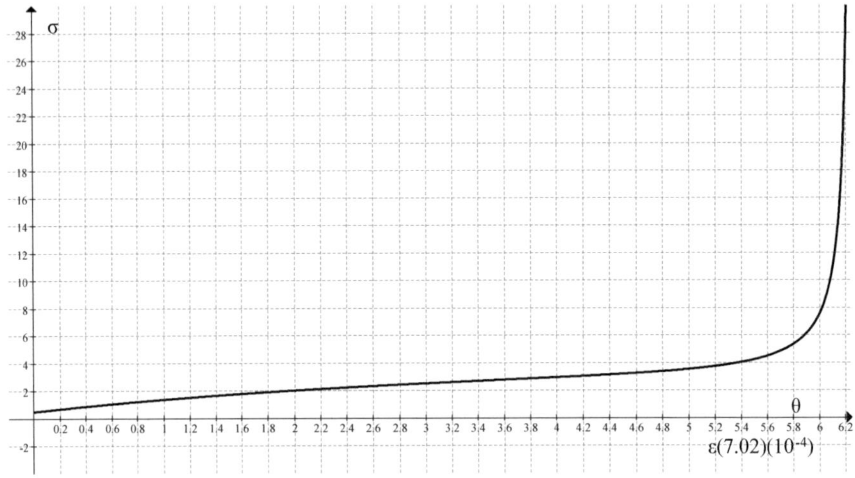

Fig. 3.20 Stress–temperature diagram (n = 0.01; $E = E_0$ exp (-0.15θ))

3.2.7 The Standard Linear Model

More complex models can be constructed using more and more elements. A complex viscoelastic rheological model will usually be of the form of the generalized Maxwell model or the generalized Kelvin chain [25–27], shown in Fig. 3.21. The generalized Maxwell model consists of N different Maxwell units in parallel, each unit with different parameter values. The absence of the isolated spring would ensure fluid-type behavior, whereas the absence of the isolated

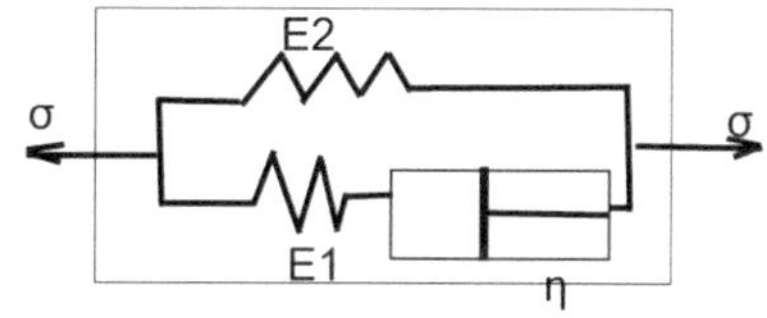

Fig. 3.21 Standard linear model

dashpot would ensure an instantaneous response. The generalized Kelvin chain consists of a chain of Kelvin units and again the isolated spring may be omitted if a fluid-type response is required. In general, the more elements one has, the more accurate a model will be in describing the response of real materials. That said, the more complex the model, the more material parameters there are which need to be evaluated by experiment—the determination of a large number of material parameters might be a difficult, if not an impossible, task. Furthermore, for a given complex model the set of material parameters is not unique, therefore the idea itself of building such model based exclusively on experimental data is a very risky proposition. The probability-based approach based on experimental data of material parameters coupled with the deterministic (even approximate) solution of a corresponding constitutive equation for a given material.

We are now in a position to look at a more complicated (and realistic) model, the standard linear model. This model consists of a spring in series with a Kelvin unit. Upon loading we expect the upper spring to stretch immediately. The dashpot then takes up the stress, transferring the load to the second spring as it slowly opens over time. Upon unloading we expect the upper spring to contract immediately and for the lower spring to slowly contract, being held back by the dashpot.

The equations for this model are, from the Fig. 3.21:

$$\begin{aligned} &\varepsilon = \varepsilon_1 + \varepsilon_2; \quad \sigma = \sigma_1 + \sigma_2; \quad \sigma = E_1 \varepsilon_1; \\ &\sigma_1 = E_2 \varepsilon_2; \quad \sigma_2 = \eta \dot{\varepsilon}_2 \end{aligned} \tag{3.20}$$

We can eliminate the four unknowns from these five equations and after simplifications we have

$$\sigma + \frac{\eta}{E_1 + E_2} \dot{\sigma} = \frac{E_1 E_2}{E_1 + E_2} \varepsilon + \frac{E_1 \eta}{E_1 + E_2} \dot{\varepsilon} \tag{3.21}$$

The creep compliance function in this case is

$$\begin{aligned} &J(t) = \frac{1}{E_1} \mathrm{e}^{-(E_2/\eta)t} + \frac{E_1 + E_2}{E_1 E_2} \left(1 - \mathrm{e}^{-(E_2/\eta)t}\right) \\ &\text{where } E = E_1; \; H = \frac{E_1 E_2}{E_1 + E_2}; \; n = \frac{\eta}{E_1 + E_2} \end{aligned} \tag{3.22}$$

The Eq. (3.20) will have a standard form now as follows:

$$n\dot{\sigma} + \sigma = En\dot{\varepsilon} + H\varepsilon \tag{3.23}$$

Stiffness parameter E in Eq. (3.23) represents the instantaneous modulus of elasticity, and H—long-term elastic modulus (prolonged modulus of elasticity). The meaning of these terms is as follows: at very slow deformation processes in Eq. (3.23) the rates $\dot{\sigma}$ and $\dot{\varepsilon}$ can be neglected compared with the values σ and ε, and then we come to the usual Hooke's law with a long-term elastic modulus $\sigma = H\varepsilon$. At very fast processes of deformation, on the contrary, the strain rate $\dot{\varepsilon}$ and the stress-rate $\dot{\sigma}$ are very high, and compared to them the strains ε and stresses σ can be ignored. In this case, we again obtain Hooke's law, but differentiated with respect to time and instantaneous modulus E $\dot{\sigma} = E\dot{\varepsilon}$.

Note that in all cases $E > H$, as evidenced by the formulas (3.21). Parameter n is the relaxation time. It is easy to see that the Kelvin–Voigt Model and the Maxwell Model are special cases of general Eq. (3.14) corresponding to parallel and serial connections of one elastic and one viscous element respectfully. In order to get the Maxwell Model the instantaneous modulus of elasticity E should be set equal to infinity, and the relaxation time n—zero, with so that the product $nE = \eta$ is finite. To obtain the Kelvin–Voigt Model (Eq. 3.15) is enough to put the long-term (prolonged) elastic modulus H to zero.

The differential constitutive relations for the Maxwell and Kelvin models were not too difficult to derive. However, even with three elements, deriving them can be a difficult task. This is because one needs to eliminate variables from a set of equations, one or more of which is a differential equation. The task is more easily accomplished using integral formulations, which are discussed below.

The constitutive relations for the four models shown in Fig. 3.22 are as follows:

$$\begin{aligned}
&1.\ \sigma + \frac{\eta}{E_1 + E_2}\dot{\sigma} = \frac{E_1 E_2}{E_1 + E_2}\varepsilon + \frac{\eta E_1}{E_1 + E_2}\dot{\varepsilon} \\
&2.\ \sigma + \frac{\eta}{E_2}\dot{\sigma} = E_1\varepsilon + \frac{\eta(E_1 + E_2)}{E_2}\dot{\varepsilon} \\
&3.\ \sigma + \frac{\eta_2}{E}\dot{\sigma} = (\eta_1 + \eta_2)\dot{\varepsilon} + \frac{\eta_1\eta_2}{E}\ddot{\varepsilon} \\
&4.\ \sigma + \frac{\eta_1 + \eta_2}{E}\dot{\sigma} = \eta_1\dot{\varepsilon} + \frac{\eta_1\eta_2}{E}\ddot{\varepsilon}
\end{aligned} \tag{3.24}$$

Generalized Kelvin Models (Fig. 3.23)

In order to solve any problems of deformation in time, one must have the Eq. (3.18) and the corresponding initial condition, which is expressed through the initial values of stresses and strains:

$$\varepsilon(0) = \varepsilon_0;\ \sigma(0) = \sigma_0 \quad \text{when}\, t = 0 \tag{3.25}$$

Fig. 3.22 Generalized Kelvin models

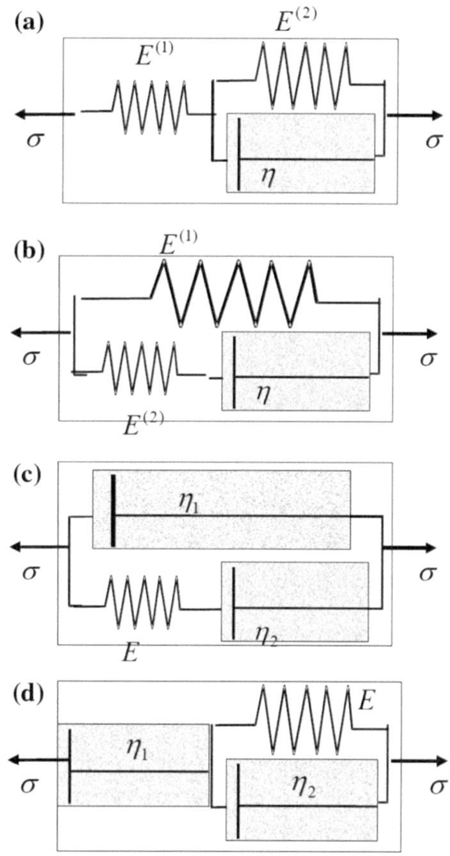

Fig. 3.23 Generalized Maxwell models

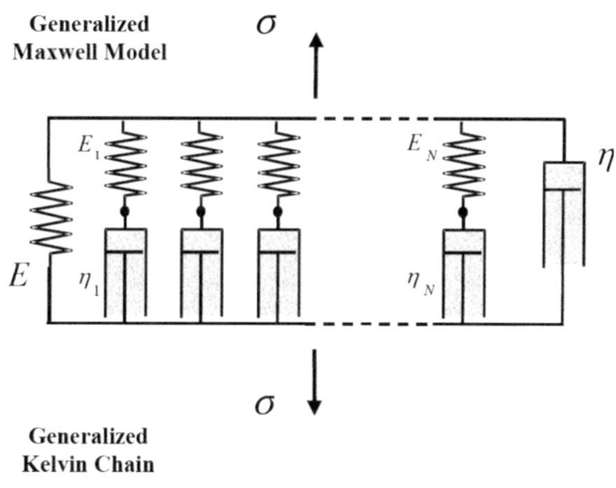

The main initial condition is the natural state of the material in which the viscous deformation element is zero. From Eqs. 3.23 and considering 3.25 we see that

$$\begin{aligned} &\varepsilon_2 = \varepsilon + \varepsilon_1 = \varepsilon - \frac{\sigma_1}{E_1} = \varepsilon - \frac{\sigma}{E_1} = \varepsilon - \frac{\sigma}{E} = 0; \\ &\text{or: } \varepsilon_0 = \frac{\sigma_0}{E} \quad \text{when } t = 0 \end{aligned} \tag{3.26}$$

3.3 Various Cases of Loading Conditions

Consider now various cases of loading material governed by Eq. 3.23.

3.3.1 Constant Load

Solution of the Eq. (3.23) with the initial condition (3.25) is as follows (Fig. 3.24):

$$\varepsilon = \frac{\sigma}{H} + \sigma\left(\frac{1}{E} - \frac{1}{H}\right)e^{-\frac{Ht}{En}} \tag{3.27}$$

Note that the value $\left(\frac{1}{E} - \frac{1}{H}\right)$ is negative, since $E > H$. Unlike the Kelvin–Voigt model, there is an initial strain $\varepsilon_0 = \frac{\sigma_0}{E}$ appearing instantly at the moment stress application. Over time, the deformation asymptotically approaches the magnitude of $\frac{\sigma}{H}$ (see Fig. 3.25).

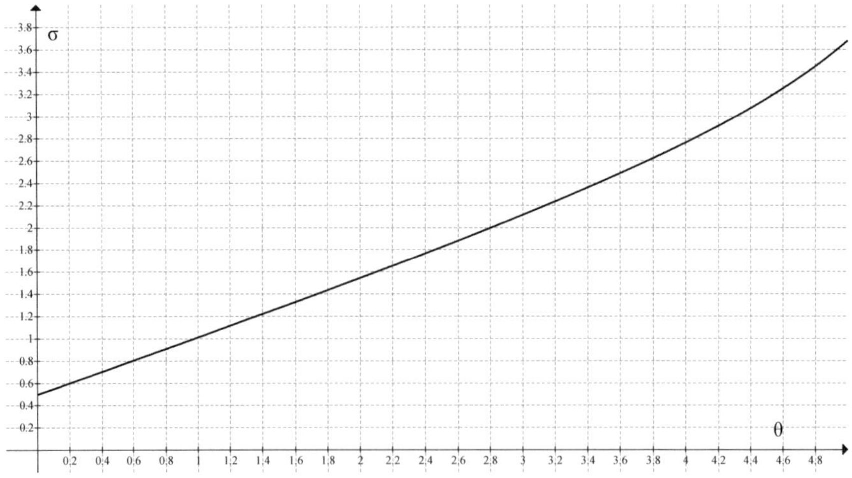

Fig. 3.24 Stress–temperature diagram (constant load)

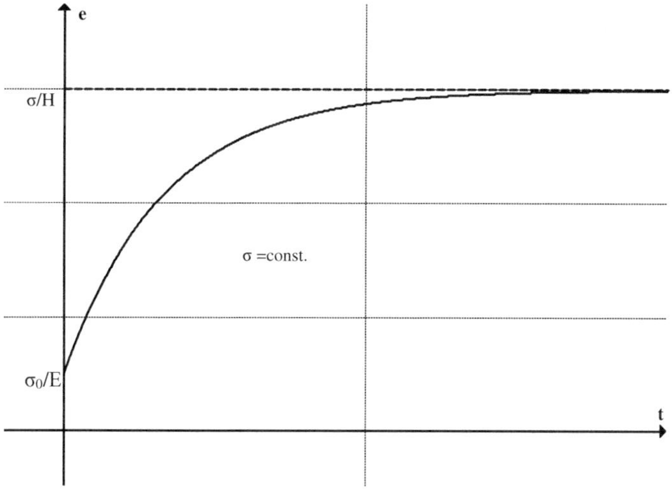

Fig. 3.25 Standard linear model (σ = const.)

3.3.2 Unloading

Suppose the material is unloaded at time $\tau = t$. The constitutive law (3.23), with zero stress, reduces to

$$\varepsilon = Ce^{-\frac{Ht}{En}} \tag{3.28}$$

Suppose at the time $t = t_0$ the value of ε is ε_0 and the stress $\sigma = \sigma_0$. At the same time t, due to the instantaneous stress reduction by the amount of σ_0 the deformation will decrease to a value:

$$\varepsilon(t_0) = \varepsilon_0 - \frac{\sigma_0}{E} \tag{3.29}$$

Later deformation will vary smoothly by the law (3.28). The constant C must be determined from the initial conditions. The final result is as follows:

$$\varepsilon = \left(\varepsilon_0 - \frac{\sigma_0}{E}\right)e^{-\frac{H(t-t_0)}{En}} \tag{3.30}$$

Over time, the deformation after unloading vanish to zero (see Fig. 3.26), i.e., to the natural state of the material. Thus the residual deformation vanishes when $t \to \infty$.

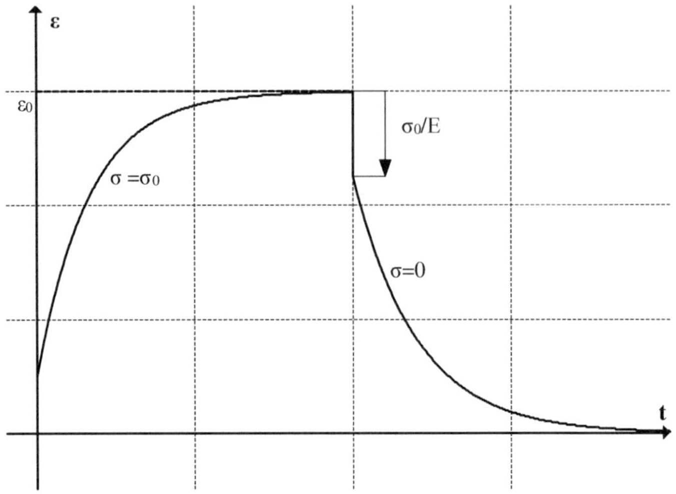

Fig. 3.26 Standard linear model (unloading)

3.3.3 Load Increases with Time Uniformly

Consider next different a priory given simple loading conditions that will allow us later to understand better the creep and recovery response of a material affected by high temperature. Lets say stress varies according to law $\sigma = vt$, where v—a constant velocity of stress increase. Substituting this expression for σ in the Eq. (3.23) we find a general solution:

$$\varepsilon(t) = Ce^{-\frac{Ht}{En}} + \frac{vt}{H} + \frac{vn}{H} - \frac{vnE}{H^2} \tag{3.31}$$

With a natural initial condition $\varepsilon(0) = 0$ constant C gets the value $C = -\frac{vn}{H} + \frac{vnE}{H^2}$ and formula (3.31) becomes:

$$\varepsilon(t) = \frac{vn}{H}\left(1 - \frac{E}{H}\right)\left(1 - e^{-\frac{Ht}{En}}\right) + \frac{vt}{H} \tag{3.32}$$

The strain tends to increase (with $t \to \infty$) to the value that is different from the elastic deformation $\varepsilon = vt/H$ (with long-term modulus of elasticity H) by an amount equal to $\frac{vn}{H^2}(E - H)$.

We can get also the stress–strain function by substituting time $t = \sigma/v = \hat{\sigma}$ in Eq. (3.32)

$$\varepsilon(\widehat{\sigma}) = \frac{vn}{H}\left(1 - \frac{E}{H}\right)\left(1 - \mathrm{e}^{-\frac{H\widehat{\sigma}}{En}}\right) + \frac{v\widehat{\sigma}}{H}$$

3.3.4 Load Decreases with Time Uniformly

Suppose at some point in time t is measured from the point where "zero load" begins and the material is unloaded linearly $\sigma = \sigma_0 - v_1 t$. Substituting this expression into (3.18) and solving it with respect to ε, we obtain

$$\varepsilon = C\mathrm{e}^{-\frac{Ht}{En}} - \frac{v_1 t}{H} - \frac{v_1 n}{H} + \frac{v_1 n E}{H^2} + \frac{\sigma_0}{H} \tag{3.33}$$

Suppose that at the start of unloading as a result of previous loadings the strain had reached the value ε_0. Then the constant C can be determined from the initial condition $\varepsilon(0) = \varepsilon_0$. The final solution (3.33) is now

$$\varepsilon = \varepsilon_0 \mathrm{e}^{-\frac{H}{En}} - \frac{v_1 t}{H} + \left(\frac{\sigma_0}{H} + \frac{v_1 n E}{H^2} - \frac{v_1 n}{H}\right)\left(1 - \mathrm{e}^{-\frac{Ht}{En}}\right) \tag{3.34}$$

Strain rate at the initial unloading is

$$\dot{\varepsilon} = \frac{\sigma_0 - H\varepsilon_0}{En} - \frac{v_1}{E} \tag{3.35}$$

If the loading process had followed the previous Eq. (3.35) until $t = 0$, then $\sigma_0 > H\varepsilon_0$. Therefore, the first term on the right (3.35) is positive and the initial strain rate $\dot{\varepsilon}$ for sufficiently low speed unloading will also be positive. This means that the strain can grow for a while after the start of unloading. The same phenomenon occurs with other methods of loading material until unloading.

3.3.5 Permanent Deformations in Time

Assuming $\varepsilon(0) = \varepsilon_0 =$ const., we obtain from Eq. (3.23):

$$\sigma = H\varepsilon_0 + (\sigma_0 - H\varepsilon_0)\mathrm{e}^{-t/n} \tag{3.36}$$

Comparing this expression with the solution of Maxwell deformation model, we note that as time t increases the stress according to the Eq. (3.36) strive not to zero but to a constant $H\,\varepsilon_0$ (see Fig. 3.27). Undoubtedly, the solution (3.36) better corresponds to the behavior of most materials subjected to time-constant

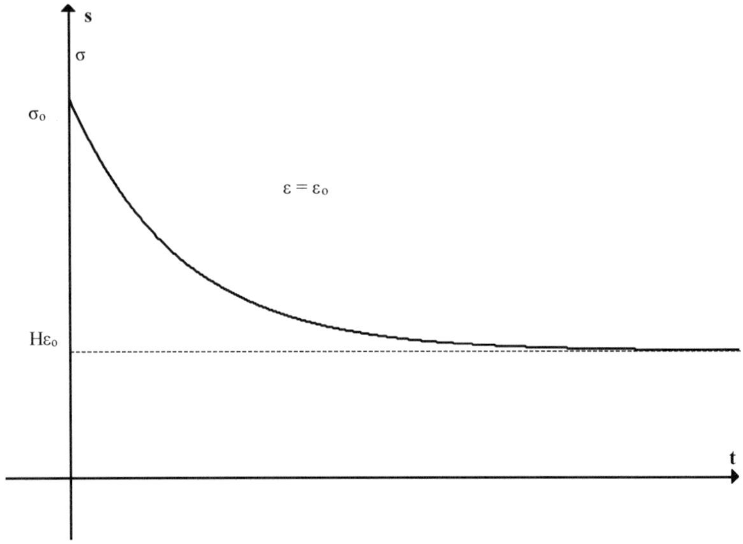

Fig. 3.27 Permanent deformations in time

deformation. Respectively, if the initial state of the material at time $t = 0$ is natural (the strain $\varepsilon(0) = \varepsilon_0$; or $\sigma_0 = E\varepsilon_0$) and applied instantly, we have now instead of (3.36):

$$\sigma = H\varepsilon_0 + (E - H)\,\varepsilon_0 e^{-t/n} \tag{3.37}$$

If the initial stress $\sigma_0 = H\varepsilon_0$, then based on Eq. (3.31) the stress relaxation is absent (Fig. 3.27).

3.3.6 Linearly Increasing (Decreasing) Deformations in Time

This case is very important for the future high temperature creep and recovery analysis, since the strain ε is linearly proportional to the dimensionless temperature θ. Therefore, more details in analysis are provided below including the application of substitution method (see above) that is used often in solving many non-invariant and nonlinear integral-type creep deformations problems. So, if the strain is linearly increasing ($\varepsilon = wt$), where w is a constant strain rate, Eq. (3.23) has the form

$$n\dot{\sigma} + \sigma = Enw + Hwt \tag{3.38}$$

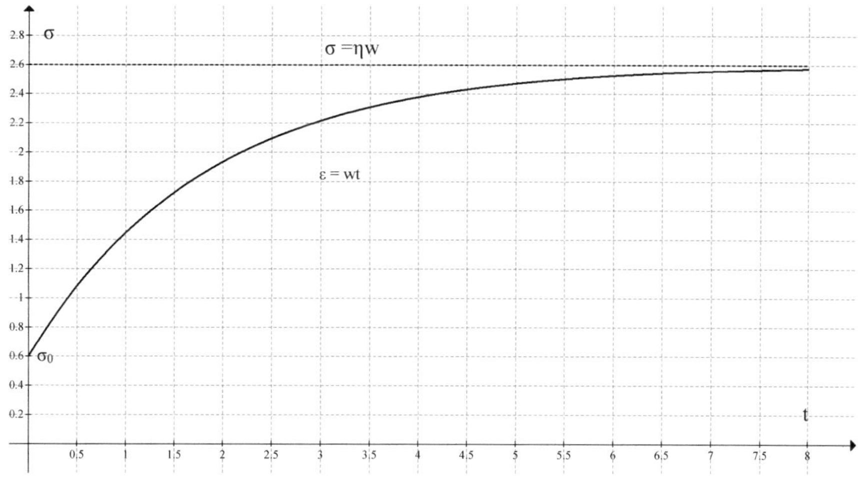

Fig. 3.28 Stress–time diagram

Its solution is:

$$\begin{aligned}&\sigma = Hwt + \sigma_0 e^{-t/n} + n(E - H)\left[1 - e^{-t/n}\right]w \\ &\sigma_0 - \text{ initial stress}\end{aligned} \tag{3.39}$$

If $H = 0$ (Maxwell model), reduced formula (3.39) has a form $\sigma = \sigma_0 e^{-t/n} + nwE\left[1 - e^{-t/n}\right]$. The limit stress value is $\sigma = nwE = \eta w$ (see Fig. 3.28).

If the strain is linearly decreasing ($\varepsilon = -wt$), where w is a constant strain rate, Eq. (3.23) has the form (initial stress σ_0).

$$\begin{aligned}&\sigma = -Hwt + \sigma_0 e^{-t/n} - n(E - H)\,w(1 - e^{-t/n}) \\ &\text{where } \sigma_0 - \text{initial stress value}\end{aligned} \tag{3.40}$$

If initial natural condition is zero ($\varepsilon_0 = \sigma_0/E = 0$), Eq. (3.40) is reduced as follows:

$$\sigma = -Hwt - n(E - H)\,w(1 - e^{-t/n}) \tag{3.41}$$

At zero long-term modulus $H = 0$, i.e., for the Maxwell model, the expression (3.40) gets as follows:

$$\sigma = \sigma_0 e^{-t/n} - \eta w(1 - e^{-t/n}) \tag{3.42}$$

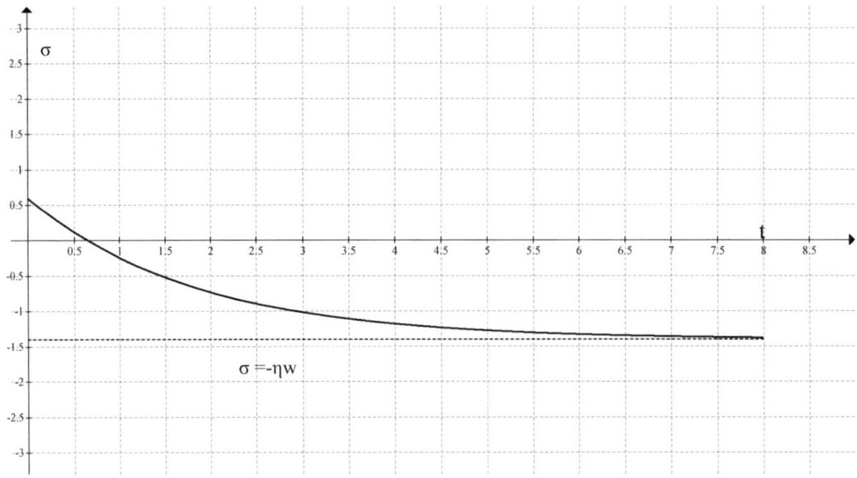

Fig. 3.29 Uniform strain increases (Maxwell model)

where $\eta = nE$—the coefficient of viscosity. After long period of time σ converges here to a constant value $-\eta w$ (see Fig. 3.28). If $\sigma_0 = nw(E - H)$ then the stress function (3.42) becomes a linear function of time (Fig. 3.29).

Uniform change deformations occur, e.g., when testing samples in hydraulic presses with a constant feed rate of the piston. When unloading at a constant negative strain rate, in some cases may be some increase in stress in the initial period of unloading process similar to the phenomenon of increasing strains in the initial period of unloading process with a constant rate of stress reduction.

3.4 Substitution Method

Substituting $t = 0.01 + 0.02\theta - 0.0016\theta^2$ (the inverse function of $\theta(t)$) in Eq. (3.23) we obtain the following equivalent differential equation:

$$\begin{aligned}
&n\dot{\sigma} + \sigma = En\dot{\varepsilon} + H\varepsilon = E_0A[\exp(-0.15\theta)]\{w + \frac{mw}{n}\theta[0.02 - 0.0032\theta]\}\\
&H = mE;\\
&E = E_0\exp(-0.15\theta)\\
&\frac{d\sigma}{d\theta} = -\frac{1}{n}\sigma[0.02 - 0.0032\theta] + E_0A[\exp(-0.15\theta)]\{w + \frac{mw}{n}\theta[0.02 - 0.0032\theta]\}
\end{aligned} \tag{3.43}$$

3.4.1 Linear Strain Increase

The solutions of Eq. (3.43), Standard Linear Model (SLM), for different rates "w" of strain increase are presented below (Figs. 3.30, 3.31 and 3.32):

$$\mathrm{d}y/\mathrm{d}x = -(1/n)\,y(0.02 - 0.0032x) + E_0A(\exp(-0.15x))(w + (mw/n)\,x(0.02 - 0.0032x))$$

$$w = 0.1; \quad n = 0.01; \quad m = H/E = 0.5$$

$$\mathrm{d}(y3)/\mathrm{d}(x) = -(1/0.01) * y3 * (0.02 - 0.0032 * x) + 2.9 * 7.02 * (\exp(-0.15 * (x))^\wedge 1) * (0.5 + (0.5 * 0.5/0.01) * x * (0.02 - 0.0032 * x))$$

$$y3(0) = 0$$

Calculated values of DEQ variables

	Variable	Initial value	Minimal value	Maximal value	Final value
1	x	0	0	8	8
5	$y3$	0	0	7.299647	7.299647

Differential equations

$$\mathrm{d}(y3)/\mathrm{d}(x) = -(1/0.01) * y3 * (0.02 - 0.0032 * x) + 2.9 * 7.02 * (\exp(-0.15 * (x))^\wedge 1) * (0.1 + (0.1 * 0.5/0.01) * x * (0.02 - 0.0032 * x))$$

Model: $y3 = -0.013 + 1.685 * x - 0.35 * x^\wedge 2 + 0.031 * x^\wedge 3$

Variable	Value
$a0$	−0.0130265
$a1$	1.685332
$a2$	−0.3505818
$a3$	0.0311403

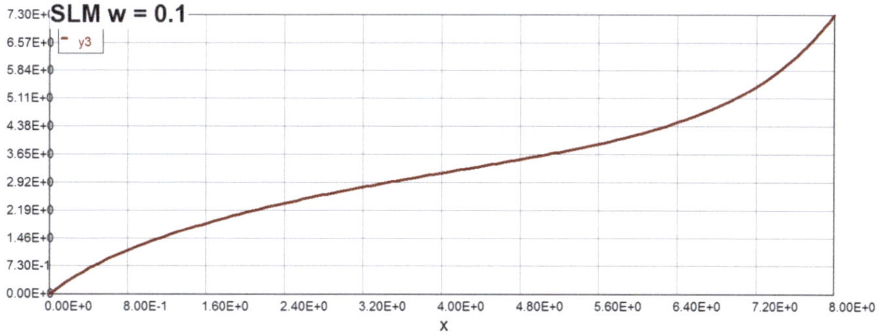

Fig. 3.30 Standard linear model (SLM) (linear strain increase w = 0.1)

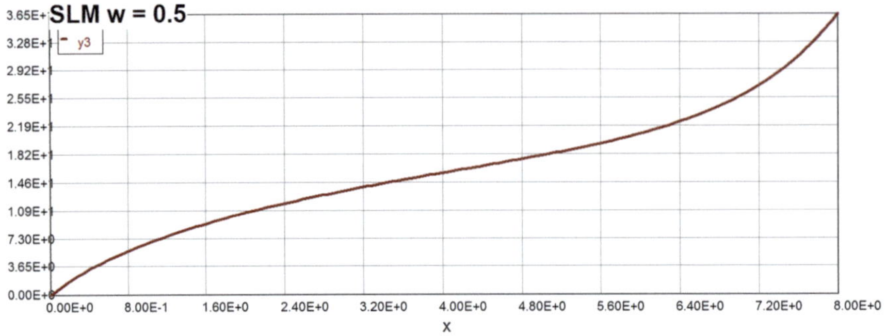

Fig. 3.31 Standard linear model (SLM) (linear strain increase $w = 0.5$)

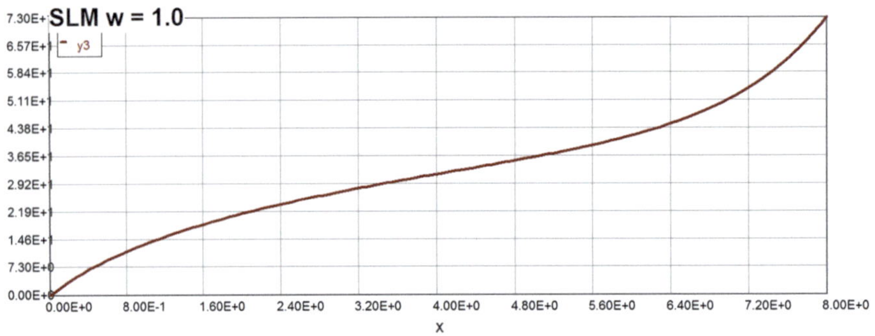

Fig. 3.32 Standard linear model (SLM) (linear strain increase $w = 1.0$)

$w = 0.5$; $n = \eta/E = 0.01$; $m = 0.5$

Calculated values of DEQ variables

	Variable	Initial value	Minimal value	Maximal value	Final value
1	x	0	0	8	8
5	$y3$	0	0	36.49823	36.49823

Differential equations

4 $\mathrm{d}(y3)/\mathrm{d}(x) = -(1/0.01) * y3 * (0.02 - 0.0032 * x) + 2.9 * 7.02 * (\exp(-0.15 * (x))^\wedge 1) * (0.5 + (0.5 * 0.5/0.01) * x * (0.02 - 0.0032 * x))$

Model: $y3 = -0.065 + 8.427 * x - 1.753 * x\hat{\ }2 + 0.156 * x\hat{\ }3$

Variable	Value
$a0$	−0.0651322
$a1$	8.426661
$a2$	−1.752909
$a3$	0.1557014

$w = 1.0; n = 0.01; m = 0.5$

Calculated values of DEQ variables

	Variable	Initial value	Minimal value	Maximal value	Final value
1	x	0	0	8	8
5	$y3$	0	0	72.99647	72.99647

Differential equations

$$4 \quad d(y3)/d(x) = -(1/0.01) * y3 * (0.02 - 0.0032 * x) + 2.9 * 7.02 * (\exp(-0.15 * (x))\hat{\ }1) * (1.0 + (1.0 * 0.5/0.01) * x * (0.02 - 0.0032 * x))$$

Model: $y3 = -0.13 + 16.85 * x - 3.5 * x\hat{\ }2 + 0.311 * x\hat{\ }3$

Variable	Value
$a0$	−0.1302647
$a1$	16.85332
$a2$	−3.505818
$a3$	0.3114028

The summary of results in case of constant rates of strain increases are presented in Fig. 3.33.

3.4.2 *Linear Strain Decrease*

The solutions of Eq. (3.38) for different rates "*w*" of strain decrease are presented below (Figs. 3.34, 3.35 and 3.36):

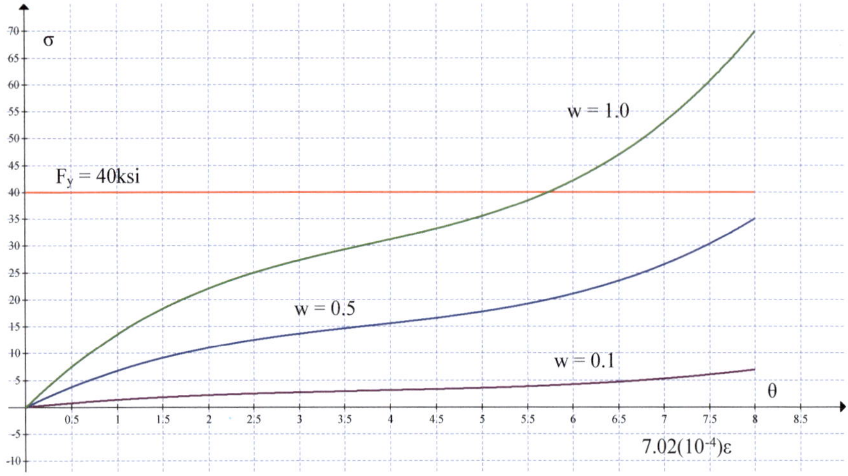

Fig. 3.33 Standard linear model (SLM) (summary of results)

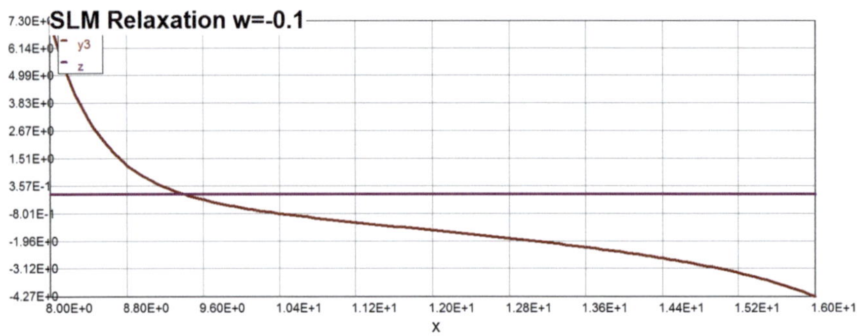

Fig. 3.34 Standard linear model (SLM) (linear strain decrease $w = 0.1$)

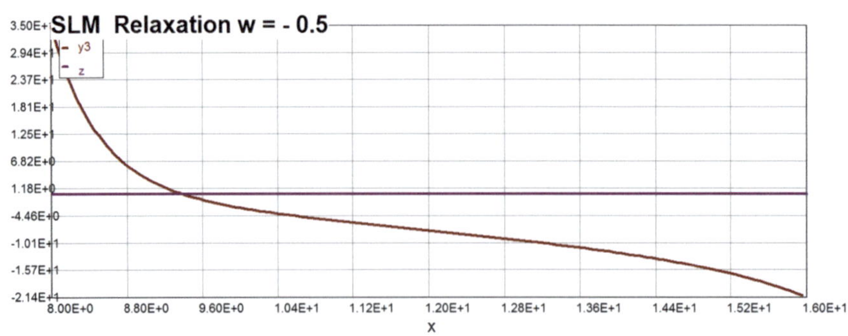

Fig. 3.35 Standard linear model (SLM) (linear strain decrease $w = 0.5$)

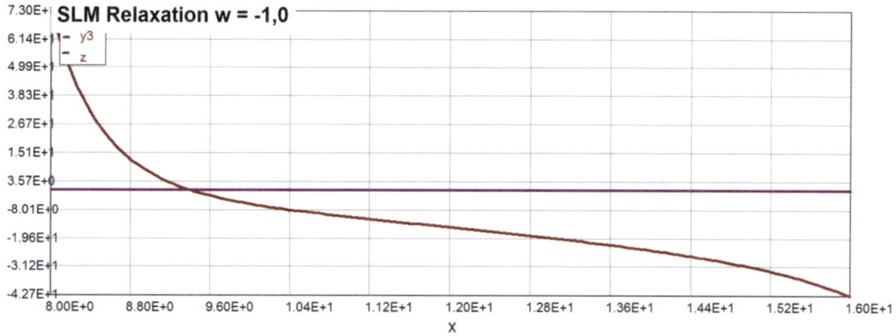

Fig. 3.36 Standard linear model (SLM) (linear strain decrease w = 1.0)

$$dy/dx = -(1/n)\,y(0.02 - 0.0032x) + E_0A0.3(\exp(0.15x))(-w - (mw/n)\,x(0.02 - 0.0032x))$$

$$\begin{aligned}
&n\dot{\sigma} + \frac{1}{n}\sigma[0.02 - 0.0032\theta] = E\dot{\varepsilon} + \frac{1}{n}H\varepsilon\\
&E_0A0.3[\exp(0.15\theta)]\{-w - \frac{mw}{n}\theta[0.02 - 0.0032\theta]\}\\
&H = mE; E = E_0\exp(-0.15\theta)\\
&\frac{d\sigma}{d\theta} = -\frac{1}{n}\sigma[0.02 - 0.0032\theta] + 0.3E_0A[\exp(0.15\theta)]\{-w - \frac{mw}{n}\theta[0.02 - 0.0032\theta]\}\\
&\sigma(0) = \sigma_0\,\text{ksi}
\end{aligned} \quad (3.44)$$

Calculated values of DEQ variables

	Variable	Initial value	Minimal value	Maximal value	Final value
1	x	8	8	16	16
2	y	50	3.218843	50	4.37016

Differential equations

$$1\ d(y)/d(x) = -100 * y * (0.02 + 0.0032 * (x - 8)) + 2.9 * 7.02 * 0.3 * (\exp(0.15 * (x - 8)))\hat{}1$$

Model: $y3 = 522.9 - 167.9 * x + 20 * x\hat{}2 - 1.047 * x\hat{}3 + 0.0203 * x\hat{}4$

Variable	Value
$a0$	522.8858
$a1$	−167.8642
$a2$	19.99072

(continued)

(continued)

Variable	Value
$a3$	−1.047406
$a4$	0.0203215

Model: $y3 = 2497 - 801 * x + 95.3 * x\hat{}2 - 4.99 * x\hat{}3 + 0.0967 * x\hat{}4$

Variable	Value
$a0$	2497.472
$a1$	−801.0649
$a2$	95.30641
$a3$	−4.988553
$a4$	0.0966723

Calculated values of DEQ variables

	Variable	Initial value	Minimal value	Maximal value	Final value
1	x	8	8	16	16
2	y	50	3.218843	50	4.37016

Differential equations

1 $d(y)/d(x) = -100 * y * (0.02 + 0.0032 * (x - 8)) + 2.9 * 7.02 * 0.3 * (\exp(0.15 * (x - 8)))\hat{}1$

Model: $y3 = 5228.8 - 1678.6 * x + 199.9 * x\hat{}2 - 10.474 * x\hat{}3 + 0.2032 * x\hat{}4$

Variable	Value
$a0$	5228.858
$a1$	−1678.642
$a2$	199.9072
$a3$	−10.47406
$a4$	0.2032149

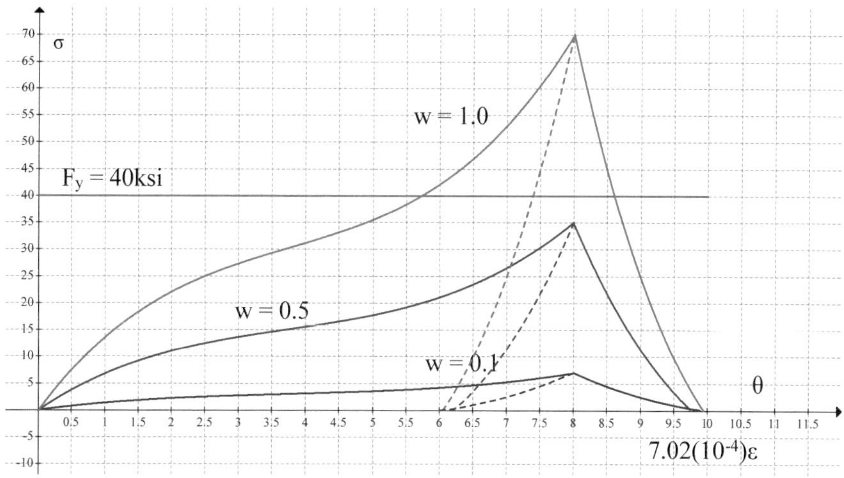

Fig. 3.37 Standard linear model (SLM) (summary of results)

The summary of results in case of constant rates of strain decreases are presented in Fig. 3.37.

3.5 Harmonic Variation of the Stress and Strain

Creep and stress relaxation tests are convenient for studying material response at long times (minutes to days), but less accurate at shorter times (seconds and less). Dynamic tests, in which the stress (or strain) resulting from a sinusoidal strain (or stress) is measured, are often well suited for filling out the "short-time" range of material response. When a viscoelastic material is subjected to a sinusoidal varying stress, a steady state will eventually be reached in which the resulting strain is also sinusoidal, having the same angular frequency but retarded in phase by an angle φ; this is analogous to the delayed strain observed in creep experiments. The strain lags the stress by the phase angle δ, and this is true even if the strain rather than the stress is the controlled variable.

If the stress varies sinusoidal $\sigma = \sigma_0 \sin\omega t$, then a solution of the Eq. (3.23) for deformations can be sought in the form $\varepsilon = \varepsilon_0 \sin(\omega t + \varphi)$. The parameter φ is a phase angle relative to the phase of deformation vibrations stress fluctuations. Introducing ε in the form $\varepsilon = \mathrm{A} \sin\omega t + \mathrm{B} \cos\omega t$, where $A = \varepsilon_0 \cos\varphi$; $B = \varepsilon_0 \sin\varphi$ and substituting these expressions into (3.23), we obtain a system of algebraic equations for determining the values of A and B:

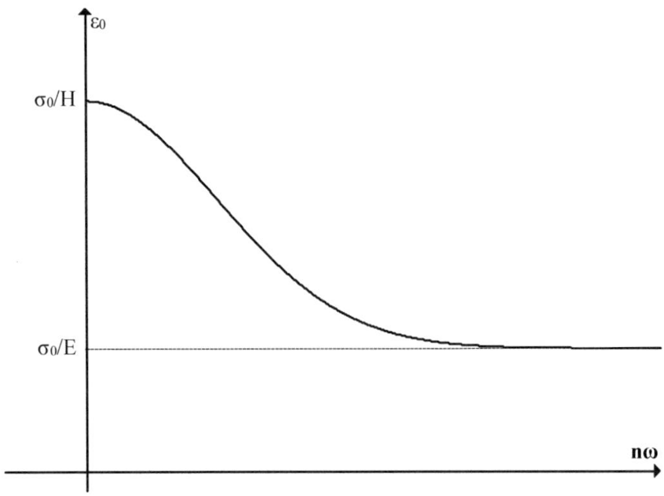

Fig. 3.38 Amplitude ε_0 versus dimensionless frequency $n\omega$

$$\begin{cases} HA - EnB = \sigma_0 \\ En\omega A + HB = n\omega\sigma_0 \end{cases} \quad \mathrm{A} = \frac{\sigma_0(H + En^2\omega^2)}{H^2 + E^2n^2\omega^2}; \; B = \frac{-\sigma_0(E - H)n\omega}{H^2 + E^2n^2\omega^2}$$
$$\varepsilon_0 = \sqrt{A^2 + B^2} = \sigma_0\sqrt{\frac{1 + n^2\omega^2}{H^2 + E^2n^2\omega^2}}; \; \tan\varphi = \frac{B}{A} = -\frac{(E - H)n\omega}{(H + En^2\omega^2)} \tag{3.45}$$

The amplitude ε_0 is changing from σ_0/H to σ_0/E, when the frequency ω is changing from zero to infinity. Similarly, ε_0 is dependent on the relaxation time and n (Fig. 3.38).

Phase angle φ is zero for values of $n\omega$ equal to zero and infinity. The maximum value of the absolute value of the phase angle φ is at $\omega v = H/E$, then $\tan \varphi = (H - E)/2HE$. Note that the angle φ is always negative, this means that the deformation delayed relative to change in stress. In particular cases, schemes of work material of formula (3.39) take the following form

(a) non-relaxing for material (model Kelvin–Voigt model): $E = \infty$; $n = 0$ and $En = \eta$, then

$$\varepsilon_0 = \frac{\sigma_0}{\sqrt{H^2 + \eta^2\omega^2}}; \; \tan\varphi = -\frac{\eta\omega}{H} \tag{3.46}$$

(b) relaxing material (Maxwell model) $H = 0$ then we obtain:

$$\varepsilon_0 = \frac{\sigma_0\sqrt{1+n^2\omega^2}}{En\omega}; \; \tan\varphi = -\frac{1}{n\omega} \tag{3.47}$$

3.6 Arbitrary Law of Variation of Stress and Strain

If the stress varies with time under arbitrary law, then by solving the Eq. (3.18) with respect to deformations we obtain the formula

$$\varepsilon = \mathrm{e}^{-\frac{Ht}{En}}\left[\frac{1}{nE}\int_0^t (\sigma + n\dot{\sigma})\mathrm{e}^{\frac{H\tau}{En}}\mathrm{d}\tau + C\right] \tag{3.48}$$

An arbitrary constant C in (3.42) is obtained now using the initial condition $\varepsilon(0)$ and the result is as follows:

$$\varepsilon = \varepsilon(0)\mathrm{e}^{-\frac{Ht}{En}} + \left[\frac{1}{nE}\int_0^t (\sigma + n\dot{\sigma})\mathrm{e}^{-\frac{H(t-\tau)}{En}}\mathrm{d}\tau\right] \tag{3.49}$$

This expression (3.49) can be simplified by integration by parts of the second term in the brackets. The formula (3.49) takes the form:

$$\varepsilon(t) = \left[\varepsilon(0) - \frac{\sigma(0)}{E}\right]\mathrm{e}^{-\frac{Ht}{En}} + \frac{\sigma(t)}{E} + \left[\frac{E-H}{nE^2}\int_0^t \sigma(\tau)\mathrm{e}^{-\frac{H(t-\tau)}{En}}\mathrm{d}\tau\right] \tag{3.50}$$

If at the initial time $t = 0$, the material is in its natural state, then $\varepsilon(0) = \frac{\sigma(0)}{E}$, and formula (3.50) can be simplified again:

$$\varepsilon(t) = \frac{\sigma(t)}{E} + \left[\frac{E-H}{nE^2}\int_0^t \sigma(\tau)\mathrm{e}^{-\frac{H(t-\tau)}{En}}\mathrm{d}\tau\right] \tag{3.51}$$

In general case we can put:

$$\begin{gathered}\varepsilon(t) = \frac{1}{E}\left[\sigma(t) + \frac{E-H}{nE}\int_0^t \sigma(\tau)\mathrm{e}^{-\frac{H(t-\tau)}{En}}\mathrm{d}\tau\right] \\ \text{where } K(t-\tau) = \frac{E-H}{nE}\mathrm{e}^{-\frac{H(t-\tau)}{En}}\end{gathered} \tag{3.52}$$

Integral Eq. (3.52) is equivalent to the differential Eq. (3.23). Formula (3.52) considers all load changes which took place before the considered time t, so that the function expressed by it, is called hereditary stress–strain function. The function

$K(t - \tau)/E$ is the influence function (*creep compliance function*, *memory function* or simply "creep function") of elementary impulse $\sigma(\tau)\mathrm{d}\tau$ that is applied to the element (sample) at time τ for the deformation seen in the other time $t > \tau$. This kind of influence function corresponds to the material whose properties do not change over time, or as we call it, time invariant material. This follows from the fact that any addition of the same value to t and τ does not change the value of $K(t - \tau)$.

Similarly, from Eq. (3.23) by solving it with respect to σ one can obtain an expression for the stress at any given strain law in time. Under natural initial conditions $\sigma(0) = E\,\varepsilon(0)$, we obtain:

$$\begin{aligned} \sigma(t) &= E\left[\varepsilon(t) - \frac{E - H}{nE}\int_0^t \varepsilon(\tau)\mathrm{e}^{-\frac{(t-\tau)}{n}}\mathrm{d}\tau\right] \\ \text{where } R(t - \tau) &= \frac{E - H}{nE}\mathrm{e}^{-\frac{(t-\tau)}{n}} \end{aligned} \quad (3.53)$$

Influence function R describes the deformations that had occurred at time τ on the stresses that are occurring at time t.

It is, as well as the influence function $K(t - \tau)$, a single variable function (the time difference $t - \tau$).

Equation (3.53) can be considered as the solution of the integral Eq. (3.51) with respect to the stresses. In this case $K(t - \tau)$ is the kernel of the integral Eq. (3.51) and $R(t - \tau)$—its resolvent. In turn, the formula (3.45) is a solution of an integral Eq. (3.53) with respect to deformations. Equation (3.51) is an integral equation of Volterra second type with respect to σ. Special simplified cases (modified Eqs. 3.51 and 3.53) are as follows:

1. In case of the Maxwell model at $H = 0$:

$$\begin{aligned} \varepsilon(t) &= \frac{1}{E}\left[\sigma(t) + \frac{1}{n}\int_0^t \sigma(\tau)\mathrm{d}\tau\right] \\ \sigma(t) &= E\left[\varepsilon(t) - \frac{1}{n}\int_0^t \varepsilon(\tau)\mathrm{e}^{-\frac{(t-\tau)}{n}}\mathrm{d}\tau\right] \end{aligned} \quad (3.54)$$

Here, the kernel is a constant: $K = 1/n$

2. In the case of the Kelvin–Voigt model, putting $E = \infty$; $n = 0$ and $En = \eta$, we obtain:

$$\varepsilon(t) = \frac{1}{\eta}\left[\int_0^t \sigma(\tau)\mathrm{e}^{-\frac{H(t-\tau)}{\eta}}\mathrm{d}\tau\right] \quad (3.55)$$

The latter formula means that the stress does not depend on previous stress and strain values, and they are determined directly from the current time τ values.

3.7 Impulse Stress Function

Now we examine the response to a sudden load. Equation (3.23) allows also finding the solution of an impact force (stress) action on a given structural element at a given time t. We will say that the force action is instant if:

$$\begin{aligned} &\sigma(t) = 0 \;\text{ if }\; t \neq t_0 \\ &\sigma(t) = 1 \;\text{ if }\; t = t_0 \\ &\int_{t_0-\alpha}^{t_0+\alpha} \sigma(t)\mathrm{d}t = I = \text{const} \\ &\text{and} - \alpha \text{ very small} \end{aligned} \tag{3.56}$$

After substituting Eq. (3.55) onto (3.23), we have

$$\varepsilon(t) = \frac{I}{E}K(t - t_0) = I\frac{E - H}{E^2 n}\mathrm{e}^{-\frac{H(t-t_0)}{En}} \tag{3.57}$$

The deformations $\varepsilon(t)$ are equal to zero if $t < t_0$, and if $t = t_0$, then:

$$\varepsilon(t = t_0) = \frac{I}{E}K(0) = I\frac{E - H}{E^2 n} \tag{3.58}$$

Therefore, the deformations have a jump (3.58) at $t = t_0$, and after that they are decreasing to zero when time $t \to \infty$. Differentiating Eq. (3.57) ones, we will get the deformation rate at the time of impact force application (it has also a jump):

$$\dot{\varepsilon}(t = t_0) = I\frac{E - H}{E^3 n^2}H \tag{3.59}$$

3.8 Linearly Hereditary Creep

As seen in the previous sections, linear viscoelasticity can be stated in terms of mechanical models constructed from linear springs and dashpots. These models generate constitutive relations that are ordinary differential equations. However, integral equations could be used as well, and this integral approach is also used as a starting point for viscoelastic theory.

Integrals are summing operations, and this view of viscoelasticity takes the response of the material at time t to be the sum of the responses to excitations imposed at all previous times.

The ability to sum these individual responses requires the material to obey a more general statement of linearity than we have invoked previously, specifically that the response to a number of individual excitations be the sum of the responses that would have been generated by each excitation acting alone. Mathematically, if the stress due to a strain $\varepsilon_1(t)$ is $\sigma(\varepsilon_1)$ and that due to a different strain $\varepsilon_2(t)$ is $\sigma(\varepsilon_2)$, then the stress due to both strains is $\sigma(\varepsilon_1 + \varepsilon_2) = \sigma(\varepsilon_1) + \sigma(\varepsilon_2)$.

Combining this with the condition for multiplicative scaling, we have as a general statement of linear viscoelasticity: $\sigma(a\varepsilon_1 + b\varepsilon_2) = a\ \sigma(\varepsilon_1) + b\ \sigma(\varepsilon_2)$.

The "Boltzman Superposition Integral" statement of linear viscoelastic response follows from this definition. Consider the stress $\sigma_1(t)$ at time t due to the application of a small strain $\Delta\varepsilon_1$ applied at a time τ_1 previous to t; this is given directly from the definition of the relaxation modulus as $\sigma_1(t) = E_{\mathrm{rel}}(t - \tau_1)\ \Delta\varepsilon_1$. Similarly, the stress $\sigma_2(t)$ due to a strain increment $\Delta\varepsilon_1$ applied at a different time τ_2 is $\sigma_2(t) = E_{\mathrm{rel}}(t - \tau_2)\ \Delta\varepsilon_2$. If the material is linear, the total stress at time t due to both strain increments together can be obtained by superposition of these two individual effects: $\sigma(t) = \sigma_1(t) + \sigma_2(t) = E_{\mathrm{rel}}(t - \tau_1)\ \Delta\varepsilon_1 + E_{\mathrm{rel}}(t - \tau_2)\ \Delta\varepsilon_2$. As the number of applied strain increments increases so as to approach a continuous distribution, this becomes

$$\sigma(t) = \sum_i \sigma_i(t) = \sum_i E_{\mathrm{rel}}(t - \tau_{\mathrm{i}})\Delta\varepsilon_{\mathrm{i}}$$

$$\sigma(t) = \int_{-\infty}^{t} E_{\mathrm{rel}}(t - \tau)\mathrm{d}\varepsilon = \int_{-\infty}^{t} E_{\mathrm{rel}}(t - \tau)\frac{\mathrm{d}\varepsilon}{\mathrm{d}\tau}\mathrm{d}\tau$$

Consider now some of the models.

3.8.1 Infinite Number of Series-Connected Standard Linear Models

Let us take a complex model with an infinite number of series-connected standard linear models.

For each component of the model takes place here the relationship between stress and strain (3.51) (Fig. 3.39).

$$\varepsilon_i(t) = \frac{1}{E}\left[\sigma(t) + \frac{E_i - H_i}{n_i E_i}\int_0^t \sigma(\tau)\,\mathrm{e}^{-\frac{H_i(t-\tau)}{E_i n_i}}\mathrm{d}\tau\right] \tag{3.60}$$

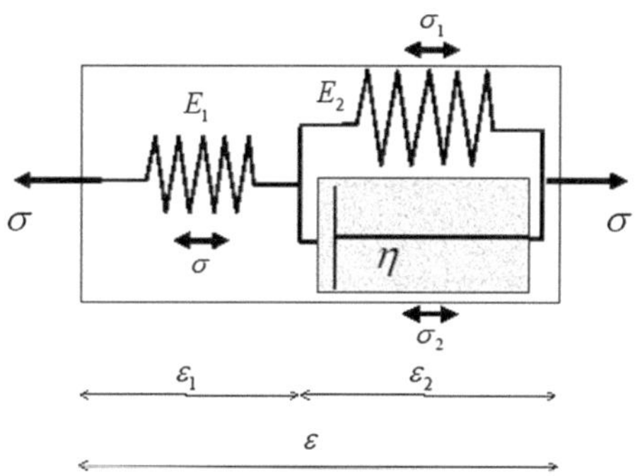

Fig. 3.39 Model of series-connected components

where i—the index relating to the component model and with σ—the stress is the same for all finite models of series-connected components. Summing up all the strains ε_i of the model, we obtain a model with a finite number of components of model is equal to k.

$$\varepsilon(t) = \sum_{i=1}^{k} \varepsilon_i(t) = \sigma(t)\sum_{i=1}^{k}\frac{1}{E_i} + \int_0^t \sigma(\tau)\sum \frac{E_i - H_i}{E_i^2 n_i} e^{-\frac{H_i(t-\tau)}{E_i n_i}} \mathrm{d}\tau \tag{3.61}$$

Equation (3.61) can be rewritten now in standard form similar to (3.51) as follows:

$$\begin{gathered}\varepsilon(t) = \frac{1}{E}\left[\sigma(t) + \frac{E - H}{nE}\int_0^t \sigma(\tau)\mathrm{e}^{-\frac{H(t-\tau)}{En}}\mathrm{d}\tau\right] \\ \text{where } K(t-\tau) = E\sum_{i=1}^{k}\frac{E_i - H_i}{n_i E_i^2}\mathrm{e}^{-\frac{H_i(t-\tau)}{E_i n_i}} \\ \frac{1}{E} = \sum_{i=1}^{k}\frac{1}{E_i} \text{ and } \frac{1}{H} = \sum_{i=1}^{k}\frac{1}{H_i}\end{gathered} \tag{3.62}$$

Thus, the kernel K of the integral equation is the sum of exponential functions with different delay times n_i that could be considered as an independent variable of a function f $(1/n_i)$. The constant values of $1/E_i$ and $1/H_i$ shall be considered infinitesimal (when the number k of components of standard models tends to infinity), because their respective sums must be finite.

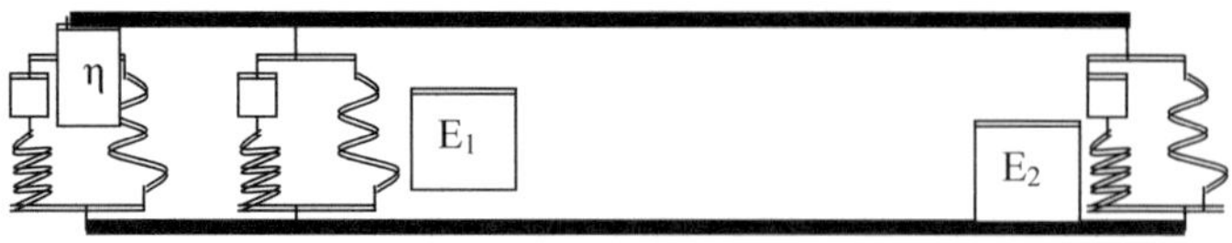

Fig. 3.40 Model of parallel-connected components

3.8.2 Infinite Number of Parallel-Connected Standard Linear Models

Let us take a complex model with an infinite number of parallel-connected standard linear models (see Fig. 3.40).

In this model, we express the stress through deformation by using formula (3.62).

$$\sigma_i(t) = E_i\varepsilon(t) + \frac{H_i - E_i}{n_i}\int_0^t \varepsilon(\tau)\mathrm{e}^{-\frac{(t-\tau)}{n_i}}\mathrm{d}\tau \quad (3.63)$$

Strain $\varepsilon(t)$ is the same for all components elementary models whose parameters are assigned indices $i = 1, 2, \ldots, k$. Full stress is the sum of stress components models.

$$\sigma_i(t) = \sum_{i=1}^{k} E_i\varepsilon(t) + \int_0^t \varepsilon(\tau)\sum_{i=1}^{k}\frac{H_i - E_i}{n_i}\mathrm{e}^{-\frac{(t-\tau)}{n_i}}\mathrm{d}\tau \quad (3.64)$$

Equation (3.64) can be rewritten now in standard form similar:

$$\begin{aligned}\sigma(t) &= E[\varepsilon(t) - \int_0^t \varepsilon(\tau)\,R(t-\tau)]\mathrm{d}\tau \\ E &= \sum_{k} E_i;\ R(t-\tau) = \frac{1}{E}\sum_{i=1}^{k}\frac{E_i - H_i}{n_i}\mathrm{e}^{-\frac{t-\tau}{n_i}} = \frac{1}{E}\sum_{i=1}^{k} f(1/n_i)\mathrm{e}^{-\frac{t-\tau}{n_i}}\end{aligned} \quad (3.65)$$

3.9 Retardation and Relaxation Spectrum

Generalized models can contain many parameters and will exhibit a whole array of relaxation and retardation times. The response of real materials can be modeled by allowing for a number of different retardation times of different orders of magnitude.

To proceed to an infinite number of components models, one should place them in the ascending order of their relaxation times n_i and assume that $E_i = E(n)\mathrm{d}n$ and

$H = H(n)\mathrm{d}n$, where n is considered a continuously changing variable. In the limit as $N \to \infty$, letting sums in (3.65) be replaced by integrals and changing the dummy variable $1/n$ to α, the integral Eq. (3.57) has a form:

$$\overline{\sigma}(t) = \sigma(t) + \int_0^t \int_0^\infty f(\alpha)[1 - \mathrm{e}^{-\alpha t}]K(t - \tau)\mathrm{d}\tau \mathrm{d}\alpha \tag{3.66}$$

The representation (3.66) allows for a continuous retardation time, in contrast to the discrete times of the model (3.63). The function $f(\alpha)$ is called the **retardation spectrum** of the model. Different responses can be modeled by simply choosing different forms for the retardation spectrum.

After substitution time t and retardation time t_{ri} by the dimensionless temperature θ and variable $\alpha_{\mathrm{ri}} = 1/\theta_{\mathrm{ri}}$ respectfully we have now

$$\overline{\sigma}(\theta) = \sigma(\theta) + \int_0^\theta \int_0^\infty f(\alpha)[1 - \mathrm{e}^{-\alpha\theta}]K[t(\theta) - \tau(\tilde{\theta})]\tau'(\tilde{\theta})\mathrm{d}\tilde{\theta}\mathrm{d}\alpha \tag{3.67}$$

A similar analysis can be carried out for the Generalized Maxwell model. For two Maxwell elements in parallel, the constitutive equation can be shown to be

$$\sigma + \frac{E_2\eta_1 + E_1\eta_2}{E_1E_2}\dot{\sigma} + \frac{\eta_1\eta_2}{E_1E_2}\ddot{\sigma} = (\eta_1 + \eta_2)\dot{\varepsilon} + \frac{E_2 + E_1}{E_1E_2}\eta_1\eta_2\ddot{\varepsilon} \tag{3.68}$$

Consider the case of specified strain, so that this is a second-order differential equation in $\sigma(t)$. The homogeneous solution is

$$\sigma_h(t) = C_1\mathrm{e}^{-\frac{t}{t_{r1}}} + C_2\mathrm{e}^{-\frac{t}{t_{r2}}}$$

For a constant strain ε_0, the full solution is

$$\sigma(t) = \varepsilon_0\left[E_1\mathrm{e}^{-t/t_{r1}} + E_2\mathrm{e}^{-t/t_{r2}}\right]$$

Thus, whereas the single Maxwell unit has a single relaxation time, this model has two relaxation times, which are the eigenvalues of the differential constitutive equation. The term inside the square brackets is evidently the relaxation modulus of the model.

By considering a model with an indefinite number of Maxwell units in parallel, each with vanishingly small elastic modulus ΔE_i, one has the expression analogous to (3.59), and $\varphi(\alpha)$ is called the **relaxation spectrum** of the model.

$$\overline{\sigma}(\theta) = \sigma(\theta) + \int_0^{\theta}\int_0^{\infty} \varphi(\alpha)[1 - \mathrm{e}^{-\alpha\theta}]R[t(\theta) - \tau(\tilde{\theta})]\tau'(\tilde{\theta})\mathrm{d}\tilde{\theta}\mathrm{d}\alpha \tag{3.69}$$

3.10 Long-Term Modulus of Elasticity

The stress–strain relationship derived from a consideration of various viscoelastic models can be expressed in an integral form. In this case the kernel K is a linear combination of a finite or infinite number of exponential functions:

$$\begin{gathered}\varepsilon(t) = \frac{1}{E(t)}\left[\sigma(t) + \int_0^t \sigma(\tau)K(t - \tau)\mathrm{d}\tau\right] \\ \text{where } K(t - \tau) = \sum_i A_i \mathrm{e}^{-\alpha_i(t-\tau)}\end{gathered} \tag{3.70}$$

In Eq. (3.62) function $[K(t - \tau)]/E$ represents the effect of the impact load applied at time τ on the deformation at time t. This function must be defined for positive values of the argument $t - \tau$ and is usually a monotonically decreasing function asymptotically approaching zero.

For negative values of the argument $t - \tau$ function of demand $K(t - \tau)$ is identically zero, since the load impulses only affect subsequent deformation, and not the previous ones. Let's substitute the independent variable τ under the integral sign in (3.70) on with $t - \tau = u$. Then the Eq. (3.70) can be rewritten as:

$$\begin{gathered}\varepsilon(t) = \frac{1}{E(t)}\left[\sigma(t) + \int_{t_0}^t \sigma(t - u)\, K(u)\mathrm{d}u\right] \\ \text{where } K(u) = \sum_i A_i \mathrm{e}^{-\alpha_i u}\end{gathered} \tag{3.71}$$

Consider the case of a constant load when the stress σ up to time t_0 is equal to zero, and after this time has a constant value, then for all $t > t_0$. Equation (3.71) has the form

$$\varepsilon(t) = \frac{\sigma_0}{E}\left[1 + \int_0^{t - t_0} K(u)\mathrm{d}u\right] \tag{3.72}$$

Immediately after the instant application of stress σ_0 we have $\varepsilon = \sigma_0/E$. Consequently E represents the instantaneous modulus of elasticity. The quantity E, obviously, must always be greater than zero. Strain rate $\dot{\varepsilon}(t)$ after instantaneous application constant stress can be obtained by differentiating Eq. (3.72) with respect to time t.

$$\dot{\varepsilon}(t) = \frac{\sigma_0}{E} K(t - t_0) \text{ and if } t_0 = 0, \text{ then } K(t) = \frac{E\dot{\varepsilon}(t)}{\sigma_0} \tag{3.73}$$

Thus $K(t)/E$ is a function of rate of change in deformation after constant stress (σ_0 = 1) application to the sample that has a natural initial conditions. Hence, this is a simple method of experimental determination of function $K(t)$ based on standard creep tests under constant load. From Eq. (3.72) it follows that in the limit as $t \to \infty$ the strain $\varepsilon(\infty)$ is equal to:

$$\varepsilon(\infty) = \frac{\sigma_0}{E}\left[1 + \int_0^{\infty} K(u)\mathrm{d}u\right]$$

The quantity

$$H = E/\left[1 + \int_0^{\infty} K(u)\mathrm{d}u\right] \tag{3.74}$$

can be called long-term modulus of elasticity that is the ratio of stress to strain after a very long constant load application. Long-term modulus is finite, if the integral $\int_0^{\infty} K(u)\mathrm{d}u =_t \overset{\lim}{\to} \infty \int_0^t K(u)\mathrm{d}u$ has a finite limit. Otherwise—the long-term modulus of elasticity is zero. An analogous definition of long-term modulus of elasticity is used in creep buckling problems [28–30].

3.11 The Resolvent of Linear Equation Hereditary Deformation

We seek a solution of Eq. (3.53) in the form:

$$\frac{1}{E}\sigma(t) = \left[\varepsilon(t) - \int_0^t \varepsilon(\tau)R(t - \tau)\mathrm{d}\tau\right] \tag{3.75}$$

The function R is a resolution of the integral Eq. (3.70) and can be considered as a function of the influence of strain at time τ on the voltage at the time t. Function R (same as a function K) must be positive monotonically decreasing function of its argument. For negative values of the argument function R is identically zero. Substitute solution (3.53) in Eq. (3.70). After elementary algebraic manipulations, we have:

$$\left[\int_0^t \varepsilon(\tau)K(t-\tau)\mathrm{d}\tau - \int_0^t \varepsilon(\tau)R(t-\tau)\mathrm{d}\tau\right] = \int_0^t\int_0^\tau \varepsilon(s)R(t-s)K(t-\tau)\mathrm{d}s\mathrm{d}\tau \tag{3.76}$$

Since the function $\varepsilon(\tau)$—arbitrary function, then you can put that $\varepsilon(\tau) = \delta(t)$ ($\delta(t)$ is delta-function) and then from Eq. (3.76) we obtain:

$$K(t) - R(t) = \int_0^t R(\tau)K(t-\tau)\mathrm{d}\tau \tag{3.77}$$

Equation (3.77) can be regarded as an integral equation with respect to functions R. In order to solve this equation the one-sided Laplace transform can be used on both sides of equation. Further, from the formula (3.77) is also evident that $K(0) = R(0)$ and $K(t) > R(t)$.

The simplest kernel integral Eq. (3.70) is the decaying exponential function $K(u) = C\mathrm{e}^{-au}$, where $u = t - \tau$. This kernel is not fully consistent with experimental data, as when $t = 0$, it has a finite value. However, it still gives us an idea and that in a simple and understandable form, the principal qualitative phenomena of linear creep deformation. In particular, as will be shown below, this corresponds to the kernel of the differential dependence of the standard linear model of a viscoelastic material, which was previously shown in Eq. (3.23). The integral Eq. (3.70) in this case has a form:

$$E\varepsilon(t) = \sigma(t) + \int_0^t C\sigma(u)\,\mathrm{e}^{-au}\mathrm{d}u \tag{3.78}$$

Let's take the derivative of Eq. (3.78) with respect to time:

$$E\dot{\varepsilon}(t) = \dot{\sigma}(t) + C\sigma(t) - \int_0^t aC\sigma(u)\,\mathrm{e}^{-au}\mathrm{d}u \tag{3.79}$$

Multiply Eq. (3.78) by "a" and add it to the Eq. (3.79). Then the integral terms disappear and remain differential equation of the form:

$$E\dot{\varepsilon}(t) + Ea\varepsilon(t) = \dot{\sigma}(t) + (C+a)\,\sigma(t) \tag{3.80}$$

If in Eq. (3.80) is set $C + a = 1/n$ and $Ea = H/n$, we obtain the differential equation given above (3.23). Resolution of (3.78) kernel can be found using the rules of the Laplace transform and the inverse transformations. The final result in this case has the form $R(u) = C\mathrm{e}^{-(a+C)u}$.

3.12 Examples

Example 1 Consider two Kelvin units in series, as in the generalized Kelvin chain; the first unit has properties E_1, η_1 and the second unit has properties E_1, η_1. The constitutive equation is given by $\sigma + \frac{\eta_1+\eta_2}{E_1+E_2}\dot{\sigma} = \frac{E_1E_2}{E_1+E_2}\varepsilon + \frac{E_2\eta_1+E_1\eta_2}{E_1+E_2}\dot{\varepsilon} + \frac{\eta_1\eta_2}{E_1+E_2}\ddot{\varepsilon}$. Considering the characteristic equation, show that the eigenvalues are $\lambda_1 = -E_1/\eta_1$ and $\lambda_2 = -E_2/\eta_2$ and find the homogeneous solution.

Solution:

The homogeneous equation is of the form $\frac{\eta_1\eta_2}{E_1+E_2}\ddot{\varepsilon} + \frac{E_2\eta_1+E_1\eta_2}{E_1+E_2}\dot{\varepsilon} + \frac{E_2E_1}{E_1+E_2}\varepsilon = 0$. $\varepsilon(t) = C_1 e^{-t/t_{r1}} + C_2 e^{-t/t_{r2}}$; $n_1 = t_{r1} = \eta_1/E_1$; $n_2 = t_{r2} = \eta_2/E_2$.

The characteristic equation is $n_1n_2\lambda^2 + (n_1 + n_2)\lambda + 1 = 0$, therefore: $\lambda_1 = -n_1$ $\lambda_2 = -n_2$.

The homogeneous solution is $\varepsilon(t) = C_1 e^{-t/n_1} + C_2 e^{-t/n_2}$.

Example 2 Consider now a constant load $\sigma = \sigma_0$. Show that the particular solution (see Example 1) is $\varepsilon_p = \frac{E_1+E_2}{E_1E_2}\sigma_0$.

Solution:

$$\sigma_0 = \frac{E_1E_2}{E_1+E_2}\varepsilon + \frac{E_2\eta_1+E_1\eta_2}{E_1+E_2}\dot{\varepsilon} + \frac{\eta_1\eta_2}{E_1+E_2}\ddot{\varepsilon}$$

$$\text{Therefore: } \varepsilon_p = \frac{E_1+E_2}{E_1E_2}\sigma_0$$

Example 3 One initial condition of the problem (see Example 1) is that $\varepsilon(0) = 0$. The second condition results from the fact that only the dashpots react at time $t = 0$. Find the second initial condition.

Solution:

For a constant load $\sigma = \sigma_0$, the full solution is

$$\varepsilon(t) = \sigma_0\left[\frac{1}{E_1}(1 - e^{-t/n_1}) + \frac{1}{E_2}(1 - e^{-t/n_2})\right]$$

Differentiating this equation with respect to time ones we have

$$\dot{\varepsilon}(t) = \sigma_0\left[\frac{1}{E_1}\frac{1}{n_1}e^{-t/n_1} + \frac{1}{E_2}\frac{1}{n_2}e^{-t/n_2}\right] \text{ and } \dot{\varepsilon}(t=0) = \sigma_0\frac{\eta_1+\eta_2}{\eta_2\eta_1}$$

Example 4 Consider the two Maxwell units in parallel.

For two Maxwell elements in parallel, the constitutive equation can be shown to be $\sigma + \frac{E_1\eta_2+E_2\eta_1}{E_1E_2}\dot{\sigma} + \frac{\eta_1\eta_2}{E_1E_2}\ddot{\sigma} = (\eta_1+\eta_2)\dot{\varepsilon} + \frac{E_1+E_2}{E_1E_2}\eta_1\eta_2\ddot{\varepsilon}$. Considering the characteristic equation, find the eigenvalues are λ_1 and λ_2 and the homogeneous solution.

Solution:

From this constitutive equation, the differential equation to be solved is of the form $\sigma_0 = (\eta_1 + \eta_2)\dot{\varepsilon} + \frac{E_1+E_2}{E_1E_2}\eta_1\eta_2\ddot{\varepsilon}$. The characteristic equation is:

$$\frac{E_1+E_2}{E_1E_2}\eta_1\eta_2\lambda^2 + (\eta_1+\eta_2)\lambda = 0$$

Therefore $\lambda_1 = 0;\ \lambda_2 = -\frac{\eta_1+\eta_2}{(E_1+E_2)\eta_1\eta_2}E_1E_2 = -\frac{\eta_1+\eta_2}{E_1+E_2}\frac{1}{n_{r1}n_{r2}} = -\frac{(1+\eta_2/\eta_1)}{(1+E_2/E_1)}\frac{1}{n_{r2}}$

And the eigenvalues are

$$\lambda_1 = 0;\ \lambda_2 = -\frac{(\eta_1+\eta_2)E_1E_2}{(\eta_1\eta_2)\,(E_1+E_2)}$$

Hence the homogeneous solution is $\varepsilon(t) = C_1 + C_2 e^{-t/n_{r2}}$ where $n_{r2} = -1/\lambda_2$.

Example 5 Consider now a constant stress $\sigma = \sigma_0$ (see Example 4). Using the condition that only the springs react at time $t = 0$, show that the particular solution is $\sigma_0\, t/(\eta_1 + \eta_2)$.

Solution:

$$\sigma_0 = (\eta_1+\eta_2)\dot{\varepsilon} + \frac{E_1+E_2}{E_1E_2}\eta_1\eta_2\ddot{\varepsilon}$$

$$\varepsilon_p = At \Rightarrow A = \sigma_0/(\eta_1+\eta_2) \Rightarrow \varepsilon_p = \sigma_0 t/(\eta_1+\eta_2)$$

The complete solution is given by

$$\varepsilon(t) = \sigma_0\left[\frac{1}{E_1+E_2}e^{-t/t_r} - \frac{t_r}{\eta_1+\eta_2}(1 - e^{-t/t_r}) + \frac{t}{\eta_1+\eta_2}\right];\ t_r = n_{r2}$$

While homogeneous equations are relatively easy to solve, the Lagrange method *(variation of parameters)* allows the calculation of the coefficients of the general solution of the inhomogeneous equation, and thus the complete general solution of the inhomogeneous equation can be determined. From the homogeneous solution $\varepsilon(t) = C_1 + C_2 e^{-t/n_{r2}}$ where $n_{r2} = -1/\lambda_2$ we obtain $u_1 = 1$ and $u_2 = e^{-t/n}$. The Wronskian of these two functions is

$$\begin{vmatrix} 1 & e^{-t/n_{r2}} \\ 0 & -1/n_{r2}e^{-t/n_{r2}} \end{vmatrix} = -[1/n_{r2}]e^{-t/n_{r2}} \neq 0$$

Because the Wronskian is nonzero, the two functions are linearly independent, so this is in fact the general solution for the homogeneous differential equation. We seek functions $C_1(t)$ and $C_2(t)$ so that $C_1(x)u_1 + C_2(x)i_2$ is a general solution of the non-homogeneous equation. We need only calculate the integrals. That is,

$$C_1(t) = -\sigma_0[1/n_{r2}] \int e^{-t/n_{r2}} dt = \sigma_0 e^{-t/n_{r2}} + A_1$$

$$C_2(t) = -\sigma_0[1/n_{r2}] \int e^{-2t/n_{r2}} dt = \frac{1}{2}\sigma_0 e^{-t/n_{r2}} + A_2$$

A_1 and A_2 are constants of integration. Adding a constant to $C_1(t)$ or $C_2(t)$ does not change the value of $L[u(t)]$ because the extra term is just a linear combination of u_1 and u_2, which is a solution of operator L by definition. Substituting $C_1(t)u_1$ and $C_2(t)u_2$ into ordinary differentiation equation (ODE) we have after algebraic simplifications

$$\varepsilon(t) = \sigma_0 \left[\frac{1}{E_1 + E_2} e^{-t/t_r} - \frac{t_r}{\eta_1 + \eta_2} (1 - e^{-t/t_r}) + \frac{t}{\eta_1 + \eta_2} \right]$$

Example 6 Consider two Maxwell units in parallel, as in the generalized Maxwell model. The constitutive equation is the same as in Example 4. Consider now a constant load ε_0. Show that the homogeneous solution is, analogous to Example 4:

Solution:

The homogeneous equation is $\sigma + \frac{E_1\eta_2 + E_2\eta_1}{E_1E_2}\dot{\sigma} + \frac{\eta_1\eta_2}{E_1E_2}\ddot{\sigma} = 0$. Solution:

$$1 + \frac{E_1\eta_2 + E_2\eta_1}{E_1E_2}\lambda + \frac{\eta_1\eta_2}{E_1E_2}\lambda^2 = 0;$$

$$n_1n_2\lambda^2 + (n_2 + n_1)\lambda + 1 = 0 \Rightarrow \lambda_1 = -1/n_1;\ \lambda_2 = -1/n_2$$

$$n_1 = \eta_1/E_1; \quad n_2 = \eta_2/E_2$$

Therefore, the homogeneous solution is

$$\sigma_h = C_1 e^{-t/n_1} + C_2 e^{-t/n_2}$$

Example 7 Show that for a constant strain ε_0, (see Example 6), the full solution is

$$\sigma(t) = \varepsilon_0 \left[E_1 e^{-t/n_1} + E_2 e^{-t/n_2} \right]$$

Solution:

Using again the Lagrange method we can show that

$$C_1(t) = -\varepsilon_0[1/n_1] \int e^{-2t/n_1} dt = \frac{1}{2}\varepsilon_0 e^{-2t/n_1} + A_1$$

$$C_2(t) = -\varepsilon_0[1/n_2] \int e^{-2t/n_2} dt = \frac{1}{2}\varepsilon_0 e^{-2t/n_{r2}} + A_2$$

The complete solution is given by

$$\sigma(t) = \varepsilon_0 \left[E_1 \mathrm{e}^{-t/n_1} + E_2 \mathrm{e}^{-t/n_2} \right]$$

References

1. Ashby MF, Jones DRH (1980). Engineering materials 1: an introduction to their properties and applications. Pergamon Press
2. Frost HJ, Ashby MF (1982). Deformation-mechanism maps: the plasticity and creep of metals and ceramics. Pergamon Press
3. Krajcinovic D (1996) Damage mechanics. Elsevier
4. Suresh S (1998) Fatigue of materials, 2nd edn. Cambridge University Press
5. ASTM D2990 Plastic creep testing
6. ASTM D1598—02 (2009) Standard test method for time-to-failure of plastic pipe
7. ASTM D2837—13e1 (2009) Standard test method for obtaining hydrostatic design basis for thermoplastic pipe materials
8. ASTM D 4065 (2009) Dynamic mechanical analysis (DMA) D 4065, D 5279 and D 4440
9. ASTM D7028—07 (2015) Standard test method for glass transition temperature (DMA Tg) of polymer matrix composites
10. Knauss WG, Emri I (1987) Volume change and nonlinearly thermo-viscoelasticity constitution of polymers. Polym Eng Sci 27(1):86–100
11. O'Connell PA, McKenna GB (1997) Large deformation response of polycarbonate: time–temperature, time–aging time, and time–strain superposition. Polym Eng Sci 37(9):1485–1495
12. Brostow W (2000) Time–stress correspondence in viscoelastic materials: an equation for the stress and temperature shift factor. Mater Res Innovations 3:347–351
13. Yen SC, Williamson FL (1990) Accelerated characterization of creep response of an off-axis composite material. Compos Sci Technol 38:103–118
14. Luo W, Yang T, An Q (2001) Time–temperature–stress equivalence and its application to nonlinear viscoelastic materials. Acta Mech Solida Sin 14(3):195–199
15. Findley WN, Lai JSY (1967) A modified superposition principle applied to creep of nonlinear viscoelastic material under abrupt change in state of combined stress. Trans Soc Rheol 11 (3):361–380
16. Schapery RA (1969) On the characterization of nonlinear viscoelastic materials. Polym Eng Sci 9(4):295–310
17. Basar Y (2000) Nonlinear continuum mechanics of solids. Springer
18. Roylance D (2001). Engineering viscoelasticity. Massachusetts Institute of Technology, Cambridge, MA 02139, pp 8–11
19. Gutierrez-Lemini D (2013) Engineering viscoelasticity. Springer, p. 88
20. Tschoegl NW (1989) The phenomenological theory of linear viscoelastic behavior, pp 119–126
21. Razdolsky L (2012) Structural fire load. McGraw-Hill Publishing Co., USA
22. NISTIR 7563 (2009) Best practice guidelines for structural fire resistance design of concrete and steel buildings. U.S. Department of Commerce National Institute of Standards and Technology, February 2009
23. Bland DR (1960) The theory of linear viscoelasticity. Pergamon, Oxford
24. Carcione JM (2007) Wave fields in real media. Theory and numerical simulation of wave propagation in anisotropic, anelastic and porous media, 2nd edn. Elsevier, Amsterdam
25. Christensen RM (1982) Theory of viscoelasticity, 2nd edn. Academic Press, New York
26. Flugge W (1975) Viscoelasticity. Springer-Verlag, New York

27. Tschoegl NW (1989) The phenomenological theory of linear viscoelastic behavior. Springer Verlag, Heidelberg
28. Shanley FR (1960) Weight-strength analysis of aircraft structures. Dover, NY
29. Gerard G (1962) A unified theory of creep buckling of columns. J Aerosp Sci Technol
30. Gerard George (1962) Theory of creep buckling of perfect plates and shells. J Aerosp Sci 29 (9):1087–1090

Chapter 4
Phenomenological Time Variant Nonlinear Creep Models

Notations

k	The thermal conductivity that has dimensions (W/m K or J/m s K)
T	Temperature in K
Q	Activation energy
R	Ideal gas constant
T_o	Ambient temperature
t	Time
$t^* = t_r$	Creep rupture (failure) time
$\vec{v}(u; v; w)$	Velocity vector
D	Diffusion coefficient (m^2/s)
p	Pressure
v	Kinematic viscosity $v = \mu/\rho$
θ	Dimensionless temperature
τ	Dimensionless time
h	Height of the compartment (m)
a	Thermal diffusivity (m^2/s)
Time	$t = \frac{h^2}{a}\tau$ (s)
Temperature	$T = \frac{RT_*^2}{E}\theta + T_*$ (K), where T_* = 600 K Base line temperature
Coordinates	$\bar{x} = x/h$ and $\bar{z} = z/h$ "x" and "z"—dimensionless coordinates
Velocities	$\bar{u} = \frac{v}{h}u$ (m/s) and $\bar{w} = \frac{v}{h}w$ (m/s) Horizontal and vertical components velocity accordingly; v—kinematic viscosity [m^2/sec]; "u" and "w"—dimensionless velocities
$J(t, t')$	Compliance function (often also called the creep function)
T_M	Melting point of the metal
$\varepsilon(t)$	Strain
$\sigma(t)$	Stress
$\bar{\sigma}(t) = E(t)\varepsilon(t)$	Instantaneous (elastic) stress
σ_{eq}	Equivalent stress

L. Razdolsky, *Probability Based High Temperature Engineering*,
DOI 10.1007/978-3-319-41909-1_4

σ^*	Ultimate stress
ε_e	Instantaneous (elastic) strain
ε_c	Creep strain
ϕ	Resistance factor
ε_T	Thermal expansion due to high temperature effect
$K(t, t') = -\partial J(t, t')/\partial t'$	Retardation function (memory function)
$R(t, t')$	Relaxation function (also called the relaxation modulus)
P_{LM}	Larson–Miller parameter
P_D	Dorn Parameter
PMH	Mason–Haferd Parameter
Z	Zener–Holloman Parameter
E	Modulus of Elasticity
α	Material property parameter (MPP)
n	Relaxation time
$\delta = 1$	Frank–Kamenetsky parameter

4.1 Introduction

In general, the mechanical properties and performance of materials change with increasing temperatures. Some properties and performance, such as elastic modulus and strength decrease with increasing temperature. Others, such as ductility, increase with increasing temperature. It is important to note that atomic mobility is related to diffusion which can be described using Ficks Law: $D = D_0[\exp(-E/RT)]$, where D is the diffusion rate, D_0 is a constant, Q is the activation energy for atomic motion, R is the universal gas constant (8.314 J/mol K) and T is the absolute temperature. Thus, diffusion-controlled mechanisms of creep that is considered in this book will have significant effect on high temperature mechanical properties and performances. For example, dislocation climb, concentration of vacancies, new slip systems, and grain boundary sliding all are diffusion-controlled and will affect the behavior of materials at high temperatures. In addition, corrosion or oxidation mechanisms, which are diffusion-rate dependent, will have an effect on the life time of materials at high temperatures.

Various components of power plants, steam generators/turbines, rotors, etc., operate at high temperatures under significant stresses. For example, pipes in steam turbine power plants carry steam at temperatures up to 650 °C and pressures of about 25 MPa. Temperatures can go to as high as 1500 °C in jet engines and initiate creep deformation even in the turbine blades. The structures/components need to be designed against excessive creep distortion/failure within the expected operating life of the component and it is thus vital to understand creep behavior in metals in order to ensure functionality of these engineering components without threat to catastrophic failures that may lead to loss of lives and economic implications in investments.

Creep occurs in metals by dislocation and atomic diffusion mechanisms. As the temperatures increase, these creep mechanisms and effects become significant and thus leads to failure which may become catastrophic if not properly checked.

Creep tests are thus carried out to investigate the creep properties of the materials at high temperatures using the specified standards including the British and American standards. This information can thus be used to develop creep resistant alloys that inhibit creep failure.

A number of properties can be deduced from the uniaxial creep curve. These are the duration of the stages, the value of minimum creep rate, the time to fracture and the strain value before fracture [1].

The shape of the creep curve and the duration of the creep stages depend strongly on the stress and temperature values. The dependencies on stress and temperature are of primary interest to an engineer designing some structure or machine.

In order to obtain mechanical properties of the material, series of creep tests are usually performed for different stress and temperature values. From the resulting families of creep curves one can obtain the minimum creep rate versus stress curve, the minimum creep rate versus temperature curve, the creep rate versus time curve, and the stress versus time to fracture curve (long term strength curve). The ranges of stress and temperature should be selected according to the ranges expected in the structure.

Two additional forms of the time-dependent stress–strain behavior are creep recovery and stress relaxation. Creep recovery is usually observed when after a certain period of time the load is spontaneously removed. After unloading the strain drops about the value ε_{el} (recovery of the elastic strain). Then the strain slowly decreases down to the permanent (irrecoverable) value.

Stress relaxation is observed when the strain is held constant in time (ε = const). A uniaxial specimen is instantaneously deformed to the strain value $\varepsilon_{el} = \sigma/E$, where E is the Young's modulus. During the test the load is continuously decreased in such a way that the initial strain remains constant. A threshold of the initial stress (strain) exists below which the relaxation is not observable.

Conceptually a creep test is rather simple: apply a force to a test specimen and measure its dimensional change over time with exposure to a relatively high temperature. If a creep test is carried to its conclusion (that is, fracture of the test specimen), often without precise measurement of its dimensional change, then this is called a stress rupture test. Although conceptually quite simple, creep tests in practice are more complicated. Temperature control is critical (fluctuation must be kept to <0.1–0.5 °C). Resolution and stability of the extensometer is an important concern (for low creeping materials, displacement resolution must be on the order of 0.5 μm).

Environmental effects can complicate creep tests by causing premature failures unrelated to elongation and thus must either mimic the actual use conditions or be controlled to isolate the failures to creep mechanisms. Uniformity of the applied stress is critical if the creep tests are too interpreted.

In addition to creep and relaxation, many different tests under variable loading, and/or strain conditions are discussed in the literature. Examples for the creep curves under stepwise loading are presented in [2, 3] among others. In this case the creep test starts under a certain value of the load. After reaching steady-state creep rate the load is rapidly increased (decreased) and kept constant over a period of time (holding time). Such tests allow analyzing transient creep effects, e.g., the duration of primary creep after the rapid change of loading. Furthermore, they indicate that the steady-state creep rate in the current loading step depends not only on the value of the applied stress but also on the loading history (e.g., the number of previous stress cycles, the holding time, etc.).

In principle, the creep deformation should be linked to an applied stress. Thus, as the specimen elongates the cross sectional area decreases and the load need to be decreased to maintain a constant stress. In practice, it is simpler to maintain a constant load. When reporting creep test results the initial applied stress is used. Note that In general the effect of the uniformity of the applied stress only really manifests itself in the tertiary region, which is beyond the region of interest in the secondary region. The effects of increasing temperature or increasing stress are to raise the levels and shapes of the strain time curves. Note that for isothermal tests, the shapes of the curves for increasing stress may change from dominant steady-state to sigmoid with little steady state to dominant primary. Similar trends are seen for iso stress tests and increasing temperature.

The above definitions of creep and relaxation are related to the case of uniaxial homogeneous stress states realized in standard material testing. Under "creep in structures" we understand time-dependent changes of strain and stress states taking place in structural components or systems as a consequence of external loading and temperature.

Examples of these changes include progressive deformations, relaxation and redistribution of stresses, local reduction of material strength, etc. Furthermore, the strain and stress states are inhomogeneous and multiaxial in most cases. The scope of "creep modeling for structural analysis" is to develop a tool which allows simulating the time-dependent behavior in engineering structures up to the critical state of creep rupture.

4.2 Volterra Equation and Creep Constitutive Law

The integral equation problem is to find the solution to

$$h(x)f(x) = g(x) + \lambda \int_a^b K(x, y)f(y)\mathrm{d}y \tag{4.1}$$

We are given functions $h(x)$, $g(x)$, $K(x; y)$, and wish to determine $f(x)$. The quantity λ is a parameter, which may be complex in general. The function $K(x; y)$ is called the kernel of the integral equation.

We shall assume that $h(x)$ and $g(x)$ are defined and continuous on the interval $a < x < b$, and that the kernel is defined and continuous on $a < x < b$ and $a < y < b$. Here we will concentrate on the problem for real variables x and y. The functions may be complex valued, although we will sometimes simplify the discussion by considering real functions. However, many of the results can be generalized in fairly obvious ways, such as relaxation to piecewise continuous functions, and generalization to multiple dimensions.

There are many resources for further reading on this subject. Some of the popular ones among physicists include the "classic" texts by [4].

Integral Equations of the Second Kind

Referring back to Eq. 4.1, if $h(x) \neq 0$ for $a < x < b$ we may rewrite the problem in a form with $h(x) = 1$

$$f(x) = g(x) - \lambda \int_a^x K(x, y) f(y) \mathrm{d}y \tag{4.2}$$

This is referred to as a linear integral equation of the second kind or as a Fredholm equation of the second kind. It defines a linear transformation from function f to function g. A value of λ for which the homogeneous equation has nontrivial solutions is called an eigenvalues of the equation (or, of the kernel). We are interested in the problem of inverting this linear transformation given g what is f? As it is a linear transformation, it should not be surprising that the techniques are analogous with those familiar in matrix equations.

Degenerate Kernels

Definition (Degenerate Kernel): We can write the kernel in the form

$$K(x, y) = \sum_{i-1}^{n} a_i(x) b_i(y) \tag{4.3}$$

The kernel K is called degenerate. We may assume that functions $a_i(x)$ is linearly independent. Otherwise we could reduce the number of terms in the sum to use only independent functions. Likewise we may assume that $b_i(x)$ is linearly independent functions.

The notion of a degenerate kernel is important due to two facts

1. Any continuous function $K(x; y)$ can be uniformly approximated by polynomials in a closed interval. That is, the polynomials are "complete" on a closed bounded interval $[a, b]$.
2. The solution of the integral equation for degenerate kernels is easy (at least formally) by exploiting the Dirichlet series expansion of the compliance function $K(x, y)$.

If the kernel is degenerate, we have shown (see Chap. 2) that the solution may be obtained by transforming the problem to that of solving a system of linear equations.

4.2.1 Volterra Equations

Integral equations of the form $f(x) = g(x) - \lambda \int_a^x K(x,y)f(y)\mathrm{d}y$ are called Volterra equations of the second kind. Often an exact closed solution is elusive, and we resort to approximate methods. For example, one common approach is the iterative solution. Notice that if the kernel is independent of x, $K(x;\ y) = K(y)$, then the solution to the Volterra equation is simple. This may be solved with various approaches (see below).

The Volterra equation readily lends itself to a numerical approach to solution on a grid (or "mesh" or "lattice"). We note first that (absorbing the λ-factor into the definition of K for convenience). This suggests building up a solution at arbitrary x by stepping along a grid starting at $x = a$. To carry out this method, we start by dividing the interval $[a, x]$ into N steps. Let us note that we don not even have to explicitly solve a system of linear equations, as we did for Fredholm equation with a degenerate kernel.

4.2.2 Creep Compliance Function and Material Property Parameters

The values of the compliance function are influenced by many factors, which may be divided into intrinsic and extrinsic. The intrinsic factors are those that become fixed when the material is cast; they include the molecular composition parameters, such as the aggregate fraction, the elastic modulus of composed components, as well as the design strength. The extrinsic factors are those that can be changed externally; they include temperature change (including the recovery histories), time when thermal loading begins, degree of inelastic deformations, etc. The mathematical expressions for the compliance function and the influencing factors in case of fire will be discussed in more detail below but in the meantime it will be assumed that the compliance function $J(t,\ t')$ is known, being given either by an analytical formula, or a graph, or a table of values. The corresponding creep constitutive equation in general form is as follows:

$$\bar{\sigma}(t) = E(t)\varepsilon(t) = \sigma(t) + \int_0^t K(t,t')\sigma(t')\mathrm{d}t' \tag{4.4}$$

In Eq. (4.4), $K(t, t') = -\partial J(t, t')/\partial t'$. Geometrically, this equation means that the stress history is decomposed into vertical strips each of which is considered as an impulse function of stress (Dirac δ-function). Thus, $K(t, t')$ represents the strain at time t caused by a unit stress impulse at time t' and is called the stress impulse memory function.

The principle of superposition may be equivalently expressed in terms of the relaxation function, $R(t, t')$ (also called the relaxation modulus), which represents the uniaxial stress σ at time t caused by a unit constant axial strain imposed at time t' and held constant afterwards. Imaging the strain history $\varepsilon(t)$ to be decomposed into small strain increments $\mathrm{d}\varepsilon(t')$ imposed at times t', the principle of superposition means that the responses to these increments, given as $R(t, t')\,\mathrm{d}\varepsilon(t')$, may be superimposed. This yields the constitutive relation of viscoelasticity in the form

$$\frac{1}{E(t)}\sigma(t) = \varepsilon(t) - \int\limits_0^t R(t,t')\varepsilon(t')\mathrm{d}t' \tag{4.5}$$

When the stress history is given, Eq. (4.5) represents a Volterra integral equation for the strain history $\varepsilon(t)$. By solving this equation for the strain history specified as a step function (a constant unit strain imposed at age t'), one may calculate the stress histories for various t' (relaxation curves), and thus obtain the relaxation function. For realistic forms of $J(t, t')$, this solution must be carried out numerically. Conversely, Eq. (4.5) represents a Volterra integral equation for $\varepsilon(t)$. By solving this equation for the stress history in the form of a step function, i.e., a constant unit stress applied at age t', one may calculate the individual creep curves, which together define the compliance function. Equation (4.5) is said to be the resolvent of Eq. (4.4) and vice versa. Functions $J(t, t')$ and $R(t, t')$, called the kernels of the integral equations, are complementary to each other, and if one of them is specified the other one follows.

Multiaxial generalization of all the preceding relations is obtained easily, by virtue of the fact that the material is essentially isotropic based on the hypothesis of linearity (principle of superposition), Eq. (4.4) is generalized as

$$\bar{\sigma}(t) = \mathbf{B}\left[\sigma(t) + \int\limits_0^t K(t,t')\sigma(t')\mathrm{d}t'\right] \tag{4.6}$$

The multiaxial stress–strain relations may also be written without matrix symbolism, as separate relations for the volumetric components and for the deviatory components of the stress and strain tensors. These equations are similar to Eqs. (4.4) and (4.5), the uniaxial compliance function $J_v(t, t')$ being replaced by the volumetric compliance function $J_v(t, t') = 3\,(1-2\nu)\,J(t, t')$ and by the deviatory compliance function $J_d(t, t') = 2(1 + \nu)\,J(t \cdot t')$.

An important property of the compliance function is that it is a function of two variables, the current time, t, and the time at thermal loading t'. It is a salient characteristic of material that the compliance function cannot be considered as a function of one variable, i.e., the time-lag $t - t'$, as is customary in classical viscoelasticity. The time variance is a considerable obstacle to analytical solutions of structural problems, and necessitates that most real problems have to be solved by numerical methods. As a consequence of numerical methods application discussed earlier in Chap. 2 the corresponding differential and integral equations describing the phenomenon of creep must be written in dimensionless form. Scale analysis is a very powerful tool used in the phenomenological creep theory for the reducing the number of the extrinsic parameters that are influencing the complicity of the compliance function. It allows also identify some "unknown" parameters that will be called later on as random parameters when the probability based approach of creep deformations will be presented. Finally, it will allow solving the corresponding applied deterministic and probabilistic creep deformation problems in closed form. Since the application of high temperature creep analysis is connected in this book with the structural fire resistance, the temperature–time relationships are taken from the previously developed by author analysis for each fire severity scenario [5].

4.3 Thermal Analysis

Thermal analysis should be conducted to obtain the temperature field of the structural system. Thermal diffusivity analysis (fire protected structural elements) is also presented in [5].

Continuity Equation

Continuity equation is also called the conservation of mass, which ensures that the total mass is conserved; in other words, the total mass of a fluid system is completely accounted for. In three dimensions, the continuity equation is

$$\frac{\partial \rho}{\partial t} + \nabla \cdot (\rho \vec{v}) = 0, \tag{4.7}$$

where ρ is the density, t is time, v is the velocity vector, and ∇ is the gradient operator.

Momentum Conservation Equation

The principal of conservation of linear momentum means the total force generated by the momentum transfer in each direction is balanced by the rate of change of momentum in each direction and is given by

$$\rho\left(\frac{\partial \vec{v}}{\partial t} + \vec{v} \cdot \nabla \vec{v}\right) = -\nabla p + \nabla S_{ij} + \vec{f}, \tag{4.8}$$

where $\vec{f}$ is the body force vector and S_{ij} is the Cauchy stress tensor and p is pressure.

Energy Conservation Equation

The energy balance can be obtained as heat entering the control volume by convection and heat entering the control volume by diffusion is equal to heat exiting the control volume by convection and heat exiting the control volume by diffusion and rate of change of energy within the control volume. Energy conservation equation is given by

$$c_p\rho\frac{\partial T}{\partial t} = \mathrm{div}(\lambda \mathrm{grad} T - c_p\rho\vec{v}\nabla T) + Qze^{-E/RT} - \frac{e\sigma A_v(T^4 - T_0^4)}{V}, \tag{4.9}$$

where c_p is heat capacity; T is temperature, and $Qze^{-E/RT}$ is the internal heat generation.

These equations (conservation of energy, mass and momentum) are also presented in dimensionless form [5]. Now let us introduce the dimensionless parameters and variables in conservation of energy, mass, and momentum equations [3]. For example, the temperature–time dimensionless function is defined as the solution of simplified differential equations

Dimensionless Temperature–Time Function (see Fig. 4.1). Fast Fire [5] $\gamma = 0.025$.

Differential equations

1 $d(y0)/d(t) = 20 * (1 - y2) * \exp(y0/(1 + 0.1 * y0)) - 0.233 * y0\wedge 4$

2 $d(y2)/d(t) = 1.0 * (1 - y2) * \exp(y0/(1 + 0.1 * y0))$

3 $d(y1)/d(t) = 0.5 * (1 - y1)\wedge 1.0 * \exp(y/(1 + 0.1 * y))$

4 $d(y)/d(t) = (1) * 20 * (1 - y1)\wedge 1.0 * \exp(y/(1 + 0.1 * y)) - 0.233 * y\wedge 4$

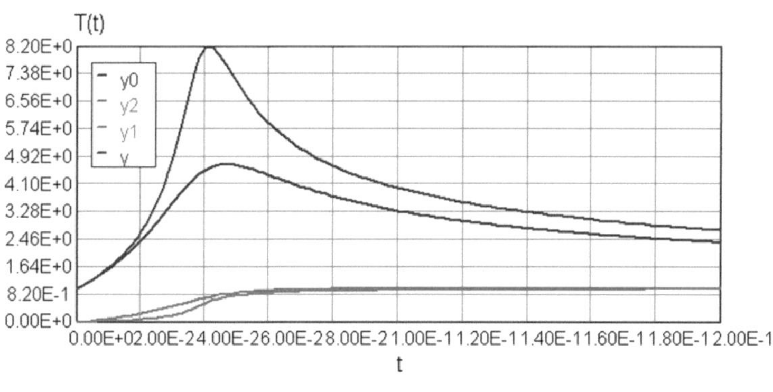

Fig. 4.1 Dimensionless temperature–time curves

Calculated values of DEQ variables

	Variable	Initial value	Minimal value	Maximal value	Final value
1	t	0	0	0.2	0.2
2	y	1	1	8.449013	3.197677
3	$y0$	1	1	7.041326	2.601609
4	$y1$	0	0	0.9251493	0.9251493
5	$y2$	0	0	0.9727725	0.9727725

Model $T(t) : y = \theta(\tau) = \varphi(\tau)$
$= 20.75 * \tau + 722.2 * \tau\hat{}2 - 30370 * \tau\hat{}3 + 523600 * \tau\hat{}4$

Variable	Value
$a0$	0
$a1$	20.75091
$a2$	722.2399
$a3$	−3.037E + 04
$a4$	5.236E + 05

We will need also the inverse function θ^{-1} and its first derivative: (Fig. 4.2).

Model $t1 = \tau(\theta) = \varphi(\theta)$
$= 0.0405 * \theta - 0.01126 * \theta\hat{}2 + 0.001462 * \theta\hat{}3 - 0.00006868 * \theta\hat{}4$

Variable	Value
$a0$	0.0016539
$a1$	0.0405241
$a2$	−0.0112592
$a3$	0.0014618
$a4$	−6.868E-05

$$\tau(\theta) = 0.0405 * \theta - 0.01126 * \theta\hat{}2 + 0.001462 * \theta\hat{}3 - 0.00006868 * \theta\hat{}4 \tag{4.10}$$

$$\mathrm{d}\tau/\mathrm{d}\theta = 0.0405 - 0.02252 * \theta\hat{}1 + 0.004386 * \theta\hat{}2 - 0.0002747 * \theta\hat{}3 \tag{4.11}$$

$$\theta = \frac{E}{RT_*^2}[T - T_*] \quad \text{and} \quad \tau = \frac{a}{h^2}t. \tag{4.12}$$

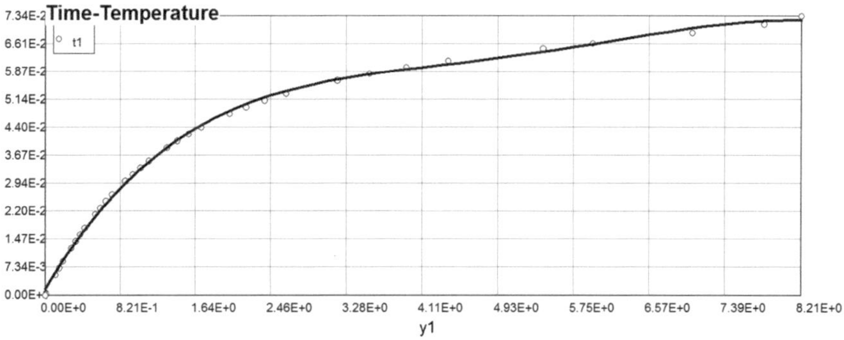

Fig. 4.2 Dimensionless time–temperature curve

Here: $T_* = 600$ K is the base line temperature (onset of creep deformations); *E*—activation energy; *R*—universal gas constant; *h*—linear dimension; *a*—thermal diffusivity.

In the differential equations describing the phenomenon of creep, the invariance of time is determined by the fact that the coefficients of the functions σ and ε and their derivatives—are constants. In the integral relations of σ and ε the invariance of time is represented by the fact that the creep functions are the functions of the difference of the current time t and time of application elementary external load τ. Meanwhile frequently there are materials whose properties over time change noticeably. These include primarily concrete, which hardens after its manufacture for quite a long time, as well as, materials that change their properties due to other factors, such as exposure to high temperature. If the change in properties of the material happens all the time in the direction of decreasing its ability to deform over time, this change is called as the aging properties of the material. Accompanying material aging often decreases its instant (elastic) deformability, i.e., increases its rigidity. In contrast, exposure to high temperature leads to an increase in the ability to deform during the time (increased its instantaneous elastic deformability), and thus, reduce stiffness. Creep deformation is important not only in systems where high temperatures are endured such as nuclear power plants, jet engines and heat exchangers, but also in the design of many other structural systems.

Creep can occur in polymers and metals which are considered viscoelastic materials. When a polymeric material is subjected to an impacted force, the response can be modeled using the Kelvin–Voigt model. In this model, the material is represented by a Hookean spring and a Newtonian dashpot in parallel. In the simplest case of differential relationships between stresses and strains account the variability of the material properties can be produced by replacing the constant coefficients of the corresponding differential equations by variable coefficients, i.e., time dependent. For example, standard, basic simplified linear differential equation of deformation (see Eq. (1.1) from Chap. 1) can be generalized to the case of variables material properties as follows (instant and long-term modulus of elasticity and the relaxation time is assumed to be functions of time):

$$n(t)E(t)\dot{\varepsilon}+H(t)\varepsilon=\sigma+n(t)\dot{\sigma} \tag{4.13}$$

General solution of Eq. (4.13) for strains (natural initial conditions $\varepsilon(0)=[\sigma(0)/E(0)]$) is Volterra integral equation of the second kind

$$\varepsilon(t)E(t)=\sigma(t)+\int_0^t\sigma(\tau)K(t,\tau)\mathrm{d}\,\tau \tag{4.14}$$

where

$$\mathrm{K}(t,\tau)=E(t)\frac{E(\tau)-H(\tau)+n(\tau)\dot{E}(\tau)}{n(\tau)E^2(\tau)}\mathrm{e}^{-\int_\tau^t\frac{H(\tau)\mathrm{d}(\tau)}{E(\tau)n(\tau)}}$$

In Eq. (4.14), for $\tau < t$, $K(t, \tau)$ is a creep function (kernel of the integral equation). Similarly, a solution of Eq. (4.14) for stresses can be written in the form

$$\sigma(t)=E(t)[\varepsilon(t)-\int_0^t\varepsilon(\tau)R(t,\tau)\mathrm{d}\,\tau] \tag{4.15}$$

where

$$R(t,\tau)=\frac{1}{E(t)}\frac{E(\tau)-H(\tau)+n(\tau)\dot{E}(\tau)}{n(\tau)}\mathrm{e}^{-\int_\tau^t\frac{\mathrm{d}(\tau)}{n(\tau)}}$$

Equation (4.15), for $\tau < t$, $R(t, \tau)$ is a relaxation function (resolvent of kernel $K(t, \tau)$).

Substituting (4.15) into (4.14) we obtain the integral type relationship between relaxation modulus and creep compliance function after some simplifications and integration by parts

$$K(t-\tau)-R(t)=\int_0^t K(t-\tau)R(\tau)\mathrm{d}\,\tau=\int_0^t K(\tau)\,R(t-\tau)\mathrm{d}\,\tau \tag{4.16}$$

4.3.1 Principle of Superposition

As a consequence of creep, the stress in redundant structures usually varies with time even if the thermal load is constant. The calculation of creep caused by variable stress is greatly facilitated by the principle of superposition. This principle is usually assumed to apply to material within the service stress range, and its use in design is permitted by contemporary structural codes and recommendations of engineering societies. The principle of superposition, which is equivalent to the hypothesis of linearity of the constitutive equation that relates the stress and strain histories, states that the response to a sum of two stress (or strain) histories is the sum of the responses to each of them taken separately. According to this principle, the strain caused by stress history $\sigma(t)$ may be obtained by decomposing the history into small increments $d\sigma(t')$ applied at times t', and summing the corresponding strains which equal $d\sigma(t')[J(t, t')]$. It may be noted that the proportionality property for creep under constant stress appears to have a broader applicability than the principle of superposition. It may be also noted that a certain simple non-linear generalization of the principle of superposition extends the applicability range significantly.

The principle of superposition over predicts the magnitude of creep recovery. For intuitive understanding of the reason for reduced recovery consider the curves of creep without the elastic strain term (see Fig. 4.3).

For recovery the stress is decreased. The reduced creep recovery is often regarded as one manifestation of the phenomenon of adaptation in creep. In a broader sense, this phenomenon also means that after a longer period of creep under a relatively small compressive stress well within the linear range, the response to any second stress change (both a decrease (recovery) and an increase) is generally stiffer than predicted from the principle of superposition, as if the material has "adapted" to the previous stress.

The experimental support for this broader interpretation of the adaptation phenomenon is, however, ambiguous. The present theory predicts the additional creep resulting from a second stress increase will always be larger than that predicted by the principle of superposition.

To generalize the present formulation for creep at temperature changes, it is necessary to add not only thermal expansion, but also the stress induced by thermal expansion, which represent manifestations of the transitional thermal creep.

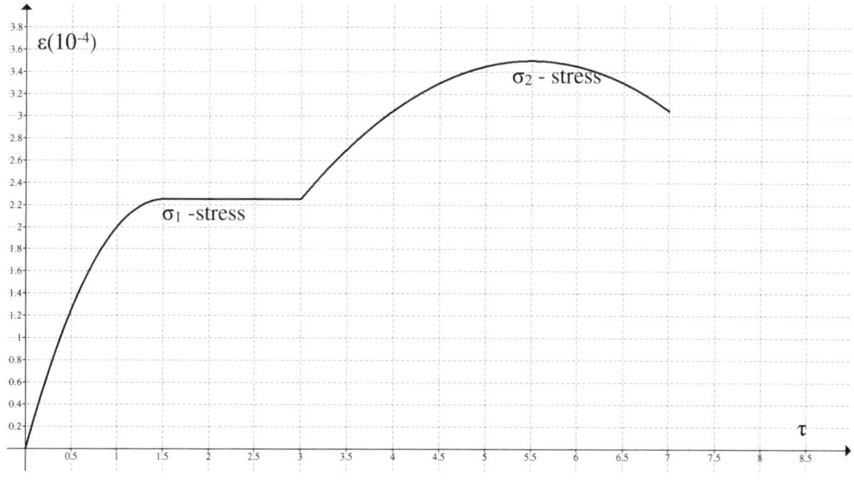

Fig. 4.3 Principle of superposition

The mathematical description of these phenomena is a separate problem, and can be accomplished in the manner described below.

4.3.2 Rate-Type Creep Law and Rheological Model

The goal in engineering design for creep is to predict the behavior over the long-time term. To this end there are three key methods: stress rupture, minimum strain rate versus time to failure, and temperature compensated time. No matter which method is used, two important rules of thumb must be borne in mind: (1) test time must be at least 10 % of design time and (2) creep and/or failure mechanism must not change with time, temperature, or stress.

Stress rupture test

This is the "brute force method" is which a large number of tests are run at various stresses and temperatures to develop plots of applied stress versus time to failure. While it is relatively easy to use these plots to provide estimates of stress rupture life within the range of stresses and lives covered by the test data, extrapolation of the data can be problematic when the failure mechanism changes as a function of time or stress as shown by the "knee" in Fig. 4.4.

Minimum strain rate versus time to failure

This type of relation is based on the observation that strain is the macroscopic manifestation of the cumulative creep damage. As such, it is implied that failure will occur when the damage in the material in form of creep cavities and cracks resulting from coalesced creep cavities reaches a critical level. This critical level of

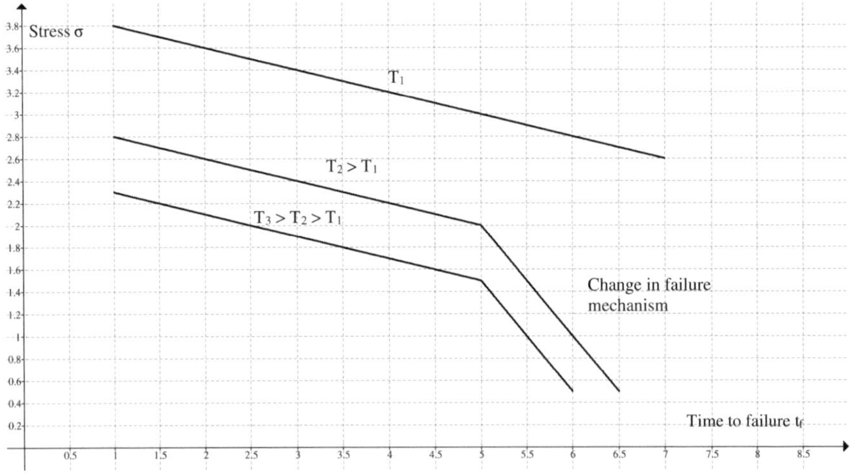

Fig. 4.4 Stress–time to failure diagram

damage is manifested as the failure which can be predicted from the minimum strain rate and the time to failure such that $\dot{\varepsilon}_{\min} t_f = C \approx \varepsilon_f$. This relationship is, known as the Monkman–Grant relation, should give a slope of −1 on a log-log plot of σ versus t_f regardless of temperature or applied stress for a particular material. It then becomes a simple matter to predict a time to failure either by measuring the minimum strain at a given stress and temperature or predicting the minimum strain rate for the given temperature and stress once the constants are determined. Having found the minimum strain rate, the time to failure can be found from the Monkman–Grant plot for the particular material.

Temperature compensated time

In these methods, a higher temperature is used at the same stress so as to cause a shorter time to failure such that temperature is traded for time. In this form of accelerated testing it is assumed that the failure mechanism does not change and hence, is not a function of temperature or time. In addition, assumptions can be made that Q is stress and temperature independent. Two of the more well-known relations are Sherby–Dorn and Larson–Miller.

In the Sherby-Dorn method, θ is the temperature compensated time such that

$$P_{\mathrm{SD}} = \log t_f - \operatorname{loge}\left[\frac{Q}{RT}\right] \tag{4.17}$$

In this method, a number of tests are run at various temperatures and stresses to determine the times to failure. A "universal" plot is then made of the stress as a function of P_{SD}. The allowable stress for a combination of time to failure and temperature (i.e., P_{SD}) can then be determined from the curve. The application of the Sherby–Dorn relation is as follows. For example, the stress-P_{SD} relation for a certain aluminum–magnesium alloy is found to be

$$\sigma = f(P_{SD}) = -11.3P_{SD} - 124 \ (25 \le \sigma \le 85 \text{ MPa}) \tag{4.18}$$

Example 4.1 Data: Activation energy, Q, is 150.5 kJ/mol; R = 8.314 J/mol K, t_f = 2000 h

Determine the allowable stress to give 2000 h life at 200 °C (473 K).

Solution:

P_{SD} is calculated from Eq. (4.17): P_{SD} = −13.21. Substituting this value of P_{SD} into Eq. (4.18) gives an allowable stress of 25.27 MPa [3.66 ksi].

Creep tests are carried out to investigate the creep properties of the materials at high temperatures using the specified standards including the British and American standards. This information can thus be used to develop creep resistant alloys that inhibit creep failure. Ductility increases with increasing temperature while elastic modulus (E) and strength decrease with increasing temperature.

High temperatures have a significant effect on materials even for moderate heating rates. The internal structural changes and defects develop, owing to physical and chemical phenomena. Macroscopically, damages are principally material is used by irreversible strains and variations of strength and elastic modulus. The important magnitude of transient thermal creep helps avoiding excessive damage during heating. This effect is, however, restricted to the heating phase and does not appear during cooling, leading to irrecoverable strains [6, 7].

Researchers are developing new solutions to improve the properties of conventional materials. Innovative processing techniques such as rapid solidification and powder metallurgy have produced some interesting alloys with improved properties, especially in room and elevated temperature strength.

Creep strength is important in considering the life of structures to be exposed to elevated temperatures and stresses for extended periods of time. Typically, in locations where creep is a potential problem, strict restrictions are placed on the allowable stresses in order to maintain the creep strain well below creep rupture. This often results in over designed structures to minimize the stresses. However, for instance, in aircraft there is a trade-off between weight and over designing the structure to minimize creep. The creep design criterion is specified for the lifetime of the structure.

It is unrealistic to perform creep tests for the exact service life of a structural system because that would take many hours of real fire test of uninterrupted creep testing for a single test; and it would not have any significant scientific meaning, because the next real fire scenario will be different anyway! There are many recognized methods for extrapolating creep data. Two major types of extrapolation techniques have emerged: linear extrapolation of the steady-state regime, and parametric models. The following paragraphs describe each of these different techniques.

Data from creep tests should extend at least 1–10 % of actual lifetimes to give meaningful extrapolation. Linear extrapolation of the creep strain along the steady-state creep rate is considered valid for no more than one to two orders of magnitude, although this method is generally unacceptable because it does not account for the possibility of entering tertiary creep. Using the abridged method of extrapolation, tests are performed at several stress levels at the anticipated maximum temperature for a given fire scenario. The creep strain versus time (at constant temperature) results in a family of creep curves, where each curves representing a different stress level. Extrapolation of the creep rate out to the design life (fire rating) will provide an estimate to the actual creep life, but will not predict if creep rupture will occur prior to reaching the design life, specifically, if the real fire duration exceeds the fire rating time.

Acceleration of creep tests is sometimes necessary to shorten the length of the tests to reasonable times. The mechanical acceleration method uses stress levels significantly higher than the anticipated service stresses to reach the limiting strain faster. The design life is again determined by developing a family of creep curves at constant stress and temperature and extrapolating the appropriate steady-state creep rate out to the design life. Accelerated tests can also use the thermal acceleration method in which the temperatures in the tests are significantly higher than the anticipated fire temperatures. A plot of the stress versus time for a constant creep strain and temperature will develop a family of creep curves from which extrapolation out to the design lifetime can be made [8]. As with the abridged method above, neither of these accelerated test methods will predict if creep rupture will occur prior to reaching the design life time. In addition, increasing the temperature or the stress significantly can change which creep deformation mechanism is active, which makes the creep data developed from these tests relatively meaningless to the actual service environment conditions.

Because of uncertainties in extrapolating creep data to determine the design lifetime, parametric models have been developed.

There is more certainty to extrapolating data using a parametric model, because many different test conditions are superimposed to create a master curve. The guidelines for extrapolating data using a parametric method allow an additional order of magnitude extrapolation with more certainty than simple linear extrapolation allows.

The creep test specimen is usually kept in a furnace lined with refractory materials for high temperature tests. Temperature control is essential and ought to be nearly constant because any slight variations may present difficulties in differentiating between the specimen movement due to thermal expansion and that due to creep. A small temperature variation would thus result in the prediction of more or less life of a material from the actual life depending on the variation. Temperature measurements are usually carried out by thermocouples in contact with the specimen. For reliable temperature measurements, thermocouples should be shielded from direct radiation from the furnace walls.

We briefly discuss now some creep damage equations that are rather prescriptive in nature and based exclusively on test results that are empirically connecting the rapture life (time to rapture) and temperature (or creep rate). The results are limited to the type of the material to be tested and testing methodology and instrumentation equipment used. Several creep parameters are utilized to correlate creep data to be presented in a format that gives more meaning in design. From the experimental data, various curves can be plotted known as master curves [9, 10].

The most common and universally recognized parametric model developed is the Larson–Miller method. This model assumes that the steady-state regime follows power law creep. Larson and Miller developed this time–temperature relationship for prediction rupture and creep stresses.

It should be noted that as (4.19) is nonlinear, the computation of the damage is an iterative process.

Larson–Miller creep parameter [11]

In the Larson–Miller method, θ, is the temperature compensated time such that

$$P_{\text{LM}} = \log e \left[\frac{Q}{RT}\right] = \left[\log t_f + (\log \theta = C)\right] \tag{4.19}$$

P_{LM} is the Larson–Miller parameter, Q is assumed to a function of stress only, and C is a constant of ~ 20 for most materials. In this method, a number of tests are run at various temperatures and stresses to determine the times to failure and activation energy. A "universal" plot is then made of the stress as a function of P_{LM}

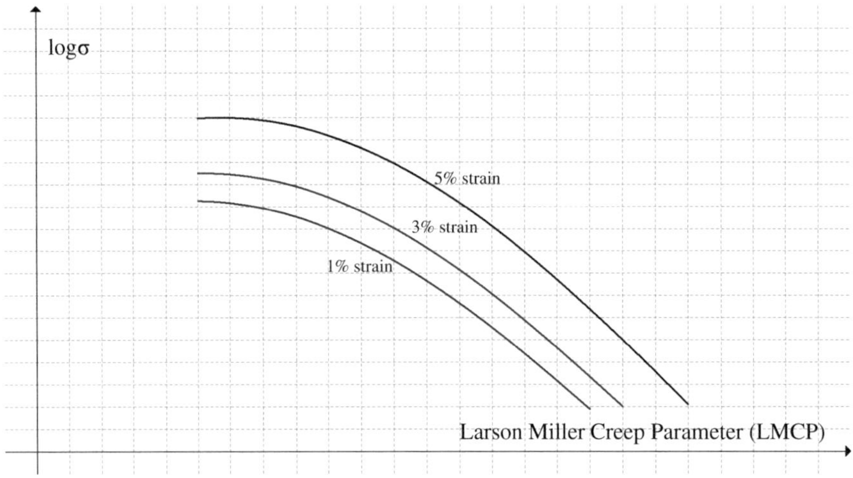

Fig. 4.5 Larson–Miller parameter

(see Fig. 4.5). The allowable stress for a combination of time to failure and temperature (i.e., P_{LM}) can then from the curve.

t_f-time to either creep rupture or to a given creep strain level and C = material constant

The Larson–Miller parameter gives better consistence with deformation process occurring at lower temperatures. It also offers better results with interpolation and extrapolation. The L.C.M.P. values for the different materials at various temperatures and stresses can be obtained from experimental data.

Whichever time is used for determining the constant will determine the lifetime extrapolated for this method. In addition, minimum creep rate can be substituted for time, allowing the prediction of creep rates instead of lifetimes.

Dorn Parameter (PD) [12].

This correlation assumes creep to be thermally activated.

The Dorn Parameter is normally expressed as

$$P_{\mathrm{D}} = te^{-E/RT} \tag{4.20}$$

Where: E—activation Energy; T—temperature; t—time; P_D—Dorn Parameter.

Generally, empirical relationship between $\dot{\varepsilon}$ and t_r can be given by Volkmann and Grant equation [13]:

$$\log(t_{\mathrm{r}}) + m\log(\dot{\varepsilon}) = B \tag{4.21}$$

where t_r-rapture life; $\dot{\varepsilon}$—steady-state creep rate; m and B are constants $0.77 < m < 0.93$ for aluminum, copper, iron, and titanium alloy

$0.48 < B < 1.3$ for aluminum, copper, iron, and titanium alloys.

Manson—Haferd Parameter (P_{MH}) [14].

Manson and Haferd developed a linear time–temperature relation for the extrapolation of creep and stress rupture data, as an improvement to the Larson–Miller method. This was developed to eliminate errors introduced in the Larson–Miller method by assuming $C = 20$ for the Larson–Miller constant. The Manson–Haferd method assumes the same starting point of steady-state creep dominated by power law behavior. However, the Manson–Haferd parameter determines two constants in lieu of the one proposed by Larson and Miller. The Manson–Haferd parameter is more reliable for data prediction at higher temperatures. It explains complex deformation pattern controlled by several mechanisms. It is expressed as

$$P_{\mathrm{MH}} = \frac{T - T_a}{\log t - log\, t_{\mathrm{a}}}, \tag{4.22}$$

where T_a—constant temperature; t_a-constant time; PMH = Mason Haferd Parameter

As with the Larson–Miller method, the time can either be the time to creep rupture or the time to a given creep strain level. The Manson–Hanford parameter is derived graphically from the intersection point of extrapolated isostress lines when plotted on a log of minimum creep rate versus absolute temperature (instead of the inverse of temperature, which is used in the Larson–Miller parameter). The intersection point will identify the constants T_a and t_a. By plotting the Manson–Hanford parameter versus stress, all creep data will collapse onto a single master curve.

The equation of this curve can be determined by curve fitting, which yields an equation relating time to a given percent creep, temperature, and stress. The Manson–Haferd parameter can be used for a variety of different materials and times to either a certain percent creep strain or creep rupture. The Manson–Haferd method tends to be more conservative by underestimating the creep extrapolations, as compared to the Larson–Miller method.

Zener–Hollomon Parameter (Z) [15].

The correlation between the flow stress (σ), temperature (T), and strain rate $\dot{\varepsilon}$, particularly at high temperatures, could be expressed by an Arrhenius type equation. Further, the effects of temperature and strain rate on deformation behavior could be represented by Zener–Holloman Parameter (Z) in an exponent type equation [15]. These are mathematically expressed as

$$Z = \dot{\varepsilon}\mathrm{e}^{-Q/RT} = f(\sigma) = \begin{cases} A'\sigma^{n'} \\ A''\exp(\beta\sigma) \\ A[\sinh(\alpha\sigma)]^{n} \end{cases} \tag{4.23}$$

Here, R is the universal gas constant; T is the absolute temperature in K; $\dot{\varepsilon}$ represents the strain rate and A; A'; A''; n; n'; β and $\alpha \approx \beta/n'$ are material constants. In the above equations, the Zener–Hollomon parameter Z is temperature compensated strain rate [16] and Q is the deformation activation energy.

Structural steels with minimum yield strength of 1380 MPa are often referred as ultra high strength steels (UHSS). The applications of these steels are in critical cases such as pressure vessels, aircraft undercarriages, rocket motor casings, turbine motors. In addition to their high strength to weight ratio, these steels must have good ductility, toughness, and fatigue resistance.

The material that used in this study is a medium carbon low alloy steel with ultra high strength that is micro-alloyed with vanadium. This steel is produced by electro slag refining (ESR).

This parametric relationship is aimed at determining the activation energy for deformation, flow stress and dynamic recrystallization. The flow stress increases with decrease in temperature and increases in strain rate. The flow curve typically shows work hardening region followed by dynamic softening due to recovery/ recrystallization. The work hardening is predominant at lower temperatures and higher strain rates. On the contrary, the extent of dynamic softening is more at higher temperatures and lower strain rates. This is due to the fact that higher temperatures and lower strain rates offer higher mobility to the grain boundary and longer time for nucleation and growth of dynamically recrystallized grains. Zener–Hollomon parameter Z is mostly applicable for pure metals and accurate for secondary creep.

Phenomenological constitutive equationsSimple phenomenological constitutive equations were proposed during the first thirty years of previous century [17]. The Norton-Bailey creep law was mostly used in practice.

$$\dot{\varepsilon} = N\sigma^n, \tag{4.24}$$

where N and n are material properties following from uniaxial tests; ε, σ denote strains and stresses, respectively, the dot is the derivation with respect to the time. The Norton-Bailey law is a suitable description for the stationary creep. This relation can be modified for the primary creep introducing an explicit dependence on the time t or for the tertiary creep introducing a creep rupture variable ω and defining a creep rupture law [10].

Numerous examples of creep equations and additional creep rupture equation are presented in the literature. One of the first was introduced by Rabotnov [10] assuming only creep rupture coupling. The starting point of this approach is the creep equation (4.24) modified as follows:

$$\dot{\varepsilon} = \frac{3}{2} K\sigma_{\mathrm{eq}}^{n-1} \frac{s}{(1-\omega^r)^k} \tag{4.25}$$

Here $(0 \leq \omega \leq 1)$ denotes a creep rupture variable; r, k are material parameters. For the creep rupture evolution law an expression similar to the Norton law can be formulated

$$\dot{\omega} = B\left[\sigma_{\mathrm{eq}}^{\omega}\right]^m \frac{1}{(1-\omega^r)^k} \tag{4.26}$$

Here m, k are material parameters. The Rabotnov creep rupture equations enable an extension of the classical creep equation for stationary creep to the tertiary creep. The advantage of this approach is the simple form easy for handling. The main problems are connected with the neglecting of primary creep and the formal introduction of the creep rupture parameter ω that leads to a range of this parameter from 0 (undamaged state) to 1 (fully damaged state) and which seems to be not realistic. In addition, the creep rupture law does not reflect the different damage behavior, for instance, resulting from tensile, or compressive loads.

One possible variation of the Rabotnov creep rupture equation is given by the introduction of other expressions for the equivalent stress σ_{eq}. With the help of the modification of the equivalent stress controlled not only by the deviatory stresses can be introduced. A suitable generalized equivalent stress expression was proposed, for instance, in [18].

It should be noted that as Eq. (4.25) is nonlinear, the computation of the creep rupture (failure) time is an iterative process. The integral type constitutive creep law can also reflect a nonlinear relationship between stress and stain by including the Norton–Bailey power law into right hand side integral Eq. (4.2)

$$\bar{\sigma}(t) = E(t)\varepsilon(t) = \sigma(t) + \int_0^t K(t, t')\sigma^n(t')\mathrm{d}t' \tag{4.27}$$

The material parameter N from Eq. (4.24) is included in kernel function $K(t, t')$ and n is a new dimensionless parameter. The solution of the integral Eq. (4.27) (assuming that it exists) is a function $\sigma = f(\varepsilon, t)$ that has only one absolute maximum value σ^* (ultimate stress) at a particular point (ε^*, t^*). Therefore, this time value $t^* = t_r$ could serve as an engineering phenomenological creep rupture (failure) time. It should be also noted that Eq. (4.27) has a nonlinear relationship between stress and stain even without Norton–Bailey power law, because the kernel $K(t, t')$ must include the Arrhenius law in case of thermal creep process (exponential function of temperature), and the temperature in turn is a linear function of strain ε. Therefore, the integral type stress–strain relationship is nonlinear. Although the understanding of non-linear methods for thermal creep is much more limited than it is for the linear methods, some noteworthy advances have been made and will now be briefly reviewed.

4.4 Nonlinearity Due to Temperature Variation

This is the main source of nonlinearity under constant loads in the service stress range.

Its source is twofold

1. The cross coupling of creep with thermal expansion, which causes that these strains are not additive and need to be taken into account in the form of, for example, stress-induced thermal expansion (or contraction).
2. Cracking or tensile strain-softening, which inevitably accompanies the changes of temperature and is caused by the delay due to thermal diffusion. These effects may be analyzed numerically in a step-by-step fashion.

A rather simple analysis of these effects was made by the author who employed the thermo-adjusted effective modulus method, introducing corrections to the instantaneous effective modulus on the basis of approximate histories of temperature dependency. A particular difficulty arises with strain-softening when the time steps are decreased substantially beyond the shortest relaxation time, as is necessary when long-time creep is analyzed.

It was found that an arbitrary increase of the time step is made possible by making six major changes in general integral type constitutive creep law (4.27) which is analogous to the algorithm for the Maxwell chain model (see Chap. 3) and is based on the assumption that the instantaneous effective modulus E (θ) is obtained from simple tensile creep test results.

4.4.1 *Relationship Between Modulus of Elasticity and Temperature*

The empirical relationship between modulus of elasticity and temperature is taken from [19] and presented in tabular form (see Table 4.1)

$$\frac{E(T)}{E(20)} = 1 + \left[\frac{T}{2000\left(\ln\frac{T}{1100}\right)}\right] \text{if: } 0\ {}^{\circ}\text{C} < 600\ {}^{\circ}\text{C}$$

$$\frac{E(T)}{E(20)} = \frac{690\left[1 - \frac{T}{1000}\right]}{T - 53.5} \text{if: } 600\ {}^{\circ}\text{C} < 1000\ {}^{\circ}\text{C}$$

$$\rho(\theta) = \exp(-0.15\theta)$$

The temperature at which the proportion of the yield stress at elevated temperature is considered to have dropped to zero differs from that of the modulus of elasticity.

Table 4.1 Function $\rho(\theta) = \exp(-0.15\theta)$

θ	0	1	2	3	4	5	6	7	8	9	10
T (°C)	300	360	420	480	540	600	660	720	780	840	900
$\rho(\theta)$	0.885	0.84	0.78	0.71	0.62	0.50	0.39	0.29	0.21	0.14	0.08

The mechanical properties of steel vary with temperature, generally reducing in strength as the temperature of the steel increases. Steel has a limited strength, meaning that at a certain temperature the strength of the steel member will eventually reduce to zero.

The temperature at which the member being analyzed will fail is determined using the formula for the variation of the yield stress of steel or modulus of elasticity [19].

This implies that the factor affecting the steel strength is the parameter $\rho(\theta) = \exp(-0.15\theta)$—the ratio of the design stiffness of the member under the design temperature load for fire specified to the design stiffness of the member at room temperature.

The design capacity of the steel section is based on the yield stress and the cross-sectional area of the beam, assuming the cross section of the structural element remains constant and a uniform temperature is maintained throughout the steel. This only occurs with four-sided exposure, as with three-sided exposure to a fire, there will be significant temperature differences across the cross section of the steel that has to be accounted for the temperature gradient in the steel.

Finally, the changes to the integral type constitutive creep law (4.27) are as follows:

Let us substitute the independent variables t and t' with some new functions of temperature in integral Eq. (4.27) and rewrite this equation in dimensionless form

$$\begin{aligned} &\bar{\sigma}(\theta) = E(\theta)\varepsilon(\theta) = \sigma(\theta) + \int_0^\theta L[\theta,\tau]\sigma^n(\tau)\mathrm{d}\tau \\ &\varepsilon(\theta) = A\theta;\ E(\theta) = E_0\exp(-a\theta);\ t = \varphi(\theta);\ t' = \varphi(\tau) \\ &L[\theta,\tau] = K[\varphi(\theta),\varphi(\tau)]\varphi' \\ &\quad a \text{ — modulus of elasticity deterioration parameter} \end{aligned} \tag{4.28}$$

Functions $\varphi(\theta)$ and $\varphi'(\theta)$ in Eq. (4.28) are inverse functions of temperature–time functions (Eqs. (4.10) and (4.11) respectfully) that characterize fire severity scenario: Very Fast; Fast; Medium, or Slow fire [5]. Deterioration parameter “a” characterizes reduction of modulus of elasticity with temperature increase [13]; parameter A is a constant ($A = 7.02\ (10^{-4})$ for regular steel material) and n is the creep stress exponent.

The Arrhenius law must be included in our new creep function $L[\varsigma(\theta), \tau(\theta')]$, therefore the integral Eq. (4.28) has a form

$$\bar{\sigma}(\theta) = E(\theta)\varepsilon(\theta) = \sigma(\theta) + \delta\int_0^\theta \exp\left[\frac{\tau}{1+\beta\tau}\right]L[\theta,\tau]\sigma^n(\tau)\mathrm{d}\tau \tag{4.29}$$

Here: $\delta = 1$—Frank–Kamenetsky parameter in our case [3].

The usual form of Arrhenius law ($\exp\left[-\frac{E}{RT}\right]$ where E—activation energy, R—universal gas constant, and T—temperature in K) in Eq. (4.25) is substituted by the approximate Frank-Kamenetsky expression $\delta\exp\left[\frac{\theta}{1+\beta\theta}\right]$, where θ—dimensionless temperature and $\beta = RT_*/E$—dimensionless parameter. The base line temperature $T_* = 600$ K ≈ 300 °C is the onset creep deformations temperature (assumed here for steel structures).

The kernel of Eq. (4.29) must comply with inequality

$$\left[\int_0^{\theta_{\max}}\int_0^{\theta_{\max}}\left\{\left|\exp\left[\frac{\tau}{1+\beta\tau}\right]L[\theta,\tau]\right|\right\}^2 \mathrm{d}\tau\mathrm{d}\theta\right]^{-1/2} \geq 1 \qquad (4.30)$$
$$0<\tau<\theta<\theta_{\max}$$

If the function $\exp\left[\frac{\tau}{1+\beta\tau}\right]L[\theta,\tau]$ is square-integrable for $0 < \theta < \theta_{\max}$ and $0 < \tau < \theta_{\max}$, while $E(\theta)\varepsilon(\theta)$ is square-integrable for $0 < \theta < \theta_{\max}$ then the solution of Eq. (4.29) is mean-square convergent for $\delta = 1$. Obviously, inequality (4.30) can be satisfied if the interval $0 < \theta < \theta_{\max}$ is small. However, if this interval is too big then it should be divided into "N" subintervals and the strip method (see Chap. 2) must be applied.

The dimensionless variables (temperature and time θ and τ) that are introduced above substantially compress real values of T and t, so that the solution of Eq. (4.29) can be obtained for much bigger temperatures range without subdividing the total temperature interval into smaller subintervals. For instance, the real temperature interval [300 °C; 1200 °C] is compressed into dimensionless interval [$0 < \theta < 15$], and real-time interval [0; 14,400 s.] into dimensionless interval [0; 0.22]. This in turn practically eliminates one of the biggest numerical computation problem connected with subdividing the small time steps into even smaller steps when solving the Eq. (4.29) by using the very popular finite differences method.

The solution of any Volterra equation of the second kind can be approximated by a solution of degenerate integral equation in the mean-square (and certain other) metrics to any degree of accuracy (see Chap. 2). The degenerate kernel approximates the kernel of the initial Volterra equation in one sense or another. An IE (4.26) with a degenerate kernel has the following form:

$$\bar{\sigma}(\theta) = E(\theta)\varepsilon(\theta) = \sigma(\theta) + \int_0^{\theta}\exp\left[\frac{\tau}{1+\beta\tau}\right]\sum_{i=1}^{N} a_i[\varphi(\theta)]b_i[\varphi(\tau)]\varphi'(\tau)\sigma^n(\tau)\mathrm{d}\tau \qquad (4.31)$$

Here one may assume that functions $a_i(x)$ (and $b_i(y)$) are linearly independent (otherwise, the number of terms in Eq. (4.31)) can be reduced). If $n = 1$ then the

Eq. (4.31) is linear integral e Volterra equation of the second kind, and if $n > 1$, then the equation is nonlinear.

Consider first the linear Eq. (4.28), e.g., $n = 1$.

4.4.2 Approximate Solutions of Linear Volterra Integral Equation

The main purpose of the present section is to investigate best approximate solutions, i.e., solutions in the sense of least squares, and to establish the applicability of these approximate methods to much more complex problem of thermally activated nonlinear creep constitutive integral type equations by using the strip method that has been discussed in Chap. 2. Since thermal creep constitutive equations usually cannot be solved explicitly, so it is required to obtain at least approximate solutions. There are numerous numerical methods which have been focusing on the solution of linear integral equations. For example, Tricomi in his book [1], introduced the classical method of successive approximations for integral equations. Variation iteration method [20] and was effective and convenient for solving integral equations. Taylor expansion approach was presented for solving integral equations by Kanwal [21]. In addition, Babolian et al. [22] solved some integral equations systems by using the orthogonal triangular basis functions. Jafari et al. [23] applied Gauss–Legendre method to find numerical solution system of linear integral equations. Moreover, some different valid methods for solving this kind of equations have been developed. The application of these methods (modified and specifically tailored for solving general type of integral type constitutive creep equations) is presented below. The mathematical background of these methods is presented in Chap. 2 and the detailed applications to the nonlinear creep deformations problem is presented in the corresponding section below.

Consider first the method of moments (MoM) method that is often used for obtaining the maximum value of unknown function and can be applied to linear operators describing different physical phenomenal.

4.4.3 Method of Moments

The method of moments (MoM) is a general solution method that is widely used in all of engineering. A Fourier series approximation to a periodic time function has a similar solution process as the MoM solution for creep constitutive law.

The method of moments (MoM) is a technique used to solve the Eq. (4.31) for the creep stress.

This method as applied to integral Volterra equations of the second kind consists of an approximation which replaces the kernel of integral Eq. (4.31) by a degenerate kernel

$$K_N(\theta,\tau) = \sum_{n=1}^{N} a_n(\theta) b_n(\tau) \varphi'(\tau) \tag{4.32}$$

Followed by the solution of the degenerate integral Eq. (4.31)

$$\begin{aligned} &\sigma(\theta) + \int_0^{\theta} \exp\left[\frac{\tau}{1+\beta\tau}\right] K_N(\theta,\tau)\sigma(\tau)\mathrm{d}\tau = f(\theta) \\ &f(\theta) = \theta \mathrm{e}^{-a\theta}; \quad a = 0.15 \text{ (steel structure)} \end{aligned} \tag{4.33}$$

The step-by-step procedure in this case is as follows:

Expand stress function $\sigma(\theta)$ into a series with unknown expansion coefficients

$$\sigma(\theta) = f(\theta) - \sum_{i=1}^{N} A_i a_i(\theta) \tag{4.34}$$

Define weighting (testing) functions. Perform a testing (or weighting) procedure to obtain a set of N linear equations (to solve for N unknown coefficients). The selection of the testing functions to be the complex conjugates of the expansion functions is referred to as Galerkin method. We get to choose, but generally select the complex conjugates of the expansion functions.

The constants A_i are obtained by multiplying each side of Eq. (4.33) by the testing (basis) function and integrating over a dimensionless temperature interval [0, 8]. The coefficients are found from the system of linear algebraic equations

$$\begin{aligned} &A_i + \lambda \sum_{j=1}^{N} A_i b_{i,j} = f_i \\ &\text{where } b_{i,j} = \int_0^{\theta_{\max}} a_i(x) b_j(x) \mathrm{d}x; \; f_i = \int_0^{\theta_{\max}} f(x) b_i(x) \mathrm{d}x \\ &\Delta(\lambda) = \begin{vmatrix} 1+\lambda b_{1,1} & \lambda b_{1,2} & \lambda b_{1,3} \ldots\ldots\ldots\ldots\ldots & \lambda b_{1,N} \\ \lambda b_{2,1} & 1+\lambda b_{2,2} & \lambda b_{2,3} \ldots\ldots\ldots\ldots\ldots & \lambda b_{2,N} \\ \ldots & \ldots & \ldots & \ldots \\ \lambda b_{N,1} & \lambda b_{N,2} & \lambda b_{N,3} \ldots\ldots\ldots\ldots\ldots & 1+\lambda b_{N,N} \end{vmatrix} \end{aligned} \tag{4.35}$$

If a system (4.35) has a unique solution, Eq. (4.33) is uniquely solvable as well. The values $\lambda \neq 0$ (there is not more than N such values) for which the determinant of the system (4.35) is zero are eigenvalues. The conditions for solvability of a degenerate integral Eq. (4.33) are given by the Fredholm alternative. If $f = 0$, a degenerate integral Eq. (4.33) is a Volterra equation of the first kind; in order that it is solvable it is necessary and sufficient that the function "f" is presentable as a linear combination of the functions a_i. The importance of degenerate integral equations in the general theory of Fredholm/Volterra equations is based on the fact that the solution of any Fredholm/Volterra equation of the second kind can be approximated by solutions of degenerate integral equations in the mean-square metrics. Solving (4.33) is reduced to solving a system of linear algebraic equations. The degenerate kernel $K_N(\theta, \tau)$ may be found from the kernel $K(\theta, \tau)$ in several ways, e.g., by expanding the kernel into a Taylor series or a Fourier series. The method of degenerate kernels may be applied to systems of integral equations of the type (4.28), to multidimensional equations with relatively simple domains of integration and to certain nonlinear equations [24–26].

4.4.4 Galerkin Method (Linear Volterra Equation)

Galerkin Method: 1st Approximation [basic function $\sigma(\theta) = A\ (\theta \exp(-0.15\theta))$]

Consider now the Galerkin method that is often used for obtaining the maximum value of unknown function and can be applied to linear and nonlinear operators.

The linear Volterra integral equations are usually difficult to solve analytically and in many cases the solution must be approximated. Therefore, in recent years several numerical approaches have been proposed. The numerical methods usually transform the integral equation into a linear system that can be solved by direct or iterative methods [27].

In comparison to the other methods, this approach has some advantages. For example, this method is not iterative and it solves the problem directly, without need of any initial guess. The Galerkin method works much more efficiently if the type of the stress–strain function affected by thermal creep process is known in advance from some other physical evidence (for instance, from similar tests results). For metal matrix materials (MMM) one can assume that the stress–strain diagram in case of fire temperature effect should have some similarity with the elastic–plastic nonlinear behavior of steel materials, so let us select the stress–temperature (strain) basic function as follows:

$$\sigma(\theta) = A_0[\theta \exp(-0.15\theta)] \tag{4.36}$$

We have selected the stress–temperature function identical to the instantaneous stress function simply because it is the first approximation of unknown function $\sigma(\theta)$ when sequential method is employed (see below).

Let us assume also that the kernel in Eq. (4.33) is presented as

$$K_N(\theta,\tau) = \sum_{i=1}^{N} \exp\{-\alpha_i[\varphi(\theta) - \varphi(\tau)]\}\varphi'(\tau) \tag{4.37}$$

Here α_i—material property parameters (MPP) that approximately can be estimated from simple tension creep test and their values are reciprocal of retardation times (corresponding temperatures from Eq. (4.10)). Substituting (4.37) into (4.33) and taking $N = 1$ we have after integration ($\alpha_1 = 0.333$ is assumed)

$$\begin{aligned}\int_0^8 \theta e^{-0.15\theta}(\theta e^{-0.15\theta})d\theta = A_0 \int_0^8 (\theta e^{-0.15\theta})^2 d\theta + \\ + A_0 \int_0^8 (\theta e^{-0.15\theta})[e^{-0.333\varphi(\theta)}]d\theta \int_0^\theta (\tau e^{-\tau})e^{\frac{\tau}{1+0.1\tau}}e^{0.333\varphi(\tau)}\varphi'(\tau)d\tau\end{aligned} \tag{4.38}$$

The second integral from 0 to θ in the right hand site of Eq. (4.38) is the solution of a dummy differential equation

$$\begin{aligned}d(K)/d(t) = {} & (\exp(t/(1+0.1*t)))*(\exp(0.333*m))*(0.0405-0.02252*t^\wedge 1 \\ & +0.004386*t^\wedge 2-0.0002747*t^\wedge 3)*(t*\exp(-t))\end{aligned} \tag{4.39}$$

Initial condition is $K(0) = 0$, and explicit equation for m is

$$m = \varphi(\tau) = (0.0405*t - 0.01126{}^*t^\wedge 2 + 0.001462*t^\wedge 3 - 0.00006868*t^\wedge 4) \tag{4.40}$$

The solution of ODE (4.39) and (4.40) can be obtained easy by using any popular mathematical software such as POLYMATH for example. The step-by-step integration process is presented below in Example 4.2

Example 4.2 Data: $0 < \theta < 8$ (see Fig. 4.1); $\sigma = A\theta[\exp(-0.15\theta)]$; $\alpha_1 = 0.333$

Equation (4.38) can be rewritten as follows: $f1 = A[\,R1 + H1]$. Here $f1$ is the value of left hand site of Eq. (4.38); $R1$ and $H1$ are the value of the first and second terms of the right hand site sum of Eq. (4.38).

The dummy function $K(\theta)$—solution of Eq. (4.39)—becomes an integrand part of $H1$.

The simple computer code in this case is as follows:

Differential equations

1 $d(f1)/d(t) = t^\wedge 2 * (\exp(-0.3 * t))$
2 $d(H1)/d(t) = (\exp(-0.333 * (z))) * (t^\wedge 1) * (\exp(-0.15 * t)) * G011$
3 $d(R1)/d(t) = t^\wedge 2 * (\exp(-0.3 * t))$
4 $d(K)/d(t) = (\exp(t/(1 + 0.1 * t))) * (\exp(0.333 * z)) * (z1) * (t^\wedge 1) * (\exp(-0.15 * t))$

Explicit equations

1 $m = \theta(\tau) = (0.0405 * t - 0.01126 * t^\wedge 2 + 0.001462 * t^\wedge 3 - 0.00006868 * t^\wedge 4)$
2 $m1 = \varphi'(\tau) = (0.0405 - 0.02252 * t^\wedge 1 + 0.004386 * t^\wedge 2 - 0.0002747 * t^\wedge 3)$

After integration we have (using POLYMATH software).

Calculated values of DEQ variables

	Variable	Initial value	Minimal value	Maximal value	Final value
1	f01	0	0	31.87343	31.873
2	G011	0	0	1.660922	1.661
3	H011	0	0	31.87343	31.873
4	R011	0	0	8.944085	8.9440
5	t	0	0	8	8
6	m	0	0	0.0705907	0.0706
7	m1	0.0405	0.0003976	0.0405	0.0004

5. Finally we have

$$A = \frac{31.873}{31.873 + 8.944} = 0.781$$

$$\boldsymbol{\sigma}(\boldsymbol{\theta}{=}0) = 0;\ \boldsymbol{\sigma}(\boldsymbol{\theta}{=}8) = 0.781(2.4) = 1.874$$

$$\boldsymbol{\sigma}(\boldsymbol{\theta}{=}8) = 2.40 - \text{stress without creep}$$

The stress–temperature–strain diagram is presented below.

From Fig. 4.6 we can define the resistance factor ϕ as a ratio of creep stress value at the end of monotonically increased temperature (see Fig. 4.1) and the temperature stress value (without creep effect—$\delta = 0$ in Eq. (4.29)) due to combined effect of hardening and dynamic softening processes. The first approximation is $\phi = 1.874/2.4 = 0.781$.

Let us present now the second approximation of Galerkin method.

2nd Approximation: [basic function σ = $A\theta[\exp(-0.15\theta)] + B\theta[\exp(-0.2\theta)]$

The computations are very similar to the 1st approximation; therefore we present just the results.

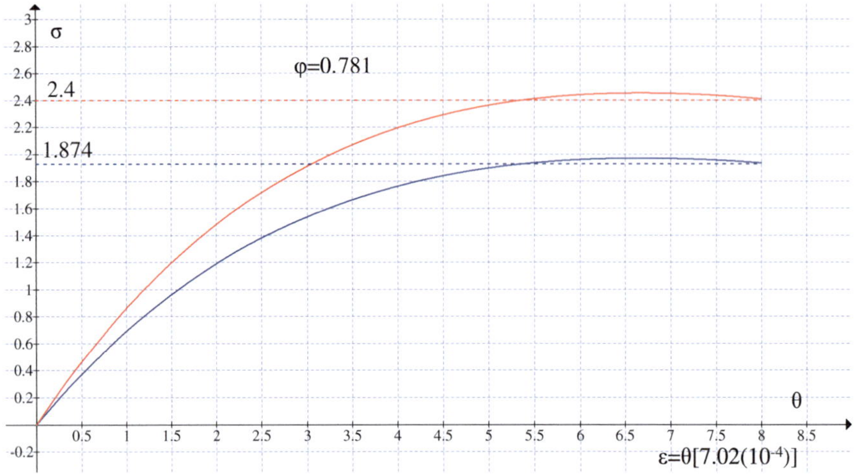

Fig. 4.6 Stress–temperature (strain) diagrams. First approximation

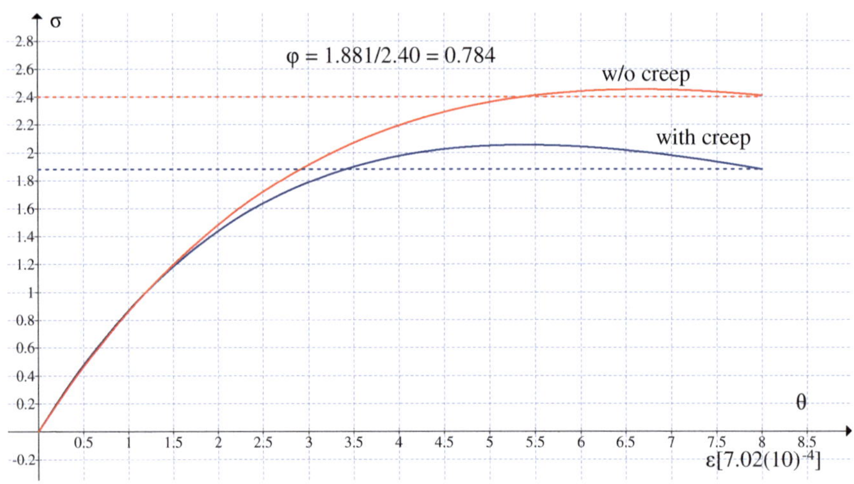

Fig. 4.7 Stress–temperature (strain) diagrams. Galerkin second approximation

$$\theta e^{-0.15\theta} = \sigma(\theta) + e^{-0.333\varphi(\theta)} \int_0^\theta e^{\frac{\tau}{1+0.1\tau}} e^{0.333\varphi(\tau)} \varphi'(\tau)\sigma(\tau)d\tau$$

$$m = \varphi(\theta) = \mathbf{0.0405 * \theta - 0.01126 * \theta\text{^}2 + 0.001462 * \theta\text{^}3 - 0.00006868 * \theta\text{^}4}$$

$$m1 = \varphi'(\theta) = \mathbf{0.0405 - 0.02252 * \theta\text{^}1 + 0.004386 * \theta\text{^}2 - 0.0002747 * \theta\text{^}3}$$

$$K(\theta, \tau) = e^{-0.333\varphi(\theta)} e^{\frac{\tau}{1+0.1\tau}} e^{0.333\varphi(\tau)} \varphi'(\tau)$$

$$\sigma(\theta) = A\theta e^{-0.15\theta} + B\theta[\exp(-0.2\theta)]$$

$$\text{I)} \int_0^8 [\theta^2 e^{-0.3\theta})d\theta = \int_0^8 \{A\theta e^{-0.15\theta} + B\theta[\exp(-0.2\theta)]\}\theta e^{-0.15\theta}d\theta$$

$$+ \int_0^8 e^{-0.333m}\theta e^{-0.15\theta}d\theta \int_0^\theta e^{\frac{\tau}{1+0.1\tau}} e^{0.333m}[m1]\{A\tau e^{-0.15\tau} + B\tau[\exp(-0.2\tau)]\}d\tau$$

and dummy differential equation is:

$$\frac{dK1(\theta)}{d\tau} = e^{\frac{\tau}{1+0.1\tau}} e^{0.333m}[m1]\{\tau e^{-0.15\tau}\}$$

$$\frac{dK2(\theta)}{d\tau} = e^{\frac{\tau}{1+0.1\tau}} e^{0.333m}[m1]\{\tau[\exp(-0.2\tau)]\}$$

$$\text{II)} \int_0^8 [\theta^2 e^{-0.35\theta})d\theta = \int_0^8 \{A\theta e^{-0.15\theta} + B\theta[\exp(-0.2\theta)]\}\theta e^{-0.2\theta}d\theta$$

$$+ \int_0^8 e^{-0.333m}\theta e^{-0.2\theta}d\theta \int_0^\theta e^{\frac{\tau}{1+0.1\tau}} e^{0.333m}[m1]\{A\tau e^{-0.15\tau} + B\tau[\exp(-0.2\tau)]\}d\tau$$

and initial condition $K_1(\tau = 0) = K_2(\tau = 0) = 0$.

$m = \varphi(\theta) = \mathbf{0.0405 * \theta - 0.01126 * \theta^\wedge 2 + 0.001462 * \theta^\wedge 3 - 0.00006868 * \theta^\wedge 4}$

$m1 = \varphi'(\theta) = \mathbf{0.0405 - 0.02252 * \tau^\wedge 1 + 0.004386 * \tau^\wedge 2 - 0.0002747 * \tau^\wedge 3}]$

After integration, we have

Calculated values of DEQ variables

	Variable	Initial value	Minimal value	Maximal value	Final value
2	f1	0	0	31.87343	31.87343
3	f2	0	0	24.74852	24.74852
6	H11	0	0	31.87343	31.87343
7	H12	0	0	24.74852	24.74852
8	H21	0	0	24.74852	24.74852
9	H22	0	0	19.37801	19.37801
10	K1	0	0	1.660922	1.660922
11	K2	0	0	0.3087773	0.3087773
13	R11	0	0	8.944085	8.944085

(continued)

(continued)

	Variable	Initial value	Minimal value	Maximal value	Final value
14	R12	0	0	2.632417	2.632417
15	R21	0	0	6.547027	6.547027
16	R22	0	0	1.99845	1.99845
17	t	0	0	8	8
18	m	0	0	0.0705907	0.0705907
21	$m1$	0.0405	0.0003976	0.0405	0.0003976

Differential equations

1 $\mathrm{d}(f1)/\mathrm{d}(t) = t^2 * (\exp(-0.3 * t))$

2 $\mathrm{d}(f2)/\mathrm{d}(t) = t^2 * (\exp(-0.35 * t))$

3 $\mathrm{d}(\mathrm{K1})/\mathrm{d}(t) = (\exp(t/(1+0.1 * t))) * (\exp(0.333 * z)) * (z1) * t * (\exp(-0.15 * t))$

4 $\mathrm{d}(\mathrm{K2})/\mathrm{d}(t) = (\exp(\mathrm{t}/(1+0.1 * t))) * (\exp(0.333 * z)) * (z1) * (\exp(-0.2 * t))$

5 $\mathrm{d}(\mathrm{R11})/\mathrm{d}(t) = (\exp(-0.333 * (z))) * t * (\exp(-0.15 * t)) * K1$

6 $\mathrm{d}(\mathrm{R12})/\mathrm{d}(t) = \exp(-0.333 * (z))) * t * (\exp(-0.15 * t)) * K2$

7 $\mathrm{d}(\mathrm{R21})/\mathrm{d}(t) = \exp(-0.333 * (z)) * t * (\exp(-0.2 * t)) * K1$

8 $\mathrm{d}(\mathrm{R22})/\mathrm{d}(t) = \exp(-0.333 * (z)) * t * (\exp(-0.2 * t)) * K2$

9 $\mathrm{d}(H11)/\mathrm{d}(t) = (\exp(-0.3 * t)) * (t^2)$

10 $\mathrm{d}(H12)/\mathrm{d}(t) = (\exp(-0.35 * t)) * (t^2)$

11 $\mathrm{d}(H21)/\mathrm{d}(t) = (\exp(-0.35 * t)) * (t^2)$

12 $\mathrm{d}(H22)/\mathrm{d}(t) = (\exp(-0.4 * t)) * (t^2)$

Explicit equations

1 $m = (0.0405 * t - 0.01126 * t^\wedge 2 + 0.001462 * t^\wedge 3 - 0.00006868 * t^\wedge 4)$

2 $m1 = (0.0405 - 0.02252 * t^\wedge 1 + 0.004386 * t^\wedge 2 - 0.0002747 * t^\wedge 3)$

$$\begin{cases} 40.814A + 27.38B = 31.87 \\ 31.3A + 21.38B = 24.75 \end{cases}$$

$A = 0.239; B = 0.808$ $\boldsymbol{\sigma}(\boldsymbol{\theta}{=}0) = 0$; $\boldsymbol{\sigma}(\boldsymbol{\theta}{=}8) = 1.881$

$\boldsymbol{\sigma}(\boldsymbol{\theta}{=}8) = 2.40$ – temperature stress without creep

Resistance factor $\phi = 1.881/2.40 = 0.784$

40.814	27.38	31.87
31.3	21.38	24.75

Linear equations solution

	Variable	Value
1	*x*1	0.239
2	*x*2	0.808

The equations

[1] $40.814 \cdot x1 + 27.38 \cdot x2 = 31.87$
[2] $31.3 \cdot x1 + 21.38 \cdot x2 = 24.75$

4.5 Replacement of the Integral Equation with Finite System of Linear Algebraic Equations

Let us say that the integral Eq. (4.31) of the second kind is given. Integral entering into this equality we can replace by means of any formula of the approximate integration with some simple type the expression which is not containing a sign of integral. Really, any linear formula of the approximate integration looks like

$$\int_a^b \psi(x)\mathrm{d}x = \sum_{k=1}^{n} A_k\psi(x_k) + \rho \tag{4.41}$$

where A_k and x_k are constants for the given interval and for the given formula of approximate integration method, and ρ—an error. Thus it is usual $A_k > 0$ and the $\sum_{k=1}^{n} A_k = b - a$.

We will be using here the Gauss–Legendre method and the A_k and x_k—constants are as follows:

$$x_k = a + (b - a)x_k^{(n)}; \quad A_k = (b - a)A_k^{(n)}, \tag{4.42}$$

where $x_k^{(n)}$—Gaussian points (roots of polynomial Legendre) and $A_k^{(n)}$—Gaussian factors made for an interval [0, 1]. After application of the formula (4.42) to the integral in left hand side of the Eq. (4.41) we come to equality

$$\varphi(x) - \lambda\sum_{k=1}^{n} A_k K(x, x_k)\varphi(x_k) = f(x) + \lambda\rho(x) \tag{4.43}$$

In particular, believing in equality (4.43) it is consecutive $x = x_1; x_2 \ldots . x_n$ we come to following system of the equations with which satisfy numbers $\varphi(x_i)$-values of required function in points x_i.

$$\varphi(x_i) - \lambda \sum_{k=1}^{n} A_k K(x_i, x_k)\varphi(x_k) = f(x_i) + \lambda\rho(x_i) \tag{4.44}$$

Rejecting in the right hand part of equality of (4.44) small number $\rho(x_i)$, which precise value is not known to us, we shall receive following system n the equations with n unknowns—$\varphi(x_1)$; $\varphi(x_2)$;…., $\varphi(x_n)$. The Eq. (4.44) can be presented in the developed type as follows:

$$\begin{cases} \varphi(x_1)[1 - \lambda A_1 K(x_1, x_1)] - \lambda\varphi(x_2)A_2K(x_1, x_2) - \ldots - \lambda]\varphi(x_n)A_nK(x_1, x_n) = f(x_1) \\ -\lambda\varphi(x_1)A_1K(x_2, x_1) + \varphi(x_2)[1 - \lambda A_2K(x_2, x_2)] - \ldots - \lambda\varphi(x_n)A_nK(x_2, x_n) = f(x_2) \\ \ldots\ldots\ldots\ldots\ldots\ldots\ldots\ldots\ldots\ldots\ldots\ldots\ldots\ldots\ldots\ldots\ldots\ldots\ldots\ldots \\ -\lambda\varphi(x_1)A_1K(x_n, x_1) - \lambda\varphi(x_2)A_2K(x_n, x_2) - \ldots + \varphi(x_n)[1 - \lambda A_nK(x_n, x_n)] = f(x_n) \end{cases} \tag{4.45}$$

4.5.1 Derivation of Two-Point Gauss Quadrature Rule

The two-point Gauss quadrature rule is an extension of the trapezoidal rule approximation where the arguments of the function are not predetermined as *a* and *b*, but as unknowns x_1 and x_2. So in the two-point Gauss quadrature rule, the integral is approximated as

$$I = \int_a^b f(x)\mathrm{d}x \approx c_1 f(x_1) + c_2 f(x_2) \tag{4.46}$$

We can state here (without proof) that the above four simultaneous nonlinear equations have only one acceptable solution

$$c_1 = c_2 = \frac{b-a}{2}$$

$$x_1 = \frac{b-a}{2}\left[-\frac{1}{\sqrt{3}}\right] + \frac{b+a}{2}; \quad x_2 = \frac{b-a}{2}\left[\frac{1}{\sqrt{3}}\right] + \frac{b+a}{2}$$

In our case: $a = 0$; $b = 8$; $c_1 = c_2 = 4.0$; $x_1 = 8(0.2113) = 1.69$; $x_2 = 8$ $(0.7887) = 6.31$.

Applying the Gaussian quadrature rule then results in the following approximation:

$$I = \int_a^b f(x)\mathrm{d}x \approx \frac{b-a}{2}[f(x_1) + f(x_2)] \tag{4.47}$$

Solving this system of Eqs. (4.45), we find approximations for the values of $\tilde{\varphi}(x_1); \tilde{\varphi}(x_2); \ldots \tilde{\varphi}(x_n)$ of unknown function $\varphi(x_1);\ \varphi(x_2);\ \ldots \varphi(x_n)$. From these values, the approximate value of the function can be found with the help of a method of interpolation. In this special case, it is best to get this value—$\tilde{\varphi}(x)$—based on equality (4.44), discarding $\rho(x)$ and replacing $\varphi(x)$ to $\tilde{\varphi}(x)$. Then $\tilde{\varphi}(x)$ can be written as:

$$\tilde{\varphi}(x) = f(x) + \lambda \sum_{k=1}^{n} A_k K(x, x_k) \tilde{\varphi}(x_k) \tag{4.48}$$

Obviously, the accuracy of the result obtained by replacing the integral Eq. (4.2) with the systems of linear Eqs. (4.45), will be higher than a smaller error we make by replacing the integral by the sum. Accurate error estimation, is presented in [1, 2], however, the method itself has been applied much earlier in the fundamental work of Fredholm and Hilbert.

Example 4.3 Data from Example 4.2

$$\theta = [0; 8] \quad \text{Kernel type}: \quad K(\theta, \theta_k) = \left[e^{\theta_k/(1+0.1\theta_k)}\left(e^{-[\varphi(\theta)-\varphi(\theta_k)]\alpha}\right)\right]\varphi'(\theta_k)$$
$$\alpha = 0.333 \quad f(\theta) = \theta\rho(\theta) = \theta\exp(-0.15\theta)$$

Applying the Gauss–Legendre quadrature rule we have

$$\begin{aligned}
\tilde{\sigma}(\theta) &= f(\theta) - 8\sum_{k=1}^{2}\frac{1}{2}\tilde{\sigma}(\theta_k)\left[e^{\theta_k/(1+0.1\theta_k)}\right]\left[e^{-[\varphi(\theta)-\varphi(\theta_k)]\alpha)}\right]\varphi'(\theta)\text{where:} f(\theta)\\
&= 7.02\left(10^{-4}\right)E_0\theta[R(\theta)];\ \alpha = 0.333\theta_1 = 1.69; \theta_2 = 6.31; \varphi(\theta) = m\\
&= \mathbf{0.0405{*}\theta - 0.01126 * \theta\hat{}2 + 0.001462 * \theta\hat{}3 - 0.00006868 * \theta\hat{}4} m1\\
&= \varphi'(\theta) = \mathbf{0.0405 - 0.02252 * \theta\hat{}1 + 0.004386 * \theta\hat{}2 - 0.0002747 * \theta\hat{}3} K_{1,1}\\
&= \frac{8}{2}e^{1.69/1.17}\left(e^{-0.333(0)}\right)(1.69) = 16.96(0.0136) = 0.231 K_{1,2}\\
&= \frac{8}{2}\left[e^{1.69/1.17}\left(e^{-0.333(-0.0657+0.0382)}\right)\right](0.0136) = 0.1857 = 0 K_{2,1}\\
&= \frac{8}{2}\left[e^{6.31/1.631}\left(e^{-0.333(0.0657-0.0428)}\right)\right]0.004 = 0.760 K_{2,2}\\
&= \frac{8}{2}e^{6.31/1.631}\left(e^{-0.333(7.887)} - e^{-0.333(0)}\right)0.004 = 0.766; f_1(\theta_1)\\
&= A(1.31); f_2(\theta_2) = A(2.45)
\end{aligned}$$

The system of linear equations is

$$\begin{cases}\sigma(\theta_1)\left[1 + \mathrm{K}_{1,1}\right] + \sigma(\theta_2)\mathrm{K}_{1,2} = \mathrm{f}_1\\ \sigma(\theta_1)\mathrm{K}_{2,1} + \sigma(\theta_2)\left[1 + \mathrm{K}_{2,2}\right] = \mathrm{f}_2\end{cases} \quad \begin{cases}\sigma(\theta_1)1.231 + \sigma(\theta_2)0 = 1.31\\ \sigma(\theta_1)0.76 + \sigma(\theta_2)1.766 = 2.45\end{cases}$$

Linear equations solution is

	Variable	Value
1	$x1$	1.064
2	$x2$	0.929

Finally, the approximate closed form of stress–temperature (strain) relationship is as follows (Fig. 4.8 and Table 4.2):

$$\tilde{\sigma}(\theta) = f(\theta) - \sum_{k=1}^{2} \frac{1}{2}\tilde{\sigma}(\theta_k)\left[\mathrm{e}^{-[\varphi(\theta)-\varphi(\theta_k)]\alpha}\right]\varphi'(\theta_k)$$

$$= 7.02\left(10^{-4}\right)E_0 \begin{bmatrix} \theta\left[\mathrm{e}^{-0.15\theta}\right] - \frac{1}{2}\{(1.064)\left[\mathrm{e}^{1.69/1.169}\left(\mathrm{e}^{-[\varphi(\theta)-\varphi(1.69)]0.333}\right)\right]\varphi'(1.69) \\ +0.929\left[\mathrm{e}^{6.31/1.631}((\mathrm{e}^{-[\varphi(\theta)-\varphi(6.31)]0.333})\right]\varphi'(6.31)\} \end{bmatrix}$$

Now we repeat the computations for MPP $\alpha = 0.1$.

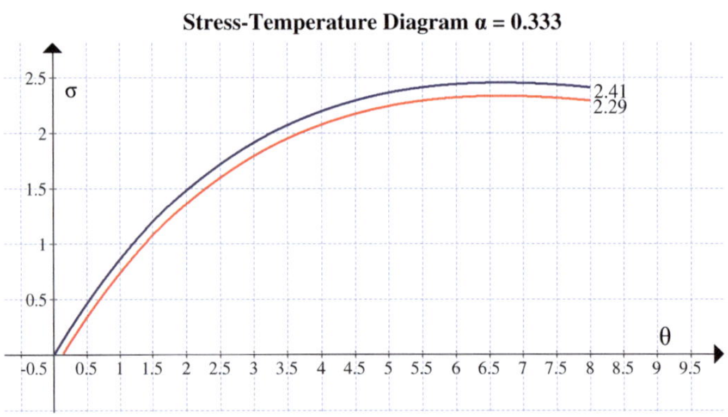

Fig. 4.8 Stress–temperature diagram, Gauss–Legendre method

Table 4.2 Stress–temperature data: Gauss–Legendre Method ($\alpha = 0.33$)

θ	0	1.0	2.0	3.0	4.0	5.0	6.0	7.0	8.0
$\sigma(\theta)$ w/o creep	0	0.86	1.48	1.91	2.20	2.36	2.44	2.45	2.41
$\sigma(\theta)$ w/creep	0	0.74	1.36	1.79	2.08	2.24	2.32	2.33	2.29

$\alpha = 0.1 \quad R_E(\theta) = \rho(\theta) = \exp(-0.15\theta)$

$$\tilde{\sigma}(\theta) = f(\theta) - 8\sum_{k=1}^{2}\frac{1}{2}\tilde{\sigma}(\theta_k)\left[e^{\theta_k/(1+0.1\theta_k)}\right]\left[e^{-[\varphi(\theta)-\varphi(\theta_k)]\alpha)}\right]\varphi'(\theta)$$

where : $f(\theta) = 7.02(10^{-4})E_0\theta[R(\theta)]$

$\theta_1 = 1.69;\ \theta_2 = 6.31;\ \varphi(\theta) = m = \mathbf{0.0405*\theta - 0.01126*\theta\hat{}2 + 0.001462*\theta\hat{}3 - 0.00006868*\theta\hat{}4}$

$m1 = \varphi'(\theta) = \mathbf{0.0405 - 0.02252*\theta\hat{}1 + 0.004386*\theta\hat{}2 - 0.0002747*\theta\hat{}3}$

$$K_{1,1} = \frac{8}{2}e^{1.69/1.17}\left(e^{-0.1(0)}\right)z1(1.69) = 16.96(0.0136) = 0.231$$

$$K_{1,2} = \frac{8}{2}\left[e^{1.69/1.17}\left(e^{-0.1(-0.0657+0.0428)}\right)\right](0.0136) = 0.1857 = 0$$

$$K_{2,1} = \frac{8}{2}\left[e^{6.31/1.631}\left(e^{-0.1(0.0657-0.0428)}\right)\right]0.004 = 0.749$$

$$K_{2,2} = \frac{8}{2}e^{6.31/1.631}\left(e^{-0.1(0)}\right)0.004 = 0.766;\ f_1(\theta_1) = A(1.31);\ f_2(\theta_2) = A(2.45)$$

Linear equations solution

	Variable	Value
1	$x1$	1.064
2	$x2$	0.936

The equations

1. $1.231 \cdot x1 = 1.31$
2. $0.749 \cdot x1 + 1.766 \cdot x2 = 2.45$

Both methods above have shown relatively good correlating results with respect to maximum values of thermal creep stress. However, the sequential approximation method (see below) provides better qualitative results of creep behavior at high temperature range and at the same time allows analyzing some very important types of nonlinear creep deformations. It is worthwhile to mention also that this method provides much faster convergence of the iteration process since the result is presented in alternating series form.

4.6 Successive Approximation (Sequential Approximation Method)

The method of successive approximations is as follows. We seek a solution of Eq. (4.2) in form of a series in powers of the parameter λ.

$$\begin{aligned} y(x) &= y_0(x) + \lambda y_1(x) + \lambda^2 y_2(x) + \ldots \\ K_1(x,t) &= K(x,t) \\ K_2(x,t) &= \int_0^x K(x,\tau)K_1(t,\tau)\mathrm{d}\tau \\ &\ldots\ldots\ldots\ldots\ldots\ldots\ldots\ldots\ldots \end{aligned} \tag{4.49}$$

Substituting this series in Eq. (4.2), we find.

$$\lambda y_0(x) + \lambda y_1(x) + \ldots = f(x) + \lambda \int_0^x K(x,\tau)[y_0(\tau) + \lambda y_1(\tau) + \cdots]\mathrm{d}\tau \tag{4.50}$$

Equating the coefficients of equal powers of λ, we obtain.

$$\begin{aligned} y_0(x) &= f(x) \\ y_1(x) &= \int_0^x K(x,\tau)y_0(\tau)\mathrm{d}\tau \\ y_2(x) &= \int_0^x K(x,\tau)y_1(\tau)\mathrm{d}\tau \\ &\ldots\ldots\ldots\ldots\ldots\ldots\ldots\ldots \end{aligned} \tag{4.51}$$

From these equations we can determine successively all functions $y_i(x)$. If we introduce the so-called integrated kernel, you can get the following recursive relation for the kernel of the original Eq. (4.2).

$$\begin{aligned} K_1(x,t) &= K(x,t) \\ K_2(x,t) &= \int_0^x K(x,\tau)K_1(t,\tau)\mathrm{d}\tau \\ K_3(x,t) &= \int_0^x K(x,\tau)K_2(t,\tau)\mathrm{d}\tau \\ &\ldots\ldots\ldots\ldots\ldots\ldots\ldots\ldots\ldots\ldots \end{aligned} \tag{4.52}$$

The power series (4.50) can now be written in the form.

$$
\begin{aligned}
y(x) &= f(x) + \lambda \int_0^x K_1(x,\tau) f(\tau) d\tau + \lambda^2 \int_0^x K_2(x,\tau) f(\tau) \mathrm{d}\tau + \cdots \\
&= f(x) + \lambda \int_0^x \Gamma(x,\tau,\lambda) f(\tau) \mathrm{d}\tau
\end{aligned}
\tag{4.53}
$$

We now denote $\Gamma(x,t,\lambda)$ function in parentheses integrand on the right side of Eq. (4.53). The function $\Gamma(x,t,\lambda)$ is called the resolvent of the Eq. (4.2). We can prove, assuming the kernel $K(x, t)$ is limited $|K(x, t)| < M$, that the series (4.49) and (4.53) will converge uniformly and the function $y(x)$ is a solution of Eq. (4.2) if λ satisfies the condition.

$$
|\lambda| < \left[\int_0^a \int_0^a K^2(x,t) \mathrm{d}x \mathrm{d}t \right]^{-1/2}
\tag{4.54}
$$

The step-by-step computations and computer code are presented below.

Example 4.4 Data: see Example 4.2.

$\sigma_0 = \theta \mathrm{e}^{-0.15\theta}$; $\lambda = -1$; $\alpha = 0.333$
Substituting $\sigma_0 = \theta \mathrm{e}^{-0.15\theta}$ into Eq. (4.56) we have

$$
\theta \mathrm{e}^{-0.15\theta} = \sigma(\theta) + \mathrm{e}^{-0.333\varphi(\theta)} \int_0^\theta \mathrm{e}^{\frac{\tau}{1+0.1\tau}} \mathrm{e}^{0.333\varphi(\tau)} \varphi'(\tau) \sigma(\tau) \mathrm{d}\tau
$$

$m = \varphi(\theta) = \mathbf{0.0405{*}\theta - 0.01126{*}\theta\hat{}2 + 0.001462{*}\theta\hat{}3 - 0.00006868{*}\theta\hat{}4}$

$m1 = \varphi'(\theta) = \mathbf{0.0405 - 0.02252{*}\theta\hat{}1 + 0.004386{*}\theta\hat{}2 - 0.0002747{*}\theta\hat{}3}$

$$
\sigma_1 = \mathrm{e}^{-0.333\varphi(\theta)} \int_0^\theta e^{\frac{\tau}{1+0.1\tau}} \mathrm{e}^{0.333m} m1 \tau \mathrm{e}^{-0.15\tau} \mathrm{d}\tau
$$

$$
\sigma_2 = \mathrm{e}^{-0.333\varphi(\theta)} \int_0^\theta e^{\frac{\tau}{1+0.1\tau}} \mathrm{e}^{0.333m} m1(\sigma_1) \mathrm{d}\tau
$$

$$
\sigma_3 = \mathrm{e}^{-0.333\varphi(\theta)} \int_0^\theta \mathrm{e}^{\frac{\tau}{1+0.1\tau}} \mathrm{e}^{0.333m} m1(\sigma_2) \mathrm{d}\tau
$$

……………………………………

$$
\sigma(x) = \sigma_0 + \lambda\sigma_1 + \lambda^2\sigma_2 + \cdots
$$

Substituting integral in the right hand side of Eq. (4.2) with the solution of the corresponding differential equation (ODE) we have four (4) ODE equations and twelve (12) explicit equations. Therefore, we have four sequential approximations. The computer code using POLYMATH software is as follows:

Differential equations

1 $d(Z11)/d(t) = (\exp(t/(1+0.1*t)))*(\exp(0.333*m))*m1*(Z0^\wedge p)$
2 $d(Z22)/d(t) = (\exp(t/(1+0.1*t)))*(\exp(0.333*m))*m1*(Z1^\wedge p)$
3 $d(Z33)/d(t) = (\exp(t/(1+0.1*t)))*(\exp(0.333*m))*m1*(Z2^\wedge p)$
4 $d(Z44)/d(t) = (\exp(t/(1+0.1*t)))*(\exp(0.333*m))*m1*(Z3^\wedge p)$

Explicit equations

1 $Z0 = (t*\exp(-0.15*t))$
2 $m = (0.0405*t - 0.01126*t^\wedge 2 + 0.001462*t^\wedge 3 - 0.00006868*t^\wedge 4)$
3 $A = (\exp(-0.333*n))$
4 $Z1 = A*Z11$
5 $Z2 = A*Z22$
6 $Z3 = A*Z33$
7 $Z4 = A*Z44$
8 $Z = Z0 - Z1 + Z2 - Z3 + Z4$
9 $m1 = (0.0405 - 0.02252*t^\wedge 1 + 0.004386*t^\wedge 2 - 0.0002747*t^\wedge 3)$
10 $R = t*(\exp(-0.15*t))$
11 step = if $(t< = 8)$ then$(Z0 == t*\exp(-0.15*t))*(1)$ else $(Z0 == Z)$
12 $p = 1$

Calculated values of DEQ variables

	Variable	Initial value	Minimal value	Maximal value	Final value
1	A	1	0.9767674	1	0.9767674
2	m	0	0	0.0705907	0.0705907
3	$m1$	0.0405	0.0003976	0.0405	0.0003976
4	p	1	1	1	1
5	R	0	0	2.45253	2.409554
6	Step	1	1	1	1
7	Step1	0	0	0	0
8	t	0	0	8	8
9	Z	0	0	1.97477	1.231449
10	Z0	0	0	2.45253	2.409554
11	Z1	0	0	1.622335	1.622335
12	Z11	0	0	1.660922	1.660922

(continued)

(continued)

	Variable	Initial value	Minimal value	Maximal value	Final value
13	Z2	0	0	0.5473212	0.5473212
14	Z22	0	0	0.5603394	0.5603394
15	Z3	0	0	0.1245904	0.1245904
16	Z33	0	0	0.1275538	0.1275538
17	Z4	0	0	0.0214986	0.0214986
18	Z44	0	0	0.0220099	0.0220099

We can see now from Fig. 4.5 that maximum thermal creep stress value ($\sigma = 1.96$) is very close to the same stress value that was obtained above by using Galerkin method ($\sigma = 2.055$). However the stress–temperature diagram is quite different in a high temperature range $4.8 < \theta < 8$ (588 °C < T < 780 °C). We have to underline here that this is so far the linear Volterra integral equation of the second kind (Fig. 4.9).

The numerical solution of Eq. (4.31) can be approximated now by using polynomial regression method (Fig. 4.10).

Model $Z = 0.0459 + 0.872 * t\^{}0.0971 * t\^{}2 - 0.00183 * t\^{}3 + 0.000312 * t\^{}4$

Variable	Value
$a0$	0.0459
$a1$	0.872
$a2$	−0.0971
$a3$	−0.00183
$a4$	0.000312

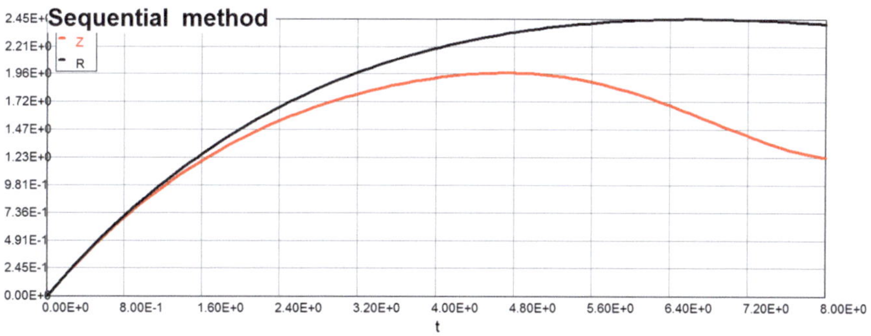

Fig. 4.9 Stress–temperature diagram, sequential method ($\alpha = 0.333$)

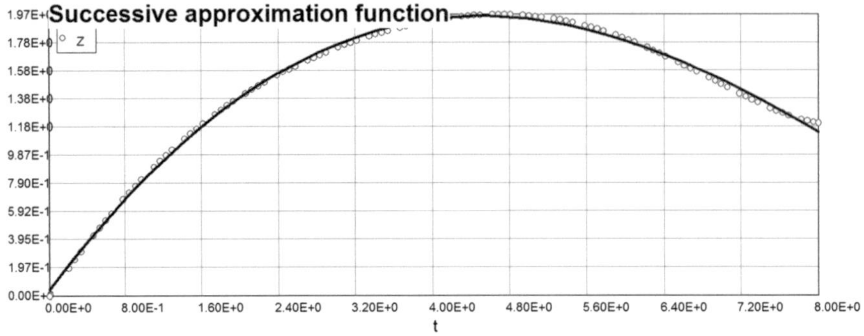

Fig. 4.10 Successive approximation of stress–strain function

$$\sigma(\theta) = 0.0459 + 0.872 * \theta - 0.0971 * \theta^\wedge 2 - 0.00183 * \theta^\wedge 3 + 0.000312 * \theta^\wedge 4 \tag{4.55}$$

The resistance factor can be defined as $\phi = \mathbf{1.971/2.29} = \mathbf{0.86}$.

Example 4.5 Data: see Example 4.1.

$\sigma_0 = \theta e^{-0.15\theta}$; $\lambda = -1$; $\alpha = 0.1$

The computer code in this case is

```
t(0) = 0
t(f) = 8
Z0 = (t * exp(−0.15 * t)) * (1)
A = (exp(−0.1 * n))
d(Z11)/d(t) = (exp(t/(1 + 0.1 * t))) * (exp(0.1 * m)) * m1 * Z0^1
Z1 = A * Z11
d(Z22)/d(t) = (exp(t/(1 + 0.1 * t))) * (exp(0.1 * m)) * m1 * Z1^1
Z2 = A*Z22
d(Z33)/d(t) = (exp(t/(1 + 0.1 * t))) * (exp(0.1 * m)) * m1 * Z2^1
Z3 = A * Z33
Z = Z0 − Z1 + Z2 − Z3
Z11(0) = 0
Z22(0) = 0
Z33(0) = 0
m = (0.0405 * t − 0.01126 * t^2 + 0.001462 * t^3 − 0.00006868 * t^4)
m1 = (0.0405 − 0.02252 * t^1 + 0.004386 * t^2 − 0.0002747 * t^3)
R = t * (exp(−0.15 * t)) Calculated values of DEQ variables
```

	Variable	Initial value	Minimal value	Maximal value	Final value
1	*A*	1	0.449329	1	0.449329
2	*m*	0	0	0.0705907	0.0705907

(continued)

(continued)

	Variable	Initial value	Minimal value	Maximal value	Final value
3	$m1$	0.0405	0.0003976	0.0405	0.0003976
4	R	0	0	2.452527	2.409554
5	t	0	0	8	8
6	Z	0	0	2.114516	1.788328
7	$Z0$	0	0	2.452527	2.409554
8	$Z1$	0	0	0.7368736	0.7354935
9	$Z11$	0	0	1.636871	1.636871
10	$Z2$	0	0	0.1306919	0.1306919
11	$Z22$	0	0	0.2908601	0.2908601
12	$Z3$	0	0	0.0164245	0.0164245
13	$Z33$	0	0	0.0365534	0.0365534

The numerical solution of Eq. (4.31) can be approximated now by using polynomial regression method. The results are presented below (see Fig. 4.11 and Eq. 4.56).

Resistance factor $\phi = 1.788/2.4 = 0.745$ Model: $Z = 0.0325 + 0.907 * t - 0.116 * t$^2 + 0.00307 * t^3 + 0.0000786 * t^4

Variable	Value	95 % confidence
$a0$	0.0325102	0.0143249
$a1$	0.9070876	0.0239788
$a2$	−0.116173	0.0119181
$a3$	0.0030717	0.0022035
$a4$	7.862E-05	0.0001349

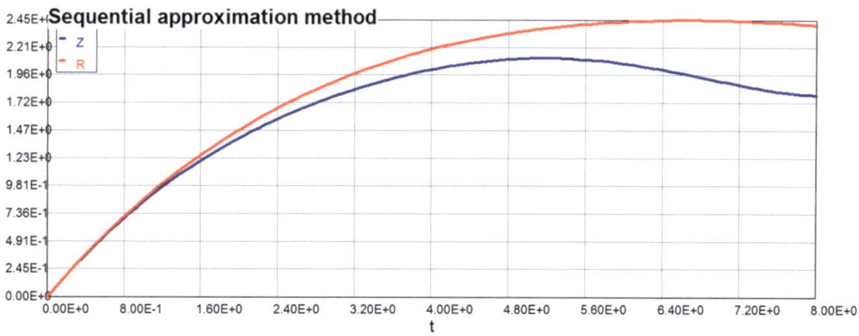

Fig. 4.11 Stress–temperature diagram, sequential method ($\alpha = 0.1$)

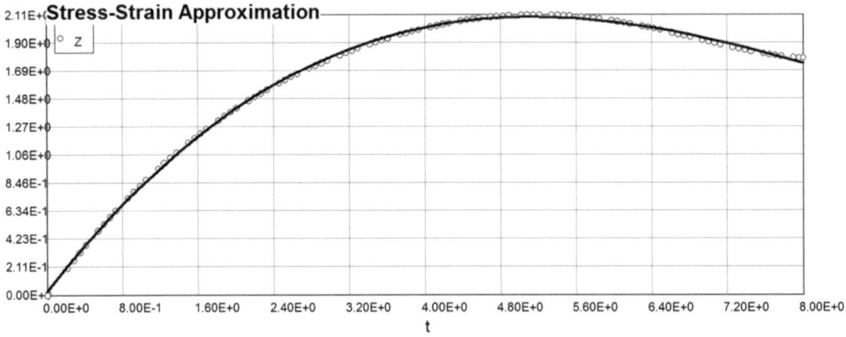

Fig. 4.12 Successive approximation of stress–strain function ($\alpha = 0.1$)

$$\sigma(\theta) = 0.0325 + 0.907 * \theta - 0.116 * \theta \wedge 2 + 0.00307 * \theta \wedge 3 + 0.0000786 * \wedge \theta 4 \tag{4.56}$$

Resistance factor $\phi = 2.11/2.4 = 0.879$ (Fig. 4.12).

Example 4.5 Data: see Example 4.2.

$\sigma_0 = \theta e^{-0.15\theta}$; $\lambda = -1$; $\alpha = 0.01$

The computer code in this case is

Differential equations

1 $d(Z11)/d(t) = (\exp(t/(1 + 0.1 * t))) * (\exp(0.01 * m)) * m1 * Z0$
2 $d(Z22)/d(t) = (\exp(t/(1 + 0.1 * t))) * (\exp(0.01 * m)) * m1 * Z1$
3 $d(Z33)/d(t) = (\exp(t/(1 + 0.1 * t))) * (\exp(0.01 * m)) * m1 * Z2$

Explicit equations

1 $Z0 = t * \exp(-0.15 * t)$
2 $A = \exp(-0.01 * t)$
3 $Z1 = A * Z11$
4 $Z2 = A * Z22$
5 $Z3 = A * \ Z33$
6 $Z = Z0 \ - \ Z1 + Z2 - Z3$
7 $m = (0.0405 * t - 0.01126 * t^\wedge + 0.001462 * t^\wedge 3 - 0.00006868 * t^\wedge 4)$
8 $m1 = (0.0405 - 0.02252 * t^\wedge 1 + 0.004386 * t^\wedge 2 - 0.0002747 * t^\wedge 3)$
9 $R = t * (\exp(-0.15 * t))$

Calculated values of DEQ variables

	Variable	Initial value	Minimal value	Maximal value	Final value
1	*A*	1	0.9231163	1	0.9231163
2	*m*	0	0	0.0705907	0.0705907
3	*m*1	0.0405	0.0003976	0.0405	0.0003976
4	*R*	0	0	2.45253	2.409554
5	*t*	0	0	8	8
6	*Z*	0	0	1.988681	1.280742
7	*Z*0	0	0	2.45253	2.409554
8	*Z*1	0	0	1.502534	1.502534
9	*Z*11	0	0	1.627676	1.627676
10	*Z*2	0	0	0.4757539	0.4757539
11	*Z*22	0	0	0.5153781	0.5153781
12	*Z*3	0	0	0.102031	0.102031
13	*Z*33	0	0	0.1105289	0.1105289

Model: *Z* = *a*0 + *a*1 * *t* + *a*2 * *t*^2 + *a*3 * *t*^3 + *a*4 * *t*^4

Variable	Value
*a*0	0.044659
*a*1	0.8751279
*a*2	−0.0992168
*a*3	−0.0012786
*a*4	0.0002878

Resistance factor $\phi = 1.99/2.4 = 0.839$ (Figs. 4.13 and 4.14).

Similar computations of linear relationship between thermal creep stresses and dimensionless temperature (strain) are obtained for different MPP and presented in Table 4.4.

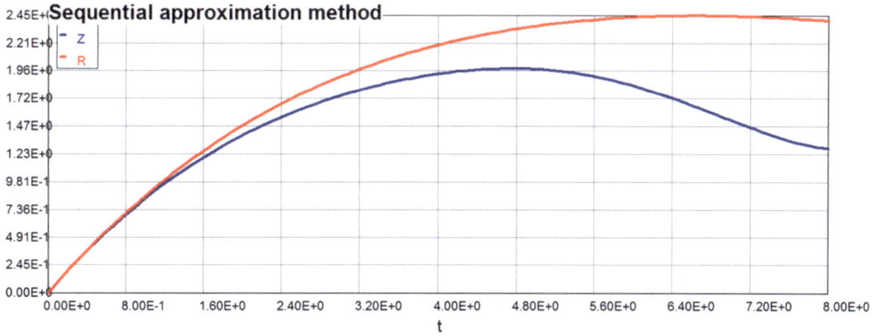

Fig. 4.13 Stress–temperature diagram, sequential method (α = 0.01)

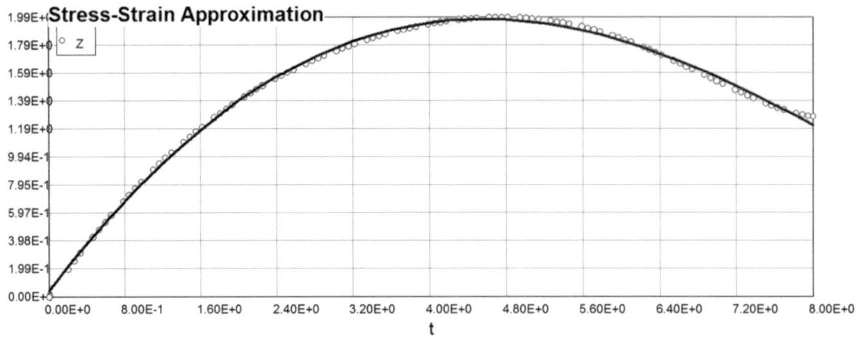

Fig. 4.14 Successive approximation of stress–strain function ($\alpha = 0.01$)

4.7 Nonlinear Integral Type Creep Constitutive Law

Nowadays, numerical methods are widely used in many engineering areas, as for instance mechanical, civil, aeronautical, or in our case electrical engineering. They have helped many engineers to solve efficiently the problems and challenges, which are inherent to their work, since a huge number of analyses and studies can be performed with the help of computers. This has also helped to reduce considerably the economic cost typically associated to trial and error procedures. Usually, a researcher on numerical modeling aspires to find the method, which can solve any generic problem. Unfortunately, this is still an open problem. According to the type of problem one method can be more suitable than others, for instance finite element methods (FEM) are typically used for the analysis of 3-D enclosed structures, but they exhibit less performance than other methods when simulating for instance open problems or planar structures. This section is devoted to exhibit several numerical examples and practical applications of the theoretical results developed in previous sections. The solutions to all the presented nonlinear creep deformation problems are obtained within the framework of discretization and linearization of constitutive creep integral equations (CIE) through the method of moments (MoM); Galerkin Method, and method of successive approximation. Each technique is illustrated by practical examples (with step-by-step procedures) and the results comparison is presented at the end. The dimensionless numerical integration strategy is utilized to study the effects of material property parameters (MPP) on the nonlinear material response. Advantageously, the usage of the dimensionless numerical integration strategy permits to implement not only deterministic but the probability based approach to the creep deformation problem by using the deterministic results as a stochastic data for further applied probability methods of analysis.

In addition to allow the derivation of several relevant scientific conclusions, this study has also a practical impact. Indeed, it has been used to ascertain the good behavior of all the tools needed in the development of new structural engineering materials. Corresponding Computer Codes are provided, which are being implemented in the framework of this section. The objective of this software is to show the ability of the use of some practical forms of nonlinear creep integral equation together with the simple approximate solutions to analyze combine effect of MPP and structural system on fire resistance technology.

The steady-state or minimum strain rate is often used as a design tool. For example, what is the stress needed to produce a minimum strain rate of 10^{-6} m/m/h or what is the stress needed to produce a minimum strain rate of 10^{-7} m/m/h . An Arrhenius-type rate model is used to include the effect of temperature in the model (Eq. 4.2) such that

$$\dot{\varepsilon} = N\sigma^n \exp\left(-\frac{Q}{RT}\right), \tag{4.57}$$

where n is the stress exponent, Q is the activation energy for creep, R is the universal gas constant and T is the absolute temperature.

To determine the various constants in Eq. (4.57) a series of isothermal and iso stress tests are required. For isothermal tests, the exponential function of Eq. (4.57) becomes a constant resulting in the Norton–Bailey creep law (see Eq. 4.24) $\dot{\varepsilon} = N\sigma^n$ that can be rewritten in integral form

$$\varepsilon_{cr} = N\int\limits_0^t \sigma^n \mathrm{d}t \tag{4.58}$$

Combining Eq. (4.58) with general creep constitutive law (4.2) we have

$$\bar{\sigma} = E\varepsilon = \sigma(t) + \int\limits_0^t K(t,\tau)\sigma^n \mathrm{d}\tau \tag{4.59}$$

Parameter N in Eq. (4.59) is absorbed now by kernel $K(t, \tau)$. Applying the substitution method (see above) to Eq. (4.59) we have in case of thermal creep process

$$\begin{aligned}
&f(\theta) = \theta \mathrm{e}^{-0.15\theta} = \sigma(\theta) + \mathrm{e}^{-\alpha\varphi(\theta)} \int_0^\theta e^{\frac{\tau}{1+0.1\tau}} \mathrm{e}^{\alpha\varphi(\tau)} \varphi'(\tau)\sigma^n(\tau)\mathrm{d}\tau \\
&m = \varphi(\theta) = \mathbf{0.0405{*}\theta - 0.01126{*}\theta\hat{}2 + 0.001462{*}\theta\hat{}3 - 0.00006868{*}\theta\hat{}4} \\
&m1 = \varphi'(\theta) = \mathbf{0.0405 - 0.02252{*}\theta\hat{}1 + 0.004386{*}\theta\hat{}2 - 0.0002747{*}\theta\hat{}3}
\end{aligned} \tag{4.60}$$

Obviously, if n = 1 we have the linear Volterra equation of the second kind and "n" and "α" in Eq. (4.59) are material parameters. We have selected again only one parameter "α" in Eq. (4.60), because it will be called later on (in probability based approach) a random parameter.

4.7.1 Strip Method and Stress Function Linearization

The dimensionless temperature interval [0, 8] has been partitioned by integer numbers 0, 1, 2… 8. In each two adjoining subintervals $[i-1, i]$ and $[i, i+1]$ the kernel $K(\theta, s)$ remains unchanged and the stress from creep deformation is equal at partitioned point i. Thus the stress–temperature (strain) diagram is composed of the stress diagram form previous interval $[i-1, i]$ and additional stress from the next interval $[i, i+1]$ as it has been stated earlier in strip method application (see Chap. 2). That in turn ensures the continuity of the whole stress–strain diagram. This type of contingency of stresses and strains at partitioned points guarantees the convergence of approximate solution (4.2) in L_2 space.

Substituting the function σ^n in Eq. (4.60) by a linear part of Taylor series expansion at every partitioning point θ_i we have

$$\begin{aligned}
\sigma_i(\theta) &= \theta \mathrm{e}^{-0.15\theta} - \mathrm{e}^{-\alpha\varphi(\theta)} \int_{\theta_i}^{\theta} \mathrm{e}^{\frac{\tau}{1+0.1\tau}} \mathrm{e}^{\alpha\varphi(\tau)} \varphi'(\tau)[\sigma^n|_{\sigma_i} + n\sigma^{n-1}|_{\sigma_i}(\sigma - \sigma_i)]d\tau \\
&= \theta \mathrm{e}^{-0.15\theta} - \mathrm{e}^{-\alpha\varphi(\theta)} \int_{\theta_i}^{\theta} [\mathrm{e}^{\frac{\tau}{1+0.1\tau}} \mathrm{e}^{\alpha\varphi(\tau)} \varphi'(\tau)][\sigma^n|_{\sigma_i}]d\tau \\
&+ \mathrm{e}^{-\alpha\varphi(\theta)} \int_{\theta_i}^{\theta} [\mathrm{e}^{\frac{\tau}{1+0.1\tau}} \mathrm{e}^{\alpha\varphi(\tau)} \varphi'(\tau)][n\sigma^{n-1}|_{\sigma_i}(\sigma_i)]\mathrm{d}\tau \\
&- \mathrm{e}^{-\alpha\varphi(\theta)} \int_{\theta_i}^{\theta} [\mathrm{e}^{\frac{\tau}{1+0.1\tau}} \mathrm{e}^{\alpha\varphi(\tau)} \varphi'(\tau)][n\sigma^{n-1}|_{\sigma_i}]\sigma \mathrm{d}\tau \\
& i = 0, 1, 2 \ldots N; \quad \theta_i < \theta < \theta_{i+1}
\end{aligned} \tag{4.61}$$

If $n = 2$ Eq. (4.61) is reduced to

$$\begin{aligned}
\sigma_i(\theta) = \theta \mathrm{e}^{-0.15\theta} + \mathrm{e}^{-\alpha\varphi(\theta)}(\sigma_i^2) \int_{\theta_i}^{\theta} [\mathrm{e}^{\frac{\tau}{1+0.1\tau}} \mathrm{e}^{\alpha\varphi(\tau)} \varphi'(\tau)]\mathrm{d}\tau \\
- \mathrm{e}^{-\alpha\varphi(\theta)}[2\sigma_i] \int_{\theta_i}^{\theta} [\mathrm{e}^{\frac{\tau}{1+0.1\tau}} \mathrm{e}^{\alpha\varphi(\tau)} \varphi'(\tau)]\sigma \mathrm{d}\tau
\end{aligned} \tag{4.62}$$

If $n = 3$ Eq. (4.61) is reduced to

$$\begin{aligned}
\sigma_i(\theta) &= \theta \mathrm{e}^{-0.15\theta} - b^3 z + c^3 z - \mathrm{e}^{-\alpha\varphi(\theta)} c \int_{\theta_i}^{\theta} [\mathrm{e}^{\frac{\tau}{1+0.1\tau}} \mathrm{e}^{\alpha\varphi(\tau)} \varphi'(\tau)]\sigma d\tau \\
&= \theta \mathrm{e}^{-0.15\theta} + 2c^3 z - \mathrm{e}^{-\alpha\varphi(\theta)} 3c^2 \int_{\theta_i}^{\theta} [\mathrm{e}^{\frac{\tau}{1+0.1\tau}} \mathrm{e}^{\alpha\varphi(\tau)} \varphi'(\tau)]\sigma \mathrm{d}\tau \\
c &= [\sigma|_{\sigma_i}]; b = [\sigma|_{\sigma_i}]; z = \int_{\theta_i}^{\theta} [\mathrm{e}^{\frac{\tau}{1+0.1\tau}} \mathrm{e}^{\alpha\varphi(\tau)} \varphi'(\tau)]\sigma \mathrm{d}\tau
\end{aligned} \tag{4.63}$$

The solution (stress function σ) of linear integral Eq. (4.61) is defined on each subinterval $[\sigma_i, \sigma_{i+1}]$ and corresponding dimensionless temperature subinterval $[\theta_i, \theta_{i+1}]$, $i = 0, 1, 2\ldots.. N$. Therefore, now we are able to compute the stress function values at each partitioning point $i = 0, 1, 2\ldots N$ and then to obtain the whole stress–temperature (strain) function $\sigma(\theta)$ by using any best-to-fit regression method. Stress function continuity can be achieved by equating the values of two adjoining functions σ_i and σ_{i+1} at the partitioning point i. The application of a combined linearization and strip method is illustrated below in Example 4.6.

Example 4.6 Data: see Example 4.2.

$\sigma_0 = \theta e^{-0.15\theta}$; $\lambda = -1$; $n = 2$; $\alpha = 0.1$

The computer code in this case is

```
t(0) = 0
t(f) = 0.5
y0 = (t * exp(−0.15 * t)) + (c^2) * z
A = ((exp(−0.1 * m)))
d(y11)/d(t) = 2 * c * (exp(t/(1 + 0.1 * t))) * ((exp(0.1 * m)) * m1 * (y0)
y1 = A * y11
d(y22)/d(t) = 2 *c * (exp(t/(1 + 0.1 * t))) * ((exp(0.1 * m)) * m1 * (y1)
y2 = A * y22
d(y33)/d(t) = 2 * c * (exp(t/(1 + 0.1 * t))) * ((exp(0.1 * m)) * m1 * (y2)
y3 = A * y33
d(y44)/d(t) = 2 * c * (exp(t/(1 + 0.1 * t))) * ((exp(0.1 * m)) * m1 * (y3)
y4 = A * y44
d(y55)/d(t) = 2 * c * (exp(t/(1 + 0.1 * t))) * ((exp(0.1 * m)) * m1 * (y4)
y5 = A * y55
d(y66)/d(t) = 2 * c * (exp(t/(1 + 0.1 * t))) * ((exp(0.1 * m)) * m1 * (y5)
y6 = A * y66
y = y0 − y1 + y2 − y3 + y4 − y5 + y6
c = 0
d(z)/d(t) = (exp(t/(1 + 0.1 * t))) * ((exp(0.10 * m)) * m1
z(0) = 0
y11(0) = 0
y22(0) = 0
y33(0) = 0
y44(0) = 0
y55(0) = 0
y66(0) = 0
m = (0.0405 * t − 0.01126 * t^2 + 0.001462 * t^3 − 0.00006868 * t^4) *
0 + 0.02 * t − 0.01
m1 = (0.0405 − 0.02252 * t^1 + 0.004386 * t^2 − 0.0002747 * t^3) * 0 + 0.02
R = t * (exp(−0.15 * t))
```

Calculated values of DEQ variables [0 < t < 0.5]

	Variable	Initial value	Minimal value	Maximal value	Final value
1	A	1.001001	1	1.001001	1
2	c	0	0	0	0
3	m	−0.01	−0.01	0	0
4	$m1$	0.02	0.02	0.02	0.02
5	R	0	0	0.4638717	0.4638717
6	t	0	0	0.5	0.5
7	y	0	0	0.4638717	0.4638717
8	$y0$	0	0	0.4638717	0.4638717
9	$y1$	0	0	0	0
10	$y11$	0	0	0	0
11	$y2$	0	0	0	0
12	$y22$	0	0	0	0
13	$y3$	0	0	0	0
14	$y33$	0	0	0	0
15	$y4$	0	0	0	0
16	$y44$	0	0	0	0
17	$y5$	0	0	0	0
18	$y55$	0	0	0	0
19	$y6$	0	0	0	0
20	$y66$	0	0	0	0
21	z	0	0	0.0128521	0.0128521

Calculated values of DEQ variables [0.5 < t < 1.0]

	Variable	Initial value	Minimal value	Maximal value	Final value
1	A	1	0.9990005	1	0.9990005
2	c	0.464	0.464	0.464	0.464
3	m	0	0	0.01	0.01
4	$m1$	0.02	0.02	0.02	0.02
5	R	0.4638717	0.4638717	0.860708	0.860708
6	t	0.5	0.5	1	1
7	y	0.4638717	0.4638717	0.8523454	0.8523454
8	$y0$	0.4638717	0.4638717	0.8650623	0.8650623
9	$y1$	0	0	0.0128249	0.0128249
10	$y11$	0	0	0.0128377	0.0128377
11	$y2$	0	0	0.0001086	0.0001086
12	$y22$	0	0	0.0001087	0.0001087
13	$y3$	0	0	6.407E−07	6.407E−07

(continued)

(continued)

	Variable	Initial value	Minimal value	Maximal value	Final value
14	*y*33	0	0	6.413E−07	6.413E−07
15	*y*4	0	0	2.892E−09	2.892E−09
16	*y*44	0	0	2.895E−09	2.895E−09
17	*y*5	0	0	1.056E−11	1.056E−11
18	*y*55	0	0	1.057E−11	1.057E−11
19	*y*6	0	0	3.235E−14	3.235E−14
20	*y*66	0	0	3.238E−14	3.238E−14
21	*z*	0	0	0.0202246	0.0202246

Similar computations of nonlinear relationship between thermal creep stresses and dimensionless temperature (strain) are obtained for different MMP α and presented in Tables 4.3 and 4.4.

The convergence of solution of Eq. (4.2) is deteriorating with an increase in the stress exponent n. In this case it is necessary to use the strip method (see Chap. 2), thus reducing the length of an interval, and replacing it with a finite number of

Table 4.3 Stress–temperature data: Gauss–Legendre Method ($\alpha = 0.1$)

θ	0	1.0	2.0	3.0	4.0	5.0	6.0	7.0	8.0
$\sigma(\theta)$ w/o creep	0	0.86	1.48	1.91	2.20	2.36	2.44	2.45	2.41
$\sigma(\theta)$ w/creep	0	0.74	1.36	1.79	2.22	2.24	2.32	2.33	2.29

Table 4.4 Linear creep deformations

θ	0	1	2	3	4	5	6	7	8	α_i	ϕ
$f(\theta)$	0	0.86	1.48	1.76	2.05	2.24	2.35	2.4	2.41		
$\sigma(\theta)$	0	0.838	1.394	1.74	1.931	1.965	1.802	1.476	1.207	0.01	0.877
$\sigma(\theta)$	0	0.838	1.394	1.74	1.931	1.965	1.802	1.477	1.208	0.1	0.877
$\sigma(\theta)$	0	0.838	1.394	1.74	1.931	1.966	1.803	1.477	1.209	0.2	0.877
$\sigma(\theta)$	0	0.838	1.394	1.74	1.932	1.966	1.803	1.478	1.210	0.333	0.877
$\sigma(\theta)$	0	0.838	1.395	1.741	1.932	1.966	1.804	1.479	1.211	0.5	0.877
$\sigma(\theta)$	0	0.838	1.395	1.741	1.933	1.967	1.805	1.481	1.213	0.75	0.877
$\sigma(\theta)$	0	0.838	1.395	1.742	1.933	1.968	1.806	1.483	1.216	1.0	0.877
$\sigma(\theta)$	0	0.840	1.402	1.756	1.954	1.995	1.844	1.538	1.283	10	0.877
$\sigma(\theta)$	0	0.849	1.44	1.828	2.054	2.133	2.053	1.858	1.68	100	0.952
$\sigma(\theta)$	0	0.859	1.474	1.896	2.164	2.306	2.349	2.319	2.244	1000	1.0

Table 4.5 Nonlinear creep stress data ($n = 2$)

θ	0	1	2	3	4	5	6	7	8	α_i	ϕ
$\sigma(\theta)$	0	0.861	1.435	1.757	1.871	1.878	1.468	0.784	0.351	100	0.84
$\sigma(\theta)$	0	0.861	1.417	1.714	1.813	1.968	1.502	0.763	0.23	10	0.88
$\sigma(\theta)$	0	0.861	1.411	1.701	1.788	1.656	1.194	0.5	0.138	1.0	0.87
$\sigma(\theta)$	0	0.861	1.410	1.776	1.863	1.727	1.257	0.546	0.16	0.1	0.91
$\sigma(\theta)$	0	0.861	1.4	1.682	1.768	1.62	1.126	0.461	0.158	0.01	0.862

Table 4.6 Nonlinear creep stress data ($n = 3$)

θ	0	1	2	3	4	5	6	7	8	α_i	ϕ
$f(\theta)$	0	0.86	1.48	1.76	2.05	2.24	2.35	2.4	2.41		
$\sigma(\theta)$	0	0.861	1.408	1.626	1.589	1.26	0.626	0.208	0.12	0.01	0.92
$\sigma(\theta)$	0	0.861	1.408	1.627	1.690	1.262	0.628	0.208	0.12	0.1	0.91
$\sigma(\theta)$	0	0.861	1.408	1.628	1.591	1.264	0.63	0.209	0.121	1.0	0.92
$\sigma(\theta)$	0	0.861	1.416	1.649	1.621	1.299	0.664	0.216	0.126	10	0.94
$\sigma(\theta)$	0	0.861	1.467	1.813	1.89	1.672	1.135	0.437	0.231	100	0.92

subintervals. Continuity of solution of integral Eq. (4.2) carried out by equating the two adjacent subintervals solutions at their partitioning point. The practical application of the strip method in case of the stress exponent $n = 2$ and $n = 3$ as well as analysis of the solution in this case is described in detail in Example 4.7 and the data is shown in Tables 4.5 and 4.6 respectfully.

4.7.2 Galerkin Method (Nonlinear Creep Law)

$$
\begin{aligned}
&\sigma(\theta) = \theta e^{-0.15\theta} - e^{-\alpha\varphi(\theta)} \int_{\theta_i}^{\theta} e^{\frac{\tau}{1+0.1\tau}} e^{\alpha\varphi(\tau)} \varphi'(\tau)[\sigma^n] d\tau \\
&\tilde{\sigma}(\theta) = A\theta e^{-0.15\theta} \\
&A\int_0^8 \theta^2 e^{-0.3\theta} d\theta = \int_0^8 \theta^2 e^{-0.3\theta} d\theta \\
&\quad - A^n \int_0^8 e^{-\alpha\varphi(\theta)} \theta e^{-0.15\theta} d\theta \int_0^{\theta} e^{\frac{\tau}{1+0.1\tau}} e^{\alpha\varphi(\tau)} \varphi'(\tau) \tau^n e^{-0.15n\tau} d\tau
\end{aligned}
\tag{4.64}
$$

Denote now

$$\int_0^8 \theta^2 e^{-0.3\theta} d\theta = 31.87; \quad B = \int_0^8 e^{-\alpha\varphi(\theta)} \theta e^{-0.15\theta} z d\theta$$
$$\frac{dz}{d\tau} = e^{\frac{\tau}{1+0.1\tau}} e^{\alpha\varphi(\tau)} \varphi'(\tau) \tau^n e^{-0.15n\tau}$$
$$z(0) = 0 \tag{4.65}$$

We have

$$31.87(1 - A) = B[A^n] \tag{4.66}$$

Computer Code is as follows:

```
t(0) = 0
t(f) = 8
A = ((exp(−0.1 * m))) * t * ((exp(−0.15 * t)))
d(z)/d(t) = (t^n) * (exp(t/(1 + 0.1 * t))) * ((exp(0.1 * m))) * m1 *
((exp(−0.15 * n * t)))
d(z1)/d(t) = A * z
d(z)/d(t) = (exp(t/(1 + 0.1 * t))) * ((exp(0.10 * m))) * m1
z(0) = 0
z1(0) = 0
m = (0.0405 * t − 0.01126 * t^2 + 0.001462 * t^3 − 0.00006868 * t^4)
m1 = (0.0405 − 0.02252 * t^1 + 0.004386 * t^2 − 0.0002747 * t^3)
n = 2
B = z1_{t = 8}
```

Calculated values of DEQ variables

	Variable	Initial value	Minimal value	Maximal value	Final value
1	A	0	0	2.43612	2.392604
2	m	0	0	0.0705907	0.0705907
3	$m1$	0.0405	0.0003976	0.0405	0.0003976
4	n	2	2	2	2
5	t	0	0	8	8
6	z	0	0	3.735222	3.735222
7	$z1$	0	0	18.60611	18.60611

$31.87(1-A) = 18.606[A^2\ ]\ n = 2$

$A^2 + 1.713A - 1.713 = 0 \rightarrow A = 0.7076 \rightarrow \sigma = 0.7076\ \theta[\exp(-0.15\theta)]$

If $n = 3$, then

Calculated values of DEQ variables

	Variable	Initial value	Minimal value	Maximal value	Final value
1	*A*	0	0	2.436116	2.392604
2	*m*	0	0	0.0705907	0.0705907
3	*m*1	0.0405	0.0003976	0.0405	0.0003976
4	*n*	3	3	3	3
5	*t*	0	0	8	8
6	*z*	0	0	8.734284	8.734284
7	*z*1	0	0	40.87779	40.87779

$$A^3 + 1.283\text{A} - 1.283 = 0 \rightarrow A = 0.715 \rightarrow \sigma = 0.715\ \theta[\exp(-0.15\theta)]$$

Calculated values of NLE variables

	Variable	Value	*f*(*x*)	Initial Guess
1	*y*	0.7150458	3.117E-10	1.25 (0 < *y* < 2.5)

Calculated values of DEQ variables

	Variable	Initial value	Minimal value	Maximal value	Final value
1	*A*	0	0	2.436141	2.392604
2	*m*	0	0	0.0705907	0.0705907
3	*m*1	0.0405	0.0003976	0.0405	0.0003976
4	*n*	4	4	4	4
5	*t*	0	0	8	8
6	*z*	0	0	20.67413	20.67413
7	*z*1	0	0	92.50077	92.50077

Differential equations

1 $d(z)/d(t) = (t \wedge n) * (\exp(t/(1 + 0.1 * t))) * ((\exp(0.1 * m))) * m1 * ((\exp(-0.15 * n * t)))$
2 $\mathrm{d}(z1)/\mathrm{d}(t) = A * z$

Explicit equations

1 $m = (0.0405 * t - 0.01126 * t \wedge 2 + 0.001462 * t \wedge 3 - 0.00006868 * t \wedge 4)$
2 $A = ((\exp(-0.1 * m))) * t * ((\exp(-0.15 * t)))$
3 $m1 = (0.0405 - 0.02252 * t \wedge 1 + 0.004386 * t \wedge 2 - 0.0002747 * t \wedge 3)$
4 $n = 4$

$$A^4 + 2.9A - 2.9 = 0 \rightarrow A = 0.834 \rightarrow \sigma = 0.834\ \theta[\exp(-0.15\theta)]$$

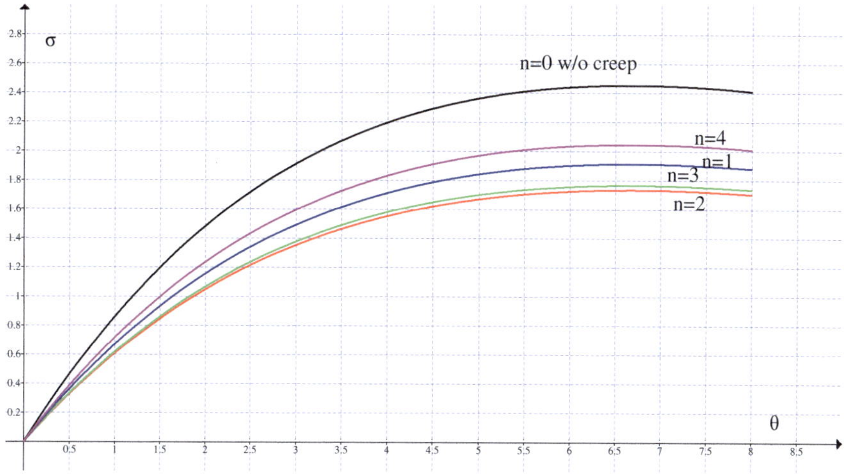

Fig. 4.15 Nonlinear creep: Galerkin method

Calculated values of NLE variables

	Variable	Value	$f(x)$	Initial guess
1	y	0.8335404	1.439E-10	1.25 ($0 < y < 2.5$)

Nonlinear equations

1	$f(y) = y\text{^}4 + 2.9 * y - 2.9 = 0$

If $n = 1$, then Calculated values of DEQ variables (Fig. 4.15).

	Variable	Initial value	Minimal value	Maximal value	Final value
1	A	0	0	2.436135	2.392604
2	m	0	0	0.0705907	0.0705907
3	$m1$	0.0405	0.0003976	0.0405	0.0003976
4	n	1	1	1	1
5	t	0	0	8	8
6	z	0	0	1.636871	1.636871
7	$z1$	0	0	8.961484	8.961484

$$31.87(1 - A) = 8.96[A]\ n = 1 \rightarrow A = 0.78 \rightarrow \sigma = 0.78\ \theta[\exp(-0.15\theta)]$$

4.7.3 Method of Moments

The method of moments (MoM) is a general solution method that is widely used in all of engineering. A Fourier series approximation to a periodic time function has a similar solution process as the MoM solution for creep constitutive law.

The method of moments (MoM) is a technique used to solve Eq. (4.31) for the temperature creep stress. The step-by-step procedure in this case is as follows:

1. Expand the kernel $K(\theta,\ \tau)$ into a series

$$K_N(\theta,\tau)=\sum_{n=1}^{N}a_n[\varphi(\theta)]b_n[\varphi(\tau)]\varphi'(\tau) \tag{4.67}$$

2. Expand stress function $\sigma(\theta)$ into a series with unknown expansion coefficients

$$\sigma(\theta)=f(\theta)-\sum_{i=1}^{N}A_if_i(\theta) \tag{4.68}$$

3. Substitute Eqs. (4.67) and (4.68) into the degenerate integral Eq. (4.31)

$$\begin{aligned}&L(\theta)\equiv f(\theta)-\sum_{i=1}^{N}A_if_i(\theta)=f(\theta)-\sum_{n=1}^{N}a_n(\theta)\int_0^{\theta}\exp\left[\frac{\tau}{1+\beta\tau}\right]b_n(\tau)\left[f(\tau)-\sum_{i=1}^{N}A_if_i(\tau)\right]^n\varphi'(\tau)\mathrm{d}\,\tau\\&\sum_{i=1}^{N}A_if_i(\theta)+\sum_{n=1}^{N}a_n(\theta)\int_0^{\theta}\exp\left[\frac{\tau}{1+\beta\tau}\right]b_n(\tau)\left[f(\tau)-\sum_{i=1}^{N}A_if_i(\tau)\right]^n\varphi'(\tau)\mathrm{d}\,\tau\\&f(\theta)=\theta\mathrm{e}^{-a\theta};\ a=0.15\ \text{(steel structure)};\ f_i(\theta)=\theta^i\mathrm{e}^{-a\theta}\end{aligned} \tag{4.69}$$

Since the function $f_i(\theta)$ is an arbitrary function, Eq. (4.31) must be orthogonal to any $f_j(\theta)$

$$L(\theta)\cdot f_i(\theta)=\int_0^8 L(\theta)f_i(\theta)\mathrm{d}\,\theta=0\quad i=1,2,\ldots N \tag{4.70}$$

4. Define weighting (basis) functions.

The selection of the basis functions to be orthogonal to the nonlinear operator $L(\theta)$ is referred to as Galerkin method. We get to choose, but generally select the basis functions from the expansion (4.68).

5. The constants A_i are obtained by multiplying each side of Eq. (4.70) by the testing (basis) function and integrating over a dimensionless temperature interval [0,8]. The unknowns A_i are found from the algebraic equations. Assuming for instance i = 1 and n = 2 we have (see Example 4.9):

Example 4.9

$$A_1\int_0^8 f_1^2 \mathrm{d}\theta+\int_0^8\left\{a_1[\varphi(\theta)]\int_0^{\theta}\exp\left[\frac{\tau}{1+\beta\tau}\right]b_1[\varphi(\tau)]\varphi'(\tau)[f(\tau)-A_1f_1(\tau)]^2\mathrm{d}\tau\right\}f_1\mathrm{d}\theta=0$$

$$f_1(\theta)=A_1\theta\exp(-0.15\theta)$$

Introducing the dummy differential equations (as it has been done above) the computer Code in this case is written as follows:

$d(y1)/d(t) = (t \wedge 2) * (\exp(t/(1+0.1*t))) * ((\exp(0.1*m))) * m1 * ((\exp(-0.30*t)))$
$d(y2)/d(t) = 2*(t \wedge 3) * (\exp(t/(1+0.1*t))) * m1 * ((\exp(0.1*m))) * ((\exp(-0.45*t)))$
$d(y3)/d(t) = (t \wedge 3) * (\exp(t/(1+0.1*t))) * ((\exp(0.1*m))) * m1 * ((\exp(-0.45*t)))$
$d(c)/d(t) = (\exp(-0.3*t)) * t \wedge 2$
$d(c1)/d(t) = y1 * (\exp(-0.1*m)) * t * (\exp(-0.15*t))$
$d(c2)/d(t) = y2 * (\exp(-0.1*m)) * t * (\exp(-0.15*t))$
$d(c3)/d(t) = y3 * (\exp(-0.1*m)) * \mathrm{t} * (\exp(-0.15*t))$
$y1(0) = 0$
$y2(0) = 0$
$y3(0) = 0$
$c1(0) = 0$
$c2(0) = 0$
$c3(0) = 0$
$c(0) = 0$

Explicit equations

$m = (0.0405*t - 0.01126*t \wedge 2 + 0.001462*t \wedge 3 - 0.00006868*t \wedge 4)$
$m1 = (0.0405 - 0.02252*t \wedge 1 + 0.004386*t \wedge 2 - 0.0002747*t \wedge 3)$
$n = 2$
$[\mathrm{cA} = c1 - Ac2 + c3A^2]_{t=\theta=8}$

Calculated values of DEQ variables

	Variable	Initial value	Minimal value	Maximal value	Final value
1	c	0	0	31.87343	31.87343
2	$c1$	0	0	18.60611	18.60611
3	$c2$	0	0	81.75558	81.75558
4	$c3$	0	0	40.87779	40.87779
5	m	0	0	0.0705907	0.0705907
6	$m1$	0.0405	0.0003976	0.0405	0.0003976

(continued)

(continued)

	Variable	Initial value	Minimal value	Maximal value	Final value
7	n	2	2	2	2
8	t	0	0	8	8
9	$y1$	0	0	3.735222	3.735222
10	$y2$	0	0	17.46857	17.46857
11	$y3$	0	0	8.734284	8.734284

$31.87A = 18.61 - 81.76A + 40.88A^2$ $40.88\,A^2 - 113.63A + 18.61 = 0$

The nonlinear (quadratic) algebraic equation in this case is (using POLYMATH software)

$f(y) = 40.88 * y\wedge 2 - 113.63 * y + 18.61$

$y(0) = 0.8$

$y(\max) = 2.5$

$y(\min) = 0$

Calculated values of NLE variables

	Variable	Value	$f(x)$	Initial guess
1	y	0.1747654	3.175E-10	1.25 ($0 < y < 2.5$)

Nonlinear equations

$$1 \quad f(y) = 40.88 * y^2 - 113.63 * y + 18.61 = 0$$

$$40.88A^2 - 113.63A + 18.61 = 0 \rightarrow A = 0.175 \text{ and } \sigma = 0.825\theta[\exp(-0.15\theta)].$$

Compare with Galerkin Method result $\sigma = 0.7076\ \theta[\exp(-0.15\theta)]$.

Example 4.10 Data: $\varepsilon_{all} = 0.0042$; $n = 3$

Find: σ_{all} & t_f

Solution

$\varepsilon_{all} = 0.0042 = 7.02(\theta)(10^{-4})$; Therefore $\theta = 6$. From Fig. 4.7 (Galerkin method) we have: $\sigma_{all} = 1.744\ (2.9)7.02 = 35.5$ ksi [245 MPa]; $t_f = m\ (\theta = 6) = 0.0644 \rightarrow (0.0644/0.22)4 = 1.17$ h.

Thus the allowable stress is 35.5 ksi [245 MPa] and the fire duration is 1.17 h to failure.

4.6 Reduction of the Volterra integral equation with the degenerate kernels to an equivalent system of first order differential equations

Consider the degenerate kernel (4.67) where a_i and b_i are two systems of linearly independent functions. Substituting Eq. (4.67) into Eq. (4.31) we have

$$\begin{aligned}\bar{\sigma}(\theta) &= E(\theta)\varepsilon(\theta) = \sigma(\theta) + \sum_{i=1}^{N}\int_{0}^{\theta} a_i[\varphi(\theta)]b_i[\varphi(\tau)]\varphi'(\tau)\sigma^n(\tau)d\tau \\ &= \sigma(\theta) + \sum_{i=1}^{N} a_i[\varphi(\theta)]\int_{0}^{\theta} b_i[\varphi(\tau)]\varphi'(\tau)\sigma^n(\tau)\mathrm{d}\tau = \sigma(\theta) + \sum_{i=1}^{N} a_i[\varphi(\theta)]Z_i \\ \frac{dZ_i}{\mathrm{d}\tau} &= b_i[\varphi(\tau)]\varphi'(\tau)\sigma^n(\tau); \quad Z_i(0) = 0\end{aligned} \tag{4.71}$$

The degenerate kernels (with new independent variables θ, τ) may be written as

$$K(\theta,\tau) = \sum_{i=1}^{N} \exp[-\alpha_i\varphi(\theta)]\exp[\alpha\varphi(\tau)] \tag{4.72}$$

The expansion in Eq. (4.72) represents a series of real exponentials, called the Dirichlet series (also called the Prony series). Here $t = \varphi(\theta)$ and $t' = \varphi(\tau)$ are substitution time functions, called the reduced times and α_i is called MPP—material property parameter (retardation temperature/time in Eq. (4.2) or relaxation temperature/time in Eq. (4.3)). In general, function $t = \varphi(\theta)$ can be obtained for a given fire severity scenario [5] (for instance, by the method of least squares). As for α_i parameters, however, they cannot be calculated from measured creep data but must be suitably chosen in advance. The choice of α_i cannot be arbitrary but must satisfy certain conditions. The values of α_i must not be spaced too sparsely, and they must cover the entire temperature–time range of interest, in particular, the smallest α_i must be such that $\alpha_i < 3\ \alpha_{\min}$ and the largest α_i must be such that

$\alpha_i > 0.5\ \alpha_{\max}$, in which $\alpha_{\min}$ and $\alpha_{\max}$ are the smallest and the largest time delay after temperature load application for which the response is of interest.

The α_i values that give a close fit of given $K(t, t')$ data are not unique. Equally good fits of the given compliance function data can be obtained for many possible choices of α_i values which are spaced in the temperature–time scale and cover the entire temperature–time range of interest.

The Dirichlet series expansion should be regarded only as an approximation to the kernel function, motivated by computational convenience, rather than as a fundamental law.

The expansion contains unnecessarily many material parameters defining all the compliance or relaxation functions. Therefore, for all practical purposes in case of fire we will limit the range of MPP to the interval $10^{-3} < \alpha_i < 10^5$.

As already mentioned, the purpose of the Dirichlet series expansion and degenerate kernel form is to convert a constitutive equation of an integral type to one of a differential type.

For a temperature–time dependant material, this conversion is somewhat simpler not only for linear creep stress–strain integral type relationship but also for a nonlinear when the stress exponent $n > 1$. Equation (4.31) with $K(\theta, \tau)$ given by Eq. (4.72) may be rewritten as

$$\begin{aligned}
&\sigma(\theta) = E(\theta)\varepsilon(\theta) - \sum_{i=1}^{N} a_i[\varphi(\theta)]Z_i \\
&\frac{dZ_i}{d\tau} = b_i[\varphi(\tau)]\varphi'(\tau)\sigma^n(\tau); \quad Z_i(0) = 0 \\
&b_i[\varphi(\tau)] = (\exp(t)/(1+0.1t))[((\exp(\alpha_i m)))]m1 \\
&a_i[\varphi(\tau)] = [(\exp(-\alpha_i m))] \\
&m = (0.0405\tau - 0.01126\tau\hat{}2 + 0.001462\tau\hat{}3 - 0.00006868\tau\hat{}4) \\
&m1 = (0.0405 - 0.02252\tau\hat{}1 + 0.004386\tau\hat{}2 - 0.0002747\tau\hat{}3)
\end{aligned} \tag{4.73}$$

Nonlinear creep behavior is essential component in constructing any phenomenological model when the high temperature load is present. At the same time it is necessary to analyze the effect of nonlinearity and retaining more than one term of the series expansion (4.73) on the convergence of the solution of the integral Eq. (4.31). Computational technique for different stress exponent numbers "n" with two terms of the series expansion (4.72) as well as the Computer Code and step-by-step calculations procedure is presented below via Examples 4.11–4.14. Summary of results are presented below in tabular form (see Table 4.1) and in graph form (see Fig. 4.1).

Example 4.11 Data: $0 < \theta < 8$; $\alpha_1 = 0.1$ & $\alpha_2 = 0.5$; $n = 1$; $N = 2$

$$\begin{aligned}
\bar{\sigma}(\theta) = \mathrm{E}(\theta)\varepsilon(\theta) &= \sigma(\theta) + \sum_{\mathrm{i}=1}^{N} a_i[\varphi(\theta)] \int_0^\theta b_i[\varphi(\tau)]\varphi'(\tau)\sigma^n(\tau)\mathrm{d}\tau \\
&= \sigma(\theta) + a_1 Z1 + a_2 Z2; \ \frac{\mathrm{d}Z_i}{\mathrm{d}\tau} = b_i[\varphi(\tau)]\varphi'(\tau)\sigma^n(\tau); \quad Z_i(0) = 0
\end{aligned}$$

$$\begin{aligned}
&b_1[\varphi(\tau)] = (\exp(\mathrm{t})/(1 + 0.1\mathrm{t})))[((\exp(0.1 * m)))] * m1 \\
&b_2[\varphi(\tau)] = (\exp(\mathrm{t})/(1 + 0.1\mathrm{t})))[((\exp(0.5 * m)))] * m1 \\
&a_1[\varphi(\tau)] = [((\exp(-0.1 * m)))]; \ a_2[\varphi(\tau)] = [((\exp(-0.5 * m)))]
\end{aligned}$$

The computer code is as follows:

$$\bar{\sigma}(\theta) = E(\theta)\varepsilon(\theta) = \sigma(\theta) + \sum_{i=1}^{N} \int_0^{\theta} a_i[\varphi(\theta)]b_i[\varphi(\tau)]\varphi'(\tau)\sigma^n(\tau)\mathrm{d}\tau$$

$$\sigma(\theta) = \theta \mathrm{e}^{-0.15\theta} - \exp[(-\alpha m)] \int_0^{\theta} \exp[(\alpha m)]exp()m1\sigma^n(\tau)\mathrm{d}\,\tau| \cdot \alpha\, m[m1]$$

$$\sigma'(\theta) = \mathrm{e}^{-0.15\theta}[1 - 0.15\theta] - \exp()m1\sigma^n(\theta) - \alpha\, m[m1]\exp[(-\alpha m)]$$

$$\sigma'(\theta) = \sigma(\theta)\alpha\, m[m\,1] + \mathrm{e}^{-0.15\theta}[1 - 0.15\theta] - \exp()m1\sigma^n(\theta) - \theta \mathrm{e}^{-0.15\theta}\alpha\, m[m1]$$

$$\sigma(0) = 0;$$

$$\exp() = (\exp(t)/(1\ +\ 0.1t)))$$

Differential equations

1 $\mathrm{d}(Z1)/\mathrm{d}(t) = (\exp(t/(1 + 0.1 * t))) * (\exp(0.1 * m)) * m1 * (Z \wedge n)$

2 $\mathrm{d}(Z2)/\mathrm{d}(t) = (\exp(t/(1 + 0.1 * t))) * (\exp(0.5 * m)) * m1 * (Z \wedge n)$

Explicit equations

1 $m = (0.0405 * (t) - 0.01126 * (t) \wedge 2 + 0.001462 * (t) \wedge 3 - 0.00006868 * (t) \wedge 4)$

2 $m1 = (0.0405 - 0.02252 * (t) \wedge 1 + 0.004386 * (t) - 0.0002747 * (t) \wedge 3)$

3 $n = 1$

4 $A2 = \exp(-0.5 * m)$

5 $A1 = \exp(-0.10 * m)$

6 $Z = t * (\exp(-0.15 * t)) - A1 * Z1 - A2 * Z2$

Calculated values of DEQ variables (Fig. 4.16).

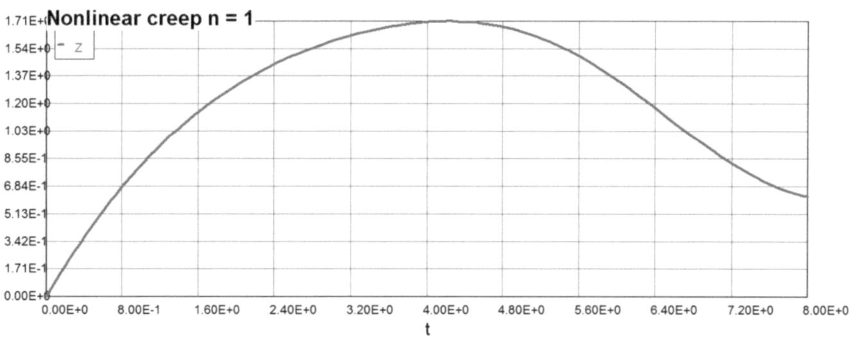

Fig. 4.16 Stresstemperature-strain function

	Variable	Initial value	Minimal value	Maximal value	Final value
1	$A1$	1	0.9929658	1	0.9929658
2	$A2$	1	0.9653203	1	0.9653203
3	m	0	0	0.0705907	0.0705907
4	$m1$	0.0405	0.0003976	0.0405	0.0003976
5	n	1	1	1	1
6	t	0	0	8	8
7	Z	0	0	1.710496	0.6238547
8	$Z1$	0	0	0.9010917	0.9010917
9	$Z2$	0	0	0.9229535	0.9229535

Model: $Z = a0 + a1 * t + a2 * t\text{^}2 + a3*t\text{^}3 + a4 * t\text{^}4$

Variable	Value
$a0$	0.0736719
$a1$	0.7884001
$a2$	−0.0682128
$a3$	−0.0102645
$a4$	0.0009255

$$\sigma(\theta) = 0.0737 + 0.788(\theta) - 0.0682 * \theta\text{^}2 - 0.0103 * \theta\text{^}3 + 0.00092 * \theta\text{^}4 \quad (4.74)$$

Example 4.12 Data: $0 < \theta < 8$; $\alpha_1 = 0.1$ & $\alpha_2 = 0.5$; $n = 2$

Computations are similar to Example 4.

Differential equations

1 $d(Z1)/d(t) = (\exp(t/(1+0.1*t)))*(\exp(0.1*m))*m1*(Z \wedge n)$

2 $d(Z2)/d(t) = (\exp(t/(1+0.1*t)))*(\exp(0.5*m))*m1*(Z \wedge n)$

Explicit equations

1 $m = (0.0405*(t) - 0.01126*(t) \wedge 2 + 0.001462*(t) \wedge 3 - 0.00006868*(t) \wedge 4)$

2 $m1 = (0.0405 - 0.02252*(t) \wedge 1 + 0.004386*(t) \wedge 2 - 0.0002747*(t) \wedge 3)$

3 $n = 2$

4 $A2 = \exp(-0.5*m)$

5 $A1 = \exp(-0.10*m)$

6 $Z = t*(\exp(-0.15*t)) - A1*Z1 - A2*Z2$

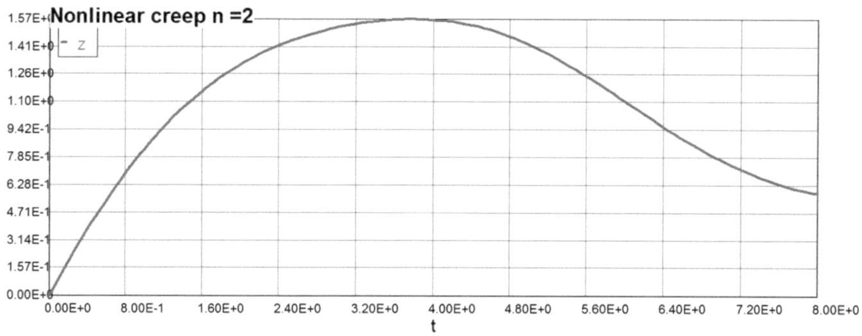

Fig. 4.17 Stress-temperature-strain function (n = 2)

Calculated values of DEQ variables (Fig. 4.17).

	Variable	Initial value	Minimal value	Maximal value	Final value
1	*A*1	1	0.9929658	1	0.9929658
2	*A*2	1	0.9653203	1	0.9653203
3	*m*	0	0	0.0705907	0.0705907
4	*m*1	0.0405	0.0003976	0.0405	0.0003976
5	*n*	2	2	2	2
6	*t*	0	0	8	8
7	*Z*	0	0	1.570386	0.5836825
8	*Z*1	0	0	0.9215698	0.9215698
9	*Z*2	0	0	0.9435044	0.9435044

Model: $Z = a0 + a1 * t + a2 * t\wedge 2 + a3 * t\wedge 3 + a4 * t\wedge 4$

Variable	Value
*a*0	0.0468408
*a*1	0.8920563
*a*2	−0.1248912
*a*3	−0.0042307
*a*4	0.0008622

$$\sigma(\theta) = 0.0468 + 0.892(\theta) - 0.125*\theta\wedge 2 - 0.00423*\theta\wedge 3 + 0.00086*\theta\wedge 4 \quad (4.75)$$

Example 4.13 Data: $0 < \theta < 8$; $\alpha_1 = 0.1$ & $\alpha_2 = 0.5$; $n = 1.5$

Differential equations

$$1 \quad d(Z1)/d(t) = (\exp(t/(1+0.1*t)))*(\exp(0.1*m))*m1*(Z \wedge n)$$
$$2 \quad d(Z2)/d(t) = (\exp(t/(1+0.1*t)))*(\exp(0.5*m))*m1*(Z \wedge n)$$

Explicit equations

$$1 \quad m = (0.0405*(t) - 0.01126*(t)\wedge 2 + 0.001462*(t)\wedge 3 - 0.00006868*(t)\wedge 4)$$
$$2 \quad m1 = (0.0405 - 0.02252*(t)\wedge 1 + 0.004386*(t)\wedge 2 - 0.0002747*(t)\wedge 3)$$
$$3 \quad n = 1.5$$
$$4 \quad A2 = \exp(-0.5*m)$$
$$5 \quad A1 = \exp(-0.10*m)$$
$$6 \quad Z = t*(\exp(-0.15*t)) - A1*Z1 - A2*Z2$$

Calculated values of DEQ variables (Fig. 4.18).

	Variable	Initial value	Minimal value	Maximal value	Final value
1	$A1$	1	0.9929658	1	0.9929658
2	$A2$	1	0.9653203	1	0.9653203
3	m	0	0	0.0705907	0.0705907
4	$m1$	0.0405	0.0003976	0.0405	0.0003976
5	n	1.5	1.5	1.5	1.5
6	t	0	0	8	8
7	Z	0	0	1.639287	0.5887536
8	$Z1$	0	0	0.9188983	0.9188983
9	$Z2$	0	0	0.9409992	0.9409992

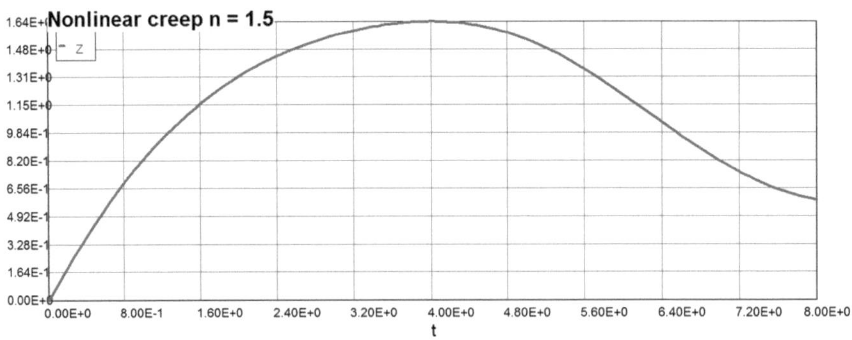

Fig. 4.18 Stress-temperature-strain function (n = 1.5)

Model: $Z = a0 + a1 * t + a2 * t\text{^}2 + a3 * t\text{^}3 + a4*t\text{^}4$

Variable	Value
$a0$	0.0647465
$a1$	0.8280175
$a2$	−0.0869317
$a3$	−0.0095334
$a4$	0.0010487

$$\sigma(\theta) = 0.0647 + 0.828(\theta) - 0.0869{*}\theta^\wedge 2 - 0.00953{*}\theta^\wedge 3 + 0.00105{*}\theta^\wedge 4 \quad (4.76)$$

Example 4.14 Data: $0 < \theta < 8$; $\alpha_1 = 0.1$ & $\alpha_2 = 0.5$; $n = 3.0$

Differential equations

1 $\mathrm{d}(Z1)/\mathrm{d}(t) = (\exp(t/(1 + 0.1 * t))) * (\exp(0.1 * m)) * m1 * (Z \wedge n)$

2 $\mathrm{d}(Z2)/\mathrm{d}(t) = (\exp(t/(1 + 0.1 * t))) * (\exp(0.5 * m)) * m1 * (Z \wedge n)$

Explicit equations (Fig. 4.19).

1 $m = (0.0405 * (t) - 0.01126 * (t) \wedge 2 + 0.001462 * (t) \wedge 3 - 0.00006868 * (t) \wedge 4)$

2 $m1 = (0.0405 - 0.02252 * (t) \wedge 1 + 0.004386 * (t) \wedge 2 - 0.0002747 * (t) \wedge 3)$

3 $n = 3$

4 $A2 = \exp(-0.5 * m)$

5 $A1 = \exp(-0.10 * m)$

6 $Z = t * (\exp(-0.15 * \mathrm{t})) - A1 * Z1 - A2 * Z2$

Model: $Z = a0 + a1 * t + a2 * t\text{^}2 + a3 * t\text{^}3 + a4 * t\text{^}4$

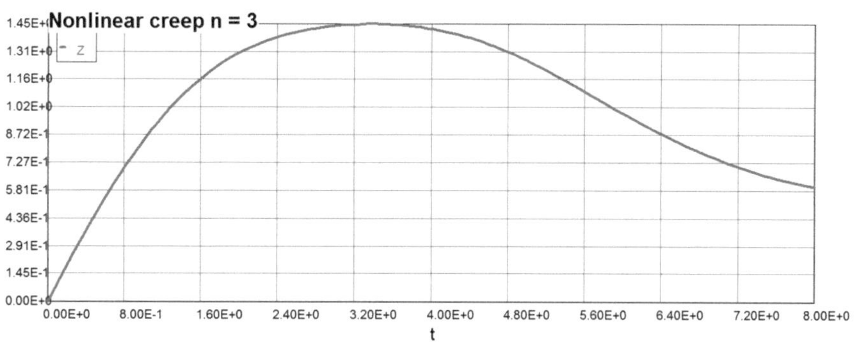

Fig. 4.19 Stress-temperature-strain function (n = 3)

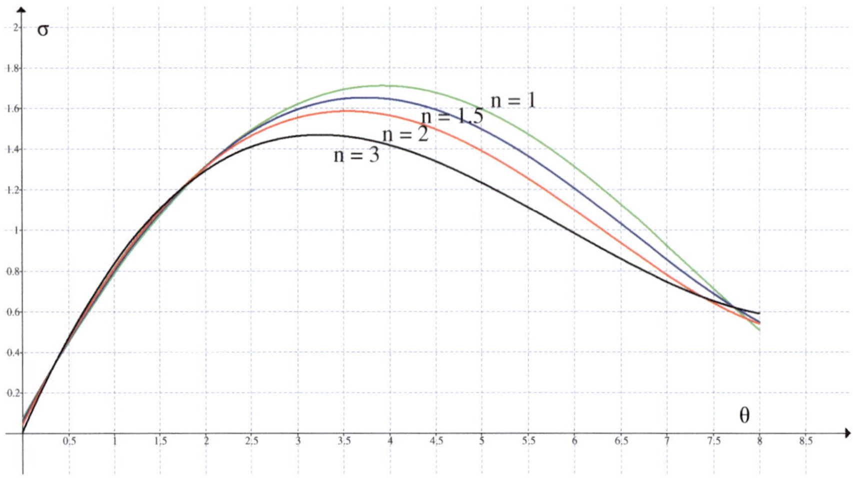

Fig. 4.20 Stress-temperature-strain functions: degenerate kernel

Variable	Value
$a0$	0.0071387
$a1$	1.033211
$a2$	−0.2196109
$a3$	0.0124993
$a4$	−5.614E-06

$$\sigma(\theta) = 0.00714 + 1.033(\theta) - 0.22{*}\theta{}^\wedge 2 + 0.0125{*}\theta{}^\wedge 3 \tag{4.77}$$

It can be seen from the Fig. 4.20 and Tables 4.7 and 4.8, the increase in non-linearity (stress exponent numbers n) reduces the maximum thermal stresses as well

Table 4.7 Creep stress data (n = 2)

θ	0	1	2	3	4	5	6	7	8	α_i	ϕ
$f(\theta)$	0	0.86	1.48	1.76	2.05	2.24	2.35	2.4	2.41		
n = 2											
θ	0	1	2	3	4	5	6	7	8	α_i	ϕ
$\sigma(\theta)$	0	0.847	1.392	1.692	1.820	1.766	1.504	1.160	0.943	0.01	0.80
$\sigma(\theta)$	0	0.847	1.392	1.692	1.820	1.766	1.504	1.161	0.944	0.1	0.80
$\sigma(\theta)$	0	0.847	1.392	1.694	1.822	1.769	1.508	1.166	0.950	1.0	0.80
$\sigma(\theta)$	0	0.848	1.398	1.707	1.841	1.795	1.544	1.217	1.007	10	0.80
$\sigma(\theta)$	0	0.852	1.431	1.782	1.956	1.955	1.781	1.545	1.380	100	0.873

Table 4.8 Creep stress data ($n = 3$)

θ	0	1	2	3	4	5	6	7	8	α_i	ϕ
$f(\theta)$	0	0.86	1.48	1.76	2.05	2.24	2.35	2.4	2.41		
n = 3											
θ	0	1	2	3	4	5	6	7	8	α_i	ϕ
$\sigma(\theta)$	0	0.851	1.382	1.623	1.677	1.554	1.275	0.996	0.832	0.01	0.714
$\sigma(\theta)$	0	0.852	1.382	1.623	1.678	1.554	1.275	0.997	0.833	0.1	0.714
$\sigma(\theta)$	0	0.847	1.382	1.624	1.680	1.556	1.279	1.0	0.839	1.0	0.714
$\sigma(\theta)$	0	0.848	1.388	1.637	1.698	1.580	1.312	1.049	0.893	10	0.714
$\sigma(\theta)$	0	0.855	1.419	1.720	1.822	1.745	1.540	1.348	1.224	100	0.803

as shifts the graph to the left (i.e., decreases the allowable stresses and the strain at which this maximum is reached).

4.8 Comparison of the Numerical Solution Obtained by Different Approximate Methods

Precision of calculations using the method of replacing the kernel of the integral Eq. (4.31) to degenerate kernel (for simplicity we will call this method “degenerate method”) represented in the form of the expansion (4.53) of course depends on the number of members held in this series. In applying the linearization method in conjunction with the method of strips (for simplicity we will call this method “linearization method”) set out above, this approximation kernel’s accuracy of the integral Eq. (4.31) is not required, because the entire interval of integration was divided into a finite number of small subintervals. On the other hand, we would like to compare the solution of the creep constitutive integral equation obtained by these two different methods while only one MPP parameter is considered. Therefore only one member is held in the series (4.58) and we cannot expect that the results (stress–temperature diagram) will coincide on the whole range of dimensionless temperature. In fact the correlated computations have shown that the creep stress values are very close in the primary and secondary creep development areas and differ substantially in the decay area of the stress–temperature (stain) diagram. In order to improve this situation we need to keep at least 10 members of the expansion of (4.58). However, holding a large number of terms in the expansion series will cover virtually the entire spectrum of mechanical properties of materials (MPP) of interest to us, consequently the maximum creep stress will be close the

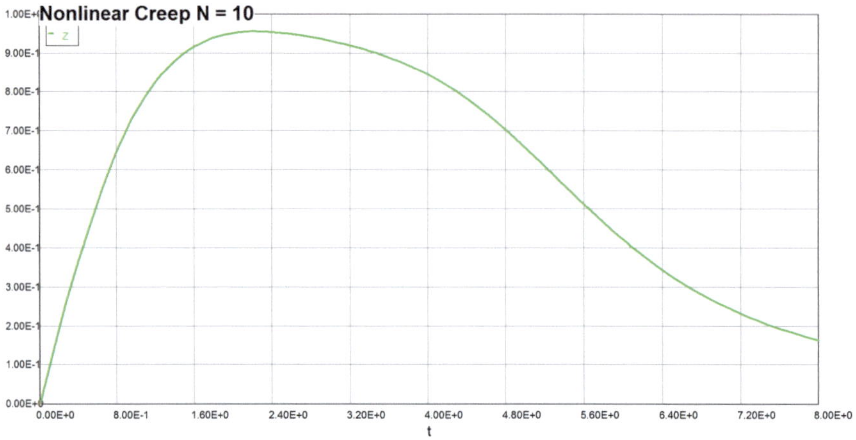

Fig. 4.21 Stress–temperature (strain) diagram (N = 10)

mean value. It should also be noted that the retention of such a large number of terms in the expansion series does not create any significant additional difficulties from a computational point of view, due to the similarity of computing operating procedures.

Reduction of creep integral type equation to an equivalent system of differential equations (with N = 10 in expansion series 4.49) and the corresponding computer code in this case is as follows:

Differential equations

1 $d(Z1)/d(t) = (\exp(t/(1+0.1*t)))*(\exp(1*m))*m1*(Z\wedge n)$

2 $d(Z2)/d(t) = (\exp(t/(1+0.1*t)))*(\exp(0.5*m))*m1*(Z\wedge n)$

3 $d(Z3)/d(t) = (\exp(t/(1+0.1*t)))*(\exp(0.1*m))*m1*(Z\wedge n)$

4 $d(Z4)/d(t) = (\exp(t/(1+0.1*t)))*(\exp(0.01*m))*m1*(Z\wedge n)$

5 $d(Z5)/d(t) = (\exp(t/(1+0.1*t)))*(\exp(0.15*m))*m1*(Z\wedge n)$

6 $d(Z6)/d(t) = (\exp(t/(1+0.1*t)))*(\exp(0.2*m))*m1*(Z\wedge n)$

7 $d(Z7)/d(t) = (\exp(t/(1+0.1*t)))*(\exp(0.3*m))*m1*(Z\wedge n)$

8 $d(Z8)/d(t) = (\exp(t/(1+0.1*t)))*(\exp(0.75*m))*m1*(Z\wedge n)$

9 $d(Z9)/d(t) = (\exp(t/(1+0.1*t)))*(\exp(10*m))*m1*(Z\wedge n)$

10 $d(Z10)/d(t) = (\exp(t/(1+0.1*t)))*(\exp(100*m))*m1*(Z\wedge n)$

Explicit equations

1 $m = (0.0405 * (t) - 0.01126 * (t) \wedge 2 + 0.001462 * (t) \wedge 3 - 0.00006868 * (t) \wedge 4)$

2 $m1 = (0.0405 - 0.02252 * (t) \wedge 1 + 0.004386 * (t) \wedge 2 - 0.0002747 * (t) \wedge 3)$

3 $n = 2$

4 $A10 = \exp(-100 * m)$

5 $A1 = \exp(-1 * m)$

6 $A2 = \exp(-0.5 * m)$

7 $A3 = \exp(-0.1 * m)$

8 $A4 = \exp(-0.01 * m)$

9 $A5 = \exp(-0.15 * m)$

10 $A6 = \exp(-0.2 * m)$

11 $A7 = \exp(-0.3 * m)$

12 $A8 = \exp(-0.75 * m)$

13 $A9 = \exp(-10 * m)$

14 $Z = t * (\exp(-0.15 * t)) - A1 * Z1 - A2 * Z2 - A3 * Z3 - A4 * Z4 - A5 * Z5 - A6 * Z6 - A7 * Z7 - A8 * Z8 - A9 * Z9 - A10 * Z10$

Calculated values of DEQ variables (Fig. 4.21, Tables 4.9 and 4.10).

	Variable	Initial value	Minimal value	Maximal value	Final value
1	A1	1	0.9318432	1	0.9318432
2	A10	1	0.0008596	1	0.0008596
3	A2	1	0.9653203	1	0.9653203
4	A3	1	0.9929658	1	0.9929658
5	A4	1	0.9992943	1	0.9992943
6	A5	1	0.9894673	1	0.9894673
7	A6	1	0.985981	1	0.985981
8	A7	1	0.9790454	1	0.9790454
9	A8	1	0.948434	1	0.948434
10	A9	1	0.4936605	1	0.4936605
11	m	0	0	0.0705907	0.0705907
12	m1	0.0405	0.0003976	0.0405	0.0003976
13	n	2	2	2	2
14	t	0	0	8	8
15	Z	0	0	0.955553	0.1642291
16	Z1	0	0	0.2612134	0.2612134
17	Z10	0	0	78.46712	78.46712
18	Z2	0	0	0.2543317	0.2543317
19	Z3	0	0	0.2489625	0.2489625
20	Z4	0	0	0.2477707	0.2477707

(continued)

(continued)

	Variable	Initial value	Minimal value	Maximal value	Final value
21	*Z5*	0	0	0.2496271	0.2496271
22	*Z6*	0	0	0.2502936	0.2502936
23	*Z7*	0	0	0.2516321	0.2516321
24	*Z8*	0	0	0.2577486	0.2577486
25	*Z9*	0	0	0.4244781	0.4244781

Table 4.9 Creep stress data: comparison of the numerical solution (n = 2)

θ	0	1	2	3	4	5	6	7	8	α_i	
$\sigma(\theta)$	0	0.838	1.395	1.742	1.933	1.968	1.806	1.483	1.216	1.0	Linear creep $n = 1$
$\sigma(\theta)$	0	0.861	1.411	1.701	1.788	1.656	1.194	0.5	0.138	1.0	Linear. method
$\sigma(\theta)$	Check	0.847	1.392	1.694	1.822	1.769	1.508	1.166	0.950	1.0	Degen. method

Table 4.10 Creep stress data: comparison of the numerical solution (n = 3)

θ	0	1	2	3	4	5	6	7	8	α_i	
$\sigma(\theta)$	0	0.861	1.467	1.813	1.89	1.672	1.135	0.437	0.231	100	Linear. method
$\sigma(\theta)$	check	0.855	1.419	1.720	1.822	1.745	1.540	1.348	1.224	100	Degen. method
$\sigma(\theta)$	0	0.861	1.408	1.628	1.591	1.264	0.63	0.209	0.121	1.0	Linear. method
$\sigma(\theta)$	check	0.847	1.382	1.624	1.680	1.556	1.279	1.0	0.839	1.0	Degen. method
$\sigma(\theta)$	0	0.861	1.408	1.627	1.690	1.262	0.628	0.208	0.12	0.1	Linear. method
$\sigma(\theta)$	check	0.852	1.382	1.623	1.678	1.554	1.275	0.997	0.833	0.1	Degen. method

4.9 Stress–Strain Unload

A continuous function $\sigma(\theta)$ is one-to-one (or it is strictly increasing) therefore the stress–strain unloading function should be strictly decreasing (with no local maxima or minima). Thus the graph of unloading function $g(\theta)$ can be obtained from the graph of $\sigma(\theta)$ by reflecting the graph across the line $\theta = 8$. This can be simply evaluated by shifting the function by c units towards negative x-axis or in transformation terms $f(x) \rightarrow f(x + c)$, finding the mirror image of this new curve along $x = 0$, i.e., $f(x + c) \rightarrow f(-x+c)$ and then re-shifting the new curve by c units towards right, i.e., $f(-x+c) \rightarrow f(2c - x)$. Conclusively, the mirror image of any curve $y = f(x)$ about $x = c$ would be given by $y = f(2c - x)$. Therefore from mathematical point of view the independent variable (θ or s) in differential Eq. (4.31) has to be substituted by 8- θ (or 8-s) and the positive direction of axis has to be substituted by negative. The initial condition obviously is $\sigma_0\ (\theta) = \sigma_{\max}(\theta)$. The solution of the corresponding differential equation is as follows ($\alpha = 0.33$):

$$\begin{aligned}
&\bar{\sigma}(\theta) = \mathrm{E}(\theta)\varepsilon(\theta) = \sigma(\theta) + \sum_{i=1}^{N} \int_0^{\theta} a_i[\varphi(\theta)]b_i[\varphi(\tau)]\varphi'(\tau)\sigma^n(\tau)\mathrm{d}\tau \\
&\sigma(\theta) = \theta \mathrm{e}^{-0.15\theta} - \exp[(-\alpha m)] \int_0^{\theta} \exp[(-\alpha m)] \exp() m\,1 \sigma^n(\tau)\mathrm{d}\tau| \cdot \alpha[m\,1] \\
&\sigma' = \theta \mathrm{e}^{-0.15\theta}[1 - 0.15\theta] - \exp() m 1 \sigma^n(\theta)[m\,1] \exp[(-\alpha m)] \int_0^{\theta} F(\tau)\mathrm{d}\tau \\
&\sigma'(\theta) = -\sigma(\theta)\alpha[m\,1] + \mathrm{e}^{-0.15\theta}[1 - 0.15\theta] - \exp() m 1 \sigma^n(\theta) + \theta \mathrm{e}^{-0.15\theta}\alpha[m\,1] \\
&\sigma(0) = 0; \\
&\exp() = (\exp(t)/(1 + 0.1t)))
\end{aligned} \tag{4.78}$$

Differential equations two differ. Equations y → up and z → dawn; but the same m1

Equivalent diff. equation for one MPP (n = 2;3)

1 $\mathrm{d}(y)/\mathrm{d}(t) = -0.33 * y * m1 + (\exp(-0.15 * t)) * (1 - 0.15 * t)$
$- (\exp(t/(1.0 + 0.1 * t))) * m1 * y \wedge n + 0.33 * m1 * t * (\exp(-0.15 * t))$

2 $\mathrm{d}(z)/\mathrm{d}(t) = -0.33 * y * m1 + (\exp(-0.15 * (8 - t))) * (1 - 0.15 * (8 - t))$
$- (\exp((8 - t)/(1.0 + 0.1 * (8 - t)))) * m1 * y \wedge n + 0.33 * m1 * (8 - t) * (\exp(-0.15 * (8 - t)))$

Explicit equations

1 $n = 2$

2 $m1 = 0.0405 - 0.02252 * t \wedge 1 + 0.004386 * t \wedge 2 - 0.0002747 * t \wedge 3$

3 $m = (0.0405 * t - 0.01126 * t \wedge 2 + 0.001462 * t \wedge 3 - 0.00006868 * t \wedge 4)$

Calculated values of DEQ variables

	Variable	Initial value	Minimal value	Maximal value	Final value
1	m	0	0	0.0578259	0.0578259
2	$m01$	−10	−10	−0.0648732	−0.0712832
3	$m1$	0.0405	0.0030152	0.0405	0.0030152
4	n	2	2	2	2
5	t	0	0	4	4
6	y	0	0	1.820251	1.820251
7	z	1.82	0.2879122	1.82	0.5010474

The residual stress value (from computer output) is: $\sigma_{res} = 0.288$ (Fig. 4.22). Model of stress–temperature diagram (loading part)

Variable	Value
$a0$	−0.0034631
$a1$	1.021777
$a2$	−0.1795741
$a3$	0.0079214
$a4$	0.0004135

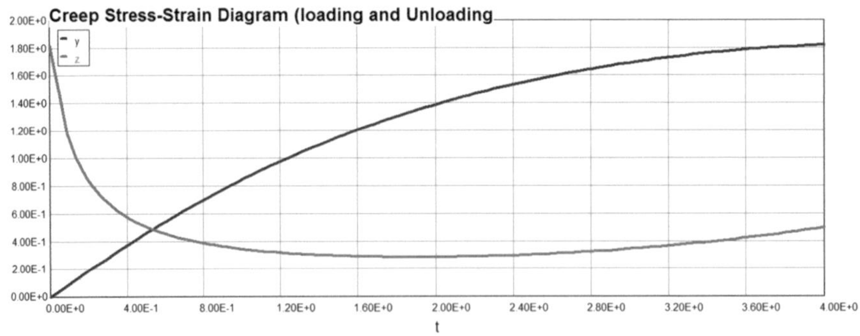

Fig. 4.22 Creep stress curves (loading and unloading)

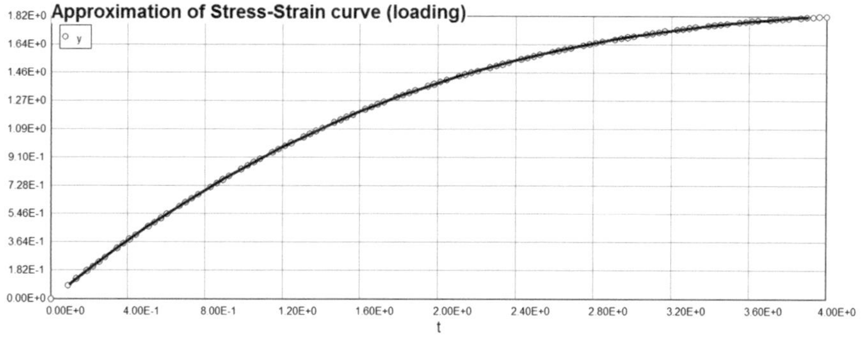

$$\sigma_{\mathbf{up}} = -\mathbf{0.00346} + \mathbf{1.0218} * \boldsymbol{\theta} - \mathbf{0.18} * \boldsymbol{\theta}^{\wedge}\mathbf{2} + \mathbf{0.008} * \boldsymbol{\theta}^{\wedge}\mathbf{3} + \mathbf{0.0004} * \boldsymbol{\theta}^{\wedge}\mathbf{4} \quad (4.79)$$

Model of stress–temperature diagram (unloading part)

Variable	Value
$a0$	1.326
$a1$	−2.197
$a2$	1.593
$a3$	−0.475
$a4$	0.0506

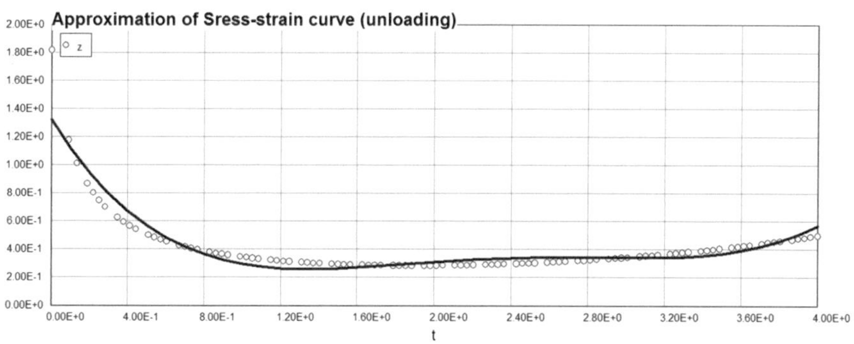

$$\sigma_{\mathbf{dw}} = \mathbf{1.326} - \mathbf{2.197} * \boldsymbol{\theta} + \mathbf{1.593} * \boldsymbol{\theta}^{\wedge}\mathbf{2} - \mathbf{0.475} * \boldsymbol{\theta}^{\wedge}\mathbf{3} + \mathbf{0.0506} * \boldsymbol{\theta}^{\wedge}\mathbf{4} \quad (4.80)$$

We can see from Fig. 4.23 that there is a residual stress at the end of unloading diagram, but the residual strain is zero. This is because the main assumption in all previous examples was that loading and unloading processes follows the same

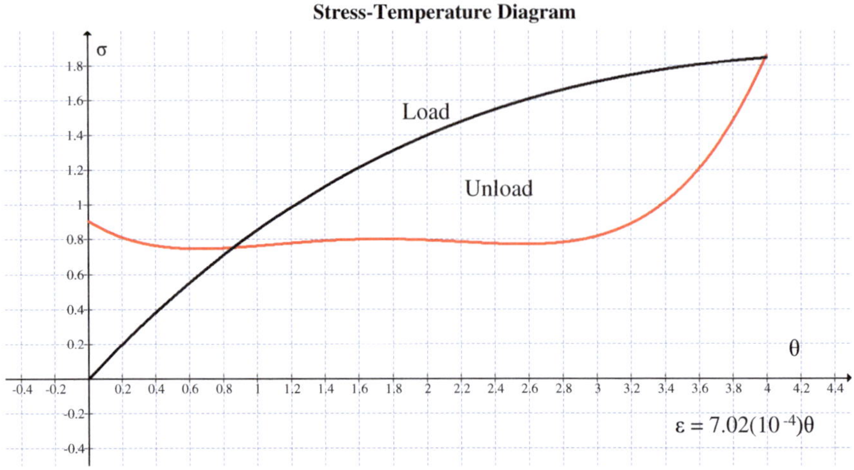

Fig. 4.23 Creep stress—temperature diagram

temperature–time curve and the reduction (increase) of modulus of elasticity also follows the same Eq. (4.28). However, it is a well-known fact that the decay phase of any design fire is quite different from the heating (developing) phase. In fact the heating part of fire is controlled by standard temperature–time curve [28], but the decay period is described by the linear decrease in temperature per unit time. In case of real fire scenario (see Fig. 4.1) it will be approximated by different analytical expressions of function "*m*" and "*m*1". Since we agreed to consider only one MPP parameter at the time, the integral form of creep constitutive law can be substituted by the first order ordinary differential equation (ODE), which is considerably easier to analyze. Equation (4.31) will remain unchanged except functions "*m*" and "*m*1" that we obtain now from Eq. (4.9) for decay period (see Fig. 4.1). The computer code in this case is (Fig. 4.24).

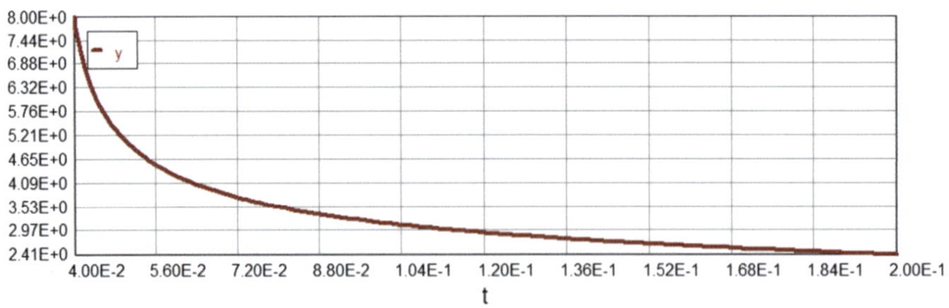

Fig. 4.24 Temperature decreasing part

Differential equations

$$d(y1)/d(t) = 0.5 * (1 - y1) \wedge 1.0 * \exp(y/(1 + 0.1 * y))$$
$$d(y)/d(t) = (1) * 20 * (1 - y1) \wedge 1.0 * \exp(y/(1 + 0.1 * y)) - 0.233 * y \wedge 4$$
$$t(0) = 0.04$$
$$t(f) = 0.2$$
$$y1(0) = 0.925$$
$$y(0) = 8.0$$

Calculated values of DEQ variables

	Variable	Initial value	Minimal value	Maximal value	Final value
1	*t*	0.04	0.04	0.2	0.2
2	*y*	8	2.411669	8	2.411669
3	*y*1	0.925	0.925	0.9738659	0.9738659

$$\textbf{Model}: \theta = 60 - 2.98 * 10^{\wedge}3 * \tau + 6.45 * 10^{\wedge}4 * \tau^{\wedge}2 - 7.27 * 10^{\wedge}5 * \tau^{\wedge}3 + 4.48 * 10^{\wedge}6 * \tau^{\wedge}4 - 1.42 * 10^{\wedge}7 * \tau^{\wedge}5 + \; + 1.84 * 10^{\wedge}7 * \tau^{\wedge}6 \quad (4.81)$$

The inverse function θ^{-1} is:

Model: *t* = *a*0 + *a*1 * y + *a*2 * *y*^2 + *a*3 * y^3 + *a*4 * *y*^4 + *a*5 * *y*^5 + *a*6 * *y*^6 + *a*7 * *y*^7 + *a*8 * *y*^8

Variable	Value
*a*0	7.092698
*a*1	−10.01532
*a*2	6.394174
*a*3	−2.359154
*a*4	0.5452611
*a*5	−0.08040 8
*a*6	0.0073643
*a*7	−0.0003822
*a*8	8.594E−06

$$\mathbf{Model: m} = \tau = 7.09 - \mathbf{10} * \theta + \mathbf{6.394} * \theta^\wedge 2 - \mathbf{2.359} * \theta^\wedge 3 + \mathbf{0.545} * \theta^\wedge 4 - \mathbf{0.0804} * \theta^\wedge 5 + \mathbf{0.00736} * \theta^\wedge 6 - 0.000382 * \theta^\wedge 7 + \mathbf{8.594} * 10^\wedge - 6 * \theta^\wedge 8 \tag{4.82}$$

$$\mathbf{m01} = -10 + 12.79 * \theta^\wedge 1 - 7.077 * \theta^\wedge 2 + 2.18 * \theta^\wedge 3 - 0.402 * \theta^\wedge 4 + 0.04416 * \theta^\wedge 5 - 0.002674 * \theta^\wedge 6 + 0.0000687 * \theta^\wedge 7 \tag{4.83}$$

Differential equations

1 $d(y11)/d(t) = -0.33 * y11 * (-m01) + (\exp(-0.15 * (t))) * (1 - 0.15 * (t)) - (\exp((t)/(1.0 + 0.1 * (t)))) * (-m01) * y11 \wedge n + 0.33 * (-m01) * (8 - t) * (\exp(-0.15 * (t)))$

$y11(0) = 1.82$

$t(0) = 2.41$

$t(f) = 4$

Calculated values of DEQ variables

	Variable	Initial value	Minimal value	Maximal value	Final value
2	$m01$	−0.2278211	−0.2278211	−0.06486	−0.0712832
3	$m1$	0.007856	0.0030152	0.007856	0.0030152
4	n	2	2	2	2
5	t	2.41	2.41	4	4
7	y11	1.82	0.3316012	1.82	0.37174

Differential equations

1 $d(y)/d(t) = -0.33 * y * m1 + (\exp(-0.15 * t)) * (1 - 0.15 * t) - (\exp(t/(1.0 + 0.1 * t))) * m1 * y \wedge n + 0.33 * m1 * t * (\exp(-0.15 * t))$

2 $d(y11)/d(t) = 1 * (-0.33 * y11 * (-m01) + (\exp(-0.15 * (8 - t))) * (1 - 0.15 * (8 - t)) - (\exp((8 - t)/(1.0 + 0.1 * (8 - t)))) * (-m01) * y11 \wedge n + 0.33 * (-m01) * (8 - t) * (\exp(-0.15 * (8 - t))))$

Explicit equations

1 $n = 2$

2 $m1 = 0.0405 - 0.02252 * t \wedge 1 + 0.004386 * t \wedge 2 - 0.0002747 * t \wedge 3$

3 $m = (0.0405 * t - 0.01126 * t \wedge 2 + 0.001462 * t \wedge 3 - 0.00006868 * t \wedge 4)$

4 $m01 = -10 + 12.79 * t \wedge 1 - 7.077 * t \wedge 2 + 2.18 * t \wedge 3 - 0.402 * t \wedge 4 + 0.04416 * t \wedge 5 - 0.002674 * t \wedge 6 + 0.0000687 * t \wedge 7$

Model of stress–temperature diagram (unloading part) (Fig. 4.25).

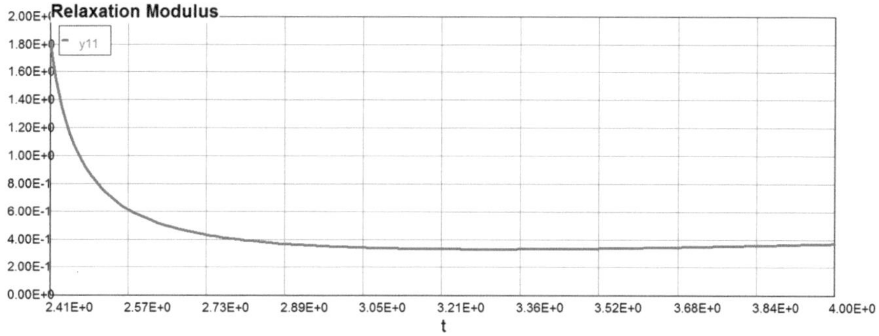

Fig. 4.25 Relaxation modulus (n = 2)

Model: $y11 = 316.3 - 384.7 * t + 174.7 * t\hat{}2 - 35.05 * t\hat{}3 + 2.623 * t\hat{}4$

Variable	Value
$a0$	316.3428
$a1$	−384.7456
$a2$	174.6771
$a3$	−35.0503
$a4$	2.623023

$$\sigma_{\mathbf{rlx}} = \mathbf{316.3 - 384.7} * \boldsymbol{\theta} + \mathbf{174.7} * \boldsymbol{\theta}\hat{}\mathbf{2} - \mathbf{35.05} * \boldsymbol{\theta}\hat{}\mathbf{3} + \mathbf{2.623} * \boldsymbol{\theta}\hat{}\mathbf{4} \quad (4.84)$$

Let us analyze now the relaxation modulus dependence from exponential stress parameter n and how it affects the residual stresses. The results are self explanatory therefore they are presented in graphical form and illustrate quantitative effect on residual stresses. All other data are not changed (Figs 4.26 and 4.27).

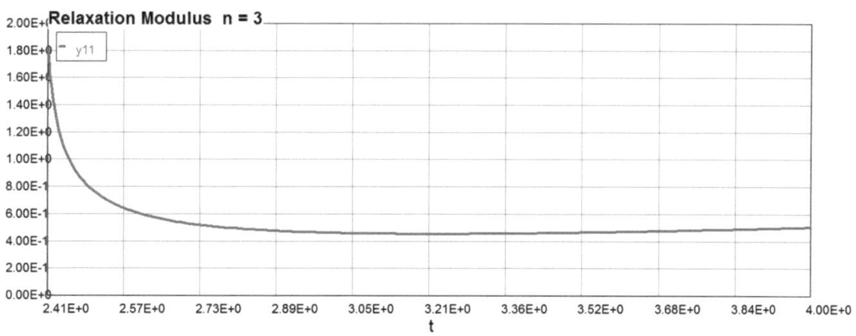

Fig. 4.26 Relaxation modulus (n = 3)

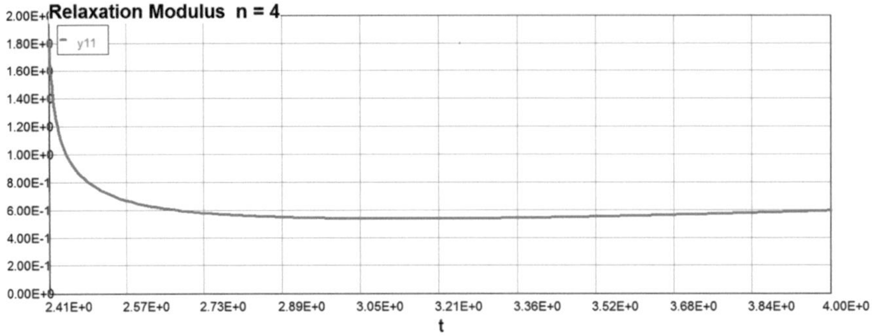

Fig. 4.27 Relaxation modulus (n = 4)

Calculated values of DEQ variables

	Variable	Initial value	Minimal value	Maximal value	Final value
2	$m01$	−0.2278211	−0.2278211	−0.0648617	−0.0712832
4	n	3	3	3	3
5	t	2.41	2.41	4	4
7	$y11$	1.82	0.4555582	1.82	0.5060631

Model: $y11 = a0 + a1 * t + a2 * t\text{^}2 + a3 * t\text{^}3 + a4 * t\text{^}4$

Variable	Value	95 % confidence
$a0$	301.9382	48.62396
$a1$	−369.257	62.6926
$a2$	168.5539	30.03641
$a3$	−33.98863	6.33916
$a4$	2.555125	0.4973982

$$\sigma_{\mathbf{rlx}} = -\mathbf{384.7} * \boldsymbol{\theta} + \mathbf{174.7} * \boldsymbol{\theta}\text{^}\mathbf{2} - \mathbf{35.05} * \boldsymbol{\theta}\text{^}\mathbf{3} + * \boldsymbol{\theta}\text{^}\mathbf{4} \qquad \text{(4.84correct)}$$

$$\mathbf{n = 4}$$

Calculated values of DEQ variables (Fig. 4.28)

	Variable	Initial value	Minimal value	Maximal value	Final value
1	m	0.0503533	0.0503533	0.0578259	0.0578259
2	$m01$	−0.2278211	−0.2278211	−0.0648607	−0.0712832
3	$m1$	0.007856	0.0030152	0.007856	0.0030152

(continued)

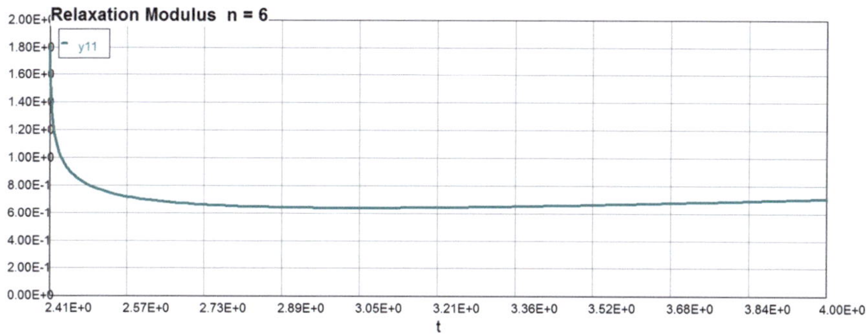

Fig. 4.28 Relaxation modulus (n = 6)

(continued)

	Variable	Initial value	Minimal value	Maximal value	Final value
4	n	4	4	4	4
5	t	2.41	2.41	4	4
6	$y11$	1.82	0.5377373	1.82	0.5957399

Model: $y11 = a0 + a1 * t + a2 * t\wedge 2 + a3 * t\wedge 3 + a4 * t\wedge 4$

Variable	Value	95 % confidence
$a0$	284.9743	52.44922
$a1$	−349.3612	67.72147
$a2$	159.8576	32.48534
$a3$	−32.302	6.863065
$a4$	2.432713	0.5389692

$$\boldsymbol{\sigma_{rlx} = -384.7 * \theta + 174.7 * \theta\wedge 2 - 35.05 * \theta^\wedge 3 + * \theta^\wedge 4} \quad (4.84\text{correct})$$

$$\mathbf{n = 6}$$

Calculated values of DEQ variables

	Variable	Initial value	Minimal value	Maximal value	Final value
1	m	0.0503533	0.0503533	0.0578259	0.0578259
2	$m01$	−0.2278211	−0.2278211	−0.0648602	−0.0712832
3	$m1$	0.007856	0.0030152	0.007856	0.0030152
4	n	6	6	6	6

(continued)

(continued)

	Variable	Initial value	Minimal value	Maximal value	Final value
5	*t*	2.41	2.41	4	4
6	*y*11	1.82	0.641552	1.82	0.7082142

Model: $y11 = a0 + a1 * t + a2 * t\hat{}2 + a3 * t\hat{}3 + a4 * t\hat{}4$

Variable	Value	95 % confidence
*a*0	258.1833	56.53482
*a*1	−317.3477	73.14711
*a*2	145.6102	35.15143
*a*3	−29.49292	7.438095
*a*4	2.22576	0.5849385

$$\boldsymbol{\sigma_{rlx} = -384.7 * \theta + 174.7 * \theta\hat{}2 - 35.05 * \theta\hat{}3 + * \theta\hat{}4} \quad (4.84\text{correct})$$

$$\mathbf{n = 8}$$

Calculated values of DEQ variables

	Variable	Initial value	Minimal value	Maximal value	Final value
1	*m*	0.0503533	0.0503533	0.0578259	0.0578259
2	*m*01	−0.2278211	−0.2278211	−0.0648613	−0.0712832
3	*m*1	0.007856	0.0030152	0.007856	0.0030152
4	*n*	8	8	8	8
5	*t*	2.41	2.41	4	4
7	*y*11	1.82	0.7054143	1.82	0.775645

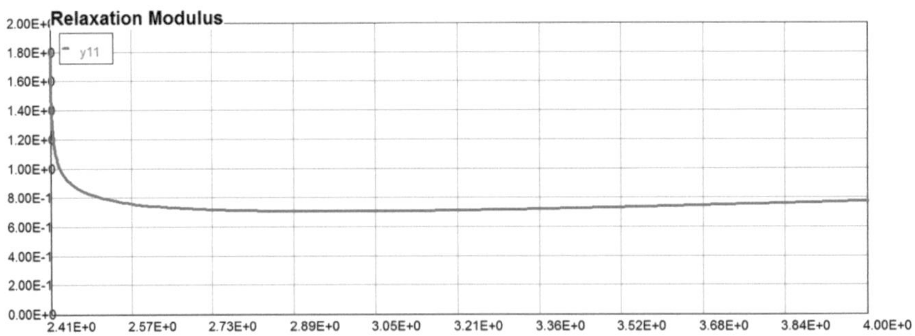

Model: $y11 = a0 + a1 * t + a2 * t\hat{}2 + a3 * t\hat{}3 + a4 * t\hat{}4$

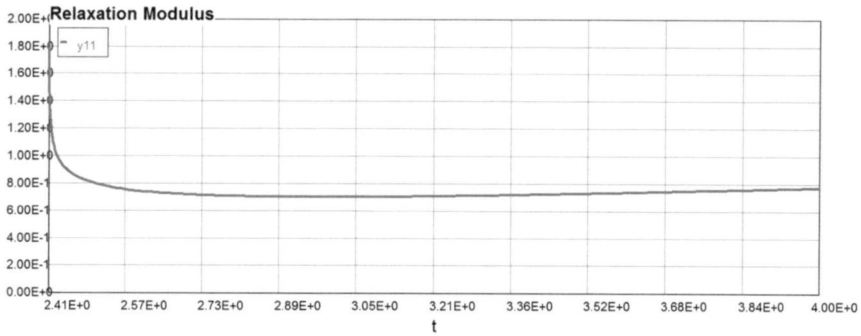

Fig. 4.29 Relaxation modulus: residual stress

Variable	**Value**	**95 % confidence**
$a0$	235.2392	57.30992
$a1$	−289.2497	74.19151
$a2$	132.7941	35.6663
$a3$	−26.90558	7.548448
$a4$	2.030738	0.5936326

$$\boldsymbol{\sigma_{\mathrm{rlx}} = 10.92 - 7.35 * \theta + 1.984 * \theta^\wedge 2 - 0.242 * \theta^\wedge 3 + 0.0109 * \theta^\wedge 4} \qquad (4.85)$$

One can see from Fig. 4.29 that there is a residual stress at the end of unloading diagram.

4.9.1 Ordinary Differential Equation (ODE) of Creep Constitutive Law

Consider now the general case of degenerate creep function: $K(\theta, \tau) = a(\theta)b(\tau)$, where a and b—arbitrary functions. Equation (4.78) has the form as follows:

$$\begin{aligned} &\bar{\sigma}(\theta) = E(\theta)\varepsilon(\theta) = \sigma(\theta) + a[\varphi(\theta)] \int_0^\theta (\exp(\tau)/(1+0.1\tau)))\mathrm{b}[\varphi(\tau)]\varphi'(\tau)\sigma^n(\tau)\mathrm{d}\tau \\ &\bar{\sigma}(\theta) = \sigma(\theta) + a[\varphi(\theta)]Z \\ &\frac{dZ}{\mathrm{d}\tau} = b[\varphi(\tau)]\varphi'(\tau)\sigma^n(\tau); \quad Z(0) = 0 \end{aligned} \qquad (4.86)$$

Integral type of creep constitutive law can be substituted by the ODE. Differentiating first equation of (4.86) we have

$$
\begin{aligned}
\bar{\sigma}(\theta) &= E(\theta)\varepsilon(\theta) = \sigma(\theta) + a[\varphi(\theta)] \int_0^\theta (\exp(\tau)/(1+0.1\tau))\mathrm{b}[\varphi(\tau)]\varphi'(\tau)\sigma^n(\tau)\mathrm{d}\tau \\
\frac{\mathrm{d}\bar{\sigma}(\theta)}{\mathrm{d}\theta} &= \frac{\mathrm{d}\sigma(\theta)}{\mathrm{d}\theta} + a[\varphi(\theta)](\exp(\theta)/(1+0.1\theta))\mathrm{b}[\varphi(\theta)]\varphi'(\theta)\sigma^n(\theta) \\
&\quad + \frac{da[\varphi(\theta)]}{\mathrm{d}\theta} \int_0^\theta (\exp(\tau)/(1+0.1\tau))\mathrm{b}[\varphi(\tau)]\varphi'(\tau)\sigma^n(\tau)\mathrm{d}\tau
\end{aligned}
\tag{4.87}
$$

Multiplying first equation of (4.87) by $\mathrm{da}[\varphi(\theta)]/\mathrm{d}\theta$ and second equation by $a\{\varphi(\theta)]$ and subtracting first from second equation we have

$$
\begin{aligned}
a[\varphi(\theta)]\frac{\mathrm{d}\bar{\sigma}(\theta)}{\mathrm{d}\theta} &= a[\varphi(\theta)]\frac{\mathrm{d}\sigma(\theta)}{\mathrm{d}\theta} + \{a[\varphi(\theta)]\}^2(\exp(\theta)/(1+0.1\theta))\mathrm{b}[\varphi(\theta)]\varphi'(\theta)\sigma^n(\theta) \\
&\quad + \frac{\mathrm{da}[\varphi(\theta)]}{\mathrm{d}\theta}\bar{\sigma}(\theta) - \frac{\mathrm{da}[\varphi(\theta)]}{\mathrm{d}\theta}\sigma(\theta) \\
\frac{\mathrm{d}\sigma(\theta)}{\mathrm{d}\theta} &= \frac{\mathrm{d}\bar{\sigma}(\theta)}{\mathrm{d}\theta} + \frac{1}{a[\varphi(\theta)]}\frac{\mathrm{da}[\varphi(\theta)]}{\mathrm{d}\theta}\sigma(\theta) - a[\varphi(\theta)](\exp(\theta)/(1+0.1\theta))\mathrm{b}[\varphi(\theta)]\varphi'(\theta)\sigma^n(\theta) \\
&\quad - \frac{1}{a[\varphi(\theta)]}\frac{\mathrm{da}[\varphi(\theta)]}{\mathrm{d}\theta}\bar{\sigma}(\theta); \quad \bar{\sigma}(\theta) = E(\theta)\varepsilon(\theta); \quad \sigma(0) = 0
\end{aligned}
\tag{4.88}
$$

The solution of Eq. (4.88) exists if $a[\varphi\ (\theta)]$ and $b[\varphi\ (\theta)]$ are monotonically increasing (decreasing) positive functions.

Consider the particular case of the Eq. (4.88) when the cooling process (temperature decrease with time) runs much faster than it is prescribed in the case of fast fire by formulas (4.10) and (4.11). In other words, we assume that the function $m = \varphi(\theta) = \theta$ *and* $m1 = \varphi'(\theta) = 1$, that is, the fire decay process occurs in a linear fashion. This approach is adopted in a number of regulatory standards for fire safety [29–31].

Equation (4.88) in this case (loading and unloading) is as follows:

$$
\begin{aligned}
\frac{\mathrm{d}\sigma}{\mathrm{d}\theta} &= -\frac{1}{(\mathrm{e}^{-0.33(8-\theta)})}\left(0.33\mathrm{e}^{-0.33(8-\theta)} + \mathrm{e}^{-(8-\theta)}\left[\exp\left(\frac{(8-\theta)}{1+0.1(8-\theta)}\right)\right][\mathrm{e}^{-0.66(8-\theta)}]\right)\sigma \\
&\quad + \frac{1}{(\mathrm{e}^{-0.33(8-\theta)})}\left\{0.33\mathrm{e}^{-0.33(8-\theta)}\left((8-\theta)\mathrm{e}^{-0.15(8-\theta)}\right) + (\mathrm{e}^{-0.33(8-\theta)})\mathrm{e}^{-0.15(8-\theta)}(1-0.15(8-\theta))\right\} \\
\sigma(0) &= 1.82
\end{aligned}
\tag{4.89}
$$

$$\begin{aligned}\frac{d\sigma}{d\theta} &= -\frac{1}{(e^{-0.33(\theta)})}\left(0.33e^{-0.33(\theta)} + e^{-(\theta)}\left[\exp\left(\frac{(\theta)}{1+0.1(\theta)}\right)\right][e^{-0.66(\theta)}]\right)\sigma \\ &+ \frac{1}{(e^{-0.33(\theta)})}\left\{0.33e^{-0.33(\theta)}\left((\theta)e^{-0.15(\theta)}\right) + (e^{-0.33(\theta)})e^{-0.15(\theta)}(1-0.15(\theta))\right\} \\ \sigma(0) &= 0\end{aligned} \tag{4.90}$$

The solution of Eqs. (4.89) and (4.90) and the corresponding computer code is

Calculated values of DEQ variables

	Variable	Initial value	Minimal value	Maximal value	Final value
1	*m*1	−9.872806	−9.872806	−0.0648881	−0.9038336
2	*n*	3	3	3	3
3	*t*	0.01	0.01	8	8
5	*Y*	1.228	0.9401084	1.257237	0.9401084
8	*Y*1	0	0	1.256987	1.228217

Differential equations

1	d(*Y*)/d(*t*) = (−(1/((exp(−0.33 * (8−*t*)))) * (0.33 * (exp(−0.33 * (8−*t*))) * *Y* + (exp(−(8 − *t*))) * (exp(((8 − *t*))/(1.0 + 0.1*(8 − t)))) * ((exp(−0.66 * (8 − *t*)))^1)) * *Y*^*n*) + (1/((exp(−0.33 * (8 − *t*))))) * (0.33 * (exp(−0.33 * (8 − t))) * (8 − t) * (exp(−0.15 * (8−*t*))) + (exp(−0.15 * (8−t))) * ((exp(−0.33 * (8−t)))) * (1−0.15 * (8−t))))
2	d(*Y*1)/d(*t*) = (−(1/((exp(−0.33 * (*t*)))) * (0.33 * (exp(−0.33 * (*t*))) * *Y*1 + (exp(−(*t*))) * (exp(((*t*))/(1.0 + 0.1*(*t*)))) * ((exp(−0.66 * (*t*)))^1))*Y1^*n*) + (1/((exp(−0.33 * (*t*))))) * (0.33 * (exp(−0.33 * (*t*))) * (*t*) * (exp(−0.15 * (*t*))) + (exp(−0.15 * (*t*))) * ((exp(−0.33 * (t)))) * (1−0.15 * (*t*))))

Explicit equations (Fig. 4.30).

1	*n* = 3

Model: *Y* = *a*0 + *a*1 * *t* + *a*2 * *t*^2 + *a*3 * *t*^3 + *a*4 * *t*^4

Variable	Value
*a*0	1.230423
*a*1	−0.0158938
*a*2	0.0224413
*a*3	−0.0055378
*a*4	0.0003007

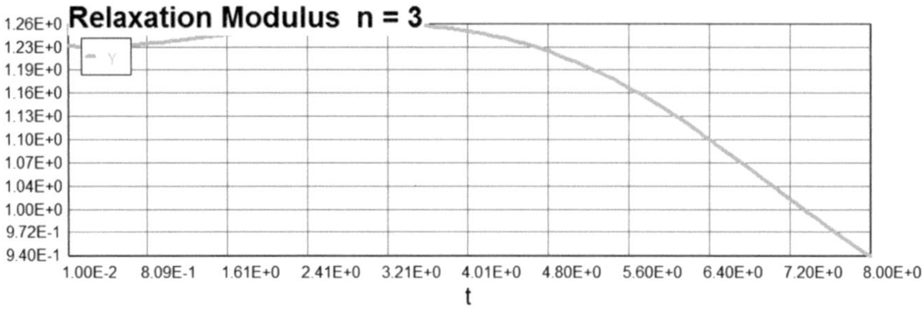

Fig. 4.30 Relaxation modulus (ODE method)

$$\sigma_{\mathbf{rlx}} = \mathbf{1.23 - 0.0159} * \theta + \mathbf{0.0224} * \theta^\wedge \mathbf{2 - 0.0055} * \theta^\wedge \mathbf{3 + 0.0003} * \theta^\wedge \mathbf{4} \quad (4.91)$$

Model: *Y*1 = *a*0 + *a*1 * *t* + *a*2 * *t*^2 + *a*3 * *t*^3 + *a*4 * *t*^4

Variable	**Value**
*a*0	0.0549013
*a*1	0.9941044
*a*2	−0.317813
*a*3	0.0444852
*a*4	−0.0022609

$$\sigma_{\mathbf{up}} = \mathbf{0.055 + 0.994} * \theta - \mathbf{0.318} * \theta^\wedge \mathbf{2 + 0.0445} * \theta^\wedge \mathbf{3 - 0.00226} * \theta^\wedge \mathbf{4} \quad (4.92)$$

n = 2

Calculated values of DEQ variables

	Variable	**Initial value**	**Minimal value**	**Maximal value**	**Final value**
1	*m*1	−9.872806	−9.872806	−0.0648959	−0.9038336
2	*n*	2	2	2	2
3	*t*	0.01	0.01	8	8
5	*Y*	1.357	0.9300436	1.357972	0.9300436
8	*Y*1	0	0	1.35668	1.31975

Model: *Y*1 = *a*0 + *a*1 * *t* + *a*2 * *t*^2 + *a*3 * *t*^3 + *a*4 * *t*^4

Variable	**Value**
*a*0	0.0566054
*a*1	0.9001886
*a*2	−0.2532709

(continued)

(continued)

Variable	**Value**
*a*3	0.03287
*a*4	−0.0016078

$$\boldsymbol{\sigma_{up} = 0.0566 + 0.90 * \theta - 0.253 * \theta^\wedge 2 + 0.0329 * \theta^\wedge 3 - 0.0016 * \theta^\wedge 4} \quad (4.93)$$

Model: $Y = a0 + a1 * t + a2 * t^\wedge 2 + a3 * t^\wedge 3 + a4 * t^\wedge 4$

Variable	**Value**
*a*0	1.355353
*a*1	−0.0643995
*a*2	0.0459639
*a*3	−0.0094684
*a*4	0.0004862

$$\boldsymbol{\sigma_{rlx} = 1.355 - 0.0644 * \theta + 0.0459 * \theta^\wedge 2 - 0.0097 * \theta^\wedge 3 + 0.00049 * \theta^\wedge 4} \quad (4.94)$$

The instantaneous stresses (due to modulus of elasticity changes—temperature effect without creep and loading/unloading process follows the same curve) are: $\bar{\sigma} = (\theta)e^{-0.15(\theta)}$ (See Fig. 4.31). Finally, the combined stress–strain graph (loading and unloading case) is presented below.

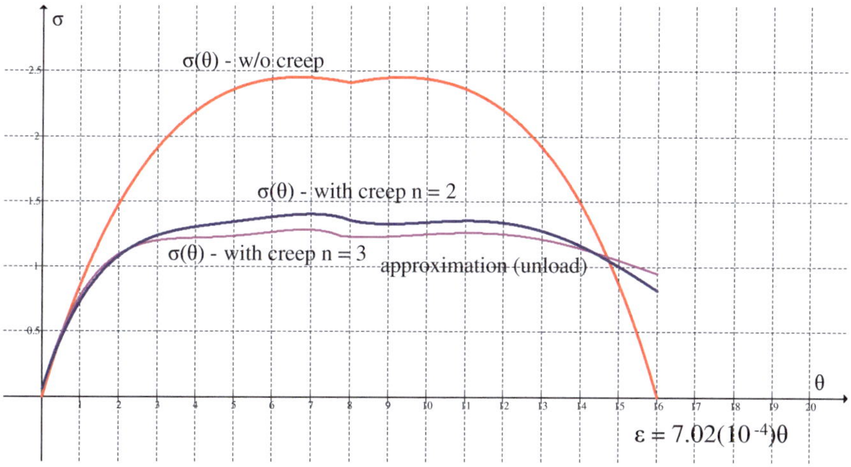

Fig. 4.31 Creep stress diagram (loading and unloading)

One can see from Fig. 4.31 that there is a residual stress at the end of unloading diagram. The computer output of unloading part of stress–strain diagram is approximated by the "best-to-fit" forth power polynomial (see Eq. 4.94).

References

1. Tricomi FG (1982) Integral equations. Dover, New York
2. ASM Handbook, Vol. 8: Mechanical testing and evaluation (2000) ASM International
3. Dowling NE (2006) Mechanical behavior of materials: engineering methods for deformation, fracture, and fatigue, 3rd edn. Prentice Hall, Englewood Cliffs
4. Mathews and Walker, Courant and Hilbert, Whittaker and Watson, and Mathews and Walker " Mathematical Methods of Physics" ... Vladimirov "Equations of mathematical physics" Jan 23, 2011 ... R. L. WALKER ... Addison-Wesley Publishing Company, Inc
5. Razdolsky L (2012) Structural fire load. Theory and principals. McGraw-Hill, New York
6. Guo JS, Waldron P (2000) Development of the stiffness damage test (SDT) for characterization of thermally loaded concrete. Mater Struct 33:483–491
7. Rilem TC (1998) 129-MHT. Test methods for mechanical properties of concrete at high temperatures. Recommendations: Part 7: transient creep for service and accident conditions. Mater Struct 31:290
8. Collins JA (1981) Failure of materials in mechanical design (pp. 31–32, 435–478). Wiley, New York
9. Hertzberg RW, Vinci RP, Hertzberg JL (2012) Deformation and fracture mechanics of engineering materials, 5th edn. Wiley, New York
10. Rabotnov YN (1969) Creep problems in structural members (North-Holland, Amsterdam, 1969), trans. from Russ. Polzuchest' elementov konstrukcij (Nauka, Moskva, 1966)
11. Larson FR, Miller J (1952) A time-temperature relationship for rupture and creep stresses. Trans ASME 74:765–775
12. Dorn JE (1956) The spectrum of activation energies for creep. Creep recovery. American Society for Metals, pp 255–283
13. Volkmann P (1998) Proceedings of the international conference numbers, functions, equations '98, Leaflets in Mathematics, ed. Pales, Z. (Janus Pannonius University), pp 150, 151
14. Manson SS, Haferd AM (1953) A linear "lime-temperature relation for extrapolation of creep and stress rupture data. NACA Technical Note 2890
15. Sellars CM, Tegart WJM (1966) Acta Metallurgica 14:1136–1138
16. Zener C, Hollomon H (1944) J Appl Phys 15:22
17. Odqvist FKG, Hult J (1962) Kriechfestigkeitmetallischer Werkstoffe. Springer, Berlin
18. Altenbach H, Zolochevsky A (1966) // Engng Frac Mech 54:75
19. SAA (2002) Structural design actions. AS 1170.1:2002. Standards Association of Australia
20. Lan X (2007) Variational iteration method for solving integral equations. Comput Math Appl 54:1071–1078
21. Kanwal RP, Liu KC (1989) A Taylor expansion approach for solving integral equations. Int J Math Educ Sci Technol 2:411–414
22. Babolian E, Masouri Z, Hatamzadeh-Varmazyar S (2009) A direct method for numerically solving integral equations system using orthogonal triangular functions. Int J Ind Math 2:135–145
23. Jafari H, Hosseinzadeh H, Mohamadzadeh S (2010) Numerical solution of system of linear integral equations by using Legendre wavelets. Int J Open Probl Comput Sci Math 5:63–71
24. Kantorovich LV, Krylov VI (1958) Approximate methods of higher analysis, Noordhoff (Translated from Russian)

25. Maleknejad K, Aghazadeh N (2005) Numerical solution of Volterra integral equations of the second kind with convolution kernel by using Taylor-series expansion method. Appl Math Comput 161:915–922
26. Baker CTH (2000) A perspective on the numerical treatment of Volterra equations. J Comput Appl Math 125(1–2):217–249
27. Baker CTH (1977) The numerical treatment of integral equations. Clarendon Press, Oxford, pp Chapter 4
28. SFPE (2011) SFPE standard on calculating fire exposure to structural elements
29. SFPE (2004) Engineering guide: fire exposures to structural elements, Bethesda, Md., Society of Fire Protection Engineers
30. CEN; EN 1991-1-2 (2002) Actions on Structures, Part 1-2–Actions on structures exposed to fire, CEN Central Secretariat, Brussels
31. Kunio K, Takashi S (1963) Estimation of fire temperature-time curve in rooms. Building Research Institute, Ministry of Construction, Japanese Government June 1963

Chapter 5
Transient Engineering Creep of Materials Under Various Fire Conditions

Notation

k	The thermal conductivity that has the dimensions W/m K or J/m s K
T	Temperature
d	Thickness in the direction of heat flow
c	Specific heat capacity
Q	Activation energy
R	Ideal gas constant
P_l	Losses of heat due to thermal radiation
e	Emissivity factor
σ	Stefan–Boltzmann constant (σ = 5.6703 (10^{-8}) W/m^2 K^4)
T_o	Ambient temperature
A_v	Area of openings
c_p	Average specific heat at constant pressure
t	Time
$\vec{v}(u; v; w)$	Velocity vector
D	Diffusion coefficient (m^2/s)
p	Pressure
v	Kinematic viscosity
θ	Dimensionless temperature
τ	Dimensionless time
h	Height of the compartment (m)
a	Thermal diffusivity (m^2/s)
Time	$t = \frac{h^2}{a}\tau$ (s)
Temperature	$T = \frac{RT_*^2}{E}\theta + T_*$ (K), where T_* = 600 °K is the base line temperature

L. Razdolsky, *Probability Based High Temperature Engineering*,
DOI 10.1007/978-3-319-41909-1_5

Coordinates	$\bar{x} = x/h$ and $\bar{z} = z/h$—"x" and "z"—dimensionless coordinates
Velocities	$\bar{u} = \frac{v}{h}u$(m/s) and $\bar{w} = \frac{v}{h}w$ (m/s)—horizontal and vertical components velocity accordingly; v–Kinematic viscosity (m^2/s); "u" and "w"—dimensionless velocities
$Pr = v/a$	Prandtl number
$Fr = \frac{gh^3}{va}$	Froude number
g	Gravitational acceleration
$Le = a/D = Sc/Pr$	The Lewis number
$Sc = v/D$	The Schmidt number
$\bar{\beta} = \frac{RT_*}{E}$	Dimensionless parameter
$\bar{\gamma} = \frac{c_pRT_*^2}{QE}$	Dimensionless parameter
$P = \frac{e\sigma K_v(\beta T_*)^3 h}{\lambda}$	Thermal radiation dimensionless coefficient
$\sigma = 5.67\ (10^{-8})$ (W/m^2 K^4)	Stefan–Boltzmann constant
$K_v = A_o\ h/V$	Dimensionless opening factor
A_o	Total area of vertical and horizontal openings
$\delta = \left(\frac{E}{RT_*^2}\right)Qz\left(\exp\left(-\frac{E}{RT_*}\right)\right)$	Frank-Kamenetskii's parameter
$C = [1 - P(t)/P_o]$	Concentration of the burned fuel product in the fire compartment
$\bar{W} = \frac{v}{h}W$	Vertical component of gas velocity
$\bar{U} = \frac{v}{h}U$	Horizontal component of gas velocity
$b = L/h$, "L" and "h"	Length (width) and height of fire compartment accordingly
W; U	Dimensionless velocities
R_n	Nominal strength
S_i	Nominal load
φ	Resistance factor
γ	Load factor
R_c	Characteristic value for the resistance
A	Cross-sectional area
I	Moment of inertia
W	Total weight
G_c	Characteristic value for the permanent load
S	Characteristic value for the variable load
ψ_1	Partial safety factor for the permanent load
ψ_2	Partial safety factor for the variable load
β	Reliability index

S	Probability space
A	Set of outcomes (events) to which a probability is assigned
$P(E_2 \mid E_1)$	Conditional probability
$\Phi^*(.)$	Denotes the cumulative distribution function of standard normal distribution $\Phi^*(z) = \frac{1}{\sqrt{2\pi}} \int_{-\infty}^{z} e^{-\frac{z^2}{2}} dz$
μ_A, μ_B, σ_A, σ_B	Are mean and standard deviation of A and B, respectively
$J(t, t')$	Compliance function
T_M	Melting point of the metal matrix material
$\varepsilon(t)$	Strain
$\sigma(t)$	Stress
$\bar{\sigma}(t) = E(t)\varepsilon(t)$	Instantaneous stress
ε_e	Instantaneous (elastic) strain
ε_c	Creep strain
ε_T	Thermal expansion due to high-temperature effect
$K(t, t') = -\partial J(t, t')/\partial t'$	Retardation function (memory function)
$R(t, t')$	Relaxation function (also called the relaxation modulus)
$M(\theta)$	Bending moment
$V(\theta)$	Shear force
$P(\theta)$	Axial force
$y(\theta)$	Deformation
P_f	Probability of failure
P_{rel}	Reliability
σ	ψ
ε	ψ
θ	Dimensionless temperature
κ	Curvature
τ	Dimensionless time
ω	Frequency
$S(\omega)$	Spectral density
μ	Poisson coefficient
D	Diffusion rate (Flick's law)
η	Viscosity parameter of the material
E	Modulus of elasticity
$n = \eta/E$	Relaxation time
n	Power law exponent
α_i	Material property parameter
L	Span (spring spacing)
k_0	Subgrade modulus

5.1 Introduction

5.1.1 Overview

The most significant input to a fire engineering analysis is the choice of fire scenarios and related design fires, since they underpin the ensuing fire safety design and its authority approval.

Underplaying the design fires could result in inadequate protection and the unacceptable possibility of major fire incidents; being too conservative and design requirements could become unrealistic, unachievable, and uneconomic.

Fundamental to the process of fire safety engineering is the proper evaluation of fire hazards and the resulting developments of fire scenarios. These are then translated into design fires which provide the "load" in any structural and fire engineering analysis. This could be argued as the most critical part of the overall process embodied in structural fire resistance analysis and design.

In fire safety engineering for the built environment it would seem that the process of hazard analysis, development of fire scenarios, and choice of design fires for structural fire engineering calculations are somewhat problematic. A review by Johnson [1] highlighted that part of the problem may stem from the lack of clarity in many guidelines documents on matters of:

1. Definition of what constitutes "worst credible scenario".
2. How to address issues of redundancy and uncertainty?
3. No proper consideration of "high challenge fires" or extreme events.
4. Very limited direction on sensitivity analysis.

A more productive approach would seem to be to develop a robust methodology for the development of fire scenarios and design fires that could be agreed by the structural fire engineering profession, regulators, and certifiers and used on all projects. This could be enshrined in guidelines documents and codes or practice and become the way of improving fire engineering resistance design solutions. Key to this improved, systematic approach is a more structured identification of all credible fire scenarios and their classification into those for standard design, those for testing design beyond standard limits, and those which are extreme events and beyond design. Further consideration of the required expertise amongst those involved in scenario planning is also an issue. As a result, structural and fire protection engineers are involved in a wide range of projects from office, high rise buildings to more complex projects such as airframe structural system design, aerospace structures, airports, and road tunnels.

Structural codes worldwide are moving from prescriptive to performance-based approaches. Performance-based codes establish fire resistance design objectives and leave the means for achieving those objectives to the designer. One of the main advantages of performance-based designs is that the most recent rheological creep constitutive models can be used by practicing engineers, inevitably leading to innovative and cost-effective designs. Prescriptive codes are easy to use, and regulatory officials can quickly determine whether a design follows code requirements. However, the application of prescriptive methods in many modern-design structural systems is very questionable. By assuming a worst-case but realistic natural fire scenario and calculating the heat transfer to the steel, the load-bearing capacity of the steel members can be checked at high temperatures, and requirements for fire protection, if any, can be judged in a rational manner.

Performance-based fire resistance design is now implemented and accepted in many countries. The design methodology has key advantages over prescriptive-based design. Structural behavior in fire depends on a number of variables. These include material deterioration at elevated temperature and reduction of stiffness of the structural system as a whole.

The energy- and mass balance equations can be used to determine the actual thermal exposure and fire duration. This is known as the *natural fire method*. This method allows the combustion characteristics of the fire, the ventilation and wind effects, and the thermal properties of the material to be considered. It is the most rigorous means of determining temperature–time function that affects fire resistance capability of the structural system. The rapid growth of computing power and the corresponding maturing of computational fluid dynamics (CFD) have led to the development of CFD-based field models applied to fire research problems. The use of CFD models has allowed the description of fires in complex geometries and the incorporation of a wide variety of physical phenomena. The differential equations are solved numerically by dividing the physical space where the fire is to be simulated into a large number of rectangular cells. Within each cell, the gas velocity, temperature, and so on are assumed to be uniform, changing only with time. The accuracy with which the fire dynamics can be simulated depends on the number of cells that can be incorporated into the simulation structural analysis; therefore, the simplifications and approximations of the temperature–time relationship structural fire load are absolutely essential.

5.1.2 Temperature–Time Function

The analytical approach in the structural fire engineering field typically comprises thermal and subsequent structural analyses. When designing structures for thermal fire load, the first step is to calculate the temperature field and then the ultimate strength capacity, based on the temperatures assessed. This is possible by using the simplified (but conservative) design method or the more sophisticated global analysis and design in accordance with the structural code requirements. The simplification (where it is possible for the determination of thermal fire load only) is the key element of the methodology proposed here. The overall system of conservation of energy, mass, and momentum equations that are analyzed here is similar to the Fire Dynamics Simulator (FDS) model [2]. However, the limitations and simplifications are different because they are concentrated on a narrowly focused problem: temperature—time fire load. For example, the large eddy simulation technique, the mixture fraction combustion model, active fire protection systems, and so on are not needed in the case of assessing creep deformations and further down the probability-based structural fire resistance. The FDS model solves the conservation equations of mass, momentum, and energy using the finite-difference method and the solution is updated in time on a three-dimensional (3D) rectilinear grid. However, the thermal radiation is computed using a finite-volume technique. The method proposed in this book uses the spatial averaging of variables; therefore, it is similar to the two-zone method in this respect. Consequently, this method has an intermediate position between the FDS and two-zone methods.

Since the existence of heat within a substance is caused by molecular action, the greater the molecular activity, the more intense is the heat. *Conduction* is heat transfer by means of molecular agitation within a material without any motion of the material as a whole. The energy in this case will be transferred from the higher speed particles to the slower ones with a net transfer of energy to the slower ones. *Convection* is heat transfer by mass motion of a fluid such as air when heated is caused to move away from the source of heat, carrying energy with it.

Radiated heat is one of the major sources of fire spread. This method of heat transmission is known as *radiation of heat waves*. Heat and light waves are similar in nature, but they differ in length per cycle. Heat waves are longer than light waves, and they are sometimes called *infrared rays*. Radiated heat will travel through space in all directions.

Compartment fire development can be described as being comprised of four stages: incipient, growth, fully developed, and decay. Flashover is not a stage of development, but simply a rapid transition between the growth and fully developed stages.

The overall structural fire resistance (SFR) design process can be separated into the activities illustrated in the following flowchart that emphasizes the fact that the ultimate strength (SFR) and overall stability of a structure very much depend on the assessed design fire scenario.

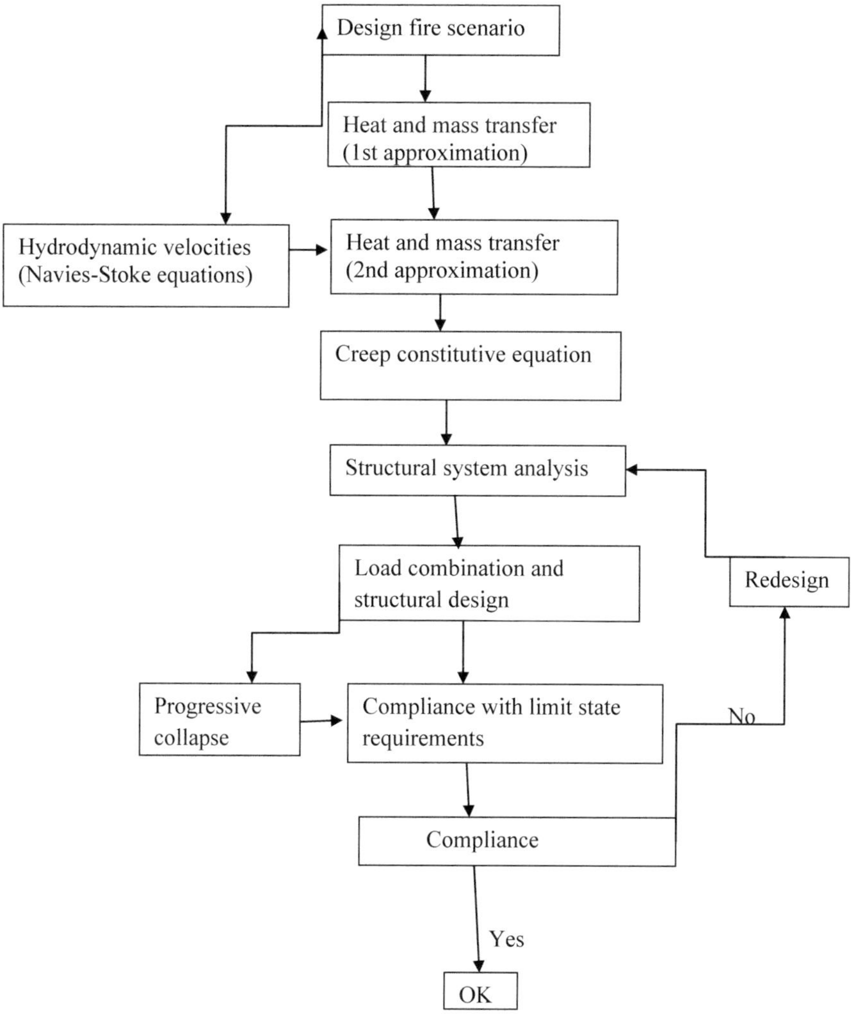

Consider now nonlinear parabolic differential equations representing the energy and mass conservation laws [3]:

Energy Conservation Law

$$\frac{\partial \theta}{\partial \tau} + Pr\left(U\frac{\partial \theta}{\partial x} + W\frac{\partial \theta}{\partial z}\right) = \delta(1 - C)^k \exp\left(\frac{\theta}{1 + \beta\theta}\right) - P\theta^4 \tag{5.1}$$

Mass Conservation Law

$$\frac{\partial C}{\partial \tau} + Pr\left(U\frac{\partial C}{\partial x} + W\frac{\partial C}{\partial z}\right) = \gamma\delta(1 - C)^k \exp\left(\frac{\theta}{1 + \beta\theta}\right), \tag{5.2}$$

where "*U*" and "*W*"—horizontal and vertical velocities, respectively, that should be obtained from the Navier–Stokes equations (see Chap. 3).

Thermal input data are as follows: heat transfer coefficients, thermal conductivity and thermal capacity of the air–gas mixture, heat effect (chemical reaction heat effect), activation energy, and so on. The thermal analysis comprises a determination of the temperature field versus time under design consideration. Let us now consider the spatial averaging of temperature and combustion rate (the unsteady process of chemical reaction). The detailed explanation of additional assumptions and conditions specifically tailored for obtaining temperature–time relationship versus fire severity scenario is presented in [4]. Equations (5.1) and (5.2) are simplified further [5]:

$$\frac{\partial \theta}{\partial \tau} = \delta(1 - C)^k \exp\left(\frac{\theta}{1 + \beta\theta}\right) - P\theta^4 \tag{5.3}$$

$$\frac{\partial C}{\partial \tau} = \gamma\delta(1 - C)^k \exp\left(\frac{\theta}{1 + \beta\theta}\right) \tag{5.4}$$

Equations (5.3) and (5.4) describe the unsteady combustion process at any temperature level. However, as stated in Frank-Kamenetskii [6], the parameter δ is calculated for the temperature close to the flashover point. In our case it is the onset temperature of creep deformations process, therefore the incipient stage of fire development is omitted.

The post-flashover stage of the fire (the fully developed fire) is the most important one from creep stresses computational point of view as well as from structural design point of view. As was mentioned above the base line temperature for creep effect on structural design is assumed as 300 °C, therefore the metal matrix material (MMM) is uniformly preheated to this temperature level. The initial condition for differential equations (5.3) and (5.4) is

$$\theta(\tau = 0) = 0; \quad C(\tau = 0) = 0 \tag{5.5}$$

Based on the Society of Five Protection Engineers (SFPE) guide [7] and Swedish fire curves [8, 9] for post-flashover realistic fire exposure, we can standardize fires as shown in Table 5.1.

The temperature–time function for a given fire scenario is the solution of Eqs. (5.3) and (5.4). However, in case of a developed fire, mathematical modeling of the physical and chemical transformations of real materials is known to have only with a small degree of confidence. At the same time, based on many full fire test results data, one can expect that curtain parameters, such as maximum

Table 5.1 Fire severity

Category	Fuel load L (MJ/m^2)	Max. temperature T_{max} (K)	Max. dimensionless temperature θ_{max}	Parameter "γ" from Eq. (5.4)
Ultra fast	$500 < L < 700$	$1020 < T_{max} < 1300$	$7.0 < \theta_{max} < 11.67$	$0 < \gamma < 0.05$
Fast	$300 < L < 500$	$880 < T_{max} < 1020$	$4.67 < \theta_{max} < 7.0$	$0.05 < \gamma < 0.175$
Medium	$100 < L < 300$	$820 < T_{max} < 880$	$3.67 < \theta_{max} < 4.67$	$0.175 < \gamma < 0.275$
Slow	$50 < L < 100$	$715 < T_{max} < 820$	$1.92 < \theta_{max} < 3.67$	$0.275 < \gamma < 1.0$

Note If fuel load $L > 700$, select $\gamma = 0$

temperature, type of temperature–time function, and so on are well known. On the other hand, some other parameters [e.g., parameter γ from Eq. (5.4)] are known with some degree of approximation.

From a physical point of view, this parameter characterizes the ratio of heat loss during the development stage of a fire (incipient and free-burning) divided by total energy released (heat rate) [10], that is, $\gamma = \frac{c_p R T_*^2}{QE}$. If, for example, the heat rate of a chemical reaction is large, then parameter γ is small. Therefore, parameter γ has a bounded variation between 0 and 1. It is also important to underline here that for any given value of parameter γ from the interval [0; 1], only one solution of Eqs. (5.3) and (5.4) exists, and the temperature–time function in this case has only one maximum value. It can be seen by observation (see below) that this maximum temperature value increases when parameter γ decreases from 1 to 0. On the other hand, the maximum gas temperature in a real fire compartment and the fuel load defines the category of fire severity (see Table 5.1); therefore, there is a correlation between the fire severity category and the value of parameter γ. Therefore, in our case, any solution of differential equations (5.3) and (5.4) is a function of two independent variables: t (time) and γ from an interval [0; 1]. Now, in order to select the solution that is needed, an additional condition has to be imposed. Each fire severity category (see Table 5.1) is defined by a corresponding maximum (averaged up in space) gas temperature T_{max}. The mathematical model of a real fire in a compartment now can be formulated as follows:

1. For each fixed number of γ from the interval [0; 1], find the discrete number of solutions of differential equations (5.3) and (5.4)—temperature–time curves—collection of functions.
2. Find the maximum values of temperature from this collection of functions.
3. Select the temperature–time curve from a given family of curves in order to apply in the future the method of change of variables in solving integral-type creep constitutive equation.
4. Obviously, all solutions of differential equations have to be obtained in dimensionless forms (temperature θ and time τ) and then should be transferred into real temperatures and time variables (see "Notation").

Now we can set up a typical computational problem from mathematical modeling of a real fire in a compartment for each category of fire exposure from Table 5.1. Each case of fire exposure is presented below by the numerical solutions of Eqs. (5.3) and (5.4) (using the simple mathematical software POLYMATH) in tabular and analytical forms.

Very Fast Fire; Temperature Increase (1022 K < T < 1305 K)

Data $T_* = 600$ K; $0 < \tau < 0.2$; $0 < \gamma < 0.05$

Select $\gamma = 0.01$

Differential equations (5.3) and (5.4) are rewritten as an input for the POLYMATH software:

1. d(y)/d(t) = 20 * (1 − y1)^1.0 * exp(y/(1 + 0.1 * y)) − 0.233 * y^4
2. d(y1)/d(t) = 0.2 * (1 − y1)^1.0 * exp(y/(1 + 0.1 * y))

"y" is the dimensionless temperature θ with the corresponding parameter $\gamma = 0.01$, $y1$ is the concentration of the burned fuel product C in the fire compartment. Solving differential equations (5.3) and (5.4) with initial conditions (5.5), we have

Calculated values of DEQ variables

	Variable	Initial value	Minimal value	Maximal value	Final value
4	t	0	0	0.2	0.2
5	y	0	0	*9.81*	*4.71*
7	$y1$	0	0	0.792	0.792

Tabulated data and additional assumptions are presented in [4]. For the final temperature–time graph, see Fig. 5.1.

In case of Very Fast Fire, we have (development stage): time interval $0 < \tau < 0.0744$; temperature increasing interval: $0 < \theta < 9.81$. For the solution of creep deformation of structural element (or system) at high-temperature effects from fire, we need to examine separately the monotonically increasing and monotonically decreasing portion of the curve (see Fig. 5.1). We will call increasing part—temperature loading condition, and decreasing part—temperature unloading condition. Using again POLYMATH software we approximate the increasing part of the curve (see Fig. 5.1) by analytical expression as follows (see Fig. 5.2 and Eq. 5.6):

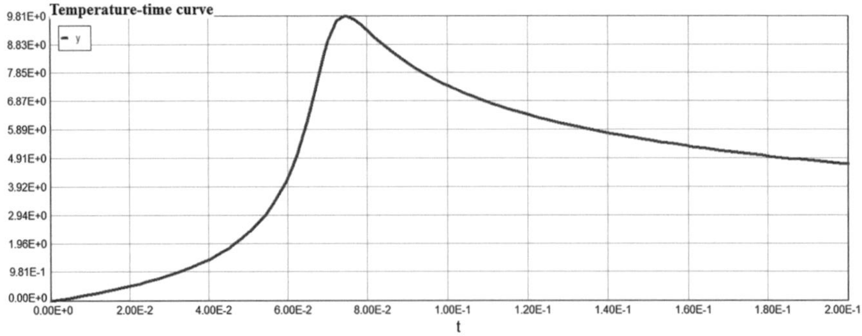

Fig. 5.1 Temperature–time curve. Very fast fire

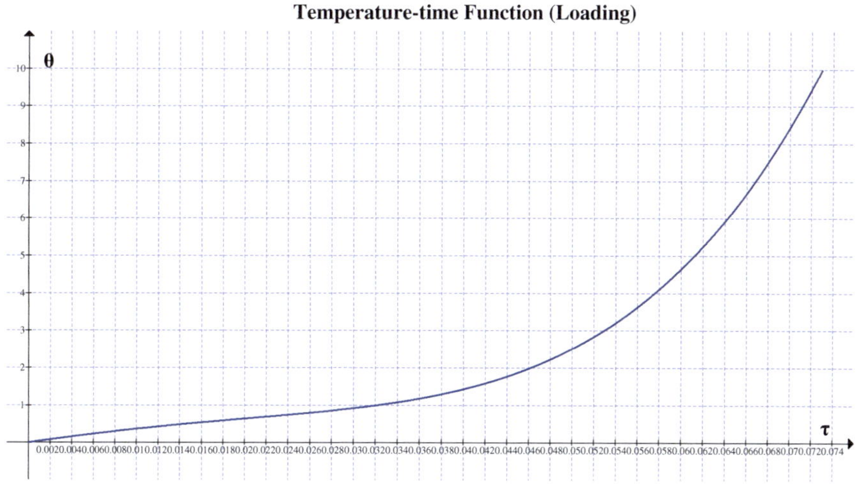

Fig. 5.2 Temperature–time curve (increasing stage)

$$\textbf{Model}: \theta = 40.86 * \tau - 337.2 * \tau\hat{}2 - 16{,}060 * \tau\hat{}3 + 530{,}400 * \tau\hat{}4 \tag{5.6}$$

Variable	Value
$a0$	0.0
$a1$	40.860
$a2$	−337.2
$a3$	−1.606E+04
$a4$	5.304E+05

First derivative of Eq. (5.6) with respect to dimensionless time τ is

$$\dot{\theta} = 40.86 - 674.4 * \tau\hat{}1 - 48{,}180 * \tau\hat{}2 + 2{,}121{,}600 * \tau\hat{}3$$

Therefore, the strain rate can be computed as follows:

$$\begin{aligned} \dot{\varepsilon} &= 7.02\left(10^{-4}\right)[40.86 - 674.4 * \tau\hat{}1 - 48{,}180 * \tau\hat{}2 + 2{,}121{,}600 * \tau\hat{}3] \\ &= 0.0287 - 0.4734 * \tau\hat{}1 - 33.822 * \tau\hat{}2 + 1489.36 * \tau\hat{}3 \end{aligned} \tag{5.7}$$

The graph of strain rate function $\varepsilon'(\tau)$ (see Fig. 5.3).

One can see from Fig. 5.3 that the strain rate is approximately constant up to $\tau = 0.03$ in this case (real time $t = (h^2/a_0) * \tau$ or 36 min) and represents the secondary phase of creep deformation process. It is shown in detailed form later in this chapter, when change of variable method is applied to the solution of creep constitutive equation, that the inverse time–temperature function and it derivative

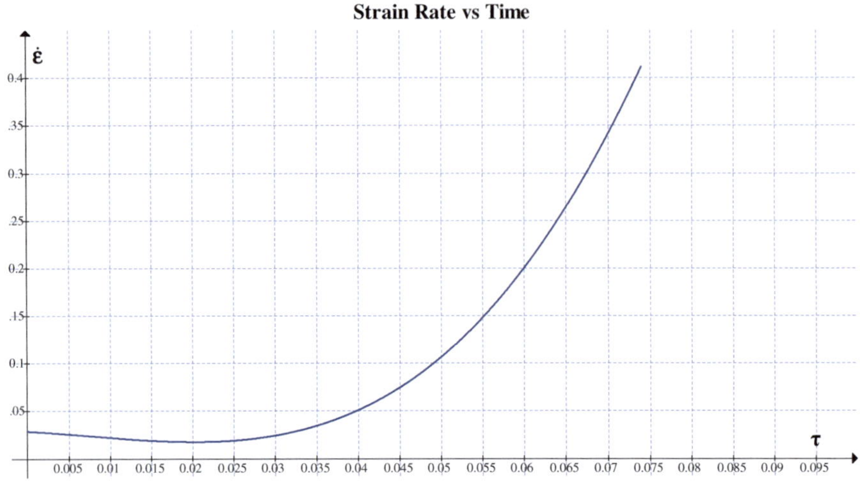

Fig. 5.3 Strain rate–time diagram. Increasing temperature stage ($0 < \tau < 0.0744$)

characterizes all three stages of general creep process: primary; secondary and tertiary. The inverse dimensionless temperature–time function θ^{-1} is

Model:
τ = 0.00245 + 0.0375 * θ − 0.00934 * θ^2 + 0.00104 * θ^3 − 0.000041 * θ^4

Variable	Value
*a*0	0.0024484
*a*1	0.0375392
*a*2	−0.0093426
*a*3	0.0010392
*a*4	−4.104E−05

The graph of inverse function $\tau(\theta)$ (see Fig. 5.4).
Finally,

$$m = \tau(\theta) = 0.00245 + 0.0375 * \theta - 0.00934 * \theta^\wedge 2 + 0.00104 * \theta^\wedge 3 - 0.000041 * \theta^\wedge 4 \tag{5.8}$$

$$m1 = \tau'(\theta) = 0.0375 - 0.01868 * \theta + 0.00312\, \theta^\wedge 2 - 0.000164 * \theta^\wedge 3 \tag{5.9}$$

Equations (5.8) and (5.9) are used below when we consider the problem of creep during monotonically increasing thermal load. From Fig. 5.4 (inverse strain rate vs. dimensionless time) one can see that in this fire scenario case the temperature range $0 < \theta < 2$ represents (approximately) the primary stage of creep; the temperature

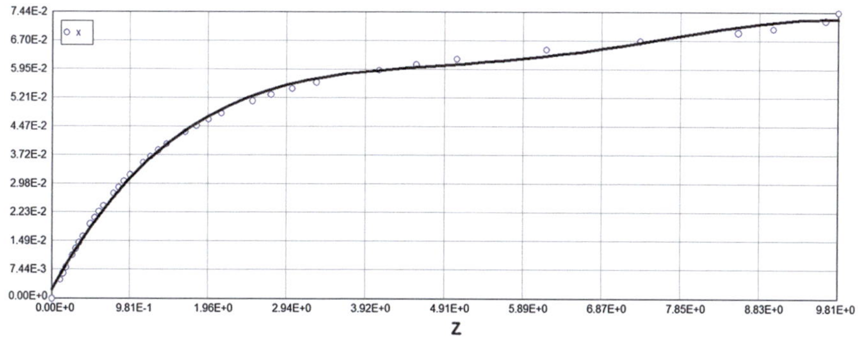

Fig. 5.4 The inverse function $\tau(\theta)$. Temperature increase stage

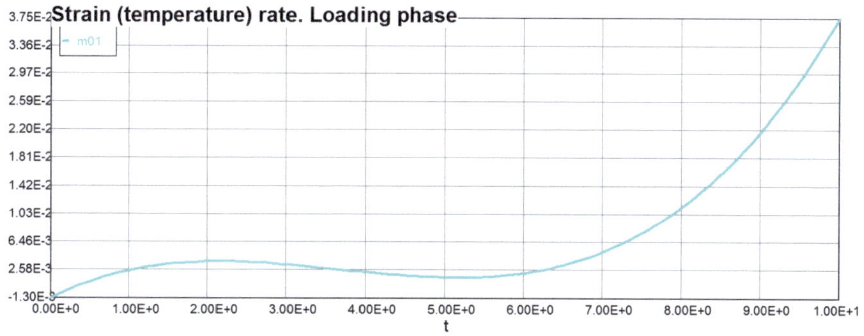

Fig. 5.5 Strain rate versus dimensionless temperature

range $2 < \theta < 6$—secondary stage and the temperature range $6 < \theta < 10$—tertiary stage (Fig. 5.5).

Let us consider now the temperature decreasing part of real fire curve (see Fig. 5.1).

Very Fast Fire; Temperature Decrease

Data $T_* = 600$ K; $0 < \gamma < 0.05$. Select $\gamma = 0.01$

In case of Very Fast Fire we have: time interval $0.0744 < \tau < 0.2$; temperature decreasing interval: $9.69 > \theta > 4.71$. Using again POLYMATH software we approximate the decreasing part of the curve (see Fig. 5.1) by analytical expression as follows:

Model:

$\theta = 37.95 - 718.4 * \tau + 6282.6 * \tau\text{^}2 - 25{,}400 * \tau\text{^}3 + 38{,}980 * \tau\text{^}4$

Variable	Value
$a0$	37.94618
$a1$	−718.4317
$a2$	6282.564
$a3$	−2.54E+04
$a4$	3.898E+04

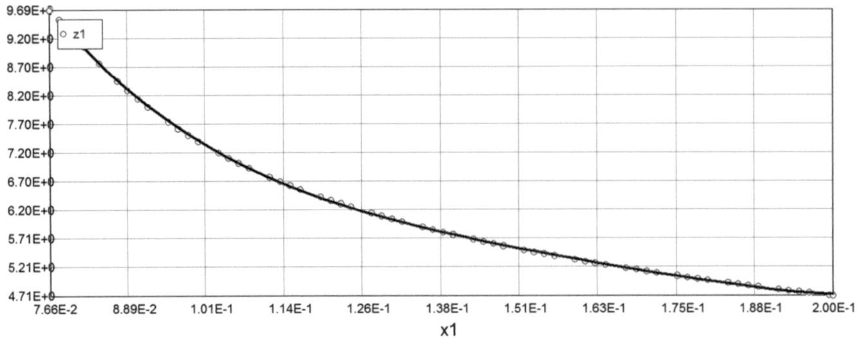

Fig. 5.6 Temperature–time graph of decay stage

The dimensionless temperature–time graph of fire decay stage (see Fig. 5.6).

The nonlinear approximation of the inverse temperature–time decay function is provided by the model as follows:

Model:

$\tau = 1.62 - 0.668 * \theta + 0.114 * \theta\text{^}2 - 0.00885 * \theta\text{^}3 + 0.000262 * \theta\text{^}4$

Variable	Value
$a0$	1.619
$a1$	−0.6677056
$a2$	0.1136394
$a3$	−0.008849
$a4$	0.0002619

The graph of inverse function θ^{-1} (see Fig. 5.7).

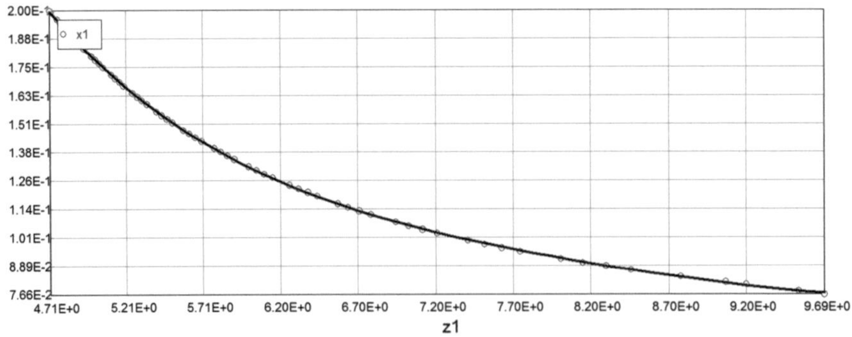

Fig. 5.7 Graph of inverse function θ^{-1}

Finally, functions $m0$ and $m01$ for fire decay stage are $(4.71 < \theta < 9.69; 0.0766 < \tau < 0.2)$.

$$m0 = \tau(\theta) = 1.62 - 0.668 * \theta + 0.114 * \theta^\wedge 2 - 0.00886 * \theta^\wedge 3 + 0.000262 * \theta^\wedge 4 \tag{5.10}$$

$$m01 = \tau'(\theta) = -0.668 + 0.228 * \theta - 0.0266\,\theta^\wedge 2 - 0.00105 * \theta^\wedge 3 \tag{5.11}$$

Equations (5.10) and (5.11) are used below when we discuss the problem of creep deformations during monotonically decreasing thermal load (fire decay phase).

5.2 Creep Constitutive Equation. Very Fast Fire (VFF)

Analysis of creep in engineering structures requires the formulation and the solution of an initial boundary value problem including the equilibrium equations and the constitutive assumptions. Equations describing the kinematics of three-dimensional solids or plane structural systems as well as equilibrium equations of mechanics of media are presented in monographs and textbooks, e.g., [11–13]. In what follows we discuss constitutive equations for the description of creep behavior in case of fire. The starting point of the engineering creep theory is the introduction of the inelastic strain, the creep constitutive law, the flow rule, the equivalent stress, and internal state variables. One of the most important and fundamental questions is that of the definition of the inelastic strain and the decomposition of the total strain into elastic and irreversible parts within the material description. From the theoretical point of view this is still a subject of many discussions within the nonlinear continuum mechanics, e.g., [14, 15].

In engineering mechanics, these concepts are often introduced based on intuitive assumptions, available experimental data and applications. Therefore, a lot of formulations of uniaxial and multiaxial creep equations can be found in the literature. In what follows some of them will be discussed. First, let us recall several assumptions usually made in the creep mechanics [16, 17].

The assumption of infinitesimal strains allows neglecting the difference between the true stresses and strains and the engineering stresses and strains. Creep equations in the geometrical nonlinear case (finite strains) are discussed in the monograph [18], for example. Finite strain equations based on rheological models are presented in the monographs [19, 20]. The linearized equations of creep continuum mechanics can be used in the majority of engineering applications because structures are usually designed such that the displacements and strains arising as a consequence of the applied loading do not exceed the prescribed small values. The exception is the case of thin-walled shells, where geometrical nonlinearities must be considered even if strains are infinitesimal.

Consider general uniaxial creep constitutive model (see Chap. 4). The primary variables associated with creep deformation and creep rate are stress, temperature, and time. Much of the early work characterizing creep behavior has been aimed at fitting empirical equations to a general form where it has been typically assumed that the functions for stress, temperature, and time are separable into a product as $\sigma = f(T)\ J(t,\ t_1)$. For variable stress and temperature loading conditions for rate formulation must be adopted. The creep strain rate equation is often cast in either a time hardening, or strain hardening formulation.

One of the most critical factors determining the structural integrity of elevated temperature components is their creep behavior. Creep at high temperature can lead to micro-cracking and ultimate fracture and, therefore, is one of the main mechanisms that limits the component life.

Creep properties are generally determined by means of a test in which a constant load or stress is applied to a specimen and the resulting strain is recorded as a function of time. After the initial instantaneous strain, a decelerating strain rate stage (transient primary creep) leads to a steady minimum creep rate (MCR), which is finally followed by an accelerating stage (tertiary creep) that ends to fracture at a rupture time. During primary creep, the decreasing slope of the creep curve is attributed to strain hardening. Secondary stage creep is explained in terms of a balance between strain hardening, softening, and damage processes, resulting in a nearly constant creep rate. The tertiary stage is attributed to the appearance of internal or external damage processes coupled with softening processes, resulting in a decrease in the resistance to load or a significant increase in the net sectional stress.

The rupture life, t_R, at a given temperature and stress level is generally obtained when it is necessary to evaluate the response of a material for using in short-life situations, such as for rocket engine (t_R = 100 s) or a turbine blade in a military aircraft engine (t_R = 100 h). In such short-life situations, the major question is whether the component will or will not fail, rather than how much it will deform. As a result, the details of the creep-time curve are not of central importance to the engineering problem.

Over the past several decades, a range of methods have been developed for the prediction and evaluation of creep resistance. The most widely used methods include constitutive equations and parametric correlations [21–26].

In this chapter, variations of modulus of elasticity, stress hardening exponent with temperature are assumed to be known and based on data from standard tensile tests. In addition, standard tensile test results at time variable high temperatures should be analyzed, since creep plays an important role in the determination of the stress–strain response at high temperatures. The stress–strain curve for the material under investigation at a given temperature–time curve and design fire scenario, is then predicted using a general rheological model which combines time-independent material properties parameters (MPP); temperature-dependent modulus of elasticity and time-dependent creep data. The predicted stress–strain curves should be then compared with the experimental curves.

Research into relaxation of residual stresses subject to mechanical loading has followed a much different path than thermal relaxation. For thermal relaxation, temperature and exposure time are the primary parameters, while for mechanical loading the important factors are temperature, maximum and minimum applied stress, loading frequency, hold time, and number of applied cycles. The thermo-mechanical relaxation can be described as a two-stage process, in which the first stage is a shakedown of the initial residual stresses described by a monotonic stress–strain law, and a second stage that is slower and is described by a recrystallization process related to the continued temperature decrease (see Fig. 5.1) and the tendency for the material to return to its original shape when unloaded.

Over time, the stress and temperature decrease, while the residual stress remains constant.

The stress–strain response of a material is often described using a differential equation of the form (temperature-dependent modulus of elasticity is assumed to be [11]): $E = E_0[\exp(-0.15\theta)]$.

The differential equations describing creep constitutive law is as follows (see Chap. 4).

5.2.1 Creep Constitutive Equation. Equivalent ODE Method

$$\begin{aligned}
\bar{\sigma}(\theta) &= E(\theta)\varepsilon(\theta) = \sigma(\theta) + \sum_{i=1}^{N}\int_{0}^{\theta} a_i[\varphi(\theta)]b_i[\varphi(\tau)]\varphi'(\tau)\sigma^n(\tau)d\tau \\
\sigma'(\theta) &= -\sigma(\theta)\varphi'(\theta) + \theta e^{-0.15\theta}[1-0.15\theta] - a[\varphi(\theta)]b[\varphi(\theta)]\varphi'(\theta)\sigma^n(\theta) + \varphi'(\theta)\theta e^{-0.15\theta} \\
b_1[\varphi(\tau)] &= (\exp(t)/(1+0.1t))[((\exp(\alpha_1 m)))]m1 \\
a_1[\varphi(\tau)] &= [(\exp(-\alpha_1 m))]m1 \\
m &= \varphi(\theta) = 0.00245 + 0.0375 * \theta - 0.00934 * \theta\hat{}2 + 0.00104 * \theta\hat{}3 - 0.000041 * \theta\hat{}4 \\
m1 &= \varphi'(\theta) = 0.0375 - 0.01868 * \theta + 0.00312\theta\hat{}2 - 0.000164 * \theta\hat{}3 \\
\sigma(\theta = 0) &= 0 \\
\sigma'(\theta) &= -\sigma(\theta)\alpha m1 + e^{-0.15\theta}[1-0.15\theta] - (\exp(t)/(1+0.1t))m1\sigma^n(\theta) + m1\alpha\theta e^{-0.15\theta}
\end{aligned} \tag{5.12}$$

Consider now general uniaxial creep constitutive model for monotonically increasing part of real fire temperature load (equivalent differential equation with one MPP $\alpha = 0.33$):

$$\begin{aligned}
d(\sigma)/d(\theta) &= -0.33 * \sigma * m1 + (\exp(-0.15 * \theta)) * (1 - 0.15 * \theta) \\
&\quad - (\exp(\theta/(1.0 + 0.1 * \theta))) * m1 * \sigma^n + 0.33 * m1 * \theta * (\exp(-0.15 * \theta))
\end{aligned} \tag{5.13}$$

The temperature range is $[0 < \theta < 10]$ and the dimensionless time duration for increasing part of fully developed fire stage is $[0 < \tau < 0.0744]$ (see Fig. 5.1). It is interesting to note that the formula (5.13) of the creep retardation contains only the first derivative of the function $m1$, not the function m itself.

The creep differential equation in case of the temperature decrease (decay fire stage), as mentioned above, can be easily obtained from the above Eq. (5.13) by changing the temperature range to $[10 > \theta > 4.71]$ (see Fig. 5.1); the time interval to [0.0744 < τ < 0.2]; the initial conditions to σ (θ = 10) from loading stage and function $m1$ to the corresponding function $m01$. The differential equation of creep in this case takes the form:

$$\begin{aligned} \mathrm{d}(\sigma)/\mathrm{d}\sigma(\theta) = & -0.33\sigma(-m01) + (\exp(-0.15(\theta)))[1 - 0.15(\theta)] \\ & - (\exp((\theta)/(1.0 + 0.1(\theta))))(-m01)\sigma\hat{}n + 0.33(-m01)(\theta)(\exp(-0.15(\theta))) \end{aligned} \tag{5.14}$$

A typical computational problem from mathematical modeling of a real fire for this category of fire exposure (loading and unloading) is as follows (the computer code and the numerical solutions of Eqs. (5.13) and (5.14) using the simple mathematical software POLYMATH):

$$\begin{aligned} \bar{\sigma}(\theta) &= E(\theta)\varepsilon(\theta) = \sigma(\theta) + \sum_{i=1}^{N} \int_0^{\theta} a_i[\varphi(\theta)]b_i[\varphi(\tau)]\varphi'(\tau)\sigma^n(\tau)\mathrm{d}\tau \\ \sigma'(\theta) &= -\sigma(\theta)\varphi'(\theta) + \theta e^{-0.15\theta}[1 - 0.15\theta] - a[\varphi(\theta)]b[\varphi(\theta)]\varphi'(\theta)\sigma^n(\theta) + \varphi'(\theta)\theta e^{-0.15\theta} \\ b_1[\varphi(\tau)] &= (\exp(t)/(1 + 0.1t))[((\exp(\alpha_1 m)))]m1 \\ a_1[\varphi(\tau)] &= [(\exp(-\alpha_1 m))]m1 \\ m &= \varphi(\theta) = 0.00245 + 0.0375 * \theta - 0.00934 * \theta\hat{}2 + 0.00104 * \theta\hat{}3 - 0.000041 * \theta\hat{}4 \\ m1 &= \varphi'(\theta) = 0.0375 - 0.01868 * \theta + 0.00312\theta\hat{}2 - 0.000164 * \theta\hat{}3 \\ \sigma(\theta = 0) &= 0 \\ \sigma'(\theta) &= -\sigma(\theta)\alpha m1 + e^{-0.15\theta}[1 - 0.15\theta] - (\exp(\theta)/(1 + 0.1\theta))m1\sigma^n(\theta) + m1\alpha\theta e^{-0.15\theta} \end{aligned} \tag{5.15}$$

Differential equations

Loading

$$\begin{aligned} \mathrm{d}(y)/\mathrm{d}(t) = & -0.33 * y * m1 + (\exp(-0.15 * t)) * (1 - 0.15 * t) \\ & - (\exp(t/(1.0 + 0.1 * t))) * m1 * y\hat{}n + 0.33 * m1 * t * (\exp(-0.15 * t)) \end{aligned}$$

Explicit equations

$$\begin{aligned} n &= 2 \\ m1 &= \tau'(\theta) = 0.0375 - 0.01868 * t + 0.00312 * t\hat{}2 - 0.000164 * t\hat{}2 \end{aligned} \tag{5.16}$$

Solution for loading stage:

Differential equation

1 $\mathrm{d}(\sigma)/\mathrm{d}(\theta) = -0.33 * \sigma * m1 + (\exp(-0.15 * \theta)) * (1 - 0.15 * \theta)$
$- (\exp(\theta/(1.0 + 0.1 * \theta))) * m1 * \sigma\hat{}n + 0.33 * m1 * \theta * (\exp(-0.15 * \theta))$

Explicit equations

1 $n = 2$

2 $m1 = \tau = 0.0375 - 0.01868 * \theta + 0.00312 * \theta\hat{}2 - 0.000164 * \theta\hat{}3$

Initial condition for loading part obviously is $\sigma(0) = y(0) = 0$. Initial condition for unloading part will be the final stress value $y(f)$ from the solution of loading part. Therefore, first we have to solve the creep constitutive equation for loading stage and then for unloading.

Solution of Eq. (5.15) is:

Calculated values of DEQ variables

	Variable	Initial value	Minimal value	Maximal value	Final value
2	$m1$	0.0375	−0.0013	0.0375	−0.0013
3	n	2	2	2	2
4	θ	0	0	10	10
5	σ	0	0	1.867	0.5350

Finally, the stress–strain diagram (temperature increase stage of fully developed fire) is presented in Fig. 5.8.

The analytical expression is given below by nonlinear polynomial approximation:

$$\textbf{Model}: \sigma = 0.1 + 0.78 * \theta - 0.06 * \theta\hat{}2 - 0.0089 * \theta\hat{}3 + 0.000755 * \theta\hat{}4 \quad (5.17)$$

Variable	Value
$a0$	0.0996562
$a1$	0.7801044
$a2$	−0.0600009
$a3$	−0.0089423
$a4$	0.0007548

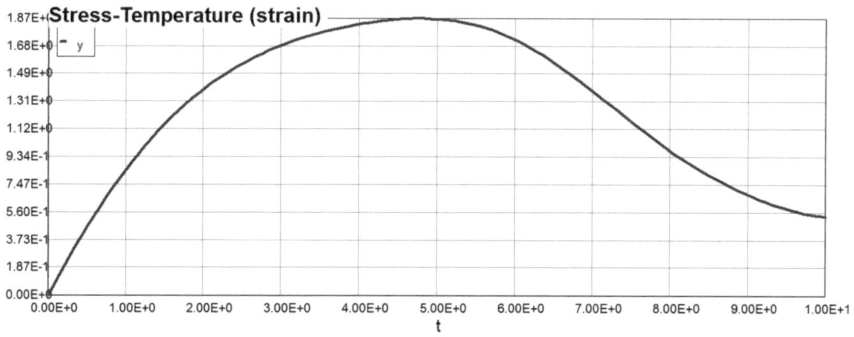

Fig. 5.8 Stress–temperature–strain diagram. Very fast fire (loading stage)

Relaxation Modulus

Relaxation of residual stresses subject to thermal loading condition can be described by the creep differential equation as follows:

Unloading (recovery process)

Differential equation

$$\begin{aligned} d(y)/d(t) = & -0.33 * y * (-m01) + (\exp(-0.15 * (t))) * (1 - 0.15 * (t)) \\ & - (\exp((t)/(1.0 + 0.1 * (t)))) * (-m01) * y^{\wedge}n \\ & + 0.33 * (-m01) * (t) * (\exp(-0.15 * (t))) \end{aligned}$$

Explicit equations

$n = 2$; $10 > \theta > 4.71$; $0.0744 < \tau < 0.2$

$$m01 = \tau'(\theta) = -0.668 + 0.228 * \theta - 0.0266\,\theta^{\wedge}2 - 0.00105 * \theta^{\wedge}3 \tag{5.18}$$

$y(0) = \sigma(\theta = 10)_{\text{loading}} = 0.535$

Solution for unloading stage is:

Calculated values of DEQ variables

	Variable	Initial value	Minimal value	Maximal value	Final value
2	$m01$	−0.2939285	−2.098	−0.2939285	−2.098
4	n	2	2	2	2
5	t	4.71	4.71	10	10
7	y	0.535	0.0669744	0.535	0.0669744

Differential equations

$$\begin{aligned} 1 \quad d(y)/d(t) = & 1 * (-0.33 * y * (-m01) + (\exp(-0.15 * (t))) * (1 - 0.15 * (t)) \\ & - (\exp((t)/(1.0 + 0.1 * (t)))) * (-m01) * y^{\wedge}n \\ & + 0.33 * (-m01) * (t) * (\exp(-0.15 * (t)))) \end{aligned}$$

Explicit equations

1 $n = 2$

2 $m01 = -0.668 + 0.228 * t - 0.0266 * t^{\wedge}2 - 0.00105 * t^{\wedge}3$

See Fig. 5.9.

The analytical expression for relaxation modulus is given below by nonlinear approximation:

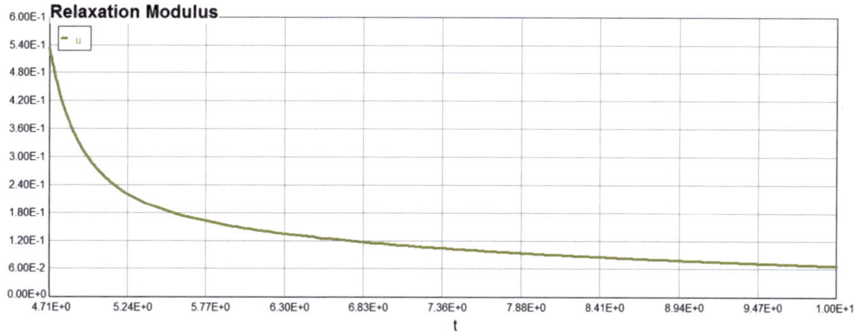

Fig. 5.9 Relaxation modulus. Unloading stage

Model: $y = a0 + a1 * t + a2 * t\char`^2 + a3 * t\char`^3 + a4 * t\char`^4$

Variable	Value
$a0$	18.57414
$a1$	−9.690931
$a2$	1.891751
$a3$	−0.1625901
$a4$	0.0051838

$$\textbf{Model}: \sigma = 18.57 - 9.69 * \theta + 1.892 * \theta^\wedge 2 - 0.162 * \theta^\wedge 3 + 0.00518 * \theta^\wedge 4 \tag{5.19}$$

The combined graph of both processes (loading and unloading) is presented in Fig. 5.10.

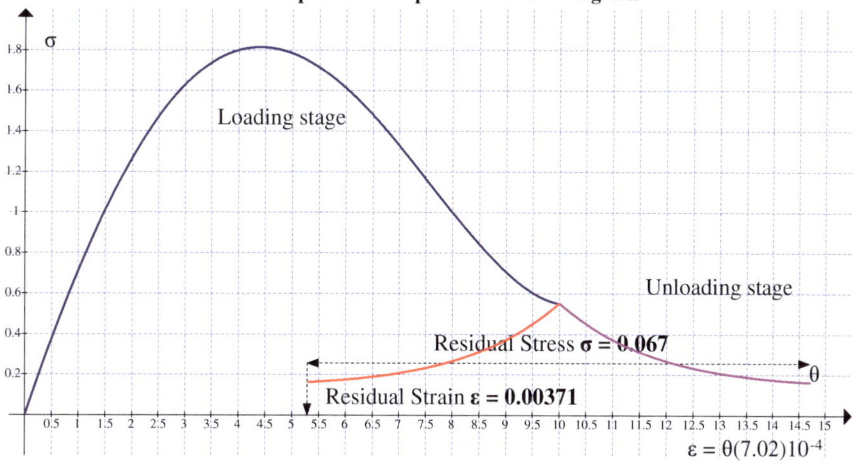

Fig. 5.10 Creep stress–temperature–strain diagram

The maximum design temperature of the stress–strain diagram presented above is $\theta = 10$ ($T = 1200$ °K). We had analyzed (see Chap. 4) some creep damage equations that are connecting the rapture life (time to rapture) and temperature (or creep rate). Several creep parameters are utilized to correlate creep data to be presented in a format that gives more meaning in design. The most common and universally recognized parametric model developed is the Larson–Miller method [21, 22] Larson and Miller developed the time–temperature relationship for prediction rupture and creep stresses. This critical level of damage is manifested as the failure which can be predicted from the minimum strain rate and the time to failure that by measuring the minimum strain at a given stress and temperature or obtaining the minimum strain rate for the given temperature and stress. Graphically, it can be illustrated as follows (see Fig. 5.11).

Example 5.1 Assuming (conservatively) that the fracture point is located at the maximum stress σ_{max} (i.e., allowable design temperature in this case is equal to $\theta = 5$ or $T = 900$ °K). Obtain the minimum strain and maximum allowable stress. The process of thermal loading can be described as follows:

Loading case: **$0 < \theta < 5$; $0 < \tau < 0.0375$**

Calculated values of DEQ variables

	Variable	Initial value	Minimal value	Maximal value	Final value
1	$m1$	0.0375	0.00158	0.0375	0.0016
2	n	2	2	2	2
3	θ	0	0	5	5
4	σ	0	0	1.867	1.863

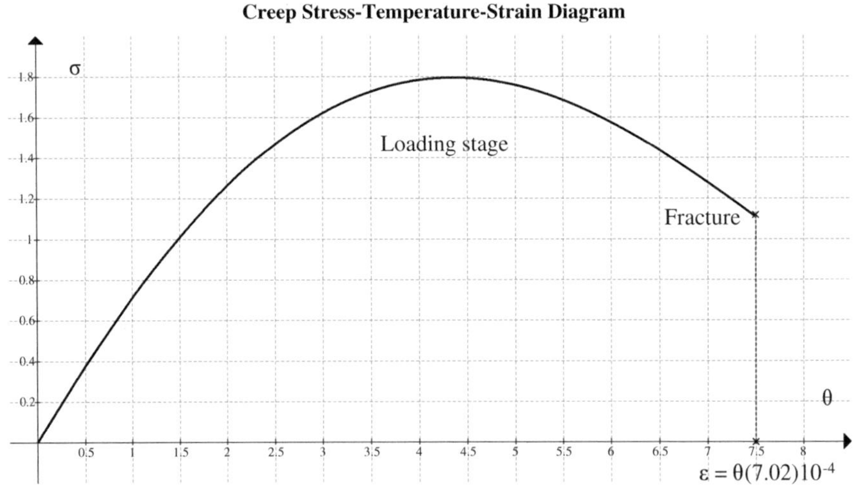

Fig. 5.11 Creep–stress–temperature–strain diagram and fracture point

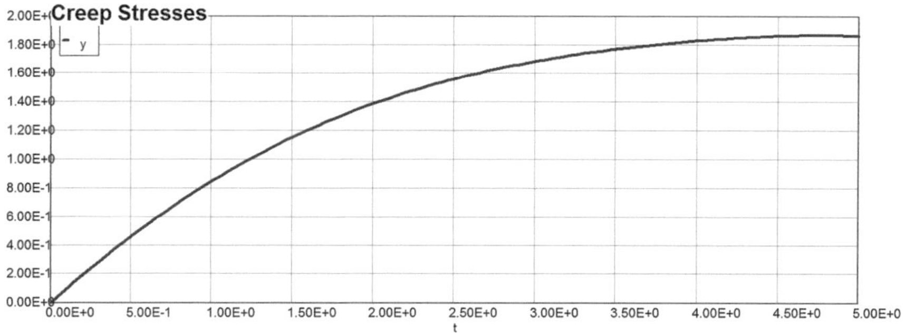

Model: $\sigma = 1.05 * \theta - 0.208 * \theta$^2 + 0.0168 * θ^3 − 0.00044 * θ^4

Variable	Value
$a0$	0
$a1$	1.050542
$a2$	−0.208
$a3$	0.0168
$a4$	−0.00044

1. Find the strain first: $\varepsilon = 7.02\ (10^{-4})\theta = 7.02\ (10^{-4})5 = 0.00351$
2. Obtain the allowable stress value: σ = 1.05 * θ − 0.208 * θ^2 + 0.0168 * θ^3 − 0.00044 * θ^4 = 1.05 * (5) − 0.208 * 5^2 + 0.0168 * 5^3 − 0.00044 * 5^4 = 1.875 [σ(7.02)2.9 = 38.2 ksi = 263.4 MPa]

5.2.2 *The Functional Dependencies of Creep Stresses and Strains from Material Properties Parameters (MPP)*

Parameters characterizing the mechanical response in the primary and secondary regions of creep deformations demonstrate pronounced changes at some high temperatures of fully developed fire scenarios (which imply that the creep processes at low and high values of dimensionless MPP parameters α_i are to be distinguished). It has been assumed (in all analysis above) that the material parameter $\alpha = 0.33$ and for practical purposes we have limited to the range of these α_i from the interval $[10^{-3} < \alpha_i < 10^5]$. We will consider now just the temperature increasing part (loading stage) of creep deformation process. Since all computations are similar for different α in the case of $\alpha = 0.33$, the intermediate computations are omitted and the final results are presented in tabular form. The main goal of this section is to create a statistical data base, since α_i will be called a random parameter, for the probability-based method of creep analysis.

Consider now Eq. (5.12) and $\alpha = 0.001$. Again, in case of VFF we have

$$\begin{aligned} d(y)/d(t) = &-\alpha_i * y * m1 + (\exp(-0.15 * t)) * (1 - 0.15 * t) \\ &- (\exp(t/(1.0 + 0.1 * t))) * m1 * y\hat{}n + \alpha_i * m1 * t * (\exp(-0.15 * t)) \end{aligned}$$

Solution for loading stage:
α = 0.001

Differential equation

$$\begin{aligned} 1 \quad d(\sigma)/d(\theta) = &-0.001 * \sigma * m1 + (\exp(-0.15 * \theta)) * (1 - 0.15 * \theta) \\ &- (\exp(\theta/(1.0 + 0.1 * \theta))) * m1 * \sigma\hat{}n + 0.001 * m1 * \theta * (\exp(-0.15 * \theta)) \end{aligned}$$

Calculated values of DEQ variables

	Variable	Initial value	Minimal value	Maximal value	Final value
1	*m*1	0.0375	0.0079932	0.1463	0.1463
2	*n*	3	3	3	3
3	*r*	0	0	2.45251	2.231302
4	*t*	0	0	10	10
5	*y*	0	0	1.572992	0.0009539

Differential equations

$$\begin{aligned} 1 \quad d(y)/d(t) = &-0.001 * y * m1 + (\exp(-0.15 * t)) * (1 - 0.15 * t) \\ &- (\exp(t/(1.0 + 0.1 * t))) * m1 * y\hat{}n + 0.001 * m1 * t * (\exp(-0.15 * t)) \end{aligned}$$

Explicit equations

1 $n = 3$

2 $m1 = 0.0375 - 0.01868 * t + 0.00312 * t\hat{}2 - 0.000164 * t\hat{}2$

3 $r = t * (\exp(-0.15 * t))$

The upper curve in Fig. 5.12 represents the stress–temperature function when the creep function in Eq. (5.12) is identically equal to zero, i.e., there is no combined effect of creep and temperature, but the deformation occurs just due to temperature and modulus of elasticity changes. The lower curve in Fig. 5.12 represents the combined effect of engineering creep, modulus of elasticity, and continuously changed

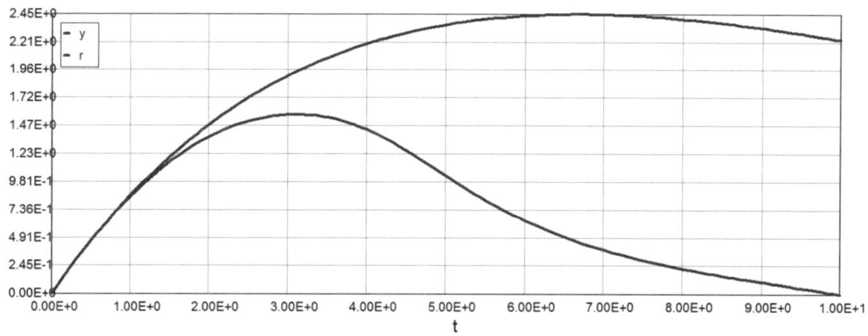

Fig. 5.12 Stress–temperature diagrams

temperature on stress and deformation. The analytical expression for the stress–temperature function (lower curve) is given below by nonlinear polynomial approximation:

Model: $y = a0 + a1 * \mathrm{t} + a2 * t\hat{}2 + a3 * t\hat{}3 + a4 * t\hat{}4$

Variable	Value
$a0$	0
$a1$	1.428
$a2$	−0.3811
$a3$	0.0337
$a4$	−0.00097

$$\textbf{Model}: \sigma = 1.428 * \theta - 0.381 * \theta\hat{}2 + 0.0337 * \theta\hat{}3 - 0.00097 * \theta\hat{}4 \tag{5.20}$$

Similar computations are provided below for the following MPP values: **α = 0.01; 0.1; 0.2; 0.33; 0.5; 1.0; 2.0; 3.0; 5.0; 10; 10^2; 10^3; 10^4; 10^5**. The results are summarized in the tabular form (see Table 5.2) as well as in analytical approximations. The creep stress exponent is assumed to be $n = 3$ for these types of analyses.

α = 0.01

Calculated values of DEQ variables

	Variable	Initial value	Minimal value	Maximal value	Final value
1	$m1$	0.0375	0.0079932	0.1463	0.1463
2	n	3	3	3	3
3	t	0	0	10	10
4	y	0	0	1.573011	0.0049354

Differential equations

1 $d(y)/d(t) = -0.01 * y * m1 + (\exp(-0.15 * t)) * (1 - 0.15 * t) - (\exp(t/(1.0 + 0.1 * t))) * m1 * y\wedge n + 0.01 * m1 * t * (\exp(-0.15 * t))$

Explicit equations

1 $n = 3$

2 $m1 = 0.0375 - 0.01868 * t + 0.00312 * t\wedge 2 - 0.000164 * t^2$

See Fig. 5.13.

Model: $y = a0 + a1 * t + a2 * t\wedge 2 + a3 * t\wedge 3 + a4 * t\wedge 4$

Variable	Value
$a0$	0.1
$a1$	1.428
$a2$	−0.381
$a3$	0.0337
$a4$	−0.00097

$$\textbf{Model}: \sigma = 1.428 * \theta - 0.381 * \theta\wedge 2 + 0.0337 * \theta\wedge 3 - 0.00097 * \theta\wedge 4 \quad (5.21)$$

α = 0.1

Calculated values of DEQ variables

	Variable	Initial value	Minimal value	Maximal value	Final value
1	$m1$	0.0375	0.0079932	0.1463	0.1463
2	n	3	3	3	3
3	t	0	0	10	10
4	y	0	0	1.573204	0.0417526

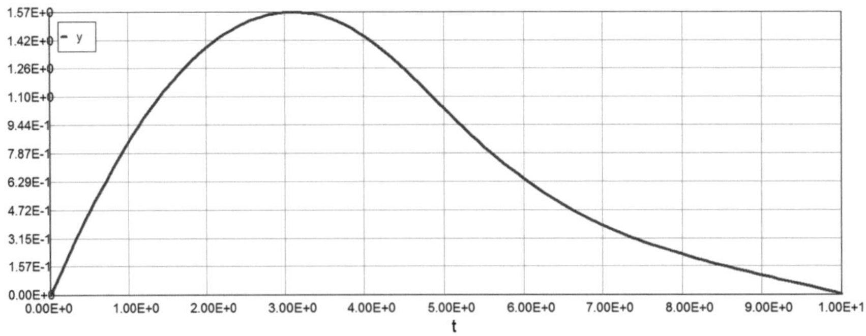

Fig. 5.13 Stress–temperature diagram α = 0.01

Differential equations

1 $d(y)/d(t) = -0.1 * y * m1 + (\exp(-0.15 * t)) * (1 - 0.15 * t) - (\exp(t/(1.0 + 0.1 * t))) * m1 * y^{\wedge}n + 0.1 * m1 * t * (\exp(-0.15 * t))$

Explicit equations

1 $n = 3$

2 $m1 = 0.0375 - 0.01868 * t + 0.00312 * t^{\wedge}2 - 0.000164 * t^{\wedge}2$

See Fig. 5.14.

Model: $y = a0 + a1 * t + a2 * t^{\wedge}2 + a3 * t^{\wedge}3 + a4 * t^{\wedge}4$

Variable	Value
$a0$	−0.166
$a1$	1.425
$a2$	−0.38
$a3$	0.0335
$a4$	−0.001

$$\textbf{Model}: \sigma = 1.425 * \theta - 0.38 * \theta^{\wedge}2 + 0.0335 * \theta^{\wedge}3 - 0.001 * \theta^{\wedge}4 \quad (5.22)$$

α = 0.2

Calculated values of DEQ variables

	Variable	Initial value	Minimal value	Maximal value	Final value
1	$m1$	0.0375	0.0079932	0.1463	0.1463
2	n	3	3	3	3
3	t	0	0	10	10
4	y	0	0	1.573417	0.0761399

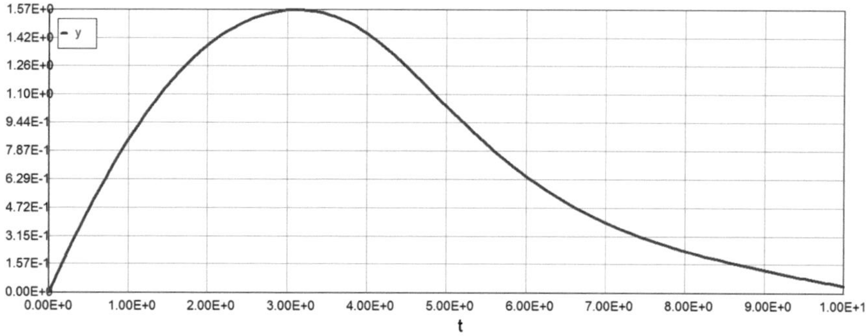

Fig. 5.14 Stress–temperature diagram α = 0.1

Differential equations

1 $d(y)/d(t) = -0.2 * y * m1 + (\exp(-0.15 * t)) * (1 - 0.15 * t) - (\exp(t/(1.0 + 0.1 * t))) * m1 * y^\wedge n + 0.2 * m1 * t * (\exp(-0.15 * t))$

Explicit equations

1 $n = 3$

2 $m1 = 0.0375 - 0.01868 * t + 0.00312 * t^\wedge 2 - 0.000164 * t^\wedge 2$

See Fig. 5.15.

Model: $y = a0 + a1 * t + a2 * t^\wedge 2 + a3 * t^\wedge 3 + a4 * t^\wedge 4$

Variable	Value
$a0$	−0.1655119
$a1$	1.423881
$a2$	−0.3787287
$a3$	0.0332697
$a4$	−0.0009323

Model: $\sigma = 0 + 1.424 * \theta - 0.379 * \theta^\wedge 2 + 0.0332 * \theta^\wedge 3 - 0.00093 * \theta^\wedge 4$ (5.23)

α = 0.33

Calculated values of DEQ variables

	Variable	Initial value	Minimal value	Maximal value	Final value
1	$m1$	0.0375	0.0079932	0.1463	0.1463
2	n	3	3	3	3
3	t	0	0	10	10
4	y	0	0	1.573695	0.1118357

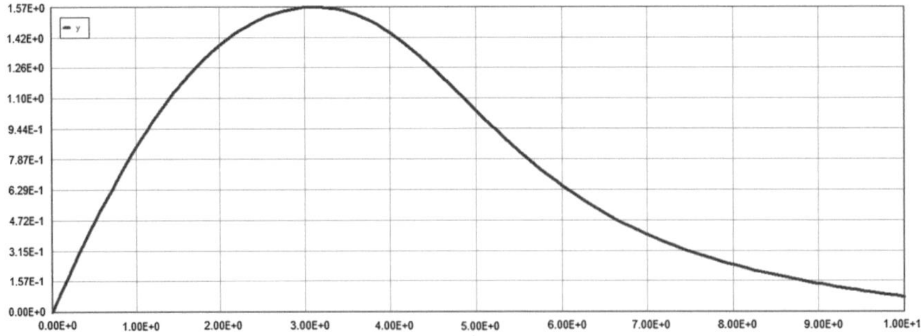

Fig. 5.15 Stress–temperature diagram α = 0.2

Differential equations

1 $d(y)/d(t) = -0.33 * y * m1 + (\exp(-0.15 * t)) * (1 - 0.15 * t) - (\exp(t/(1.0 + 0.1 * t))) * m1 * y^\wedge n + 0.33 * m1 * t * (\exp(-0.15 * t))$

Explicit equations

1 $n = 3$

2 $m1 = 0.0375 - 0.01868 * t + 0.00312 * t^\wedge 2 - 0.000164 * t^\wedge 2$

See Fig. 5.16.

Model: $y = a0 + a1 * t + a2 * t^\wedge 2 + a3 * t^\wedge 3 + a4 * t^\wedge 4$

Variable	Value
$a0$	−0.1653686
$a1$	1.423448
$a2$	−0.3783983
$a3$	0.0331841
$a4$	−0.000923

$$\textbf{Model}: \sigma = 1.423 * \theta - 0.378 * \theta^\wedge 2 + 0.0332 * \theta^\wedge 3 - 0.00093 * \theta^\wedge 4 \quad (5.24)$$

α = 0.5

Calculated values of DEQ variables

	Variable	Initial value	Minimal value	Maximal value	Final value
1	$m1$	0.0375	0.007989	0.1463	0.1463
2	n	3	3	3	3
3	t	0	0	10	10
4	y	0	0	1.57388	0.1471535

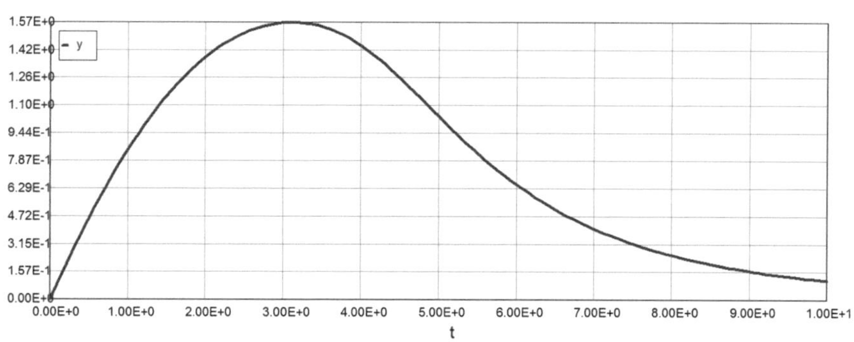

Fig. 5.16 Stress–temperature diagram α = 0.33

Differential equations

1 $d(y)/d(t) = -0.5 * y * m1 + (\exp(-0.15 * t)) * (1 - 0.15 * t) - (\exp(t/(1.0 + 0.1 * t))) * m1 * y^{\wedge}n + 0.5 * m1 * t * (\exp(-0.15 * t))$

Explicit equations

1 $n = 3$

2 $m1 = 0.0375 - 0.01868 * t + 0.00312 * t^{\wedge}2 - 0.000164 * t^{\wedge}2$

See Fig. 5.17.

Model: $y = a0 + a1 * t + a2 * t^{\wedge}2 + a3 * t^{\wedge}3 + a4 * t^{\wedge}4$

Variable	Value
$a0$	−0.1708483
$a1$	1.430736
$a2$	−0.3811908
$a3$	0.0335784
$a4$	−0.0009379

Model: $\sigma = 0 + 1.431 * \theta - 0.381 * \theta^{\wedge}2 + 0.0336 * \theta^{\wedge}3 - 0.00094 * \theta^{\wedge}4$ (5.25)

α = 1.0

Calculated values of DEQ variables

	Variable	Initial value	Minimal value	Maximal value	Final value
1	$m1$	0.0375	0.0079932	0.1463	0.1463
2	n	3	3	3	3
3	t	0	0	10	10
4	y	0	0	1.575117	0.2121058

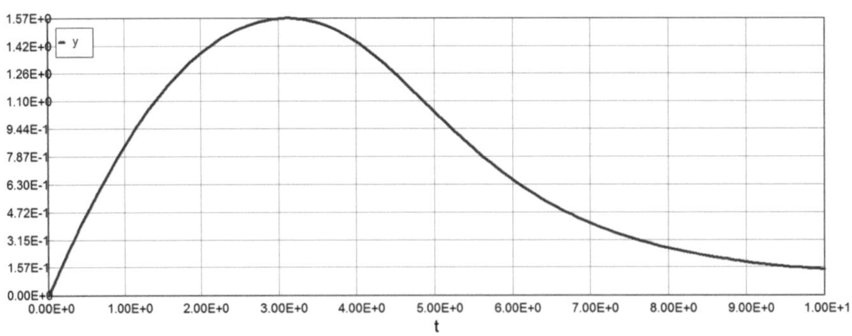

Fig. 5.17 Stress–temperature diagram $\alpha = 0.5$

Differential equations

1 $d(y)/d(t) = -1 * y * m1 + (\exp(-0.15 * t)) * (1 - 0.15 * t) - (\exp(t/(1.0 + 0.1 * t))) * m1 * y^{\wedge}n + 1 * m1 * t * (\exp(-0.15 * t))$

Explicit equations

1 $n = 3$

2 $m1 = 0.0375 - 0.01868 * t + 0.00312 * t^{\wedge}2 - 0.000164 * t^{\wedge}2$

See Fig. 5.18.

Model: $y = a0 + a1 * t + a2 * t^{\wedge}2 + a3 * t^{\wedge}3 + a4 * t^{\wedge}4$

Variable	Value
$a0$	−0.1688143
$a1$	1.431877
$a2$	−0.3828975
$a3$	0.0339888
$a4$	−0.0009562

$$\textbf{Model}: \sigma = 1.432 * \theta - 0.383 * \theta^{\wedge}2 + 0.034 * \theta^{\wedge}3 - 0.00096 * \theta^{\wedge}4 \qquad (5.26)$$

α = 2.0

Calculated values of DEQ variables

	Variable	Initial value	Minimal value	Maximal value	Final value
1	$m1$	0.0375	0.0079888	0.1463	0.1463
2	n	3	3	3	3
3	t	0	0	10	10
4	y	0	0	1.577074	0.2820312

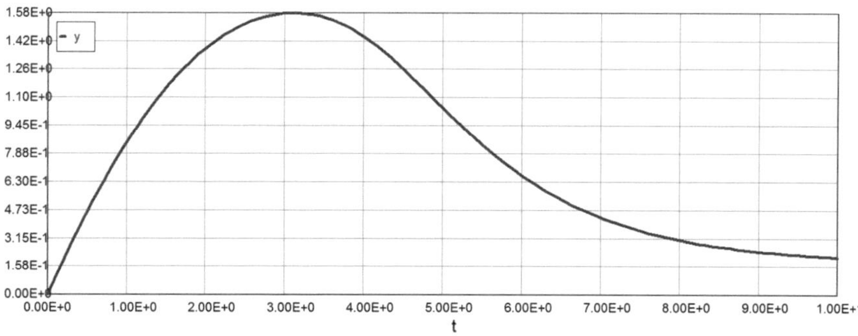

Fig. 5.18 Stress–temperature diagram α = 1.0

Differential equations

1 $d(y)/d(t) = -2 * y * m1 + (\exp(-0.15 * t)) * (1 - 0.15 * t) - (\exp(t/(1.0 + 0.1 * t))) * m1 * y^\wedge n + 2 * m1 * t * (\exp(-0.15 * t))$

Explicit equations

1 $n = 3$

2 $m1 = 0.0375 - 0.01868 * t + 0.00312 * t^\wedge 2 - 0.000164 * t^\wedge 2$

See Fig. 5.19.

Model: $y = a0 + a1 * t + a2 * t\wedge 2 + a3 * t\wedge 3 + a4 * t\wedge 4$

Variable	Value
$a0$	−0.1800632
$a1$	1.452742
$a2$	−0.3934394
$a3$	0.0359411
$a4$	−0.0010591

Model: $\sigma = 0 + 1.453 * \theta - 0.393 * \theta^\wedge 2 + 0.0359 * \theta^\wedge 3 - 0.00106 * \theta^\wedge 4$ (5.27)

α = 3.0

Calculated values of DEQ variables

	Variable	Initial value	Minimal value	Maximal value	Final value
1	$m1$	0.0375	0.0079932	0.1463	0.1463
2	n	3	3	3	3
3	t	0	0	10	10
4	y	0	0	1.579293	0.3266515

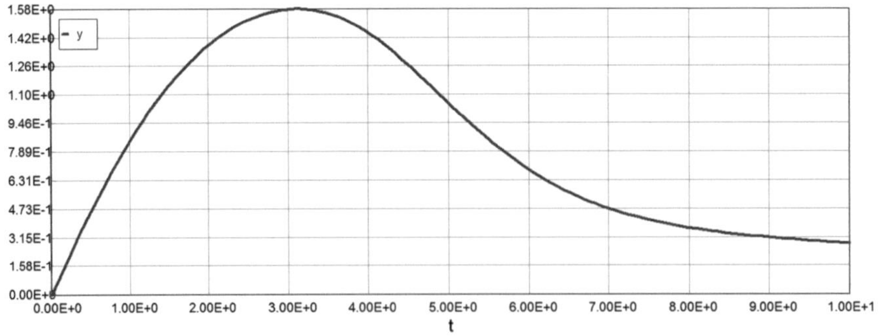

Fig. 5.19 Stress–temperature diagram α = 2.0

Differential equations

1 $d(y)/d(t) = -3 * y * m1 + (\exp(-0.15 * t)) * (1 - 0.15 * t) - (\exp(t/(1.0 + 0.1 * t))) * m1 * y^\wedge n + 3 * m1 * t * (\exp(-0.15 * t))$

Explicit equations

1 $n = 3$

2 $m1 = 0.0375 - 0.01868 * t + 0.00312 * t^\wedge 2 - 0.000164 * t^\wedge 2$

See Fig. 5.20.

Model: $y = a0 + a1 * t + a2 * t^\wedge 2 + a3 * t^\wedge 3 + a4 * t^\wedge 4$

Variable	Value
$a0$	−0.1762943
$a1$	1.452532
$a2$	−0.3955017
$a3$	0.0366488
$a4$	−0.0011049

Model: $\sigma = 0 + 1.453 * \theta - 0.396 * \theta^\wedge 2 + 0.0366 * \theta^\wedge 3 - 0.0011 * \theta^\wedge 4$ (5.28)

α = 5.0

Calculated values of DEQ variables

	Variable	Initial value	Minimal value	Maximal value	Final value
1	$m1$	0.0375	0.0079932	0.1463	0.1463
2	n	3	3	3	3
3	t	0	0	10	10
4	y	0	0	1.583371	0.3886332

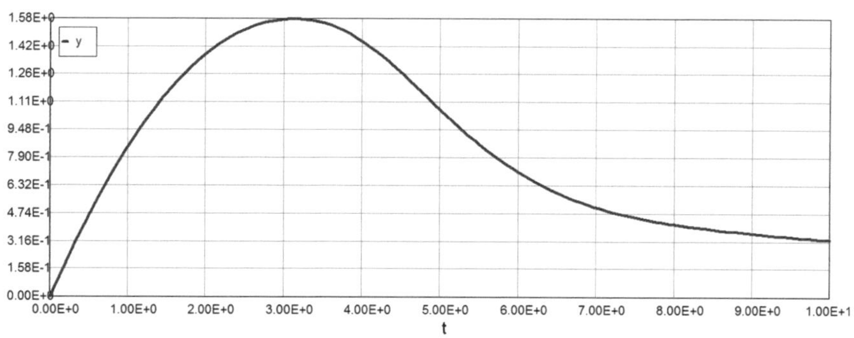

Fig. 5.20 Stress–temperature diagram α = 3.0

Differential equations

$$1 \quad d(y)/d(t) = -5 * y * m1 + (\exp(-0.15 * t)) * (1 - 0.15 * t) - (\exp(t/(1.0 + 0.1 * t))) * m1 * y^{\wedge}n + 5 * m1 * t * (\exp(-0.15 * t))$$

Explicit equations

$$1 \quad n = 3$$

$$2 \quad m1 = 0.0375 - 0.01868 * t + 0.00312 * t^{\wedge}2 - 0.000164 * t^{\wedge}2$$

See Fig. 5.21.

Model: $y = a0 + a1 * t + a2 * t^{\wedge}2 + a3 * t^{\wedge}3 + a4 * t^{\wedge}4$

Variable	Value
$a0$	−0.1774674
$a1$	1.458105
$a2$	−0.4001859
$a3$	0.0379626
$a4$	−0.0011896

$$\textbf{Model}: \sigma = 0 + 1.458 * \theta - 0.4 * \theta^{\wedge}2 + 0.038 * \theta^{\wedge}3 - 0.00119 * \theta^{\wedge}4 \quad (5.29)$$

α = 10

Calculated values of DEQ variables

	Variable	Initial value	Minimal value	Maximal value	Final value
1	$m1$	0.0375	0.0079888	0.1463	0.1463
2	n	3	3	3	3
3	t	0	0	10	10
4	y	0	0	1.593417	0.4857851

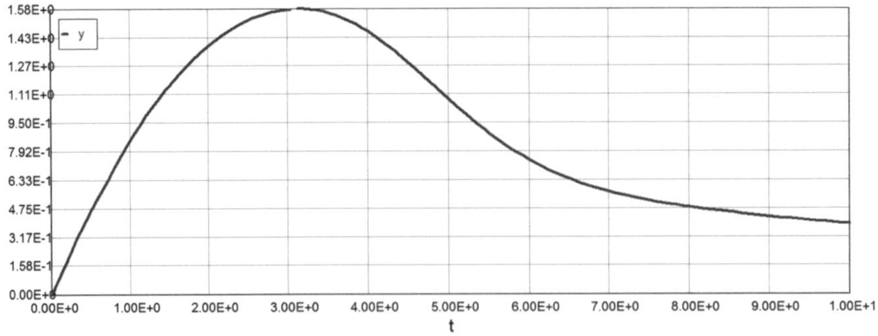

Fig. 5.21 Stress–temperature diagram α = 5

Differential equations

1 $d(y)/d(t) = -10 * y * m1 + (\exp(-0.15 * t)) * (1 - 0.15 * t) - (\exp(t/(1.0 + 0.1 * t))) * m1 * y^{\wedge}n + 10 * m1 * t * (\exp(-0.15 * t))$

Explicit equations

1 $n = 3$

2 $m1 = 0.0375 - 0.01868 * t + 0.00312 * t^{\wedge}2 - 0.000164 * t^{\wedge}2$

See Fig. 5.22.

Model: $y = a0 + a1 * t + a2 * t^2 + a3 * t^3 + a4 * t^4$

Variable	Value
$a0$	−0.1781578
$a1$	1.45727
$a2$	−0.4016061
$a3$	0.0391376
$a4$	−0.0012838

$$\textbf{Model}: \sigma = 1.457 * \theta - 0.402 * \theta^{\wedge}2 + 0.0391 * \theta^{\wedge}3 - 0.00128 * \theta^{\wedge}4 \quad (5.30)$$

α = 100

Calculated values of DEQ variables

	Variable	Initial value	Minimal value	Maximal value	Final value
1	$m1$	0.0375	0.0079895	0.1463	0.1463
2	n	3	3	3	3
3	t	0	0	10	10
4	y	0	0	1.724496	0.951412

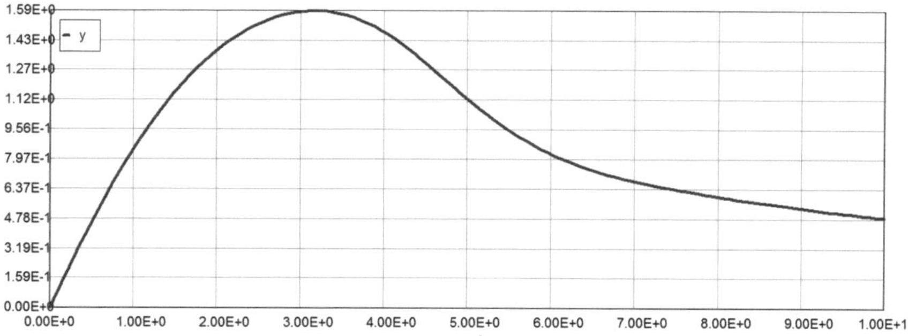

Fig. 5.22 Stress–temperature diagram α = 10

Differential equations

$$1\quad d(y)/d(t) = -100 * y * m1 + (\exp(-0.15 * t)) * (1 - 0.15 * t) - (\exp(t/(1.0 + 0.1 * t))) * m1 * y^\wedge n + 100 * m1 * t * (\exp(-0.15 * t))$$

Explicit equations

$$1\quad n = 3$$

$$2\quad m1 = 0.0375 - 0.01868 * t + 0.00312 * t^\wedge 2 - 0.000164 * t^\wedge 2$$

See Fig. 5.23.

Model: $y = a0 + a1 * t + a2 * t^\wedge 2 + a3 * t^\wedge 3 + a4 * t^\wedge 4$

Variable	Value
$a0$	0
$a1$	1.271
$a2$	−0.304
$a3$	0.0276
$a4$	−0.00088

$$\textbf{Model}: \sigma = 1.271 * \theta - 0.304 * \theta^\wedge 2 + 0.0276 * \theta^\wedge 3 - 0.00088 * \theta^\wedge 4 \quad (5.31)$$

α = 1000

Calculated values of DEQ variables

	Variable	Initial value	Minimal value	Maximal value	Final value
1	$m1$	0.0375	0.0079966	0.1463	0.1463
2	n	3	3	3	3
3	r	0	0	2.45248	2.231302
4	t	0	0	10	10
5	y	0	0	2.106468	1.610935

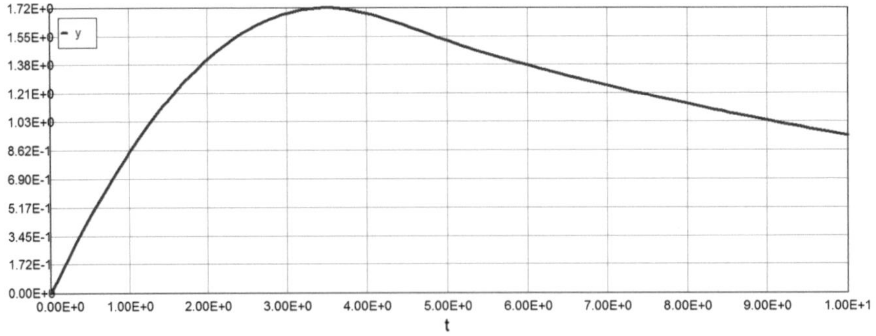

Fig. 5.23 Stress–temperature diagram α = 100

Differential equations

1 $d(y)/d(t) = -1000 * y * m1 + (\exp(-0.15 * t)) * (1 - 0.15 * t) - (\exp(t/(1.0 + 0.1 * t))) * m1 * y^\wedge n + 1000 * m1 * t * (\exp(-0.15 * t))$

Explicit equations

1 $n = 3$

2 $m1 = 0.0375 - 0.01868 * t + 0.00312 * t^\wedge 2 - 0.000164 * t^\wedge 2$

3 $r = t * (\exp(-0.15 * t))$

See Fig. 5.24.

Model: $y = a0 + a1 * t + a2 * t\wedge 2 + a3 * t\wedge 3 + a4 * t\wedge 4$

Variable	Value
$a0$	0
$a1$	1.045
$a2$	−0.169
$a3$	0.0099
$a4$	−0.000182

Model: $\sigma = 1.045 * \theta - 0.169 * \theta^\wedge 2 + 0.0099 * \theta^\wedge 3 - 0.000182 * \theta^\wedge 4$ (5.32)

α = 10,000

Calculated values of DEQ variables

	Variable	Initial value	Minimal value	Maximal value	Final value
1	$m1$	0.0375	0.0079956	0.1463	0.1463
2	n	3	3	3	3
3	r	0	0	2.452492	2.231302
4	t	0	0	10	10
5	y	0	0	2.384092	2.094868

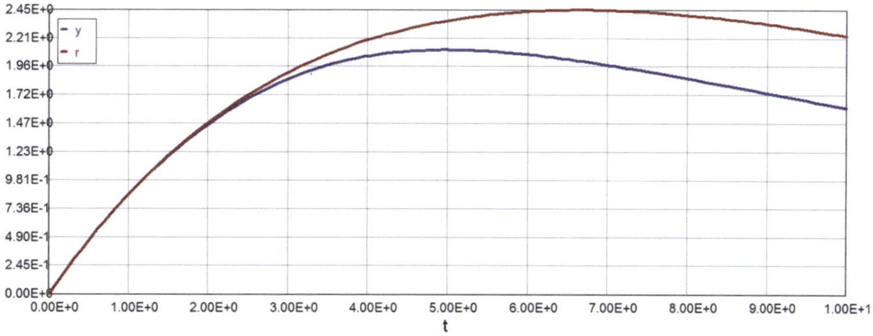

Fig. 5.24 Stress–temperature diagram α = 1000

Differential equations

1 $d(y)/d(t) = -10{,}000 * y * m1 + (\exp(-0.15 * t)) * (1 - 0.15 * t) - (\exp(t/(1.0 + 0.1 * t))) * m1 * y^{\wedge}n + 10{,}000 * m1 * t * (\exp(-0.15 * t))$

Explicit equations

1 $n = 3$
2 $m1 = 0.0375 - 0.01868 * t + 0.00312 * t^{\wedge}2 - 0.000164 * t^{\wedge}2$
3 $r = t * (\exp(-0.15 * t))$

See Fig. 5.25.
Model: $y = a0 + a1 * t + a2 * t^{\wedge}2 + a3 * t^{\wedge}3 + a4 * t^{\wedge}4$

Variable	Value
$a0$	0.0
$a1$	0.99
$a2$	−0.141
$a3$	0.00814
$a4$	−0.000184

$$\textbf{Model}: \sigma = 0.99 * \theta - 0.141 * \theta^{\wedge}2 + 0.00814 * \theta^{\wedge}3 - 0.000184 * \theta^{\wedge}4 \quad (5.33)$$

α = 100,000
Calculated values of DEQ variables

	Variable	Initial value	Minimal value	Maximal value	Final value
1	$m1$	0.0375	0.0079938	0.1463	0.1463
2	n	3	3	3	3
3	r	0	0	2.452498	2.231302
4	t	0	0	10	10
5	y	0	0	2.444621	2.215169

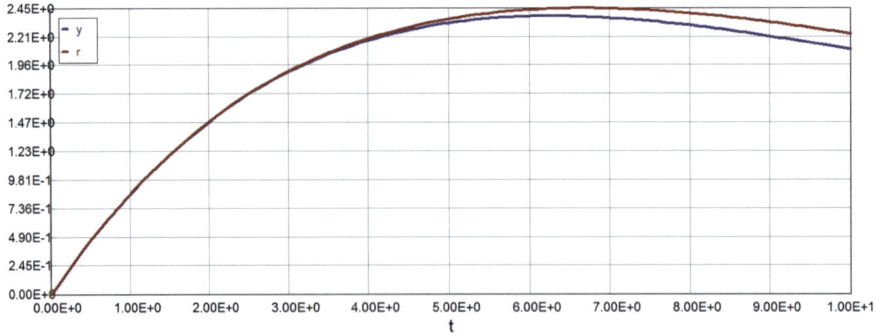

Fig. 5.25 Stress–temperature diagram $\alpha = 10{,}000$

Differential equations

1 $d(y)/d(t) = -100{,}000 * y * m1 + (\exp(-0.15 * t)) * (1 - 0.15 * t) - (\exp(t/(1.0 + 0.1 * t))) * m1 * y^\wedge n + 100{,}000 * m1 * t * (\exp(-0.15 * t))$

Explicit equations

1 $n = 3$

2 $m1 = 0.0375 - 0.01868 * t + 0.00312 * t^\wedge 2 - 0.000164 * t^\wedge 2$

3 $r = t * (\exp(-0.15 * \mathrm{t}))$

See Fig. 5.26.

Model: $y = a0 + a1 * t + a2 * t\text{^}2 + a3 * t\text{^}3 + a4 * t\text{^}4$

Variable	Value
$a0$	0.0
$a1$	0.988
$a2$	−0.141
$a3$	0.00863
$a4$	−0.000219

$$\textbf{Model}: \sigma = 0.988 * \theta - 0.141 * \theta^\wedge 2 + 0.00863 * \theta^\wedge 3 - 0.000219 * \theta^\wedge 4 \quad (5.34)$$

The results are summarized in Table 5.2.

The data in Table 5.3 and the corresponding analytical expressions of stress–strain functions (see Eqs. 5.20–5.34) allow making the following conclusions regarding the impact of material parameter α_i on creep stress diagram.

1. The range of variation of the parameter α_i, although limited, covers a fairly wide range of values, namely, from $10^{-4} < \alpha_i < 10^5$. The creep stresses are limited,

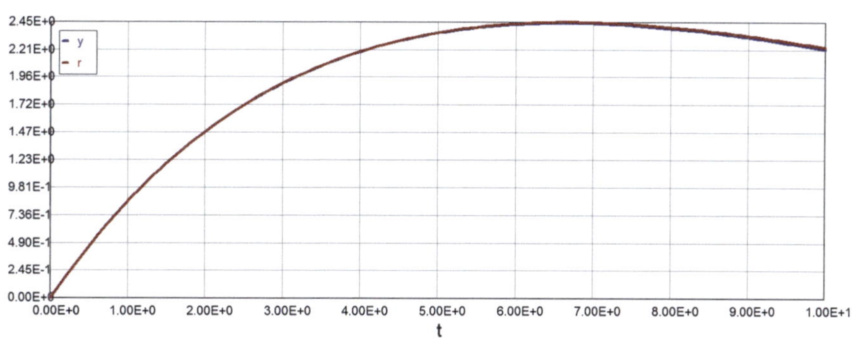

Fig. 5.26 Stress–temperature diagram $\alpha = 100{,}000$

Table 5.2 Stress–temperature data (very fast fire)

$\alpha\|\theta$		1	2	3	4	5	6	7	8	9	10
0.001	0	0.85	1.378	1.57	1.438	1.035	0.645	0.388	0.224	0.107	0.001
0.01	0	0.85	1.378	1.57	1.438	1.035	0.645	0.388	0.225	0.109	0.005
0.1	0	0.85	1.378	1.571	1.439	1.036	0.647	0.392	0.233	0.126	0.042
0.2	0	0.85	1.378	1.572	1.439	1.037	0.65	0.397	0.243	0.144	0.076
0.33	0	0.85	1.378	1.572	1.44	1.038	0.652	0.403	0.255	0.165	0.112
0.5	0	0.852	1.378	1.572	1.44	1.04	0.656	0.411	0.269	0.189	0.147
1.0	0	0.852	1.379	1.573	1.442	1.045	0.667	0.433	0.307	0.244	0.212
2.0	0	0.852	1.379	1.575	1.446	1.054	0.688	0.473	0.368	0.315	0.282
3.0	0	0.852	1.380	1.577	1.45	1.063	0.708	0.509	0.414	0.363	0.327
5.0	0	0.852	1.381	1.580	1.458	1.080	0.746	0.569	0.484	0.43	0.389
10	0	0.852	1.384	1.589	1.477	1.122	0.826	0.68	0.598	0.537	0.486
100	0	0.855	1.417	1.691	1.692	1.525	1.379	1.255	1.143	1.042	0.951
1000	0	0.859	1.467	1.856	2.055	2.106	2.067	1.977	1.861	1.736	1.611
10,000	0	0.860	1.480	1.906	2.177	2.327	2.382	2.368	2.305	2.21	2.095
100,000	0	0.861	1.481	1.912	2.193	2.358	2.433	2.441	2.398	2.319	2.215
r	0	0.861	1.482	1.913	2.195	2.362	2.439	2.450	2.41	2.333	2.231

Table 5.3 Maximum creep stress versus stress exponent

n	1	3	5	6	8	10
σ						
σ_{max}	1.802	1.575	1.416	1.361	1.283	1.231

both at the bottom and the top. For example, when α_i tends to infinity (10^5 in our case) the maximum creep stress approaches the value of stress at which creep of the material does not affect the temperature stresses. On the other hand, when α_i approaches zero (10^{-5} in this case) the maximum creep stress tends to a constant value. This relationship is presented graphically in Fig. 5.

2. As mentioned above, the parameter α_i represents the inverse of the retardation time, which is interconnected by Eq. 5.15. Therefore, large values of α_i values correspond to very small retardation times and vice versa.
3. As mentioned above, the parameter α_i cannot be uniquely obtained based on experimental data on creep material, and hence can be selected arbitrarily. Based on this parameter α_i will be considered in the following chapters as a random variable, and the data in Table 5.2 thus constitute a statistical basis on which to apply probabilistic methods for solving relevant creep engineering problems.
4. The maximum creep stress value depends not only on the mechanical parameters of the material, as well as by nonlinear stress relationship (see Eq. 5.12). This functional dependence for some fixed values of the MPPs is analyzed in the next section of this chapter.
5. Temperature–time curves, of course, are different for each design fire scenario, so the construction of the stress–strain diagram is necessary for all cases of fire severity.

5.2.3 Functional Dependencies of Creep Stresses and Strains from the Stress Exponent "n"

Let us now consider the question of constructing a stress–strain diagram, depending on the creep stress exponent n. The values of the stress exponents have been semi-theoretically explained [23]. Previous studies [24, 25] have revealed that the deformation mechanisms that operate during high-temperature creep of metals and alloys are characterized by the effective stress exponent n_e in the creep rate equation, i.e., the value of n has been observed to be equal to be between 3.0 and 6.0 for dislocation power law creep. Some researchers reported that the modified stress exponent, n, approached to the low values, which were explained by the classical creep theory [17, 26].

However, for many practical materials, the values of n in Eq. (5.12) are much larger than n = 3–5.

Some alloys exhibit a very large stress exponent ($n > 10$), and this has typically been explained by introducing a "threshold stress," σ_{th}, below which creep cannot be measured (so $2 \leq n \leq 10$).

Since our goal is to introduce the phenomenological engineering creep theory (regardless of any particular conventional or nonconventional creep mechanism), it will be assumed here that the stress exponent, n, is a real number from the closed interval [1, 10]. Obviously, $n = 1$ corresponds to linear creep theory.

Let us consider again the creep differential constitutive law (5.12) for instance, with fixed value $\alpha = 1.0$ and variable n from [1, 10]. The computational process is similar to the above analysis with variable MPP; therefore any intermediate calculations are omitted. The computer code and the results are as follows:

Consider Eq. (5.12) and $\alpha = 1.0$. In case of VFF, we have

$$\begin{aligned} d(y)/d(t) = & -\alpha_i * y * m1 + (\exp(-0.15 * t)) * (1 - 0.15 * t) \\ & - (\exp(t/(1.0 + 0.1 * t))) * m1 * y^\wedge n + \alpha_i * m1 * t * (\exp(-0.15 * t)) \end{aligned}$$

Solution for this loading stage:

n = 1 α = 1

Calculated values of DEQ variables

	Variable	Initial value	Minimal value	Maximal value	Final value
1	$m1$	0.0375	0.0079914	0.1463	0.1463
2	n	1	1	1	1
3	r	0	0	2.452476	2.231302
4	t	0	0	10	10
5	y	0	0	1.802403	0.0099515

Differential equations

1 $\mathrm{d}(y)/\mathrm{d}(t) = -1 * y * m1 + (\exp(-0.15 * t)) * (1 - 0.15 * t)$
$- (\exp(t/(1.0 + 0.1 * t))) * m1 * y^{\wedge}n + 1 * m1 * t * (\exp(-0.15 * t))$

Explicit equations

1 $n = 1$

2 $m1 = 0.0375 - 0.01868 * t + 0.00312 * t^{\wedge}2 - 0.000164 * t^{\wedge}2$

3 $r = t * (\exp(-0.15 * \mathrm{t}))$

See Fig. 5.27.
Model: $y = a0 + a1 * t + a2 * t\text{^}2 + a3 * t\text{^}3 + a4 * t\text{^}4$

Variable	Value
$a0$	0
$a1$	1.186
$a2$	−0.184
$a3$	−0.005
$a4$	0.0012

$$\textbf{Model}: \sigma = 1.186 * \theta - 0.184 * \theta^{\wedge}2 - 0.005 * \theta^{\wedge}3 - 0.0012 * \theta^{\wedge}4 \qquad (5.35)$$

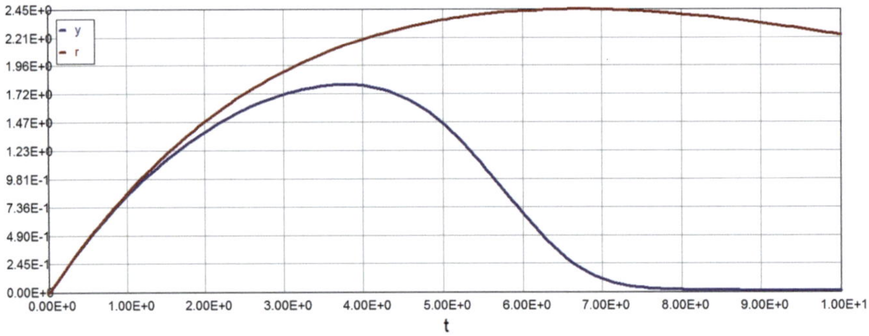

Fig. 5.27 Stress–temperature diagram $n = 1$

n = 3
Calculated values of DEQ variables

	Variable	Initial value	Minimal value	Maximal value	Final value
1	$m1$	0.0375	0.0079932	0.1463	0.1463
2	n	3	3	3	3
3	r	0	0	2.45251	2.231302
4	t	0	0	10	10
5	y	0	0	1.575117	0.2121058

Differential equations

1 $\mathrm{d}(y)/\mathrm{d}(t) = -1 * y * m1 + (\exp(-0.15 * t)) * (1 - 0.15 * t) - (\exp(t/(1.0 + 0.1 * t))) * m1 * y\hat{}n + 1 * m1 * t * (\exp(-0.15 * t))$

Explicit equations

1 $n = 3$
2 $m1 = 0.0375 - 0.01868 * t + 0.00312 * t\hat{}2 - 0.000164 * t\hat{}2$
3 $r = t * (\exp(-0.15 * \mathrm{t}))$

See Fig. 5.28.
Model: $y = a0 + a1 * t + a2 * t\hat{}2 + a3 * t\hat{}3 + a4 * t\hat{}4$

Variable	Value
$a0$	0
$a1$	1.432
$a2$	−0.383
$a3$	0.034
$a4$	−0.000956

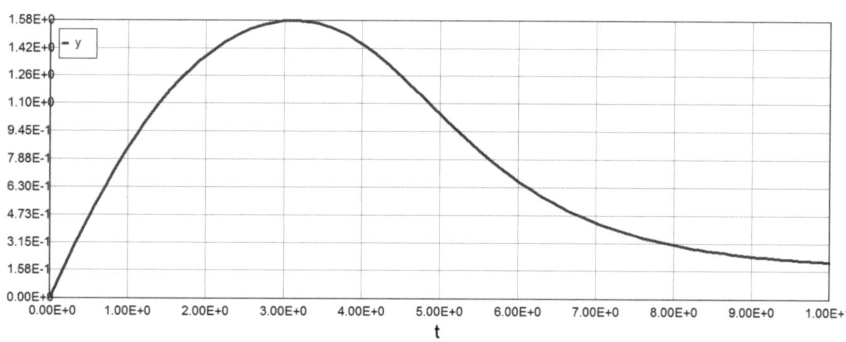

Fig. 5.28 Stress–temperature diagram $n = 3$

$$\textbf{Model}: \sigma = 1.432 * \theta - 0.383 * \theta^\wedge 2 + 0.034 * \theta^\wedge 3 - 0.000956 * \theta^\wedge 4 \quad (5.36)$$

n = 5
Calculated values of DEQ variables

	Variable	Initial value	Minimal value	Maximal value	Final value
1	$m1$	0.0375	0.007989	0.1463	0.1463
2	n	5	5	5	5
3	r	0	0	2.45253	2.231302
4	t	0	0	10	10
5	y	0	0	1.416329	0.3855116

Differential equations

1 $\mathrm{d}(y)/\mathrm{d}(t) = -1 * y * m1 + (\exp(-0.15 * t)) * (1 - 0.15 * t)$
$- (\exp(t/(1.0 + 0.1 * t))) * m1 * y^\wedge n + 1 * m1 * t * (\exp(-0.15 * t))$

Explicit equations

1 $n = 5$
2 $m1 = 0.0375 - 0.01868 * t + 0.00312 * t^\wedge 2 - 0.000164 * t^\wedge 2$
3 $r = t * (\exp(-0.15 * \mathrm{t}))$

See Fig. 5.29.

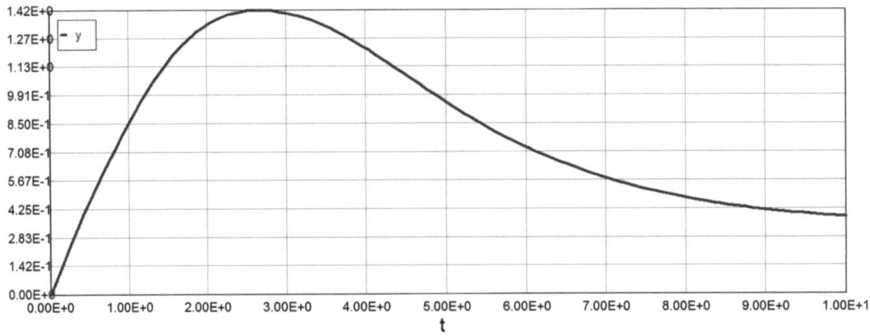

Fig. 5.29 Stress–temperature diagram n = 5

Model: $y = a0 + a1 * t + a2 * t\text{^}2 + a3 * t\text{^}3 + a4 * t\text{^}4$

Variable	Value
$a0$	0
$a1$	1.380
$a2$	−0.410
$a3$	0.0435
$a4$	−0.00157

$$\textbf{Model}: \sigma = 1.38 * \theta - 0.410 * \theta\hat{}2 + 0.0435 * \theta\hat{}3 - 0.00157 * \theta\hat{}4 \tag{5.37}$$

n = 6

Calculated values of DEQ variables

	Variable	Initial value	Minimal value	Maximal value	Final value
1	$m1$	0.0375	0.007989	0.1463	0.1463
2	n	6	6	6	6
3	r	0	0	2.45253	2.231302
4	t	0	0	10	10
5	y	0	0	1.361207	0.4477313

Differential equations

1 $\mathrm{d}(y)/\mathrm{d}(t) = -1 * y * m1 + (\exp(-0.15 * t)) * (1 - 0.15 * t) - (\exp(t/(1.0 + 0.1 * t))) * m1 * y\hat{}n + 1 * m1 * t * (\exp(-0.15 * t))$

Explicit equations

1 $n = 6$

2 $m1 = 0.0375 - 0.01868 * t + 0.00312 * t\hat{}2 - 0.000164 * t\hat{}2$

3 $r = t * (\exp(-0.15 * \mathrm{t}))$

See Fig. 5.30.

Model: $y = a0 + a1 * t + a2 * t\text{^}2 + a3 * t\text{^}3 + a4 * t\text{^}4$

Variable	Value
$a0$	0
$a1$	1.330
$a2$	−0.404
$a3$	0.0442
$a4$	−0.00166

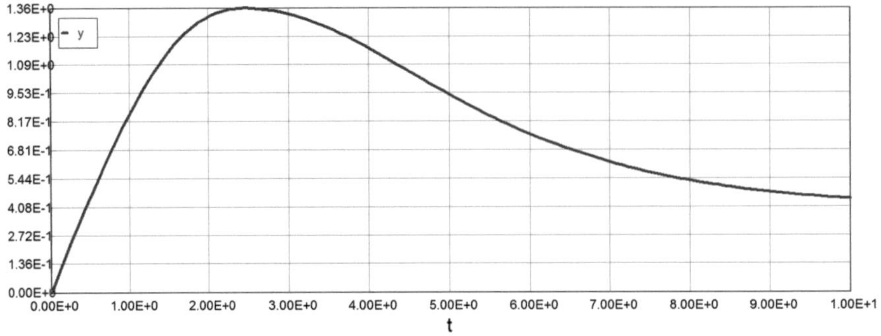

Fig. 5.30 Stress–temperature diagram $n = 6$

$$\textbf{Model}: \sigma = 1.33 * \theta - 0.404 * \theta\hat{}2 + 0.0442 * \theta\hat{}3 - 0.00166 * \theta\hat{}4 \tag{5.38}$$

n = 8

Calculated values of DEQ variables

	Variable	Initial value	Minimal value	Maximal value	Final value
1	$m1$	0.0375	0.007989	0.1463	0.1463
2	n	8	8	8	8
3	r	0	0	2.45253	2.231302
4	t	0	0	10	10
5	y	0	0	1.282874	0.5407249

Differential equations

1 $\mathrm{d}(y)/\mathrm{d}(t) = -1 * y * m1 + (\exp(-0.15 * t)) * (1 - 0.15 * t) - (\exp(t/(1.0 + 0.1 * t))) * m1 * y\hat{}n + 1 * m1 * t * (\exp(-0.15 * t))$

Explicit equations

1 $n = 8$

2 $m1 = 0.0375 - 0.01868 * t + 0.00312 * t\hat{}2 - 0.000164 * t\hat{}2$

3 $r = t * (\exp(-0.15 * \mathrm{t}))$

See Fig. 5.31.

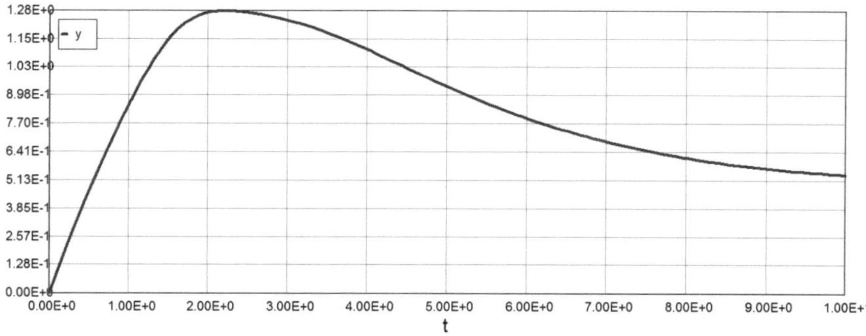

Fig. 5.31 Stress–temperature diagram $n = 8$

Model: $y = a0 + a1 * t + a2 * t\text{^}2 + a3 * t\text{^}3 + a4 * t\text{^}4$

Variable	Value
$a0$	0
$a1$	1.24
$a2$	−0.384
$a3$	0.0435
a4	−0.0017

$$\textbf{Model}: \sigma = 1.24 * \theta - 0.384 * \theta^\wedge 2 + 0.0435 * \theta^\wedge 3 - 0.0017 * \theta^\wedge 4 \quad (5.39)$$

n = 10

Calculated values of DEQ variables

	Variable	Initial value	Minimal value	Maximal value	Final value
1	$m1$	0.0375	0.0079969	0.1463	0.1463
2	n	10	10	10	10
3	r	0	0	2.452485	2.231302
4	t	0	0	10	10
5	y	0	0	1.231285	0.6065921

Differential equations

1 $\mathrm{d}(y)/\mathrm{d}(t) = -1 * y * m1 + (\exp(-0.15 * t)) * (1 - 0.15 * t) - (\exp(t/(1.0 + 0.1 * t))) * m1 * y^\wedge n + 1 * m1 * t * (\exp(-0.15 * t))$

Explicit equations

1 $n = 10$

2 $m1 = 0.0375 - 0.01868 * t + 0.00312 * t^\wedge 2 - 0.000164 * t^\wedge 2$

3 $r = t * (\exp(-0.15 * \mathrm{t}))$

See Fig. 5.32.

Model: $y = a0 + a1 * t + a2 * t$^2 $+ a3 * t$^3 $+ a4 * t$^4

Variable	Value
$a0$	0.0
$a1$	1.173
$a2$	−0.367
$a3$	0.0424
$a4$	−0.0017

$$\textbf{Model}: \sigma = 1.173 * \theta - 0.367 * \theta^\wedge 2 + 0.0424 * \theta^\wedge 3 - 0.0017 * \theta^\wedge 4 \quad (5.40)$$

See Fig. 5.33.

Approximately, we have $\sigma_{max} = 1.8 - 0.0634n$

n = 1 α = 100

Calculated values of DEQ variables

	Variable	Initial value	Minimal value	Maximal value	Final value
1	$m1$	0.0375	0.0079926	0.1463	0.1463
2	n	1	1	1	1
3	r	0	0	2.452514	2.231302
4	t	0	0	10	10
5	y	0	0	2.006567	0.9001359

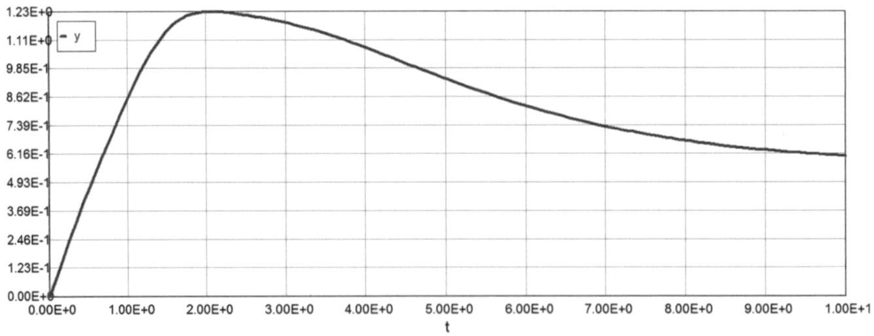

Fig. 5.32 Stress–temperature diagram $n = 10$

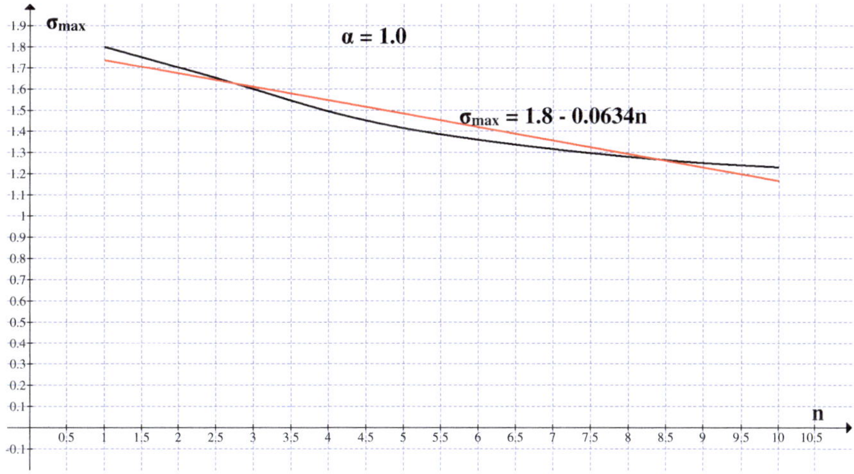

Fig. 5.33 Stress–temperature diagram $\alpha = 1$

Differential equations

$$1 \quad d(y)/d(t) = -100 * y * m1 + (\exp(-0.15 * t)) * (1 - 0.15 * t) - (\exp(t/(1.0 + 0.1 * t))) * m1 * y^{\wedge}n + 100 * m1 * t * (\exp(-0.15 * t))$$

Explicit equations

1 $n = 1$

2 $m1 = 0.0375 - 0.01868 * t + 0.00312 * t^{\wedge}2 - 0.000164 * t^{\wedge}2$

3 $r = t * (\exp(-0.15 * \mathrm{t}))$

See Fig. 5.34.

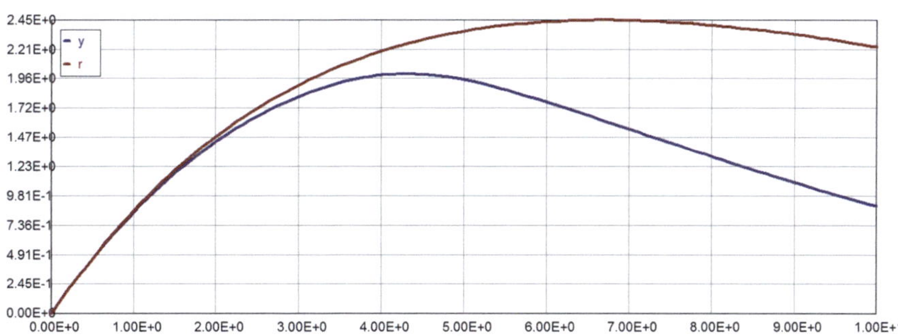

Fig. 5.34 Stress–temperature diagram $n = 1$

Model: $y = a0 + a1 * t + a2 * t\text{^}2 + a3 * t\text{^}3 + a4 * t\text{^}4$

Variable	Value
$a0$	0
$a1$	1.0144
$a2$	−0.146
$a3$	0.00315
$a4$	0.000223

Model: $\sigma = 1.0144 * \theta - 0.146 * \theta\text{^}2 + 0.00315 * \theta\text{^}3 - 0.000223 * \theta\text{^}4$ (5.41)

n = 3

Calculated values of DEQ variables

	Variable	Initial value	Minimal value	Maximal value	Final value
1	$m1$	0.0375	0.0079895	0.1463	0.1463
2	n	3	3	3	3
3	r	0	0	2.452519	2.231302
4	t	0	0	10	10
5	y	0	0	1.724496	0.951412

Differential equations

1 $$\mathrm{d}(y)/\mathrm{d}(t) = -100 * y * m1 + (\exp(-0.15 * t)) * (1 - 0.15 * t) - (\exp(t/(1.0 + 0.1 * t))) * m1 * y\text{^}n + 100 * m1 * t * (\exp(-0.15 * t))$$

Explicit equations

1 $n = 3$

2 $m1 = 0.0375 - 0.01868 * t + 0.00312 * t\text{^}2 - 0.000164 * t\text{^}2$

3 $r = t * (\exp(-0.15 * \mathrm{t}))$

See Fig. 5.35.

Model: $y = a0 + a1 * t + a2 * t\text{^}2 + a3 * t\text{^}3 + a4 * t\text{^}4$

Variable	Value
$a0$	0
$a1$	1.271
$a2$	−0.304
$a3$	0.0276
$a4$	−0.000886

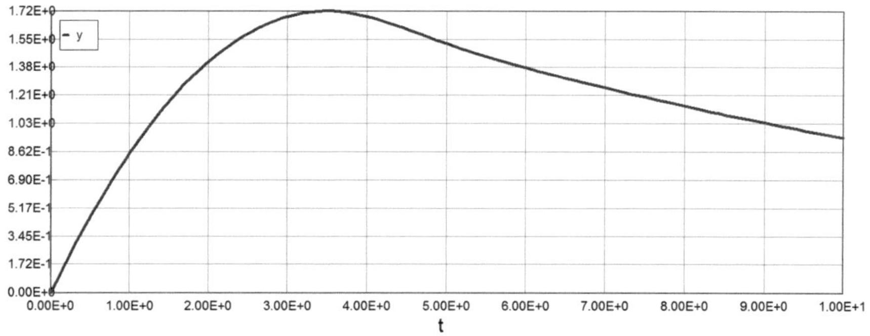

Fig. 5.35 Stress–temperature diagram $n = 3$

$$\textbf{Model}: \sigma = 1.271 * \theta - 0.304 * \theta^\wedge 2 + 0.0276 * \theta^\wedge 3 - 0.000886 * \theta^\wedge 4 \quad (5.42)$$

n = 5

Calculated values of DEQ variables

	Variable	Initial value	Minimal value	Maximal value	Final value
1	$m1$	0.0375	0.0079898	0.1463	0.1463
2	n	5	5	5	5
3	r	0	0	2.452483	2.231302
4	t	0	0	10	10
5	y	0	0	1.523301	0.9677567

Differential equations

1 $\mathrm{d}(y)/\mathrm{d}(t) = -100 * y * m1 + (\exp(-0.15 * t)) * (1 - 0.15 * t) - (\exp(t/(1.0 + 0.1 * t))) * m1 * y^\wedge n + 100 * m1 * t * (\exp(-0.15 * t))$

Explicit equations

1 $n = 5$

2 $m1 = 0.0375 - 0.01868 * t + 0.00312 * t^\wedge 2 - 0.000164 * t^\wedge 2$

3 $r = t * (\exp(-0.15 * \mathrm{t}))$

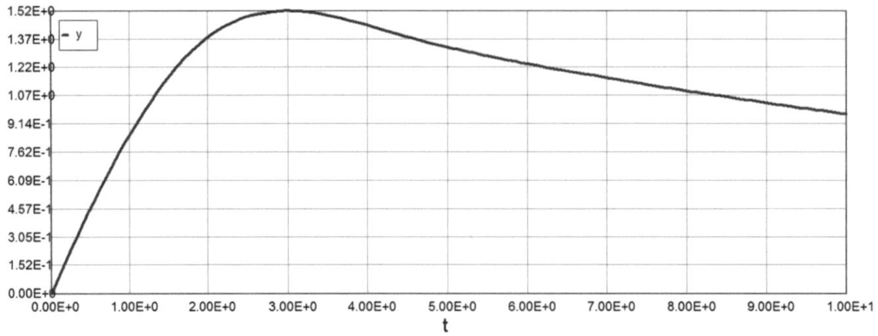

Fig. 5.36 Stress–temperature diagram $n = 5$

See Fig. 5.36.

Model: $y = a0 + a1 * t + a2 * t^2 + a3 * t^3 + a4 * t^4$

Variable	Value
$a0$	0
$a1$	1.286
$a2$	−0.354
$a3$	0.0380
$a4$	-0.00144

$$\textbf{Model}: \sigma = 1.286 * \theta - 0.354 * \theta^{\wedge}2 + 0.038 * \theta^{\wedge}3 - 0.00144 * \theta^{\wedge}4 \quad (5.43)$$

n = 6

Calculated values of DEQ variables

	Variable	Initial value	Minimal value	Maximal value	Final value
1	$m1$	0.0375	0.0079898	0.1463	0.1463
2	n	6	6	6	6
3	r	0	0	2.452436	2.231302
4	t	0	0	10	10
5	y	0	0	1.452124	0.9723881

Differential equations

1 $$\mathrm{d}(y)/\mathrm{d}(t) = -100 * y * m1 + (\exp(-0.15 * t)) * (1 - 0.15 * t) - (\exp(t/(1.0 + 0.1 * t))) * m1 * y^{\wedge}n + 100 * m1 * t * (\exp(-0.15 * t))$$

Explicit equations

1 $n = 6$

2 $m1 = 0.0375 - 0.01868 * t + 0.00312 * t^\wedge 2 - 0.000164 * t^\wedge 2$

3 $r = t * (\exp(-0.15 * \mathrm{t}))$

See Fig. 5.37.

Model: $y = a0 + a1 * t + a2 * t^\wedge 2 + a3 * t^\wedge 3 + a4 * t^\wedge 4$

Variable	Value
$a0$	0
$a1$	1.26
$a2$	−0.358
$a3$	0.0397
$a4$	−0.00155

$$\textbf{Model}: \sigma = 1.26 * \theta - 0.358 * \theta^\wedge 2 + 0.0397 * \theta^\wedge 3 - 0.00155 * \theta^\wedge 4 \quad (5.44)$$

n = 8

Calculated values of DEQ variables

	Variable	Initial value	Minimal value	Maximal value	Final value
1	$m1$	0.0375	0.00799	0.1463	0.1463
2	n	8	8	8	8
3	r	0	0	2.452467	2.231302
4	t	0	0	10	10
5	y	0	0	1.35083	0.9785441

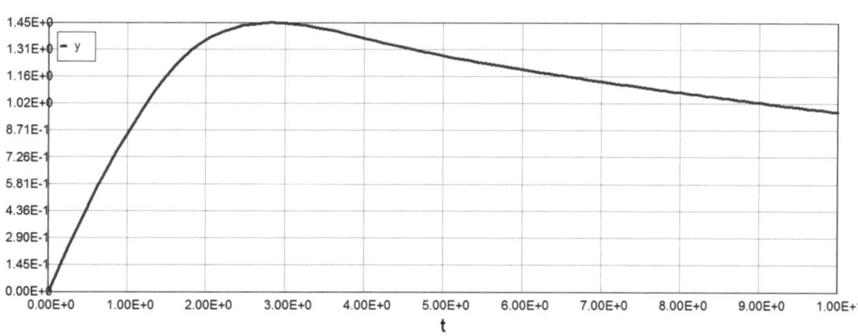

Fig. 5.37 Stress–temperature diagram $n = 6$

Differential equations

1 $d(y)/d(t) = -100 * y * m1 + (\exp(-0.15 * t)) * (1 - 0.15 * t) - (\exp(t/(1.0 + 0.1 * t))) * m1 * y\hat{}n + 100 * m1 * t * (\exp(-0.15 * t))$

Explicit equations

1 $n = 8$

2 $m1 = 0.0375 - 0.01868 * t + 0.00312 * t\hat{}2 - 0.000164 * t\hat{}2$

3 $r = t * (\exp(-0.15 * \mathrm{t}))$

See Fig. 5.38.
Model: $y = a0 + a1 * t + a2 * t\hat{}2 + a3 * t\hat{}3 + a4 * t\hat{}4$

Variable	Value
$a0$	0
$a1$	1.20
$a2$	−0.353
$a3$	0.0406
a4	−0.00164

$$\textbf{Model}: \sigma = 1.2 * \theta - 0.353 * \theta\hat{}2 + 0.0406 * \theta\hat{}3 - 0.00164 * \theta\hat{}4 \quad (5.45)$$

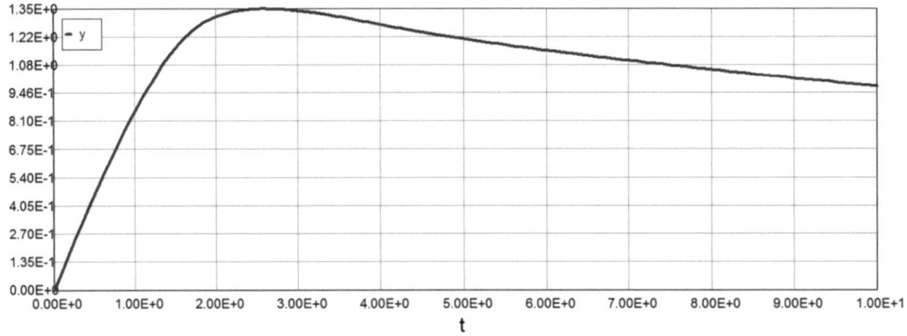

Fig. 5.38 Stress–Temperature diagram $n = 8$

n = 10

Calculated values of DEQ variables

	Variable	Initial value	Minimal value	Maximal value	Final value
1	$m1$	0.0375	0.0079913	0.1463	0.1463
2	n	10	10	10	10
3	r	0	0	2.452486	2.231302
4	t	0	0	10	10
5	y	0	0	1.284375	0.9824526

Differential equations

1 $\mathrm{d}(y)/\mathrm{d}(t) = -100 * y * m1 + (\exp(-0.15 * t)) * (1 - 0.15 * t) - (\exp(t/(1.0 + 0.1 * t))) * m1 * y\hat{\ }n + 100 * m1 * t * (\exp(-0.15 * t))$

Explicit equations

1 $n = 10$

2 $m1 = 0.0375 - 0.01868 * t + 0.00312 * t\hat{\ }2 - 0.000164 * t\hat{\ }2$

3 $r = t * (\exp(-0.15 * \mathrm{t}))$

See Fig. 5.39.

Model: $y = a0 + a1 * t + a2 * t\hat{\ }2 + a3 * t\hat{\ }3 + a4 * t\hat{\ }4$

Variable	Value
$a0$	0.0
$a1$	1.148
$a2$	−0.344
$a3$	0.0403
$a4$	−0.00165

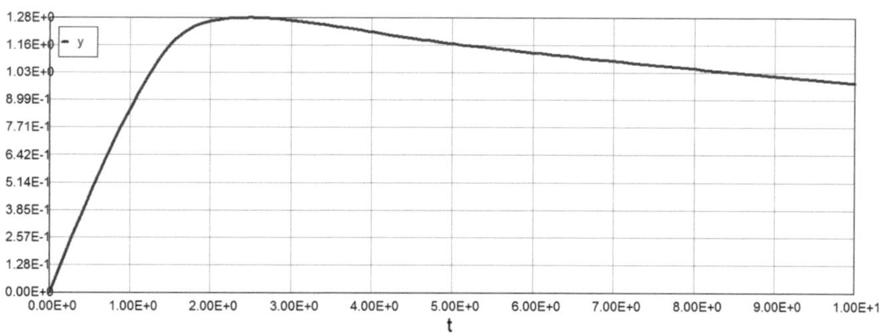

Fig. 5.39 Stress–temperature diagram $n = 10$

$$\textbf{Model}: \sigma = 1.148 * \theta - 0.344 * \theta^\wedge 2 + 0.0403 * \theta^\wedge 3 - 0.00165 * \theta^\wedge 4 \tag{5.46}$$

See Fig. 5.40 and Table 5.4.

$$\begin{aligned} &\text{Approximately we have} \\ &\sigma_{\max} = -0.08\text{n} + 2 \end{aligned} \tag{5.47}$$

Thus, the main points of the calculation of creep stresses and strains can be summarized as follows:

1. For each type of fire severity it is necessary to obtain the temperature–time function using the appropriate differential equation (5.12).
2. We need to find an analytical expression of the inverse function θ^{-1} and its first derivative of m and m1 at a monotonically increasing function of temperature. It should also be noted that the first derivative of the function θ determines creep strain rate (up to a constant factor).
3. Similar calculations (see p. 2) should be performed for the interval monotonically decreasing temperature.
4. Applying the method of change of variables in the integral equation of creep (at a monotonically increasing function of temperature) and then reducing it to the corresponding differential equations of the first order, we find the approximate analytical creep stress–temperature (strain) function.

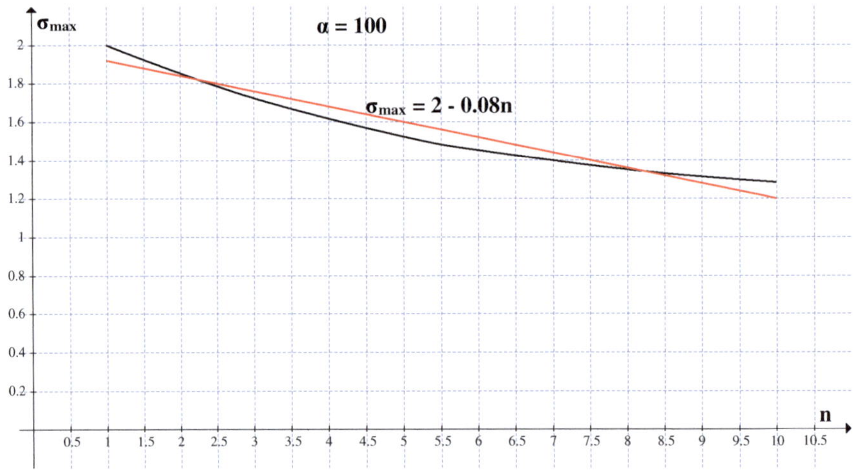

Fig. 5.40 Maximum creep stress versus stress exponent $\alpha = 100$

Table 5.4 Maximum creep stress versus stress exponent

n	1	3	5	6	8	10	α
σ							
σ_{max}	2.0	1.724	1.523	1.452	1.351	1.284	100
σ_{max}	1.802	1.575	1.416	1.361	1.283	1.231	1

5. Similar calculations (see p. 4) should be performed for the interval monotonically decreasing temperature.
6. At the end, the combined graph of both processes (loading and unloading) should be presented for each fire severity case.

Due to the fact that all the calculations required obtaining a stress–strain diagram for other cases of fire severity is very similar to the case of a Very Fast Fire (VFF), some intermediate computations will be omitted.

5.3 Creep Stress–Strain Diagram. Fast Fire

All subsequent calculations for this case of fire severity are presented in a step-by-step procedural form in accordance with the main points of the above.

5.3.1 *Temperature–Time Function*

Fast Fire; Temperature Increase (882 K < T < 1022 K)

Data $T_* = 600$ K; $0 < \tau < 0.2$; $0.05 < \gamma < 0.175$

Select $\gamma = 0.05$

Differential Eqs. (5.3) and (5.4) are rewritten as an input for "Polymath" software:

```
1  d(y0)/d(t) = 20 * (1 - y2) * exp(y0/(1 + 0.1 * y0)) - 0.157 * y0^4
2  d(y2)/d(t) = 5.5 * (1 - y2) * exp(y0/(1 + .1 * y0))
3  d(y1)/d(t) = 1.0 * (1 - y1)^1.0 * exp(y/(1 + .1 * y))
4  d(y)/d(t) = (1) * 20 * (1 - y1)^1.0 * exp(y/(1 + .1 * y)) - 0.157 * y^4,
```

where

"y" is the dimensionless temperature "θ" with the corresponding parameter "$\gamma = 0.05$".

"y0" is the dimensionless temperature "θ" with the corresponding parameter "$\gamma = 0.175$".

"y1" is the concentration of the product of the first-order chemical reaction with "$\gamma = 0.05$".

"y2" is the concentration of the product of the first-order chemical reaction with "$\gamma = 0.175$".

Calculated values of DEQ variables

	Variable	Initial value	Minimal value	Maximal value	Final value
1	*t*	0	0	0.2	0.2
2	*y*	0	0	7.916172	3.026519
3	*y*0	0	0	4.121423	2.629209
4	*y*1	0	0	0.9812653	0.9812653
5	*y*2	0	0	0.9987195	0.9987195

Tabulated solution of Eqs. (5.3) and (5.4) is presented in [4] and the graphs are shown in Fig. 5.41.

5.3.2 Analytical Expression of the Inverse Function θ^{-1} and Its First Derivative

Based on the tabulated data shown above, the final approximation of the dimensionless temperature–time curve can be presented as follows:

Monotonically increasing temperature–time function: $(0 < \theta < 8)$; $(0 < \tau 1 < 0.0763)$

$$\textbf{Model}: \theta = 39.35 * \tau - 688.9 * \tau^\wedge 2 + 6463.5 * \tau^\wedge 3 + 197000 * \tau^\wedge 4 \quad (5.48)$$

Variable	Value
*a*0	0.0
*a*1	39.35
*a*2	−688.9
*a*3	6463.5
*a*4	1.97E+05

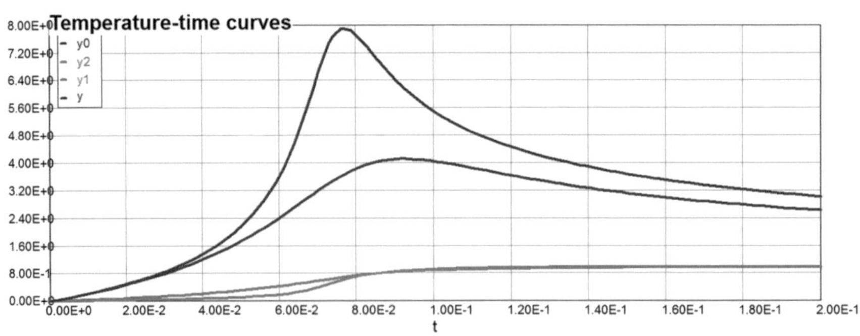

Fig. 5.41 Dimensionless time–temperature curves

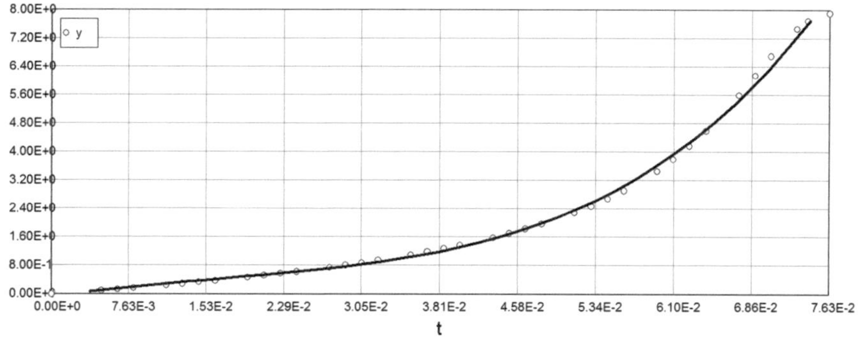

Fig. 5.42 Monotonically increasing temperature–time function

See Fig. 5.42.

The inverse function θ^{-1} and its first derivative (monotonically increasing functions) are

$$\textbf{Model}: \mathbf{m} = \tau = 0.00154 + 0.0411 * \theta - 0.0113 * \theta^\wedge 2 + 0.00147 * \theta^\wedge 3 \\ - 0.0000697 * \theta^\wedge 4$$

Variable	Value
*a*0	0.00154
*a*1	0.0411
*a*2	−0.0113
*a*3	0.00147
*a*4	−6.97E−05

See Fig. 5.43.

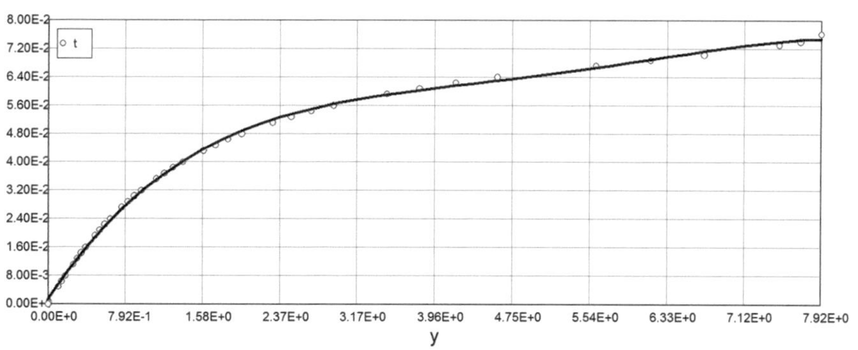

Fig. 5.43 Monotonically increasing inverse function θ^{-1}

$$\mathbf{m1} = \tau' = 0.0411 - 0.0226 * \tau + 0.00441 * \tau^{\wedge}2 - 0.000279 * \tau^{\wedge}3 \qquad (5.50)$$

Monotonically decreasing temperature–time function: $(8 > \theta > 3)$; $(0.07 < \tau 1 <$ 0;0.2)

Model:

$\theta = 40.58 - 822 * \tau1 + 7094 * \tau1^{\wedge}2 - 27{,}920 * \tau1^{\wedge}3 + 41{,}560 * \tau1^{\wedge}4$

Variable	Value
*a*0	40.5781
*a*1	−822.036
*a*2	7094.178
*a*3	−2.792E+04
*a*4	4.156E+04

See Fig. 5.44.

The inverse function $\theta^{-1}(\tau 1)$:

Model:

$t1 = 1.072 - 0.577 * x + 0.132 * x^{\wedge}2 - 0.0137 * x^{\wedge}3 + 0.000531 * x^{\wedge}4$

Variable	Value
*a*0	1.071476
*a*1	−0.5771325
*a*2	0.1318912
*a*3	−0.0136581
*a*4	0.0005312

See Fig. 5.45.

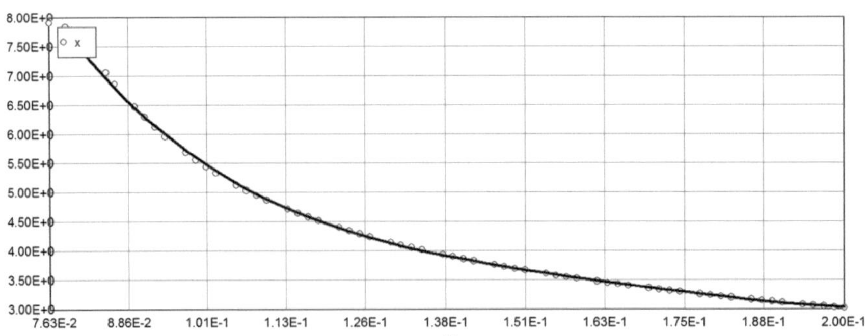

Fig. 5.44 Monotonically decreased temperature–time function

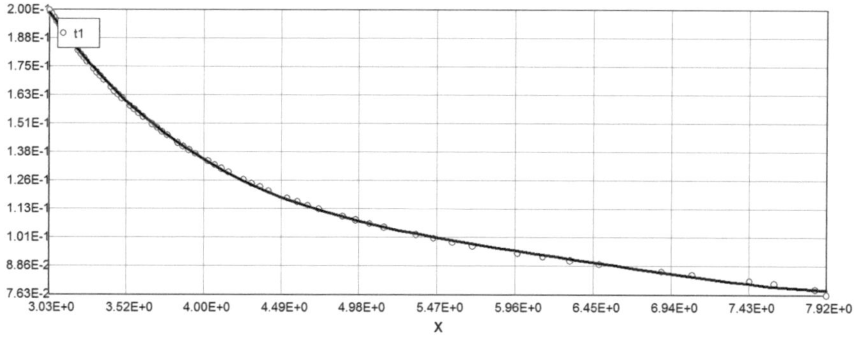

Fig. 5.45 Monotonically decreasing inverse function $\theta^{-1}(\tau 1)$

$$\textbf{Model} : \textbf{m0} = \tau 1 = 1.072 - 0.577 * \theta + 0.132 * \theta^\wedge 2 - 0.0137 * \theta^\wedge 3 + 0.000531 * \theta^\wedge 4$$

The first derivative of inverse function $\theta^{-1}(\tau 1)$:

$$\textbf{m01} = \tau' = -0.577 + 0.264 * \theta - 0.0441 * \theta^\wedge 2 - 0.00212 * \theta^\wedge 3 \tag{5.52}$$

5.3.3 Creep Constitutive Equation. Equivalent ODE Method

Monotonically increasing temperature–time function

The stress–strain response of a material is described using differential equation of the form:

$$\begin{aligned}
&\sigma'(\theta) = -\sigma(\theta)\alpha m1 + \mathrm{e}^{-0.15\theta}[1 - 0.15\theta] - (\exp(t)/(1+0.1t))m1\sigma^n(\theta) + m1\alpha\theta\mathrm{e}^{-0.15\theta}\\
&m = \varphi(\theta) = 0.00154 + 0.0411 * \theta - 0.0113 * \theta^\wedge 2 + 0.00147 * \theta^\wedge 3 - 0.0000697 * \theta^\wedge 4\\
&m1 = \varphi'(\theta) = 0.0411 - 0.0226 * \theta + 0.00441 * \theta^\wedge 2 - 0.000279 * \theta^\wedge 3\\
&\sigma(\theta = 0) = 0; \quad \alpha = 0.33
\end{aligned} \tag{5.53}$$

Consider now general uniaxial creep constitutive model for monotonically increasing part of real fire temperature load (equivalent differential equation with one MPP $\alpha = 0.33$):

$$\begin{aligned}
\mathrm{d}(\sigma)/\mathrm{d}(\theta) = &-0.33 * \sigma * m1 + (\exp(-0.15 * \theta)) * (1 - 0.15 * \theta) - (\exp(\theta/(1.0 + 0.1 * \theta))) * m1 * \sigma^n\\
&+ 0.33 * m1 * \theta * (\exp(-0.15 * \theta))
\end{aligned} \tag{5.54}$$

The temperature range is $[0 < \theta < 8]$ and the dimensionless time duration for increasing part of fully developed fire stage is $[0 < \tau < 0.0763]$.

Calculated values of DEQ variables

	Variable	Initial value	Minimal value	Maximal value	Final value
1	$m1$	0.0411	−0.000308	0.0411	−0.000308
2	n	2	2	2	2
3	t	0	0	8	8
4	z	0	0	1.797189	0.9668527

Differential equation

$$1 \quad d(z)/d(t) = -0.33 * z * m1 + (\exp(-0.15 * t)) * (1 - 0.15 * t) - (\exp(t/(1.0 + 0.1 * t))) * m1 * z^n + 0.33 * m1 * t * (\exp(-0.15 * t))$$

See Fig. 5.46.

Model: $\sigma = 0.89 * \theta - 0.0973 * \theta$^2 $- 0.0076 * \theta$^3 $+ 0.00095 * \theta$^4

Variable	Value
$a0$	0.041
$a1$	0.89
$a2$	−0.0973
$a3$	−0.0076
$a4$	0.00095

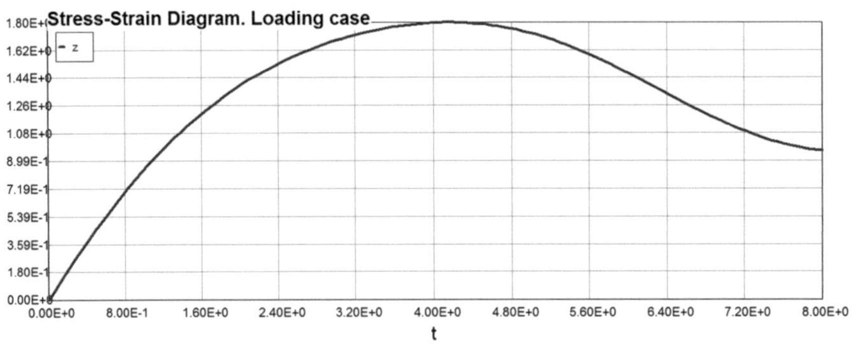

Fig. 5.46 Stress–strain diagram. Fast fire

Monotonically decreasing temperature–time function

Calculated values of DEQ variables

	Variable	Initial value	Minimal value	Maximal value	Final value
1	m	0.0540709	0.0540709	0.0705907	0.0705907
2	$m01$	−0.23914	−2.37284	−0.23914	−2.37284
3	$m1$	0.0049971	0.0003976	0.0049971	0.0003976
4	n	2	2	2	2
5	t	3	3	8	8
6	y	0	0	0.4927664	0.3907289
7	$y11$	0.967	0.0935758	0.967	0.0935758
8	z	0.967	0.967	0.967	0.967

Differential equations

1 $\mathrm{d}(y)/\mathrm{d}(t) = -0.33 * y * m1 + (\exp(-0.15 * t)) * (1 - 0.15 * t) - (\exp(t/(1.0 + 0.1 * t))) * m1 * y^\wedge n + 0.33 * m1 * t * (\exp(-0.15 * t))$

2 $\mathrm{d}(z)/\mathrm{d}(t) = 0 * (-0.33 * z * (m1) + (\exp(-0.15 * (8 - t))) * (1 - 0.15 * (8 - t)) - (\exp((8 - t)/(1.0 + 0.1 * (8 - t)))) * (m1) * z^\wedge n + 0.33 * (m1) * (8 - t) * (\exp(-0.15 * (8 - t))))$

3 $\mathrm{d}(y11)/\mathrm{d}(t) = 1 * (-0.33 * y11 * (-m01) + (\exp(-0.15 * (t))) * (1 - 0.15 * (t)) - (\exp((t)/(1.0 + 0.1 * (t)))) * (-m01) * y11^\wedge n + 0.33 * (-m01) * (t) * (\exp(-0.15 * (t))))$

Explicit equations

1 $n = 2$

2 $m1 = 0.0405 - 0.02252 * t^\wedge 1 + 0.004386 * t^\wedge 2 - 0.0002747 * t^\wedge 3$

3 $m = (0.0405 * t - 0.01126 * t^\wedge 2 + 0.001462 * t^\wedge 3 - 0.00006868 * t^\wedge 4)$

4 $m01 = -0.577 + 0.264 * t - 0.0441 * t^\wedge 2 - 0.00212 * t^\wedge 3$

Model: $y11 = a0 + a1 * t + a2 * t^\wedge 2 + a3 * t^\wedge 3 + a4 * t^\wedge 4$

Variable	Value
$a0$	12.04
$a1$	−7.913
$a2$	1.974
$a3$	−0.218
$a4$	0.0089

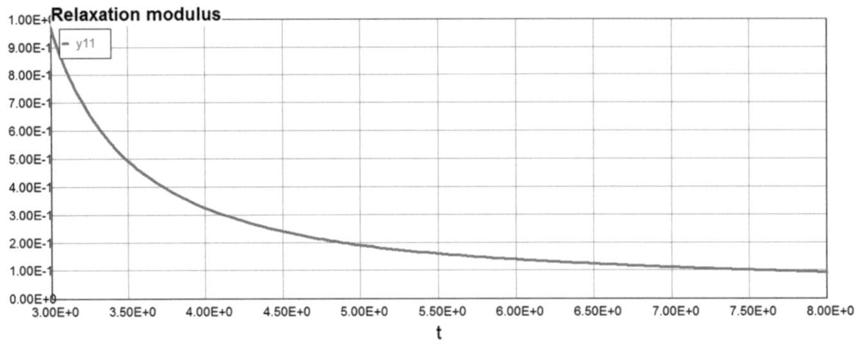

Fig. 5.47 Relaxation modulus

See Fig. 5.48.

$$\sigma = 12.04 - 7.913 * \theta + 1.974 * \theta^\wedge 2 - 0.218 * \theta^\wedge 3 + 0.0089 * \theta^\wedge 4 \qquad (5.55)$$

The combined stress–temperature–strain diagram (loading and unloading) graph is presented in Fig. 5.49.

Sinuosity of relaxation modulus curve (see Fig. 5.49) is the consequence of using polynomial regression method for approximation. The real very smooth relaxation modulus curve one can see in Fig. 5.47.

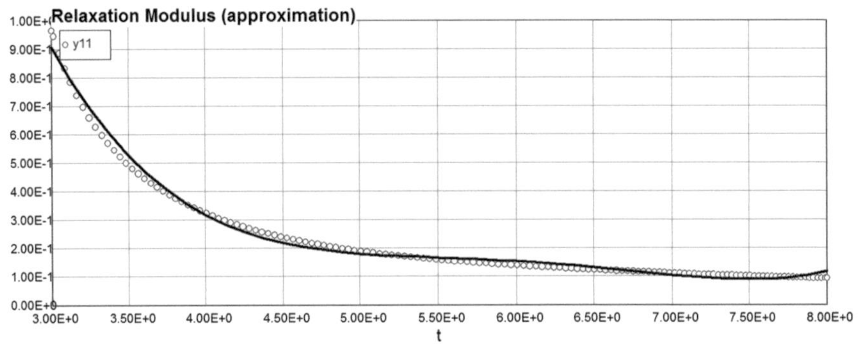

Fig. 5.48 Relaxation modulus. Approximate function

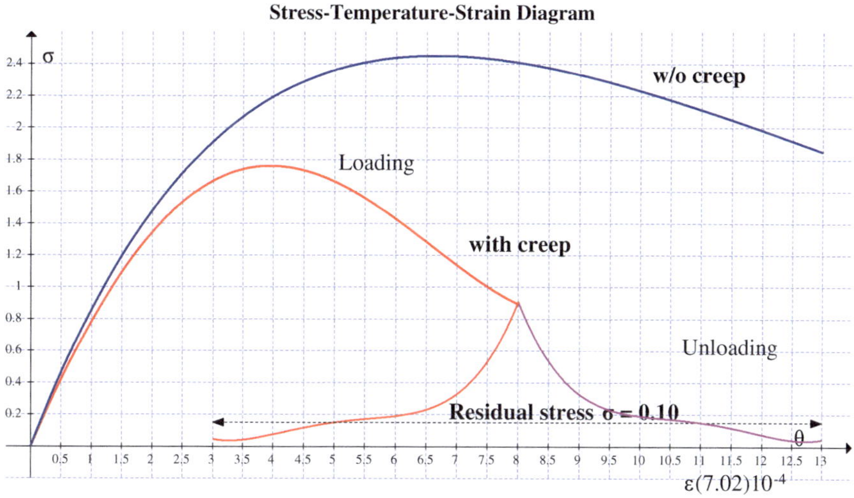

Fig. 5.49 Stress–temperature–strain diagram. Fast fire

5.3.4 *The Functional Dependencies of Creep Stresses and Strains from Material Properties Parameters (MPP)*

Consider now Eq. (5.12) and $\alpha = 0.001$. Again, in case of **FF (Fast Fire)** we have

Differential equations

$$1 \quad d(y)/d(t) = -\alpha_i * y * m1 + (\exp(-0.15 * t)) * (1 - 0.15 * t) - (\exp(t/(1.0 + 0.1 * t))) * m1 * y\hat{}n + \alpha_i * m1 * t * (\exp(-0.15 * t))$$

Explicit equations

$$n = 3$$
$$m1 = \tau(\theta) = 0.0405 - 0.02252 * t\hat{}1 + 0.004386 * t\hat{}2 - 0.0002747 * t\hat{}3 \quad (5.56)$$

Computer code and solution of Eq. (5.12) for loading stage is as follows:

α = 0.001
Calculated values of DEQ variables

	Variable	Initial value	Minimal value	Maximal value	Final value
1	*m*1	0.0411	−0.000308	0.0411	−0.000308
2	*n*	3	3	3	3
3	*r*	0	0	2.452527	2.409554

(continued)

(continued)

	Variable	Initial value	Minimal value	Maximal value	Final value
4	t	0	0	8	8
5	y	0	0	1.653557	0.8486171

Differential equations

1 $d(y)/d(t) = -0.001 * y * m1 + (\exp(-0.15 * t)) * (1 - 0.15 * t) - (\exp(t/(1.0 + 0.1 * t))) * m1 * y^\wedge n + 0.001 * m1 * t * (\exp(-0.15 * t))$

Explicit equations

1 $n = 3$

2 $r = t * (\exp(-0.15 * t))$

3 $m1 = 0.0411 - 0.0226 * t + 0.00441 * t^\wedge 2 - 0.000279 * t^\wedge 3$

See Fig. 5.50.

Model: $y = a0 + a1 * t + a2 * t$^2 $+ a3 * t$^3 $+ a4 * t$^4

Variable	Value
$a0$	0.0
$a1$	0.993
$a2$	−0.160
$a3$	0.00108
$a4$	0.000633

Model: $\sigma = 0.993 * \theta - 0.16 * \theta^\wedge 2 + 0.00108 * \theta^\wedge 3 - 0.000633 * \theta^\wedge 4$ (5.57)

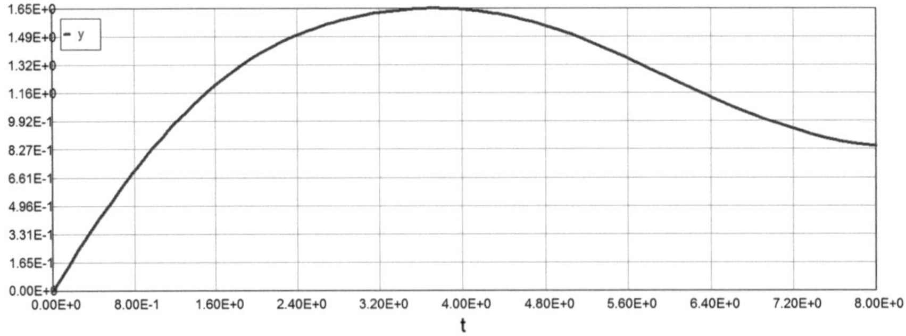

Fig. 5.50 Stress–temperature–strain diagram. Fast fire $\alpha = 0.001$

α = 0.01

Calculated values of DEQ variables

	Variable	Initial value	Minimal value	Maximal value	Final value
1	$m1$	0.0411	−0.000308	0.0411	−0.000308
2	n	3	3	3	3
3	r	0	0	2.452527	2.409554
4	t	0	0	8	8
5	y	0	0	1.653577	0.8486759

Differential equations

1 $\mathrm{d}(y)/\mathrm{d}(t) = -0.01 * y * m1 + (\exp(-0.15 * t)) * (1 - 0.15 * t) - (\exp(t/(1.0 + 0.1 * t))) * m1 * y\hat{\ }n + 0.01 * m1 * t * (\exp(-0.15 * t))$

Explicit equations

1 $n = 3$

2 $r = t * (\exp(-0.15 * t))$

3 $m1 = 0.0411 - 0.0226 * t + 0.00441 * t\hat{\ }2 - 0.000279 * t\hat{\ }3$

See Fig. 5.51.

Model: $y = a0 + a1 * t + a2 * t\hat{\ }2 + a3 * t\hat{\ }3 + a4 * t\hat{\ }4$

Variable	Value
$a0$	0
$a1$	0.992
$a2$	−0.160
$a3$	0.00108
$a4$	0.000633

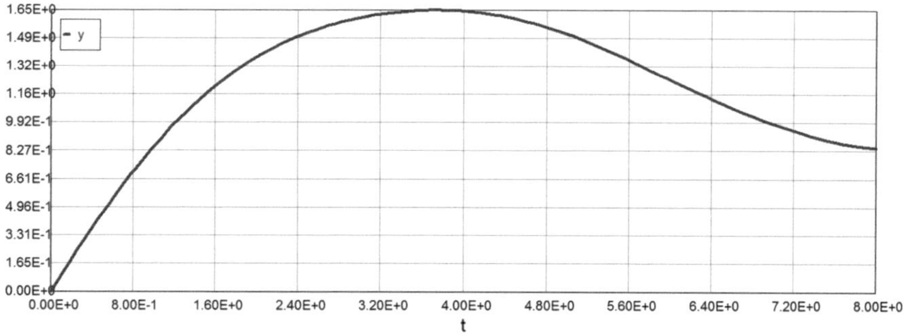

Fig. 5.51 Stress–temperature–strain diagram. Fast fire α = 0.01

$$\textbf{Model}: \sigma = 0.992 * \theta - 0.16 * \theta^\wedge 2 + 0.00108 * \theta^\wedge 3 - 0.000633 * \theta^\wedge 4 \quad (5.58)$$

α = 0.1

Calculated values of DEQ variables

	Variable	Initial value	Minimal value	Maximal value	Final value
1	*m*1	0.0411	−0.000308	0.0411	−0.000308
2	*n*	3	3	3	3
3	*r*	0	0	2.452516	2.409554
4	*t*	0	0	8	8
5	*y*	0	0	1.653602	0.8492636

Differential equations

$$1 \quad d(y)/d(t) = -0.1 * y * m1 + (\exp(-0.15 * t)) * (1 - 0.15 * t) - (\exp(t/(1.0 + 0.1 * t))) * m1 * y^\wedge n + 0.1 * m1 * t * (\exp(-0.15 * t))$$

Explicit equations

$$1 \quad n = 3$$
$$2 \quad r = t * (\exp(-0.15 * t))$$
$$3 \quad m1 = 0.0411 - 0.0226 * t + 0.00441 * t^\wedge 2 - 0.000279 * t^\wedge 3$$

See Fig. 5.52.

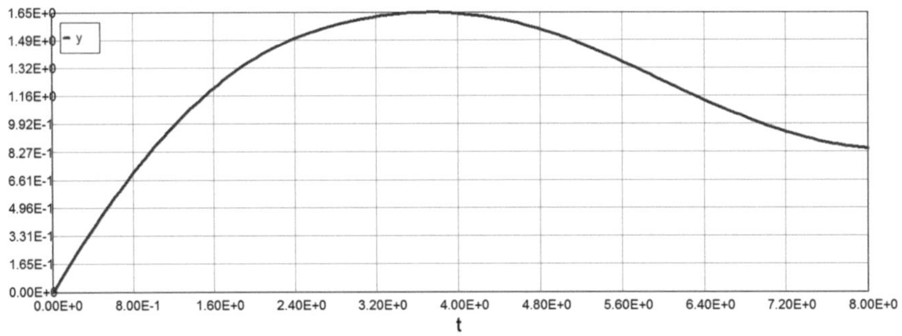

Fig. 5.52 Stress–temperature–strain diagram. Fast fire α = 0.1

Model: $y = a0 + a1 * t + a2 * t^2 + a3 * t^3 + a4 * t^4$

Variable	Value
$a0$	0.0
$a1$	0.992
$a2$	−0.16
$a3$	0.00107
$a4$	0.000633

Model: $\sigma = 0.992 * \theta - 0.16 * \theta^2 + 0.00107 * \theta^3 - 0.000633 * \theta^4$ (5.59)

α = 1

Calculated values of DEQ variables

	Variable	Initial value	Minimal value	Maximal value	Final value
1	$m1$	0.0411	−0.000308	0.0411	−0.000308
2	n	3	3	3	3
3	r	0	0	2.452516	2.409554
4	t	0	0	8	8
5	y	0	0	1.65559	0.855097

Differential equations

1 $\mathrm{d}(y)/\mathrm{d}(t) = -1 * y * m1 + (\exp(-0.15 * t)) * (1 - 0.15 * t) - (\exp(t/(1.0 + 0.1 * t))) * m1 * y^n + 1 * m1 * t * (\exp(-0.15 * t))$

Explicit equations

1 $n = 3$

2 $r = t * (\exp(-0.15 * t))$

3 $m1 = 0.0411 - 0.0226 * t + 0.00441 * t^2 - 0.000279 * t^3$

See Fig. 5.53.

Model: $y = a0 + a1 * t + a2 * t^2 + a3 * t^3 + a4 * t^4$

Variable	Value
$a0$	0.0
$a1$	0.992
$a2$	−0.16
$a3$	0.00104
$a4$	0.000635

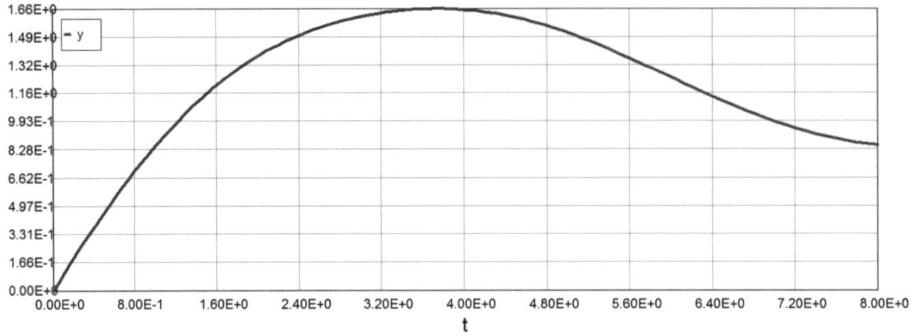

Fig. 5.53 Stress–temperature–strain diagram. Fast fire α = 1

$$\textbf{Model}: \sigma = 0.992 * \theta - 0.16 * \theta^\wedge 2 + 0.00104 * \theta^\wedge 3 + 0.000635 * \theta^\wedge 4 \quad (5.60)$$

α = 10

Calculated values of DEQ variables

	Variable	Initial value	Minimal value	Maximal value	Final value
1	$m1$	0.0411	−.000308	0.0411	−0.000308
2	n	3	3	3	3
3	r	0	0	2.452527	2.409554
4	t	0	0	8	8
5	y	0	0	1.675066	0.9095045

Differential equations

1 $\mathrm{d}(y)/\mathrm{d}(t) = -10 * y * m1 + (\exp(-0.15 * t)) * (1 - 0.15 * t) - (\exp(t/(1.0 + 0.1 * t))) * m1 * y^\wedge n + 10 * m1 * t * (\exp(-0.15 * t))$

Explicit equations

1 $n = 3$

2 $r = t * (\exp(-0.15 * t))$

3 $m1 = 0.0411 - 0.0226 * t + 0.00441 * t^\wedge 2 - 0.000279 * t^\wedge 3$

See Fig. 5.54.

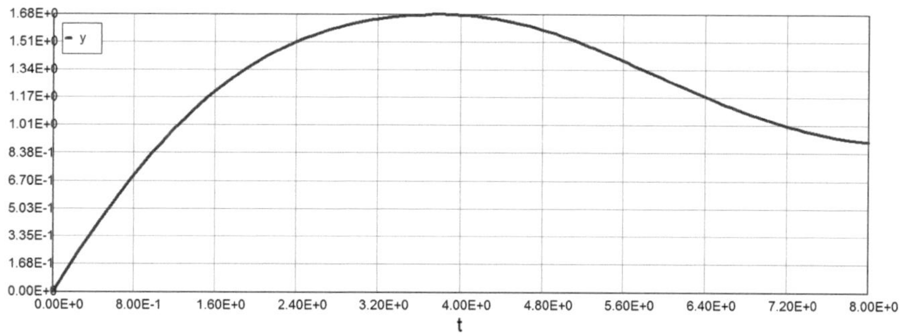

Fig. 5.54 Stress–temperature–strain diagram. Fast fire $\alpha = 10$

Model: $y = a0 + a1 * t + a2 * t\verb|^|2 + a3 * t\verb|^|3 + a4 * t\verb|^|4$

Variable	Value
$a0$	0.0
$a1$	0.993
$a2$	−0.158
$a3$	0.000792
$a4$	0.000648

$$\textbf{Model}: \sigma = 0.993 * \theta - 0.158 * \theta^\wedge 2 + 0.000792 * \theta^\wedge 3 + 0.000648 * \theta^\wedge 4 \quad (5.61)$$

α = 100

Calculated values of DEQ variables

	Variable	Initial value	Minimal value	Maximal value	Final value
1	$m1$	0.0411	−0.000308	0.0411	−0.000308
2	n	3	3	3	3
3	r	0	0	2.452498	2.409554
4	t	0	0	8	8
5	y	0	0	1.805355	1.238978

Differential equations

$$1 \quad d(y)/d(t) = -100 * y * m1 + (\exp(-0.15 * t)) * (1 - 0.15 * t) - (\exp(t/(1.0 + 0.1 * t))) * m1 * y^\wedge n + 100 * m1 * t * (\exp(-0.15 * t))$$

Explicit equations

1 $n = 3$

2 $r = t * (\exp(-0.15 * t))$

3 $m1 = 0.0411 - 0.0226 * t + 0.00441 * t^\wedge 2 - 0.000279 * t^\wedge 3$

See Fig. 5.55.

Model: $y = a0 + a1 * t + a2 * t^\wedge 2 + a3 * t^\wedge 3 + a4 * t^\wedge 4$

Variable	Value
$a0$	0.0
$a1$	0.996
$a2$	−0.148
$a3$	0.000172
$a4$	0.000641

$$\textbf{Model}: \sigma = 0.996 * \theta - 0.148 * \theta^\wedge 2 + 0.000172 * \theta^\wedge 3 + 0.000641 * \theta^\wedge 4 \quad (5.62)$$

α = 1000

Calculated values of DEQ variables

	Variable	Initial value	Minimal value	Maximal value	Final value
1	$m1$	0.0411	−0.000308	0.0411	−0.000308
2	n	3	3	3	3
3	r	0	0	2.452529	2.409554
4	t	0	0	8	8
5	y	0	0	2.12684	1.895772

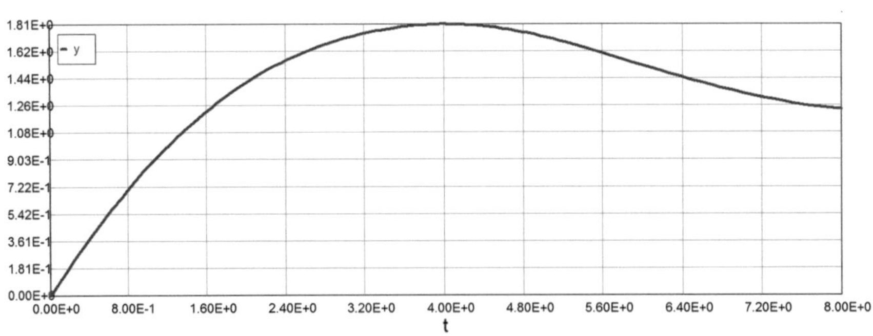

Fig. 5.55 Stress–temperature–strain diagram. Fast fire α = 100

Differential equations

1 $d(y)/d(t) = -1000 * y * m1 + (\exp(-0.15 * t)) * (1 - 0.15 * t) - (\exp(t/(1.0 + 0.1 * t))) * m1 * y^\wedge n + 1000 * m1 * t * (\exp(-0.15 * t))$

Explicit equations

1 $n = 3$

2 $r = t * (\exp(-0.15 * t))$

3 $m1 = 0.0411 - 0.0226 * t + 0.00441 * t^\wedge 2 - 0.000279 * t^\wedge 3$

See Fig. 5.56.

Model: $y = a0 + a1 * t + a2 * t^\wedge 2 + a3 * t^\wedge 3 + a4 * t^\wedge 4$

Variable	Value
$a0$	0.0
$a1$	0.986
$a2$	−0.135
$a3$	0.00337
$a4$	0.000217

Model: $\sigma = 0.986 * \theta - 0.135 * \theta^\wedge 2 + 0.00337 * \theta^\wedge 3 + 0.000217 * \theta^\wedge 4$ (5.63)

α = 10,000

Calculated values of DEQ variables

	Variable	Initial value	Minimal value	Maximal value	Final value
1	$m1$	0.0411	−0.000308	0.0411	−0.000308
2	n	3	3	3	3
3	r	0	0	2.452526	2.409554
4	t	0	0	8	8
5	y	0	0	2.384656	2.310501

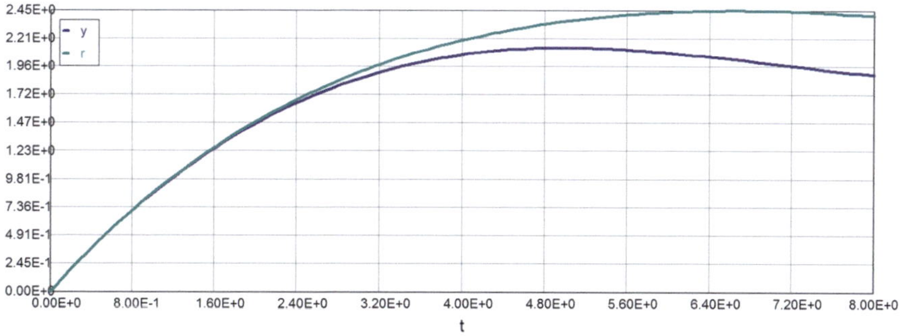

Fig. 5.56 Stress–temperature–strain diagram. Fast fire α = 1000

Differential equations

1 $d(y)/d(t) = -10{,}000 * y * m1 + (\exp(-0.15 * t)) * (1 - 0.15 * t) - (\exp(t/(1.0 + 0.1 * t))) * m1 * y^{\wedge}n + 10{,}000 * m1 * t * (\exp(-0.15 * t))$

Explicit equations

1 $n = 3$

2 $r = t * (\exp(-0.15 * t))$

3 $m1 = 0.0411 - 0.0226 * t + 0.00441 * t^{\wedge}2 - 0.000279 * t^{\wedge}3$

See Fig. 5.57.

Model: $y = a0 + a1 * t + a2 * t^{\wedge}2 + a3 * t^{\wedge}3 + a4 * t^{\wedge}4$

Variable	Value
$a0$	0.0
$a1$	0.991
$a2$	−0.142
$a3$	0.00831
$a4$	−0.000193

$$\textbf{Model}: \sigma = 0.991 * \theta - 0.142 * \theta^{\wedge}2 + 0.00831 * \theta^{\wedge}3 - 0.000193 * \theta^{\wedge}4 \quad (5.64)$$

α = 100,000

Calculated values of DEQ variables

	Variable	Initial value	Minimal value	Maximal value	Final value
1	$m1$	0.0411	−0.000308	0.0411	−0.000308
2	n	3	3	3	3
3	r	0	0	2.452514	2.409554
4	t	0	0	8	8
5	y	0	0	2.444614	2.398451

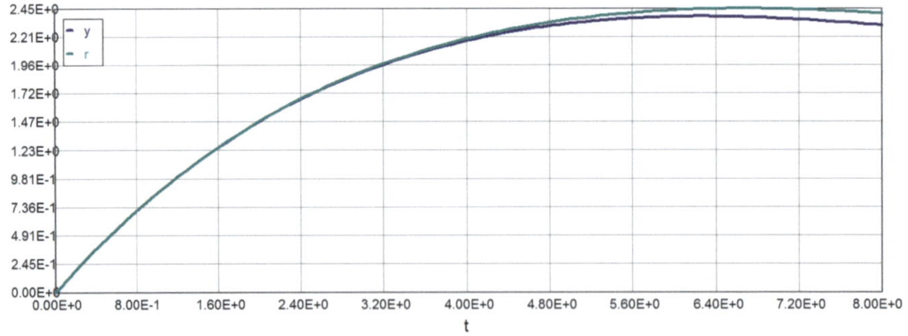

Fig. 5.57 Stress–temperature–strain diagram. Fast fire α = 10,000

Differential equations

1 $d(y)/d(t) = -100{,}000 * y * m1 + (\exp(-0.15 * t)) * (1 - 0.15 * t) - (\exp(t/(1.0 + 0.1 * t))) * m1 * y\hat{\ }n + 100{,}000 * m1 * t * (\exp(-0.15 * t))$

Explicit equations

1 $n = 3$

2 $r = t * (\exp(-0.15 * t))$

3 $m1 = 0.0411 - 0.0226 * t + 0.00441 * t\hat{\ }2 - 0.000279 * t\hat{\ }3$

See Fig. 5.58.

Model: $y = a0 + a1 * t + a2 * t\hat{\ }2 + a3 * t\hat{\ }3 + a4 * t\hat{\ }4$

Variable	Value
$a0$	0.0
$a1$	0.994
$a2$	−0.144
$a3$	0.0093
$a4$	−0.000261

Model: $\sigma = 0.994 * \theta - 0.144 * \theta\hat{\ }2 + 0.0093 * \theta\hat{\ }3 - 0.000261 * \theta\hat{\ }4$ (5.65)

See Table 5.5.

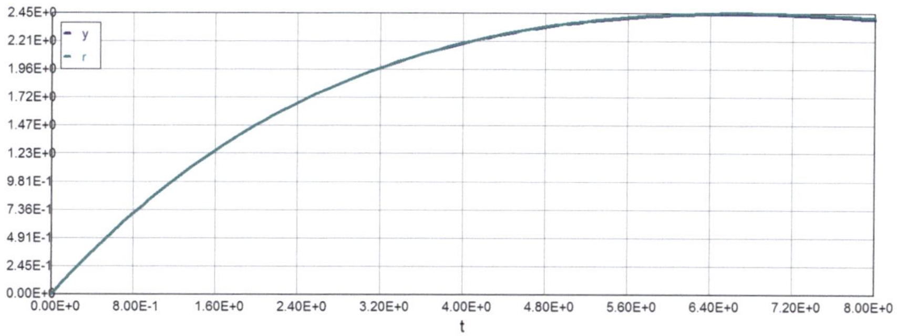

Fig. 5.58 Stress–temperature–strain diagram. Fast fire α = 100,000

Table 5.5 Stress–temperature data (fast fire)

$\alpha\|\theta$		1	2	3	4	5	6	7	8
0.001	0	0.852	1.379	1.610	1.648	1.513	1.246	0.987	0.849
0.01	0	0.852	1.379	1.610	1.648	1.513	1.246	0.987	0.849
0.1	0	0.852	1.379	1.610	1.648	1.514	1.250	0.988	0.849
1.0	0	0.852	1.379	1.611	1.650	1.516	1.250	0.993	0.855
10	0	0.852	1.385	1.625	1.671	1.543	1.286	1.042	0.910
100	0	0.855	1.418	1.713	1.805	1.723	1.530	1.349	1.239
1000	0	0.859	1.467	1.859	2.068	2.127	2.084	1.991	1.896
10,000	0	0.860	1.480	1.906	2.177	2.327	2.382	2.368	2.31
100,000	0	0.861	1.481	1.912	2.193	2.358	2.433	2.441	2.398
r	0	0.861	1.482	1.913	2.195	2.362	2.439	2.450	2.41

5.3.5 *Functional Dependencies of Creep Stresses and Strains from the Stress Exponent "n"*

n = 1 α = 100
Calculated values of DEQ variables

	Variable	Initial value	Minimal value	Maximal value	Final value
1	$m1$	0.0411	−0.000308	0.0411	−0.000308
2	n	1	1	1	1
3	r	0	0	2.452517	2.409554
4	t	0	0	8	8
5	y	0	0	2.120507	1.710478

Differential equations

1 $$\mathrm{d}(y)/\mathrm{d}(t) = -100 * y * m1 + (\exp(-0.15 * t)) * (1 - 0.15 * t) - (\exp(t/(1.0 + 0.1 * t))) * m1 * y\hat{}n + 100 * m1 * t * (\exp(-0.15 * t))$$

Explicit equations

1 $n = 1$

2 $r = t * (\exp(-0.15 * t))$

3 $m1 = 0.0411 - 0.0226 * t + 0.00441 * t\hat{}2 - 0.000279 * t\hat{}3$

See Fig. 5.59.

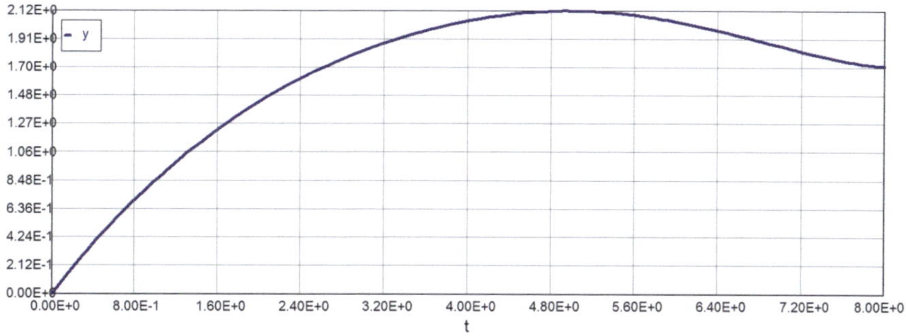

Fig. 5.59 Stress–temperature–strain diagram. Fast fire $n = 1$

Model: $y = a0 + a1 * t + a2 * t^2 + a3 * t^3 + a4 * t^4$

Variable	Value
$a0$	0.0
$a1$	0.917
$a2$	−0.106
$a3$	−0.00048
$a4$	0.000327

$$\textbf{Model}: \sigma = 0.917 * \theta - 0.106 * \theta^\wedge 2 - 0.00048 * \theta^\wedge 3 + 0.000327 * \theta^\wedge 4 \quad (5.66)$$

n = 3

Calculated values of DEQ variables

	Variable	Initial value	Minimal value	Maximal value	Final value
1	$m1$	0.0411	−0.000308	0.0411	−0.000308
2	n	3	3	3	3
3	r	0	0	2.452498	2.409554
4	t	0	0	8	8
5	y	0	0	1.805355	1.238978

Differential equations

$$1 \quad d(y)/d(t) = -100 * y * m1 + (\exp(-0.15 * t)) * (1 - 0.15 * t) - (\exp(t/(1.0 + 0.1 * t))) * m1 * y^\wedge n + 100 * m1 * t * (\exp(-0.15 * t))$$

Explicit equations

1 $n = 3$
2 $r = t * (\exp(-0.15 * t))$
3 $m1 = 0.0411 - 0.0226 * t + 0.00441 * t^\wedge 2 - 0.000279 * t^\wedge 3$

See Fig. 5.60.
Model: $y = a0 + a1 * t + a2 * t^\wedge 2 + a3 * t^\wedge 3 + a4 * t^\wedge 4$

Variable	Value
$a0$	0.0
$a1$	0.996
$a2$	−0.148
$a3$	0.000172
$a4$	0.000641

$$\textbf{Model}: \sigma = 0.996 * \theta - 0.148 * \theta^\wedge 2 + 0.000172 * \theta^\wedge 3 + 0.000641 * \theta^\wedge 4 \quad (5.67)$$

n = 5
Calculated values of DEQ variables

	Variable	Initial value	Minimal value	Maximal value	Final value
1	$m1$	0.0411	−0.000308	0.0411	−0.000308
2	n	5	5	5	5
3	r	0	0	2.452495	2.409554
4	t	0	0	8	8
5	y	0	0	1.561903	1.116931

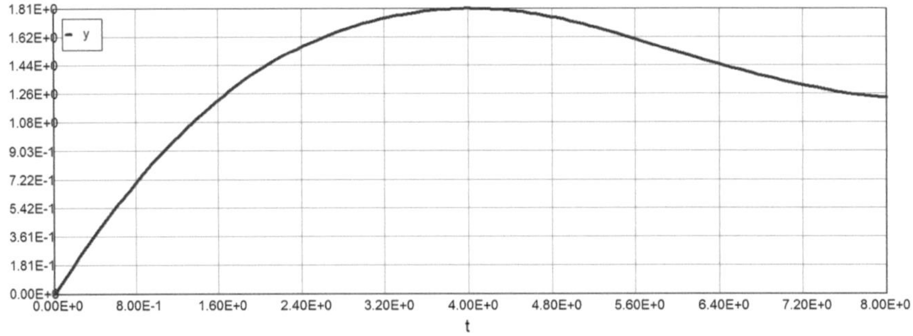

Fig. 5.60 Stress–temperature–strain diagram. Fast fire $n = 3$

Differential equations

1 $d(y)/d(t) = -100 * y * m1 + (\exp(-0.15 * t)) * (1 - 0.15 * t) - (\exp(t/(1.0 + 0.1 * t))) * m1 * y^\wedge n + 100 * m1 * t * (\exp(-0.15 * t))$

Explicit equations

1 $n = 5$

2 $r = t * (\exp(-0.15 * t))$

3 $m1 = 0.0411 - 0.0226 * t + 0.00441 * t^\wedge 2 - 0.000279 * t^\wedge 3$

See Fig. 5.61.

Model: $y = a0 + a1 * t + a2 * t^\wedge 2 + a3 * t^\wedge 3 + a4 * t^\wedge 4$

Variable	Value
$a0$	0
$a1$	1.20
$a2$	−0.296
$a3$	0.0274
$a4$	−0.00085

$$\textbf{Model}: \sigma = 1.2 * \theta - 0.296 * \theta^\wedge 2 + 0.0274 * \theta^\wedge 3 - 0.00085 * \theta^\wedge 4 \quad (5.68)$$

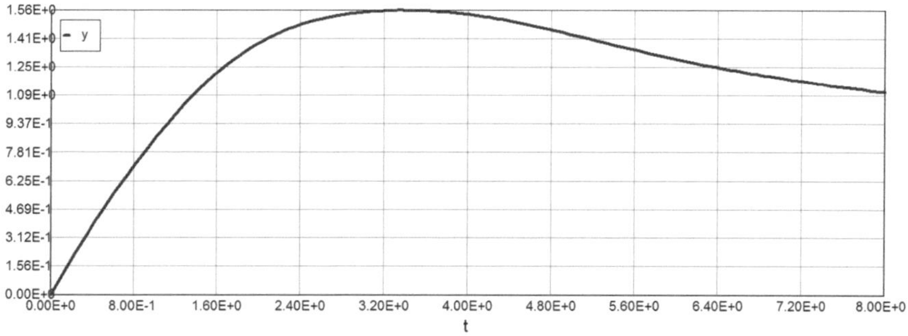

Fig. 5.61 Stress–temperature–strain diagram. Fast fire $n = 5$

n = 8

Calculated values of DEQ variables

	Variable	Initial value	Minimal value	Maximal value	Final value
1	$m1$	0.0411	−0.000308	0.0411	−0.000308
2	n	8	8	8	8
3	r	0	0	2.452497	2.409554
4	t	0	0	8	8
5	y	0	0	1.3654	1.059061

Differential equations

1 $d(y)/d(t) = -100 * y * m1 + (\exp(-0.15 * t)) * (1 - 0.15 * t) - (\exp(t/(1.0 + 0.1 * t))) * m1 * y^\wedge n + 100 * m1 * t * (\exp(-0.15 * t))$

Explicit equations

1 $n = 8$

2 $r = t * (\exp(-0.15 * t))$

3 $m1 = 0.0411 - 0.0226 * t + 0.00441 * t^\wedge 2 - 0.000279 * t^\wedge 3$

See Fig. 5.62.

Model: $y = a0 + a1 * t + a2 * t^\wedge 2 + a3 * t^\wedge 3 + a4 * t^\wedge 4$

Variable	Value
$a0$	0
$a1$	1.294
$a2$	−0.4
$a3$	0.05
$a4$	−0.00224

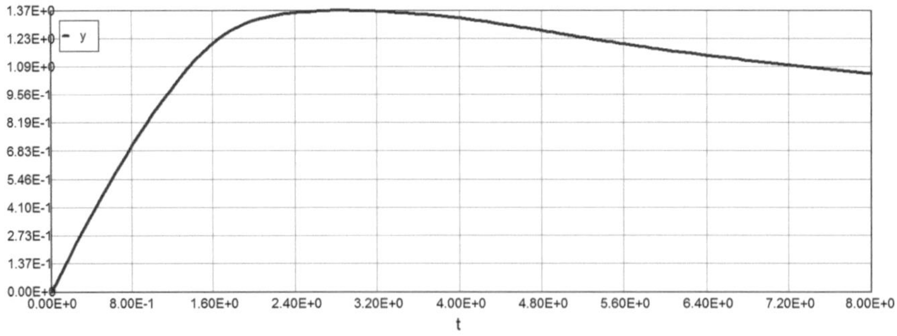

Fig. 5.62 Stress–temperature–strain diagram. Fast fire n = 8

$$\textbf{Model}: \sigma = 1.294 * \theta - 0.4 * \theta^\wedge 2 + 0.05 * \theta^\wedge 3 - 0.00224 * \theta^\wedge 4 \qquad (5.69)$$

n = 10

Calculated values of DEQ variables

	Variable	Initial value	Minimal value	Maximal value	Final value
1	$m1$	0.0411	−0.000308	0.0411	−0.000308
2	n	10	10	10	10
3	r	0	0	2.452512	2.409554
4	t	0	0	8	8
5	y	0	0	1.293494	1.041724

Differential equations

1 $\mathrm{d}(y)/\mathrm{d}(t) = -100 * y * m1 + (\exp(-0.15 * t)) * (1 - 0.15 * t) - (\exp(t/(1.0 + 0.1 * t))) * m1 * y^\wedge n + 100 * m1 * t * (\exp(-0.15 * t))$

Explicit equations

1 $n = 10$

2 $r = t * (\exp(-0.15 * t))$

3 $m1 = 0.0411 - 0.0226 * t + 0.00441 * t^\wedge 2 - 0.000279 * t^\wedge 3$

See Fig. 5.63.

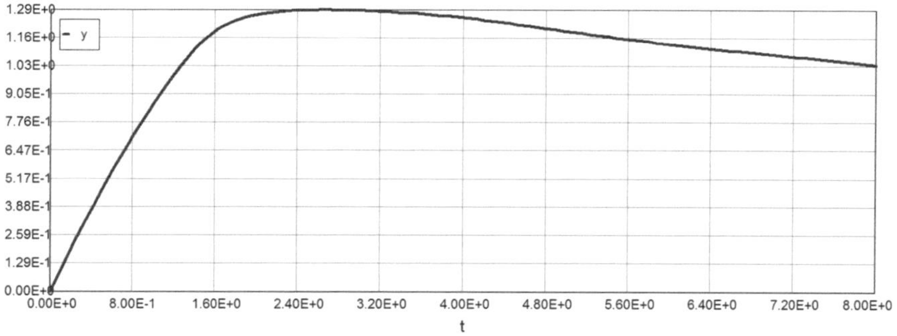

Fig. 5.63 Stress–temperature–strain diagram. Fast fire n = 10

Model: $y = a0 + a1 * t + a2 * t\text{^}2 + a3 * t\text{^}3 + a4 * t\text{^}4$

Variable	Value
$a0$	0
$a1$	1.296
$a2$	−0.423
$a3$	0.0559
$a4$	−0.00265

Table 5.6 Maximum creep stress versus stress exponent

n	1	3	5	8	10	α
σ						
σ_{max}	2.12	1.805	1.562	1.365	1.290	100

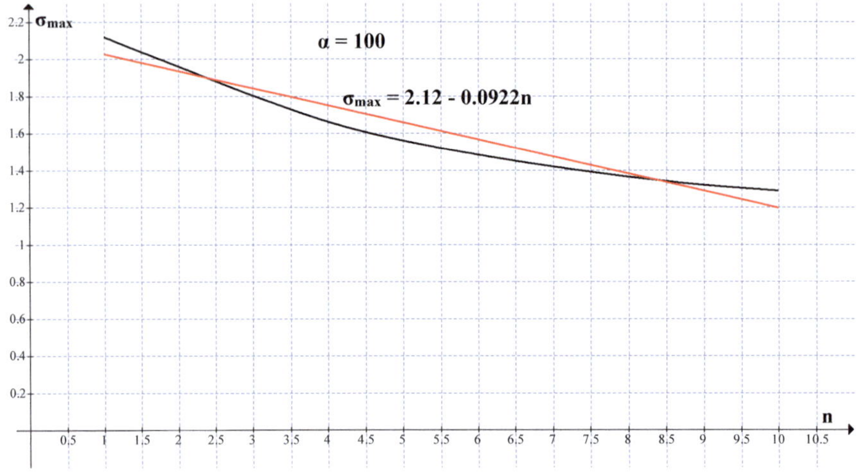

Fig. 5.64 Maximum creep stress versus stress exponent. Fast fire $\alpha = 100$

$$\textbf{Model}: \sigma = 1.296 * \theta - 0.423 * \theta^{\wedge}2 + 0.0559 * \theta^{\wedge}3 - 0.00265 * \theta^{\wedge}4 \quad (5.70)$$

See Table 5.6.
$\sigma_{max} = -0.0922n + 2.12$ (Fig. 5.64).

5.3.6 *Temperature–Time Function. Medium Fire*

Medium Fire; Temperature Increase (822 K < T < 882 K)
Data $T_* = 600$ K; $0 < \tau < 0.2$; $0.175 < \gamma < 0.275$
Select $\gamma = 0.175$

Differential equations (5.3) and (5.4) are rewritten as an input for "Polymath" software:

1 d(y0)/d(t) = 20 ∗ (1 − y2) ∗ exp(y0/(1 + 0.1 ∗ y0))^ − 0.157 ∗ y0^4
2 d(y2)/d(t)^= 5.5 ∗ (1 − y2) ∗ exp(y0/(1 + 0.1 ∗ y0))
3 d(y1)/d(t) = 3.5 ∗ (1 − y1)^1.0 ∗ exp(y/(1 + 0.1 ∗ y))
4 d(y)/d(t) = (1) ∗ 20 ∗ (1 − y1)^1.0 ∗ exp(y/(1 + 0.1 ∗ y))^ − 0.157 ∗ y^4,

where

"y" is the dimensionless temperature "θ" with the corresponding parameter "γ = 0.175".
"y0" is the dimensionless temperature "θ" with the corresponding parameter "γ = 0.275".
"y1" is the concentration of the product of the first-order chemical reaction with "γ = 0.175".
"y2" is the concentration of the product of the first-order chemical reaction with "γ = 0.275".

Calculated values of DEQ variables

	Variable	Initial value	Minimal value	Maximal value	Final value
1	*t*	0	0	0.2	0.2
2	*y*	0	0	4.121	2.629
3	*y*0	0	0	3.013	2.398
4	*y*1	0	0	0.9987	0.9987
5	*y*2	0	0	0.9993	0.9993

Tabulated solution of Eqs. (5.3) and (5.4) is presented in [4] and the graphs are shown in Fig. 5.65.

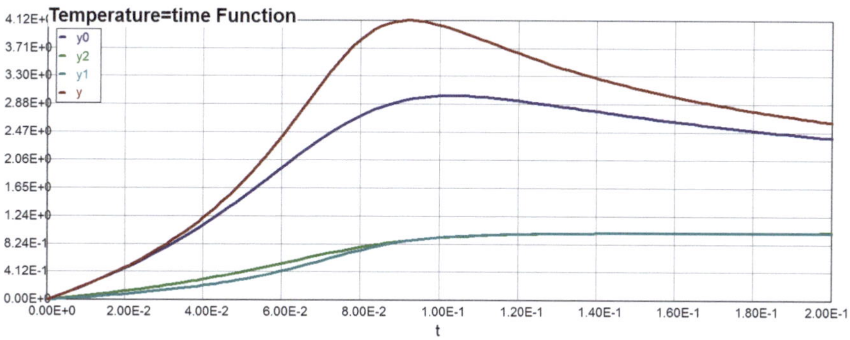

Fig. 5.65 Dimensionless temperature–time curves

5.3.7 *Analytical Expression of the Inverse Function* θ^{-1} *and Its First Derivative*

First, based on the data shown above, the "best-to-fit" (polynomial regression) approximation of the dimensionless temperature–time curve can be presented as follows:

Model: $y = a0 + a1 * t + a2 * t^2 + a3 * t^3 + a4 * t^4 + a5 * t^5$

Variable	Value
*a*0	0
*a*1	−75.87
*a*2	3910
*a*3	−4.542E+04
*a*4	2.06E+05
*a*5	−3.29E+05

$$\textbf{Model}: \theta = -75.87 * \tau + 3910 * \tau^{\wedge}2 - 45{,}420 * \tau^{\wedge}3 + 206{,}000 * \tau^{\wedge}4 - 329{,}000 * \tau^{\wedge}5 \tag{5.71}$$

Monotonically increasing temperature–time function: $(0 < \theta < 4)$; $(0 < \tau 1 < 0.0913)$

As noted above, a function of the temperature–time should be divided into two portions monotonically increasing and monotonically decreasing. Let us start with a monotonically increasing area. The range of temperature values in this case is the interval from 0 to 4 and the time interval from 0 to 0.0913. Therefore, using polynomial regression method again we have

Model: $y = a0 + a1 * t + a2 * t^2 + a3 * t^3 + a4 * t^4$

Variable	Value
*a*0	0
*a*1	42.27
*a*2	−1347
*a*3	3.603E+04
*a*4	−2.277E+05

$$\textbf{Model}: \theta = 42.27 * \tau - 1347 * \tau^{\wedge}2 + 36{,}030 * \tau^{\wedge}3 - 227{,}700 * \tau^{\wedge}4 \tag{5.72}$$

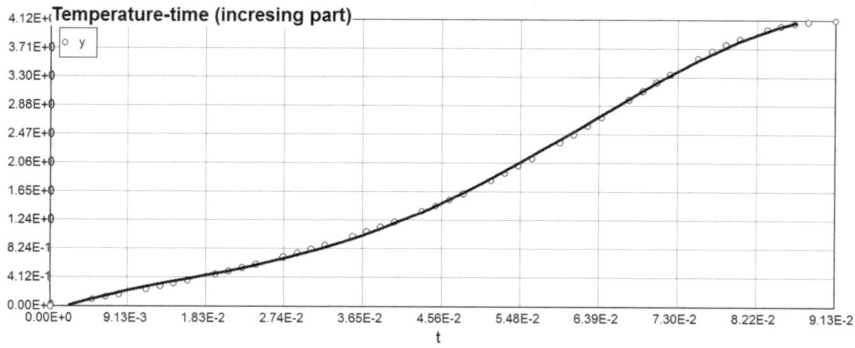

Fig. 5.66 Monotonically increasing temperature–time function

See Fig. 5.66.

The inverse function θ^{-1} and its first derivative (monotonically increasing functions) are:

Model: $t = a0 + a1 * y + a2 * y\^{}2 + a3 * y\^{}3 + a4 * y\^{}4$

Variable	Value
$a0$	0.0008285
$a1$	0.0438492
$a2$	−0.010139
$a3$	0.0004935
$a4$	0.0001537

The graph of the inverse function θ^{-1} (Fig. 5.67).

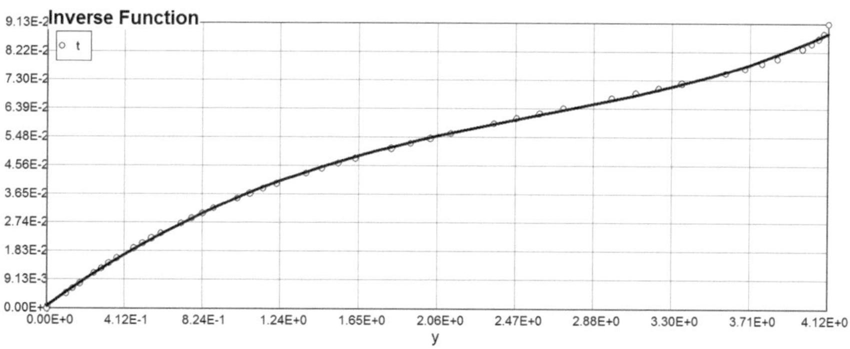

Fig. 5.67 Monotonically increasing inverse function θ^{-1}

Therefore,

$$\textbf{Model}: \mathbf{m} = \tau = 0.000928 + 0.0438 * \theta - 0.0101 * \theta\hat{\ }2 + 0.000494 * \theta\hat{\ }3 + 0.000154 * \theta\hat{\ }4$$

$$\mathbf{m1} = \tau' = 0.0438 - 0.0202 * \theta + 0.00148 * \theta\hat{\ }2 + 0.000616 * \theta\hat{\ }3 \qquad (5.74)$$

Monotonically decreasing temperature–time function: $(4 > \theta > 2.6)$; $(0.0913 < \tau 1 < 0.2)$

Performing calculations now that are similar to the previous case, we obtain corresponding formulas for a monotonically decreasing temperature range.

Temperature–time function

Model: $x = a0 + a1 * t1 + a2 * t1\hat{\ }2 + a3 * t1\hat{\ }3 + a4 * t1\hat{\ }4$

Variable	Value
*a*0	−1.5
*a*1	204.8
*a*2	−2450
*a*3	1.147E+04
*a*4	−1.914E+04

$$\textbf{Model}: \theta = -1.5 + 204.8 * \tau 1 - 2450 * \tau 1\hat{\ }2 + 11470 * \tau 1\hat{\ }3 - 19140 * \tau 1\hat{\ }4 \qquad (5.75)$$

The corresponding graph is: (Fig. 5.68).

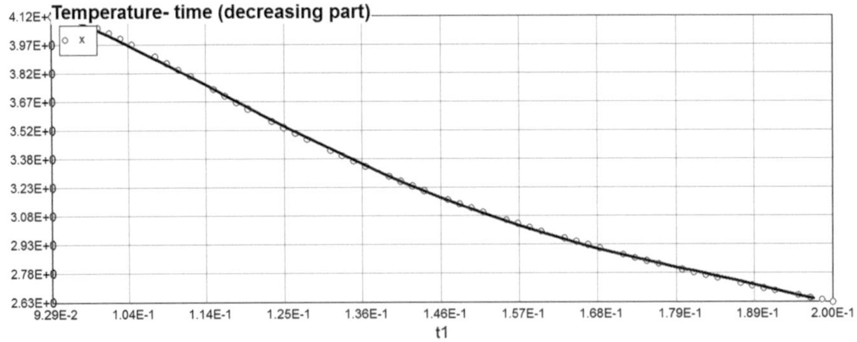

Fig. 5.68 Monotonically decreased temperature–time function

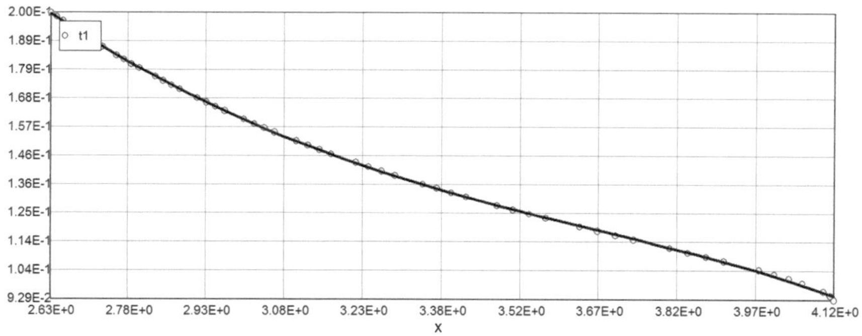

Fig. 5.69 Monotonically decreasing inverse function $\theta^{-1}(\tau 1)$

The inverse function $\theta^{-1}(\tau 1)$:

Model: $t1 = a0 + a1 * x + a2 * x\text{^}2 + a3 * x\text{^}3 + a4 * x\text{^}4$

Variable	Value
$a0$	0.79
$a1$	−0.1315
$a2$	−0.151
$a3$	0.061
$a4$	−0.0064

See Fig. 5.69.

$$\textbf{Model}: m0 = \tau 1 = 0.79 - 0.1315 * \theta - 0.151 * \theta\text{^}2 + 0.061 * \theta\text{^}3 - 0.0064 * \theta\text{^}4 \tag{5.76}$$

The first derivative of inverse function $\theta^{-1}(\tau 1)$:

$$m01 = \tau 1' = -0.1315 - 0.302 * \theta + 0.183 * \theta\text{^}2 - 0.0256 * \theta\text{^}3 \tag{5.77}$$

5.3.8 Creep Constitutive Equation. Equivalent ODE Method

Monotonically increasing temperature–time function

The stress–strain response of a material is described using differential equation of the form:

$$\sigma'(\theta) = -\sigma(\theta)\alpha m1 + e^{-0.15\theta}[1 - 0.15\theta] - (\exp(t)/(1 + 0.1t))m1\sigma^n(\theta) + m1\alpha\theta e^{-0.15\theta}$$
$$m = \varphi(\theta) = 0.00154 + 0.0411 * \theta - 0.0113 * \theta\hat{}2 + 0.00147 * \theta\hat{}3 - 0.0000697 * \theta\hat{}4$$
$$m1 = \tau' = \varphi'(\theta) = 0.0411 - 0.0226 * \theta + 0.00441 * \theta\hat{}2 - 0.000279 * \theta\hat{}3$$
$$\sigma(\theta = 0) = 0; \quad \alpha = 0.33 \tag{5.78}$$

Consider now general uniaxial creep constitutive model for monotonically increasing part of real fire temperature load (equivalent differential equation with one MPP $\alpha = 0.33$):

$$\begin{aligned} d(\sigma)/d(\theta) = &-0.33 * \sigma * m1 + (\exp(-0.15 * \theta)) * (1 - 0.15 * \theta) - (\exp(\theta/(1.0 + 0.1 * \theta))) * m1 * \sigma^n \\ &+ 0.33 * m1 * \theta * (\exp(-0.15 * \theta)) \end{aligned} \tag{5.79}$$

The temperature range is $[0 < \theta < 4]$ and the dimensionless time duration for increasing part of fully developed fire stage is $[0 < \tau < 0.0929]$. The solution of Eq. (5.79) is as follows:

Calculated values of DEQ variables

	Variable	Initial value	Minimal value	Maximal value	Final value
1	$m01$	−0.1315	−0.2764831	−0.0403539	−0.0499
2	$m1$	0.0438	0.0121118	0.0438	0.026104
3	n	2	2	2	2
4	t	0	0	4	4
5	z	0	0	1.587823	1.318068

Differential equations

1 $$\begin{aligned} d(z)/d(t) = &-1 * (-0.33 * z * (-m01) + (\exp(-0.15 * (t))) * (1 - 0.15 * (t)) \\ &- (\exp((t)/(1.0 + 0.1 * (t)))) * (-m01) * z\hat{}n \\ &+ 0.33 * (-m01) * (t) * (\exp(-0.15 * (t)))) \end{aligned}$$

The corresponding graph is: (Fig. 5.70).

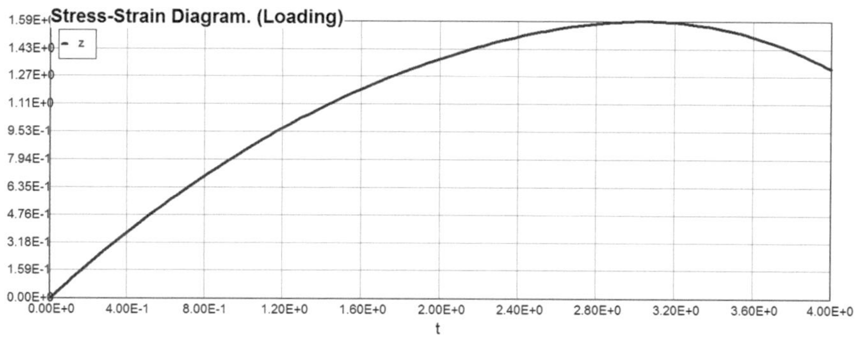

Fig. 5.70 Stress–strain diagram. Loading stage

Model: $z = a0 + a1 * t + a2 * t\text{^}2 + a3 * t\text{^}3 + a4 * t\text{^}4$

Variable	Value
$a0$	0
$a1$	1.052
$a2$	−0.234
$a3$	0.0402
$a4$	−0.0067

$$\textbf{Model}: \sigma = 1.052 * \theta - 0.234 * \theta^\wedge 2 + 0.0402 * \theta^\wedge 3 - 0.0067 * \theta^\wedge 4 \qquad (5.80)$$

Monotonically decreasing temperature–time function

Consider now general uniaxial creep constitutive model for monotonically decreasing part of real fire temperature load. Equivalent differential equation again is (5.78), but function m1 is substituted by m01 and the corresponding graph will be shifted to the end of increasing temperature interval $\theta = 4$.

Calculated values of DEQ variables

	Variable	Initial value	Minimal value	Maximal value	Final value
3	$m01$	−0.1256683	−0.1256683	−0.0403455	−0.0597487
5	n	2	2	2	2
6	t	2.63	2.63	4.12	4.12
11	z	1.32	0.7445308	1.32	0.7445308

Differential equations

$$5\quad d(z)/d(t) = -1 * (-0.33 * z * (-m01) + (\exp(-0.15 * (t))) * (1 - 0.15 * (t)) - (\exp((t)/(1.0 + 0.1 * (t)))) * (-m01) * z^{\wedge}n + 0.33 * (-m01) * (t) * (\exp(-0.15 * (t))))$$

Explicit equations

$$1\quad n = 2$$

$$3\quad m01 = -0.1315 - 0.302 * t + 0.183 * t^{\wedge}2 - 0.0256 * t^{\wedge}3$$

Initial condition
$z(0) = 1.318$

Explicit equations

$$1\quad n = 2$$

$$3\quad m01 = -0.1315 - 0.302 * \theta + 0.183 * \theta^{\wedge}2 - 0.0256 * \theta^{\wedge}3$$

Thus the relaxation modulus graph is shown in Fig. 5.71.
Using polynomial regression method again, we have

Model: $z = a0 + a1 * t + a2 * t^2 + a3 * t^3 + a4 * t^4$

Variable	Value
$a0$	32.63
$a1$	−31.87
$a2$	11.94
$a3$	−1.97
$a4$	0.12

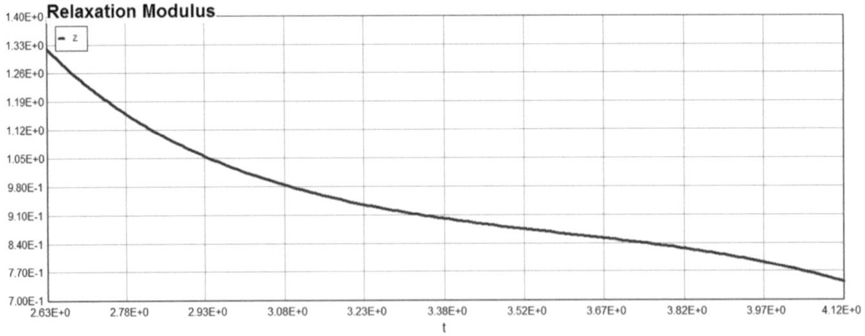

Fig. 5.71 Relaxation modulus. Fast fire

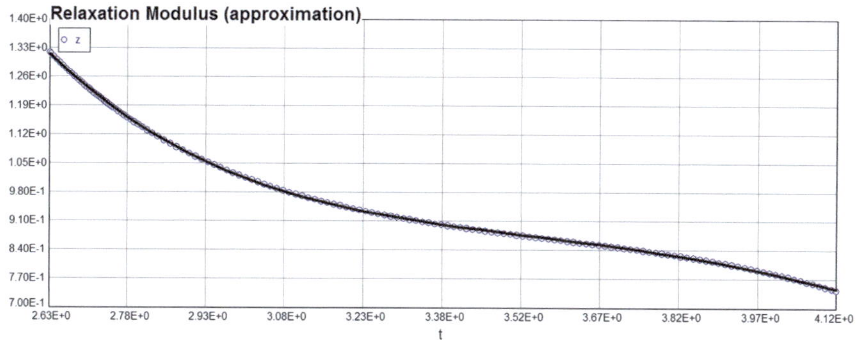

Fig. 5.72 Relaxation modulus (approximation)

See Fig. 5.72.

$$\sigma = 32.63 - 31.87 * \theta + 11.94 * \theta^\wedge 2 - 1.97 * \theta^\wedge 3 + 0.12 * \theta^\wedge 4 \tag{5.81}$$

Finally, the combined graph (loading and unloading creep stresses) is as follows: (Fig. 5.73).

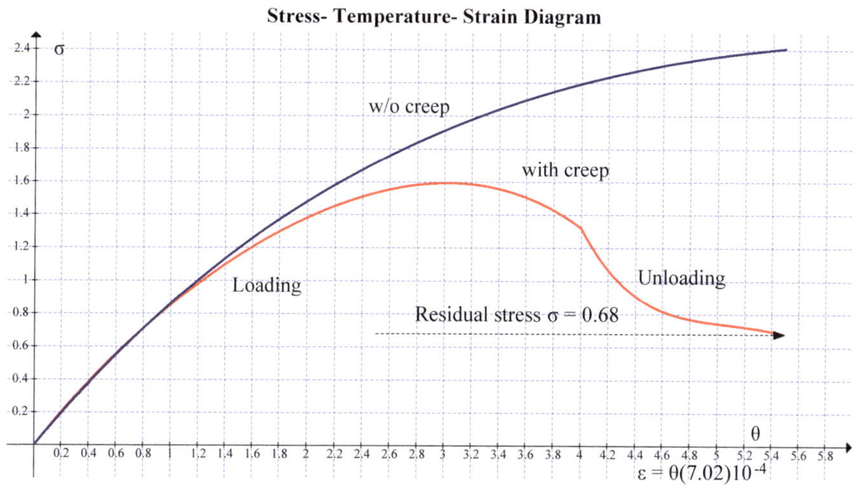

Fig. 5.73 Stress–temperature–strain diagram. Medium Fire

5.3.9 The Functional Dependencies of Creep Stresses and Strains from Material Properties Parameters (MPP)

Consider now Eq. (5.12) and $\alpha = 0.001$. Again, in case of FF, we have

Differential equations

$$\begin{aligned} \mathrm{d}(y)/\mathrm{d}(t) = & -\alpha_i * y * m1 + (\exp(-0.15 * t)) * (1 - 0.15 * t) \\ & - (\exp(t/(1.0 + 0.1 * t))) * m1 * y^\wedge n + \alpha_i * m1 * t * (\exp(-0.15 * t)) \end{aligned}$$

Explicit equations

$$\begin{aligned} n &= 3 \\ m1 &= \tau'(\theta) = 0.0405 - 0.02252 * t^\wedge 1 + 0.004386 * t^\wedge 2 - 0.0002747 * t^\wedge 3 \end{aligned} \quad (5.82)$$

Computer code and solution of Eq. (5.12) for loading stage is as follows:
α = 0.001

Calculated values of DEQ variables

	Variable	Initial value	Minimal value	Maximal value	Final value
1	$m1$	0.0438	0.0121152	0.108936	0.108936
2	n	3	3	3	3
3	r	0	0	2.439418	2.439418
4	t	0	0	6	6
5	y	0	0	1.501322	0.3841946

Differential equations

$$\begin{aligned} 1 \quad \mathrm{d}(y)/\mathrm{d}(t) = & -0.001 * y * m1 + (\exp(-0.15 * t)) * (1 - 0.15 * t) \\ & - (\exp(t/(1.0 + 0.1 * t))) * m1 * y^\wedge n + 0.001 * m1 * t * (\exp(-0.15 * t)) \end{aligned}$$

Explicit equations

1 $n = 3$
2 $r = t * (\exp(-0.15 * t))$
3 $m1 = 0.0438 - 0.0202 * t + 0.00148 * t^\wedge 2 + 0.000616 * t^\wedge 3$

See Fig. 5.74.

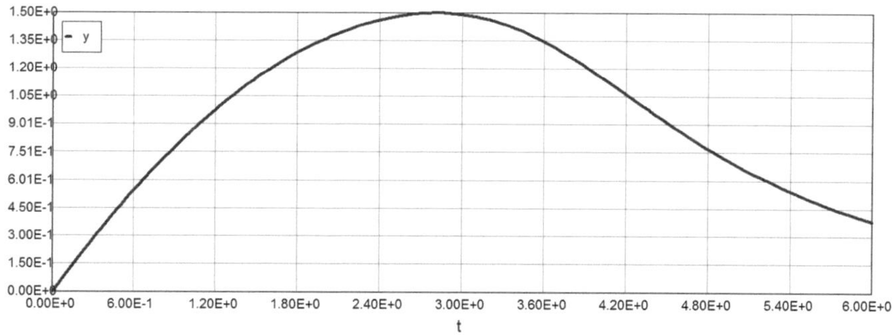

Fig. 5.74 Stress–temperature–strain diagram. Medium Fire $\alpha = 0.001$

Model: $y = a0 + a1 * t + a2 * t\wedge 2 + a3 * t\wedge 3 + a4 * t\wedge 4$

Variable	Value
$a0$	0.0
$a1$	0.86
$a2$	0.0189
$a3$	−0.0713
$a4$	0.00767

$$\textbf{Model}: \sigma = 0.86 * \theta + 0.0189 * \theta^\wedge 2 - 0.0713 * \theta^\wedge 3 + 0.00767 * \theta^\wedge 4 \quad (5.83)$$

α = 0.01

Calculated values of DEQ variables

	Variable	Initial value	Minimal value	Maximal value	Final value
1	$m1$	0.0438	0.0121152	0.108936	0.108936
2	n	3	3	3	3
3	r	0	0	2.439418	2.439418
4	t	0	0	6	6
5	y	0	0	1.501342	0.3848787

Differential equations

1 $\mathrm{d}(y)/\mathrm{d}(t) = -0.01 * y * m1 + (\exp(-0.15 * t)) * (1 - 0.15 * t) - (\exp(t/(1.0 + 0.1 * t))) * m1 * y^\wedge n + 0.01 * m1 * t * (\exp(-0.15 * t))$

Explicit equations

1 $n = 3$

2 $r = t * (\exp(-0.15 * t))$

3 $m1 = 0.0438 - 0.0202 * t + 0.00148 * t^\wedge 2 + 0.000616 * t^\wedge 3$

See Fig. 5.75.

Model: $y = a0 + a1 * t + a2 * t^2 + a3 * t^3 + a4 * t^4$

Variable	Value
$a0$	0.0
$a1$	0.861
$a2$	0.019
$a3$	−0.0713
$a4$	0.00767

$$\textbf{Model}: \sigma = 0.861 * \theta + 0.019 * \theta^\wedge 2 - 0.0713 * \theta^\wedge 3 + 0.00767 * \theta^\wedge 4 \quad (5.84)$$

α = 0.1

Calculated values of DEQ variables

	Variable	Initial value	Minimal value	Maximal value	Final value
1	$m1$	0.0438	0.0121152	0.108936	0.108936
2	n	3	3	3	3
3	r	0	0	2.439418	2.439418
4	t	0	0	6	6
5	y	0	0	1.501549	0.3916381

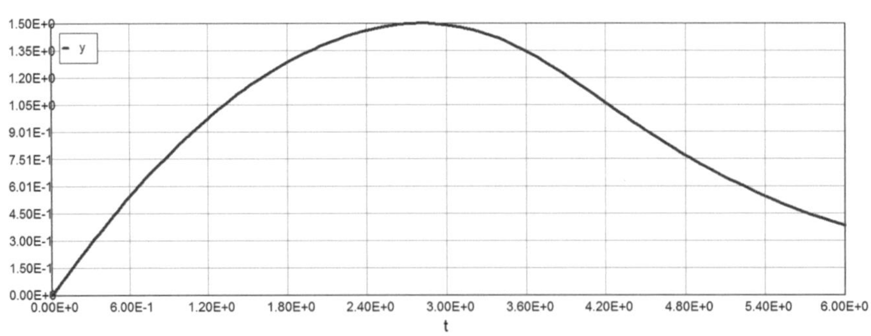

Fig. 5.75 Stress–temperature–strain diagram. Medium Fire α = 0.01

Differential equations

1 $d(y)/d(t) = -0.1 * y * m1 + (\exp(-0.15 * t)) * (1 - 0.15 * t) - (\exp(t/(1.0 + 0.1 * t))) * m1 * y^{\wedge}n + 0.1 * m1 * t * (\exp(-0.15 * t))$

Explicit equations

1 $n = 3$

2 $r = t * (\exp(-0.15 * t))$

3 $m1 = 0.0438 - 0.0202 * t + 0.00148 * t^{\wedge}2 + 0.000616 * t^{\wedge}3$

See Fig. 5.76.

Model: $y = a0 + a1 * t + a2 * t^{\wedge}2 + a3 * t^{\wedge}3 + a4 * t^{\wedge}4$

Variable	Value
$a0$	0.0
$a1$	0.861
$a2$	0.0193
$a3$	−0.0714
$a4$	0.00769

Model: $\sigma = 0.861 * \theta + 0.0193 * \theta^{\wedge}2 - 0.0714 * \theta^{\wedge}3 + 0.00769 * \theta^{\wedge}4$ (5.85)

α = 1

Calculated values of DEQ variables

	Variable	Initial value	Minimal value	Maximal value	Final value
1	$m1$	0.0438	0.0121152	0.108936	0.108936
2	n	3	3	3	3
3	r	0	0	2.439418	2.439418
4	t	0	0	6	6
5	y	0	0	1.503757	0.4520366

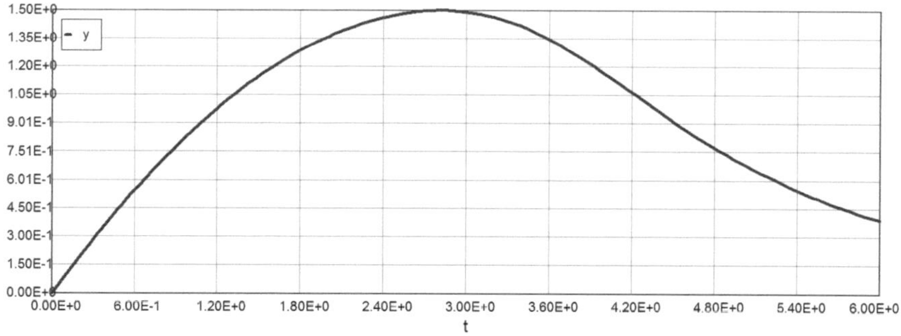

Fig. 5.76 Stress–temperature–strain diagram. Medium Fire α = 0.1

Differential equations

1 $d(y)/d(t) = -1 * y * m1 + (\exp(-0.15 * t)) * (1 - 0.15 * t) - (\exp(t/(1.0 + 0.1 * t))) * m1 * y^\wedge n + 1 * m1 * t * (\exp(-0.15 * t))$

Explicit equations

1 $n = 3$

2 $r = t * (\exp(-0.15 * t))$

3 $m1 = 0.0438 - 0.0202 * t + 0.00148 * t^\wedge 2 + 0.000616 * t^\wedge 3$

See Fig. 5.77.

Model: $y = a0 + a1 * t + a2 * t$^2 $+ a3 * t$^3 $+ a4 * t$^4

Variable	Value
$a0$	0.0
$a1$	0.86
$a2$	0.0208
$a3$	−0.0721
$a4$	0.00782

$$\textbf{Model}: \sigma = 0.86 * \theta + 0.0208 * \theta^\wedge 2 - 0.0721 * \theta^\wedge 3 + 0.00782 * \theta^\wedge 4 \quad (5.86)$$

α = 10

Calculated values of DEQ variables

	Variable	Initial value	Minimal value	Maximal value	Final value
1	$m1$	0.0438	0.0121152	0.108936	0.108936
2	n	3	3	3	3
3	r	0	0	2.439418	2.439418
4	t	0	0	6	6
5	y	0	0	1.524363	0.7534053

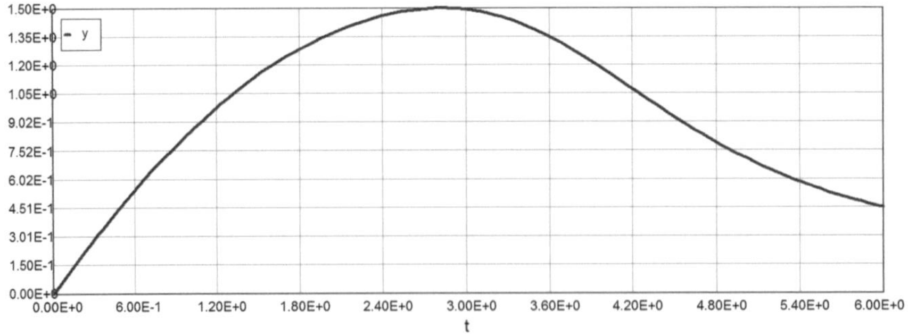

Fig. 5.77 Stress–temperature–strain diagram. Medium Fire α = 1

Differential equations

1 $d(y)/d(t) = -10 * y * m1 + (\exp(-0.15 * t)) * (1 - 0.15 * t) - (\exp(t/(1.0 + 0.1 * t))) * m1 * y^{\wedge}n + 10 * m1 * t * (\exp(-0.15 * t))$

Explicit equations

1 $n = 3$

2 $r = t * (\exp(-0.15 * t))$

3 $m1 = 0.0438 - 0.0202 * t + 0.00148 * t^{\wedge}2 + 0.000616 * t^{\wedge}3$

See Fig. 5.78.

Model: $y = a0 + a1 * t + a2 * t^2 + a3 * t^3 + a4 * t^4$

Variable	Value
$a0$	0.0
$a1$	0.905
$a2$	−0.0169
$a3$	−0.0611
a4	0.00707

$$\textbf{Model}: \sigma = 0.905 * \theta - 0.0169 * \theta^{\wedge}2 - 0.0611 * \theta^{\wedge}3 + 0.00707 * \theta^{\wedge}4 \quad (5.87)$$

α = 100

Calculated values of DEQ variables

	Variable	Initial value	Minimal value	Maximal value	Final value
1	$m1$	0.0438	0.0121148	0.108936	0.108936
2	n	3	3	3	3
3	r	0	0	2.439418	2.439418
4	i	0	0	6	6
5	y	0	0	1.668804	1.367226

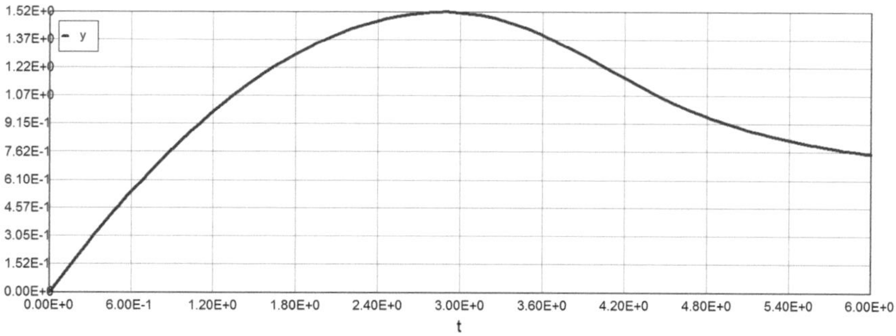

Fig. 5.78 Stress–temperature–strain diagram. Medium Fire α = 10

Differential equations

1 $d(y)/d(t) = -100 * y * m1 + (\exp(-0.15 * t)) * (1 - 0.15 * t) - (\exp(t/(1.0 + 0.1 * t))) * m1 * y \hat{} n + 100 * m1 * t * (\exp(-0.15 * t))$

Explicit equations

1 $n = 3$

2 $r = t * (\exp(-0.15 * t))$

3 $m1 = 0.0438 - 0.0202 * t + 0.00148 * t \hat{} 2 + 0.000616 * t \hat{} 3$

See Fig. 5.79.

Model: $y = a0 + a1 * t + a2 * t\hat{}2 + a3 * t\hat{}3 + a4 * t\hat{}4$

Variable	Value
$a0$	0
$a1$	1.0
$a2$	−0.120
$a3$	−0.0184
$a4$	0.00286

$$\textbf{Model}: \sigma = 1.0 * \theta - 0.012 * \theta\hat{}2 - 0.0184 * \theta\hat{}3 + 0.00286 * \theta\hat{}4 \qquad (5.88)$$

α = 1000

Calculated values of DEQ variables

	Variable	Initial value	Minimal value	Maximal value	Final value
1	$m1$	0.0438	0.0121116	0.108936	0.108936
2	n	3	3	3.	3
3	r	0	0	2.439418	2.439418
4	t	0	0	6	6
5	y	0	0	2.103128	2.065647

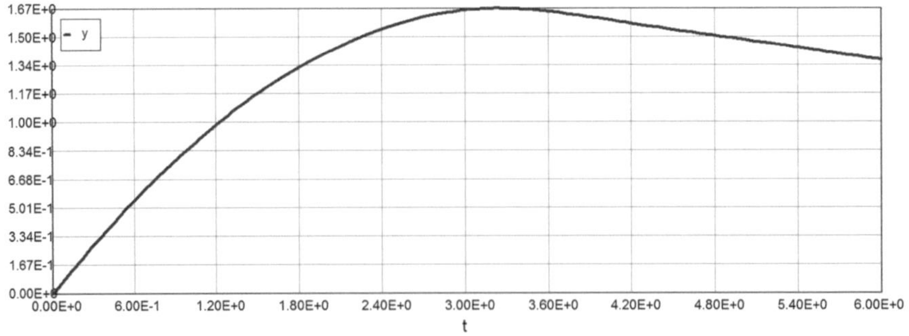

Fig. 5.79 Stress–temperature–strain diagram. Medium Fire α = 100

Differential equations

1 $d(y)/d(t) = -1000 * y * m1 + (\exp(-0.15 * t)) * (1 - 0.15 * t) - (\exp(t/(1.0 + 0.1 * t))) * m1 * y^{\wedge}n + 1000 * m1 * t * (\exp(-0.15 * t))$

Explicit equations

1 $n = 3$

2 $r = t * (\exp(-0.15 * t))$

3 $m1 = 0.0438 - 0.0202 * t + 0.00148 * t^{\wedge}2 + 0.000616 * t^{\wedge}3$

See Fig. 5.80.

Model: $y = a0 + a1 * t + a2 * t^{\wedge}2 + a3 * t^{\wedge}3 + a4 * t^{\wedge}4$

Variable	Value
$a0$	0.0
$a1$	0.987
$a2$	−0.132
$a3$	0.00172
$a4$	0.000408

Model: $\sigma = 0.987 * \theta - 0.132 * \theta^{\wedge}2 + 0.00172 * \theta^{\wedge}3 + 0.000408 * \theta^{\wedge}4$ (5.89)

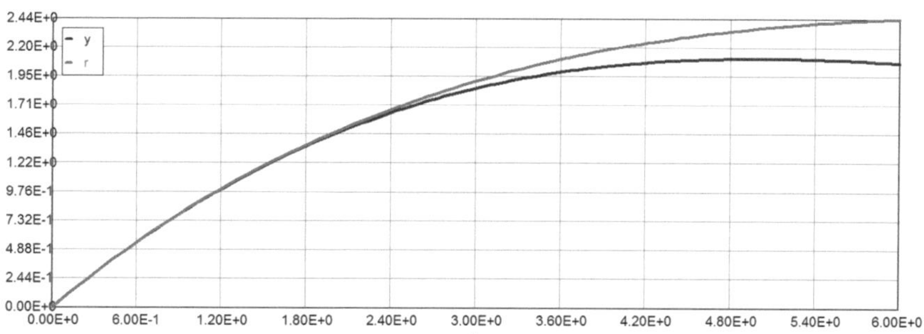

Fig. 5.80 Stress–temperature–strain diagram. Medium Fire α = 1000

α = 10,000
Calculated values of DEQ variables

	Variable	Initial value	Minimal value	Maximal value	Final value
1	$m1$	0.0438	0.0121133	0.108936	0.108936
2	n	3	3	3	3
3	r	0	0	2.439418	2.439418
4	t	0	0	6	6
5	y	0	0	2.381973	2.381973

Differential equations

1 $\mathrm{d}(y)/\mathrm{d}(t) = -10{,}000 * y * m1 + (\exp(-0.15 * t)) * (1 - 0.15 * t) - (\exp(t/(1.0 + 0.1 * t))) * m1 * y^\wedge n + 10{,}000 * m1 * t * (\exp(-0.15 * t))$

Explicit equations

1 $n = 3$

2 $r = t * (\exp(-0.15 * t))$

3 $m1 = 0.0438 - 0.0202 * t + 0.00148 * t^\wedge 2 + 0.000616 * t^\wedge 3$

See Fig. 5.81.
Model: $y = a0 + a1 * t + a2 * t\text{^}2 + a3 * t\text{^}3 + a4 * t\text{^}4$

Variable	Value
$a0$	0.0
$a1$	0.996
$a2$	−0.146
$a3$	0.0092
$a4$	−0.000265

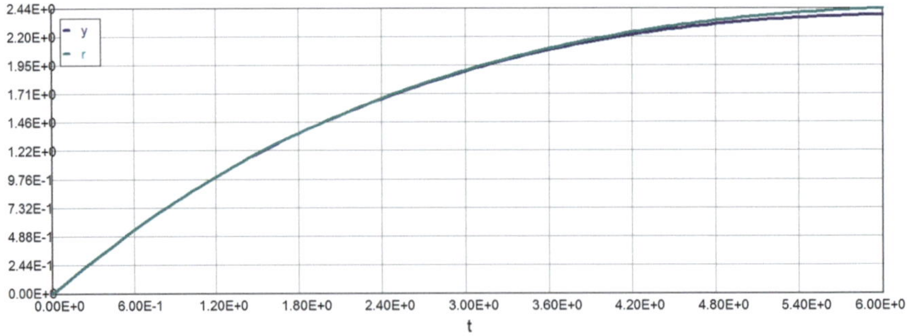

Fig. 5.81 Stress–temperature–strain diagram. Medium Fire α = 10,000

Model: $\sigma = 0.996 * \theta - 0.146 * \theta^\wedge 2 + 0.0092 * \theta^\wedge 3 - 0.000265 * \theta^\wedge 4$ (5.90)

α = 100,000

Calculated values of DEQ variables

	Variable	Initial value	Minimal value	Maximal value	Final value
1	$m1$	0.0438	0.0121141	0.108936	0.108936
2	n	3	3	3	3
3	r	0	0	2.439418	2.439418
4	t	0	0	6	6
5	y	0	0	2.433292	2.433292

Differential equations

1 $\mathrm{d}(y)/\mathrm{d}(t) = -100 * y * m1 + (\exp(-0.15 * t)) * (1 - 0.15 * t) - (\exp(t/(1.0 + 0.1 * t))) * m1 * y^\wedge n + 100 * m1 * t * (\exp(-0.15 * t))$

Explicit equations

1 $n = 1$

2 $r = t * (\exp(-0.15 * t))$

3 $m1 = 0.0438 - 0.0202 * t + 0.00148 * t^\wedge 2 + 0.000616 * t^\wedge 3$

See Fig. 5.82.

Model: $y = a0 + a1 * t + a2 * t^\wedge 2 + a3 * t^\wedge 3 + a4 * t^\wedge 4$

Variable	Value
$a0$	0.0
$a1$	0.998
$a2$	−0.147
$a3$	0.010
$a4$	−0.000317

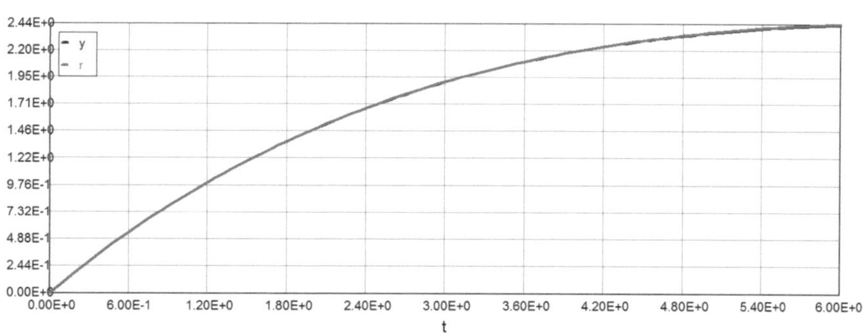

Fig. 5.82 Stress–temperature–strain diagram. Medium Fire α = 100,000

Table 5.7 Stress–temperature data (medium fire)

α|θ	0	1	2	3	4	5	6
0.001	0	0.850	1.362	1.494	1.166	0.687	0.384
0.01	0	0.850	1.362	1.494	1.166	0.688	0.385
0.1	0	0.850	1.362	1.494	1.167	0.690	0.392
1.0	0	0.850	1.363	1.497	1.175	0.715	0.452
10	0	0.851	1.370	1.520	1.246	0.904	0.735
100	0	0.854	1.412	1.659	1.606	1.485	1.367
1000	0	0.860	1.467	1.854	2.049	2.103	2.066
10,000	0	0.860	1.480	1.906	2.177	2.327	2.382
100,000	0	0.861	1.481	1.912	2.193	2.358	2.433
r	0	0.861	1.482	1.913	2.195	2.362	2.439

$$\textbf{Model}: \sigma = 0.998 * \theta - 0.147 * \theta\hat{}2 + 0.01 * \theta\hat{}3 - 0.000317 * \theta\hat{}4 \qquad (5.91)$$

See Table 5.7

5.3.10 *Functional Dependencies of Creep Stresses and Strains from the Stress Exponent "n"*

n = 1 α = 100
Calculated values of DEQ variables

	Variable	Initial value	Minimal value	Maximal value	Final value
1	$m1$	0.0438	0.0121132	0.108936	0.108936
2	n	1	1	1	1
3	r	0	0	2.439418	2.439418
4	t	0	0	6	6
5	y	0	0	1.937033	1.725965

Differential equations

1 $d(y)/d(t) = -100 * y * m1 + (\exp(-0.15 * t)) * (1 - 0.15 * t) - (\exp(t/(1.0 + 0.1 * t))) * m1 * y\hat{}n + 100 * m1 * t * (\exp(-0.15 * t))$

Explicit equations

1 $n = 1$

2 $r = t * (\exp(-0.15 * t))$

3 $m1 = 0.0438 - 0.0202 * t + 0.00148 * t\hat{}2 + 0.000616 * t\hat{}3$

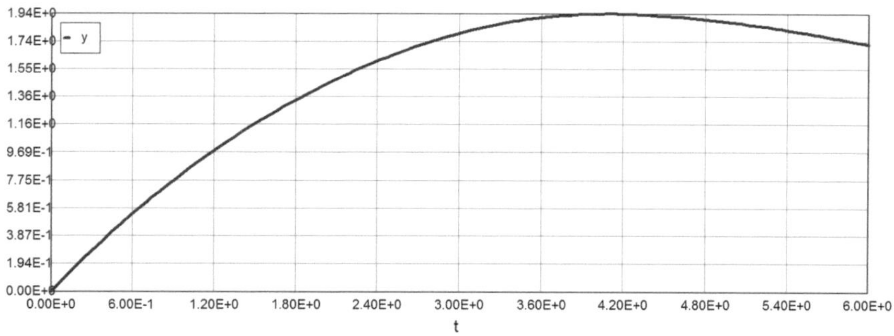

Fig. 5.83 Stress–temperature–strain diagram. Medium Fire $n = 1$

See Fig. 5.83.

Model: $y = a0 + a1 * t + a2 * t\text{^}2 + a3 * t\text{^}3 + a4 * t\text{^}4$

Variable	Value
$a0$	0.0
$a1$	0.927
$a2$	−0.0913
$a3$	−0.0103
$a4$	0.00127

$$\textbf{Model}: \sigma = 0.927 * \theta - 0.0913 * \theta^\wedge 2 - 0.0103 * \theta^\wedge 3 + 0.00127 * \theta^\wedge 4 \quad (5.92)$$

n = 3

Calculated values of DEQ variables

	Variable	Initial value	Minimal value	Maximal value	Final value
1	$m1$	0.0438	0.0121148	0.108936	0.108936
2	n	3	3	3	3
3	r	0	0	2.439418	2.439418
4	t	0	0	6	6
5	y	0	0	1.668804	1.367226

Differential equations

1 $\mathrm{d}(y)/\mathrm{d}(t) = -100 * y * m1 + (\exp(-0.15 * t)) * (1 - 0.15 * t) - (\exp(t/(1.0 + 0.1 * t))) * m1 * y^\wedge n + 100 * m1 * t * (\exp(-0.15 * t))$

Explicit equations

1 $n = 3$
2 $r = t * (\exp(-0.15 * t))$
3 $m1 = 0.0438 - 0.0202 * t + 0.00148 * t^\wedge 2 + 0.000616 * t^\wedge 3$

See Fig. 5.84.
Model: $y = a0 + a1 * t + a2 * t^\wedge 2 + a3 * t^\wedge 3 + a4 * t^\wedge 4$

Variable	Value
$a0$	0
$a1$	1.0
$a2$	−0.120
$a3$	−0.0184
$a4$	0.00286

$$\textbf{Model}: \sigma = 1.0 * \theta - 0.12 * \theta^\wedge 2 - 0.0184 * \theta^\wedge 3 + 0.00286 * \theta^\wedge 4 \quad (5.93)$$

n = 5
Calculated values of DEQ variables

	Variable	Initial value	Minimal value	Maximal value	Final value
1	$m1$	0.0438	0.0121149	0.108936	0.108936
2	n	5	5	5	5
3	r	0	0	2.439418	2.439418
4	t	0	0	6	6
5	y	0	0	1.489246	1.233896

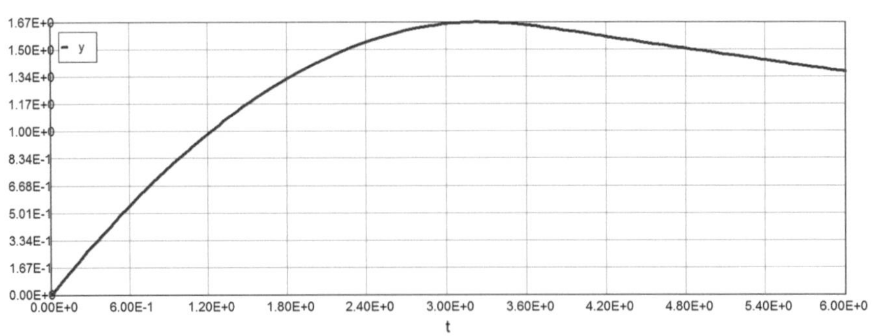

Fig. 5.84 Stress–temperature–strain diagram. Medium Fire $n = 3$

Differential equations

1 $d(y)/d(t) = -100 * y * m1 + (\exp(-0.15 * t)) * (1 - 0.15 * t) - (\exp(t/(1.0 + 0.1 * t))) * m1 * y^{\wedge}n + 100 * m1 * t * (\exp(-0.15 * t))$

Explicit equations

1 $n = 5$

2 $r = t * (\exp(-0.15 * t))$

3 $m1 = 0.0438 - 0.0202 * t + 0.00148 * t^{\wedge}2 + 0.000616 * t^{\wedge}3$

See Fig. 5.85.

Model: $y = a0 + a1 * t + a2 * t^{\wedge}2 + a3 * t^{\wedge}3 + a4 * t^{\wedge}4$

Variable	Value
$a0$	0
$a1$	1.202
$a2$	−0.287
$a3$	0.0176
$a4$	0.000496

$$\textbf{Model}: \sigma = 1.202 * \theta - 0.287 * \theta^{\wedge}2 + 0.0176 * \theta^{\wedge}3 + 0.000496 * \theta^{\wedge}4 \quad (5.94)$$

n = 8

Calculated values of DEQ variables

	Variable	Initial value	Minimal value	Maximal value	Final value
1	$m1$	0.0438	0.0121149	0.108936	0.108936
2	n	8	8	8	8
3	r	0	0	2.439418	2.439418
4	t	0	0	6	6
5	y	0	0	1.33335	1.14968

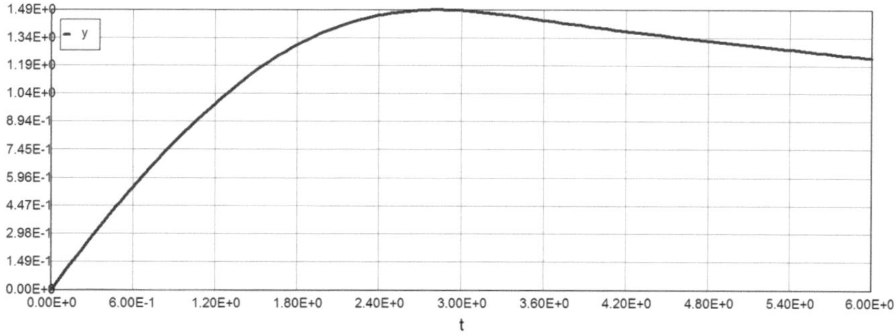

Fig. 5.85 Stress–temperature–strain diagram. Medium Fire $n = 5$

Differential equations

1 $\mathrm{d}(y)/\mathrm{d}(t) = -100 * y * m1 + (\exp(-0.15 * t)) * (1 - 0.15 * t) - (\exp(t/(1.0 + 0.1 * t))) * m1 * y^\wedge n + 100 * m1 * t * (\exp(-0.15 * t))$

Explicit equations

1 $n = 8$

2 $r = t * (\exp(-0.15 * t))$

3 $m1 = 0.0438 - 0.0202 * t + 0.00148 * t^\wedge 2 + 0.000616 * t^\wedge 3$

See Fig. 5.86.

Model: $y = a0 + a1 * t + a2 * t^\wedge 2 + a3 * t^\wedge 3 + a4 * t^\wedge 4$

Variable	Value
$a0$	0
$a1$	1.373
$a2$	−0.457
$a3$	0.0606
$a4$	−0.0028

$$\textbf{Model}: \sigma = 1.373 * \theta - 0.457 * \theta^\wedge 2 + 0.0606 * \theta^\wedge 3 - 0.0028 * \theta^\wedge 4 \quad (5.95)$$

n = 10

Calculated values of DEQ variables

	Variable	Initial value	Minimal value	Maximal value	Final value
1	$m1$	0.0438	0.0121138	0.108936	0.108936
2	n	10	10	10	10
3	r	0	0	2.439418	2.439418
4	t	0	0	6	6
5	y	0	0	1.271732	1.120468

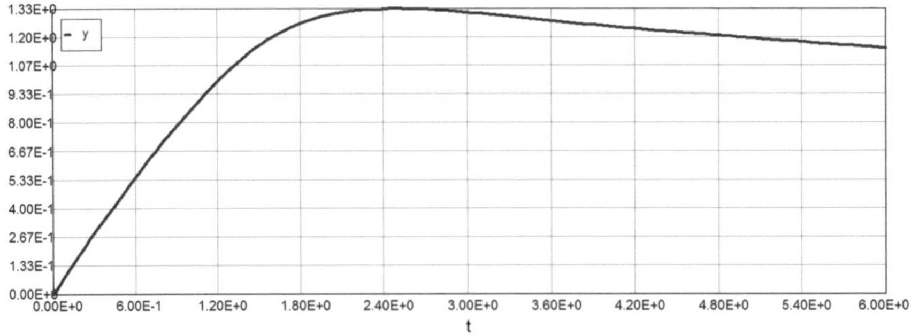

Fig. 5.86 Stress–temperature–strain diagram. Medium Fire n = 8

Differential equations

1 $d(y)/d(t) = -100 * y * m1 + (\exp(-0.15 * t)) * (1 - 0.15 * t) - (\exp(t/(1.0 + 0.1 * t))) * m1 * y^{\wedge}n + 100 * m1 * t * (\exp(-0.15 * t))$

Explicit equations

1 $n = 10$

2 $r = t * (\exp(-0.15 * t))$

3 $m1 = 0.0438 - 0.0202 * t + 0.00148 * t^{\wedge}2 + 0.000616 * t^{\wedge}3$

See Fig. 5.87.

Model: $y = a0 + a1 * t + a2 * t^{\wedge}2 + a3 * t^{\wedge}3 + a4 * t^{\wedge}4$

Variable	Value
$a0$	0
$a1$	1.422
$a2$	−0.516
$a3$	0.0768
$a4$	−0.00411

$$\textbf{Model}: \sigma = 1.422 * \theta - 0.516 * \theta^{\wedge}2 + 0.0768 * \theta^{\wedge}3 - 0.00411 * \theta^{\wedge}4 \quad (5.96)$$

See Fig. 5.88 and Table 5.8.

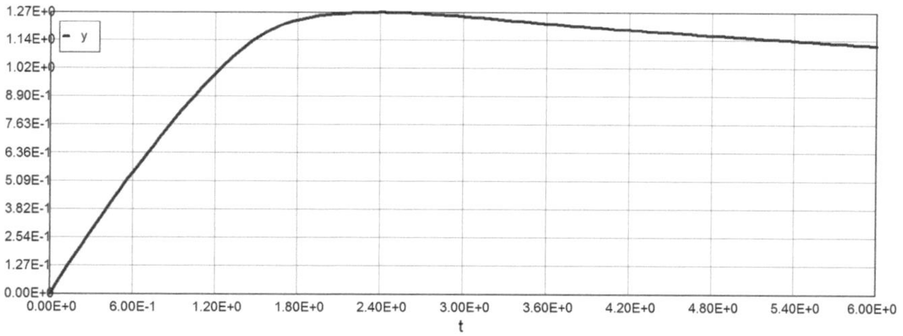

Fig. 5.87 Stress–temperature–strain diagram. Medium Fire $n = 10$

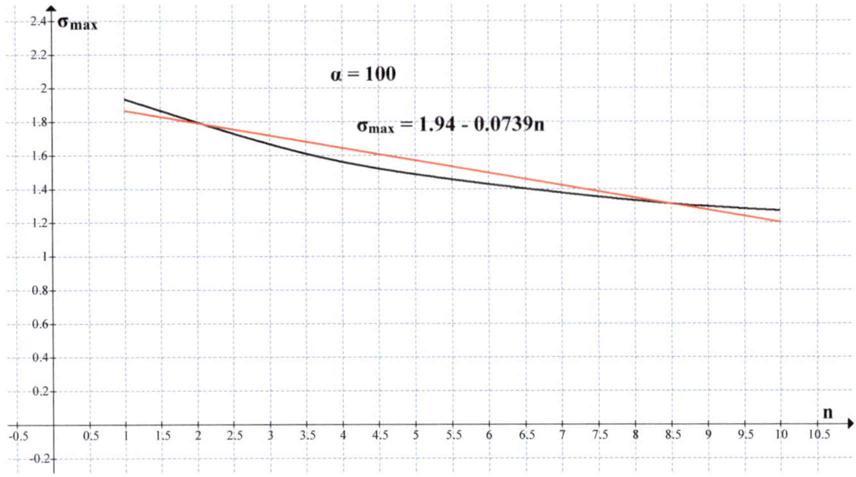

Fig. 5.88 Maximum creep stress versus stress exponent. Medium fire $\alpha = 100$

Table 5.8 Maximum creep stress versus stress exponent

n	1	3	5	8	10	α
σ						
σ_{max}	1.937	1.669	1.489	1.333	1.272	100

5.4 Stress–Strain Diagram. Slow Fire

All subsequent calculations for this case of fire severity are presented in a step-by-step procedural form in accordance with the main points of the above.

5.4.1 Temperature–Time Function

Slow Fire; Temperature Increase (715 K < T < 822 K)

Data $T_* = 600$ K; $0 < \tau < 0.2$; $0.275 < \gamma < 1$

Select $\gamma = 0.275$

Differential equations (5.3) and (5.4) are rewritten as an input for "Polymath" software:

1 $d(y0)/d(t) = 20 * (1 - y2) * \exp(y0/(1 + 0.1 * y0)) - 2.53 * 0 - 0.157 * y0\hat{}4$

2 $d(y2)/d(t) = 20 * (1 - y2) * \exp(y0/(1 + .1 * y0))$

3 $d(y1)/d(t) = 5.5 * (1 - y1)\hat{}1.0 * \exp(y/(1 + .1 * y))$

4 $d(y)/d(t) = 20 * (1 - y1)\hat{}1.0 * \exp(y/(1 + .1 * y)) - 2.53 * 0 - 0.157 * y\hat{}4,$

where

"y" is the dimensionless temperature "θ" with the corresponding parameter "$\gamma = 0.275$".
"$y0$" is the dimensionless temperature "θ" with the corresponding parameter "$\gamma = 1.0$".
"$y1$" is the concentration of the product of the first-order chemical reaction with "$\gamma = 0.275$".
"$y2$" is the concentration of the product of the first-order chemical reaction with "$\gamma = 1.0$".

Calculated values of DEQ variables

	Variable	Initial value	Minimal value	Maximal value	Final value
4	*t*	0	0	0.2	0.2
5	*y*	0	0	*3.01*	*2.4*
6	*y*0	0	0	0.9848585	0.9788831
7	*y*1	0	0	0.9993049	0.9993049
8	*y*2	0	0	0.9998429	0.9998429

Differential equations

1 $d(y0)/d(t) = 20 * (1 - y2) * \exp(y0/(1 + .1 * y0)) - 0.157 * y0^\wedge 4$

2 $d(y2)/d(t) = 20 * (1 - y2) * \exp(y0/(1 + .1 * y0))$

3 $d(y1)/d(t) = 5.5 * (1 - y1)^\wedge 1.0 * \exp(y/(1 + .1 * y))$

4 $d(y)/d(t) = 20 * (1 - y1)^\wedge 1.0 * \exp(y/(1 + .1 * y)) - 0.157 * y^\wedge 4$

Tabulated solution of Eqs. (5.3) and (5.4) is presented in [4] and the graphs are shown in Fig. 5.89.

Now based on the data shown above, the "best-to-fit" (polynomial regression) approximation of the dimensionless temperature–time curve can be presented as follows:

Model: $y = a0 + a1 * t + a2 * t^\wedge 2 + a3 * t^\wedge 3 + a4 * t^\wedge 4$

Variable	Value
*a*0	0.0
*a*1	9.924
*a*2	779.8
*a*3	−7836
*a*4	2.006E+04

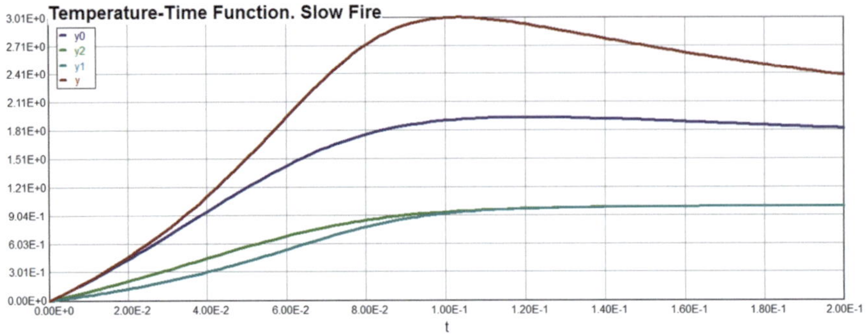

Fig. 5.89 Dimensionless temperature–time curve

$$\theta(\tau) = 9.924\tau + 779.8\tau^\wedge 2 - 7836\,\tau^\wedge 3 + 20060\tau^\wedge 4 \tag{5.97}$$

5.4.2 *Analytical Expression of the Inverse Function* θ^{-1} *and Its First Derivative*

Monotonically increasing temperature–time function: $(0 < \theta < 3)$; $(0 < \tau 1 < 0.104)$

As noted above, a function of the temperature–time should be divided into two portions monotonically increasing and monotonically decreasing. Let us start with a monotonically increasing area. The range of temperature values in this case is the interval from 0 to 3 and the time interval from 0 to 0.104. Therefore, using polynomial regression method again we have

Model: $y = a0 + a1 * t + a2 * t\verb|^|2 + a3 * t\verb|^|3 + a4 * t\verb|^|4$

Variable	Value
$a0$	0
$a1$	18.08
$a2$	137.7
$a3$	4376.9
$a4$	−4.56E+04

$$\textbf{Model}: \theta = 18.08 * \tau + 137.7 * \tau^\wedge 2 + 4376.9 * \tau^\wedge 3 - 45600 * \tau^\wedge 4 \tag{5.98}$$

See Fig. 5.90.

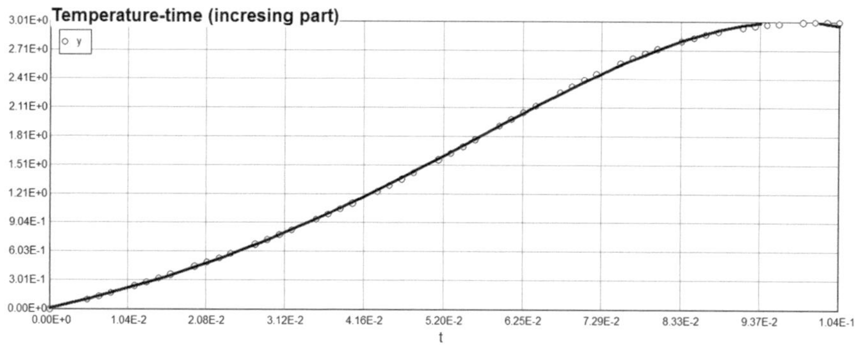

Fig. 5.90 Monotonically increasing temperature–time function

The inverse function θ^{-1} and its first derivative (monotonically increasing functions) are:

Model: $t = a0 + a1 * y + a2 * y\text{^}2 + a3 * y\text{^}3 + a4 * y\text{^}4$

Variable	Value
$a0$	0
$a1$	0.0351
$a2$	0.0104
$a3$	−0.0124
$a4$	0.00286

The graph of the inverse function θ^{-1} is shown in Fig. 5.91.

Therefore, the analytical expression of the inverse function and its first derivative is as follows:

$$\textbf{Model} : \mathbf{m} = \tau = 0.351 * \tau + 0.0104 * \theta^\wedge 2 - 0.0124 * \theta^\wedge 3 + 0.00286 * \theta^\wedge 4 \tag{5.99}$$

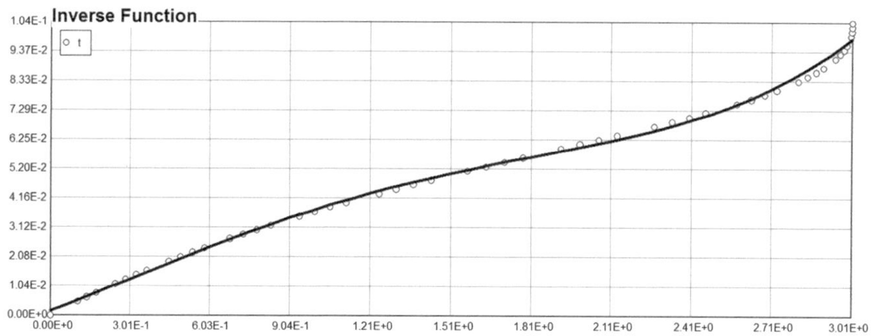

Fig. 5.91 Monotonically increasing inverse function θ^{-1}

$$\mathbf{m1} = \tau' = 0.351 + 0.0208 * \theta^\wedge 1 + 0.0372 * \theta^\wedge 2 + 0.01144 * \theta^\wedge 3 \quad (5.100)$$

Monotonically decreasing temperature–time function: $(3 > \theta > 2.41)$; $(0.104 < \tau 1 < 0.2)$.

Performing calculations that are similar to the previous case, we obtain corresponding formulas for a monotonically decreasing temperature range. The temperature–time relationship in this case is very close to linear, therefore the function itself and the inverse function will be approximated by linear function.

Temperature–time function

Model: $x = a0 + a1 * t1$

Variable	Value
$a0$	3.756
$a1$	−6.908

$$\mathbf{Model} : \theta = 3.756 - 6.908 * \tau 1 \quad (5.101)$$

The corresponding graph is: (Fig. 5.92)

The inverse function $\theta^{-1}(\tau 1)$:

Model: $t1 = a0 + a1 * x$

Variable	Value
$a0$	0.542
$a1$	−0.144

The corresponding graph is: (Fig. 5.93)

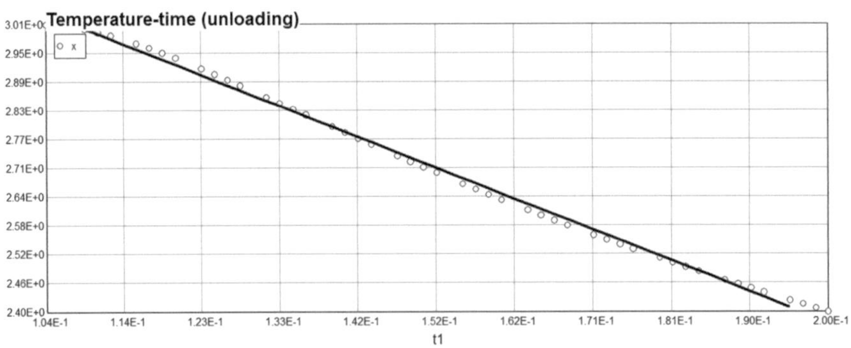

Fig. 5.92 Monotonically decreasing temperature–time function

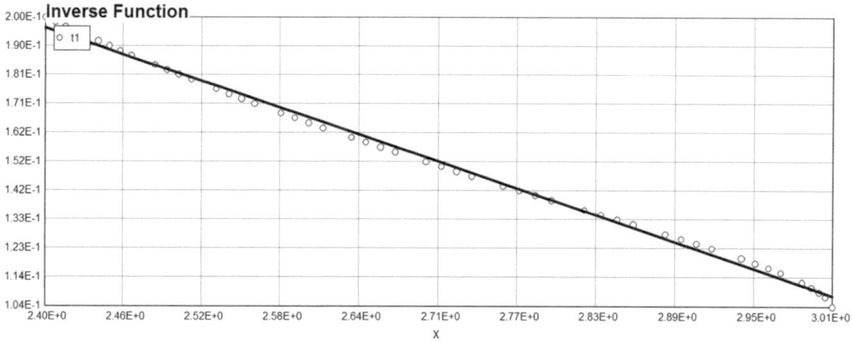

Fig. 5.93 Monotonically decreasing inverse function $\theta^{-1}(\tau 1)$

$$\textbf{Model}: \tau 1 = 0.542 - 0.144 * \theta \tag{5.102}$$

Finally,

$$\mathbf{m0} = \tau 1 = 0.542 - 0.144 * \theta \tag{5.103}$$

The first derivative of inverse function $\theta^{-1}(\tau 1)$:

$$\mathbf{m01} = \tau 1' = -0.144 \tag{5.104}$$

5.4.3 Creep Constitutive Equation. Equivalent ODE Method

Monotonically increasing temperature–time function

The stress–strain response of a material is described using differential equation of the form

$$\begin{aligned}
&\sigma'(\theta) = -\sigma(\theta)\alpha m1 + e^{-0.15\theta}[1 - 0.15\theta] - (\exp(t)/(1+0.1t))m1\sigma^n(\theta) + m1\alpha\theta e^{-0.15\theta} \\
&\mathbf{m} = \tau = 0.351 * \theta + 0.0104 * \theta^\wedge 2 - 0.0124 * \theta^\wedge 3 + 0.00286 * \theta^\wedge 4 \\
&\mathbf{m1} = \tau' = 0.351 + 0.0208 * \theta^\wedge 1 + 0.0372 * \theta^\wedge 2 + 0.01144 * \theta^\wedge 3 \\
&\sigma(\theta = 0) = 0; \quad \alpha = 0.33
\end{aligned} \tag{5.105}$$

Consider now general uniaxial creep constitutive model for monotonically increasing part of real fire temperature load (equivalent differential equation with one MPP $\alpha = 0.33$):

$$d(\sigma)/d(\theta) = -0.33 * \sigma * m1 + (\exp(-0.15 * \theta)) * (1 - 0.15 * \theta) - (\exp(\theta/(1.0 + 0.1 * \theta))) * m1 * \sigma^n + 0.33 * m1 * \theta * (\exp(-0.15 * \theta)) \quad (5.106)$$

The temperature range is $[0 < \theta < 3]$ and the dimensionless time duration for increasing part of fully developed fire stage is $[0 < \tau < 0.104]$. The solution of Eq. (5.106) is as follows:

Calculated values of DEQ variables

	Variable	Initial value	Minimal value	Maximal value	Final value
1	m	0	0	1.04346	1.04346
2	$m1$	0.351	0.351	1.05708	1.05708
3	n	2	2	2	2
4	t	0	0	3	3
5	z	0	0	0.7269596	0.3155685

Differential equations

1 $d(z)/d(t) = -0.33 * z * (m1) + (\exp(-0.15 * (t))) * (1 - 0.15 * (t)) - (\exp((t)/(1.0 + 0.1 * (t)))) * (m1) * z\hat{}n + 0.33 * (m1) * (t) * (\exp(-0.15 * (t)))$

Explicit equations

1 $n = 2$

2 $m = 0.351 * t + 0.0104 * t\hat{}2 - 0.0124 * t\hat{}3 + 0.00286 * t\hat{}4$

3 $m1 = 0.351 + 0.0208 * t\hat{}1 + 0.0372 * t\hat{}2 + 0.01144 * t\hat{}34$

The corresponding graph is: (Fig. 5.94)

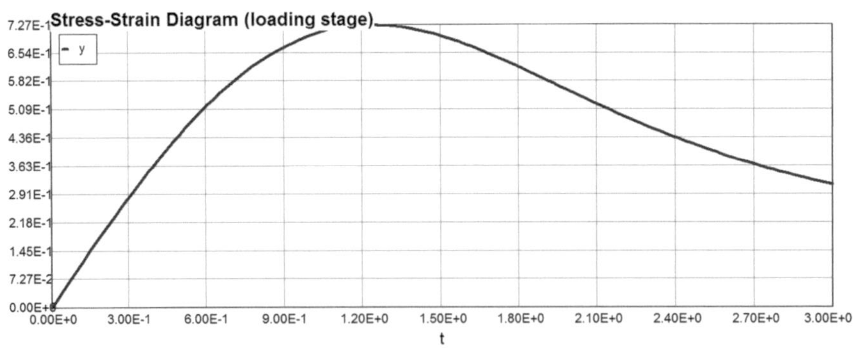

Fig. 5.94 Stress–temperature–strain diagram. Slow fire

Using polynomial regression method again, we have
Model: $z = a0 + a1 * t + a2 * t\^2 + a3 * t\^3 + a4 * t\^4$

Variable	Value
$a0$	0
$a1$	1.248
$a2$	−0.545
$a3$	−0.00797
$a4$	0.0216

Model: $\sigma = 1.248 * \theta - 0.545 * \theta\hat{}2 - 0.00797 * \theta\hat{}3 + 0.0216 * \theta\hat{}4$ (5.107)

Monotonically decreasing temperature–time function

Consider now general uniaxial creep constitutive model for monotonically decreasing part of real fire temperature load. Equivalent differential equation again is (5.13), but function m1 is substituted by $m01$and the corresponding graph will be shifted to the end of increasing temperature interval $\theta = 3$.

Calculated values of DEQ variables

	Variable	Initial value	Minimal value	Maximal value	Final value
1	$m01$	−0.144	−0.144	−0.144	−0.144
2	n	2	2	2	2
3	t	2.41	2.41	6	6
4	z	0.316	0.191514	0.4793088	0.191514

Differential equations

5 $$\mathrm{d}(z)/\mathrm{d}(t) = 1 * (-0.33 * z * (-m01) + (\exp(-0.15 * (t))) * (1 - 0.15 * (t)) - (\exp((t)/(1.0 + 0.1 * (t)))) * (-m01) * z^n + 0.33 * (-m01) * (t) * (\exp(-0.15 * (t))))$$

Explicit equations

1 $n = 2$

2 $m01 = -0.144$

Thus the relaxation modulus graph is shown in Fig. 5.95.

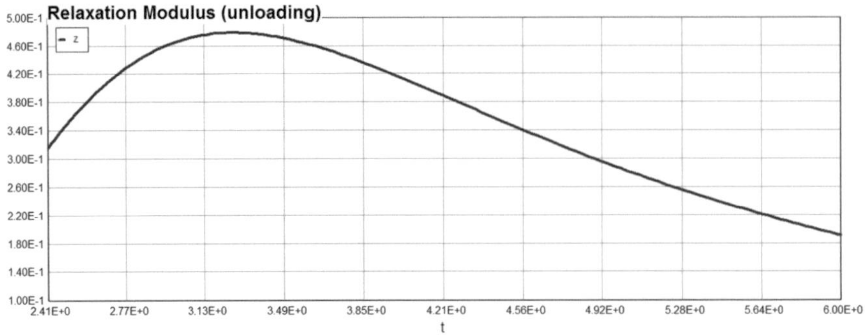

Fig. 5.95 Relaxation modulus. Slow fire

Using polynomial regression method again, we have
Model: $z = -4.86 + 4.759 * t - 1.524 * t\^2 + 0.208 * t\^3 - 0.0104 * t\^4$

Variable	Value
$a0$	−4.86
$a1$	4.759
$a2$	−1.524
$a3$	0.208
$a4$	−0.0104

$$\textbf{Model}: \sigma = -4.86 + 4.759 * \theta - 1.524 * \theta\^2 + 0.208 * \theta\^3 - 0.0104 * \theta\^4 \quad (5.108)$$

Finally, the combined graph (loading and unloading creep stresses) is as follows: (Fig. 5.96)

Temperature decrease rate in case of Slow Fire severity (see Fig. 5.96) is very small, therefore (as it has been shown in Chap. 3) the creep stresses are increasing (for some period of time) while the fire decay stage had already started, i.e., the relaxation process has a delay period. We do not see this effect in any previously discussed fire severity scenarios, since the temperature decrease rate is more rapid.

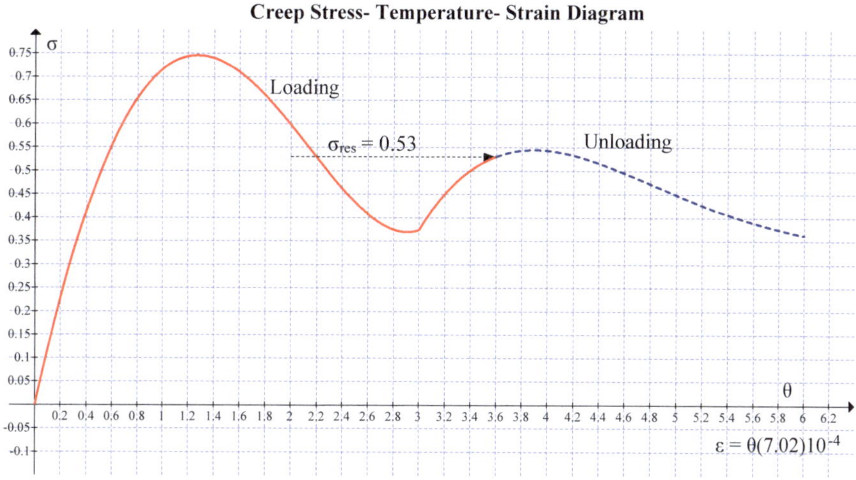

Fig. 5.96 Stress–temperature–strain diagram. Slow fire

5.4.4 *The Functional Dependencies of Creep Stresses and Strains from Material Properties Parameters (MPP)*

Consider now Eq. (5.12) and $\alpha = 0.001$. Again, in case of FF we have
Differential equations

$$1\quad d(y)/d(t) = -\alpha_i * y * m1 + (\exp(-0.15 * t)) * (1 - 0.15 * t) - (\exp(t/(1.0 + 0.1 * t))) * m1 * y\hat{}n + \alpha_i * m1 * t * (\exp(-0.15 * t))$$

Explicit equations

$$n = 3$$
$$m1 = \tau'(\theta) = 0.351 + 0.0208 * \theta\hat{}1 + 0.0372 * \theta\hat{}2 + 0.01144 * \theta\hat{}3 \qquad (5.109)$$

Computer code and solution of Eq. (5.12) for loading stage is as follows:

α = 0.001
Calculated values of DEQ variables

	Variable	Initial value	Minimal value	Maximal value	Final value
1	$m1$	0.351	0.351	1.76156	1.76156
2	n	3	3	3	3
3	r	0	0	2.195247	2.195247

(continued)

(continued)

	Variable	Initial value	Minimal value	Maximal value	Final value
4	t	0	0	4	4
5	y	0	0	0.7828498	0.2219964

Differential equations

1 $d(y)/d(t) = -0.001 * y * m1 + (\exp(-0.15 * t)) * (1 - 0.15 * t) - (\exp(t/(1.0 + 0.1 * t))) * m1 * y^\wedge n + 0.001 * m1 * t * (\exp(-0.15 * t))$

Explicit equations

1 $n = 3$
2 $r = t * (\exp(-0.15 * t))$
3 $m1 = 0.351 + 0.0208 * t^\wedge 1 + 0.0372 * t^\wedge 2 + 0.01144 * t^\wedge 3$

See Fig. 5.97.

Model: $y = a0 + a1 * t + a2 * t^\wedge 2 + a3 * t^\wedge 3 + a4 * t^\wedge 4$

Variable	Value
$a0$	0
$a1$	1.585
$a2$	−0.974
$a3$	0.212
$a4$	−0.0157

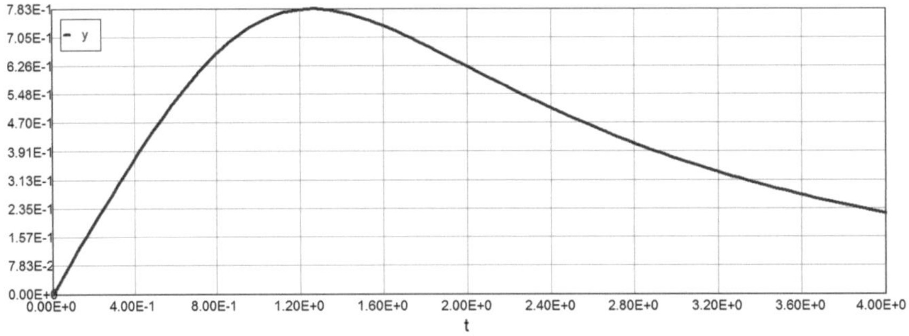

Fig. 5.97 Stress–temperature–strain diagram. Slow fire $\alpha = 0.001$

$$\textbf{Model}: \sigma = 1.585 * \theta - 0.974 * \theta^\wedge 2 + 0.212 * \theta^\wedge 3 - 0.0157 * \theta^\wedge 4 \quad (5.110)$$

α = 0.01
Calculated values of DEQ variables

	Variable	Initial value	Minimal value	Maximal value	Final value
1	$m1$	0.351	0.351	1.76156	1.76156
2	n	3	3	3	3
3	r	0	0	2.195247	2.195247
4	t	0	0	4	4
5	y	0	0	0.7830289	0.2278537

Differential equations

1 $\mathrm{d}(y)/\mathrm{d}(t) = -0.01 * y * m1 + (\exp(-0.15 * t)) * (1 - 0.15 * t) - (\exp(t/(1.0 + 0.1 * t))) * m1 * y^\wedge n + 0.01 * m1 * t * (\exp(-0.15 * t))$

Explicit equations

1 $n = 3$
2 $r = t * (\exp(-0.15 * t))$
3 $m1 = 0.351 + 0.0208 * t^\wedge 1 + 0.0372 * t^\wedge 2 + 0.01144 * t^\wedge 3$

See Fig. 5.98.

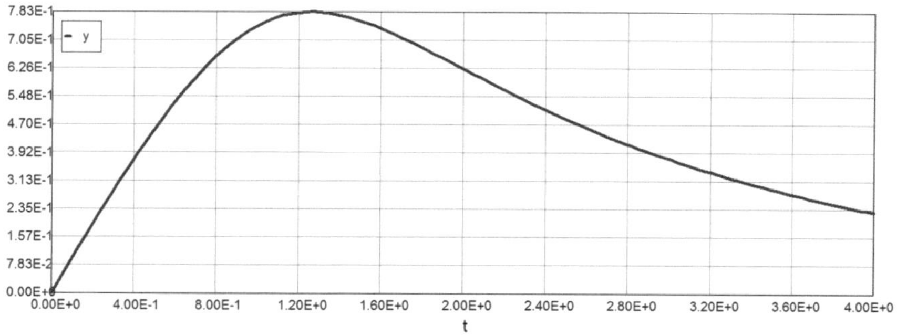

Fig. 5.98 Stress–temperature–strain diagram. Slow fire α = 0.01

Model: $y = a0 + a1 * t + a2 * t^2 + a3 * t^3 + a4 * t^4$

Variable	Value
$a0$	0
$a1$	1.585
$a2$	−0.974
$a3$	0.212
$a4$	−0.0157

$$\textbf{Model}: \sigma = 1.585 * \theta - 0.974 * \theta^{\wedge}2 + 0.212 * \theta^{\wedge}3 - 0.0157 * \theta^{\wedge}4 \tag{5.111}$$

α = 0.1

Calculated values of DEQ variables

	Variable	Initial value	Minimal value	Maximal value	Final value
1	$m1$	0.351	0.351	1.76156	1.76156
2	n	3	3	3	3
3	r	0	0	2.195247	2.195247
4	t	0	0	4	4
5	y	0	0	0.7848064	0.2758059

Differential equations

1 $\mathrm{d}(y)/\mathrm{d}(t) = -0.1 * y * m1 + (\exp(-0.15 * t)) * (1 - 0.15 * t) - (\exp(t/(1.0 + 0.1 * t))) * m1 * y^{\wedge}n + 0.1 * m1 * t * (\exp(-0.15 * t))$

Explicit equations

1 $n = 3$

2 $r = t * (\exp(-0.15 * t))$

3 $m1 = 0.351 + 0.0208 * t^{\wedge}1 + 0.0372 * t^{\wedge}2 + 0.01144 * t^{\wedge}3$

See Fig. 5.99.

Model: $y = a0 + a1 * t + a2 * t^2 + a3 * t^3 + a4 * t^4$

Variable	Value
$a0$	0
$a1$	1.583
$a2$	−0.973
$a3$	0.214
$a4$	−0.0159

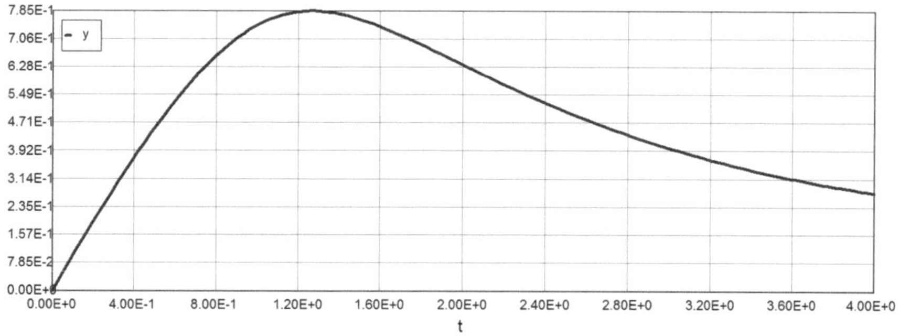

Fig. 5.99 Stress–temperature–strain diagram. Slow fire $\alpha = 0.1$

$$\textbf{Model}: \sigma = 1.583 * \theta - 0.973 * \theta\hat{}2 + 0.214 * \theta\hat{}3 - 0.0159 * \theta\hat{}4 \quad (5.112)$$

$\alpha = 1$

Calculated values of DEQ variables

	Variable	Initial value	Minimal value	Maximal value	Final value
1	$m1$	0.351	0.351	1.76156	1.76156
2	n	3	3	3	3
3	r	0	0	2.195247	2.195247
4	t	0	0	4	4
5	y	0	0	0.8038112	0.4764483

Differential equations

1 $\mathrm{d}(y)/\mathrm{d}(t) = -1 * y * m1 + (\exp(-0.15 * t)) * (1 - 0.15 * t) - (\exp(t/(1.0 + 0.1 * t))) * m1 * y\hat{}n + 1 * m1 * t * (\exp(-0.15 * t))$

Explicit equations

1 $n = 3$

2 $r = t * (\exp(-0.15 * t))$

3 $m1 = 0.351 + 0.0208 * t\hat{}1 + 0.0372 * t\hat{}2 + 0.01144 * t\hat{}3$

See Fig. 5.100.

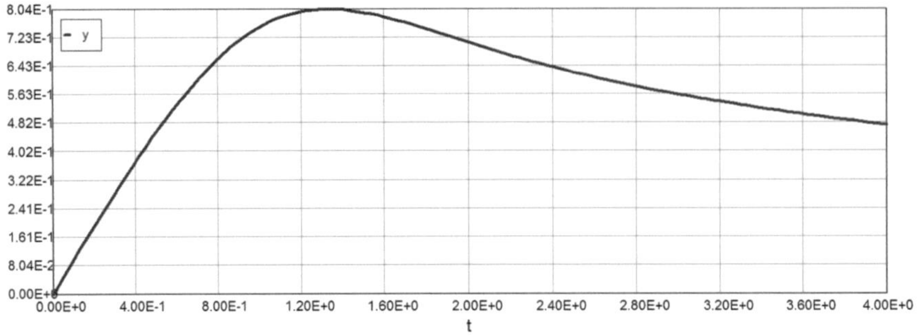

Fig. 5.100 Stress–temperature–strain diagram. Slow fire $\alpha = 1$

Model: $y = a0 + a1 * t + a2 * t\text{^}2 + a3 * t\text{^}3 + a4 * t\text{^}4$

Variable	Value
$a0$	0
$a1$	1.538
$a2$	−0.924
$a3$	0.211
$a4$	−0.0168

$$\textbf{Model}: \sigma = 1.538 * \theta - 0.924 * \theta\hat{}2 + 0.211 * \theta\hat{}3 - 0.0168 * \theta\hat{}4 \qquad (5.113)$$

$\alpha = 10$

Calculated values of DEQ variables

	Variable	Initial value	Minimal value	Maximal value	Final value
1	$m1$	0.351	0.351	1.76156	1.76156
2	n	3	3	3	3
3	r	0	0	2.195247	2.195247
4	t	0	0	4	4
5	y	0	0	1.015137	0.9082207

Differential equations

1 $\mathrm{d}(y)/\mathrm{d}(t) = -10 * y * m1 + (\exp(-0.15 * t)) * (1 - 0.15 * t) - (\exp(t/(1.0 + 0.1 * t))) * m1 * y\hat{}n + 10 * m1 * t * (\exp(-0.15 * t))$

Explicit equations

1 $n = 3$

2 $r = t * (\exp(-0.15 * t))$

3 $m1 = 0.351 + 0.0208 * t^\wedge 1 + 0.0372 * t^\wedge 2 + 0.01144 * t^\wedge 3$

See Fig. 5.101.

Model: $y = a0 + a1 * t + a2 * t^2 + a3 * t^3 + a4 * t^4$

Variable	Value
$a0$	0
$a1$	1.249
$a2$	−0.505
$a3$	0.0801
$a4$	−0.00427

Model: $\sigma = 1.249 * \theta - 0.505 * \theta^\wedge 2 + 0.0801 * \theta^\wedge 3 - 0.00427 * \theta^\wedge 4$ (5.114)

α = 100

Calculated values of DEQ variables

	Variable	Initial value	Minimal value	Maximal value	Final value
1	$m1$	0.351	0.351	1.76156	1.76156
2	n	3	3	3	3
3	r	0	0	2.195247	2.195247
4	t	0	0	4	4
5	y	0	0	1.562581	1.549283

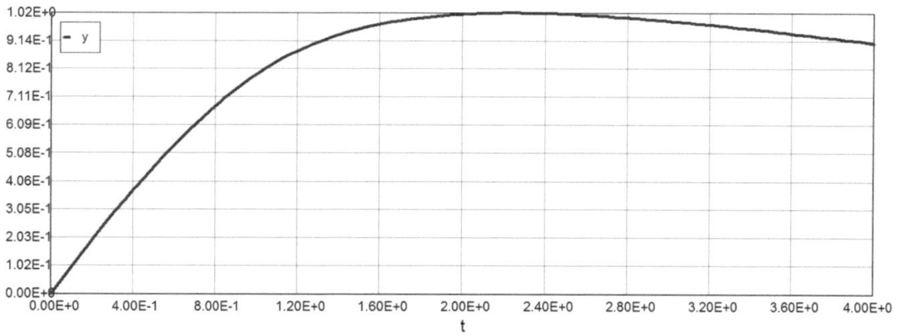

Fig. 5.101 Stress–temperature–strain diagram. Slow fire α = 10

Differential equations

1 $d(y)/d(t) = -100 * y * m1 + (\exp(-0.15 * t)) * (1 - 0.15 * t) - (\exp(t/(1.0 + 0.1 * t))) * m1 * y^{\wedge}n + 100 * m1 * t * (\exp(-0.15 * t))$

Explicit equations

1 $n = 3$
2 $r = t * (\exp(-0.15 * t))$
3 $m1 = 0.351 + 0.0208 * t^{\wedge}1 + 0.0372 * t^{\wedge}2 + 0.01144 * t^{\wedge}3$

See Fig. 5.102.
Model: $y = a0 + a1 * t + a2 * t^{\wedge}2 + a3 * t^{\wedge}3 + a4 * t^{\wedge}4$

Variable	Value
$a0$	0
$a1$	1.0
$a2$	−0.142
$a3$	−0.0193
$a4$	0.00406

$$\textbf{Model}: \sigma = 1.0 * \theta - 0.142 * \theta^{\wedge}2 - 0.0193 * \theta^{\wedge}3 + 0.00406 * \theta^{\wedge}4 \tag{5.115}$$

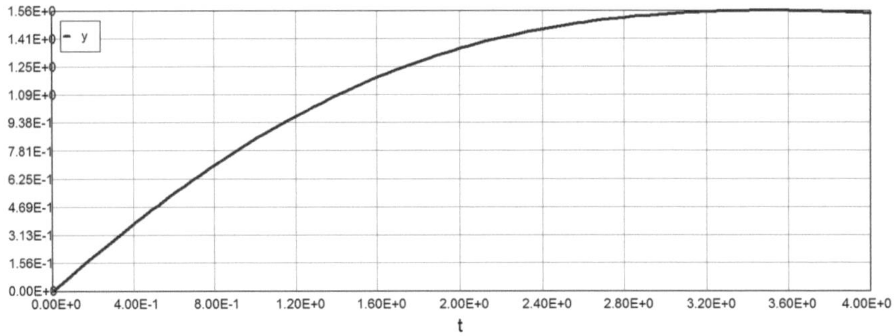

Fig. 5.102 Stress–temperature–strain diagram. Slow fire α = 100

α = 1000

Calculated values of DEQ variables

	Variable	Initial value	Minimal value	Maximal value	Final value
1	$m1$	0.351	0.351	1.76156	1.76156
2	n	3	3	3	3
3	r	0	0	2.195247	2.195247
4	t	0	0	4	4
5	y	0	0	2.046145	2.046145

Differential equations

1 $\mathrm{d}(y)/\mathrm{d}(t) = -1000 * y * m1 + (\exp(-0.15 * t)) * (1 - 0.15 * t) - (\exp(t/(1.0 + 0.1 * t))) * m1 * y^\wedge n + 1000 * m1 * t * (\exp(-0.15 * t))$

Explicit equations

1 $n = 3$

2 $r = t * (\exp(-0.15 * t))$

3 $m1 = 0.351 + 0.0208 * t^\wedge 1 + 0.0372 * t^\wedge 2 + 0.01144 * t^\wedge 3$

See Fig. 5.103.

Model: $y = a0 + a1 * t + a2 * t^\wedge 2 + a3 * t^\wedge 3 + a4 * t^\wedge 4$

Variable	Value
$a0$	0.0
$a1$	0.995
$a2$	−0.141
$a3$	0.00463
$a4$	9.509E−05

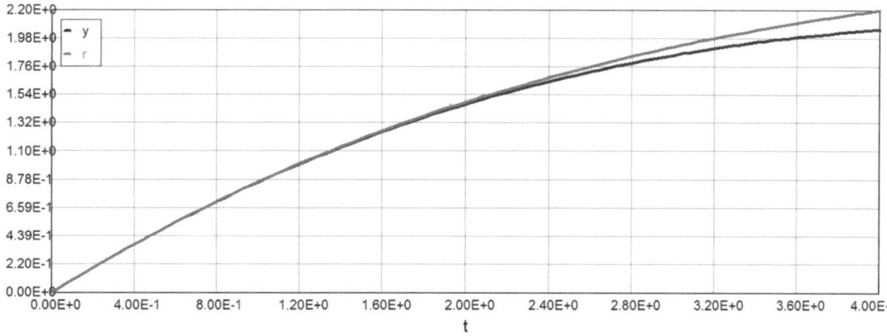

Fig. 5.103 Stress–temperature–strain diagram. Slow fire α = 1000

$$\textbf{Model}: \sigma = 0.995 * \theta - 0.141 * \theta\hat{\ }2 + 0.00463 * \theta\hat{\ }3 + 0.000095 * \theta\hat{\ }4 \quad (5.116)$$

α = 10,000

Calculated values of DEQ variables

	Variable	Initial value	Minimal value	Maximal value	Final value
1	*m*1	0.351	0.351	1.76156	1.76156
2	*n*	3	3	3	3
3	*r*	0	0	2.195247	2.195247
4	*t*	0	0	4	4
5	*y*	0	0	2.177276	2.177276

Differential equations

1 $\mathrm{d}(y)/\mathrm{d}(t) = -10{,}000 * y * m1 + (\exp(-0.15 * t)) * (1 - 0.15 * t) - (\exp(t/(1.0 + 0.1 * t))) * m1 * y\hat{\ }n + 10{,}000 * m1 * t * (\exp(-0.15 * t))$

Explicit equations

1 $n = 3$

2 $r = t * (\exp(-0.15 * t))$

3 $m1 = 0.351 + 0.0208 * t\hat{\ }1 + 0.0372 * t\hat{\ }2 + 0.01144 * t\hat{\ }3$

See Fig. 5.104.

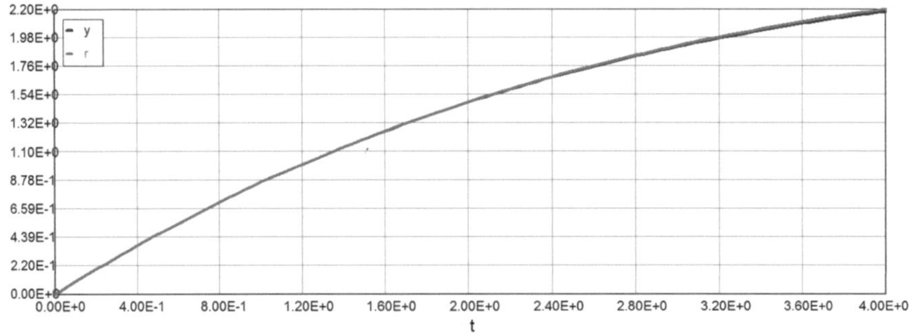

Fig. 5.104 Stress–temperature–strain diagram. Slow fire α = 10,000

Table 5.9 Stress–temperature data (slow fire)

$\alpha \vert \theta$		1	2	3	4
0.001	0	0.746	0.624	0.372	0.222
0.01	0	0.746	0.625	0.375	0.228
0.1	0	0.747	0.634	0.40	0.276
1.0	0	0.753	0.709	0.562	0.476
10	0	0.793	1.011	0.987	0.908
100	0	0.847	1.353	1.545	1.549
1000	0	0.859	1.465	1.849	2.046
10,000	0	0.860	1.480	1.906	2.177
r	0	0.861	1.482	1.913	2.195

Model: $y = a0 + a1 * t + a2 * t\text{^}2 + a3 * t\text{^}3 + a4 * t\text{^}4$

Variable	Value
$a0$	0.0
$a1$	0.999
$a2$	−0.148
$a3$	0.0102
$a4$	−0.000378

$$\textbf{Model}: \sigma = 0.999 * \theta - 0.148 * \theta^\wedge 2 + 0.0102 * \theta^\wedge 3 - 0.000378 * \theta^\wedge 4 \quad (5.117)$$

See Table 5.9.

5.4.5 *Functional Dependencies of Creep Stresses and Strains from the Stress Exponent "n"*

n = 1 α = 100
Calculated values of DEQ variables

	Variable	Initial value	Minimal value	Maximal value	Final value
1	$m1$	0.351	0.351	1.76156	1.76156
2	n	1	1	1	1
3	r	0	0	2.195247	2.195247
4	t	0	0	4	4
5	y	0	0	1.870543	1.870543

Differential equations

1 $d(y)/d(t) = -100 * y * m1 + (\exp(-0.15 * t)) * (1 - 0.15 * t) - (\exp(t/(1.0 + 0.1 * t))) * m1 * y^\wedge n + 100 * m1 * t * (\exp(-0.15 * t))$

Explicit equations

1 $n = 1$

2 $r = t * (\exp(-0.15 * t))$

3 $m1 = 0.351 + 0.0208 * t^\wedge 1 + 0.0372 * t^\wedge 2 + 0.01144 * t^\wedge 3$

See Fig. 5.105.
Model: $y = a0 + a1 * t + a2 * t^\wedge 2 + a3 * t^\wedge 3 + a4 * t^\wedge 4$

Variable	Value
$a0$	0.0
$a1$	0.988
$a2$	−0.154
$a3$	0.00586
$a4$	2.47E−05

$$\textbf{Model}: \sigma = 0.988 * \theta - 0.154 * \theta^\wedge 2 + 0.00586 * \theta^\wedge 3 + 0.0000247 * \theta^\wedge 4 \tag{5.118}$$

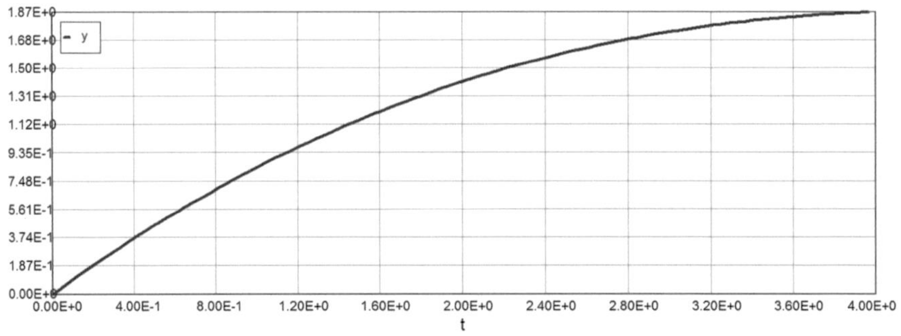

Fig. 5.105 Stress–temperature–strain diagram. Slow fire $n = 1$

n = 3

Calculated values of DEQ variables

	Variable	Initial value	Minimal value	Maximal value	Final value
1	$m1$	0.351	0.351	1.76156	1.76156
2	n	3	3	3	3
3	r	0	0	2.195247	2.195247
4	t	0	0	4	4
5	y	0	0	1.562581	1.549283

Differential equations

1 $d(y)/d(t) = -100 * y * m1 + (\exp(-0.15 * t)) * (1 - 0.15 * t) - (\exp(t/(1.0 + 0.1 * t))) * m1 * y^{\wedge}n + 100 * m1 * t * (\exp(-0.15 * t))$

Explicit equations

1 $n = 3$

2 $r = t * (\exp(-0.15 * t))$

3 $m1 = 0.351 + 0.0208 * t^{\wedge}1 + 0.0372 * t^{\wedge}2 + 0.01144 * t^{\wedge}3$

See Fig. 5.106.

Model: $y = a0 + a1 * t + a2 * t^{\wedge}2 + a3 * t^{\wedge}3 + a4 * t^{\wedge}4$

Variable	Value
$a0$	0
$a1$	1.0
$a2$	−0.142
$a3$	−0.0193
$a4$	0.00406

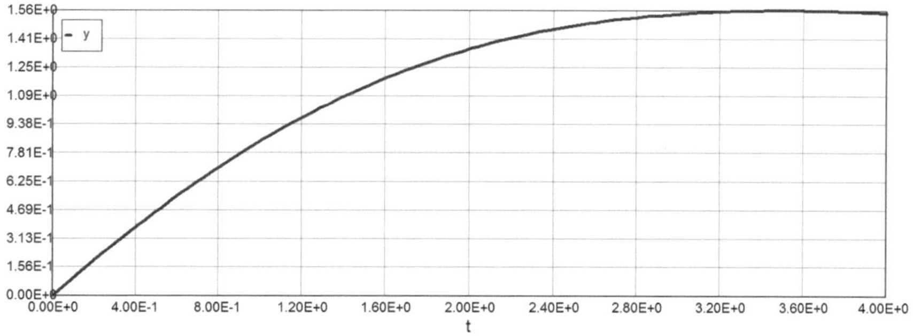

Fig. 5.106 Stress–temperature–strain diagram. Slow fire $n = 3$

$$\textbf{Model}: \sigma = 1.0 * \theta - 0.142 * \theta\hat{}2 - 0.0193 * \theta\hat{}3 + 0.00406 * \theta\hat{}4 \qquad (5.119)$$

n = 5

Calculated values of DEQ variables

	Variable	Initial value	Minimal value	Maximal value	Final value
1	$m1$	0.351	0.351	1.76156	1.76156
2	n	5	5	5	5
3	r	0	0	2.195247	2.195247
4	t	0	0	4	4
5	y	0	0	1.39242	1.366659

Differential equations

1 $d(y)/d(t) = -100 * y * m1 + (\exp(-0.15 * t)) * (1 - 0.15 * t) - (\exp(t/(1.0 + 0.1 * t))) * m1 * y\hat{}n + 100 * m1 * t * (\exp(-0.15 * t))$

Explicit equations

1 $n = 5$

2 $r = t * (\exp(-0.15 * t))$

3 $m1 = 0.351 + 0.0208 * t\hat{}1 + 0.0372 * t\hat{}2 + 0.01144 * t\hat{}3$

See Fig. 5.107.

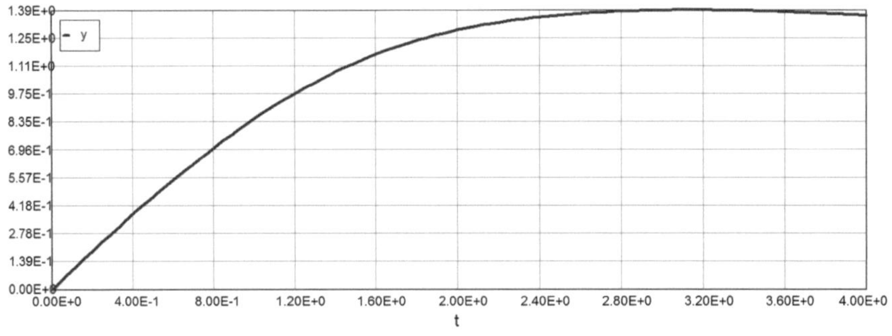

Fig. 5.107 Stress–temperature–strain diagram. Slow fire $n = 5$

Model: $y = a0 + a1 * t + a2 * t^2 + a3 * t^3 + a4 * t^4$

Variable	Value
$a0$	0
$a1$	1.06
$a2$	−0.176
$a3$	−0.0267
$a4$	0.00658

$$\textbf{Model}: \sigma = 1.06 * \theta - 0.176 * \theta^{\wedge}2 - 0.0267 * \theta^{\wedge}3 + 0.00658 * \theta^{\wedge}4 \quad (5.120)$$

n = 8

Calculated values of DEQ variables

	Variable	Initial value	Minimal value	Maximal value	Final value
1	$m1$	0.351	0.351	1.76156	1.76156
2	n	8	8	8	8
3	r	0	0	2.195247	2.195247
4	t	0	0	4	4
5	y	0	0	1.263959	1.237717

Differential equations

1 $$d(y)/d(t) = -100 * y * m1 + (\exp(-0.15 * t)) * (1 - 0.15 * t) - (\exp(t/(1.0 + 0.1 * t))) * m1 * y^{\wedge}n + 100 * m1 * t * (\exp(-0.15 * t))$$

Explicit equations

1 $n = 8$

2 $r = t * (\exp(-0.15 * t))$

3 $m1 = 0.351 + 0.0208 * t^{\wedge}1 + 0.0372 * t^{\wedge}2 + 0.01144 * t^{\wedge}3$

See Fig. 5.108.

Model: $y = a0 + a1 * t + a2 * t^2 + a3 * t^3 + a4 * t^4$

Variable	Value
$a0$	0
$a1$	1.163
$a2$	−0.279
$a3$	−0.00466
$a4$	0.00550

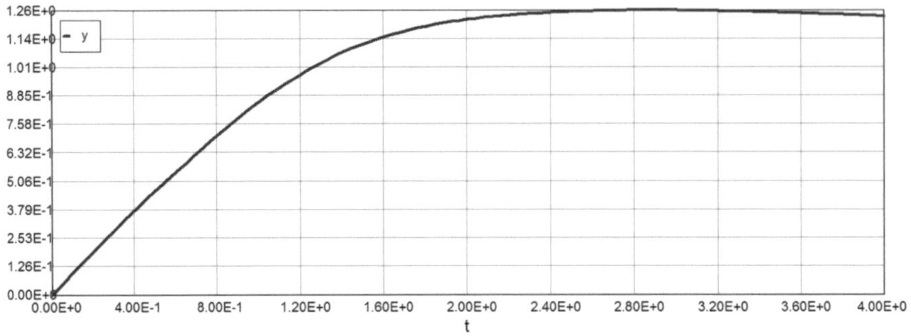

Fig. 5.108 Stress–temperature–strain diagram. Slow fire n = 8

$$\textbf{Model}: \sigma = 1.163 * \theta - 0.279 * \theta^\wedge 2 - 0.00466 * \theta^\wedge 3 + 0.0055 * \theta^\wedge 4 \quad (5.121)$$

n = 10

Calculated values of DEQ variables

	Variable	Initial value	Minimal value	Maximal value	Final value
1	$m1$	0.351	0.351	1.76156	1.76156
2	n	10	10	10	10
3	r	0	0	2.195247	2.195247
4	t	0	0	4	4
5	y	0	0	1.215626	1.191612

Differential equations

1 $\mathrm{d}(y)/\mathrm{d}(t) = -100 * y * m1 + (\exp(-0.15 * t)) * (1 - 0.15 * t) - (\exp(t/(1.0 + 0.1 * t))) * m1 * y^\wedge n + 100 * m1 * t * (\exp(-0.15 * t))$

Explicit equations

1 $n = 10$

2 $r = t * (\exp(-0.15 * t))$

3 $m1 = 0.351 + 0.0208 * t^\wedge 1 + 0.0372 * t^\wedge 2 + 0.01144 * t^\wedge 3$

See Fig. 5.109.

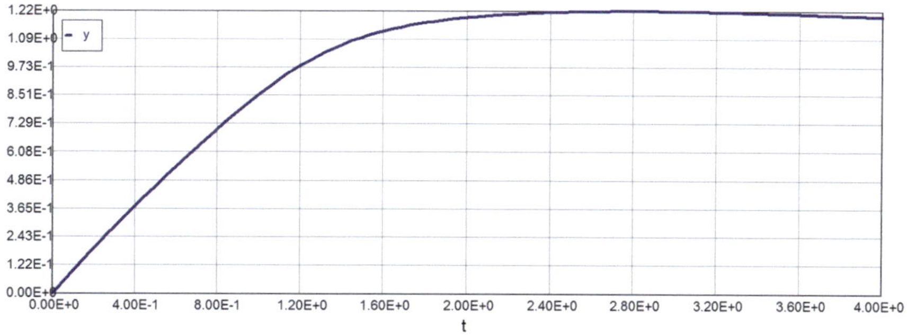

Fig. 5.109 Stress–temperature–strain diagram. Slow fire $n = 10$

Model: $y = a0 + a1 * t + a2 * t^2 + a3 * t^3 + a4 * t^4$

Variable	Value
$a0$	0
$a1$	1.217
$a2$	−0.341
$a3$	0.0126
$a4$	0.00403

$$\textbf{Model}: \sigma = 1.217 * \theta - 0.341 * \theta^{\wedge}2 + 0.0126 * \theta^{\wedge}3 + 0.00403 * \theta^{\wedge}4 \quad (5.122)$$

See Fig. 5.110 and Table 5.10.

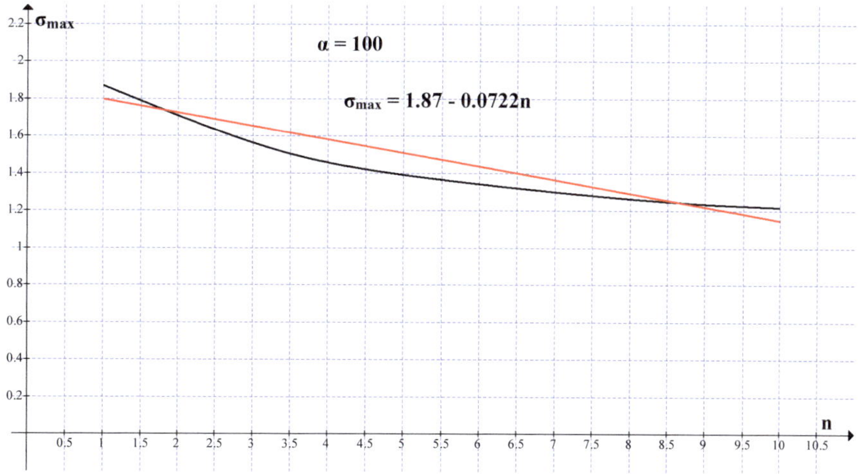

Fig. 5.110 Maximum creep stress versus stress exponent. Slow fire $\alpha = 100$

Table 5.10 Maximum creep stress versus stress exponent

n	1	3	5	8	10	α
σ						
σ_{max}	1.87	1.563	1.392	1.264	1.220	100

References

1. Johnson PF (2007) Use of worst case scenarios in fire safety design. NFPA Annual Meeting, Boston
2. National Institute of Standards and Technology (NIST) (2008) Fire dynamics simulator (Version 5) technical reference guide, vol. 1: mathematical model. NIST Special Publication 1018-5. NIST, Gaithersburg
3. Razdolsky L, Petrov A, Shtessel E (1977) Critical conditions of local ignition in a large medium with convective heat transfer. In: Physics of combustions and explosions. Academy of Science, Moscow
4. Razdolsky L (2012) Structural fire load. Theory and principals. McGraw-Hill, New York
5. Razdolsky L (2009) Mathematical modeling of fire dynamics. In: Proceedings of the World Congress on Engineering and Computer Science. WCE, London
6. Frank-Kamenestkii DA (1969) Diffusion and heat transfer in chemical kinetics. Plenum Press, New York
7. SFPE (2011) Engineering guide: fire exposures to structural elements. Society of Fire Protection Engineers, Quincy, MA
8. Magnusson SE, Thelandersson S (1970) Temperature-time curves of complete process of fire development in enclosed spaces. Acta Polytech Scand
9. Society of Fire Protection Engineers (1998) The SFPE engineering guide to performance-based fire protection analysis and design. SFPE, Quincy
10. Razdolsky L (2009) Structural fire loads in a high-rise building design. In: Proceedings of the SFPE engineering technology conference. NFPA-Publisher, Scottsdale
11. Eringen AC (1999) Microcontinuum field theories. Vol. I: foundations and solids. Springer, New York
12. Eringen AC (2001) Microcontinuum field theories. Vol. II: fluent media. Springer, New York
13. Lai WM, Rubin D, Krempl E (1993) Introduction to continuum mechanics. Pergamon Press, Oxford
14. Bertram A (2003) Finite thermoplasticity based on isomorphismus. Int J Plast 19:2027–2050
15. Bertram A (2005) Elasticity nad plasticity of large deformations. Springer, Berlin
16. Betten J (2005) Creep mechanics. Springer, Berlin
17. Odqvist FKG (1974) Mathematical theory of creep and creep rupture. Oxford University Press, Oxford
18. Billington EW (1986) Introduction to the Mechanics and Physics of Solids. Adam Hilger, Bristol
19. Krausz AS, Krausz K (1996) Unified constitutive laws of plastic deformation. Academic press, San Diego
20. AISC 13th Ed
21. Larson FR, Miller J (1952) A time—temperature relationship for rupture and creep stresses. Trans ASME 74:765–771

22. Hertzberg RW (1996) Deformation and fracture mechanics of engineering materials, 4th edn. Wiley, Hoboken
23. Weertman J (1957) Steady-state creep through dislocation climb. J Appl Phys 28(3):362–364
24. Blum W (1993) High-temperature deformation and creep of crystalline solids. In: Cahn RW et al (eds) Materials science and technology, vol 6. VCH, Weinheim, pp 360–405
25. Evans RW, Wilshire B (1993) Introduction to creep. Institute of Materials, London, pp 45–66
26. Sherby OD, Burke PM (1968) Mechanical behavior of crystalline solid at elevated temperature. Prog Mater Sci 13:340–347

Chapter 6
Anisotropic Structural Plates—Anisotropic Materials and Composite Structures

Notation

k	The thermal conductivity that has the dimensions W/m K or J/m s K
T	Temperature
d	Thickness in the direction if heat flow
c	Specific heat capacity
Q	Activation energy
R	Ideal gas constant
P_1	Losses of heat due to thermal radiation
e	Emissivity factor
σ	Stefan–Boltzmann constant (σ = 5.6703 (10^{-8}) W/m^2 K^4)
T_o	Ambient temperature
A_v	Area of openings
c_p	Average specific heat at constant pressure
t	Time
$\vec{v}(u;\,v;\,w)$	Velocity vector
D	Diffusion coefficient (m^2/s)
p	Pressure
v	Kinematic viscosity
θ	Dimensionless temperature
τ	Dimensionless time
h	Height of the compartment (m)
a	Thermal diffusivity (m^2/s)
Time	$t = \frac{h^2}{a}\tau$ (s)

L. Razdolsky, *Probability Based High Temperature Engineering*,
DOI 10.1007/978-3-319-41909-1_6

Temperature	$T = \frac{RT_*^2}{E}\theta + T_*$ (K), where T_* = 600 K is the base line temperature
Coordinates	$\bar{x} = x/h$ and $\bar{z} = z/h$—"x" and "z"—Dimensionless coordinates
Velocities	$\bar{u} = \frac{v}{h}u$ (m/s) and $\bar{w} = \frac{v}{h}w$ (m/s)—Horizontal and vertical components velocity accordingly; v—kinematic viscosity (m^2/s); "u" and "w"—dimensionless velocities
$\Pr = v/a$	Prandtl number
$\mathrm{Fr} = \frac{gh^3}{va}$	Froude number
g	Gravitational acceleration
$\mathrm{Le} = a/D = \mathrm{Sc}/\Pr$	The Lewis number
$\mathrm{Sc} = v/D$	The Schmidt number
$\bar{\beta} = \frac{RT_*}{E}$	Dimensionless parameter
$\bar{\gamma} = \frac{c_p RT_*^2}{QE}$	Dimensionless parameter
$P = \frac{e\sigma K_v(\beta T_*)^3 h}{\lambda}$	Thermal radiation dimensionless coefficient
$\sigma = 5.67\ (10^{-8})\ (\mathrm{W/m^2\ K^4})$	Stefan-Boltzman constant
$K_v = A_o h/V$	Dimensionless opening factor
A_o	Total area of vertical and horizontal openings
$\delta = \left(\frac{E}{RT_*^2}\right)Qz\left(\exp\left(-\frac{E}{RT_*}\right)\right)$	Frank-Kamenetskii's parameter
$C = [1 - P(t)/P_o]$	Concentration of the burned fuel product in the fire compartment
$\bar{W} = \frac{v}{h}W$	Vertical component of gas velocity
$\bar{U} = \frac{v}{h}U$	Horizontal component of gas velocity
$b = L/h$, "L" and "h"	Length (width) and height of fire compartment accordingly
W; U	Dimensionless velocities
R_n	Nominal strength
S_i	Nominal load
φ	Resistance factor
γ	Load factor
R_c	Characteristic value for the resistance
A	Cross-sectional area
I	Moment of inertia
W	Total weight
G_c	Characteristic value for the permanent load
S	Characteristic value for the variable load
ψ_1	Partial safety factor for the permanent load

ψ_2	Partial safety factor for the variable load
β	Reliability index
S	Probability space
A	Set of outcomes (events) to which a probability is assigned
$P(E_2\|E_1)$	Conditional probability
Φ^* (.)	Denotes the cumulative distribution function of standard normal distribution $\Phi^*(z) = \frac{1}{\sqrt{2\pi}}\int_{-\infty}^{z} e^{-\frac{z^2}{2}}dz$
μ_A, μ_B, σ_A, σ_B	Mean and standard deviation of A and B, respectively
$J(t, t')$	Compliance function
T_M	Melting point of the metal matrix material
$\varepsilon(t)$	Strain
$\sigma(t)$	Stress
$\bar{\sigma}(t) = E(t)\varepsilon(t)$	Instantaneous stress
ε_e	Instantaneous (elastic) strain
ε_c	Creep strain
ε_T	Thermal expansion due to high temperature effect
$K(t, t') = -\partial J(t, t')/\partial t'$	Retardation function (memory function)
$R(t, t')$	Relaxation function (also called the relaxation modulus)
$M(\theta)$	Bending moment
$V(\theta)$	Shear force
$P(\theta)$	Axial force
$y(\theta)$	Deformation
P_f	Probability of failure
P_{rel}	Reliability
σ	Ψ
ε	Ψ
θ	Dimensionless temperature
κ	Curvature
τ	Dimensionless time
ω	Frequency
$S(\omega)$	Spectral density
μ	Poisson coefficient
D	Diffusion rate (Flick's law)
η	*Viscosity* parameter of the material
E	Modulus of elasticity
$n = \eta/E$	Relaxation time
n	Power law exponent
α_i	Material property parameter
L	Span (spring spacing)
k_0	Subgrade modulus

6.1 Introduction

In recent years, various constitutive models have been proposed to describe fundamental features of an anisotropy, structure, and rate dependence (e.g. [1–5]). Different approaches have been used to capture the various rate-dependent phenomena, such as strain-rate effects, creep, relaxation, and accumulated effects. These constitutive models include empirical models, rheological models, and general stress–strain–time models that are based on theories of viscoelasticity. Viscous–plastic models are easily adaptable to numerical implementation in a general purpose finite element framework, as they are often formulated in incremental form. Most of the rate-dependent models were developed based on the Perzyna's [6, 7] overstress theory. This approach has become a preferred basis for the further development of viscoelastic models. However, determination of model input parameters for overstress models is difficult (see e.g. [4]), and strictly speaking not feasible in practical context due to the very low loading rates required in the laboratory tests. As a consequence, the input values require calibration via parametric studies, which limits practical adaptation, and furthermore, the values for the input parameters are not necessarily unique [8, 9]. It is commonly thought that a consequence of the overstress theory is that it lacks the capability to model tertiary creep, i.e., the acceleration of the creep process [8], but this problem can be overcome by introducing some damage or deteriorations law in the formulation. However, it is only possible to model stress relaxation if the stress state lies outside the current static yield surface.

The elastic and creep parts of the strain in the model are combined with an additive law, expressing the total strain rate as combination of elastic and creep components. New alloying combinations are introducing different precipitate phases, which offer promising mechanical properties over a wider range of temperatures. Another solution involves the addition of extrinsic reinforcement, either in the form of dispersions, particulates, whiskers, or continuous fibers. The addition of a reinforcement phase can help to overcome these current limitations in aluminum alloys and expand the practical service parameters for aluminum. The latter solution to improving properties resulted in the development of metal matrix composites (MMC). Metal matrix composites typically involve a relatively ductile metal matrix that transfers load to a high-strength reinforcing phase. The metal matrix is designed to absorb energy, which provides impact resistance, ductility, toughness, and plastic deformation. The role of the matrix depends on the form of the reinforcement, but in general, the matrix transfers load to the reinforcing phase. The reinforcement phase is designed to be the load-bearing phase that increases the strength, stiffness, and creep resistance of the metal matrix. Reinforcements can be effective in many different forms, and the strengthening mechanisms are dependent upon the morphology of the reinforcing phase. The final form of extrinsic reinforcement used in metal matrix composites is continuous filaments. Continuously reinforced metals are designed such that the fibers are oriented in the direction of the applied loading. This takes greatest advantage of the strength of the fibers.

The metal matrix absorbs energy and transfers essentially the entire load to the fibers. These composites display high degrees of anisotropy, depending on the orientation of the fibers, and typical fiber volume fractions range from 0.30 to greater than 0.50. High residual stresses in the matrix due to the large mismatch in coefficients of thermal expansion also aid in strengthening in some cases [10, 11]. High strength is retained at high temperatures almost to the melting temperature of the metal matrix in the direction of the length of the fibers. One of the most obvious effects of the reinforcement on the matrix properties is observed in the mechanical properties. Whiskers increase the modulus of elasticity; the actual amount of the increase in modulus is determined by the alignment and the volume fraction of the whiskers. The yield or flow stress and the ultimate tensile strength experience up to 60 % increase over the unreinforced alloy. The creep properties of whisker and particulate reinforced MMCs have been studied [12, 13]. Elevating the temperatures and shortening the times too much will lead to two problems when trying to predict actual creep properties. First, test temperatures significantly higher than the anticipated service conditions during a real fire event can cause the changes in the material and its mechanical properties as previously mentioned in this section. This can put the material into a different creep deformation regime, which may produce erroneous data because it does not accurately reproduce the deformation mechanisms active under design fire conditions. This can also cause large changes in the precipitates which results in a nonrepresentative material. Second, elevating the temperature and stress so that creep tests only last a few hours results in the problem of extrapolating the data from a test of less than one hour to a service life of many hours [14, 15]. The reliability of extrapolating such data is questionable, for the reasons given above. Creep in MMCs is more complex than creep in a monolithic alloy. The mechanisms described for unreinforced alloys can occur in both constituent phases. Creep will have a far smaller effect in the reinforcement phase than in the matrix because the melting temperature is much higher. Thus, at temperatures below 300 °C, the creep effects can be considered negligible. As a result, the reinforcement can be described as creep-resistant, and the aluminum matrix creep-compliant. The extent of primary creep is dependent on the volume fraction of the reinforcement phase due to load redistribution between fibers and matrix during creep. As the matrix creeps, the stress in the matrix decreases. The fibers deform elastically, so the stress in the fibers increases, resulting in an overall decrease in the creep rate as compared to the unreinforced alloy (see Examples below). Other local mechanisms also occur which combine to create the overall creep response of the composite. Reinforced phases increase the creep resistance of the MMC matrix by obstructing dislocation, atom, or vacancy motion. However, the reinforcement phase may not be capable of impeding other deformation mechanisms active within different temperature or stress regimes. Reinforcement phases can also introduce new creep mechanisms which do not occur in monolithic alloys.

6.2 Stress, Strain, and Deformations in Solids. Constitutive Relations

It is physically observed that the deformation and motion of the particles in a continuous body are in some way related to the forces applied to the body. Alternately, we can say that the strain (a measure of deformation per unit length) is related to the stress (force per unit area) applied to the body. This relationship essentially provides specific information about the characteristics of the specific material that the body is made of [16–18]. In general, the 9 components of the stress tensor $[\sigma]$ can be related to the 9 components of the strain tensor. Due to the symmetry of both tensors ($\sigma_{xy} = \sigma_{yx}$, etc.), there are only 6 independent stresses and strains and the number of material parameters is reduced to 36. For a linear relationship between stress and strain, one can thus write:

$$\begin{aligned} \boldsymbol{\sigma}_{xx} &= C_{11}\boldsymbol{\varepsilon}_{xx} + C_{12}\boldsymbol{\varepsilon}_{yy} + C_{13}\boldsymbol{\varepsilon}_{zz} + C_{14}\boldsymbol{\varepsilon}_{xy} + C_{15}\boldsymbol{\varepsilon}_{xz} + C_{16}\boldsymbol{\varepsilon}_{yz} \\ \boldsymbol{\sigma}_{yy} &= C_{21}\boldsymbol{\varepsilon}_{xx} + C_{22}\boldsymbol{\varepsilon}_{yy} + C_{23}\boldsymbol{\varepsilon}_{zz} + C_{24}\boldsymbol{\varepsilon}_{xy} + C_{25}\boldsymbol{\varepsilon}_{xz} + C_{26}\boldsymbol{\varepsilon}_{yz} \\ \ldots \\ \boldsymbol{\sigma}_{yz} &= C_{61}\boldsymbol{\varepsilon}_{xx} + C_{62}\boldsymbol{\varepsilon}_{yy} + C_{63}\boldsymbol{\varepsilon}_{zz} + C_{64}\boldsymbol{\varepsilon}_{xy} + C_{65}\boldsymbol{\varepsilon}_{xz} + C_{66}\boldsymbol{\varepsilon}_{yz} \end{aligned} \quad (6.1)$$

In matrix form:

$$\begin{Bmatrix} \boldsymbol{\sigma}_{xx} \\ \boldsymbol{\sigma}_{yy} \\ \boldsymbol{\sigma}_{zz} \\ \boldsymbol{\sigma}_{xy} \\ \boldsymbol{\sigma}_{xz} \\ \boldsymbol{\sigma}_{yz} \end{Bmatrix} = \begin{Bmatrix} C_{11}C_{12}C_{13}C_{14}C_{15}C_{16} \\ C_{21}C_{22}C_{23}C_{24}C_{25}C_{26} \\ C_{31}C_{32}C_{33}C_{34}C_{35}C_{36} \\ C_{41}C_{42}C_{43}C_{44}C_{45}C_{46} \\ C_{51}C_{52}C_{53}C_{54}C_{55}C_{56} \\ C_{61}C_{62}C_{63}C_{64}C_{65}C_{66} \end{Bmatrix} \begin{Bmatrix} \boldsymbol{\varepsilon}_{xx} \\ \boldsymbol{\varepsilon}_{yy} \\ \boldsymbol{\varepsilon}_{zz} \\ \boldsymbol{\varepsilon}_{xy} \\ \boldsymbol{\varepsilon}_{xz} \\ \boldsymbol{\varepsilon}_{yz} \end{Bmatrix} \quad (6.2)$$

In general, C_{ij} are functions of coordinates x, y, and z. For a homogeneous material, C_{ij} are independent of coordinates x, y, and z. For an isotropic material, C_{ij} are independent of the orientation of the coordinate axes (i.e., properties are same in all directions). For an orthotropic material, C_{ij} are different in each of the 3 coordinate directions. For an anisotropic material, C_{ij} are different in each of the 3 coordinate directions and are dependent on the orientation of the coordinate axes (i.e., properties are not the same in all directions). For the simplest solid material (linear, isotropic), one can deduce from physical observation that there are only two independent material constants that relate all stress and strain components. These are

E Young's modulus (slope of uniaxial stress–strain curve)
ν Poisson's ratio (ratio of contraction to extension strains)

These properties are typically measured in a uniaxial tensile test.

Consider a test where we apply normal tractions (stresses) in the *x*, *y*, and *z* directions simultaneously. For a linear material, we can think of this as three separate problems:

$v\varepsilon_{xx}$ = normal strain in *x* direction due to σ_{xx}+ normal strain in *x* direction due to σ_{yy}+ normal strain in *x* direction due to σ_{zz}. From Hooke's Law, the strain in the x direction for each case is:

$$\boldsymbol{\varepsilon}_{xx} = \frac{1}{E}[\boldsymbol{\sigma}_{xx} - \boldsymbol{\mu}(\boldsymbol{\sigma}_{yy} + \boldsymbol{\sigma}_{zz})]. \tag{6.3}$$

The stress in the *x* direction increases the strain while the transverse stresses cause a contraction (decrease). Doing similar computations in the *y* and *z* directions gives:

$$\begin{aligned} \boldsymbol{\varepsilon}_{yy} &= \frac{1}{E}[\boldsymbol{\sigma}_{yy} - \boldsymbol{\mu}(\boldsymbol{\sigma}_{xx} + \boldsymbol{\sigma}_{zz})] \\ \boldsymbol{\varepsilon}_{zz} &= \frac{1}{E}[\boldsymbol{\sigma}_{zz} - \boldsymbol{\mu}(\boldsymbol{\sigma}_{xx} + \boldsymbol{\sigma}_{yy})] \end{aligned}. \tag{6.4}$$

Experiments with shear tractions will show that an *xy* shear stress in the *xy* plane produces only *xy* shear strain in the *xy* plane and NO extensional strain* (e.g., the shear strain is uncoupled from the extensional strain). Thus, we obtain the following experimental observations for the shear strains:

$$\boldsymbol{\varepsilon}_{xy} = \left[\frac{1+\boldsymbol{\mu}}{E}\right]\boldsymbol{\sigma}_{xy}; \quad \boldsymbol{\varepsilon}_{xz} = \left[\frac{1+\boldsymbol{\mu}}{E}\right]\boldsymbol{\sigma}_{xz}; \quad \boldsymbol{\varepsilon}_{yz} = \left[\frac{1+\boldsymbol{\mu}}{E}\right]\boldsymbol{\sigma}_{yz} \tag{6.5}$$

The term $E/(1 + \mu)$ defines a shear modulus; *G*, relating shear strain and shear stress (similar to Young's modulus, *E*, for extensional strain). $G = E/2(1 + \mu)$.

For a linear, elastic, isotropic material, we can superimpose all of the six equations to obtain the constitutive relations [19–21]:

$$\begin{aligned} &\boldsymbol{\varepsilon}_{xx} = \frac{1}{E}[\boldsymbol{\sigma}_{xx} - \boldsymbol{\mu}(\boldsymbol{\sigma}_{yy} + \boldsymbol{\sigma}_{zz})]; \quad \boldsymbol{\varepsilon}_{yy} = \frac{1}{E}[\boldsymbol{\sigma}_{yy} - \boldsymbol{\mu}(\boldsymbol{\sigma}_{xx} + \boldsymbol{\sigma}_{zz})]; \\ &\boldsymbol{\varepsilon}_{zz} = \frac{1}{E}[\boldsymbol{\sigma}_{zz} - \boldsymbol{\mu}(\boldsymbol{\sigma}_{xx} + \boldsymbol{\sigma}_{yy})] \\ &\boldsymbol{\varepsilon}_{xy} = \left[\frac{1+\boldsymbol{\mu}}{E}\right]\boldsymbol{\sigma}_{xy}; \quad \boldsymbol{\varepsilon}_{xz} = \left[\frac{1+\boldsymbol{\mu}}{E}\right]\boldsymbol{\sigma}_{xz}; \quad \boldsymbol{\varepsilon}_{yz} = \left[\frac{1+\boldsymbol{\mu}}{E}\right]\boldsymbol{\sigma}_{yz} \end{aligned} \tag{6.6}$$

The above may be solved for the stresses in terms of the strains to obtain:

$$\begin{aligned}
\sigma_{xx} &= \frac{E}{(1+\mu)(1-2\mu)}[(1-\mu)\varepsilon_{xx} + \mu\varepsilon_{yy} + \mu\varepsilon_{zz}] \\
\sigma_{yy} &= \frac{E}{(1+\mu)(1-2\mu)}[\mu\varepsilon_{xx} + (1-\mu)\varepsilon_{yy} + \mu\varepsilon_{zz}] \\
\sigma_{zz} &= \frac{E}{(1+\mu)(1-2\mu)}[\mu\varepsilon_{xx} + \mu\varepsilon_{yy} + (1-\mu)\varepsilon_{zz}] \\
\sigma_{xy} &= \frac{E}{1+\mu}\varepsilon_{xy}; \quad \sigma_{xz} = \frac{E}{1+\mu}\varepsilon_{xz}; \quad \sigma_{yz} = \frac{E}{1+\mu}\varepsilon_{yz}..
\end{aligned} \tag{6.7}$$

Important Note: Hooke's Law in frequently written in terms of the *engineering shear strain* γ. Recall, that the engineering shear strain is defined to be twice that of the tensor shear strain; for example, $\gamma_{xy} = 2\varepsilon_{xy}$. Hence the shear stress in terms of engineering shear strain becomes:

$$\begin{aligned}
\sigma_{xy} &= \frac{E}{1+\mu} \quad \varepsilon_{xy} = \frac{E}{2(1+\mu)}\gamma_{xy} \\
\sigma_{xz} &= \frac{E}{2(1+\mu)}\gamma_{xz} = G\gamma_{xz}; \quad \sigma_{yz} = G\gamma_{yz}; \quad \sigma_{xy} = G\gamma_{xy}
\end{aligned} \tag{6.8}$$

For a linear, elastic, homogeneous material we note that the extensional strains and stresses are uncoupled from the shear strains and stresses.

Experimentally, we observe that a temperature increase, ΔT, produces a uniform expansion but no shear and the expansion is proportional to a material constant α (coefficient of thermal expansion). The additional strain due to heating is thus: $\varepsilon_{xx} = \varepsilon_{yy} = \varepsilon_{zz} = \alpha\Delta T$.

$$\begin{aligned}
\varepsilon_{xx} &= \frac{1}{E}[\sigma_{xx} - \mu(\sigma_{yy} + \sigma_{zz})] + \alpha\Delta T \\
\varepsilon_{yy} &= \frac{1}{E}[\sigma_{yy} - \mu(\sigma_{xx} + \sigma_{zz})] + \alpha\Delta T \\
\varepsilon_{zz} &= \frac{1}{E}[\sigma_{zz} - \mu(\sigma_{xx} + \sigma_{yy})] + \alpha\Delta T \\
\varepsilon_{xy} &= \left[\frac{1+\mu}{E}\right]\sigma_{xy}; \quad \varepsilon_{xz} = \left[\frac{1+\mu}{E}\right]\sigma_{xz}; \quad \varepsilon_{yz} = \left[\frac{1+\mu}{E}\right]\sigma_{yz}.
\end{aligned} \tag{6.9}$$

It should be noted that the first term in the extensional strain terms above (the [ ] term) is due to elastic behavior of the material (i.e., it has Young's modulus in it). The second part is due to thermal strain. We can separate the *total strain* into *elastic* and *thermal strains components*: $\varepsilon_{x,\text{tot}} = \varepsilon_{x,\text{el.}} + \varepsilon_{\text{therm.}}$

Recall, in the above, that shear strains have no thermal component.

6.3 Principal Stresses and Stress Invariants

At every point in a stressed body there are at least three planes, called *principal planes*, with normal vectors, called *principal directions*, where the corresponding stress vector is perpendicular to the plane n, i.e., parallel or in the same direction as the normal vector n, and where there are no normal shear stresses τ. The three stresses normal to these principal planes are called *principal stresses*.

The components σ_{ij} of the stress tensor depend on the orientation of the coordinate system at the point under consideration. However, the stress tensor itself is a physical quantity and as such, it is independent of the coordinate system chosen to represent it. There are certain invariants associated with every tensor which are also independent of the coordinate system. For example, a vector is a simple tensor of rank one. In three dimensions, it has three components. The value of these components will depend on the coordinate system chosen to represent the vector, but the magnitude of the vector is a physical quantity (a scalar) and is independent of the Cartesian coordinate system chosen to represent the vector. Similarly, every second rank tensor (such as the stress and the strain tensors) has three independent invariant quantities associated with it. One set of such invariants are the principal stresses of the stress tensor, which are just the eigenvalues of the stress tensor. Their direction vectors are the principal directions or eigenvectors.

$$T^{(n)} = \lambda\vec{n} = \sigma_n\vec{n}$$
$$\text{or: } (\sigma_{ij} - \lambda\delta_{ij})n_j = 0$$

This is a homogeneous system, i.e. equal to zero, of three linear equations. To obtain a nontrivial (nonzero) solution, the determinant matrix of the coefficients must be equal to zero, i.e., the system is singular. Thus,

$$\left|\sigma_{ij} - \lambda\delta_{ij}\right| = \begin{vmatrix} \sigma_{11} - \lambda & \sigma_{12} & \sigma_{13} \\ \sigma_{21} & \sigma_{22} - \lambda & \sigma_{23} \\ \sigma_{31} & \sigma_{32} & \sigma_{33} - \lambda \end{vmatrix}. \tag{6.10}$$

Expanding the determinant leads to the *characteristic equation*

$$\left|\sigma_{ij} - \lambda\,\delta_{ij}\right| = -\lambda^3 + I_1\lambda^2 - I_2\lambda + I_3 = 0, \tag{6.11}$$

where

$$\begin{aligned} I_1 &= \sigma_{kk} = \sigma_{11} + \sigma_{22} + \sigma_{33} \\ I_2 &= \begin{vmatrix} \sigma_{22}\sigma_{23} \\ \sigma_{32}\sigma_{33} \end{vmatrix} + \begin{vmatrix} \sigma_{11}\sigma_{13} \\ \sigma_{31}\sigma_{33} \end{vmatrix} + \begin{vmatrix} \sigma_{11}\sigma_{12} \\ \sigma_{21}\sigma_{22} \end{vmatrix} = \frac{1}{2}\left(\sigma_{ii}\sigma_{jj} - \sigma_{ij}\sigma_{ji}\right). \\ I_3 &= \det(\sigma_{ij}). \end{aligned} \tag{6.12}$$

The characteristic equation has three real roots λ_i, i.e., not imaginary due to the symmetry of the stress tensor. The $\sigma_1 = \max(\lambda_1, \lambda_2, \lambda_3)$, $\sigma_3 = \min(\lambda_1, \lambda_2, \lambda_3)$, and $\sigma_2 = I_1 - \sigma_1 - \sigma_3$ are the principal stresses, functions of the eigenvalues λ_i. The eigenvalues are the roots of the Cayley–Hamilton theorem. The principal stresses are unique for a given stress tensor. Therefore, from the characteristic equation, the coefficients $I_{1,}$ I_2 and I_3, called the first, second, and third *stress invariants*, respectively, have always the same value regardless of the coordinate system's orientation.

For each eigenvalues, there is a nontrivial solution for n_j in the equation $\left|\sigma_{ij} - \lambda\delta_{ij}\right| = 0$. These solutions are the principal directions or eigenvectors defining the plane where the principal stresses act. The principal stresses and principal directions characterize the stress at a point and are independent of the orientation.

A coordinate system with axes oriented to the principal directions implies that the normal stresses are the principal stresses and the stress tensor is represented by a diagonal matrix:

$$\sigma_{ij} = \begin{bmatrix} \sigma_1 & 0 & 0 \\ 0 & \sigma_1 & 0 \\ 0 & 0 & \sigma_1 \end{bmatrix}. \tag{6.13}$$

The principal stresses can be combined to form the stress invariants $I_{1,}$ I_2 and I_3, The first and third invariants are the trace and determinants, respectively, of the stress tensor. Thus,

$$\begin{aligned} I_1 &= \sigma_1 + \sigma_2 + \sigma_3 \\ I_2 &= \sigma_1\sigma_2 + \sigma_2\sigma_3 + \sigma_3\sigma_1 \\ I_3 &= \sigma_1\sigma_2\sigma_3 \end{aligned} \quad . \tag{6.14}$$

Because of its simplicity, the principal coordinate system is often useful when considering the state of the elastic medium at a particular point. Principal stresses are often expressed in the following equation for evaluating stresses in the x and y directions or axial and bending stresses on a part [22–24]. The principal normal stresses can then be used to calculate the von Mises stress and ultimately the safety factor and margin of safety.

$$\sigma_1, \sigma_2 = \frac{\sigma_x + \sigma_y}{2} \pm \sqrt{\left(\frac{\sigma_x - \sigma_y}{2}\right)^2 + \tau_{xy}^2}. \tag{6.15}$$

Using just the part of the equation under the square root is equal to the maximum and minimum shear stress for plus and minus. This is shown as:

$$\tau_{\max}, \tau_{\min} = \pm\sqrt{\left(\frac{\sigma_x - \sigma_y}{2}\right)^2 + \tau_{xy}^2}. \tag{6.16}$$

6.4 Maximum and Minimum Shear Stresses

The maximum shear stress or maximum principal shear stress is equal to one-half the difference between the largest and smallest principal stresses, and acts on the plane that bisects the angle between the directions of the largest and smallest principal stresses, i.e., the plane of the maximum shear stress is oriented 45° from the principal stress planes. The maximum shear stress is expressed as

$$\tau_{\max} = \frac{1}{2}|\sigma_{\max} - \sigma_{\min}| \quad \text{if } \sigma_1 \geq \sigma_2 \geq \sigma_3 \Rightarrow \tau_{\max} = \frac{1}{2}|\sigma_1 - \sigma_3|. \tag{6.17}$$

When the stress tensor is nonzero the normal stress component acting on the plane for the maximum shear stress is nonzero and it is equal to

$$\sigma_n = \frac{1}{2}(\sigma_1 + \sigma_3)$$

6.5 Stress Deviator Tensor

The stress tensor σ_{ij} can be expressed as the sum of two other stress tensors:

1. a *mean hydrostatic stress tensor* or *volumetric stress tensor* or *mean normal stress tensor*, $\sigma_{\text{aver}}\delta_{ij}$, which tends to change the volume of the stressed body; and
2. a deviatoric component called the *stress deviator tensor*, s_{ij}, which tends to distort it.

$$\begin{aligned} &\sigma_{ij} = s_{ij} + \sigma_{\text{aver}}\delta_{ij} \\ &\text{where } \sigma_{\text{aver}} = \frac{\sigma_{ij} + \sigma_{ij} + \sigma_{ij}}{3} = \frac{I_1}{3} \cdot \end{aligned} \tag{6.18}$$

The deviatoric stress tensor can be obtained by subtracting the hydrostatic stress tensor from the Cauchy stress tensor:

$$\begin{bmatrix} s_{11}s_{12}s_{13} \\ s_{21}s_{22}s_{23} \\ s_{31}s_{32}s_{33} \end{bmatrix} = \begin{bmatrix} \sigma_{11} - \sigma_{\text{aver}} & \sigma_{12} & \sigma_{13} \\ \sigma_{21} & \sigma_{22} - \sigma_{\text{aver}} & \sigma_{23} \\ \sigma_{31} & \sigma_{32} & \sigma_{33} - \sigma_{\text{aver}} \end{bmatrix}. \tag{6.19}$$

6.5.1 *Invariants of the Stress Deviator Tensor*

As it is a second-order tensor, the stress deviator tensor also has a set of invariants, which can be obtained using the same procedure used to calculate the invariants of the stress tensor. It can be shown that the principal directions of the stress deviator tensor σ_{ij} are the same as the principal directions of the stress tensor. Thus, the characteristic equation is: $\left|s_{ij} - \lambda\delta_{ij}\right| = 0$.

Because $s_{kk} = 0$, the stress deviator tensor is in a state of pure shear.

A quantity called the equivalent stress or von Mises stress is commonly used in solid mechanics. The equivalent stress is defined as

$$\sigma_e = \sqrt{\frac{1}{2}\left[(\sigma_1 - \sigma_2)^2 + (\sigma_2 - \sigma_3)^2 + (\sigma_3 - \sigma_1)^2\right]}. \tag{6.20}$$

6.6 Failure Theories

This section starts with a warning: "Failure" is a tricky term to define. Failure under load can occur due to excessive elastic deflection or due to excessive stresses—the same stress for the same material may be considered excessive in one type of loading and acceptable in another. Failure prediction theories due to excessive stresses fall into two classes: Failure when the loading is static or the number of load cycles is one or quite small, and failure due to cyclic loading when the number of cycles is large often in thousands of cycles [25–27].

Just by looking the name of the theory you will be able to formulate condition of failure in an actual case, if your concept of Principal stresses is clear. The theories along with its usability are given below.

1. *Maximum principal stress theory*—Good for brittle materials
   According to this theory when maximum principal stress induced in a material under complex load condition exceeds maximum normal strength in a simple tension test the material fails. So the failure condition can be expressed as

$$\sigma \geq \sigma_{\text{ult}}$$

2. *Maximum shear stress theory*—Good for ductile materials
   According to this theory when maximum shear strength in actual case exceeds maximum allowable shear stress in simple tension test the material case. Maximum shear stress in actual case in represented as

$$\boldsymbol{\tau}_{\text{max,act}} = \frac{\boldsymbol{\sigma}_1 - \boldsymbol{\sigma}_3}{2}$$

Maximum shear stress in simple tension case occurs at angle 45 with load, so maximum shear strength in a simple tension case can be represented as

$$\tau_{45} = \tau_{\max,\mathrm{simp}} = [\sigma_y]/2$$

Comparing these 2 quantities one can write the failure condition as

$$\sigma_1 - \sigma_3 \geq \sigma_y$$

3. *Total strain energy theory*—Good for ductile material
4. According to this theory when total strain energy in actual case exceeds total strain energy in simple tension test at the time of failure the material fails. Total strain energy in actual case is given by

$$\mathrm{T.S.E_{act}} = \frac{1}{2E}\left[\sigma_1^2 + \sigma_2^2 + \sigma_3^2 - 2\mu(\sigma_1\sigma_2 + \sigma_2\sigma_3 + \sigma_3\sigma_1)\right]$$

Total strain energy in simple tension test at the time of failure is given by $\mathrm{T.S.E_{act}} = \frac{\sigma_y^2}{2E}$. So failure condition can be simplified as

$$\frac{1}{2E}\left[\sigma_1^2 + \sigma_2^2 + \sigma_3^2 - 2\mu(\sigma_1\sigma_2 + \sigma_2\sigma_3 + \sigma_3\sigma_1)\right] \geq \frac{\sigma_y^2}{2E} \tag{6.21}$$

5. *Shear strain energy theory*—Highly recommended

According to this theory when shear strain energy in actual case exceeds shear strain energy in simple tension test at the time of failure the material fails. Shear strain energy in actual case is given by

$$\mathrm{S.S.E_{act}} = \frac{1}{12G}\left[(\sigma_1 - \sigma_2)^2 + (\sigma_2 - \sigma_3)^2 + (\sigma_3 - \sigma_1)^2\right]$$

Shear strain energy in simple tension test at the time of failure is given by $\mathrm{S.S.E_{act}} = \frac{\sigma_y^2}{6G}$. So the failure condition can be deduced as

$$\frac{1}{12G}\left[(\sigma_1 - \sigma_2)^2 + (\sigma_2 - \sigma_3)^2 + (\sigma_3 - \sigma_1)^2\right] \geq \frac{\sigma_y^2}{6G} \tag{6.22}$$

where G is shear modulus of the material.

Industrial Applications of Failure Theories

Nowadays FEA-based solvers are well integrated to use failure theories. User can specify kind of failure criterion in his solution method. Shear strain energy theory is the most commonly used method. It is a common practice to introduce Factor of Safety (F.S) while designing, in order to take care of worst loading scenario.

6.7 General Equations of Anisotropic Creep

Previous research has focused on the isotropic materials but few constitutive models have been developed for anisotropic creeping solids. When an anisotropic material is subject to creep conditions and a complex state of stress, the integral form of the constitutive creep law should exist between all the components of the stress tensor and the strain tensor. It should be mentioned that the uniaxial creep equation had been presented as

$$E(t)\varepsilon(t) = \boldsymbol{\sigma}(t) + \int_0^t K(t,\tau)\mathrm{d}\tau. \tag{6.23}$$

In the general case of an isotropic body stress–strain state can be described as follows:

$$\begin{aligned}
\varepsilon_x(t) &= \boldsymbol{\sigma}_x(t)J_1(t,t') - [\boldsymbol{\sigma}_y(t) + \boldsymbol{\sigma}_z(t)]J_2(t,t')\\
\varepsilon_y(t) &= \boldsymbol{\sigma}_y(t)J_1(t,t') - [\boldsymbol{\sigma}_z(t) + \boldsymbol{\sigma}_x(t)]J_2(t,t')\\
\varepsilon_z(t) &= \boldsymbol{\sigma}_z(t)J_1(t,t') - [\boldsymbol{\sigma}_y(t) + \boldsymbol{\sigma}_x(t)]J_2(t,t')\\
\gamma_{xy} &= \tau_{xy}J_3(t,t')\\
\gamma_{yz} &= \tau_{yz}J_3(t,t')\\
\gamma_{zx} &= \tau_{zx}J_3(t,t')
\end{aligned} \tag{6.24}$$

ε_x; ε_y; ε_z; γ_{xy}; γ_{zy} and γ_{zx}—the components of the strain tensor are functions of time;

$J_1(t, t')$, $J_2(t, t')$, and $J_3(t, t')$—experimentally obtained functional relationships. Note that the value of $1/J_1(t, t')$ is similar to the modulus of elasticity E; the ratio of $J_2(t, t')/J_1(t, t')$ similar to Poisson's ratio μ and the value of $1/J_3(t, t')$ is similar to the shear modulus G. As is well known, there is a relationship between these three variables: $G = E/2(1 + \mu)$. In our case:

$$\frac{1}{J_3(t,t')} = \frac{[J_1(t,t')]^{-1}}{2\left[1 + \frac{J_2(t,t')}{J_1(t,t')}\right]} \Rightarrow J_3(t,t') = 2[J_1(t,t') + J_2(t,t')]. \tag{6.25}$$

Let us now turn to the creep stress that are depending on the time (temperature), then we obtain the integral form of the rheological creep law for three-dimensional stress state of anisotropic material.

$$\begin{aligned}\varepsilon_x(t) = \sigma_x(t)J_1(t,t') - [\boldsymbol{\sigma}_x(t) + \boldsymbol{\sigma}_x(t)]J_2(t,t') \\ + \int_0^t [\boldsymbol{\sigma}_x(t') + \boldsymbol{\sigma}_x(t')]\frac{\partial}{\partial t'}J_2(t,t')\mathrm{d}t' \\ - \int_0^t [\boldsymbol{\sigma}_x(t')]\frac{\partial}{\partial t'}J_1(t,t')\mathrm{d}t' \quad . \end{aligned} \tag{6.26}$$

$$\ldots$$

$$\gamma_{xy} = \tau_{xy} + \int_0^t [\tau_{xy}(t')]\frac{\partial}{\partial t'}J_3(t,t')\mathrm{d}t'$$

$$\ldots$$

Denote now

$$\begin{aligned} J_1(t,t) &= \frac{1}{E(t)}; \\ \frac{J_2(t,t')}{J_1(t,t')} &= \boldsymbol{\mu}(t,t') \\ J_3(t,t) &= \frac{1}{G(t)} - \frac{\partial J_1(t,t')/\partial t'}{J_1(t,t)} = K(t,t');. \\ -\frac{\partial J_3(t,t')/\partial t'}{J_3(t,t)} &= K_c(t,t'); \\ \frac{\partial J_2(t,t')/\partial t'}{\partial J_1(t,t)/\partial t'} &= \boldsymbol{\nu}(t,t') \end{aligned} \tag{6.27}$$

Equation (6.26) are reduced now to:

$$\begin{aligned} E(t)\varepsilon_x(t) = \boldsymbol{\sigma}_x(t) - [\boldsymbol{\sigma}_y(t) + \boldsymbol{\sigma}_z(t)]\boldsymbol{\mu}(t,t) \\ + \int_0^t \{\boldsymbol{\sigma}_x(t') - v(t,t')[\boldsymbol{\sigma}_y(t') - \boldsymbol{\sigma}_z(t')]\}K(t,t')\mathrm{d}t' \\ \ldots \\ G(t)\gamma_{xy} = \tau_{xy} + \int_0^t [\tau_{xy}(t')]\frac{\partial}{\partial t'}K_c(t,t')\mathrm{d}t' \\ \ldots \end{aligned} \quad . \tag{6.28}$$

where: $G(t) = E(t)/2[1 + \boldsymbol{\mu}(t,t)]$

Formulas (6.26), (6.27), and (6.28) are written as a single line, meaning that the other two rows can be obtained by cyclic permutation of the indices x, y, and z.

The function K_c is called hereditary kernel of shear strain. Between the coefficients μ and v there is a relationship that can be easily obtained from the formula (6.28).

$$\mathbf{v}(t,t') = \boldsymbol{\mu}(t,t') + \frac{J_1(t,t')}{\partial J_1(t,t')/\partial(t')}\partial\boldsymbol{\mu}(t,t')/\partial t'. \tag{6.29}$$

This equation (6.29) shows that if μ does not depend on t', then $\mu(t) = v(t)$ and $K_c(t, t') = K(t, t')$. In order to simplify the calculations in this book we will assume that $\mu(t) = v(t) = \text{const}$.

For volume strain $\varepsilon_0 = \varepsilon_x + \varepsilon_y + \varepsilon_z$ adding the equations (6.26) we obtain

$$E_0(t)\boldsymbol{\varepsilon}_0(t) = \boldsymbol{\varepsilon}_{\text{aver}} + \int_0^t \boldsymbol{\sigma}_{\text{aver}} K(t,t')\mathrm{d}t \quad . \tag{6.30}$$

where: $\boldsymbol{\sigma}_{\text{aver}} = (\boldsymbol{\sigma}_x + \boldsymbol{\sigma}_y + \boldsymbol{\sigma}_z)/3$ and $E_0 = E(t)/[3 - 6\boldsymbol{\mu}(t,t)]$

For the sake of simplicity, the mathematical formulation of the model above is presented in triaxial stress space, which can be used only to model the response of cross-anisotropic samples subject to triaxial loading. For the sake of practical applicability we will consider a 2D material slice consisting of two materials with temperature-dependent material properties. In this case the general integral equations (6.6) will be reduced as follows (for additional assumptions and change of variables method see Chap. 4).

6.7.1 Anisotropic Creep (2D Model) with Temperature-Dependent Material Properties

Hook's Law

$$\begin{aligned}
\varepsilon_x &= \frac{\sigma_x}{E_1} - \mu_2\frac{\sigma_y}{E_2}; \quad E_1\varepsilon_x = \sigma_x - k\mu_2\sigma_y \\
\varepsilon_y &= \frac{\sigma_y}{E_2} - \mu_1\frac{\sigma_x}{E_1}; \quad E_2\varepsilon_y = \sigma_y - \frac{\mu_1\sigma_x}{k} \\
\gamma &= \frac{\tau}{G}; \quad k = \frac{E_1}{E_2} = \frac{\exp(-0.15\theta)}{\exp(-0.075\theta)} = \mathrm{e}^{-0.075\theta}
\end{aligned} \quad . \tag{6.31}$$

Creep Equations

$$E_1(\theta)\varepsilon_x(\theta) = \sigma_x(\theta) - k(\theta)\mu_2\sigma_y(\theta) + \int_0^\theta [\sigma_x^n(\tau) - k(\tau)\mu_2\sigma_y^{n1}(\tau)]K_1(\theta,\tau)\mathrm{d}\tau$$

$$\sigma_x = E_1(\theta)\varepsilon_x(\theta) + k\mu_2\sigma_y - \mathrm{e}^{-\alpha_1 m}\int_0^\theta \mathrm{e}^{\frac{\tau}{1+\beta\tau}}\mathrm{e}^{\alpha 1m(\tau)}m1\left[\sigma_x^n - k\mu_2\sigma_y^{n1}\right]\mathrm{d}\tau$$

$$\sigma_y(\theta) = E_2(\theta)\varepsilon_y(\theta) + \frac{\mu_1}{k}\sigma_x(\theta) - \mathrm{e}^{-\alpha 2m}\int_0^\theta \left[\sigma_y^{n1}(\tau) - \frac{\mu_1}{k}\sigma_x^n(\tau)\right]K_2(\theta,\tau)\mathrm{d}\tau$$

$$\sigma_y = E_2(\theta)\varepsilon_y(\theta) + \frac{\mu_1}{k}\sigma_x - \mathrm{e}^{-\alpha 2m}\int_0^\theta \mathrm{e}^{\frac{\tau}{1+\beta\tau}}\mathrm{e}^{\alpha 2m(\tau)}\left[\sigma_y^{n1}(\tau) - \frac{\mu_1}{k}\sigma_x^n(\tau)\right]m1\mathrm{d}\tau$$

$$G(\theta)[\varepsilon_y(\theta) - \varepsilon_x(\theta)] = \tau_{xy}(\theta) + \int_0^\theta \tau_{xy}(\tau)K_3(\theta,\tau)\mathrm{d}\tau$$

$$= \tau_{xy} + \mathrm{e}^{-\alpha 3m}\int_0^\theta \mathrm{e}^{\frac{\tau}{1+\beta\tau}}\mathrm{e}^{\alpha 3m(\tau)}m1(\tau_{xy})\mathrm{d}\tau$$

$$G(\theta)\gamma(\theta) = \theta G(\theta)$$

$$E_k = \frac{1}{2(1+\mu_1)}[E_2\mu_1 + E_1(1-\mu_1)] = \frac{E_2}{2(1+\mu_1)}[\mu_1 + k(1-\mu_1)] = G(\theta)$$

where: $E_2(\theta) \geq E_1(\theta)$; if $k = 1$ then $G(\theta) = \dfrac{E}{2(1+\mu)}$

$$m = \tau(\theta) = 0.00245 + 0.0375\theta - 0.009334\theta^2 + 0.001040\theta^3 - 0.0000410\theta^4$$

$$m1 = \tau'(\theta) = 0.0375 - 0.01868\theta + 0.00312\theta^2 - 0.000164\theta^3 \quad (6.32)$$

It should be noted that, if $n1 = n = 1$ we have the linear Volterra equation of the second kind, where n and $n1$—are the creep stress exponent in x and y directions respectfully (in accordance with the Norton-Bailey creep law) and "α" in Eq. (6.32) are material parameters. An Arrhenius-type rate component ($\dot{\varepsilon} = N\sigma^n \exp\left(-\frac{Q}{RT}\right)$) is used to include the effect of temperature in the model (Eq. 6.32). Function $m = \tau(\theta)$ is the inverse dimensionless temperature–time function θ^{-1} and $m1 = \tau'(\theta)$ is its first derivative (for Very Fast Fire scenario in this case [14]). We have selected again only one parameter "α" in Eq. (6.32), because it will be called later on (in probability based approach) a random parameter. The sequential method (with respect to the Poisson ratio) is used below, and the convergence of the solution is illustrated via example only. Let us remind the reader again that rigorous mathematical analyses in many cases are not presented in this book as a compromise to simplicity and transparently accurate (in a step-by-step form) structural analysis and design computations.

We seek a solution of Eq. (6.32) using the method of successive approximations.

As a first approximation we take the solution of Eq. (6.32) when both Poisson coefficients equal to zero ($\mu_1 = \mu_2 = 0$). Thus found the first approximation of the stress function is then substituted into the integrand of Eq. (6.32), whose solution is the second approximation of stress functions σ_x and σ_y. The process of computing continues until the subsequent approximation does not differ from the previous one with predetermined calculation accuracy. Let us note that in case when $\mu_1 = \mu_2 = 0$, the Eq. (6.32) automatically describes the uniaxial creep process and the results should coincide with the previous results from Chap. 4.

Calculated values of DEQ variables

	Variable	Initial value	Minimal value	Maximal value	Final value
1	A	0	0	2.430775	2.236569
2	C	0	0	4.718217	4.718217
3	m	0.00245	0.00245	0.0736126	0.07345
4	$m1$	0.0375	0.0079894	0.1463	0.1463
5	k	1	0.4723666	1	0.4723666
6	r	0	0	2.452529	2.231302
7	t	0	0	10	10
8	y	0	−0.0084432	1.799237	−0.0052673
9	Y	0	0	2.485735	2.291442
10	Z	0	0	5.077815	5.077815
11	z	0	0	2.463817	0.0054484

Differential equations

1 $d(Y)/d(t) = (\exp(t/(1.0 + 0.1 * t))) * m1 * (y - 0 * n * 0.4 * z) * (\exp(0.33 * m))$

2 $d(Z)/d(t) = (\exp(t/(1.0 + 0.1 * t))) * m1 * (z - 0 * (0.2/n) * y) * (\exp(1 * m))$

Explicit equations

1 $k = (\exp(-0.075 * t))$

2 $m = 0.00245 + 0.0375 * t - 0.00934 * t\text{^}2 + 0.00104 * t\text{^}3 - 0.000041 * t\text{^}4$

3 $A = 1 * (\exp(-0.33 * m)) * Y$

4 $y = t * (\exp(-0.15 * t)) - A + 1 * n * (-0.108 + 1.052 * t$
$+ 0.0124 * t\text{^}2 - 0.0448 * t\text{^}3 + 0.00336 * t\text{^}4) * 0.4 * 0$

$C = 1 * (\exp(-1 * m)) * Z$

$r = t * (\exp(-0.15 * t))$

$m1 = 0.0375 - 0.01868 * t + 0.00312 * t^{\wedge}2 - 0.000164 * t^{\wedge}2$

$z = t * (\exp(-0.075 * t)) - C + 0.2 * n * (-0.136 + 1.2 * t$
$\quad - 0.188 * t^{\wedge}2 - 0.0049 * t^{\wedge}3 + 0.0012 * t^{\wedge}4) * 0.2 * 0$

The best-to-fit analytical expressions can be presented (using regression method) as follows: (Fig. 6.1)

$$\textbf{Model}: y = \sigma_x = -0.136 + 1.2 * t - 0.188 * t^{\wedge}2 - 0.0049 * t^{\wedge}3 + 0.0012 * t^{\wedge}4 \quad (6.33)$$

Model: $y = a0 + a1 * t + a2 * t\char`^2 + a3 * t\char`^3 + a4 * t\char`^4$

Variable	Value
$a0$	−0.1359265
$a1$	1.195604
$a2$	−0.1874769
$a3$	−0.0048599
$a4$	0.0012085

$$\textbf{Model}: z = \sigma_y = -0.108 + 1.052 * t + 0.0124 * t^{\wedge}2 - 0.0448 * t^{\wedge}3 + 0.00336 * t^{\wedge}4 \quad (6.34)$$

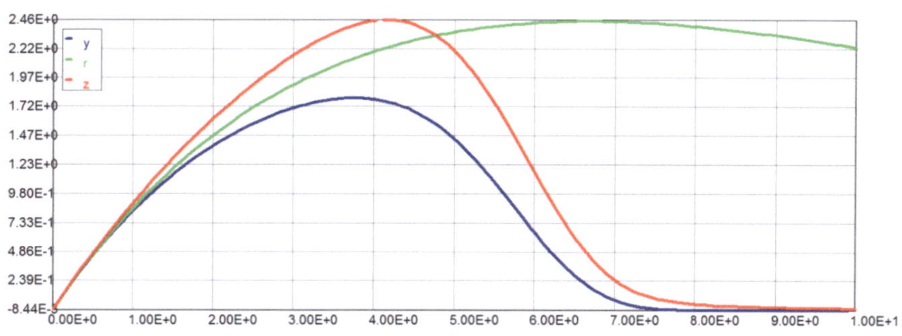

Fig. 6.1 Stress–temperature–strain diagram ($\mu_1 = \mu_2 = 0$)

Model: $z = a0 + a1 * t + a2 * t\^2 + a3 * t\^3 + a4 * t\^4$

Variable	Value
$a0$	−0.1082395
$a1$	1.051741
$a2$	0.0123743
$a3$	−0.0447911
$a4$	0.0033606

Substituting (6.12) and (6.11) into (6.10) we have (the second approximation with $n = (\exp(-0.075 * \theta))$.

Second Approximation $n = \exp(-0.075\theta)$

Calculated values of DEQ variables

	Variable	Initial value	Minimal value	Maximal value	Final value
1	A	0	0	2.456646	2.309432
2	C	0	0	4.71772	4.71772
3	m	0.00245	0.00245	0.0736129	0.07345
4	$m1$	0.0375	0.0079889	0.1463	0.1463
5	k	1	0.4723666	1	0.4723666
6	r	0	0	2.452529	2.231302
7	t	0	0	10	10
8	y	−0.0432	−0.0432	2.50528	0.0072734
9	Y	0	0	2.511763	2.366093
10	Z	0	0	5.07728	5.07728
11	z	−0.00544	−0.00544	2.647113	0.0090439

Differential equations

1 $d(Y)/d(t) = (\exp(t/(1.0 + 0.1 * t))) * m1 * (y - 1 * n * 0.4 * z) * (\exp(0.33 * m))$

2 $d(Z)/d(t) = (\exp(t/(1.0 + 0.1 * t))) * m1 * (z - 1 * (0.2/n) * y) * (\exp(1 * m))$

Explicit equations (Fig. 6.2)

1 $k = (\exp(-0.075 * t))$

2 $m = 0.00245 + 0.0375 * t - 0.00934 * t\^2 + 0.00104 * t\^3 - 0.000041 * t\^4$

3 $A = 1 * (\exp(-0.33 * m)) * Y$

4 $y = t * (\exp(-0.15 * t)) - A + 1 * n * (-0.108 + 1.052 * t$
$+ 0.0124 * t\^2 - 0.0448 * t\^3 + 0.00336 * t\^4) * 0.4 * 1$

5 $C = 1 * (\exp(-1 * m)) * Z$

6 $r = t * (\exp(-0.15 * t))$

7 $m1 = 0.0375 - 0.01868 * t + 0.00312 * t\hat{}2 - 0.000164 * t\hat{}2$

8 $z = t * (\exp(-0.075 * t)) - C + 0.2 * n * (-0.136 + 1.2 * t - 0.188 * t\hat{}2 - 0.0049 * t\hat{}3 + 0.0012 * t\hat{}4) * 0.2 * 1$

Model: $y = a0 + a1 * t + a2 * t\hat{}2 + a3 * t\hat{}3 + a4 * t\hat{}4 + a5 * t\hat{}5$

Variable	Value
$a0$	0.2276456
$a1$	0.477691
$a2$	0.5609227
$a3$	−0.2229802
$a4$	0.025753
$a5$	−0.0009558

$$y = \sigma_x = 0.228 + 0.478 * t + 0.561 * t\hat{}2 - 0.223 * t\hat{}3 + 0.0257 * t\hat{}4 - 0.00096 * t\hat{}5 \quad (6.35)$$

Model: $z = a0 + a1 * t + a2 * t\hat{}2 + a3 * t\hat{}3 + a4 * t\hat{}4 + a5 * t\hat{}5$

Variable	Value
$a0$	0.3748807
$a1$	−0.2287126
$a2$	0.8707054
$a3$	−0.2573725
$a4$	0.0256846
$a5$	−0.0008444

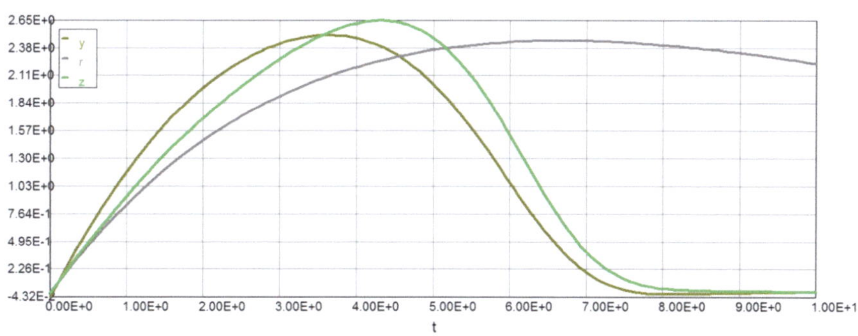

Fig. 6.2 Stress–temperature–strain diagram ($\mu_1 \neq \mu_2 \neq 0$—second approximation)

$$z = \sigma_y = 0.375 - 0.229 * t + 0.871 * t\string^2 - 0.257 * t\string^3 + 0.0257 * t\string^4 - 0.00084 * t\string^5 \quad (6.36)$$

Third Approximation $n = \exp(-0.075\theta)$

Calculated values of DEQ variables

	Variable	Initial value	Minimal value	Maximal value	Final value
1	A	0	0	2.540286	2.452133
2	C	0	0	4.700325	4.700325
3	m	0.00245	0.00245	0.0736131	0.07345
4	$m1$	0.0375	0.0079945	0.1463	0.1463
5	k	1	0.4723666	1	0.4723666
6	r	0	0	2.452502	2.231302
7	t	0	0	10	10
8	y	0.15	−0.0053859	2.604668	0.0030706
9	Y	0	0	2.597422	2.512295
10	Z	0	0	5.058559	5.058559
11	z	0.00912	0.0064863	2.667182	0.0064863

Differential equations

1 $d(Y)/d(t) = (\exp(t/(1.0 + 0.1 * t))) * m1 * (y - 1 * n * 0.4 * z) * (\exp(0.33 * m))$

2 $d(Z)/d(t) = (\exp(t/(1.0 + 0.1 * t))) * m1 * (z - 1 * (0.2/n) * y) * (\exp(1 * m))$

Explicit equations (Figs. 6.3 and 6.4)

1 $k = (\exp(-0.075 * t))$

2 $m = 0.00245 + 0.0375 * t - 0.00934 * t\string^2 + 0.00104 * t\string^3 - 0.000041 * t\string^4$

3 $A = 1 * (\exp(-0.33 * m)) * Y$

4 $y = t * (\exp(-0.15 * t)) - A + 1 * n * (0.375 - 0.229 * t + 0.871 * t\string^2 - 0.257 * t\string^3 + 0.0257 * t\string^4 - 0.00084 * t\string^5) * 0.4 * 1$

5 $C = 1 * (\exp(-1 * m)) * Z$

6 $r = t * (\exp(-0.15 * t))$

7 $m1 = 0.0375 - 0.01868 * t + 0.00312 * t\string^2 - 0.000164 * t\string^2$

8 $z = t * (\exp(-0.075 * t)) - C + 0.2 * n * (0.228 + 0.478 * t + 0.561 * t\string^2 - 0.223 * t\string^3 + 0.0257 * t\string^4 - 0.00096 * t\string^5) * 0.2 * 1$

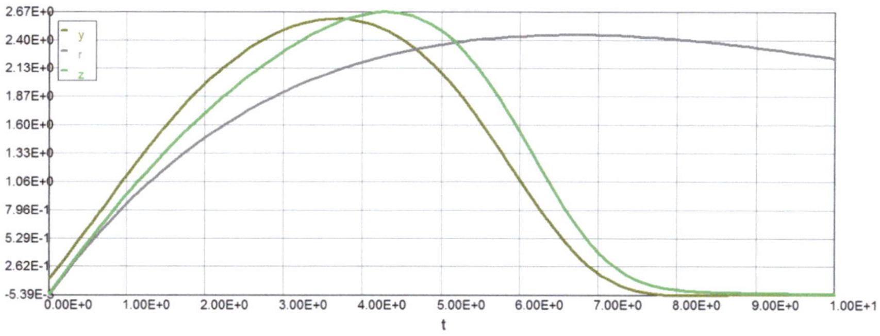

Fig. 6.3 Stress–temperature–strain diagram ($\mu_1 \neq \mu_2 \neq 0$—third approximation)

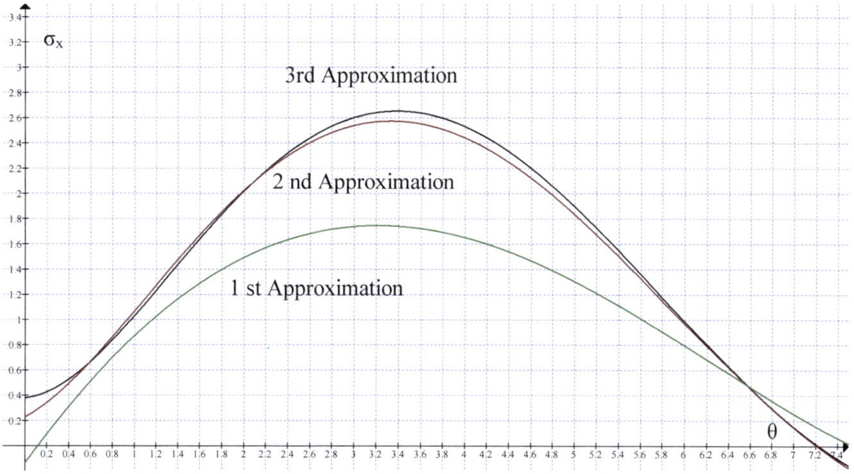

Fig. 6.4 Stress–temperature–strain diagram (σ_x—all approximations)

Model: $y = a0 + a1 * t + a2 * t\text{^}2 + a3 * t\text{^}3 + a4 * t\text{^}4 + a5 * t\text{^}5$

Variable	Value
$a0$	0.3801194
$a1$	0.0854447
$a2$	0.8206428
$a3$	−0.2866792
$a4$	0.0323315
$a5$	−0.001199

$$y = \sigma_x = 0.38 + 0.0855 * t + 0.821 * t\hat{\ }2 - 0.287 * t\hat{\ }3 + 0.0323 * t\hat{\ }4 - 0.0012 * t\hat{\ }5 \quad (6.37)$$

Model: $z = a0 + a1 * t + a2 * t\hat{\ }2 + a3 * t\hat{\ }3 + a4 * t\hat{\ }4 + a5 * t\hat{\ }5$

Variable	Value
$a0$	0.3765341
$a1$	−0.2312154
$a2$	0.8827549
$a3$	−0.2616986
$a4$	0.0262155
$a5$	−0.0008661

$$z = \sigma_y = 0.376 - 0.231 * t + 0.883 * t\hat{\ }2 - 0.262 * t\hat{\ }3 + 0.0262 * t\hat{\ }4 - 0.000866 * t\hat{\ }5 \quad (6.38)$$

One can see now from Figs. 6.5 and 6.6 that first two approximations are sufficient enough for solving Eq. (6.32).

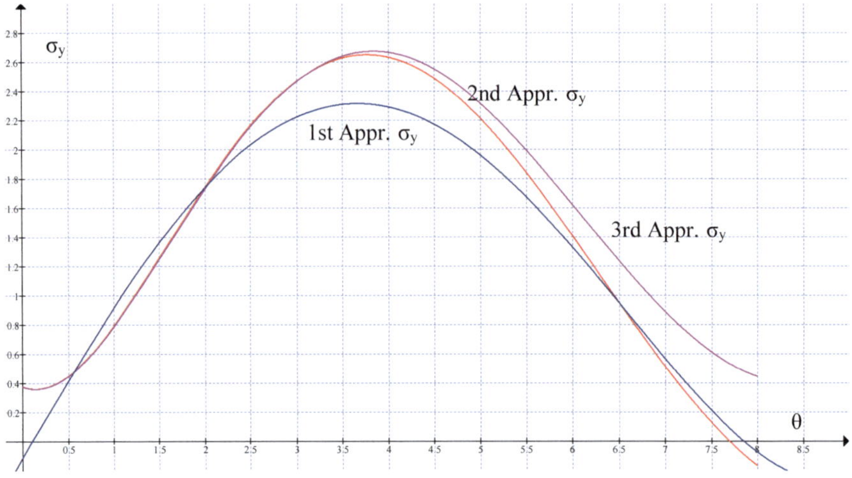

Fig. 6.5 Stress–temperature–strain diagram (σ_y—all approximations)

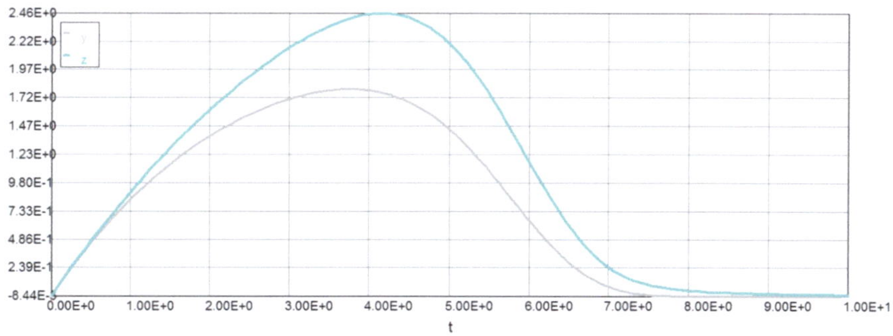

Fig. 6.6 Stress–temperature–strain diagram (σ_x; σ_y—first approximation)

6.8 High-Temperature Effect on Modulus of Elasticity Deterioration

The dimensionless integral-type creep constitutive law (6.32) has two major dimensionless parameters that can change the stress–temperature–strain diagram fundamentally. The first one is parameter "a" that characterizes the deterioration rate of modulus of elasticity in Eq. (6.32). In fact, to be more explicit we are interested to know how the ratio $k = E_1/E_2$ affects the maximum stresses in both orthogonal directions and the stress–temperature–strain diagram in general. The second one is parameter "α_i" that characterizes the material properties behavior during the continuously changing high temperature application. In fact, to be more explicit in this case we are interested to know how the stress–temperature–strain diagram changes with increase (or decrease) of "α_i" in one direction comparable to increase (or decrease) of the material property parameter (MPP) in orthogonal direction (for instance, reinforcement of composite material) including the failure region of the material. Let us start now with the high temperature effect on deterioration of modulus of elasticity. For all practical purposes we will assume here that $E_1 = \exp(-0.15\theta)$ is unchanged and $E_2 > E_1$. The solutions of Eq. (6.32) for discrete values of "k_0" are presented below (using POLYMATH software).

Example 6.1 Data: $E_2 = \exp(-0.075\theta)$; $\alpha_1 = 0.33$; $\alpha_2 = 0.99$; $k_0 = 0.15/0.075 = 2$ and $k = \exp(-0.075\theta)$.

First Approximation

Calculated values of DEQ variables

	Variable	Initial value	Minimal value	Maximal value	Final value
1	*A*	0	0	2.430775	2.236569
2	*C*	0	0	4.718215	4.718215
3	*E*2	1	0.4723666	1	0.4723666

(continued)

(continued)

	Variable	Initial value	Minimal value	Maximal value	Final value
4	k	1	0.4723666	1	0.4723666
5	m	0.00245	0.00245	0.0736127	0.07345
6	$m1$	0.0375	0.0079894	0.1463	0.1463
7	r	0	0	2.452529	2.231302
8	t	0	0	10	10
9	y	0	−0.0084432	1.799238	−0.0052673
10	Y	0	0	2.485735	2.291442
11	Z	0	0	5.074084	5.074084
12	z	0	0	2.463789	0.0054507

Differential equations

1 $\mathrm{d}(Y)/\mathrm{d}(t) = (\exp(t/(1.0 + 0.1 * t))) * m1 * (y - 0 * k * 0.4 * z) * (\exp(0.33 * m))$

2 $\mathrm{d}(Z)/\mathrm{d}(t) = (\exp(t/(1.0 + 0.1 * t)))m1 * (z - 0 * (0.2/k) * y) * (\exp(0.99 * m))$

Explicit equations

1 $E2 = (\exp(-0.075 * t))$

2 $k = (\exp(-0.075 * t))$

3 $m = 0.00245 + 0.0375 * t - 0.00934 * t^\wedge 2 + 0.00104 * t^\wedge 3 - 0.000041 * t^\wedge 4$

4 $A = (\exp(-0.33 * m)) * Y$

5 $y = t * (\exp(-0.15 * t)) - A + k * (0.375 - 0.229 * t$
$+ 0.871 * t^\wedge 2 - 0.257 * t^\wedge 3 + 0.0257 * t^\wedge 4 - 0.00084 * t^\wedge 5) * 0.4 * 0$

6 $C = (\exp(-0.99 * m)) * Z$

7 $r = t * (\exp(-0.15 * t))$

8 $m1 = 0.0375 - 0.01868 * t + 0.00312 * t^\wedge 2 - 0.000164 * t^\wedge 2$

9 $z = t * (\exp(-0.075 * t)) - C + k * (0.228 + 0.478 * t$
$+ 0.561 * t^\wedge 2 - 0.223 * t^\wedge 3 + 0.0257 * t^\wedge 4 - 0.00096 * t^\wedge 5) * 0.2 * 0$

Model: $y = a0 + a1 * t + a2 * t^\wedge 2 + a3 * t^\wedge 3 + a4 * t^\wedge 4 + a5 * t^\wedge 5$

Variable	Value
$a0$	0.2438811
$a1$	0.1228612
$a2$	0.5468612
$a3$	−0.1980324

(continued)

(continued)

Variable	Value
$a4$	0.0227155
$a5$	−0.0008527

$$y = \sigma_x = 0.244 + 0.123 * t + 0.547 * t^\wedge 2 - 0.198 * t^\wedge 3 + 0.0227 * t^\wedge 4 - 0.00085 * t^\wedge 5 \quad (6.39)$$

Model: $z = a0 + a1 * t + a2 * t^\wedge 2 + a3 * t^\wedge 3 + a4 * t^\wedge 4 + a5 * t^\wedge 5$

Variable	Value
$a0$	0.3776991
$a1$	−0.3207648
$a2$	0.9519123
$a3$	−0.2919441
$a4$	0.0308778
$a5$	−0.001091

$$z = \sigma_y = 0.378 - 0.321 * t + 0.952 * t^\wedge 2 - 0.292 * t^\wedge 3 + 0.0309 * t^\wedge 4 - 0.0011 * t^\wedge 5 \quad (6.40)$$

Second Approximation

Calculated values of DEQ variables

	Variable	Initial value	Minimal value	Maximal value	Final value
1	A	0	0	2.408151	2.118448
2	C	0	0	4.737473	4.737473
3	$E2$	1	0.4723666	1	0.4723666
4	k	1	0.4723666	1	0.4723666
5	m	0.00245	0.00245	0.0736131	0.07345
6	$m1$	0.0375	0.0079922	0.1463	0.1463
7	r	0	0	2.452517	2.231302
8	t	0	0	10	10
9	y	0.1512	−0.0108347	2.560621	−0.0065609
10	Y	0	0	2.462106	2.170424
11	Z	0	0	5.094795	5.094795
12	z	0.0488	0.002631	2.793087	0.002631

Differential equations

1 $d(Y)/d(t) = (\exp(t/(1.0 + 0.1 * t))) * m1 * (y - 1 * k * 0.4 * z) * (\exp(0.33 * m))$
2 $d(Z)/d(t) = (\exp(t/(1.0 + 0.1 * t))) * m1 * (z - 1 * (0.2/k) * y) * (\exp(0.99 * m))$

Explicit equations (Fig. 6.7)

1 $E2 = (\exp(-0.075 * t))$
2 $k = (\exp(-0.075 * t))$
3 $m = 0.00245 + 0.0375 * t - 0.00934 * t^\wedge 2 + 0.00104 * t^\wedge 3 - 0.000041 * t^\wedge 4$
4 $A = (\exp(-0.33 * m)) * Y$
5 $y = t * (\exp(-0.15 * t)) - A + k * (0.378 - 0.321 * t + 0.952 * t^\wedge 2 - 0.292 * t^\wedge 3 + 0.0309 * t^\wedge 4 - 0.0011 * t^\wedge 5) * 0.4 * 1$
6 $C = (\exp(-0.99 * m)) * Z$
7 $r = t * (\exp(-0.15 * t))$
8 $m1 = 0.0375 - 0.01868 * t + 0.00312 * t^\wedge 2 - 0.000164 * t^\wedge 2$
9 $z = t * (\exp(-0.075 * t)) - C + k * (0.244 + 0.123 * t + 0.547 * t^\wedge 2 - 0.198 * t^\wedge 3 + 0.0227 * t^\wedge 4 - 0.00085 * t^\wedge 5) * 0.2 * 1$

Model: $y = a0 + a1 * t + a2 * t^\wedge 2 + a3 * t^\wedge 3 + a4 * t^\wedge 4 + a5 * t^\wedge 5$

Variable	Value
$a0$	0.3634963
$a1$	0.1181592
$a2$	0.7932202
$a3$	−0.2805257
$a4$	0.0318167
$a5$	−0.0011849

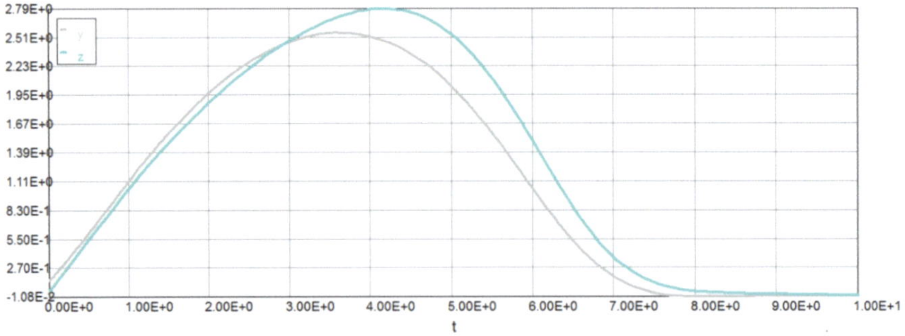

Fig. 6.7 Stress–temperature–strain diagram (σ_x; σ_y—second approximation)

$$y = \sigma_x = 0.364 + 0.118 * t + 0.793 * t^\wedge 2 - 0.280 * t^\wedge 3 + 0.0318 * t^\wedge 4 - 0.0012 * t^\wedge 5 \quad (6.41)$$

Model: $z = a0 + a1 * t + a2 * t^\wedge 2 + a3 * t^\wedge 3 + a4 * t^\wedge 4 + a5 * t^\wedge 5$

Variable	Value
$a0$	0.4033109
$a1$	−0.1899418
$a2$	0.9362042
$a3$	−0.287647
$a4$	0.0297513
$a5$	−0.0010184

$$z = \sigma_y = 0.403 - 0.190 * t + 0.936 * t^\wedge 2 - 0.288 * t^\wedge 3 + 0.0298 * t^\wedge 4 - 0.0010 * t^\wedge 5 \quad (6.42)$$

Allowable (maximum) stresses: $\sigma_x = 2.56$ (52.1 ksi); $\sigma_y = 2.79$ (56.8 ksi)

Example 6.2 Data: $E_2 = \exp(-0.05\theta)$; $\alpha_1 = 0.33$; $\alpha_2 = 0.99$; $k_0 = 0.15/0.05 = 3$ and $k = \exp(-0.1\theta)$.

It should be noted that the first approximation remains unchanged; therefore we will start with the second approximation.

Calculated values of DEQ variables

	Variable	Initial value	Minimal value	Maximal value	Final value
1	A	0	0	2.409401	2.144566
2	C	0	0	4.734461	4.734461
3	$E2$	1	0.6065307	1	0.6065307
4	k	1	0.3678794	1	0.3678794
5	m	0.00245	0.00245	0.0736129	0.07345
6	$m1$	0.0375	0.0079939	0.1463	0.1463
7	r	0	0	2.452506	2.231302
8	t	0	0	10	10
9	y	0.1512	−0.009984	2.49324	−0.0062643
10	Y	0	0	2.463603	2.197182
11	Z	0	0	5.091556	5.091556
12	z	0.0488	0.0020067	2.78123	0.0020067

Differential equations

1 $d(Y)/d(t) = (\exp(t/(1.0 + 0.1 * t))) * m1 * (y - 1 * k * 0.4 * z) * (\exp(0.33 * m))$
2 $d(Z)/d(t) = (\exp(t/(1.0 + 0.1 * t))) * m1 * (z - 1 * (0.2/k) * y) * (\exp(0.99 * m))$

Explicit equations (Fig. 6.8)

1 $E2 = (\exp(-0.05 * t))$
2 $k = (\exp(-0.1 * t))$
3 $m = 0.00245 + 0.0375 * t - 0.00934 * t^\wedge 2 + 0.00104 * t^\wedge 3 - 0.000041 * t^\wedge 4$
4 $A = (\exp(-0.33 * m)) * Y$
5 $y = t * (\exp(-0.15 * t)) - A + k * (0.378 - 0.321 * t + 0.952 * t^\wedge 2 - 0.292 * t^\wedge 3 + 0.0309 * t^\wedge 4 - 0.0011 * t^\wedge 5) * 0.4 * 1$
6 $C = (\exp(-0.99 * m)) * Z$
7 $r = t * (\exp(-0.15 * t))$
8 $m1 = 0.0375 - 0.01868 * t + 0.00312 * t^\wedge 2 - 0.000164 * t^\wedge 2$
9 $z = t * (\exp(-0.075 * t)) - C + k * (0.244 + 0.123 * t + 0.547 * t^\wedge 2 - 0.198 * t^\wedge 3 + 0.0227 * t^\wedge 4 - 0.00085 * t^\wedge 5) * 0.2 * 1$

Model: $y = a0 + a1 * t + a2 * t^\wedge 2 + a3 * t^\wedge 3 + a4 * t^\wedge 4 + a5 * t^\wedge 5$

Variable	Value
$a0$	0.3526773
$a1$	0.1601876
$a2$	0.7494802
$a3$	−0.2695805
$a4$	0.0307903
$a5$	−0.0011521

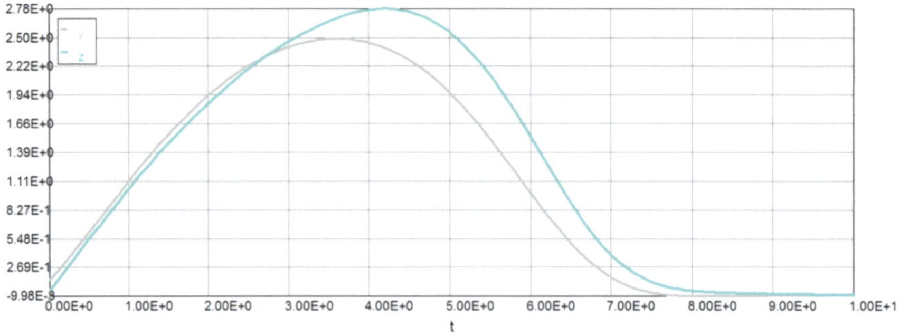

Fig. 6.8 Stress–temperature–strain diagram (σ_x; σ_y—second approximation)

$$y = \sigma_x = 0.353 + 0.160 * t + 0.750 * t^\wedge 2 - 0.270 * t^\wedge 3 + 0.0308 * t^\wedge 4 - 0.0012 * t^\wedge 5 \quad (6.43)$$

Model: $z = a0 + a1 * t + a2 * t\text{^}2 + a3 * t\text{^}3 + a4 * t\text{^}4 + a5 * t\text{^}5$

Variable	Value
$a0$	0.3991381
$a1$	−0.1639202
$a2$	0.9037879
$a3$	−0.2766069
$a4$	0.02836
$a5$	−0.0009597

$$z = \sigma_y = 0.40 - 0.164 * t + 0.904 * t^\wedge 2 - 0.277 * t^\wedge 3 + 0.0284 * t^\wedge 4 - 0.001 * t^\wedge 5 \quad (6.44)$$

Allowable (maximum) stresses: $\sigma_x = 2.49$ (50.7 ksi); $\sigma_y = 2.78$ (56.6 ksi)

Example 6.3 Data: $E_2 = \exp(-0.03\theta)$; $\alpha_1 = 0.33$; $\alpha_2 = 0.99$; $k_0 = 0.15/0.03 = 5$ and $k = \exp(-0.12\theta)$.

Calculated values of DEQ variables

	Variable	Initial value	Minimal value	Maximal value	Final value
1	A	0	0	2.410466	2.161253
2	C	0	0	4.732777	4.732777
3	$E2$	1	0.7408182	1	0.7408182
4	k	1	0.3011942	1	0.3011942
5	m	0.00245	0.00245	0.0736105	0.07345
6	$m1$	0.0375	0.0079917	0.1463	0.1463
7	r	0	0	2.452519	2.231302
8	t	0	0	10	10
9	y	0.1512	−0.0096456	2.445654	−0.0060929
10	Y	0	0	2.46472	2.214278
11	Z	0	0	5.089744	5.089744
12	z	0.0488	0.0013702	2.773993	0.0013702

Differential equations

1 $d(Y)/d(t) = (\exp(t/(1.0+0.1*t)))*m1*(y-1*k*0.4*z)*(\exp(0.33*m))$
2 $d(Z)/d(t) = (\exp(t/(1.0+0.1*t)))*m1*(z-1*(0.2/k)*y)*(\exp(0.99*m))$

Explicit equations (Fig. 6.9)

1 $E2 = (\exp(-0.03*t))$
2 $k = (\exp(-0.12*t))$
3 $m = 0.00245+0.0375*t-0.00934*t^\wedge 2+0.00104*t^\wedge 3-0.000041*t^\wedge 4$
4 $A = (\exp(-0.33*m))*Y$
5 $y = t*(\exp(-0.15*t))-A+k*(0.378-0.321*t+$
$0.952*t^\wedge 2-0.292*t^\wedge 3+0.0309*t^\wedge 4-0.0011*t^\wedge 5)*0.4*1$
6 $C = (\exp(-0.99*m))*Z$
7 $r = t*(\exp(-0.15*t))$
8 $m1 = 0.0375-0.01868*t+0.00312*t^\wedge 2-0.000164*t^\wedge 2$
9 $z = t*(\exp(-0.075*t))-C+k*(0.244+0.123*t$
$+0.547*t^\wedge 2-0.198*t^\wedge 3+0.0227*t^\wedge 4-0.00085*t^\wedge 5)*0.2*1$

Model: $y = a0 + a1 * t + a2 * t^2 + a3 * t^3 + a4 * t^4 + a5 * t^5$

Variable	Value
$a0$	0.3448446
$a1$	0.1893962
$a2$	0.7178999
$a3$	−0.2615179
$a4$	0.0300198
$a5$	−0.001127

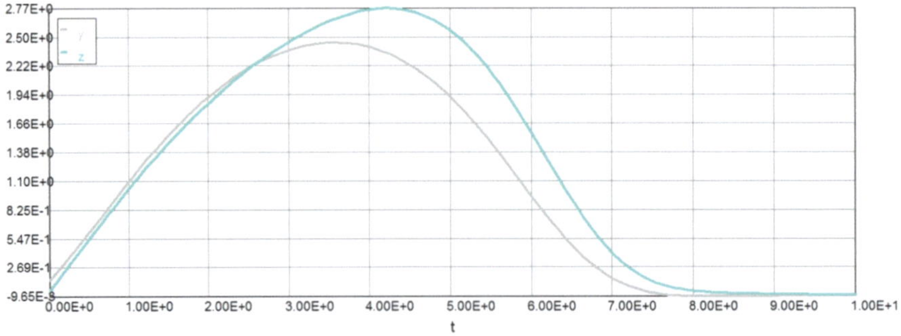

Fig. 6.9 Stress–temperature–strain diagram (σ_x; σ_y—second approximation)

$$y = \sigma_x = 0.345 + 0.189 * t + 0.718 * t^{\wedge}2 - 0.262 * t^{\wedge}3 + 0.030 * t^{\wedge}4 - 0.0011 * t^{\wedge}5 \quad (6.45)$$

Model: $z = a0 + a1 * t + a2 * t^2 + a3 * t^3 + a4 * t^4 + a5 * t^5$

Variable	Value
$a0$	0.3932188
$a1$	−0.13761
$a2$	0.8737031
$a3$	−0.2664244
$a4$	0.0270689
$a5$	−0.0009047

$$z = \sigma_y = 0.393 - 0.138 * t + 0.873 * t^{\wedge}2 - 0.266 * t^{\wedge}3 + 0.0271 * t^{\wedge}4 - 0.0009 * t^{\wedge}5 \quad (6.46)$$

Allowable (maximum) stresses: σ_x = 2.45 (49.88 ksi); σ_y = 2.77 (56.6 ksi)

Example 6.4 Data: $E_2 = \exp(-0.005\theta)$; $\alpha_1 = 0.33$; $\alpha_2 = 0.99$; $k_0 = 0.15/0.005 = 30$ and $k = \exp(-0.145\theta)$.

Calculated values of DEQ variables

	Variable	Initial value	Minimal value	Maximal value	Final value
1	A	0	0	2.411765	2.177942
2	C	0	0	4.731472	4.731472
3	$E2$	1	0.9512294	1	0.9512294
4	k	1	0.2345703	1	0.2345703
5	m	0.00245	0.00245	0.073613	0.07345
6	$m1$	0.0375	0.0079925	0.1463	0.1463
7	r	0	0	2.452515	2.231302
8	t	0	0	10	10
9	y	0.1512	−0.0092373	2.390124	−0.0059396
10	Y	0	0	2.466037	2.231377
11	Z	0	0	5.088341	5.088341
12	z	0.0488	0.0003565	2.768554	0.0003565

Differential equations

1 $d(Y)/d(t) = (\exp(t/(1.0+0.1*t)))*m1*(y-1*k*0.4*z)*(\exp(0.33*m))$
2 $d(Z)/d(t) = (\exp(t/(1.0+0.1*t)))*m1*(z-1*(0.2/k)*y)*(\exp(0.99*m))$

Explicit equations (Fig. 6.10)

1 $E2 = (\exp(-0.005*t))$
2 $k = (\exp(-0.145*t))$
3 $m = 0.00245+0.0375*t-0.00934*t^\wedge 2+0.00104*t^\wedge 3-0.000041*t^\wedge 4$
4 $A = (\exp(-0.33*m))*Y$
5 $y = t*(\exp(-0.15*t))-A+k*(0.378-0.321*t +0.952*t^\wedge 2-0.292*t^\wedge 3+0.0309*t^\wedge 4-0.0011*t^\wedge 5)*0.4*1$
6 $C = (\exp(-0.99*m))*Z$
7 $r = t*(\exp(-0.15*t))$
8 $m1 = 0.0375-0.01868*t+0.00312*t^\wedge 2-0.000164*t^\wedge 2$
9 $z = t*(\exp(-0.075*t))-C+k*(0.244+0.123*t +0.547*t^\wedge 2-0.198*t^\wedge 3+0.0227*t^\wedge 4-0.00085*t^\wedge 5)*0.2*1$

Model: $y = a0 + a1 * t + a2 * t\text{^}2 + a3 * t\text{^}3 + a4 * t\text{^}4 + a5 * t\text{^}5$

Variable	Value
$a0$	0.3359998
$a1$	0.2215092
$a2$	0.6821923
$a3$	−0.2522828
$a4$	0.0291272
$a5$	−0.0010976

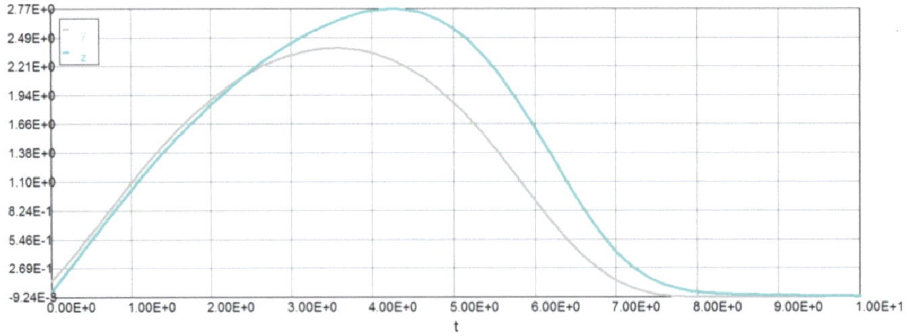

Fig. 6.10 Stress–temperature–strain diagram (σ_x; σ_y—second approximation)

$$y = \sigma_x = 0.336 + 0.222 * t + 0.682 * t^\wedge 2 - 0.252 * t^\wedge 3 + 0.0291 * t^\wedge 4 - 0.0011 * t^\wedge 5 \quad (6.47)$$

Model: $z = a0 + a1 * t + a2 * t^\wedge 2 + a3 * t^\wedge 3 + a4 * t^\wedge 4 + a5 * t^\wedge 5$

Variable	Value
$a0$	0.3879932
$a1$	−0.1135923
$a2$	0.8436114
$a3$	−0.2558922
$a4$	0.0257233
$a5$	−0.0008477

$$z = \sigma_y = 0.388 - 0.114 * t + 0.844 * t^\wedge 2 - 0.256 * t^\wedge 3 + 0.0257 * t^\wedge 4 - 0.00085 * t^\wedge 5 \quad (6.48)$$

Allowable (maximum) stresses: $\sigma_x = 2.39$ (48.66 ksi); $\sigma_y = 2.77$ (56.6 ksi).

Shear stresses are small in all these examples and do not change too much within the range of modulus of elasticity, therefore only one example is presented below.

Example 6.5 Data: $E_2 = \exp(-0.075\theta)$; $\alpha_1 = 0.33$; $\alpha_2 = 0.99$; $\alpha_3 = 1.0$; $k_0 = 0.15/0.075 = 2$ and $k = \exp(-0.075\theta)$ (see Example 6.1).

Equation (6.2) can be reduced (with respect to shear stresses) as follows:

$$G(\boldsymbol{\theta})\gamma_{xy} = \frac{E_2}{2(1+\boldsymbol{\mu}_1)}[\boldsymbol{\mu}_1 - k(1-\boldsymbol{\mu}_1)]\gamma_{xy}; \rightarrow \boldsymbol{\mu}_1 = 0.2$$

$$G(\boldsymbol{\theta})\gamma_{xy} = \frac{0.25-k}{3}\left[\boldsymbol{\sigma}_y - \frac{\boldsymbol{\sigma}_x}{k}\right]$$

$$G(\boldsymbol{\theta})[\gamma_{xy}] = \frac{0.25-k}{3}\left[\boldsymbol{\sigma}_y - \frac{\boldsymbol{\sigma}_x}{k}\right] = \boldsymbol{\tau}_{xy}(\boldsymbol{\theta}) + \int_0^{\boldsymbol{\theta}} \boldsymbol{\tau}_{xy}(\boldsymbol{\tau}) K_3(\boldsymbol{\theta}, \boldsymbol{\tau})\mathrm{d}\boldsymbol{\tau} \quad (6.49)$$

$$\boldsymbol{\tau}_{xy} = \frac{0.25-k}{3}\left[\boldsymbol{\sigma}_y - \frac{\boldsymbol{\sigma}_x}{k}\right] - \mathrm{e}^{-\alpha_3 m}\int_0^{\theta} \mathrm{e}^{\frac{\tau}{1+\beta\tau}}\mathrm{e}^{\alpha_3 m(\tau)} m1(\tau_{xy})\mathrm{d}\tau$$

Solution of Eqs. (6.2) and (6.7) is as follows (using POLYMATH software).

Calculated values of DEQ variables

	Variable	Initial value	Minimal value	Maximal value	Final value
1	A	0	0	2.407847	2.118448
2	B	0	−0.0074121	0.0399521	−0.0012417
3	C	0	0	4.737473	4.737473
4	$E2$	1	0.4723666	1	0.4723666
5	k	1	0.4723666	1	0.4723666
6	m	0.00245	0.00245	0.0736101	0.07345
7	$m1$	0.0375	0.0079922	0.1463	0.1463
8	r	0	0	2.452476	2.231302
9	t	0	0	10	10
10	X	0	−0.0079353	0.0424896	−0.0013363
11	x	0.0255744	−0.0185138	0.1002929	1.841E−05
12	Y	0	0	2.461841	2.170424
13	y	0.1512	−0.0108052	2.560621	−0.0065608
14	z	0.0488	0.002631	2.793087	0.002631
15	Z	0	0	5.094795	5.094795

Differential equations

1 $\mathrm{d}(Y)/\mathrm{d}(t) = (\exp(t/(1.0+0.1*t)))*m1*(y-1*k*0.4*z)*(\exp(0.33*m))$

2 $\mathrm{d}(Z)/\mathrm{d}(t) = (\exp(t/(1.0+0.1*t)))*m1*(z-1*(0.2/k)*y)*(\exp(0.99*m))$

3 $\mathrm{d}(X)/\mathrm{d}(t) = (\exp(t/(1.0+0.1*t)))*m1*(x)*(\exp(1*m))$

Explicit equations (Fig. 6.11)

1 $E2 = (\exp(-0.075*t))$

2 $k = (\exp(-0.075*t))$

3 $m = 0.00245+0.0375*t-0.00934*t\hat{}2+0.00104*t\hat{}3-0.000041*t\hat{}4$

4 $A = (\exp(-0.33*m))*Y$

5 $B = (\exp(-1*m))*X$

6 $y = t*(\exp(-0.15*t))-A+k*(0.378-0.321*t$
$+0.952*t\hat{}2-0.292*t\hat{}3+0.0309*t\hat{}4-0.0011*t\hat{}5)*0.4*1$

7 $C = (\exp(-0.99*m))*Z$

8 $r = t*(\exp(-0.15*t))$

9 $m1 = 0.0375-0.01868*t+0.00312*t\hat{}2-0.000164*t\hat{}2$

10 $z = t*(\exp(-0.075*t))-C+k*(0.244+0.123*t$
$+0.547*t\hat{}2-0.198*t\hat{}3+0.0227*t\hat{}4-0.00085*t\hat{}5)*0.2*1$

11 $x = 0.333*(0.25-k)*(z-y/k)-B$

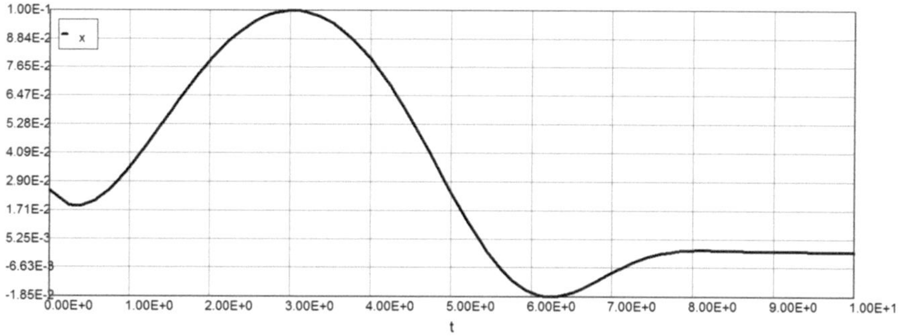

Fig. 6.11 Shear stresses

Model: $x = a0 + a1 * t + a2 * t\text{^}2 + a3 * t\text{^}3 + a4 * t\text{^}4 + a5 * t\text{^}5$

Variable	Value
$a0$	0.0140025
$a1$	−0.0064846
$a2$	0.0492076
$a3$	−0.0191358
$a4$	0.0024565
$a5$	−0.0001032

$$x = \sigma_{xy} = 0.014 - 0.0065 * t + 0.0492 * t\hat{}2 - 0.0191 * t\hat{}3 + 0.00247 * t\hat{}4 - 0.0001 * t\hat{}5 \quad (6.50)$$

Shear stresses: $\tau_{xy} = 0.1$ (2.04 ksi).

It should be noted that the maximum stress σ_y (the strongest direction of stresses) is practically unchanged in all examples above and the maximum stress σ_x (the weakest direction of stresses) is monotonically decreasing while parameter $k_0 \to \infty$. The qualitative assessment of modulus of elasticity deterioration with the temperature rise (for discrete parameters k_0) can be seen from Fig. 6.12.

The relationship between maximum creep stresses and parameters k_0 is presented below (Fig. 6.13).

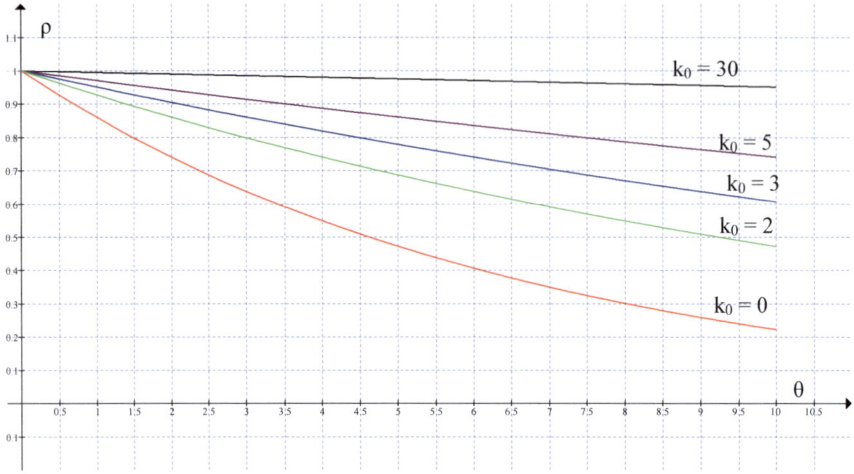

Fig. 6.12 Modulus of elasticity deterioration versus k_0

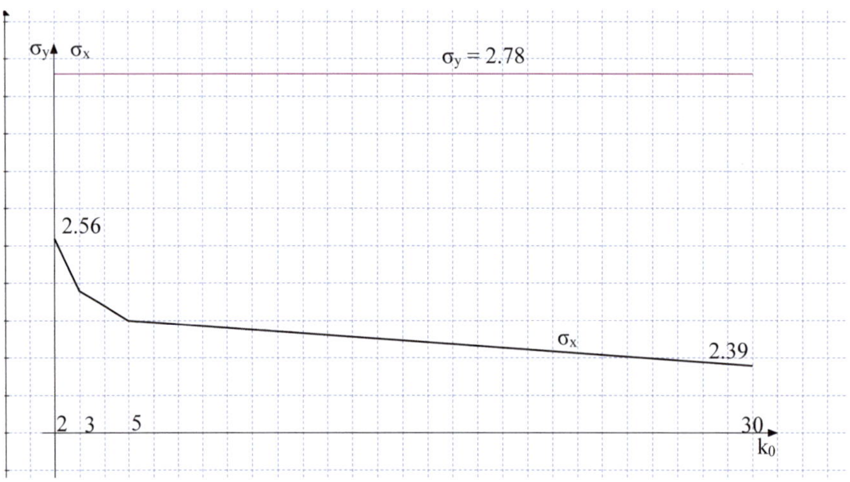

Fig. 6.13 Maximum creep stresses versus parameter k_0

Finally maximum/minimum stresses are:

$$\sigma_{1,2} = \frac{\sigma_x + \sigma_y}{2} \pm \frac{1}{2}\sqrt{(\sigma_x - \sigma_y)^2 + 4\tau_{xy}^2}$$
$$= \frac{2.56 + 2.79}{2} \pm \frac{1}{2}\sqrt{(2.56 - 2.79)^2 + 4(0.1)^2}$$

$$= 2.675 \pm 0.152$$

$$\sigma_1 = 2.827(57.55\,\text{ksi}); \quad \sigma_2 = 2.523(51.36\,\text{ksi})$$

$$\tan\alpha_1 = \frac{|\sigma_x - \sigma_y|}{\tau_{xy}} = 2.3 \Rightarrow \alpha_1 = 66.5^\circ$$

$$\tau_{\max,\min} = \pm\frac{1}{2}\sqrt{(\sigma_x - \sigma_y)^2 + 4\tau_{xy}^2} = \pm 0.152(3.09\,\text{ksi})$$

6.9 Stress–Temperature–Strain Diagram for Different MPP in Case of Fire

The response of real materials can be modeled by allowing for a number of different retardation temperature/times of different orders of magnitude, e.g. $t_R^i = \{\ldots, 10^{-2}; 10^{-1}; 1; 10^1; 10^2; \ldots\}$. In the following examples, we will choose the retardation times (temperatures) on both sides of the interval mentioned above. Thus, in one direction (e.g., along the x-axis) the retardation time is chosen large, and in the y direction—small. This combination of mechanical properties parameters (MPP) of the material (as will be shown below in the following examples) increases the value of the allowable creep stress on the one hand, and on the other—to the qualitatively different failure pattern of the material. We now fix the modulus of elasticity E_2 and parameter k_0 and proceed with computations for discrete variables $\boldsymbol{\theta}_r \sim t_R^i$. The computations below are similar to the previous examples and do not require any additional comments or explanations.

Example 6.6 Data: $E_2 = \exp(-0.075\theta)$; $\alpha_1 = 0.01$; $\alpha_2 = 10$; $\alpha_3 = 1.0$; $k_0 = 0.15/0.075 = 2$ and $k = \exp(-0.075\theta)$.

First approximation

Calculated values of DEQ variables

	Variable	Initial value	Minimal value	Maximal value	Final value
1	*A*	0	0	2.431399	2.236534
2	*B*	0	−0.0197765	0	−0.0011482
3	*C*	0	0	4.720307	4.720307
4	*E*2	1	0.4723666	1	0.4723666
5	*k*	1	0.4723666	1	0.4723666
6	*m*	0.00245	0.00245	0.0736101	0.07345
7	*m*1	0.0375	0.007997	0.1463	0.1463
8	*r*	0	0	2.452486	2.231302
9	*t*	0	0	10	10
10	*X*	0	−0.0211036	0	−0.0012358
11	*x*	0	−0.0122909	0.0024094	7.941E−05

(continued)

(continued)

	Variable	Initial value	Minimal value	Maximal value	Final value
12	Y	0	0	2.433042	2.238177
13	y	0	−0.0088626	1.798901	−0.005232
14	z	0	0	2.489237	0.0033581
15	Z	0	0	9.839195	9.839195

Differential equations

1 $d(Y)/d(t) = (\exp(t/(1.0 + 0.1 * t))) * m1 * (y - 0 * k * 0.4 * z) * (\exp(0.01 * m))$

2 $d(Z)/d(t) = (\exp(t/(1.0 + 0.1 * t))) * m1 * (z - 0 * (0.2/k) * y) * (\exp(10 * m))$

3 $d(X)/d(t) = (\exp(t/(1.0 + 0.1 * t))) * m1 * (x) * (\exp(1 * m))$

Explicit equations (Fig. 6.14)

1 $E2 = (\exp(-0.075 * t))$

2 $k = (\exp(-0.075 * t))$

3 $m = 0.00245 + 0.0375 * t - 0.00934 * t^\wedge 2 + 0.00104 * t^\wedge 3 - 0.000041 * t^\wedge 4$

4 $A = (\exp(-0.01 * m)) * Y$

5 $B = (\exp(-1 * m)) * X$

6 $y = t * (\exp(-0.15 * t)) - A + k * (0.307 - 0.131 * t + 0.83 * t^\wedge 2 - 0.255 * t^\wedge 3 + 0.0266 * t^\wedge 4 - 0.0093 * t^\wedge 5) * 0.4 * 0$

7 $C = (\exp(-10 * m)) * Z$

8 $r = t * (\exp(-0.15 * t))$

9 $m1 = 0.0375 - 0.01868 * t + 0.00312 * t^\wedge 2 - 0.000164 * t^\wedge 2$

10 $z = t * (\exp(-0.075 * t)) - C + k * (0.228 + 0.144 * t + 0.537 * t^\wedge 2 - 0.196 * t^\wedge 3 + 0.0225 * t^\wedge 4 - 0.00085 * t^\wedge 5) * 0.2 * 0$

11 $x = 0.333 * (0.25 - k) * (z - y/k) - B$

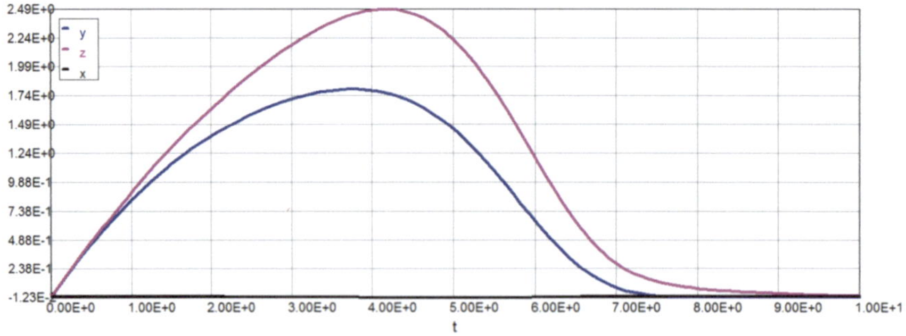

Fig. 6.14 Stress–temperature–strain diagram (σ_x; σ_y—first approximation)

Model: $y = a0 + a1 * t + a2 * t\text{^}2 + a3 * t\text{^}3 + a4 * t\text{^}4 + a5 * t\text{^}5$

Variable	Value
$a0$	0.2368994
$a1$	0.1341972
$a2$	0.5411092
$a3$	−0.19683
$a4$	0.0226039
$a5$	−0.0008489

$$y = \sigma_x = 0.237 + 0.134 * t + 0.541 * t\hat{}2 - 0.197 * t\hat{}3 + 0.0226 * t\hat{}4 - 0.00085 * t\hat{}5 \quad (6.51)$$

Model: $z = a0 + a1 * t + a2 * t\text{^}2 + a3 * t\text{^}3 + a4 * t\text{^}4 + a5 * t\text{^}5$

Variable	Value
$a0$	0.3627922
$a1$	−0.2912676
$a2$	0.9368591
$a3$	−0.2878441
$a4$	0.0304024
$a5$	−0.0010722

$$z = \sigma_y = 0.363 - 0.291 * t + 0.937 * t\hat{}2 - 0.288 * t\hat{}3 + 0.0304 * t\hat{}4 - 0.0011 * t\hat{}5 \quad (6.52)$$

Second approximation

Calculated values of DEQ variables

	Variable	Initial value	Minimal value	Maximal value	Final value
1	A	0	0	2.307525	1.709041
2	B	0	−0.0124749	0.037489	−0.0022146
3	C	0	0	4.69721	4.69721
4	$E2$	1	0.4723666	1	0.4723666
5	k	1	0.4723666	1	0.4723666
6	m	0.00245	0.00245	0.0736101	0.07345
7	$m1$	0.0375	0.0079914	0.1463	0.1463
8	r	0	0	2.452492	2.231302
9	t	0	0	10	10

(continued)

(continued)

	Variable	Initial value	Minimal value	Maximal value	Final value
10	X	0	−0.0133497	0.0398646	−0.0023834
11	x	0.0244256	−0.0203928	0.0982949	5.867E−05
12	Y	0	0	2.309065	1.710297
13	y	0.1452	−0.0207325	2.560509	−0.0156704
14	z	0.0474	−0.0040589	2.811931	−0.0040589
15	Z	0	0	9.791049	9.791049

Differential equations

1 $\mathrm{d}(Y)/\mathrm{d}(t) = (\exp(t/(1.0+0.1*t)))*m1*(y-1*k*0.4*z)*(\exp(0.01*m))$

2 $\mathrm{d}(Z)/\mathrm{d}(t) = (\exp(t/(1.0+0.1*t)))*m1*(z-1*(0.2/k)*y)*(\exp(10*m))$

3 $\mathrm{d}(X)/\mathrm{d}(t) = (\exp(t/(1.0+0.1*t)))*m1*(x)*(\exp(1*m))$

Explicit equations (Fig. 6.15)

1 $E2 = (\exp(-0.075*t))$

2 $k = (\exp(-0.075*t))$

3 $m = 0.00245+0.0375*t-0.00934*t\hat{}2+0.00104*t\hat{}3-0.000041*t\hat{}4$

4 $A = (\exp(-0.01*m))*Y$

5 $B = (\exp(-1*m))*X$

6 $y = t*(\exp(-0.15*t))-A+k*(0.363-0.291*t$
$+0.937*t\hat{}2-0.288*t\hat{}3+0.0304*t\hat{}4-0.0011*t\hat{}5)*0.4*1$

7 $C = (\exp(-10*m))*Z$

8 $r = t*(\exp(-0.15*t))$

9 $m1 = 0.0375-0.01868*t+0.00312*t\hat{}2-0.000164*t\hat{}2$

10 $z = t*(\exp(-0.075*t))-C+k*(0.237+0.134*t$
$+0.541*t\hat{}2-0.197*t\hat{}3+0.0226*t\hat{}4-0.00085*t\hat{}5)*0.2*1$

11 $x = 0.333*(0.25-k)*(z-y/k)-B$

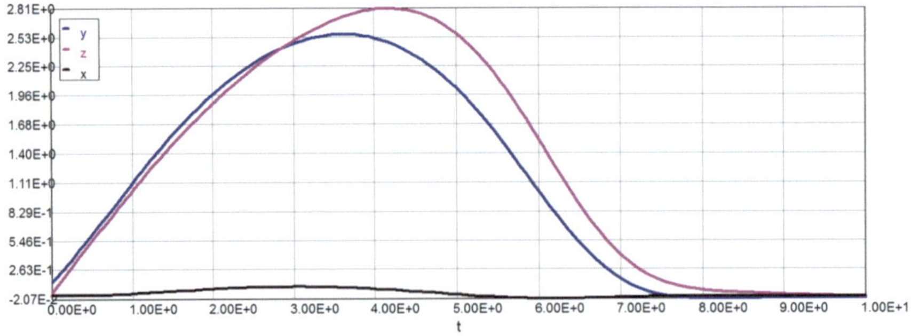

Fig. 6.15 Stress–temperature–strain diagram (σ_x; σ_y—second approximation)

Model: $y = a0 + a1 * t + a2 * t\text{^}2 + a3 * t\text{^}3 + a4 * t\text{^}4 + a5 * t\text{^}5$

Variable	Value
$a0$	0.3603427
$a1$	0.1199079
$a2$	0.7972686
$a3$	−0.2826936
$a4$	0.0321317
$a5$	−0.0011991

$$y = \sigma_x = 0.360 + 0.120 * t + 0.541 * t\text{^}2 - 0.283 * t\text{^}3 + 0.0321 * t\text{^}4 - 0.00012 * t\text{^}5 \quad (6.53)$$

Model: $z = a0 + a1 * t + a2 * t\text{^}2 + a3 * t\text{^}3 + a4 * t\text{^}4 + a5 * t\text{^}5$

Variable	Value
$a0$	0.4000137
$a1$	−0.1804668
$a2$	0.9318959
$a3$	−0.286216
$a4$	0.0295626
$a5$	−0.0010105

$$z = \sigma_y = 0.4 - 0.18 * t + 0.932 * t\text{^}2 - 0.286 * t\text{^}3 + 0.0296 * t\text{^}4 - 0.001 * t\text{^}5 \quad (6.54)$$

Example 6.7 Data: $E_2 = \exp(-0.075\theta)$; $\alpha_1 = 0.01$; $\alpha_2 = 100$; $\alpha_3 = 1.0$; $k_0 = 0.15/0.075 = 2$ and $k = \exp(-0.075\theta)$.

First approximation

Calculated values of DEQ variables

	Variable	Initial value	Minimal value	Maximal value	Final value
1	A	0	0	2.431256	2.236534
2	B	0	−0.0564415	6.007E−05	6.007E−05
3	C	0	0	4.741295	4.741295
4	$E2$	1	0.4723666	1	0.4723666

(continued)

(continued)

	Variable	Initial value	Minimal value	Maximal value	Final value
5	k	1	0.4723666	1	0.4723666
6	m	0.00245	0.00245	0.0736099	0.07345
7	$m1$	0.0375	0.0079892	0.1463	0.1463
8	r	0	0	2.452512	2.231302
9	t	0	0	10	10
10	X	0	−0.0602744	6.465E−05	6.465E−05
11	x	0	−0.0299674	0.0053566	0.0004252
12	Y	0	0	2.432904	2.238177
13	y	0	−0.008861	1.798843	−0.005232
14	z	0	−0.0176297	2.629953	−0.0176297
15	Z	0	0	7380.961	7341.587

Differential equations

1 $d(Y)/d(t) = (\exp(t/(1.0 + 0.1 * t))) * m1 * (y - 0 * k * 0.4 * z) * (\exp(0.01 * m))$

2 $d(Z)/d(t) = (\exp(t/(1.0 + 0.1 * t))) * m1 * (z - 0 * (0.2/k) * y) * (\exp(100 * m))$

3 $d(X)/d(t) = (\exp(t/(1.0 + 0.1 * t))) * m1 * (x) * (\exp(1 * m))$

Explicit equations (Fig. 6.16)

1 $E2 = (\exp(-0.075 * t))$

2 $k = (\exp(-0.075 * t))$

3 $m = 0.00245 + 0.0375 * t - 0.00934 * t^\wedge 2 + 0.00104 * t^\wedge 3 - 0.000041 * t^\wedge 4$

4 $A = (\exp(-0.01 * m)) * Y$

5 $B = (\exp(-1 * m)) * X$

6 $y = t * (\exp(-0.15 * t)) - A + k * (0.363 - 0.291 * t + 0.937 * t^\wedge 2 - 0.288 * t^\wedge 3 + 0.0304 * t^\wedge 4 - 0.0011 * t^\wedge 5) * 0.4 * 0$

7 $C = (\exp(-100 * m)) * Z$

8 $r = t * (\exp(-0.15 * t))$

9 $m1 = 0.0375 - 0.01868 * t + 0.00312 * t^\wedge 2 - 0.000164 * t^\wedge 2$

10 $z = t * (\exp(-0.075 * t)) - C + k * (0.237 + 0.134 * t + 0.541 * t^\wedge 2 - 0.197 * t^\wedge 3 + 0.0226 * t^\wedge 4 - 0.00085 * t^\wedge 5) * 0.2 * 0$

11 $x = 0.333 * (0.25 - k) * (z - y/k) - B$

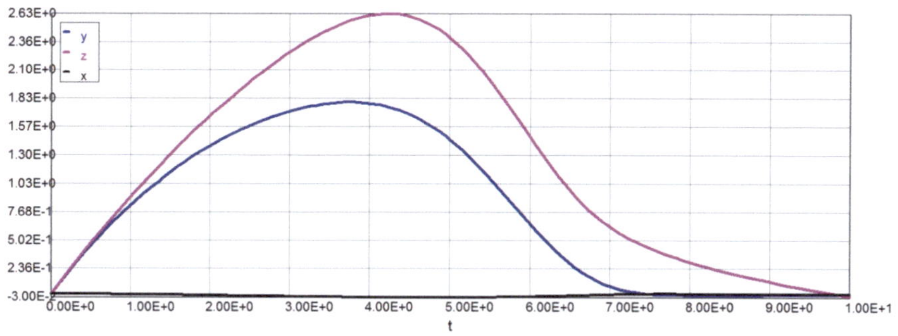

Fig. 6.16 Stress–temperature–strain diagram (σ_x; σ_y—first approximation)

Model: $y = a0 + a1 * t + a2 * t \hat{} 2 + a3 * t \hat{} 3 + a4 * t \hat{} 4 + a5 * t \hat{} 5$

Variable	Value
$a0$	0.2285023
$a1$	0.1437743
$a2$	0.5370295
$a3$	−0.1960078
$a4$	0.0225254
$a5$	−0.000846

$$y = \sigma_x \\ = 0.228 + 0.144 * t + 0.537 * t^{\wedge}2 - 0.196 * t^{\wedge}3 + 0.0225 * t^{\wedge}4 - 0.00085 * t^{\wedge}5 \quad (6.55)$$

Model: $z = a0 + a1 * t + a2 * t \hat{} 2 + a3 * t \hat{} 3 + a4 * t \hat{} 4 + a5 * t \hat{} 5$

Variable	Value
$a0$	0.3067398
$a1$	−0.1313964
$a2$	0.8300463
$a3$	−0.2554973
$a4$	0.0266284
$a5$	−0.0009273

$$z = \sigma_y \\ = 0.306 - 0.131 * t + 0.83 * t^{\wedge}2 - 0.255 * t^{\wedge}3 + 0.0266 * t^{\wedge}4 - 0.00093 * t^{\wedge}5 \quad (6.56)$$

Second approximation

Calculated values of DEQ variables

	Variable	Initial value	Minimal value	Maximal value	Final value
1	A	0	0	2.443847	2.239902
2	B	0	−0.0300594	0.0342467	−7.755E−05
3	C	0	0	4.686287	4.686287
4	$E2$	1	0.4723666	1	0.4723666
5	k	1	0.4723666	1	0.4723666
6	m	0.00245	0.00245	0.0736104	0.07345
7	$m1$	0.0375	0.007996	0.1463	0.1463
8	r	0	0	2.45248	2.231302
9	t	0	0	10	10
10	X	0	−0.032156	0.0364156	−8.346E−05
11	x	0.0191808	−0.0272754	0.0908049	0.0002642
12	Y	0	0	2.445496	2.241548
13	y	0.1224	−0.0093566	2.602704	−0.0093566
14	z	0.0456	−0.0223288	2.937081	−0.0223288
15	Z	0	0	7300.309	7256.411

Differential equations

1 $d(Y)/d(t) = (\exp(t/(1.0 + 0.1 * t))) * m1 * (y - 1 * k * 0.4 * z) * (\exp(0.01 * m))$

2 $d(Z)/d(t) = (\exp(t/(1.0 + 0.1 * t))) * m1 * (z - 1 * (0.2/k) * y) * (\exp(100 * m))$

3 $d(X)/d(t) = (\exp(t/(1.0 + 0.1 * t))) * m1 * (x) * (\exp(1 * m))$

Explicit equations (Fig. 6.17)

1 $E2 = (\exp(-0.075 * t))$

2 $k = (\exp(-0.075 * t))$

3 $m = 0.00245 + 0.0375 * t - 0.00934 * t\hat{}2 + 0.00104 * t\hat{}3 - 0.000041 * t\hat{}4$

4 $A = (\exp(-0.01 * m)) * Y$

5 $B = (\exp(-1 * m)) * X$

6 $y = t * (\exp(-0.15 * t)) - A + k * (0.306 - 0.131 * t + 0.83 * t\hat{}2 - 0.255 * t\hat{}3 + 0.0266 * t\hat{}4 - 0.00093 * t\hat{}5) * 0.4 * 1$

7 $C = (\exp(-100 * m)) * Z$

8 $r = t * (\exp(-0.15 * t))$

9 $m1 = 0.0375 - 0.01868 * t + 0.00312 * t\hat{}2 - 0.000164 * t\hat{}2$

10 $z = t * (\exp(-0.075 * t)) - C + k * (0.228 + 0.144 * t + 0.537 * t\hat{}2 - 0.196 * t\hat{}3 + 0.0225 * t\hat{}4 - 0.00085 * t\hat{}5) * 0.2 * 1$

11 $x = 0.333 * (0.25 - k) * (z - y/k) - B$

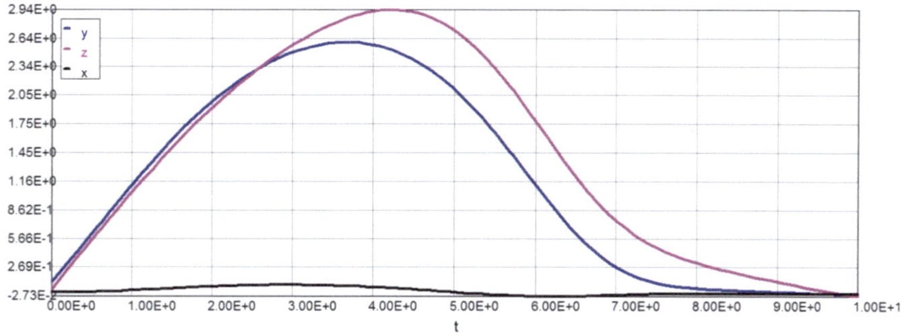

Fig. 6.17 Stress–temperature–strain diagram (σ_x; σ_y—second approximation)

Model: $y = a0 + a1 * t + a2 * t\text{^}2 + a3 * t\text{^}3 + a4 * t\text{^}4 + a5 * t\text{^}5$

Variable	Value
$a0$	0.3368873
$a1$	0.180475
$a2$	0.7571188
$a3$	−0.2700569
$a4$	0.0305787
$a5$	−0.0011357

$$\begin{aligned} y &= \sigma_x \\ &= 0.337 + 0.180 * t + 0.757 * t^\wedge 2 - 0.27 * t^\wedge 3 + 0.0306 * t^\wedge 4 - 0.0011 * t^\wedge 5 \end{aligned} \tag{6.57}$$

Model: $z = a0 + a1 * t + a2 * t\text{^}2 + a3 * t\text{^}3 + a4 * t\text{^}4 + a5 * t\text{^}5$

Variable	Value
$a0$	0.3434986
$a1$	−0.0087498
$a2$	0.8151748
$a3$	−0.2521187
$a4$	0.0257327
$a5$	−0.0008688

$$\begin{aligned} z &= \sigma_y \\ &= 0.343 - 0.00875 * t + 0.815 * t^\wedge 2 - 0.252 * t^\wedge 3 + 0.0257 * t^\wedge 4 - 0.00087 * t^\wedge 5 \end{aligned} \tag{6.58}$$

Example 6.8 Data: $E_2 = \exp(-0.075\theta)$; $\alpha_1 = 0.01$; $\alpha_2 = 1000$; $\alpha_3 = 1.0$; $k_0 = 0.15/0.075 = 2$ and $k = \exp(-0.075\theta)$.

First approximation

Calculated values of DEQ variables

	Variable	Initial value	Minimal value	Maximal value	Final value
1	*A*	0	0	2.431394	2.236534
2	*B*	0	−0.2304067	0.0125603	0.0125603
3	*C*	0	0	4.959932	4.959932
4	*E2*	1	0.4723666	1	0.4723666
5	*k*	1	0.4723666	1	0.4723666
6	*m*	0.00245	0.00245	0.0736093	0.07345
7	*m*1	0.0375	0.0079902	0.1463	0.1463
8	*r*	0	0	2.452458	2.231302
9	*t*	0	0	10	10
10	*X*	0	−0.2462148	0.0135175	0.0135175
11	*x*	0	−0.0959884	0.0146555	0.0041146
12	*Y*	0	0	2.433042	2.238177
13	*y*	0	−0.008854	1.798845	−0.005232
14	*z*	0	−0.2362664	2.935722	−0.2362664
15	*Z*	0	0	4.348E+32	3.93E+32

Differential equations

1 $d(Y)/d(t) = (\exp(t/(1.0+0.1*t)))*m1*(y-0*k*0.4*z)*(\exp(0.01*m))$

2 $d(Z)/d(t) = (\exp(t/(1.0+0.1*t)))*m1*(z-0*(0.2/k)*y)*(\exp(1000*m))$

3 $d(X)/d(t) = (\exp(t/(1.0+0.1*t)))*m1*(x)*(\exp(1*m))$

Explicit equations (Fig. 6.18)

1 $E2 = (\exp(-0.075*t))$

2 $k = (\exp(-0.075*t))$

3 $m = 0.00245+0.0375*t-0.00934*t\hat{}2+0.00104*t\hat{}3-0.000041*t\hat{}4$

4 $A = (\exp(-0.01*m))*Y$

5 $B = (\exp(-1*m))*X$

6 $y = t*(\exp(-0.15*t))-A+k*(0.306-0.131*t$
$+0.83*t\hat{}2-0.255*t\hat{}3+0.0266*t\hat{}4-0.00093*t\hat{}5)*0.4*0$

7 $C = (\exp(-1000*m))*Z$

8 $r = t*(\exp(-0.15*t))$

9 $m1 = 0.0375-0.01868*t+0.00312*t\hat{}2-0.000164*t\hat{}2$

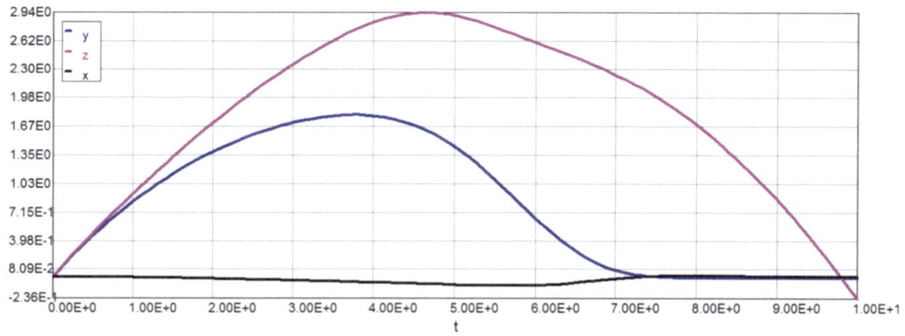

Fig. 6.18 Stress–temperature–strain diagram (σ_x; σ_y—first approximation)

10 $z = t * (\exp(-0.075 * t)) - C + k * (0.228 + 0.144 * t + 0.537 * t^\wedge 2 - 0.196 * t^\wedge 3 + 0.0225 * t^\wedge 4 - 0.00085 * t^\wedge 5) * 0.2 * 0$

11 $x = 0.333 * (0.25 - k) * (z - y/k) - B$

Model: $y = a0 + a1 * t + a2 * t$^2 $+ a3 * t$^3 $+ a4 * t$^4 $+ a5 * t$^5

Variable	Value
$a0$	0.2289081
$a1$	0.1462277
$a2$	0.5352152
$a3$	−0.1955285
$a4$	0.0224723
$a5$	−0.0008439

$$y = \sigma_x = 0.228 + 0.146 * t + 0.535 * t^\wedge 2 - 0.196 * t^\wedge 3 + 0.0225 * t^\wedge 4 - 0.00085 * t^\wedge 5 \quad (6.59)$$

Model: $z = a0 + a1 * t + a2 * t$^2 $+ a3 * t$^3 $+ a4 * t$^4 $+ a5 * t$^5

Variable	Value
$a0$	0.093095
$a1$	0.635177
$a2$	0.2324648
$a3$	−0.0859158
$a4$	0.0088596
$a5$	−0.0003263

$$z = \sigma_y = 0.0931 + 0.635 * t + 0.232 * t^\wedge 2 - 0.0859 * t^\wedge 3 + 0.00886 * t^\wedge 4 - 0.000326 * t^\wedge 5 \tag{6.60}$$

Second approximation

Calculated values of DEQ variables

	Variable	Initial value	Minimal value	Maximal value	Final value
1	A	0	0	2.405163	2.236099
2	B	0	−0.1361403	0.0249421	0.0076873
3	C	0	0	4.899476	4.899476
4	$E2$	1	0.4723666	1	0.4723666
5	k	1	0.4723666	1	0.4723666
6	m	0.00245	0.00245	0.0736098	0.07345
7	$m1$	0.0375	0.0079919	0.1463	0.1463
8	r	0	0	2.452438	2.231302
9	t	0	0	10	10
10	X	0	−0.1455801	0.0265031	0.0082731
11	x	−0.0020879	−0.0699369	0.0754045	0.0026504
12	Y	0	0	2.406792	2.237742
13	y	0.03724	−0.0533377	2.637309	−0.0533377
14	z	0.0456	−0.2525226	3.192521	−0.2525226
15	Z	0	0	4.282E+32	3.882E+32

Differential equations

1 $d(Y)/d(t) = (\exp(t/(1.0 + 0.1 * t))) * m1 * (y - 1 * k * 0.4 * z) * (\exp(0.01 * m))$
2 $d(Z)/d(t) = (\exp(t/(1.0 + 0.1 * t))) * m1 * (z - 1 * (0.2/k) * y) * (\exp(1000 * m))$
3 $d(X)/d(t) = (\exp(t/(1.0 + 0.1 * t))) * m1 * (x) * (\exp(1 * m))$

Explicit equations (Figs. 6.19, 6.20 and 6.21)

1 $E2 = (\exp(-0.075 * t))$
2 $k = (\exp(-0.075 * t))$
3 $m = 0.00245 + 0.0375 * t - 0.00934 * t^\wedge 2 + 0.00104 * t^\wedge 3 - 0.000041 * t^\wedge 4$
4 $A = (\exp(-0.01 * m)) * Y$
5 $B = (\exp(-1 * m)) * X$
6 $y = t * (\exp(-0.15 * t)) - A + k * (0.0931 + 0.635 * t + 0.232 * t^\wedge 2 - 0.0859 * t^\wedge 3 + 0.00886 * t^\wedge 4 - 0.000326 * t^\wedge 5) * 0.4 * 1$

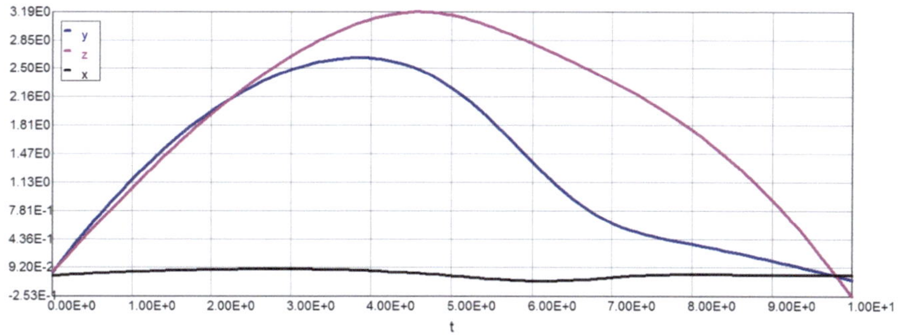

Fig. 6.19 Stress–temperature–strain diagram (σ_x; σ_y—second approximation)

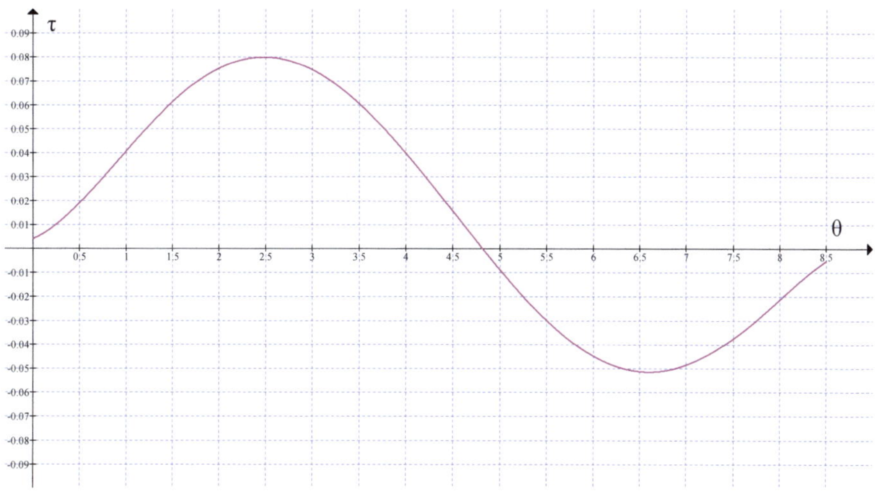

Fig. 6.20 Shear stress–temperature–strain diagram

7 $C = (\exp(-1000 * m)) * Z$

8 $r = t * (\exp(-0.15 * t))$

9 $m1 = 0.0375 - 0.01868 * t + 0.00312 * t^{\wedge}2 - 0.000164 * t^{\wedge}2$

10 $z = t * (\exp(-0.075 * t)) - C + k * (0.228 + 0.146 * t$
$+ 0.535 * t^{\wedge}2 - 0.196 * t^{\wedge}3 + 0.0225 * t^{\wedge}4 - 0.00085 * t^{\wedge}5) * 0.2 * 1$

11 $x = 0.333 * (0.25 - k) * (z - y/k) - B$

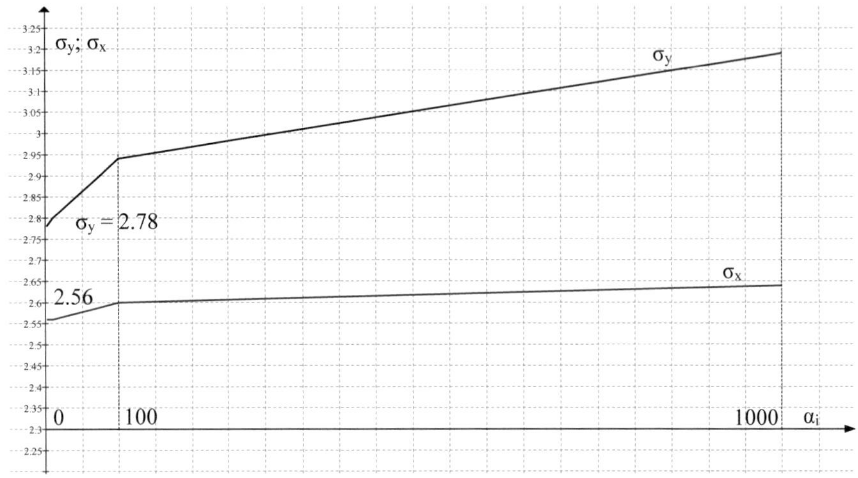

Fig. 6.21 Stresses (σ_y and σ_x)—MPP (α_i) diagram

Model: $y = a0 + a1 * t + a2 * t\hat{}2 + a3 * t\hat{}3 + a4 * t\hat{}4 + a5 * t\hat{}5$

Variable	Value
$a0$	0.2567806
$a1$	0.4415192
$a2$	0.5624329
$a3$	−0.2157032
$a4$	0.0247389
$a5$	−0.0009264

$$y = \sigma_x = 0.257 + 0.441 * t + 0.562 * t\hat{}2 - 0.216 * t\hat{}3 + 0.0247 * t\hat{}4 - 0.00093 * t\hat{}5 \tag{6.61}$$

Model: $z = a0 + a1 * t + a2 * t\hat{}2 + a3 * t\hat{}3 + a4 * t\hat{}4 + a5 * t\hat{}5$

Variable	Value
$a0$	0.1255303
$a1$	0.7304505
$a2$	0.2589966
$a3$	−0.1005632
$a4$	0.01066
$a5$	−0.0003963

$$z = \sigma_y = 0.126 + 0.73 * t + 0.259 * t^\wedge 2 - 0.1 * t^\wedge 3 + 0.0107 * t^\wedge 4 - 0.000396 * t^\wedge 5 \quad (6.62)$$

Model: $x = a0 + a1 * t + a2 * t^\wedge 2 + a3 * t^\wedge 3 + a4 * t^\wedge 4 + a5 * t^\wedge 5$

Variable	Value
$a0$	0.0041384
$a1$	0.0163532
$a2$	0.0357098
$a3$	−0.0175609
$a4$	0.0024962
$a5$	−0.0001117

$$x = \tau_{xy} = 0.0041 + 0.0164 * t + 0.0357 * t^\wedge 2 - 0.0176 * t^\wedge 3 + 0.0025 * t^\wedge 4 - 0.000112 * t^\wedge 5 \quad (6.63)$$

6.10 Nonlinear Creep Deformations

The nonlinear creep model was discussed in detail in Chap. 4, so we restrict ourselves to the specific issues associated with anisotropic material, namely the effect of varying degree of nonlinearity on the redistribution of stresses in an anisotropic medium, as well as the influence of various material parameters on the maximum values of stress and strain. Computer Code and the process of calculations similar to the case of linear creep, presented above and therefore needs no additional comment. Let us recall that the nonlinear creep in the proposed model is determined by the parameters N and $N1$ in Eq. (6.32).

Example 6.9 Data: $E_2 = \exp(-0.075\theta)$; $\alpha_1 = 0.01$; $\alpha_2 = 1.0$; $\alpha_3 = 1.0$; $k_0 = 0.15/0.075 = 2$; $n = 3$; $n1 = 1$ and $k = \exp(-0.075\theta)$.

First approximation

Calculated values of DEQ variables

	Variable	Initial value	Minimal value	Maximal value	Final value
1	A	0	0	2.230706	2.230706
2	B	0	0	7.346E−05	4.917E−05
3	C	0	0	4.521895	4.521895
4	$E2$	1	0.4723666	1	0.4723666

(continued)

(continued)

	Variable	Initial value	Minimal value	Maximal value	Final value
5	k	1	0.4723666	1	0.4723666
6	m	0.00245	0.00245	0.0736125	0.07345
7	$m1$	0.0375	0.0079888	0.1463	0.1463
8	n	3	3	3	3
9	$n1$	1	1	1	1
10	r	0	0	2.452507	2.231302
11	t	0	0	10	10
12	X	0	0	7.873E−05	5.291E−05
13	x	0	−0.0148965	0.0408754	−0.0148965
14	Y	0	0	2.232345	2.232345
15	y	0	0	1.572755	0.0005958
16	z	0	0	1.879415	0.2017703
17	Z	0	0	4.86653	4.86653

Differential equations

1 $\mathrm{d}(Y)/\mathrm{d}(t) = (\exp(t/(1.0 + 0.1 * t))) * m1 * (y^\wedge n - 0 * k * 0.4 * z^\wedge n) * (\exp(0.01 * m))$

2 $\mathrm{d}(Z)/\mathrm{d}(t) = (\exp(t/(1.0 + 0.1 * t))) * m1 * (z^\wedge n - 0 * (0.2/k) * y^\wedge n1) * (\exp(1 * m))$

3 $\mathrm{d}(X)/\mathrm{d}(t) = (\exp(t/(1.0 + 0.1 * t))) * m1 * (x^\wedge n) * (\exp(1 * m))$

Explicit equations (Fig. 6.22)

1 $E2 = (\exp(-0.075 * t))$

2 $k = (\exp(-0.075 * t))$

3 $m = 0.00245 + 0.0375 * t - 0.00934 * t^\wedge 2 + 0.00104 * t^\wedge 3 - 0.000041 * t^\wedge 4$

4 $A = (\exp(-0.01 * m)) * Y$

5 $B = (\exp(-1 * m)) * X$

6 $y = t * (\exp(-0.15 * t)) - A + k * (0.0931 + 0.635 * t$
$+ 0.232 * t^\wedge 2 - 0.0859 * t^\wedge 3 + 0.00886 * t^\wedge 4 - 0.000326 * t^\wedge 5) * 0.4 * 0$

7 $C = (\exp(-1 * m)) * Z$

8 $r = t * (\exp(-0.15 * t))$

9 $m1 = 0.0375 - 0.01868 * t + 0.00312 * t^\wedge 2 - 0.000164 * t^\wedge 2$

10 $z = t * (\exp(-0.075 * t)) - C + k * (0.228 + 0.146 * t$
$+ 0.535 * t^\wedge 2 - 0.196 * t^\wedge 3 + 0.0225 * t^\wedge 4 - 0.00085 * t^\wedge 5) * 0.2 * 0$

11 $x = 0.333 * (0.25 - k) * (z - y/k) - B$

12 $n = 3$

13 $n1 = 1$

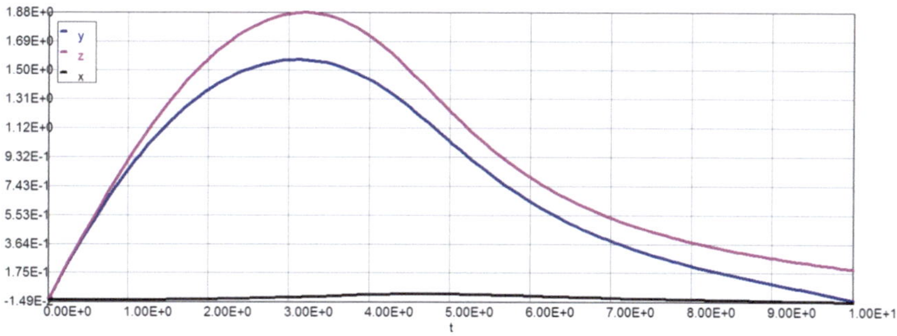

Fig. 6.22 Nonlinear creep deformations (first approximation)

Model: $y = a0 + a1 * t + a2 * t\text{^}2 + a3 * t\text{^}3 + a4 * t\text{^}4 + a5 * t\text{^}5$

Variable	Value
$a0$	0.003778
$a1$	0.9391923
$a2$	−0.0451312
$a3$	−0.0548
$a4$	0.0089022
$a5$	−0.0003916

$$\begin{aligned} y &= \sigma_x \\ &= 0.0038 + 0.939 * t - 0.0451 * t^\wedge 2 - 0.0548 * t^\wedge 3 + 0.0089 * t^\wedge 4 - 0.00039 * t^\wedge 5 \end{aligned} \tag{6.64}$$

Model: $z = a0 + a1 * t + a2 * t\text{^}2 + a3 * t\text{^}3 + a4 * t\text{^}4 + a5 * t\text{^}5$

Variable	Value
$a0$	−0.0168021
$a1$	0.9724702
$a2$	0.0511888
$a3$	−0.0913
$a4$	0.0133814
$a5$	−0.0005723

$$\begin{aligned} z &= \sigma_y \\ &= -0.0168 + 0.972 * t + 0.0512 * t^\wedge 2 - 0.0913 * t^\wedge 3 + 0.0134 * t^\wedge 4 - 0.00057 * t^\wedge 5 \end{aligned} \tag{6.65}$$

Second Approximation

Calculated values of DEQ variables

	Variable	Initial value	Minimal value	Maximal value	Final value
1	A	0	0	2.179734	2.087179
2	B	0	0	0.0003377	0.0003377
3	C	0	−0.0004692	4.255449	4.255449
4	$E2$	1	0.4723666	1	0.4723666
5	k	1	0.4723666	1	0.4723666
6	m	0.00245	0.00245	0.0736125	0.07345
7	$m1$	0.0375	0.0079908	0.1463	0.1463
8	n	3	3	3	3
9	$n1$	1	1	1	1
10	r	0	0	2.452524	2.231302
11	t	0	0	10	10
12	X	0	0	0.0003634	0.0003634
13	x	−0.0018681	−0.0018681	0.0856962	0.0024949
14	Y	0	0	2.181271	2.088713
15	y	−0.00672	−0.00672	1.97599	0.2429791
16	z	0.00076	0.00076	1.993329	0.4761333
17	Z	0	−0.0004775	4.579777	4.579777

Differential equations

1 $\mathrm{d}(Y)/\mathrm{d}(t) = (\exp(t/(1.0+0.1*t)))*m1*(y\hat{}n-1*k*0.4*z\hat{}n)*(\exp(0.01*m))$

2 $\mathrm{d}(Z)/\mathrm{d}(t) = (\exp(t/(1.0+0.1*t)))*m1*(z\hat{}n-1*(0.2/k)*y\hat{}n1)*(\exp(1*m))$

3 $\mathrm{d}(X)/\mathrm{d}(t) = (\exp(t/(1.0+0.1*t)))*m1*(x\hat{}n)*(\exp(1*m))$

Explicit equations (Fig. 6.23)

1 $E2 = (\exp(-0.075*t))$

2 $k = (\exp(-0.075*t))$

3 $m = 0.00245+0.0375*t-0.00934*t\hat{}2+0.00104*t\hat{}3-0.000041*t\hat{}4$

4 $A = (\exp(-0.01*m))*Y$

5 $B = (\exp(-1*m))*X$

6 $y = t*(\exp(-0.15*t))-A+k*(-0.0168+0.972*t$
$+0.0512*t\hat{}2-0.0913*t\hat{}3+0.0134*t\hat{}4-0.00057*t\hat{}5)*0.4*1$

7 $C = (\exp(-1*m))*Z$

8 $r = t*(\exp(-0.15*t))$

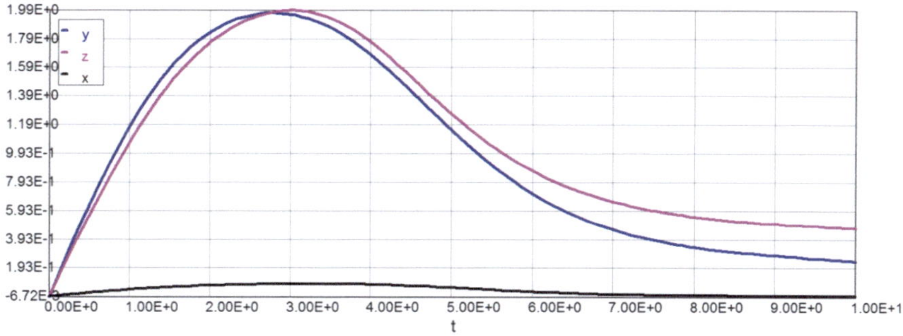

Fig. 6.23 Nonlinear creep deformations (second approximation)

9 $m1 = 0.0375 - 0.01868 * t + 0.00312 * t^{\wedge}2 - 0.000164 * t^{\wedge}2$

10 $z = t * (\exp(-0.075 * t)) - C + k * (0.0038 + 0.939 * t - 0.0451 * t^{\wedge}2 - 0.0548 * t^{\wedge}3 + 0.0089 * t^{\wedge}4 - 0.00039 * t^{\wedge}5) * 0.2 * 1$

11 $x = 0.333 * (0.25 - k) * (z - y/k) - B$

12 $n = 3$

13 $n1 = 1$

Model: $y = a0 + a1 * t + a2 * t^2 + a3 * t^3 + a4 * t^4 + a5 * t^5$

Variable	Value
$a0$	−0.0604291
$a1$	1.580236
$a2$	−0.3015318
$a3$	−0.0180808
$a4$	0.0068232
$a5$	−0.0003556

$$y = \sigma_x = -0.06 + 1.58 * t - 0.302 * t^{\wedge}2 - 0.018 * t^{\wedge}3 + 0.00682 * t^{\wedge}4 - 0.00036 * t^{\wedge}5 \tag{6.66}$$

Model: $z = a0 + a1 * t + a2 * t^2 + a3 * t^3 + a4 * t^4 + a5 * t^5$

Variable	Value
$a0$	−0.0608776
$a1$	1.35861
$a2$	−0.1548595

(continued)

(continued)

Variable	Value
*a*3	−0.0500179
*a*4	0.0098634
*a*5	−0.0004626

$$z = \sigma_y = -0.061 + 1.359 * t - 0.155 * t^\wedge 2 - 0.05 * t^\wedge 3 + 0.00986 * t^\wedge 4 - 0.00046 * t^\wedge 5 \quad (6.67)$$

Example 6.10 Data: $E_2 = \exp(-0.075\theta)$; $\alpha_1 = 0.01$; $\alpha_2 = 1.0$; $\alpha_3 = 1.0$; $k_0 = 0.15/0.075 = 2$; $n = 4$; $n1 = 1$ and $k = \exp(-0.075\theta)$.

First approximation

Calculated values of DEQ variables

	Variable	Initial value	Minimal value	Maximal value	Final value
1	*A*	0	0	2.139682	2.139682
2	*B*	0	0	4.63E−06	4.63E−06
3	*C*	0	0	4.420254	4.420254
4	*E*2	1	0.4723666	1	0.4723666
5	*k*	1	0.4723666	1	0.4723666
6	*m*	0.00245	0.00245	0.0736128	0.07345
7	*m*1	0.0375	0.0079947	0.1463	0.1463
8	*n*	4	4	4	4
9	*n*1	1	1	1	1
10	*r*	0	0	2.452501	2.231302
11	*t*	0	0	10	10
12	*X*	0	0	4.983E−06	4.983E−06
13	*x*	0	−0.0081095	0.0464286	−0.0081095
14	*Y*	0	0	2.141255	2.141255
15	*y*	0	0	1.484855	0.0916191
16	*z*	0	0	1.695421	0.3034116
17	*Z*	0	0	4.757142	4.757142

Differential equations

1 $\mathrm{d}(Y)/\mathrm{d}(t) = (\exp(t/(1.0 + 0.1 * t))) * m1 * (y^\wedge n - 0 * k * 0.4 * z^\wedge n) * (\exp(0.01 * m))$

2 $\mathrm{d}(Z)/\mathrm{d}(t) = (\exp(t/(1.0 + 0.1 * t))) * m1 * (z^\wedge n - 0 * (0.2/k) * y^\wedge n1) * (\exp(1 * m))$

3 $\mathrm{d}(X)/\mathrm{d}(t) = (\exp(t/(1.0 + 0.1 * t))) * m1 * (x^\wedge n) * (\exp(1 * m))$

Explicit equations (Fig. 6.24)

1 $E2 = (\exp(-0.075 * t))$

2 $k = (\exp(-0.075 * t))$

3 $m = 0.00245 + 0.0375 * t - 0.00934 * t^{\wedge}2 + 0.00104 * t^{\wedge}3 - 0.000041 * t^{\wedge}4$

4 $A = (\exp(-0.01 * m)) * Y$

5 $B = (\exp(-1 * m)) * X$

6 $y = t * (\exp(-0.15 * t)) - A + k * (-0.0168 + 0.972 * t + 0.0512 * t^{\wedge}2 - 0.0913 * t^{\wedge}3 + 0.0134 * t^{\wedge}4 - 0.00057 * t^{\wedge}5) * 0.4 * 0$

7 $C = (\exp(-1 * m)) * Z$

8 $r = t * (\exp(-0.15 * t))$

9 $m1 = 0.0375 - 0.01868 * t + 0.00312 * t^{\wedge}2 - 0.000164 * t^{\wedge}2$

10 $z = t * (\exp(-0.075 * t)) - C + k * (0.0038 + 0.939 * t - 0.0451 * t^{\wedge}2 - 0.0548 * t^{\wedge}3 + 0.0089 * t^{\wedge}4 - 0.00039 * t^{\wedge}5) * 0.2 * 0$

11 $x = 0.333 * (0.25 - k) * (z - y/k) - B$

12 $n = 4$

13 $n1 = 1$

Model: $y = a0 + a1 * t + a2 * t^2 + a3 * t^3 + a4 * t^4 + a5 * t^5$

Variable	Value
$a0$	−0.0479318
$a1$	1.153377
$a2$	−0.2255463
$a3$	−0.006784
$a4$	0.0038754
$a5$	−0.0002086

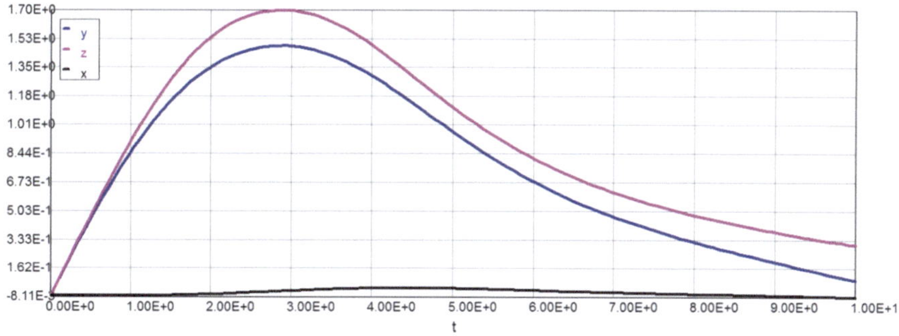

Fig. 6.24 Nonlinear creep deformations (first approximation $n = 4$)

$$y = \sigma_x = -0.048 + 1.153 * t - 0.226 * t^\wedge 2 - 0.00678 * t^\wedge 3 + 0.00387 * t^\wedge 4 - 0.00021 * t^\wedge 5 \quad (6.68)$$

Model: $z = a0 + a1 * t + a2 * t^\wedge 2 + a3 * t^\wedge 3 + a4 * t^\wedge 4 + a5 * t^\wedge 5$

Variable	Value
$a0$	−0.0891015
$a1$	1.29471
$a2$	−0.2341729
$a3$	−0.0142226
$a4$	0.0052104
$a5$	−0.0002708

$$z = \sigma_y = -0.0891 + 1.295 * t - 0.234 * t^\wedge 2 - 0.0142 * t^\wedge 3 + 0.00521 * t^\wedge 4 - 0.00027 * t^\wedge 5 \quad (6.69)$$

Second Approximation

Calculated values of DEQ variables

	Variable	Initial value	Minimal value	Maximal value	Final value
1	A	0	0	2.068744	1.902004
2	B	0	0	1.753E−05	1.753E−05
3	C	0	−0.0009595	4.061257	4.061257
4	$E2$	1	0.4723666	1	0.4723666
5	k	1	0.4723666	1	0.4723666
6	m	0.00245	0.00245	0.0736131	0.07345
7	$m1$	0.0375	0.0079894	0.1463	0.1463
8	n	4	4	4	4
9	$n1$	1	1	1	1
10	r	0	0	2.452457	2.231302
11	t	0	0	10	10
12	X	0	0	1.886E−05	1.886E−05
13	x	−0.0065035	−0.0065035	0.0757555	0.0146278
14	Y	0	0	2.070162	1.903401
15	y	−0.03564	−0.03564	1.775487	0.3974889
16	z	−0.0096	−0.0096	1.765012	0.6437031
17	Z	0	−0.0009795	4.370784	4.370784

Differential equations

1 $d(Y)/d(t) = (\exp(t/(1.0+0.1*t)))*m1*(y\hat{}n-1*k*0.4*z\hat{}n)*(\exp(0.01*m))$
2 $d(Z)/d(t) = (\exp(t/(1.0+0.1*t)))*m1*(z\hat{}n-1*(0.2/k)*y\hat{}n1)*(\exp(1*m))$
3 $d(X)/d(t) = (\exp(t/(1.0+0.1*t)))*m1*(x\hat{}n)*(\exp(1*m))$

Explicit equations (Fig. 6.25)

1 $E2 = (\exp(-0.075*t))$
2 $k = (\exp(-0.075*t))$
3 $m = 0.00245+0.0375*t-0.00934*t\hat{}2+0.00104*t\hat{}3-0.000041*t\hat{}4$
4 $A = (\exp(-0.01*m))*Y$
5 $B = (\exp(-1*m))*X$
6 $y = t*(\exp(-0.15*t))-A+k*(-0.0891+1.295*t$
$-0.234*t\hat{}2-0.0142*t\hat{}3+0.00521*t\hat{}4-0.00027*t\hat{}5)*0.4*1$
7 $C = (\exp(-1*m))*Z$
8 $r = t*(\exp(-0.15*t))$
9 $m1 = 0.0375-0.01868*t+0.00312*t\hat{}2-0.000164*t\hat{}2$
10 $z = t*(\exp(-0.075*t))-C+k*(-0.048+1.153*t$
$-0.226*t\hat{}2-0.00678*t\hat{}3+0.00387*t\hat{}4-0.00021*t\hat{}5)*0.2*1$
11 $x = 0.333*(0.25-k)*(z-y/k)-B$
12 $n = 4$
13 $n1 = 1$

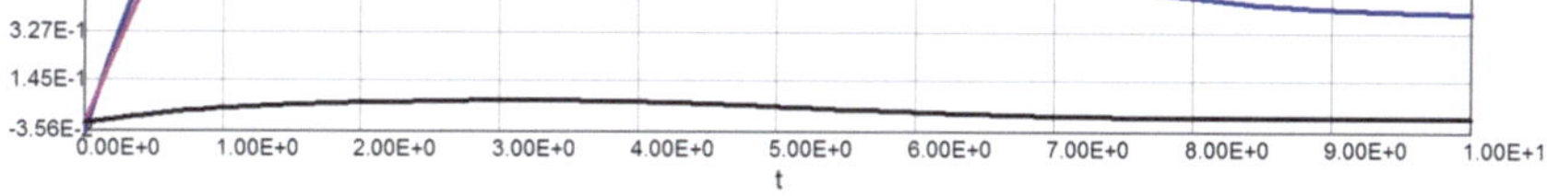

Fig. 6.25 Nonlinear creep deformations (second approximation $n = 4$)

Model: $y = a0 + a1 * t + a2 * t\text{^}2 + a3 * t\text{^}3 + a4 * t\text{^}4 + a5 * t\text{^}5$

Variable	Value
$a0$	−0.108113
$a1$	1.851691
$a2$	−0.5866825
$a3$	0.0642074
$a4$	−0.0022651
$a5$	−9.303E−06

$$\begin{aligned} y &= \sigma_x \\ &= -0.108 + 1.852 * t - 0.587 * t\hat{\ }2 + 0.0642 * t\hat{\ }3 - 0.00226 * t\hat{\ }4 - 0.000009 * t\hat{\ }5 \end{aligned} \tag{6.70}$$

Model: $z = a0 + a1 * t + a2 * t\text{^}2 + a3 * t\text{^}3 + a4 * t\text{^}4 + a5 * t\text{^}5$

Variable	Value
$a0$	−0.1191339
$a1$	1.663179
$a2$	−0.4578864
$a3$	0.0352049
$a4$	0.0006181
$a5$	−0.0001152

$$\begin{aligned} z &= \sigma_y \\ &= -0.119 + 1.663 * t - 0.458 * t\hat{\ }2 + 0.352 * t\hat{\ }3 + 0.00062 * t\hat{\ }4 - 0.000115 * t\hat{\ }5 \end{aligned} \tag{6.71}$$

Model: $z = a0 + a1 * t + a2 * t\text{^}2 + a3 * t\text{^}3 + a4 * t\text{^}4 + a5 * t\text{^}5$

Variable	Value
$a0$	−0.0608776
$a1$	1.35861
$a2$	−0.1548595
$a3$	−0.0500179
$a4$	0.0098634
$a5$	−0.0004626

Model: $z = a0 + a1 * t + a2 * t^{\wedge}2 + a3 * t^{\wedge}3 + a4 * t^{\wedge}4 + a5 * t^{\wedge}5$

Variable	Value
$a0$	−0.0608776
$a1$	1.35861
$a2$	−0.1548595
$a3$	−0.0500179
$a4$	0.0098634
$a5$	−0.0004626

Example 6.11 Data: $E_2 = \exp(-0.075\theta)$; $\alpha_1 = 0.01$; $\alpha_2 = 1.0$; $\alpha_3 = 1.0$; $k_0 = 0.15/0.075 = 2$; $n = 5$; $n1 = 1$ and $k = \exp(-0.075\theta)$.

First approximation

Calculated values of DEQ variables

	Variable	Initial value	Minimal value	Maximal value	Final value
1	A	0	0	2.064061	2.064061
2	B	0	0	2.718E−07	2.715E−07
3	C	0	0	4.338613	4.338613
4	$E2$	1	0.4723666	1	0.4723666
5	k	1	0.4723666	1	0.4723666
6	m	0.00245	0.00245	0.0736102	0.07345
7	$m1$	0.0375	0.0079901	0.1463	0.1463
8	n	5	5	5	5
9	$n1$	1	1	1	1
10	r	0	0	2.452527	2.231302
11	t	0	0	10	10
12	X	0	0	2.922E−07	2.922E−07
13	x	0	−0.002296	0.0484007	−0.002296
14	Y	0	0	2.065577	2.065577
15	y	0	0	1.414994	0.1672409
16	z	0	0	1.565348	0.3850523
17	Z	0	0	4.669279	4.669279

Differential equations

1 $\mathrm{d}(Y)/\mathrm{d}(t) = (\exp(t/(1.0 + 0.1 * t))) * m1 * (y^{\wedge}n - 0 * k * 0.4 * z^{\wedge}n) * (\exp(0.01 * m))$

2 $\mathrm{d}(Z)/\mathrm{d}(t) = (\exp(t/(1.0 + 0.1 * t))) * m1 * (z^{\wedge}n - 0 * (0.2/k) * y^{\wedge}n1) * (\exp(1 * m))$

3 $\mathrm{d}(X)/\mathrm{d}(t) = (\exp(t/(1.0 + 0.1 * t))) * m1 * (x^{\wedge}n) * (\exp(1 * m))$

Explicit equations (Fig. 6.26)

1 $E2 = (\exp(-0.075 * t))$

2 $k = (\exp(-0.075 * t))$

3 $m = 0.00245 + 0.0375 * t - 0.00934 * t^\wedge 2 + 0.00104 * t^\wedge 3 - 0.000041 * t^\wedge 4$

4 $A = (\exp(-0.01 * m)) * Y$

5 $B = (\exp(-1 * m)) * X$

6 $y = t * (\exp(-0.15 * t)) - A + k * (-0.0891 + 1.295 * t - 0.234 * t^\wedge 2 - 0.0142 * t^\wedge 3 + 0.00521 * t^\wedge 4 - 0.00027 * t^\wedge 5) * 0.4 * 0$

7 $C = (\exp(-1 * m)) * Z$

8 $r = t * (\exp(-0.15 * t))$

9 $m1 = 0.0375 - 0.01868 * t + 0.00312 * t^\wedge 2 - 0.000164 * t^\wedge 2$

10 $z = t * (\exp(-0.075 * t)) - C + k * (-0.048 + 1.153 * t - 0.226 * t^\wedge 2 - 0.00678 * t^\wedge 3 + 0.00387 * t^\wedge 4 - 0.00021 * t^\wedge 5) * 0.2 * 0$

11 $x = 0.333 * (0.25 - k) * (z - y/k) - B$

12 $n = 5$

13 $n1 = 1$

Model: $y = a0 + a1 * t + a2 * t^2 + a3 * t^3 + a4 * t^4 + a5 * t^5$

Variable	Value
$a0$	−0.0746353
$a1$	1.281349
$a2$	−0.3454275
$a3$	0.0268285
$a4$	0.0002286
$a5$	−7.176E−05

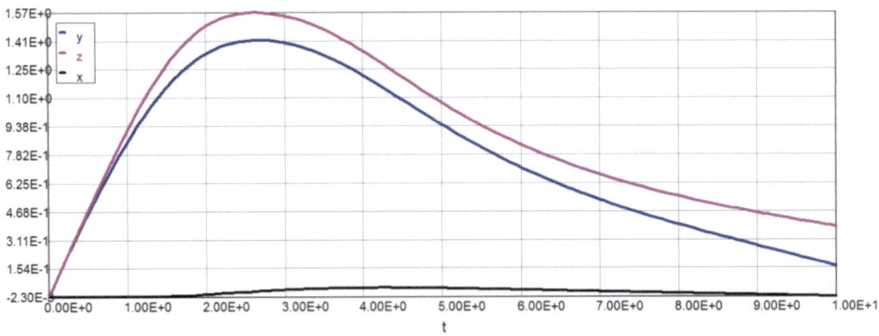

Fig. 6.26 Nonlinear creep deformations (first approximation $n = 5$)

$$y = \sigma_x = -0.0746 + 1.281 * t - 0.345 * t^\wedge 2 + 0.0268 * t^\wedge 3 + 0.000233 * t^\wedge 4 - 0.000072 * t^\wedge 5 \quad (6.72)$$

Model: $z = a0 + a1 * t + a2 * t\wedge 2 + a3 * t\wedge 3 + a4 * t\wedge 4 + a5 * t\wedge 5$

Variable	Value
$a0$	−0.1156754
$a1$	1.450165
$a2$	−0.396922
$a3$	0.0328665
$a4$	3.075E−08
$a5$	−7.215E−05

$$z = \sigma_y = -0.116 + 1.45 * t - 0.397 * t^\wedge 2 + 0.0328 * t^\wedge 3 + 0.00 * t^\wedge 4 - 0.00007 * t^\wedge 5 \quad (6.73)$$

Second Approximation

Calculated values of DEQ variables

	Variable	Initial value	Minimal value	Maximal value	Final value
1	A	0	0	2.000109	1.811564
2	B	0	0	1.036E−06	1.036E−06
3	C	0	−0.001421	4.000421	4.000421
4	$E2$	1	0.4723666	1	0.4723666
5	k	1	0.4723666	1	0.4723666
6	m	0.00245	0.00245	0.0736117	0.07345
7	$m1$	0.0375	0.0079887	0.1463	0.1463
8	n	5	5	5	5
9	$n1$	1	1	1	1
10	r	0	0	2.452528	2.231302
11	t	0	0	10	10
12	X	0	0	1.115E−06	1.115E−06
13	x	−0.0078621	−0.0078621	0.0678655	0.0254208
14	Y	0	0	2.00146	1.812895
15	y	−0.0464	−0.0464	1.633773	0.5111882
16	z	−0.01492	−0.01492	1.612898	0.7388703
17	Z	0	−0.0014529	4.305312	4.305312

Differential equations

1 $d(Y)/d(t) = (\exp(t/(1.0 + 0.1 * t))) * m1 * (y^\wedge n - 1 * k * 0.4 * z^\wedge n) * (\exp(0.01 * m))$
2 $d(Z)/d(t) = (\exp(t/(1.0 + 0.1 * t))) * m1 * (z^\wedge n - 1 * (0.2/k) * y^\wedge n1) * (\exp(1 * m))$
3 $d(X)/d(t) = (\exp(t/(1.0 + 0.1 * t))) * m1 * (x^\wedge n) * (\exp(1 * m))$

Explicit equations (Fig. 6.27)

1 $E2 = (\exp(-0.075 * t))$
2 $k = (\exp(-0.075 * t))$
3 $m = 0.00245 + 0.0375 * t - 0.00934 * t^\wedge 2 + 0.00104 * t^\wedge 3 - 0.000041 * t^\wedge 4$
4 $A = (\exp(-0.01 * m)) * Y$
5 $B = (\exp(-1 * m)) * X$
6 $y = t * (\exp(-0.15 * t)) - A + k * (-0.116 + 1.45 * t - 0.397 * t^\wedge 2 + 0.0328 * t^\wedge 3 + 0.00 * t^\wedge 4 - 0.00007 * t^\wedge 5) * 0.4 * 1$
7 $C = (\exp(-1 * m)) * Z$
8 $r = t * (\exp(-0.15 * t))$
9 $m1 = 0.0375 - 0.01868 * t + 0.00312 * t^\wedge 2 - 0.000164 * t^\wedge 2$
10 $z = t * (\exp(-0.075 * t)) - C + k * (-0.0746 + 1.281 * t - 0.345 * t^\wedge 2 + 0.0268 * t^\wedge 3 + 0.000233 * t^\wedge 4 - 0.000072 * t^\wedge 5) * 0.2 * 1$
11 $x = 0.333 * (0.25 - k) * (z - y/k) - B$
12 $n = 5$
13 $n1 = 1$

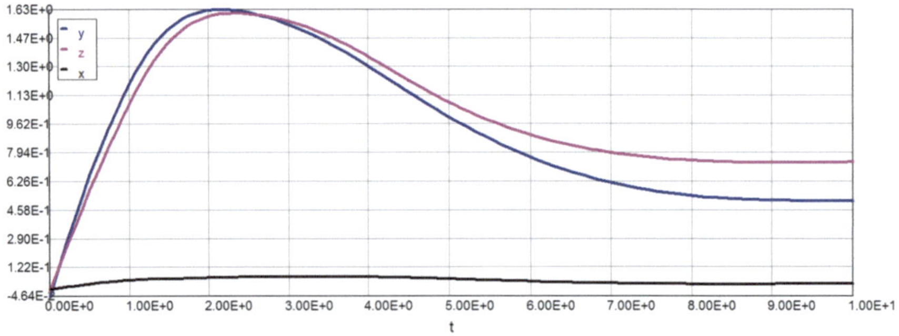

Fig. 6.27 Nonlinear creep deformations (second approximation $n = 5$)

Model: $y = a0 + a1 * t + a2 * t\text{^}2 + a3 * t\text{^}3 + a4 * t\text{^}4 + a5 * t\text{^}5$

Variable	Value
$a0$	−0.1028713
$a1$	1.922669
$a2$	−0.7167977
$a3$	0.1069766
$a4$	−0.0072846
$a5$	0.0001894

$$y = \sigma_x = -0.103 + 1.923 * t - 0.717 * t\text{^}2 + 0.107 * t\text{^}3 - 0.00729 * t\text{^}4 + 0.00019 * t\text{^}5 \quad (6.74)$$

Model: $z = a0 + a1 * t + a2 * t\text{^}2 + a3 * t\text{^}3 + a4 * t\text{^}4 + a5 * t\text{^}5$

Variable	Value
$a0$	−0.1225006
$a1$	1.765166
$a2$	−0.6040536
$a3$	0.0808737
$a4$	−0.00462
$a5$	8.92E−05

$$z = \sigma_y = -0.122 + 1.765 * t - 0.604 * t\text{^}2 + 0.0808 * t\text{^}3 - 0.00462 * t\text{^}4 - 0.00009 * t\text{^}5 \quad (6.75)$$

Example 6.12 Data: $E_2 = \exp(-0.075\theta)$; $\alpha_1 = 0.01$; $\alpha_2 = 1.0$; $\alpha_3 = 1.0$; $k_0 = 0.15/0.075 = 2$; $n = 8$; $n1 = 1$ and $k = \exp(-0.075\theta)$.

First approximation

Calculated values of DEQ variables

	Variable	Initial value	Minimal value	Maximal value	Final value
1	A	0	0	1.907976	1.907976
2	B	0	0	5.839E−11	5.839E−11
3	C	0	0	4.176237	4.176237
4	$E2$	1	0.4723666	1	0.4723666
5	k	1	0.4723666	1	0.4723666

(continued)

(continued)

	Variable	Initial value	Minimal value	Maximal value	Final value
6	m	0.00245	0.00245	0.0736126	0.07345
7	$m1$	0.0375	0.0079888	0.1463	0.1463
8	n	8	8	8	8
9	$n1$	1	1	1	1
10	r	0	0	2.452429	2.231302
11	t	0	0	10	10
12	X	0	0	6.284E−11	6.284E−11
13	x	0	0	0.0500736	0.0101485
14	Y	0	0	1.909378	1.909378
15	y	0	0	1.282263	0.3233258
16	z	0	0	1.35298	0.547428
17	Z	0	0	4.494528	4.494528

Differential equations

1 $\mathrm{d}(Y)/\mathrm{d}(t) = (\exp(t/(1.0 + 0.1 * t))) * m1 * (y\hat{}n - 0 * k * 0.4 * z\hat{}n) * (\exp(0.01 * m))$

2 $\mathrm{d}(Z)/\mathrm{d}(t) = (\exp(t/(1.0 + 0.1 * t))) * m1 * (z\hat{}n - 0 * (0.2/k) * y\hat{}n1) * (\exp(1 * m))$

3 $\mathrm{d}(X)/\mathrm{d}(t) = (\exp(t/(1.0 + 0.1 * t))) * m1 * (x\hat{}n) * (\exp(1 * m))$

Explicit equations (Fig. 6.28)

1 $E2 = (\exp(-0.075 * t))$

2 $k = (\exp(-0.075 * t))$

3 $m = 0.00245 + 0.0375 * t - 0.00934 * t\hat{}2 + 0.00104 * t\hat{}3 - 0.000041 * t\hat{}4$

4 $A = (\exp(-0.01 * m)) * Y$

5 $B = (\exp(-1 * m)) * X$

6 $y = t * (\exp(-0.15 * t)) - A + k * (-0.116 + 1.45 * t$
$- 0.397 * t\hat{}2 + 0.0328 * t\hat{}3 + 0.00 * t\hat{}4 - 0.00007 * t\hat{}5) * 0.4 * 0$

7 $C = (\exp(-1 * m)) * Z$

8 $r = t * (\exp(-0.15 * t))$

9 $m1 = 0.0375 - 0.01868 * t + 0.00312 * t\hat{}2 - 0.000164 * t\hat{}2$

10 $z = t * (\exp(-0.075 * t)) - C + k * (-0.0746 + 1.281 * t$
$- 0.345 * t\hat{}2 + 0.0268 * t\hat{}3 + 0.000233 * t\hat{}4 - 0.000072 * t\hat{}5) * 0.2 * 0$

11 $x = 0.333 * (0.25 - k) * (z - y/k) - B$

12 $n = 8$

13 $n1 = 1$

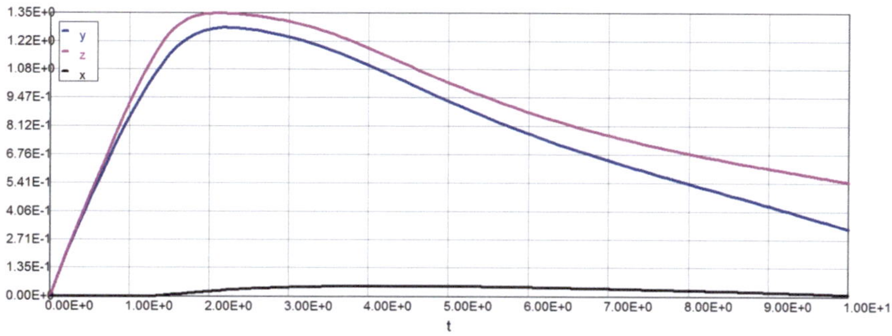

Fig. 6.28 Nonlinear creep deformations (first approximation $n = 8$)

Model: $y = a0 + a1 * t + a2 * t\text{^}2 + a3 * t\text{^}3 + a4 * t\text{^}4 + a5 * t\text{^}5$

Variable	Value
$a0$	−0.0894813
$a1$	1.415257
$a2$	−0.5068395
$a3$	0.0761069
$a4$	−0.0053556
$a5$	0.0001439

$$\begin{aligned} y &= \sigma_x \\ &= -0.0894 + 1.415 * t - 0.507 * t\text{^}2 + 0.0761 * t\text{^}3 - 0.00535 * t\text{^}4 + 0.000144 * t\text{^}5 \end{aligned} \tag{6.76}$$

Model: $z = a0 + a1 * t + a2 * t\text{^}2 + a3 * t\text{^}3 + a4 * t\text{^}4 + a5 * t\text{^}5$

Variable	Value
$a0$	−0.1101632
$a1$	1.552557
$a2$	−0.5724291
$a3$	0.0898942
$a4$	−0.0066458
$a5$	0.0001894

$$\begin{aligned} z &= \sigma_y \\ &= -0.11 + 1.552 * t - 0.572 * t\text{^}2 + 0.0898 * t\text{^}3 - 0.00665 * t\text{^}4 + 0.000189 * t\text{^}5 \end{aligned} \tag{6.77}$$

Second Approximation

Calculated values of DEQ variables

	Variable	Initial value	Minimal value	Maximal value	Final value
1	A	0	0	1.889217	1.622352
2	B	0	0	4.424E−10	4.424E−10
3	C	0	−0.0024213	3.899259	3.899259
4	$E2$	1	0.4723666	1	0.4723666
5	k	1	0.4723666	1	0.4723666
6	m	0.00245	0.00245	0.073611	0.07345
7	$m1$	0.0375	0.0079888	0.1463	0.1463
8	n	8	8	8	8
9	$n1$	1	1	1	1
10	r	0	0	2.452526	2.231302
11	t	0	0	10	10
12	X	0	0	4.761E−10	4.761E−10
13	x	−0.0065235	−0.0065235	0.0586627	0.0440343
14	Y	0	0	1.890465	1.623544
15	y	−0.044	−0.044	1.402055	0.6864175
16	z	−0.01788	−0.01788	1.376753	0.858474
17	Z	0	−0.0024829	4.196439	4.196439

Differential equations

1 $\mathrm{d}(Y)/\mathrm{d}(t) = (\exp(t/(1.0+0.1*t)))*m1*(y^\wedge n-1*k*0.4*z^\wedge n)*(\exp(0.01*m))$

2 $\mathrm{d}(Z)/\mathrm{d}(t) = (\exp(t/(1.0+0.1*t)))*m1*(z^\wedge n-1*(0.2/k)*y^\wedge n1)*(\exp(1*m))$

3 $\mathrm{d}(X)/\mathrm{d}(t) = (\exp(t/(1.0+0.1*t)))*m1*(x^\wedge n)*(\exp(1*m))$

Explicit equations (Fig. 6.29)

1 $E2 = (\exp(-0.075*t))$

2 $k = (\exp(-0.075*t))$

3 $m = 0.00245+0.0375*t-0.00934*t^\wedge 2+0.00104*t^\wedge 3-0.000041*t^\wedge 4$

4 $A = (\exp(-0.01*m))*Y$

5 $B = (\exp(-1*m))*X$

6 $y = t*(\exp(-0.15*t))-A+k*(-0.11+1.552*t$
$-0.572*t^\wedge 2+0.0898*t^\wedge 3-0.00665*t^\wedge 4+0.000189*t^\wedge 5)*0.4*1$

7 $C = (\exp(-1*m))*Z$

8 $r = t*(\exp(-0.15*t))$

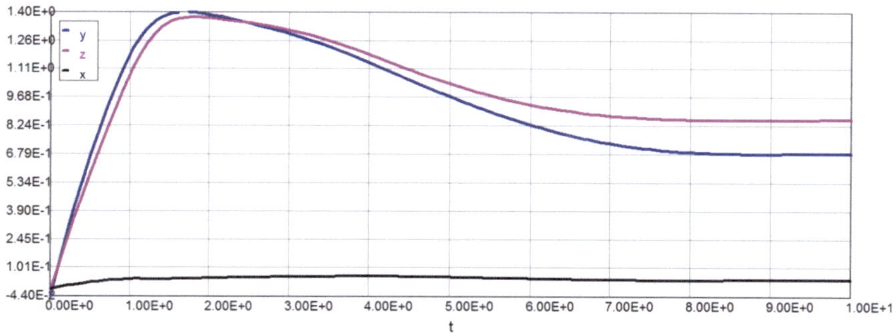

Fig. 6.29 Nonlinear creep deformations (second approximation n = 8)

9 $m1 = 0.0375 - 0.01868 * t + 0.00312 * t^\wedge 2 - 0.000164 * t^\wedge 2$

10 $z = t * (\exp(-0.075 * t)) - C + k * (-0.0894 + 1.415 * t - 0.507 * t^\wedge 2 + 0.0761 * t^\wedge 3 - 0.00535 * t^\wedge 4 + 0.000144 * t^\wedge 5) * 0.2 * 1$

11 $x = 0.333 * (0.25 - k) * (z - y/k) - B$

12 $n = 8$

13 $n1 = 1$

Model: $y = a0 + a1 * t + a2 * t$^2 + $a3 * t$^3 + $a4 * t$^4 + $a5 * t$^5

Variable	Value
$a0$	−0.0309019
$a1$	1.81085
$a2$	−0.7805769
$a3$	0.1416682
$a4$	−0.0119724
$a5$	0.0003877

$$y = \sigma_x = -0.0309 + 1.81 * t - 0.78 * t^\wedge 2 + 0.142 * t^\wedge 3 - 0.012 * t^\wedge 4 + 0.000388 * t^\wedge 5 \quad (6.78)$$

Model: $z = a0 + a1 * t + a2 * t$^2 + $a3 * t$^3 + $a4 * t$^4 + $a5 * t$^5

Variable	Value
$a0$	−0.0745918
$a1$	1.732078
$a2$	−0.7113424

(continued)

(continued)

Variable	Value
$a3$	0.1245241
$a4$	−0.0101277
$a5$	0.0003152

$$z = \sigma_y = -0.0746 + 1.732 * t - 0.711 * t^\wedge 2 + 0.125 * t^\wedge 3 + 0.0102 * t^\wedge 4 + 0.000315 * t^\wedge 5 \quad (6.79)$$

6.11 Conclusions

1. The deformation of anisotropic creeping solid is directionally dependent.
2. New constitutive models have been developed to deal with anisotropic behavior.
3. A transversely isotropic material is studied.
4. The new computational process accurately models stress–temperature–strain diagrams.

References

1. Abe F (2001) Creep rates and strengthening mechanisms in tungsten-strengthened 9Cr steels. Mat Sci Eng A319-A321:770–773
2. Altenbach H, Altenbach J, Kissing W (2004) Mechanics of composite structural elements. Springer, Berlin
3. Altenbach H (2001) A generalized limit criterion with application to strength, yielding, and damage of isotropic materials. In: Lemaitre J (ed) Handbook of materials behavior models. Academic Press, San Diego, pp 175–186
4. Becker AA, Hyde TH, Xia L (1994) Numerical analysis of creep in components. J Strain Anal 29(3):185–192
5. Bernhardi O, Mücke R (2000) A lifetime prediction procedure for anisotropic materials. Commun Numer Methods Eng 16:519–527
6. Perzyna P (1966) Fundamental problems in viscoplasticity. Adv Appl Mech 9(2):244–368
7. Robotnov Y (1969) Creep problems in structural members. North Holland Publishing Company, Amsterdam
8. Razdolsky L (2015) AIAA, Space 2015
9. Razdolsky L (2014) ASCE 2014
10. Bertram A, Olschewski J (2001) A phenomenological anisotropic creep model for cubic single crystals. In: Lemaitre J (ed) Handbook of material models. Academic Press, San Diego, pp 303–307
11. Arsenault RJ (1991) Strengthening and deformation mechanisms of discontinuous metal matrix composites. Strength of metals and alloys, 1. Freung Publishing House Ltd, London, pp 31–46

12. Bodner S, Partom Y (1975) Constitutive equations for elastic-viscoelastic strain-hardening materials. ASME J Appl Mech 42:385
13. Kaspar JW (2002) Constitutive models for engineering materials. In: Encyclopedia of physical science and technology, vol 3. bechtel.colorado.edu, pp 603–633
14. Razdolsky L. #1
15. Razdolsky L. #2
16. Bischoff-Beiermann B, Bruhns O (1994) A physically motivated set of invariants and tensor generators in the case of transverse isotropy. Int J Eng Sci 32:1531–1552
17. Weber G, Anand L (1990) Finite deformation constitutive equations and a time integrated procedure for isotropic hyperelastic and viscoelastic solids. Comput Methods Appl Mech Eng 79:173–202
18. Rice JR (1970) On the structure of stress-strain relations for time-dependent deformation of metals. J Appl Mech 37:728
19. Naumenko KD (2007) Modeling of creep for structural analysis. Springer, Berlin
20. Lemaître J (1990) Mechanics of solid materials. Cambridge University Press, Cambridge
21. Miller AA (1976) An inelastic constitutive model for monotonic, cyclic, and creep deformation: equations development and analytical procedures. ASME J Eng Mater Technol 98(2):106–112
22. Riedel H (1987) Fracture at high temperatures, materials research and engineering. Springer, Berlin
23. Boyle JT, Spence J (1983) Stress analysis for creep. Butterworth, London
24. Gibson RF (1994) Principles of composite materials. McGraw-Hill, New York
25. Altenbach H, Skrzypek J (1999) Creep and damage in materials and structures. In: CISM courses and lectures no. 399. Springer, Wien-New York
26. Krajcinovic D (1996) Damage mechanics, applied mathematics and mechanics, vol 41. North-Holland, Amsterdam
27. Lemaitre J (1996) A course on damage mechanics. Springer, Berlin

Chapter 7
Probabilistic Modeling of Creep and Stress–Strain Diagram

Notation

k	The thermal conductivity that has the dimensions W/m K or J/m s K
T	Temperature
d	Thickness in the direction of heat flow
c	Specific heat capacity
Q	Activation energy
R	Ideal gas constant
P_l	Losses of heat due to thermal radiation
e	Emissivity factor
σ	Stefan–Boltzmann constant (σ = 5.6703 (10^{-8}) W/m^2 K^4);
T_o	Ambient temperature
A_v	Area of openings
c_p	Average specific heat at constant pressure
t	Time
$\vec{v}(u; v; w)$	Velocity vector
D	Diffusion coefficient (m^2/s)
p	Pressure
v	Kinematic viscosity
θ	Dimensionless temperature
τ	Dimensionless time
h	Height of the compartment (m)
a	Thermal diffusivity (m^2/s)
Time	$t = \frac{h^2}{a}\tau(\mathrm{s})$
Temperature	$T = \frac{RT_*^2}{E}\theta + T_*$ (K), where T_* = 600 K is the base line temperature
Coordinates	$\bar{x} = x/h$ and $\bar{z} = z/h$—"x" and "z"—dimensionless coordinates

L. Razdolsky, *Probability Based High Temperature Engineering*,
DOI 10.1007/978-3-319-41909-1_7

Velocities	$\overline{u} = \frac{v}{h}u$ (m/s) and $\overline{w} = \frac{v}{h}w$ (m/s)—horizontal and vertical components' velocity accordingly; v—kinematic viscosity (m^2/s); "u" and "w"—Dimensionless velocities.
$Pr = v/a$	Prandtl number
$Fr = \frac{gh^3}{va}$	Froude number
g	Gravitational acceleration
$Le = a/D = Sc/Pr$	The Lewis number
$Sc = v/D$	The Schmidt number
$\overline{\beta} = \frac{RT_*}{E}$	Dimensionless parameter
$\overline{\gamma} = \frac{c_p RT_*^2}{QE}$	Dimensionless parameter
$P = \frac{e\sigma K_v(\beta T_*)^3 h}{\lambda}$	Thermal radiation dimensionless coefficient
$\sigma = 5.67\ (10^{-8})\ (\mathrm{W/m^2\ K^4})$	Stefan–Boltzman constant
$K_v = A_o\ h/V$	Dimensionless opening factor
A_o	Total area of vertical and horizontal openings
$\delta = \left(\frac{E}{RT_*^2}\right) Qz\left(\exp\left(-\frac{E}{RT_*}\right)\right)$	Frank-Kamenetskii's parameter
$C = [1 - P(t)/P_o]$	Concentration of the burned fuel product in the fire compartments
$\overline{W} = \frac{v}{h}W$	Vertical component of gas velocity
$\overline{U} = \frac{v}{h}U$	Horizontal component of gas velocity
$b = L/h$, L and h	Length (width) and height of fire compartment accordingly
W; U	Dimensionless velocities
R_n	Nominal strength
S_i	Nominal load
φ	Resistance factor
γ	Load factor
R_c	Characteristic value for the resistance
A	Cross-sectional area
I	Moment of inertia
W	Total weight
G_c	Characteristic value for the permanent load
S	Characteristic value for the variable load
ψ_1	Partial safety factor for the permanent load
ψ_2	Partial safety factor for the variable load
β	Reliability index
S	Probability space

A	Set of outcomes (events) to which a probability is assigned
$P(E_2\|E_1)$	Conditional probability
$\Phi^*(.)$	Denotes the cumulative distribution function of standard normal distribution $\Phi^*(z) = \frac{1}{\sqrt{2\pi}}\int_{-\infty}^{z} \mathrm{e}^{-\frac{z^2}{2}}\mathrm{d}z$
μ_A, μ_B, σ_A, σ_B	Are mean and standard deviation of A and B, respectively
$J(t, t')$	Compliance function
T_{M}	Melting point of the metal matrix material
$\varepsilon(t)$	Strain
$\sigma(t)$	Stress
$\overline{\sigma}(t) = E(t)\varepsilon(t)$	Instantaneous stress
ε_{e}	Instantaneous (elastic) strain
ε_{c}	Creep strain
ε_{T}	Thermal expansion due to high temperature effect
$K(t,t') = \partial J(t,t')/\partial t'$	Retardation function (memory function)
$R(t, t')$	Relaxation function (also called the relaxation modulus)
$M(\theta)$	Bending moment
$V(\theta)$	Shear force
$P(\theta)$	Axial force
$y(\theta)$	Deformation
P_{f}	Probability of failure
P_{rel}	Reliability
σ	Ψ
ε	Ψ
θ	Dimensionless temperature
κ	Curvature
τ	Dimensionless time
ω	Frequency
$S(\omega)$	Spectral density
μ	Poisson coefficient
D	Diffusion rate (Flick's law)
η	*Viscosity* parameter of the material
E	Modulus of elasticity
$n = \eta/E$	Relaxation time
n	Power law exponent
α_i	Material property parameter
L	Span (spring spacing)
k_0	Subgrade modulus

7.1 Introduction

It has been generally recognized that the behavior of real-life structures is determined by the aggregate effect of numerous factors which are of intrinsically stochastic nature. Therefore, dealing with issues of reliability and service life of structure, the use of probabilistic (stochastic) methods can be justified.

It appears inevitable that the structural engineering community, as well as fire protection and many other engineering communities that are ultimately responsible for life safety issues, will eventually incorporate probabilistic analysis methods to some degree. The degree to which these methods are successfully applied depends on addressing the issues and concerns discussed in this book. Certainly, one issue is to disseminate familiarity and basic understanding of the principles and assumptions made in probability-based structural design.

The field of probability-based structural fire resistance lies at the crossroads of stress analysis, fire protection and structural engineering, probability theory, thermodynamics, heat conduction theory, and advanced methods of applied mathematics. Each of these areas is covered to the extent necessary. This chapter starts from basic concepts and principles, and these are developed to more advanced levels as the text progresses. Nevertheless, some basic preparation in structural analysis/design and mathematics is expected of the reader. While selecting material for the book, the author made every effort to present both classical topics and methods and modern, or more recent, developments in the field.

7.1.1 Deterministic Approach to Structural Fire Resistance Engineering

The structural engineer ultimately is responsible to check the building structure subjected to the structural thermal load and to quantify the response of the originally proposed structural system in realistic fire scenarios in order to determine whether this response is acceptable. Strengths and weaknesses then can be clearly identified and addressed within the structural design, as appropriate. Behavior of the structural system under high thermal loading condition should be considered an integral part of the structural design process. The role of a structural engineer today involves a significant understanding of both static and thermal loading and the structures that are available to resist them. The complexity of modern structures often requires a great deal of creativity from the engineer in order to ensure that the structures support and resist the loads to which they are subjected.

At the root of the structural fire safety problem is the uncertain nature of the man-made and environmental forces that act on structures, of material properties that are changing quite rapidly under high temperature conditions, and of structural analysis procedures that, even in this computer age, are no more than models of reality. There are two issues to be of major concern and critical importance for any structural design.

1. It must be stressed that it is not possible to design for every conceivable or inconceivable event. Therefore, a threat and risk assessment can be used to quantify real risks, in order to develop suitable mitigation measures, on a project by project basis.
2. Understanding the role of structural system and its real response to fire along with the performance of fire proofing materials in real events is also key even more so as probability-based approach to structural fire resistance enhances our understanding of real structural performance. Although performance-based fire-resistant design has been available for many years, it tends to be used only by a few very knowledgeable designers. A lack of good reference books and knowledgeable design professionals has inhibited the broader use of performance-based design for fire resistance.

7.1.2 Ultimate Limit State

To satisfy the ultimate limit state, the structure must not collapse when subjected to the peak design load for which it was designed. A structure is deemed to satisfy the ultimate limit state criteria if all factored bending, shear and tensile or compressive stresses are below the factored resistances calculated for the section under consideration. The factored stresses referred to are found by applying Magnification Factors to the loads on the section. Reduction Factors are applied to determine the various factored resistances of the section.

The limit state criteria can also be set in terms of load rather than stress: using this approach, the structural element being analyzed (e.g., a beam or a column or other load bearing element, such as walls) is shown to be safe when the "Magnified" loads are less than the relevant "Reduced" resistances.

Limit state design (LSD) refers to a design method used in structural engineering. A *limit state* is a condition of a structure beyond which it no longer fulfills the relevant design criteria. The condition may refer to a degree of loading or other actions on the structure, while the criteria refer to structural integrity, fitness for use, durability, or other design requirements. A structure designed by LSD is proportioned to sustain all actions likely to occur during its design life, and to remain fit for use, with an appropriate level of reliability for each limit state. Structural codes based on LSD implicitly define the appropriate levels of reliability by their prescriptions.

7.1.3 Methods Used

Essentially, probabilistic design focuses upon the prediction of the effects of random variability. Some methods that are used to predict the random variability of an output include

- the Monte Carlo method (including Latin hypercube);
- propagation of error;
- design of experiments (DOE)
- the method of moments
- Statistical interference

Function $g(R, S)$ is commonly referred to as the performance function. The probability of failure is defined as the volume under the surface shown in the failure region where $g < 0$; $g = 0$. Volume undersurface where $g < 0$ is probability of failure. The probability of failure is defined as $P_f = P[g(R, S) \leq 0]$. Apart from applied stress and material strength, there is numerous other R, S variants, for which probabilistic analyses can be performed. The design process can be summarized here in an integrated design approach: establishing the real risks, analyzing their impact on the structure performance using the tools available to us, and developing designs to accommodate this.

7.2 Introduction to Probability Theory

7.2.1 Random Variables: Definition of a Probability

The mathematical theory of probability obtained a unified formulation in the 1930s, when Kolmogorov [1–3] introduced his axioms and defined the universal structure of a probability space. In the 1970s and 1980s, people such as Accardi [4] and George W. Mackey [5], building on ideas of von Neumann's and Segal's concerning algebras of operators, developed a unified framework, a generalized probability theory, in which classical probability theory and structural fire mechanics can be discussed together. We shall use their language in this book.

Engineering systems are designed to operate well in the face of uncertainty of characteristics of components and operating conditions. In some cases, uncertainty is introduced in the operations of the system, on purpose.

Understanding how to model uncertainty and how to analyze its effects is (or should be) an essential part of an engineer's education. Randomness is a key element of all systems we design. Building must resist the unpredictable vibrations of an earthquake or temperature load from the design fire event. What should you understand about probability? It is a complex subject that has been constructed over decades by pure and applied mathematicians. Thousands of books explore various aspects of the theory. How much do you really need to know and where do you start? The first key concept is how to model uncertainty. What do we mean by a "random experiment?" Once you understand that concept, the notion of a random variable should become transparent.

The statistician is basically concerned with drawing conclusions (or inferences) from experiments involving uncertainties. For these conclusions and inferences to be reasonably accurate, an understanding of probability theory is essential. First, we shall develop the concept of possible outcomes, which can occur or not occur. Some of these outcomes are more likely to occur than others. For example, the building with a lot of storage areas of flammable materials has a greater possibility of catching fire then an empty prefabricated steel storage building. In order to quantify the likelihood of outcome occurrence the real number can be assigned to each outcome.

This number is greater if the likelihood of an outcome is greater. This assigned number is called the *probability of an experimental outcome*. *Experiment*: This is any process of observation or procedure that (1) can be repeated (theoretically) an infinite number of times and (2) has a well-defined set of possible outcomes.

It is important to underline here that the practical meaning of probability is the ratio (frequency) of "successful" outcomes of an experiment to the total number of outcomes. It is obvious now to define the unit of measurement of probability:

The probability of a certain outcome is one (1) and the probability of an impossible outcome is zero (0). *Sample space*: This is the set of all possible outcomes of an experiment with the assigned probabilities. The probability space formalizes the notion that a physicist repeating an experiment again and again under the same general conditions, but unable to control some of the details, may expect to observe a range of different outcomes. *Probability space is complete* if and only if the sum of all probabilities is equal to 1. *Event*: This is a subset of the probability space of an experiment. When dealing with experiments that are random and well defined in purely theoretical settings, probabilities describe the statistical number of outcomes considered divided by the number of all outcomes.

The Classical Definition of Probability

This definition is for equally likely outcomes. If an experiment can produce N mutually exclusive and equally likely outcomes out of which n outcomes are favorable to the occurrence of event A, then the probability of A is denoted by $P(A)$ and is defined as the ratio n/N. Thus, the probability of A is given by $P(A) = n/N$.

This definition can be applied in a situation in which all possible outcomes and the outcomes in the events A can be counted. This definition is due to P.S. Laplace (1749–1827). The classical definition is also called the priori definition of probability.

The word "priori" is from "prior" and is used because the definition is based on the previous knowledge that the outcomes are equally likely. The probability of a certain event is a number that lies between 0 and 1. If the event does not contain any outcome, it is called an impossible event and its probability is 0. If the event is as big as the sample space, the probability of the event is 1 because in this case $P(A) = N/N = 1$. When the probability of an event is 1, it is called a "sure" or "certain" event. The classical definition of probability has always been criticized for the following reasons:

This definition assumes that the outcomes are equally likely. The term "equally likely" is almost as difficult as the word "probability" itself. Thus, the definition uses circular reasoning.

The definition is not applicable when the numbers of outcomes are not equally likely. The definition is also not applicable when the total number of outcomes is infinite or it is difficult to count the total outcomes or the outcomes favorable to the event. Like other theories, the theory of probability is a representation of probabilistic concepts in formal terms—that is, in terms that can be considered separately from their meaning. These formal terms are manipulated by the rules of mathematics and logic, and many results are interpreted or translated back into the problem domain. If one wants to explain this formulation in layman's terms, what a probability theory is all about, and how it is used in practice, one might say that this is a large number of mathematical laws and methods that allow minimizing the dependence of statistical data that is required in order to solve a particular technical problem. When it comes to practical application, however, many times it is impractical or economically not justifiable or unattainable (for example, real fire events in high-rise and super high-rise buildings) to define probabilities on the basis of the "classical method," and probability theory is very helpful in solving particular problems. At the end, of course, the results and conclusions must be verified by statistical data available, and engineering judgment is critical. In our case of obtaining the probability-based structural fire load, such a verification process represents a tieback tool to the prescriptive methods in fire protection design. Some basic principles of the probability theory are applied in a day-to-day structural and fire protection engineering practice. They are related to the concept of random phenomena, such as the law of the probability distribution of the temperature maximum, which can be a random variable or random process that crosses upward a predetermined (by an engineer) level of deterministic maximum temperature during a fire event, probabilities of structural failure, and outcomes of random experiments. In order to study random structural fire load applications, some terminology and definitions from probability theory are briefly outlined in this section [6, 7]. A summary of them follows: Outcome: The result of an experiment or occurrence of a natural phenomenon.

Random experiment: An experiment whose outcomes are not predictable in advance.
Set: A collection of individual elements in the domain D. The universal set U is defined if it contains every element in D. The null set φ is defined if it contains no element.
Event: A set of outcomes to which a probability is assigned.
Sample space: A set of all possible outcomes of an experiment, denoted by S. Every outcome of the experiment is represented by a point in S called a sample point.
Union: The union of two events A and B, which is denoted by $A \cup B$ or (A or B), is the set of all elements that belong to at least one of the sets A and B, shown in Fig. 7.1a.
Intersection: The intersection of two events A and B, which is denoted by $A \cap B$ or (A and B), is the set of elements that belong to both sets A and B, which is also referred to as a joint event of A and B, shown in Fig. 7.1b.
Complement: The complement of an event A, denoted by A, is the set containing all points in the sample space S, but not in the set A, shown in Fig. 7.1c.
Mutually exclusive events: Two events A and B are said to be mutually exclusive if they do not have common elements; that is, the intersection of A and B is a null set.

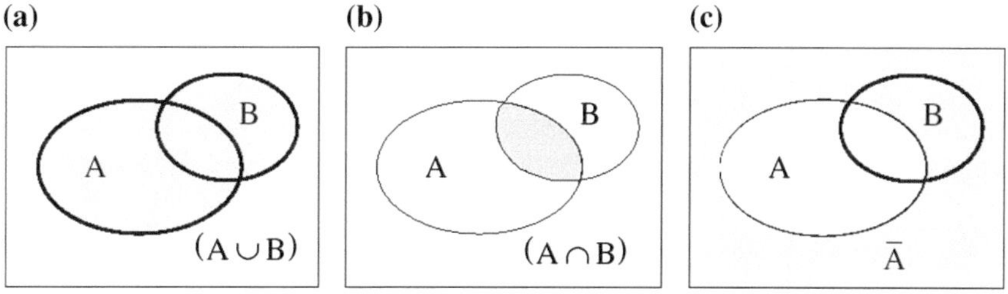

Fig. 7.1 Venn diagrams for two events: **a** union; **b** intersection; **c** complement of A

Collectively exhaustive: The events B_1, B_2, …, B_n are said to be collectively exhaustive if their union covers all the events within the entire sample space, that is, $P(B_1 \cup B_2 \cup \ldots \cup B_n) = S$ where S is the sample space. The union and intersection of the events A and B, and the complement of the event A, are shown in the Venn diagram in Fig. 7.1.
Counting: An efficient way of *counting* is necessary to handle large masses of statistical data (e.g., the level of inventory at the end of a given period or the number of production runs on a given machine in a 24 h period), and to gain an understanding of probability.

The probability of A, $P(A)$, is a number and satisfies the following three axioms [1]:

1. The probability of an event A is a number between 0 and 1, that is, $0 < P(A) < 1$.
2. For a certain event S, the probability $P(S)$ equals 1, that is, $P(A) = 1$.
3. The probability of the union of a number of mutually exclusive events, that is, intersections of null sets, is the sum of probabilities of the events, i.e.,

From these axioms it can be concluded that
Probability of null set: $P[(\varphi)] = 0$
Probability of complement: $P\left[\left(\overline{B}\right)\right] = 1 - P[(B)]$
Probability of union: $P[(A \cup B)] = P[(A)] + P[(B)] - P[(A \cap B)]$

A diagram for this situation is as follows. We see that there is some *overlap* between the events A and B. The probability of that overlap (see Fig. 7.2) portion is $P(A \cap B)$.

One other probability measure having practical importance is the conditional probability, which is denoted by $P(A|B)$. It is defined as the probability of the event A given that the event B has occurred. In probability terms, it is $P[(A|B)] = \frac{P(A \cap B)}{P(B)}$.

If the two events A and B are not related in any way they are said to be independent events. The only condition of the independence (if and only if) is as $P[(A \cap B)] = P[(A)]P[(B)]$.

Using the conditional probability definition, the total probability theorem can be derived. This is expressed as, if B_1, B_2, …, B_n are collectively exhaustive events of

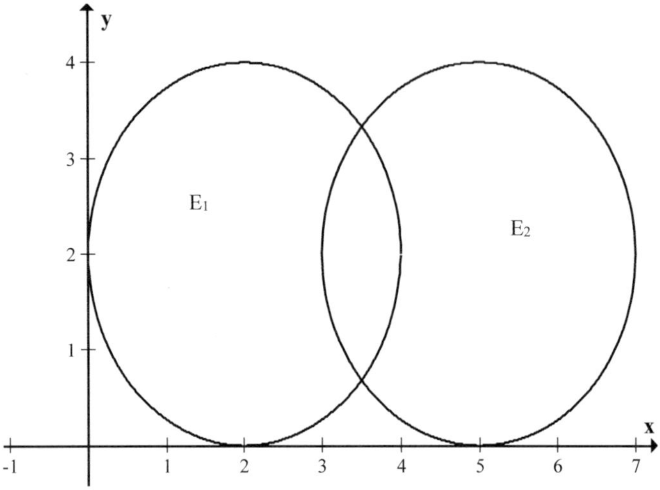

Fig. 7.2 Probability of union

the sample space S and A is an arbitrary event on S, then the total probability of A can be stated as the sum of all intersections, i.e.,

$$P(A) = \sum_{i=1}^{n} P(A \cap B_i = \sum_{i=1}^{n} P(A|B_i)P(B_i) \tag{7.1}$$

This is the statement of the total probability theorem. Using the conditional probability, the total probability theorem, and the commutative law of the events A and B, the well-known Bayes' theorem can be stated as

$$P(B_{j|}|A) = \frac{P(A|B_j)P(B_j)}{\sum_{i=1}^{n} P(A|B_i)P(B_i)} \tag{7.2}$$

Bayes' theorem helps in making decisions under uncertainties that engineers confront frequently in practice [8] provided that prior probabilistic models of uncertainties are available or determined previously by experiments.

Mutually Exclusive Events

Two (or more) events can be compatible or non-compatible (mutually exclusive events: $P[A \text{ and } B] = 0$). They are non-compatible if they cannot occur together.

For example, the structural element cannot fail and not fail at the same time; the fire cannot occur and not occur at the same time. Two (or more) events are equally possible if they are equally likely to occur. If a number of events (a) are from a *complete sample space*, (b) are non-compatible, and (c) are equally possible, then they are called random events or random variables.

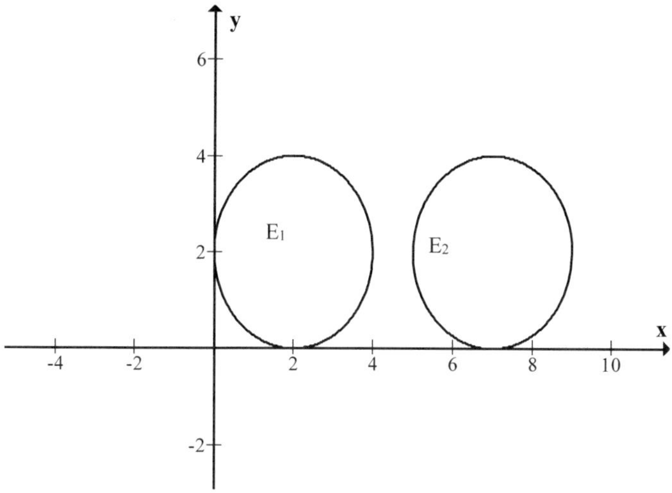

Fig. 7.3 Mutually exclusive events

If E_1 and E_2 are *mutually exclusive* events: $P(E_1 \text{ or } E_2) = P(E_1) + P(E_2)$. Our diagram for mutually exclusive events shows that there is no overlap (see Fig. 7.3).

7.3 Why Probabilistic Approach Is Needed?

Knowing the inherent risk of failure in any structural design is becoming increasingly important to both the manufacturer and the customer. Engineers and management must concern themselves with the ability to assess risk, identify parameters which drive risk, and minimize the risk given by other program constraints. Analysis of any structure using probabilistic methods provides a tool for meeting these needs. This section begins with an overview of the current structural analysis approach, expounding on its shortcomings and listing potential dilemmas in applying the approach to future structural systems designs. The general probabilistic analysis approach is then explained, with discussion on how it can address these problems.

Stochastic programming is a framework for modeling structural engineering problems that involve uncertainty. Whereas deterministic problems are formulated with known parameters, real-world problems almost invariably include some unknown parameters. When the parameters are known only within certain bounds, one approach to tackling such problems is called robust optimization. This is a field of optimization theory that deals with optimization problems in which a certain measure of robustness is sought against uncertainty that can be represented as deterministic variability in the value of the parameters of the problem itself and/ or its solution. Here, the goal is to find a solution that is feasible for all such data and optimal in some sense. Stochastic programming models are similar in style but take advantage of the fact that probability distributions governing the data are known or can be estimated. The goal here is to find some policy that is feasible for all (or almost all) the possible data instances and maximizes the expectation of some function of the decisions and the random variables. More generally, such models are formulated, solved analytically or numerically, and analyzed in order to provide useful information to a decision maker. The origins of robust optimization date back to the establishment of modern decision theory in the 1950s and use of worst case analysis and Wald's maximum model as a tool for the treatment of severe uncertainty. It became a discipline of its own in the 1970s with parallel developments in several scientific and technological fields. Over the years, it has been applied in statistics, but also in operations research, control theory [9], finance, manufacturing engineering, chemical engineering, and computer science. In engineering problems, these formulations often take the name of "robust design optimization" or "reliability based design optimization." Stochastic optimization methods generalize deterministic methods for deterministic problems.

The deterministic optimization dimensionless methods in the case of engineering creep are presented in [10], where the mathematical optimum design method has been used [11, 12].

In case of fire, let us consider a nonlinear time variant integral equation Volterra of second type representing the creep constitutive law (Eq. 7.3) with degenerate kernel (Eq. 7.4):

$$\overline{\sigma}(\theta) = E(\theta)\varepsilon(\theta) = \sigma(\theta) + \int_0^\theta L[\theta,\tau]\sigma^n(\tau)\mathrm{d}\tau$$
$$\varepsilon(\theta) = A\theta; \quad E(\theta) = E_0 \exp(-a\theta); \quad t = \varphi(\theta); \quad t' = \varphi(\tau) \tag{7.3}$$

a---modulus of elasticity deterioration parameter

$$L[\theta,\tau] = [\varphi(\theta)\varphi(\tau)]\varphi' \tag{7.4}$$

Functions $\varphi(\theta)$ and $\varphi'(\theta)$ in Eq. (4.26) are inverse functions of temperature-time functions that characterize fire severity scenario: Very Fast; Fast; Medium; or Slow fire [13]. Deterioration parameter "a" characterizes reduction of modulus of elasticity with temperature increase [13]; parameter A is a constant (A = 7.02 (10^{-4}) for regular steel material) and n is the creep stress exponent.

The Arrhenius law must be included in our new creep function (7.4); therefore, the integral Eq. (7.3) has a form

$$\overline{\sigma}(\theta) = E(\theta)\varepsilon(\theta) = \sigma(\theta) + \int_0^\theta \exp\left[\frac{\tau}{1+\beta\tau}\right] L[\theta,\tau]\sigma^n(\tau)\mathrm{d}\tau \tag{7.5}$$

For detailed explanations regarding change of variables, the Arrhenius law approximation, etc.—see Chaps. 4 and 5.

The degenerate kernels (with new independent variables θ, τ) may be written as

$$K(\theta, \tau) = \sum_{i=1}^{N} \exp[-\alpha_i \varphi(\theta)] \exp[\alpha_i \varphi(\tau)] \tag{7.6}$$

The expansion in Eq. (7.6) represents a series of real exponentials, called the Dirichlet series (also called the Prony series). Here, $t = \varphi(\theta)$ and $t' = \varphi(\tau)$ are substitution time functions, called the reduced times and α_i is called MPP—material property parameter (inverse retardation temperature/time in Eq. (7.6) or relaxation temperature/time). In general, function $t = \varphi(\theta)$ can be obtained for a given fire severity scenario [13] (for instance, by the method of least squares). As for α_i parameters, however, they cannot be calculated from measured creep data but must be suitably chosen in advance. The values of α_i must not be spaced too sparsely, and they must cover the entire temperature–time range of interest, in particular, the smallest α_i must be such that $\alpha_i < 3\ \alpha_{\min}$ and the largest α_i must be such that $\alpha_i > 0.5\ \alpha_{\max}$, in which $\alpha_{\min}$ and $\alpha_{\max}$ are the smallest and the largest time delay after temperature load application for which the response is of interest.

The α_i values that give a close fit of given $K(t, t')$ data are not unique. Equally good fits of the given compliance function data can be obtained for many possible choices of α_i values which are spaced in the temperature–time scale and cover the entire temperature–time range of interest.

The Dirichlet series expansion should be regarded only as an approximation to the kernel function, motivated by computational convenience, rather than as a fundamental law.

The expansion contains unnecessarily many material parameters defining all the compliance or relaxation functions. Therefore, for all practical purposes in case of fire we will limit the range of MPP to the interval $10^{-3} < \alpha_i < 10^5$.

As already mentioned, the purpose of the Dirichlet series expansion and degenerate kernel form is to convert a constitutive equation of an integral type to one of a differential type.

For a temperature–time-dependant material, this conversion is somewhat simpler not only for linear creep stress–strain integral type relationship but also for a nonlinear when the stress exponent $n > 1$. Equation (7.5) with $K(\theta, \tau)$ given by Eq. (7.6) may be rewritten as (for each MPP from the interval $10^{-3} < \alpha_i < 10^5$).

$$\begin{aligned} \mathrm{d}(\sigma)/\mathrm{d}(\theta) = & -\alpha * \sigma * m1 + (\exp(-0.15 * \theta)) * (1 - 0.15 * \theta) - (\exp(\theta/(1.0 + 0.1 * \theta))) \\ & * m1 * \sigma^\wedge n + \alpha * m1 * \theta * (\exp(-0.15 * \theta)) \end{aligned} \tag{7.7}$$

The initial condition is $\sigma(0) = 0$, since we are analyzing creep and high temperature deformations only. It is assumed here that initial static stresses (from dead and live loads) are much smaller then high temperature stresses during the fire event and it can be accounted for at the limit state analysis independently of creep stresses analysis by considering the corresponding load combination case. If this assumption is not acceptable, then the initial condition must be changed.

The statistical data that are required for the probability-based methods in defining and obtaining Stress–Temperature–Strain relationships are based now on discrete random variable (RV) α. Consequently, any solution of a differential type creep constitutive Eq. (7.7) is a realization of a random function σ of dimensionless temperature θ. Traditionally, structural design is based on maximum values of RV and considerably less on theory of random functions.

However, in the case of creep deformation at high temperature application it is very important to investigate both approaches: maximum random values of dimensionless allowable stresses in a given fire scenario and the applied theory of random functions. Let us start with maximum random values of dimensionless allowable stresses in the case of very fast fire. The matrix values of random maximum dimensionless allowable stresses are calculated now by solving the differential Eqs. (7.7) for each discrete RV α and presented in the following (using POLYMATH software).

7.4 Very Fast Fire: Statistical Data

Differential equations (DEQs): $\alpha = 0.001$

Calculated values of DEQ variables

	Variable	Initial value	Minimal value	Maximal value	Final value
1	m	0.00245	0.00245	0.0736123	0.07345
2	$m1$	0.0375	0.0079932	0.1463	0.1463
3	n	3	3	3	3
4	r	0	0	2.45251	2.231302
5	t	0	0	10	10
6	y	0	0	*1.572992*	0.0009539

Differential equations

1 $\mathrm{d}(y)/\mathrm{d}(t) = -0.01 * y * m1 + (\exp(-0.15 * t)) * (1 - 0.15 * t) - (\exp(t/(1.0 + 0.1 * t))) * m1 * y\^{}n + 0.01 * m1 * t * (\exp(-0.15 * t))$

Explicit equations

1 $n = 3$

2 $r = t * (\exp(-0.15 * t))$

3 $m1 = \varphi'(\theta) = 0.0375 - 0.01868 * t + 0.00312 * t\^{}2 - 0.000164 * t\^{}2$

4 $m = \varphi(\theta) = 0.00245 + 0.0375 * t - 0.00934 * t\^{}2 + 0.00104 * t\^{}3 - 0.000041 * t\^{}4$

Model: $\sigma = 1.428 * \theta - 0.381 * \theta\hat{\ }2 + 0.0337 * \theta\hat{\ }3 - 0.00097 * \theta\hat{\ }4$ (7.8)

Differential equations (DEQs): $\alpha = 0.01$

Calculated values of DEQ variables

	Variable	Initial value	Minimal value	Maximal value	Final value
1	m	0.00245	0.00245	0.0736123	0.07345
2	$m1$	0.0375	0.0079932	0.1463	0.1463
3	n	3	3	3	3
4	r	0	0	2.45251	2.231302
5	t	0	0	10	10
6	y	0	0	*1.573011*	0.0049354

Differential equations

1 $d(y)/d(t) = -0.01 * y * m1 + (\exp(-0.15 * t)) * (1 - 0.15 * t) - (\exp(t/(1.0 + 0.1 * t))) * m1 * y\hat{\ }n + 0.01 * m1 * t * (\exp(-0.15 * t))$

Explicit equations

1 $n = 3$

2 $r = t * (\exp(-0.15 * t))$

3 $m1 = 0.0375 - 0.01868 * t + 0.00312 * t\hat{\ }2 - 0.000164 * t\hat{\ }2$

4 $m = 0.00245 + 0.0375 * t - 0.00934 * t\hat{\ }2 + 0.00104 * t\hat{\ }3 - 0.000041 * t\hat{\ }4$

Model: $\sigma = 1.428 * \theta - 0.381 * \theta\hat{\ }2 + 0.0337 * \theta\hat{\ }3 - 0.00097 * \theta\hat{\ }4$ (7.9)

Differential equations (DEQs): $\alpha = 0.1$

Calculated values of DEQ variables

	Variable	Initial value	Minimal value	Maximal value	Final value
1	$m1$	0.0375	0.0079932	0.1463	0.1463
2	n	3	3	3	3
3	t	0	0	10	10
4	y	0	0	*1.573204*	0.0417526

Differential equations

1 $d(y)/d(t) = -0.1 * y * m1 + (\exp(-0.15 * t)) * (1 - 0.15 * t) - (\exp(t/(1.0 + 0.1 * t))) * m1 * y\hat{\ }n + 0.1 * m1 * t * (\exp(-0.15 * t))$

Explicit equations

1 $n = 3$
2 $m1 = 0.0375 - 0.01868 * t + 0.00312 * t^\wedge 2 - 0.000164 * t^\wedge 3$

$$\textbf{Model}: \sigma = 1.425 * \theta - 0.38 * \theta^\wedge 2 + 0.0335 * \theta^\wedge 3 - 0.001 * \theta^\wedge 4 \tag{7.10}$$

Differential equations (DEQs): $\alpha = 1$

Calculated values of DEQ variables

	Variable	Initial value	Minimal value	Maximal value	Final value
1	$m1$	0.0375	0.0079932	0.1463	0.1463
2	n	3	3	3	3
3	t	0	0	10	10
4	y	0	0	*1.575117*	0.2121058

Differential equations

1 $d(y)/d(t) = -1 * y * m1 + (\exp(-0.15 * t)) * (1 - 0.15 * t) - (\exp(t/(1.0 + 0.1 * t))) * m1 * y^\wedge n + 1 * m1 * t * (\exp(-0.15 * t))$

Explicit equations

1 $n = 3$
2 $m1 = 0.0375 - 0.01868 * t + 0.00312 * t^\wedge 2 - 0.000164 * t^\wedge 2$

$$\textbf{Model}: \sigma = 1.432 * \theta - 0.383 * \theta^\wedge 2 + 0.034 * \theta^\wedge 3 - 0.00096 * \theta^\wedge 4 \tag{7.11}$$

Differential equations (DEQs): $\alpha = 10$

Calculated values of DEQ variables

	Variable	Initial value	Minimal value	Maximal value	Final value
1	$m1$	0.0375	0.0079888	0.1463	0.1463
2	n	3	3	3	3
3	t	0	0	10	10
4	y	0	0	*1.593417*	0.4857851

Differential equations

1 $\mathrm{d}(y)/\mathrm{d}(t) = -10 * y * m1 + (\exp(-0.15 * t)) * (1 - 0.15 * t) - (\exp(t/(1.0 + 0.1 * t))) * m1 * y^\wedge n + 10 * m1 * t * (\exp(-0.15 * t))$

Explicit equations

1 $n = 3$

2 $m1 = 0.0375 - 0.01868 * t + 0.00312 * t^\wedge 2 - 0.000164 * t^\wedge 2$

$$\textbf{Model}: \sigma = 1.457 * \theta - 0.402 * \theta^\wedge 2 + 0.0391 * \theta^\wedge 3 - 0.00128 * \theta^\wedge 4 \quad (7.12)$$

Differential equations (DEQs): $\alpha = 100$

Calculated values of DEQ variables

	Variable	Initial value	Minimal value	Maximal value	Final value
1	$m1$	0.0375	0.0079895	0.1463	0.1463
2	n	3	3	3	3
3	t	0	0	10	10
4	y	0	0	*1.724496*	0.951412

Differential equations

1 $d(y)/d(t) = -100 * y * m1 + (\exp(-0.15 * t)) * (1 - 0.15 * t) - (\exp(t/(1.0 + 0.1 * t))) * m1 * y^\wedge n + 100 * m1 * t * (\exp(-0.15 * t))$

Explicit equations

1 $n = 3$

2 $m1 = 0.0375 - 0.01868 * t + 0.00312 * t^\wedge 2 - 0.000164 * t^\wedge 2$

$$\textbf{Model}: \sigma = 1.271 * \theta - 0.304 * \theta^\wedge 2 + 0.0276 * \theta^\wedge 3 - 0.00088 * \theta^\wedge 4 \quad (7.13)$$

Differential equations (DEQs): $\alpha = 1000$

Calculated values of DEQ variables

	Variable	Initial value	Minimal value	Maximal value	Final value
1	$m1$	0.0375	0.0079966	0.1463	0.1463
2	n	3	3	3	3
3	r	0	0	2.45248	2.231302
4	t	0	0	10	10
5	y	0	0	*2.106468*	1.610935

Differential equations

1 $d(y)/d(t) = -1000 * y * m1 + (\exp(-0.15 * t)) * (1 - 0.15 * t) - (\exp(t/(1.0 + 0.1 * t))) * m1 * y^\wedge n + 1000 * m1 * t * (\exp(-0.15 * t))$

Explicit equations

1 $n = 3$

2 $m1 = 0.0375 - 0.01868 * t + 0.00312 * t\wedge 2 - 0.000164 * t\wedge 2$

3 $r = t * (\exp(-0.15 * t))$

$$\textbf{Model}: \sigma = 1.045 * \theta - 0.169 * \theta\wedge 2 + 0.0099 * \theta\wedge 3 - 0.000182 * \theta\wedge 4 \quad (7.14)$$

Differential equations (DEQs): α = 10000

Calculated values of DEQ variables

	Variable	Initial value	Minimal value	Maximal value	Final value
1	$m1$	0.0375	0.0079956	0.1463	0.1463
2	n	3	3	3	3
3	r	0	0	2.452492	2.231302
4	t	0	0	10	10
5	y	0	0	*2.384092*	2.094868

Differential equations

$$1 \quad d(y)/d(t) = -10{,}000 * y * m1 + (\exp(-0.15 * t)) * (1 - 0.15 * t) - (\exp(t/(1.0 + 0.1 * t))) * m1 * y\wedge n + 10{,}000 * m1 * t * (\exp(-0.15 * t))$$

Explicit equations

1 $n = 3$

2 $m1 = 0.0375 - 0.01868 * t + 0.00312 * t\wedge 2 - 0.000164 * t\wedge 2$

3 $r = t * (\exp(-0.15 * t))$

$$\textbf{Model}: \sigma = 0.99 * \theta - 0.141 * \theta\wedge 2 + 0.00814 * \theta\wedge 3 - 0.000184 * \theta\wedge 4 \quad (7.15)$$

Differential equations (DEQs): α = 100000

Calculated values of DEQ variables

	Variable	Initial value	Minimal value	Maximal value	Final value
1	$m1$	0.0375	0.0079938	0.1463	0.1463
2	n	3	3	3	3
3	r	0	0	2.452498	2.231302
4	t	0	0	10	10
5	y	0	0	*2.444621*	2.215169

Differential equations

1 $d(y)/d(t) = -100{,}000 * y * m1 + (\exp(-0.15 * t)) * (1 - 0.15 * t) - (\exp(t/(1.0 + 0.1 * t))) * m1 * y^{\wedge}n + 100{,}000 * m1 * t * (\exp(-0.15 * t))$

Explicit equations

1 $n = 3$

2 $m1 = 0.0375 - 0.01868 * t + 0.00312 * t^{\wedge}2 - 0.000164 * t^{\wedge}2$

3 $r = t * (\exp(-0.15 * t))$

Model: $\sigma = 0.988 * \theta - 0.141 * \theta^{\wedge}2 + 0.00863 * \theta^{\wedge}3 - 0.000219 * \theta^{\wedge}4$ (7.16)

The summary of computations above is presented in Table 7.1.

Therefore, the probability that the real fire dimensionless allowable stress "**σ**" will be less than 1.57; 1.2 or 1.0 is

$$P(\boldsymbol{\sigma} < 1.57) = \Phi * \left(\frac{1.57 - 1.838}{0.369}\right) = \Phi * (-0.726) = 0.23 = 23\ \%$$
$$P(\boldsymbol{\sigma} < 1.2) = \Phi * \left(\frac{1.2 - 1.838}{0.369}\right) = \Phi * (-1.729) = 0.0418 = 4.18\ \% \quad (7.17)$$
$$P_f(\boldsymbol{\sigma} < 1.0) = \Phi * \left(\frac{1.0 - 1.838}{0.369}\right) = \Phi * (-2.27) = 0.011 = 1.1\ \%$$

The acceptable probability of failure is the criterion to which the results of the probabilistic analysis will be compared to determine if the design is acceptable.

Table 7.1 Maximum dimensionless stresses (statistical data)

Value	Maximum creep stress	Deviation	Variance	Mean value	Standard deviation	α value
1	*1.572992*	−0.265	0.0702			0.001
2	*1.573011*	−0.265	0.0702			0.01
3	*1.573204*	−0.265	0.0702			0.1
4	*1.575117*	−0.263	0.0692			1.0
5	*1.593417*	−0.245	0.060			10
6	*1.724496*	−0.113	0.0128			100
7	*2.106468*	0.268	0.0718			1000
8	*2.384092*	0.546	0.298			10,000
9	*2.444621*	0.606	0.367			100,000
Aver.	*1.838*		Total: 1.224/8 = 0.153	*1.838*	*0.391*	

Specification of this acceptable, or target, probability of failure for the total structure is a complex issue that generally will not be decided upon by the structural engineer performing the probabilistic analysis. Legal, technical, and socioeconomic considerations are involved. The agency certifying the structure should be responsible for setting this overall specification for the structure. If for instance the proposed failure probability $P_f = 0.011$ is acceptable in this case, then the reliability index β (see below) can be computed as follows []:

$\sigma_{all} = \mu_\sigma - \alpha\beta\sigma_\sigma$ ($\alpha = 1.0$ for normal probability distribution function). If $\sigma_{all} = 1.0$ (7.02) 2.9 = 20.36 ksi = 140.3 MPa; $\mu_\sigma = 1.838$ and $\sigma_\sigma = 0.391$, then $\beta_{all} = 2.143$.

It is assumed here that the maximum dimensionless stress has normal probability distribution function (pdf), since in real fire scenario there is a large number of identical independent random variables that affect the creep process. In this case, we have an aggregate effect of a large number of small random factors, the resulting random variable is normal. We can transform all the observations of normal random variable σ_{max} with mean $\mu = 1.838$ and variance $\sigma = 0.391$ to a new set of observations of another normal random variable Z with mean 0 and variance 1 using the following transformation: $Z = \sigma_{max} - \mu/\sigma$. Hence, we have the following probabilities within the interval $1.573 < \sigma_{max} < 2.44$. (To simplify notation, the fire dimensionless stress σ from creep deformations will be called simply stress σ and the subscript max will be omitted). The probability that the allowable stress "σ" will be within this interval can be computed as follows:

$$\begin{aligned} P(1.573 < \boldsymbol{\sigma} < 2.44) &= \left[\Phi * \left(\frac{2.44 - 1.838}{0.369}\right) - \Phi * \left(\frac{1.573 - 1.838}{0.369}\right)\right] \\ &= [\Phi * (1.63) - \Phi * (-0.718)] = (0.948 - 0.236) = 0.712 = 71.2\,\% \\ \Phi * (z) &= \frac{1}{\sqrt{2\pi}} \int\limits_{-\infty}^{z} e^{-\frac{z^2}{2}} dz \end{aligned} \tag{7.18}$$

The probability that the stress "σ" will be less than 1.573 is

$$P(\boldsymbol{\sigma} < 1.573) = \left[\Phi * \left(\frac{1.573 - 1.838}{0.369}\right)\right] = \Phi * (-0.718) = 0.236 = 23.6\,\%$$

If for some reason one would prefer to have 95 % probability instead of 71.2 % for the overall stresses range, then the corresponding probability problem will be formulated for the symmetrical interval (with respect to the mean value $\sigma = 1.838$) as follows: find the dimensionless stress interval "L" for a given mean value $\mu_\sigma = 1.838$ and standard deviation $\sigma_\sigma = 0.391$. The corresponding formula in this case is

$$P(|\boldsymbol{\sigma} - \sigma_m| < L) = 2\Phi * \left(\frac{L}{\boldsymbol{\sigma_\sigma}}\right) - 1 = 0.95$$

$$\text{or: } \Phi * \left(\frac{L}{\boldsymbol{\sigma_\sigma}}\right) = \frac{1 + 0.95}{2} = 0.975$$

From the integral probability function table we have

$$L/0.391 = 1.96 \text{ or } L = 0.766 \text{ and } \sigma_{\min} = 1.838 - 0.766 = 10.07$$

One can see now that the lower limit had changed substantially and it is very close to the allowable stress σ_{all} computed earlier. We have to underline here that the lower stress limit is the most important for us, since we are defining the structural fire resistance stresses that should be higher than the applied design stresses. The error function is essentially identical to the standard normal cumulative distribution function, denoted Φ^*, as they differ only by scaling and translation. Indeed,

$$\Phi * (z) = \frac{1}{\sqrt{2\pi}} \int_{-\infty}^{z} e^{-\frac{z^2}{2}} dz = \frac{1}{2}\left[1 + \text{erf}\left(\frac{z}{\sqrt{2}}\right)\right]$$

Consequently, the error function is also closely related to the Q-function, which is the tail probability of the standard normal distribution. The Q-function can be expressed in terms of the error function as

$$Q(z) = 1 - \Phi * (z) = \frac{1}{2} - \frac{1}{2}\text{erf}\left(\frac{z}{\sqrt{2}}\right)$$

The standard normal cumulative distribution function (CDF) is used more often in probability and statistics, and the error function is used more often in other branches of mathematics.

7.5 The First-Order Reliability Method (FORM)

The concept of probabilistic structural design risk assessment has been around for quite some time. In 1942, Pugsley [14] had published an article, where he proposed correlating loads and strengths with recorded structural accident rates. He states "By adopting the principle that neither design loads nor safety factors and permissible stresses should be specified arbitrarily, it will be possible to not only eliminate inadequate design, but frequently to achieve considerable economy."

In 1945, Alfred Freudenthal had published a paper [15], where he stated: "The true character of the safety factor is disclosed by the introduction of a statistical concept of physical qualities, according to which the individual properties composing strain and resistance are represented by frequency distributions, instead of by individual values... By application of the theory of probability, the concept of safety can be rationalized." Freudenthal's paper sparked international interest in structural safety and structural reliability theory. The theory was fueled by Weibull's success in 1951 [16] in developing robust statistical representations of material strength. This approach was the foundation of the First-Order Reliability Method yielding the Safety Index.

In 1967, Cornell [17] proposed a second moment format for evaluation of structural reliability. This approach generates a "safety index" calculated from the means and variances of the parameter distributions. The safety index is considered to be a measure of reliability, and is an alternative to numerically integrating the joint probability density function to determine a probability of failure. In 1973, Lind [18] demonstrated that Cornell's safety index could be used to derive safety factors on applied loads and resistance. This was a milestone; reliability analysis was at long last related to accepted (civil engineering) methods of design. Subsequent refinements were made by Hasofer and Lind [19], whose method (1974) is considered to be the foundation of probabilistic design theory. In probability-based limit states design, the structural reliability formulation is presented in such a way as to make it practical for design by engineers who may not be familiar with reliability concepts or have access to the necessary statistical data. For a typical design condition, both stress and strength can be plotted in the same horizontal axis as shown in Fig. 7.4. The mean strength, obviously, is greater than the mean applied stress. However, the overlap of PDF suggests that it is possible for strength to be less than applied stress, which is the condition for failure. This conveys the

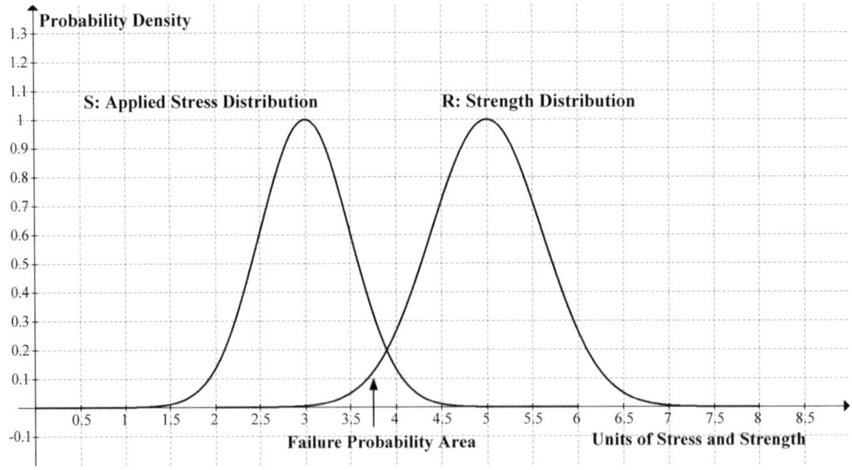

Fig. 7.4 Simplified two-dimensional formulation of failure probability

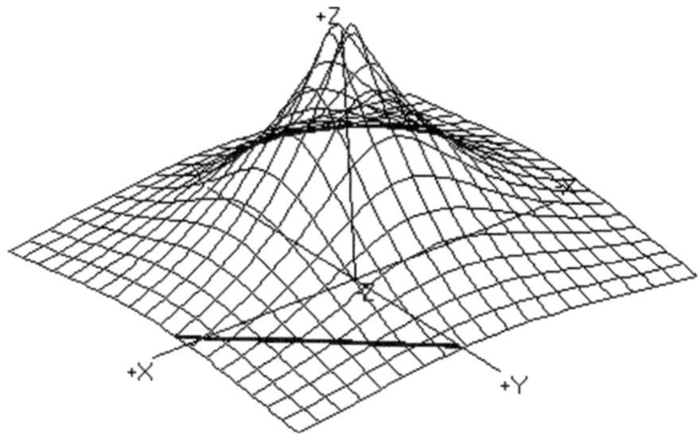

Fig. 7.5 Three-dimensional representation problems

essence of probabilistic structural analysis: there is a possibility of failure, and it is defined in the small region of overlap between the PDF. The probability of failure is defined as $P_f = P[g(R, S) \leq 0]$. The statistical variation of R and S are described by the probability density functions $f_R(r)$ and $f_S(s)$, respectively.

A technically accurate description of the stress–strength curve overlap is shown in Fig. 7.5, showing the stress and strength along the horizontal and vertical axes, respectively. The line drawn represents the scenarios where stress = strength, or $g(R, S) = R - S = 0$. This is often referred to as the "limit state" that separates the failure region ($g < 0$) from the safe region ($g > 0$). Deterministic function $g(R, S)$ is commonly referred to as the performance function. The probability of failure is defined as the volume under the surface shown in the failure region where $g < 0$.

The overlap region (volume) is quantitatively obtained from the following expression:

$$P_f = \iint_{\Omega} f_{R,S}(r, s)\mathrm{d}r\mathrm{d}s \tag{7.19}$$

$f_{R,S}(r, s)$ is the joint density function and Ω is the failure set, i.e., the set of all values of R and S such that $g(R, S) \leq 0$. If the variables R and S are statistically independent, then the joint density function is expressed as the product of individual density functions as follows:

$$P_f = \iint_{\Omega} f_R(r) f_S(s)\mathrm{d}r\mathrm{d}s \tag{7.20}$$

7.5.1 Most Probable Point Methods

There are several methods in this group, the main ones being First-Order Reliability Method (FORM) and Second-Order Reliability Method (SORM). These methods are the most complex, both mathematically and conceptually, among all

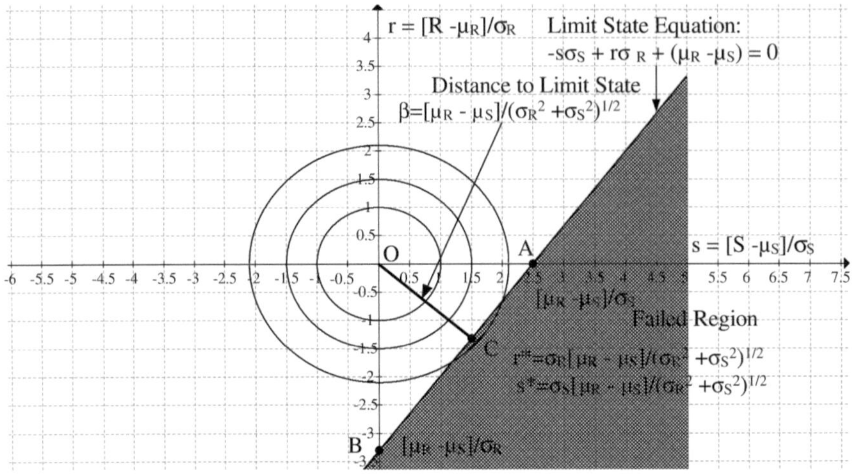

Fig. 7.6 Illustration of the two-dimensional case of a linear limit state function and standardized normally distributed variables

probabilistic analysis methods and will therefore not be described in excruciating detail here. The main steps to perform these analyses will be given and references are given for the interested reader.

In the first- and second-order reliability methods (FORM/SORM), the approach is to transform the integral (7.15) to an approximately equal integral that can be efficiently evaluated. This is done by the following steps:

1. Transform the design variable distributions into standard normal distributions. That is, transform $g(x) = 0$ into $g(u) = g(s, r) = 0$ where u is a vector of standardized, independent Gaussian variables (see Fig. 7.6).
2. Identify the most probable point (MPP), or design point C. For a given limit state function, the main contribution to failure probability comes from the region where g is closest to the origin in the transformed design variable space (u-space). The MPP is defined as the closest point to the origin in the transformed space.
3. Develop a polynomial approximation to the performance function (g-function) around the MPP. Thus, the g-function is approximated by a simply defined (quadratic) surface through that point (MPP). Compute probability of failure using the newly defined g-function and the transformed variables.

This technique is graphically depicted in Fig. 7.6 and step-by-step procedure is presented below

Step 1 *Transform Variables.*

FORM and SORM reliability approximations are carried out in the space of a set of standard, uncorrelated normal variants Y, obtained by transforming the basic variables. This transformation is dependent on the form of the probability distribution of each variable. The advantage of doing this probability transformation is to be able to exploit the superior properties of standard normal space. Specifically, the probability density in this space is symmetric (see Fig. 3.16) and it decays exponentially with the square of the distance from the origin. The transformation can be made in several ways. The specific details of the different methods for transformation are mathematically complex and lengthy and will not be given here. References [19, 20] give exhaustive details on this transformation.

Step 2 *Identify Most Probable Point.*

The MPP is the point on the limit state with the highest joint density, as can be seen in Fig. 7.6. That is, the point is most probable because it has the maximum joint probability density or largest contribution to failure. The MPP can be found by using an optimization algorithm or other iteration algorithms. The optimization begins by guessing that the MPP lies at the set of mean values of each variable involved. Then, the distance from the origin to the limit state surface is minimized subject to the constraint that the point (MPP) lays on the limit state.

Determining the MPP is at the heart of these approximate reliability methods, and many issues are involved such as dealing with correlated variables, the type of search methods used in the optimization routines, and convergence criteria.

Step 3 *Develop g-Function and Determine Failure Probability.*

The function $g(u)$ is approximated by a polynomial in the vicinity of the MPP. The first-order reliability method (FORM) estimate is $P_{\mathrm{f}} = P(g\ 0) \approx \Phi(-\beta)$, where β represents the minimum distance to the limit state. Gradients of the polynomial function are used to find the minimum distance. Several second-order (SORM) approximations are available to improve accuracy. These higher order approaches take into account the curvature of the limit state around the minimum distance point. The simplest of the SORM approximations is based on a parabolic fitting. There have been many derivatives of the second-order approach to improve upon the accuracy of the approximation as well as decrease the number of evaluation of the failure function. These can be found in reference [21].

7.5.2 *Limit State Approximation*

The load applied to the structure depicted in Fig. 7.4 is denoted by S and the resistive strength is denoted by R. In Fig. 7.6, the limit state function $g(\mathbf{x})$ has been transformed into the limit state function $g(\mathbf{u})$ by normalization of the random variables into standardized normally distributed random variables $U_i = (X - \boldsymbol{\mu}_{X_i})/\boldsymbol{\sigma}_{X_i}$ that the random variables U_{i} have zero means and unit standard deviation. Note that normal distributions for R and Q are assumed. These distributions may actually be Gaussian or may result from the transformation of a non-normal distribution to a normal distribution. The transformation of the original distribution to an equivalent normal is a critical step in the limit state approximation approach. The reliability index has the simple geometrical interpretation as the smallest distance from the line (or generally the hyper plane) forming the boundary between the safe domain and the failure domain, i.e., the domain defined by the failure event. It should be noted that this definition of the reliability index (due to Hasofer and Lind [19]) does not depend on the limit state function but rather the boundary between the safe domain and the failure domain.

The point on the failure surface with the smallest distance to origin is commonly denoted the design point or most likely failure point. It is seen that the evaluation of the probability of failure in this simple case reduces to some simple evaluations in terms of mean values and standard deviation s of the basic random variables that is the first- and second-order information.

The limit state is formulated as $g = R - S = 0$. The reduced variables are defined as

$$r = \frac{R - \boldsymbol{\mu}_R}{\boldsymbol{\sigma}_R} \quad \text{and} \quad s = \frac{S - \boldsymbol{\mu}_S}{\boldsymbol{\sigma}_S}$$

Substituting for R and S in these limit state equation yields

$$g = r\boldsymbol{\sigma}_R + \boldsymbol{\mu}_R - (s\boldsymbol{\sigma}_S + \boldsymbol{\mu}_S) = 0 \quad \text{or:} \quad g = r\boldsymbol{\sigma}_R - s\boldsymbol{\sigma}_S + (\boldsymbol{\mu}_R - \boldsymbol{\mu}_S) = 0 \tag{7.21}$$

The overlap region (volume) is quantitatively obtained from the following expression:

$$\begin{aligned} P_{\mathrm{f}} &= \iint\limits_{\Omega} f_{R,S}(r,s)\mathrm{d}r\mathrm{d}s; \\ \mu_g &= \mu_r - \mu_s \quad \text{and} \quad \sigma_g = \left(\sigma_r^2 + \sigma_s^2\right)^{1/2} \end{aligned} \tag{7.22}$$

where: $m_g; m_R$ and m_S—are mean values of random variables "g"; "R" and "S" respectfully; $\sigma_g; \sigma_R$ and σ_S—are standard deviations of random variables "g"; "R"; and "S", respectively, and $f_{R,S}(r,s)$ is the joint density function and Ω is the failure set, i.e., the set of all values of R and S such that $g(R, S) \leq 0$. If the variables R and S are statistically independent, then the joint density function is expressed as the product of individual density functions (7.16).

The event of failure is $g < 0$. The probability of failure is now given in terms of g

$$P(g < 0) = \int\limits_{-\infty}^{0} \frac{1}{\sigma_g\sqrt{2\pi}} \mathrm{e}^{-\frac{1}{2}\left(\frac{g-\mu_g}{\sigma_g}\right)^2} \mathrm{d}g$$

If we let $z = \frac{g-\mu_g}{\sigma_g}$, then $g = 0$, the upper limit of z is given by:

$$z = \frac{0-\mu_g}{\sigma_g} = -\frac{\mu_g}{\sigma_g} = -\frac{\mu_R - \mu_S}{\sqrt{\sigma_R^2 + \sigma_S^2}}$$
$$\text{and when } g \to -\infty, \quad \text{the lower limit of } z \to -\infty$$

The transformed integral becomes the probability of failure

$$P_f = \frac{1}{\sqrt{2\pi}} \int_{-\infty}^{-\frac{\mu_R-\mu_S}{\sqrt{\sigma_R^2+\sigma_S^2}}} \mathrm{e}^{-\frac{z^2}{2}} \mathrm{d}z = \frac{1}{\sqrt{2\pi}} \int_{-\infty}^{-\beta} \mathrm{e}^{-\frac{z^2}{2}} \mathrm{d}z = \Phi * (-\beta) \tag{7.23}$$

$$\text{where } \beta = \frac{\mu_z}{\sigma_z} = \frac{\mu_R - \mu_S}{\sqrt{\sigma_R^2 + \sigma_S^2}} \quad \text{and the reliability: } P_r = 1 - P_f$$

The random variable z is the standard normal variable! Therefore, the probability of failure can be found by looking in the standard normal table for the area under the curve from $-\infty$ to $z = \frac{\mu_x - \mu_y}{\sqrt{\sigma_x^2 + \sigma_y^2}}$.

This approach was developed by Cornell [17] and is a part of the First-Order Reliability Method (FORM). He named the ratio the "safety index" and denoted it as β. That is $\beta = \frac{\mu_z}{\sigma_z} = \frac{\mu_R - \mu_S}{\sqrt{\sigma_R^2 + \sigma_S^2}}$ and $P_\mathrm{f} = \Phi^*\{-\beta\}$ where Φ^* is the cumulative distribution function for a standard normal variable (look-up tables are in most statistics textbooks). This equation makes calculation of probability of failure for the normal–normal case extremely simple and fast.

The limit state Eq. (7.17) is a straight line equation, therefore, the distance between the strait line L: $Ax + By + C = 0$ and the origin is given by $d(0, L) = |C|/\sqrt{A^2 + B^2}$. Applying this equation to the limit state equation yields the following value for the distance:

$$\mathrm{d}(0, L) = \beta = |\mu_R - \mu_S| / \sqrt{\sigma_R^2 + \sigma_S^2}. \tag{7.24}$$

We have thus derived the formula for the distance from the origin to MPP (for linear limit state equation!) in the reduced coordinate space shown in Fig. (7.6).

The reliability framework for the structural design loads, load combinations, and resistance criteria in modern limit states design building standards is provided by first-order reliability methods [22] (FORM). Probability-based limit states design (or LRFD) is based on the notion that the reliability index implied by a given structural design should be equal (or exceed) a target value set by professional consensus [23] LRFD codes have preferred establishing the target reliability based on successful previous designs rather than on actuarial evaluation of the probability of exceedance, P_f [24].

Table 7.2 Target reliability Index

Code	Target reliability index
AISC, 2011	3.0–4.5
ACI, 2011	2.5–4.0
AASHTO, 2012	3.5
Eurocode, 1993	3.8 or 4.3
ISO, 2001	2.3–4.3

Also, many codes establish the performance criteria in terms of the reliability index rather than P_f because of the different connotations that the term, P_f, may carry. Because exceeding the limit state function does not necessarily imply that the structure will actually fail, many engineers have referred to P_f as the probability of limit state exceedance rather than probability of failure. Table 7.2 gives a sample of target reliability index values used for the development of various structural design codes.

7.5.2.1 Code Target Reliability Index Design or Service Life

The table reveals large differences in the target reliability index values used in different codes and different countries for seemingly similar types of limit states and materials. These differences are due to the recommended practice of extracting the target reliability index from the experience gained from the successful performance of previous designs as these experiences may differ between jurisdictions and industries. The target reliability indexes should reflect differences in the quality of the materials and construction practices as well as differences in the environmental and deterioration mechanisms whether or not these are explicitly considered in the probabilistic models utilized during the numerical evaluation of the reliability. Also, the different target reliability indexes implicitly consider the consequences of a member's failure (but not the whole structural system), the type of structures addressed by the specific code, and the risk tolerance of the engineering community and the public within a code's jurisdiction. For example, it is reasonable to have different target reliability indexes for different fire severity categories (Very Fast, Fast, Medium, and Slow fires), since the possible structural damages are quite different. The target reliability indexes are used to calibrate resistance and load factors for individual loads and combinations of loads. Specifically, the ASCE 7 (SEI/ASCE 2010) load combination factors use a "principal action companion action" format [25]. The format is based on the notion that the maximum combined load effect during a service period occurs when one time-varying load attains its maximum value while the remaining time-varying loads are at their frequent (or arbitrary-point-in-time) values [26].

Traditional structural design process is based on checking the safety of each individual member for one particular failure mode. However, structures are composed of many members each of which may fail in a different mode and the reliability of the structural system is a function of the reliability of all its components. This renders the reliability analysis quite complex requiring the implementation of advanced simulation techniques which must be linked to appropriate structural analysis. These types of analyses obviously are beyond the scope of this book. However, in order to underline the dominant affect of the reliability of the structural system as a whole, the approximate method is used here by substituting the structural system with the equivalent One Degree of Freedom (ODOF) system (see Chap. 9).

The consensus among decision makers is that acceptable levels of probability of system failure must be established based on an approximate risk analysis that includes performance criteria for the structural system (ODOF), the probability of occurrence of fire events and structural response, and the resulting consequences, which may include economic losses, or other damage. Risk is often represented in terms of the probability of failure times the consequence of failure that can be expressed in terms of the cost of failure which may include direct costs such as the cost of the structure as well as indirect costs such as users costs, economic losses, etc.

Because of the difficulties encountered in obtaining quantitative measures of loss and the complications in calculating the probability of collapse, many of the proposed performance-based design criteria have not yet been explicitly calibrated based on a quantitative assessment of risk. Instead, proposed performance-based design methods have implicitly included subjective notions of risk as perceived by the code writers based on their experience and their understanding of the public's risk tolerance as explained in [27]. ASCE Standard 7 has adopted this approach by classifying buildings into four risk categories and setting different target probabilities of failure levels for each category as shown in Table 7.2. The ASCE-7 categories are:

I. Buildings and other structures that represent a low risk to human life in the event of failure.
II. All other buildings and other structures not in Risk Categories I, III, and IV.
III. Buildings and other structures that could pose a substantial risk to human life or cause a substantial economic impact or mass disruption of everyday life in the event of failure.
IV. Buildings and other structures designated as essential facilities or the failure of which could pose a substantial hazard to the community (Table 7.3).

Table 7.3 Annualized target reliability for loads other than earthquakes

	I	II	III	IV
Failure that is not sudden and does not lead to wide-spread progression of damage	$P_f = 1.25$ (10^{-4})	$P_f = 3.0$ (10^{-5})	$P_f = 1.25$ (10^{-5})	$P_f = 5.0$ (10^{-6})
Failure that is either sudden or leads to wide-spread progression of damage	$P_f = 3.0$ (10^{-5})	$P_f = 5.0$ (10^{-6})	$P_f = 2.0$ (10^{-6})	$P_f = 7.0$ (10^{-7})
Failure that is sudden and results in wide-spread progression of damage	$P_f = 5.0$ (10^{-6})	$P_f = 7.0$ (10^{-7})	$P_f = 2.5$ (10^{-7})	$P_f = 1.0$ (10^{-7})

One can see that Annualized Target Reliability for SFL is included in Table 4.2. The philosophy of assigning these probabilities in case of fire event is rather vague and it is presented in a draft document [28]: "Annual mortality statistics in the United States provide a psychological yardstick, of sorts, in measuring and discussing risk in terms of annual frequency, although these risks are not truly comparable to building risk. For the healthy adult population, the mortality risk from cardiovascular disease and cancer is on the order of 10^{-3}/year. At the other extreme, the *de minims* risk, that risk below which society normally does not impose any regulatory guidance, is on the order of 10^{-7}/year [29]. Between these annual frequencies of 10^{-3}/year and 10^{-7}/year is a gray area in which measures to reduce risk usually are traded off against increments in cost of risk reduction. For the sake of illustrating the role of risk in PBFE in this chapter, we may take 10^{-6}/year as the upper threshold of acceptable risk (measured in terms of annual frequency) due to fires in building construction. In terms of order of magnitude, this is not inconsistent with the failure probability of building systems from other natural events. In first-generation LRFD (load and resistance factor design) [22] the target member limit state probability involving formation of the first plastic hinge was approximately 0.001 in 50 years (corresponding to a "reliability index" of about 3.0); annualized, this is on the order of 10^{-5}. The annual probability of partial or total collapse of a redundant structural frame is approximately one order of magnitude less, or on the order of 10^{-6}/year. Note that such comparisons assume that "risk" is equivalent to "annual probability or frequency." It would be tempting to assert that the acceptable risk for PBFE should be set so that the PBFE design alternatives are at least as safe as those that comply with existing prescriptive requirements. This line of thinking is analogous to that followed in first-generation LRFD, which was calibrated (in an overall reliability sense) to existing structural design practice. When applied to PBFE, however, this approach is questionable. The calibration process for structural components subjected to dead, live, wind, and snow load drew upon years of successful experience in designing for those common loads using recognized principles of structural mechanics and behavior. In contrast, fire is a low-probability event; moreover, the current fire-resistant design approach cannot be tied in any meaningful experiential way to real structural demands, behavior, or response." It should be added here that the classical definition of "probability or frequency" is not justifiable especially in case of fire events in tall and super tall buildings, therefore the indirect methods of probability theory that do not require a large amount of statistical data shell be used in this case, and only the "tie-down" statistical information must be used for verification purposes. The unit time (number of fire events per 1 year) is selected in Table 4.2 based on analogy with environmental design structural loads (wind, snow, etc.), but it is not fully specific for fire load case. For example, the Target Reliability in the aerospace and aircraft industry is based on number of flight hours or number of take off times that is logically connected with the nature of such probabilities. That leads us to idea to use the probability of failure (exceedance of target dimensionless temperature level) for a given fire severity scenario as a base of the calibration process of Annualized Target Reliability, and then by using special techniques to consider the probability of failure as a time dependent stochastic process (assuming, for example, the age dependency as an exponential stochastic process or step-function type with discrete number of times). This approach has been used in this book, and the final results are compared with the data from Table 4.2.

7.5.3 Partial Safety Factor ψ and Reliability Index β

Let us introduce now new parameter (the *partial safety factor*—load factor) $\psi = \mu_R/\mu_S$. Then the reliability index yields:

$$\beta = \frac{|\mu_R - \mu_S|}{\sqrt{\sigma_R^2 + \sigma_S^2}} = \frac{\mu_S|\psi - 1|}{\sqrt{V_R^2 + \psi^2 V_S^2}}; \tag{7.25}$$

where

$$V_R = \sigma_R/\mu_R \quad \text{and} \quad V_S = \sigma_S/\mu_S$$

If the material properties are deterministic, then $V_R = 0$ then (7.25) is reduced

$$\Psi = 1 + \beta V_S. \tag{7.26}$$

Parameter V_S is called a load variation parameter (*coefficient of variation*), and V_R is called a *material variation parameter*.

It can be seen from (7.25) and (7.26) that the reliability index β and the *partial safety factor* $\psi = \mu_R/\mu_S$ are functions of two parameters of normal PDF (mean value and standard deviation). The function $\Phi(-\beta)$ is presented in Table 4.5 [13].

Very often, in structural design, it is assumed that parameter $\beta = 3.0$ (the rule of three standards). In this case, the probability of structural failure $P_f = \Phi(-\beta) = 0.00137$.

If $\beta > 5$, the following approximation of $P_f = \Phi(-\beta)$ can be used [13]:

$$P_f = \Phi(-\beta) = \frac{1}{\sqrt{2\pi}} \frac{\beta^2 - 1}{\beta^3} e^{-\frac{\beta^2}{2}} \tag{7.27}$$

Probability-based limit states design is based on the notion that the reliability index implied by a given structural design should equal (or exceed) a target value set by professional consensus. This requirement can be met by checking safety using values $\theta_{\max,d}$ of the dimensionless temperature in structural design computations as follows [22]:

Table 7.4 Probability of Failure

$\Phi(-\beta)$	0.1	0.01	0.001	0.0001	3.2 (10^{-5})	3 (10^{-6})	2.9 (10^{-7})
β	1.28	2.32	3.15	3.77	4.0	4.5	5.0

$$\theta_{\max,\,\text{design}} = \mu_\theta + \beta\sigma_\theta = \mu_\theta(1 + \alpha\beta V_S) \tag{7.28}$$

where μ_θ—mean of θ; σ_θ—standard deviation of θ and α—sensitivity coefficient, which depends on the CDF of θ ($\alpha = 1$ for normal distribution). The coefficient of variation V_S, is defined as σ_θ/μ_θ. The quantity in the brackets is the partial safety factor. We have to underline here that the reliability index β in Eq. (7.28) reflects the combined effect of random structural fire load and statistical data of structural fire resistance due to creep deformations.

The structural system and elements are usually given in case of fire and φ = resistance factor that takes into account uncertainties in the determination of R_n (nominal strength stipulated in the material specification or code) is also given. Therefore, the nominal strength R_n of a structural element (or system) should be obtained based on a high-temperature creep effect for a given fire severity design scenario. Partial Safety factor "ψ" in structural design process usually is assumed to be equal 1.67 (if the resistance factor $\varphi = 0.9$) or 2.0 (if the resistance factor $\varphi = 0.75$) [30]. Parameter (V_R) represents the variability of the random parameter "R"–Structural Fire Resistance and can be defined based on assumed normal probability distribution (μ_R and σ_R data provided in Tables 7.4 (see above) for Very Fast fire severity case). Therefore, the approximate values of target reliability index β and structural failure probability P_f can be computed for each fire severity case, if statistical data are provided.

Example 7.1 Data: Very Fast Fire: $\mu_S = 8.77$; $\sigma_S = 1.48$; $V_S = 1.48/8.77 = 0.169$; $\psi = 1.67$. Find the reliability index β and structural failure probability P_f.

(a) From (7.28) we have $\psi = 1.67$; $\beta_{tar} = (1.67 - 1)/0.169 = 3.96$. If $\psi = 2.0$, then
$\beta_{tar} = (2.0 - 1)/0.169 = 5.92$.
From Table 7.4: If $\beta_{tar} = 3.96$ then $P_f = 4.4(10^{-5})$. If $\beta_{tar} = 5.92$ then $P_f = 1.6$ (10^{-9}) [using formula (7.27)].

$$\text{Thus: } \boldsymbol{\beta} = \begin{cases} \boldsymbol{\beta}_{\text{design}} = \boldsymbol{\beta}_{\text{computed}} & \text{if } \boldsymbol{\beta}_{\text{computed}} > \boldsymbol{\beta}_{\text{target}} \\ \boldsymbol{\beta}_{\text{design}} = \boldsymbol{\beta}_{\text{target}} & \text{if } \boldsymbol{\beta}_{\text{computed}} < \boldsymbol{\beta}_{\text{target}} \end{cases} \tag{7.29}$$

Obviously, if the target reliability index β is given, then the safety factor "ψ" can be found from (7.28). It is important to underline here that these calculated failure probabilities are based on statistical data developed from mathematical modeling of a real fire (Very Fast fire in this case) without direct consideration of high temperature creep effect. The structural fire resistance obviously is quite different from structural fire load (SFL); therefore, the full reliance on statistical data from real fires with many different architectural and structural designs and materials will not be justified. In addition, the structural failure probability is constant only when SFL is a function of maximum temperature in case of fire, however, as it was stated earlier, fire in fact is a random process and the structural resistance failure probability is a temperature and time dependant value. In this case the probability-based mathematical modeling of fire development process seems more valuable and it should be justified and verified by supporting statistical data available. That in turn leads to another interpretation of the reliability index β based on applied theory of random functions and its application to the fire development process modeling. Let us define the structural failure as the first exceedance of dimensionless stress level $\sigma = a$. Since, in applications, one is generally concerned with the statistically rare crossings of a high level of dimensionless creep stress level $\sigma = a$, it is of considerable interest to investigate the form of the first-occurrence density as function of fire duration and the creep stress–temperature–strain stationary Gaussian processes. Some simple arguments suggest a Poisson distribution of first-occurrence times [31]. The probability of exceedance of level $\sigma = a$ at least once $P_a = 1 - \mathrm{e}^{-\overline{n}_a}$. Since the reliability of the structural system must be close to 1, then:

$$\begin{aligned} &P_a \doteq 1 - \mathrm{e}^{-\overline{n}_a} \approx \overline{n}_a \\ &(\text{Note: actual}\, P_a < \overline{n}_a, \quad \text{since} \quad 1 - \mathrm{e}^{-x} < x) \end{aligned} \tag{7.30}$$

$$\overline{n}_a = T \int_0^{\infty} v f(a, v) \mathrm{d}v \tag{7.31}$$

The autocorrelation function and the formulae for the design dimensionless temperature to be equal (or exceed) a target maximum value (for detailed analysis—see this chapter below) in this fire severity case are as follows:

$$\boldsymbol{\rho}_{\boldsymbol{\sigma}}(\boldsymbol{\theta}) = (\exp(-a * \theta)) * \cos(b * \theta)$$

$$\boldsymbol{\sigma}_v^2 = K_{\dot{\sigma}}(0) = -\frac{\mathrm{d}^2}{\mathrm{d}\boldsymbol{\theta}^2} K_{\dot{\sigma}}(\boldsymbol{\theta} = 0) = \boldsymbol{\sigma}_{\boldsymbol{\sigma}}^2 (b^2 - a^2)$$

and

$$\begin{aligned} \overline{n}_a &= \boldsymbol{\theta}_{\max} \frac{\sqrt{b^2 - a^2}}{2\pi} \exp\left(-\frac{(\boldsymbol{\sigma} - \boldsymbol{\mu}_{\boldsymbol{\sigma}})^2}{2\boldsymbol{\sigma}_{\boldsymbol{\sigma}}^2} \right) = 10 \frac{\sqrt{b^2 - a^2}}{2\pi} \exp\left(-\frac{(\boldsymbol{\sigma} - \boldsymbol{\mu}_{\boldsymbol{\sigma}})^2}{2\boldsymbol{\sigma}_{\boldsymbol{\sigma}}^2} \right) \\ &= C \exp\left(-\frac{(\boldsymbol{\sigma} - \boldsymbol{\mu}_{\boldsymbol{\sigma}})^2}{2\boldsymbol{\sigma}_{\boldsymbol{\sigma}}^2} \right); \quad C = 10 \frac{\sqrt{b^2 - a^2}}{2\pi} \end{aligned} \tag{7.32}$$

Taken natural logarithms of both sides of Eq. (7.32), we have now

$$\begin{aligned}
&\ln P_{\rm a} = \ln \bar{n}_a = \ln C - \frac{(\boldsymbol{\sigma} - \boldsymbol{\mu}_\sigma)^2}{2\boldsymbol{\sigma}_\sigma^2} \\
&\text{or: } (\boldsymbol{\sigma} - \boldsymbol{\mu}_\sigma)^2 = \ln \frac{C}{P_{\rm a}} \\
&\boldsymbol{\sigma}_{\min} = \boldsymbol{\mu}_\theta - \boldsymbol{\sigma}_\sigma \sqrt{2\ln \frac{C}{P_{\rm a}}} \\
&\text{where } P_f = P_{\rm a} - \text{probability of crossing down level “a”}
\end{aligned} \tag{7.33}$$

It will be assumed now that $P_{\rm f}$ the stochastic creep stress–temperature–time function is presented as a stationary (steady time) Gaussian process, therefore the probability of structural failure is a function of minimum allowable stress and time. The minimum allowable random stress variable "$\sigma_{\min}$" has a normal probability distribution

$$f(\boldsymbol{\sigma}) = \frac{1}{\boldsymbol{\sigma}_\sigma \sqrt{2\pi}} \exp\left(-\frac{(\boldsymbol{\sigma} - \boldsymbol{\mu}_\sigma)^2}{2(\boldsymbol{\sigma}_\sigma)^2} \right)$$

The average number of the occurrences below a given level "a" for stationary Gaussian processes during temperature rise from 0 to $\theta_{\max}$ = 10 is defined as follows [13]:

$$\boldsymbol{\sigma}_{\min} = m_\sigma - \boldsymbol{\beta}_{\rm all} \boldsymbol{\sigma}_\sigma \tag{7.34}$$

$$\boldsymbol{\beta}_{\rm all} = \sqrt{2\ln \frac{C}{P_{\rm a}}} \tag{7.35}$$

Equation (7.35) in some respect similar to Eq. (7.28) that have been obtained earlier, where, in the present notation, the relation $\sigma_{\min}$ (μ_σ, σ_σ) is based on the reliability index $\beta_{\rm all}$.

It is convenient first to redefine the reliability index $\beta_{\rm all}$ as the first-occurrence stress so that the probability that the first downward crossing occurs in the temperature interval $0 < \theta < 10$. This differs from the previous definition of β; the difference is very important, since the new parameter $\beta_{\rm all}$ is based on statistical data of stress–temperature–strain diagrams for a given fire severity case (autocorrelation function) and the temperature rise interval. Defining failure as the first exceedance of $\sigma = a$, the probability $P_{\rm a}(a, T)$ of failure for stationary Gaussian processes requires only a knowledge of the mean and variance of random function σ and derivative $\dot{\boldsymbol{\sigma}}$ [32]. In addition it is required that the correlation function of the random process and its first two derivatives be known for all time. Obviously, the design temperatures on unexposed side of a structural element will be a function of type of insulation material; thickness and the thermo-diffusivity parameter of it [33]. It is important to underline here that the reliability parameter β_a is based only on stress–temperature–strain statistical data and it does not include the combination of

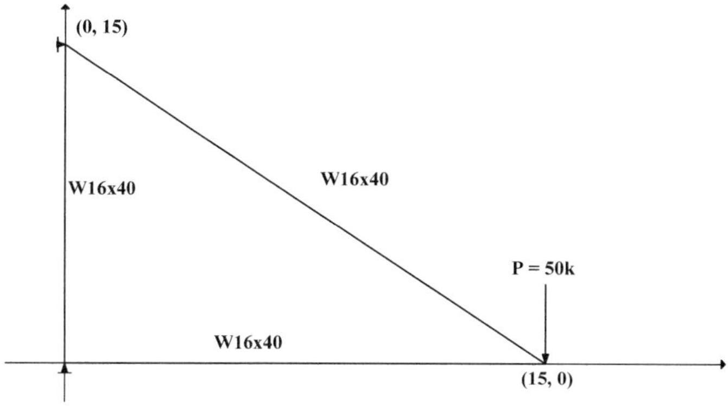

Fig. 7.7 Structural steel frame model

all structural design loads (limit state design), therefore such computations of reliability index should be considered as conservative (the same way as it has been done by author [34] with corresponding computations of structural fire load, where structural fire resistance was assumed to be a given deterministic value). Finally, we have the ability now fully implementing the FORM (SORM) methodology in obtaining the reliability index β. The step-by-step structural design computations are presented below via example.

Example 7.2 Data: The main data are taken from [] and partially reproduced here. The example consists of a steel frame depicted in Fig. 7.7, with all structural elements W 16 × 40 (cross-sectional area $A = 11.8\ \text{in}^2$; $Z_x = 73\ \text{in}^3$; $F_y = 50$ ksi, subjected to dead load (denoted $P = 50$ k) and static maximum temperature load that has normal probability distribution with parameters $\mu_T = 1000$ °F and $\sigma_T = 100$ °F. Define the performance function, reliability index β and compute probability of structural failure. The computer output in this case is as follows:

A N A L Y S I S R E S U L T S

T R A N S L A T I O N S

	Translations (ft)			Rotations (Rad)		
Node	TX	TY	TZ	RX	RY	RZ
Condition dl = dead load						
1	0.00000	0.00000	0.00000	0.00000	0.00000	−0.00029
2	0.00000	0.00000	0.00000	0.00000	0.00000	−0.00011
3	−0.00319	−0.01312	0.00000	0.00000	0.00000	−0.00108
Condition tl = temp load						
1	0.00000	0.00000	0.00000	0.00000	0.00000	0.00403
2	0.00000	0.00000	0.00000	0.00000	0.00000	0.00330
3	*0.14819*	*0.21829*	0.00000	0.00000	0.00000	0.01986

MEMBERFORCES

Condition dl = dead load						
	M33	V2	M22	V3	Axial	Torsion
	(Kip * ft)	(Kip)	(Kip * ft)	(Kip)	(Kip)	(Kip * ft)
MEMBER 1						
0 %	13.07	2.24	0.00	0.00	0.00	0.00
25 %	7.48	2.24	0.00	0.00	0.00	0.00
50 %	1.89	2.24	0.00	0.00	0.00	0.00
75 %	−3.71	2.24	0.00	0.00	0.00	0.00
100 %	−9.30	2.24	0.00	0.00	0.00	0.00
MEMBER 2						
0 %	−9.30	−0.40	0.00	0.00	87.72	0.00
25 %	−7.48	−0.40	0.00	0.00	87.72	0.00
50 %	−5.65	−0.40	0.00	0.00	87.72	0.00
75 %	−3.83	−0.40	0.00	0.00	87.72	0.00
100 %	−2.00	−0.40	0.00	0.00	87.72	0.00
MEMBER 3						
0 %	*−13.07*	−1.01	0.00	0.00	*−72.76*	0.00
25 %	−9.30	−1.01	0.00	0.00	−72.76	0.00
50 %	−5.53	−1.01	0.00	0.00	−72.76	0.00
75 %	−1.77	−1.01	0.00	0.00	−72.76	0.00
100 %	2.00	−1.01	0.00	0.00	−72.76	0.00
Condition tl = temp load						
	M33	V2	M22	V3	Axial	Torsion
	(Kip * ft)	(Kip)	(Kip * ft)	(Kip)	(Kip)	(Kip * ft)
MEMBER 1						
0 %	−214.17	−41.30	0.00	0.00	0.00	0.00
25 %	−110.92	−41.30	0.00	0.00	0.00	0.00
50 %	−7.67	−41.30	0.00	0.00	0.00	0.00
75 %	95.59	−41.30	0.00	0.00	0.00	0.00
100 %	198.84	−41.30	0.00	0.00	0.00	0.00
MEMBER 2						
0 %	198.84	11.40	0.00	0.00	42.04	0.00
25 %	147.47	11.40	0.00	0.00	42.04	0.00
50 %	96.11	11.40	0.00	0.00	42.04	0.00
75 %	44.74	11.40	0.00	0.00	42.04	0.00
100 %	-6.62	11.40	0.00	0.00	42.04	0.00
MEMBER 3						
0 %	*214.17*	13.84	0.00	0.00	*−41.30*	0.00
25 %	162.28	13.84	0.00	0.00	−41.30	0.00
50 %	110.40	13.84	0.00	0.00	−41.30	0.00
75 %	58.51	13.84	0.00	0.00	−41.30	0.00
100 %	6.62	13.84	0.00	0.00	−41.30	0.00

The deterministic performance function in this case is defined by Eq. H2-1 [35]

$$
\begin{aligned}
&g = \frac{72.76 + 0.041(\mu_{sT})}{50(11.8)} + \frac{13.07 + 0.214(12)(\mu_{sT})}{(50)73} - \mu_{\sigma R} = 0 \\
&\mu_{\sigma S} = 0.166 + 0.000773(\mu_{sT}); \quad \sigma_{\sigma S} = 0.000773(\sigma_{sT})\mu_{sT} = 1000\ {}^{\circ}\mathrm{F} \\
&\text{where } \mu_{\sigma R} = 1.838; \quad \sigma_{\sigma R} = 0.391 \rightarrow \text{see computations above} \\
&\beta = \frac{|\mu_{\sigma R} - \mu_{\sigma S}|}{\sqrt{(\sigma_{\sigma S})^2 + (\sigma_{\sigma R})^2}} = \frac{1.838 - 0.939}{\sqrt{0.0773^2 + 0.391^2}} = 2.256
\end{aligned}
\tag{7.36}
$$

The probability of structural failure and the design temperature in this case are computed as follows:

$$
\begin{aligned}
P_f &= \Phi * (-\boldsymbol{\beta}) = \Phi * (-2.256) = 0.0123; \\
T_{\text{design}} &= 1000 + (2.256)100 = 1225.6\ {}^{\circ}\mathrm{F}
\end{aligned}
\tag{7.37}
$$

7.6 Confidence Interval—Minimum Dimensionless Allowable Stress

The dimensionless minimum allowable stress σ represents the unrestraint information of input data in computerized (deterministic) analysis of limit state equation. Now, we have an opportunity to quantify our structural analysis based on probabilistic approach with respect to ranges of minimum allowable stresses.

A statistical inference is a quantifiable statement about either a population parameter or a future random variable. There are many varieties of statistical inference, but we will focus on just two of them: parameter estimation and confidence intervals. Parameter estimation is conceptually the simplest. Estimation is done by giving a single number which represents a guess at an unknown population parameter. If $\boldsymbol{\sigma}_{\min,1}; \boldsymbol{\sigma}_{\min,2}; \ldots \boldsymbol{\sigma}_{\min,n}$ is a sample of n values from a population with unknown mean μ, then we might consider using $\overline{\boldsymbol{\sigma}}_{\min}$ as an estimate of μ_{σ}. We would write: $\tilde{\boldsymbol{\mu}}_{\boldsymbol{\sigma}} = \overline{\boldsymbol{\sigma}}_{\min}$. This is not the only estimate of μ, but it makes a lot of sense. A confidence interval is an interval which has a specified (target) probability of containing an unknown population parameter $\overline{\boldsymbol{\sigma}}_{\min}$.

The sample of n values $\boldsymbol{\sigma}_{\min,1}; \boldsymbol{\sigma}_{\min,2}; \ldots \boldsymbol{\sigma}_{\min,n}$ from a population which is assumed to be normal and has an unknown mean $\overline{\boldsymbol{\sigma}}_{\min} = \text{const}$. Let us choose now a large probability α (for example, $\alpha = 0.9$; 0.95; or 0.99) that the event becomes almost certain with probability α. We have to find now such small number ε that $\mathbf{P}\{|\tilde{\boldsymbol{\mu}}_{\boldsymbol{\sigma}} - \overline{\boldsymbol{\sigma}}_{\min}| < \varepsilon\} = \boldsymbol{\alpha}$. Once the data leads to actual numbers, you will make a statement of the form "I'm 95 % confident that the value of $\overline{\boldsymbol{\sigma}}_{\min}$ lies between $\tilde{\boldsymbol{\mu}}_{\boldsymbol{\sigma}} - \varepsilon$ and $\tilde{\boldsymbol{\mu}}_{\boldsymbol{\sigma}} + \varepsilon$". The value $\overline{\boldsymbol{\sigma}}_{\min}$ can be substituted by the unbiased estimate $\tilde{\boldsymbol{\mu}}_{\boldsymbol{\sigma}}$ and the error will be not more than $\pm\varepsilon$, or:

Fig. 7.8 Confidence Interval

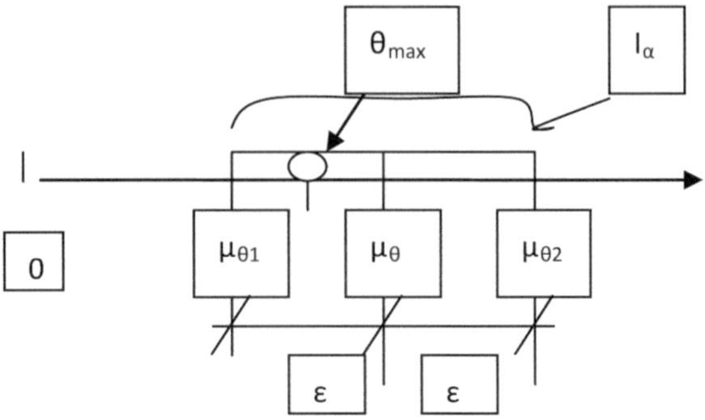

$$\mathbf{P}\{\tilde{\boldsymbol{\mu}}_\sigma - \boldsymbol{\varepsilon} < \overline{\boldsymbol{\sigma}}_{\min} < \tilde{\boldsymbol{\mu}}_\sigma + \boldsymbol{\varepsilon}\} = \boldsymbol{\alpha}. \tag{7.38}$$

Equation (7.38) means that the unknown mean value lies inside of the interval $\mathbf{I}_\alpha = (\tilde{\boldsymbol{\mu}}_\sigma - \boldsymbol{\varepsilon}; \tilde{\boldsymbol{\mu}}_\sigma + \boldsymbol{\varepsilon})$ with probability α (Fig. 7.8).

We have to underline here that an unknown mean value $\overline{\boldsymbol{\sigma}}_{\min} = \text{const}$ is not a random, however the location of $\tilde{\boldsymbol{\mu}}_\sigma$ is random and the upper and lower limits of the confidence interval are also random, since they are based on statistical data. Therefore, the probability value α is not the probability of mean value $\overline{\boldsymbol{\sigma}}_{\min}$ falling into the interval I_α, but rather the probability of the event that the random interval I_α will cover the point $\overline{\boldsymbol{\sigma}}_{\min}$. In other words, confidence interval is an interval which has a specified probability α of containing an unknown population parameter $\overline{\boldsymbol{\sigma}}_{\min}$. The procedure of obtaining small value ε based on given specified probability α is simple, if the cumulative distribution function of $\tilde{\boldsymbol{\mu}}_\sigma$ is known in advance. For example, if the population is assumed to be exactly normal to start with, then $\overline{\boldsymbol{\sigma}}_{\min}$ is automatically normally distributed. (This is not a use of the Central Limit theorem, but just the fact that the sum of n normal distributed random variables has normal probability distribution). However, the cumulative distribution function of $\tilde{\boldsymbol{\mu}}_\theta$ is not known beforehand. If one does not make the assumption that the population is exactly normal to start with, then $(\tilde{\boldsymbol{\mu}}_\sigma - \boldsymbol{\mu}_\sigma)/(\tilde{\boldsymbol{\sigma}}_\sigma/\sqrt{n})$ is approximately normal, provided n is large enough (This is precisely the Central Limit theorem.). The official standard is that n should be at least 30. The result works well for n as small as 10 in our case [sv]. In this case, with a sample of n values from a population with sample mean $\tilde{\boldsymbol{\mu}}_\sigma$ and sample standard deviation $s = \tilde{\boldsymbol{\sigma}}_\sigma$, the Central Limit theorem gives us the result that $Z = (\tilde{\boldsymbol{\mu}}_\sigma - \boldsymbol{\mu}_\sigma)/(\tilde{\boldsymbol{\sigma}}_\sigma/\sqrt{n})$ is approximately normally distributed with mean 0 and with standard deviation 1. Let us use $\tilde{\boldsymbol{\mu}}_\sigma$ for the mean of the population and use σ_σ for the standard deviation of the population. In all realistic problems, the values of $\tilde{\boldsymbol{\mu}}_\sigma$ and σ_σ are both unknown. The statistical work will be based on the sample mean $\tilde{\boldsymbol{\mu}}_\sigma$ and the sample standard deviation $\tilde{s}_\sigma$. The use of 95 % confidence intervals is most common and it will be used in this book, but there are occasions $\overline{\boldsymbol{\sigma}}_{\min}$ when we need 90 % or 99 % intervals. Clearly $\tilde{\boldsymbol{\mu}}_\sigma$, the sample mean, will be used as the estimate for $\bar{\boldsymbol{\sigma}}_{\min}$. Since $\mathrm{SD}(\tilde{\boldsymbol{\mu}}_{\boldsymbol{\sigma}}) = \tilde{\boldsymbol{\sigma}}_{\boldsymbol{\sigma}}/\sqrt{n}$, we will use $\mathrm{SE}(\tilde{\boldsymbol{\mu}}_{\boldsymbol{\sigma}}) = \tilde{\boldsymbol{\sigma}}_{\boldsymbol{\sigma}}/\sqrt{n}$ for the standard error. As a consequence of the fact that $(\tilde{\boldsymbol{\mu}}_{\boldsymbol{\sigma}} - \boldsymbol{\mu}_{\boldsymbol{\sigma}})/(\tilde{\boldsymbol{\sigma}}_{\boldsymbol{\sigma}}/\sqrt{n})$ follows the t—distribution with $n - 1$ degrees of freedom, we can given the 95 % confidence interval for $\bar{\boldsymbol{\sigma}}_{\min}$ as: $\tilde{\boldsymbol{\mu}}_{\boldsymbol{\sigma}} \pm t_{0.025,n-1}\left[s/\sqrt{n}\right]$. The t-table has been used very often in obtaining the confidence interval of the mean value. However, it is not applicable for obtaining the confidence interval of the standard deviation value (in this case the χ^2 distribution with $(n - 1)$ degrees of freedom has to be used) and this in turns makes the total process more complex. Instead the simple approximate method presented in [34] will be used throughout this book.

The step-by-step procedure in this case is as follows (for simplicity the subscript "σ" will be omitted).

1. Obtain the confidence limit (interval) I_α of mean value (μ_α) and standard deviation (σ_θ) that corresponds to a given high confidence probability "α" that is accepted (approved) by the owner or building official

$$P(|\tilde{\boldsymbol{\mu}} - \boldsymbol{\mu}| < \varepsilon_\alpha) = \boldsymbol{\alpha} \tag{7.39}$$

In case of normal probability distribution formulae (7.39) is written as

$$\begin{gathered} P(|\tilde{\boldsymbol{\mu}} - \boldsymbol{\mu}| < \varepsilon_\alpha) = 2\Phi * \left(\frac{\varepsilon_\alpha}{\boldsymbol{\sigma}_{\tilde{\boldsymbol{\mu}}}}\right) - 1 = \boldsymbol{\alpha} \\ \varepsilon_\alpha = \boldsymbol{\sigma}_{\tilde{\boldsymbol{\mu}}} \arg \Phi * \left(\frac{1+\boldsymbol{\alpha}}{2}\right); \quad t_\alpha = \arg \Phi * \left(\frac{1+\boldsymbol{\alpha}}{2}\right) \end{gathered} \tag{7.40}$$

where $\arg \Phi * (x)$ is an inverse function and the values of it can be founded from Table 7.5.

The standard deviation "$\boldsymbol{\sigma}_{\tilde{\boldsymbol{\mu}}}$" can be approximately estimated as:

$$\boldsymbol{\sigma}_{\tilde{\boldsymbol{\mu}}} = \sqrt{\frac{\boldsymbol{\sigma}_{\boldsymbol{\sigma}}^2}{n}} \tag{7.41}$$

where $\boldsymbol{\sigma}_{\boldsymbol{\sigma}}^2$ is defined from Table 7.6 and $n = 10 - 1 = 9$.

Table 7.5 Inverse Function $arg\ \Phi * (x)$

α	t_α	α	t_α	α	t_α	α	t_α
0.80	1.282	0.86	1.475	0.91	1.694	0.97	2.169
0.81	1.310	0.87	1.513	0.92	1.750	0.98	2.325
0.82	1.340	0.88	1.554	0.93	1.810	0.99	2.576
0.83	1.371	0.89	1.597	0.94	1.880	0.9973	3.0
0.84	1.404	0.90	1.643	0.95	1.96	0.999	3.29
0.85	1.439			0.96	2.053		

Table 7.6 Autocorrelation statistical data table: $\sigma_i(\theta_k)$

$\alpha\|\theta$		1	2	3	4	5	6	7	8	9	10
0.001	0	0.85	1.378	1.57	1.438	1.035	0.645	0.388	0.224	0.107	0.001
0.01	0	0.85	1.378	1.57	1.438	1.035	0.645	0.388	0.225	0.109	0.005
0.1	0	0.85	1.378	1.571	1.439	1.036	0.647	0.392	0.233	0.126	0.042
0.2	0	0.85	1.378	1.572	1.439	1.037	0.65	0.397	0.243	0.144	0.076
0.33	0	0.85	1.378	1.572	1.44	1.038	0.652	0.403	0.255	0.165	0.112
0.5	0	0.852	1.378	1.572	1.44	1.04	0.656	0.411	0.269	0.189	0.147
1.0	0	0.852	1.379	1.573	1.442	1.045	0.667	0.433	0.307	0.244	0.212
2.0	0	0.852	1.379	1.575	1.446	1.054	0.688	0.473	0.368	0.315	0.282
3.0	0	0.852	1.380	1.577	1.45	1.063	0.708	0.509	0.414	0.363	0.327
5.0	0	0.852	1.381	1.580	1.458	1.080	0.746	0.569	0.484	0.43	0.389
10	0	0.852	1.384	1.589	1.477	1.122	0.826	0.68	0.598	0.537	0.486
100	0	0.855	1.417	1.691	1.692	1.525	1.379	1.255	1.143	1.042	0.951
1000	0	0.859	1.467	1.856	2.055	2.106	2.067	1.977	1.861	1.736	1.611
10,000	0	0.860	1.480	1.906	2.177	2.327	2.382	2.368	2.305	2.21	2.095
100,000	0	0.861	1.481	1.912	2.193	2.358	2.433	2.441	2.398	2.319	2.215
r	0	0.861	1.482	1.913	2.195	2.362	2.439	2.450	2.41	2.333	2.231

Finally, the confidence limit (interval) I_β of mean value (m_γ) can be calculated now as follows:

$$I_\alpha = (\tilde{\boldsymbol{\mu}}_\sigma - t_\alpha \boldsymbol{\sigma}_{\tilde{\mu}}; \quad \tilde{\boldsymbol{\mu}}_\sigma + t_\alpha \boldsymbol{\sigma}_{\tilde{\mu}}) \tag{7.42}$$

If the confidence limit (interval) I_α of mean value (μ_σ) is less than the "original" one that was based on engineering experience/judgment, then the range of minimum allowable stresses for each "Fire Severity Case" is conservatively acceptable. Obviously, if the confidence limit (interval) I_α is larger than the "original" one, then the range of minimum allowable stresses must be decreased or the confidence probability "α" factor must be increased.

2. The confidence limit (interval) I_α of *variance* value $(\boldsymbol{\sigma}^2_{\sigma_\sigma})$ can be calculated as follows:

$$I_\alpha = (\boldsymbol{\sigma}^2_\sigma - t_\alpha \boldsymbol{\sigma}_{\sigma_\sigma}; \boldsymbol{\sigma}_\sigma + t_\alpha \boldsymbol{\sigma}_{\sigma_\sigma})$$
$$\text{where } \boldsymbol{\sigma}_{\sigma_\sigma} = \sqrt{\frac{2}{n-1}}\boldsymbol{\sigma}^2_\sigma \tag{7.43}$$

3. Therefore, the confidence limit (interval) I_α of standard deviation value $(\boldsymbol{\sigma}_{\sigma_\sigma})$ can be calculated as follows:

$$I_{\alpha}^{\sigma_{\sigma\sigma}} = \left(\sqrt{\boldsymbol{\sigma}_{\sigma}^{2} - t_{\alpha}\boldsymbol{\sigma}_{\sigma_{\sigma}}};\ \sqrt{\boldsymbol{\sigma}_{\sigma}^{2} + t_{\alpha}\boldsymbol{\sigma}_{\sigma_{\sigma}}}\right) \tag{7.44}$$

Let us apply formulas above to Fire severity cases 1—Very Fast Fire:

Example 7.3 Data: Very Fast Fire: $\alpha = 0.95$—given target confidence probability
From Table 7.1: $\mu_{\sigma R} = 1.838$; $\sigma_{\sigma R} = 0.391$.

Let us obtain now the confidence limit (interval) I_{α} of mean value (μ_{σ}) and standard deviation (σ_{σ}) that corresponds to a given high confidence probability "α" that is accepted (approved) by the owner or building official:

In case of normal probability distribution

$$\boldsymbol{\sigma}_{\boldsymbol{\mu}_{\sigma}} = \sqrt{\frac{\boldsymbol{\sigma}_{\sigma}^{2}}{n}} = \sqrt{\frac{(0.391)^{2}}{10}} = 0.124$$

$$\boldsymbol{\varepsilon}_{\alpha} = \boldsymbol{\sigma}_{\boldsymbol{\mu}_{\sigma}} \arg \Phi * \left(\frac{1+\boldsymbol{\alpha}}{2}\right) = 0.124\left[\arg \Phi * \left(\frac{1+0.95}{2}\right)\right] = 0.124(1.96) = 0.242$$

$$t_{\alpha} = \arg \Phi * \left(\frac{1+\boldsymbol{\alpha}}{2}\right) = 1.96$$

Finally, the confidence limit (interval) I_{α} for mean value is calculated as follows:

$$I_{\alpha} = (\boldsymbol{\mu}_{\sigma} - t_{\alpha}\boldsymbol{\sigma}_{\boldsymbol{\mu}_{\sigma}};\ \boldsymbol{\mu}_{\sigma} + t_{\alpha}\boldsymbol{\sigma}_{\boldsymbol{\mu}_{\sigma}}) = (1.838 - 0.242;\ 1.838 + 0.242)$$
$$= (1.596;\ 2.08)$$

The confidence limit (interval) I_{α} for standard deviation value is calculated as follows:

$$I_{\alpha} = (\boldsymbol{\sigma}_{\sigma}^{2} - t_{\alpha}\boldsymbol{\sigma}_{\sigma_{\sigma}};\ \boldsymbol{\sigma}_{\sigma}^{2} + t_{\alpha}\boldsymbol{\sigma}_{\sigma_{\sigma}}) = (0.391^{2} - (1.96)0.072;\ 0.391^{2} + (1.96)0.072)$$
$$= (0.0118;\ 0.294)$$

where $\boldsymbol{\sigma}_{\sigma_{\sigma}} = \sqrt{\frac{2}{n-1}\boldsymbol{\sigma}_{\sigma}^{2}} = \sqrt{\frac{2}{10-1}(0.391)^{2}} = 0.072$

and $I_{\alpha}^{\sigma_{\sigma\sigma}} = \left(\sqrt{\boldsymbol{\sigma}_{\sigma}^{2} - t_{\alpha}\boldsymbol{\sigma}_{\sigma_{\sigma}}};\ \sqrt{\boldsymbol{\sigma}_{\sigma}^{2} + t_{\alpha}\boldsymbol{\sigma}_{\sigma_{\sigma}}}\right);\ \sqrt{0.0118}\sqrt{0.294}) = (0.109;\ 0.542)$

The minimum allowable dimensionless stress is

$$\boldsymbol{\sigma}_{\text{min,all}} = 1.596 - 0.542 = 1.054$$

The difference in minimum allowable dimensionless stress is: $\overline{\boldsymbol{\sigma}}_{\text{min}} = 10.054/10.07 = 0.985 = 1.5\ \%$.

7.7 Structural Failures in Time

The structural system subjected to random temperature–time loading may fail when the stress in a critical member reaches a sufficiently high level. This type of failure is generally by overstress or by excessive permanent deformation rendering the structural element or system inoperative. If the allowable stress has a finite probability of crossing downwards the given level, then failure is possible, and an important problem is to find the probability that the system can operate without failure for a given time (duration of fire event). More precisely, the following problem is considered. Given a continuous and differentiable random function $\sigma(\theta)$, one wishes to find the probability that the value $\sigma = a$ will not be crossed in temperature interval $(0, \theta_{max})$. This problem is called the first-occurrence time problem and the probability density $P(a, \theta)$ is the first-occurrence density [31]. The probability of failure in $(0, \theta_{max})$ is unity if $X(0) > a$, and the probability of failure in $(0, \theta_{max})$ is (if $\sigma(\theta = 0) < a$) (Fig. 7.9):

$$P[\boldsymbol{\sigma}(\boldsymbol{\theta}_j) > a] = \int_a^{\infty} f(\boldsymbol{\sigma}|\boldsymbol{\theta}_j)\mathrm{d}\boldsymbol{\theta} \tag{7.45}$$

In the 1970s, the analysis of the response variability of stochastic structural systems received a lot of attention, consequently a new field; "Stochastic Finite Elements" was coined to stochastic mechanics. Although there have been papers on computationally expensive Monte Carlo solutions and reliability considerations, most of the studies done in stochastic finite elements have been on the second moment analysis of the response of deterministic structural systems under

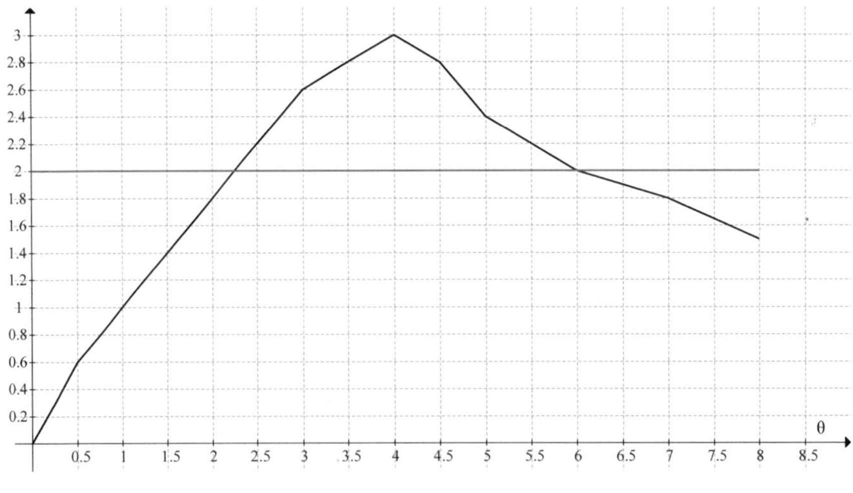

Fig. 7.9 The first-occurrence time

stochastic loading. This chapter considers the approximate solution methods for the stochastic allowable stress or strain due to stochastic nature of fire dynamic and consequently creep deformation problems. In this respect, ordinary perturbations mean-centered second-order perturbation of stress–temperature–strain relationships and ergodic expansion for the covariance function has been proposed as solution procedures. The earliest use of what would now be called *perturbation theory* was to deal with the otherwise unsolvable mathematical problem. Perturbation methods start with a simplified form of the original problem, which is *simple enough* to be solved exactly (or numerical integration results in tables), that is the integral type simplified model of creep constitutive equation (deterministic approach). The solved, but simplified problem is then "*perturbed*" to make the conditions that the perturbed solution actually satisfies closer to the real problem (real fire severity scenario), such as including the statistical dimensionless temperature–time and its first derivative dependable information. The changes that result from accommodating the perturbation, which themselves may have been simplified yet again, are used as corrections to the approximate solution. Because of simplifications introduced along every step of the way, the corrections are never perfect, and the conditions met by the corrected solution do not perfectly match the equation demanded by reality, but even one cycle of corrections often provides a remarkably better approximate answer to what the real solution should be. The partial derivatives of the stochastic processes with respect to random variables are also proportional to temperature. Structural uncertainties due to physical imperfections, model inaccuracies and system complexities are spatially distributed over the structure and can be mathematically modeled using either random variables or random processes which may be functions of time and/or space. The probabilistic study of the dynamic behavior of structural and mechanical systems requires the characterization of the random processes describing the input excitation and the structural response. A very common and powerful methodology for characterizing and describing a random process is the second moment analysis and consequently the spectral analysis, which studies random processes in the frequency domain. In particular, the use of power spectral density (PSD) functions is customary in describing stationary random processes. Definition of functions describing the spectral properties of stationary random processes is simple.

In this book, the definition of random functions is extended to general real-valued stationary random processes. These defined quantities are used in here to solve exactly and in closed-form the classical problem of computing the temperature–time variant reliability index (parameter) of the stresses and the response processes of single degree of freedom (SDOF) and both classically and non-classically damped linear elastic systems subjected to excitation from at rest initial conditions.

For the sake of simplicity and without loss of generality, the statistical data construction (based on deterministic approach of fire dynamic simulation and integral type of creep constitutive law) process of autocorrelation function; it normalization process and prove that the stationary process is ergodic are presented in details in this chapter only. All random processes considered in this book are ergodic, zero-mean processes. The application of autocorrelation theory and spectral analysis to other fire severity categories (Fast Fire; Medium Fire and Slow Fire) is very similar to the Very Fast Fire and do not require the elaborate comments.

The step-by-step procedure of the second moment analysis of the stochastic allowable stresses due to creep deformations in case of fully developed fire is as follows.

1. Create Statistics Data
   Mainstream creep analysis uses a deterministic approach to handle the compelation of required statistical data. To start with, let us say that the stochastic allowable stresses due to creep deformation process $\sigma(\theta)$ has n realizations ($\sigma_1(\theta)$, $\sigma_2(\theta)$, … $\sigma_n(\theta)$). In our case n = 10 (for each $\alpha = 10^{-3}, 10^{-2}, \ldots 10^4, 10^5$—see computations above). Instead of using reliability index β, the second moment analysis quite accurately determines the worst outcome (σ = a will be crossed downwards in the temperature interval (0, θ_{max})) and the probability of that outcome occurring in real fire scenario. This makes for a better prediction of how variability will affect each design specifically with less pessimism by targeting a slightly lower probability. Each realization function $\sigma_i(\theta)$, obviously, is just a real function (not random!) and it has a real value for any given independent value of α_k. The statistical data from are compiled and represent lows in Table 7.6.
2. In order to simplify any further computations and without loss of generality the statistical data from Table 7.6 has to be centered: the corresponding means value $\mu_\sigma(\theta_k)$ will be subtracted from the data in Table 7.6.
3. Let us find now the estimates of elements of autocorrelation matrix: the variances and correlation moments. In order to obtain the variances the following steps are required:

   1. Compute the sum of squares of matrix column elements;
   2. Divide this sum by number of rows (n = 10 in our case);
   3. In order to get the unbiased estimate multiply the result from p. 2 by the ratio $[n/n - 1]$;
   4. Repeat computations (p.p. 1, 2, 3) for each matrix column. These are the diagonal members in Correlation Matrix

   In order to obtain the correlation moments the following steps are required:

   5. Compute the sum of products of two different matrix column elements (the second moments are centered, therefore $\mu_\sigma(\theta_k) \equiv 0$);
   6. Divide this sum by number of rows (n = 10 in our case);
   7. In order to get the unbiased estimate multiply the result from p. 6 by the ratio $[n/n - 1]$;
   8. Repeat computations (p.p. 5, 6, 7) for all combinations of matrix column. These are the diagonal members in Correlation Matrix
   9. The variances are located on main diagonal of correlation matrix. The square root of these elements are representing the function: dimensionless standard deviations − dimensionless time: $\sigma_\sigma(\theta)$.

Note: The correlation function is even function, therefore only half of the matrix is required.

4. In order to the normalized cross-correlation matrix each element must be divided by product of corresponding variances
5. Finally, let us construct the normalized function (table form) by adding diagonal elements and dividing the sum on number diagonal elements [34]. For instance:

$$\boldsymbol{\rho}_{\boldsymbol{\sigma}}(\boldsymbol{\theta}_1 = 0) = (1 + 1 + \cdots + 1)/10 = 1 \tag{7.46}$$

6. Let us approximate the normalized cross-correlation function as:

$$\boldsymbol{\rho}_{\boldsymbol{\sigma}}(\boldsymbol{\theta}) = (\exp(-a * |\theta|)) * \cos(b * \theta) \tag{7.47}$$

The corresponding spectral function is [32]:

$$S(\omega) = \frac{1}{2}\frac{\sigma_{\sigma}^2}{\pi}\left[\frac{\alpha}{(\omega - \beta)^2 + \alpha^2} + \frac{\alpha}{(\omega + \beta)^2 + \alpha^2}\right] \tag{7.48}$$

7.8 Ergodicity: Very Fast Fire

The final data in this fire severity case are taken from Chap. 5 and presented below.

$$\begin{aligned} &\alpha = 0.001 \\ &\sigma = 1.428 * \theta - 0.381 * \theta\hat{}2 + 0.0337 * \theta\hat{}3 - 0.00097 * \theta\hat{}4 \end{aligned} \tag{7.49}$$

$$\begin{aligned} &\alpha = 0.01 \\ &\quad = 1.428 * \theta - 0.381 * \theta\hat{}2 + 0.0337 * \theta\hat{}3 - 0.00097 * \theta\hat{}4 \end{aligned} \tag{7.50}$$

$$\begin{aligned} &\alpha = 0.1 \\ &\sigma = 1.425 * \theta - 0.38 * \theta\hat{}2 + 0.0335 * \theta\hat{}3 - 0.001 * \theta\hat{}4 \end{aligned} \tag{7.51}$$

$$\begin{aligned} &\alpha = 0.2 \\ &\sigma = 1.424 * \theta - 0.379 * \theta\hat{}2 + 0.0332 * \theta\hat{}3 - 0.0093 * \theta\hat{}4 \end{aligned} \tag{7.52}$$

$$\begin{aligned} &\alpha = 0.33 \\ &\sigma = 1.423 * \theta - 0.378 * \theta\hat{}2 + 0.0332 * \theta\hat{}3 - 0.0093 * \theta\hat{}4 \end{aligned} \tag{7.53}$$

$$\begin{aligned}&\alpha = 0.5\\&\sigma = 0 + 1.431 * \theta - 0.381 * \theta^{\wedge}2 + 0.0336 * \theta^{\wedge}3 - 0.00094 * \theta^{\wedge}4\end{aligned} \tag{7.54}$$

$$\begin{aligned}&\alpha = 1\\&\sigma = 1.432 * \theta - 00.383 * \theta^{\wedge}2 + 00.034 * \theta^{\wedge}3 - 00.00096 * \theta^{\wedge}4\end{aligned} \tag{7.55}$$

$$\begin{aligned}&\alpha = 2\\&\sigma = 0 + 1.453 * \theta - 0.393 * \theta^{\wedge}2 + 0.0359 * \theta^{\wedge}3 - 0.00106 * \theta^{\wedge}4\end{aligned} \tag{7.56}$$

$$\begin{aligned}&\alpha = 3\\&\sigma = 0 + 1.453 * \theta - 00.396 * \theta^{\wedge}2 + 0.0366 * \theta^{\wedge}3 - 0.0011 * \theta^{\wedge}4\end{aligned} \tag{7.57}$$

$$\begin{aligned}&\alpha = 0.5\\&\sigma = 0 + 1.458 * \theta - 0.4 * \theta^{\wedge}2 + 0.038 * \theta^{\wedge}3 - 0.00119 * \theta^{\wedge}4\end{aligned} \tag{7.58}$$

$$\begin{aligned}&\alpha = 10\\&\sigma = 1.457 * \theta - 0.402 * \theta^{\wedge}2 + 0.0391 * \theta^{\wedge}3 - 0.00128 * \theta^{\wedge}4\end{aligned} \tag{7.59}$$

$$\begin{aligned}&\alpha = 100\\&\sigma = 1.271 * \theta - 0.304 * \theta^{\wedge}2 + 0.0276 * \theta^{\wedge}3 - 0.00088 * \theta^{\wedge}4\end{aligned} \tag{7.60}$$

$$\begin{aligned}&\alpha = 1000\\&\sigma = 1.045 * \theta - 0.169 * \theta^{\wedge}2 + 0.0099 * \theta^{\wedge}3 - 0.000182 * \theta^{\wedge}4\end{aligned} \tag{7.61}$$

$$\begin{aligned}&\alpha = 10,000\\&\sigma = 0.99 * \theta - 0.141 * \theta^{\wedge}2 + 0.00814 * \theta^{\wedge}3 - 0.000184 * \theta^{\wedge}4\end{aligned} \tag{7.62}$$

$$\begin{aligned}&\alpha = 100,000\\&\sigma = 0.988 * \theta - 0.141 * \theta^{\wedge}2 + 0.00863 * \theta^{\wedge}3 - 0.000219 * \theta^{\wedge}4\end{aligned} \tag{7.63}$$

From the above, one can see that the steady time random function has an ergodic character, therefore only *one* random realization function is sufficient enough in order to obtain the correlation function in this case. The summary of all stress–temperature–strain relationships graphically are presented in Fig. 7.10.

Let us choose the random realization function that corresponds to α = 100.

$$\begin{aligned}&\alpha = 100\\&\sigma = 1.271 * \theta - 0.304 * \theta^{\wedge}2 + 0.0276 * \theta^{\wedge}3 - 0.00088 * \theta^{\wedge}4\end{aligned} \tag{7.64}$$

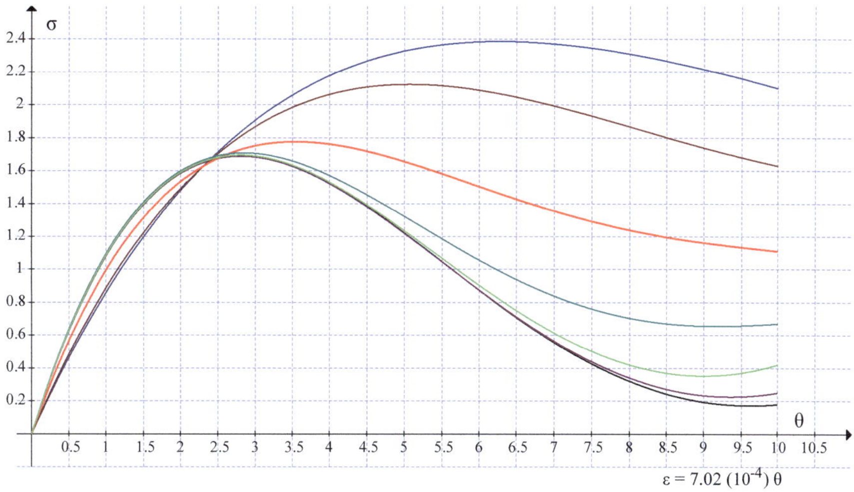

Fig. 7.10 Random realization functions

The mean value can be calculated now as follows:

$$\boldsymbol{\mu}_{\boldsymbol{\sigma}} = \frac{1}{10}\int_{0}^{10} \boldsymbol{\sigma}(\boldsymbol{\theta})\mathrm{d}\boldsymbol{\theta}$$

$$= \frac{1}{10}\int_{0}^{10} (\mathbf{1.271} * \boldsymbol{\theta} - \mathbf{0.304} * \boldsymbol{\theta}^\wedge \mathbf{2} + \mathbf{0.0276} * \boldsymbol{\theta}^\wedge \mathbf{3} - \mathbf{0.00088} * \boldsymbol{\theta}^\wedge \mathbf{4})\mathrm{d}\boldsymbol{\theta} = \frac{13.66}{10} = 1.37 \tag{7.65}$$

In order to calculate the correlation function in this case the chosen function (7.64) has to be centered (see Table 7.7). Again, after using the POLYMATH software we have:

Model: $z = a0 + a1 * t + a2 * t^\wedge 2 + a3 * t^\wedge 3 + a4 * t^\wedge 4$

Variable	Value
$a0$	−1.354451
$a1$	1.087588
$a2$	−0.2713503
$a3$	0.0271354
$a4$	−0.000996

Table 7.7 Centered correlation function (7.64)

$\alpha \vert \theta$	0	1	2	3	4	5	6	7	8	9	10
100	0	0.855	1.417	1.691	1.692	1.525	1.379	1.255	1.143	1.042	0.951
σ^*	−1.37	−0.515	0.047	0.321	0.322	0.155	0.009	−0.115	−0.227	−0.328	−0.419

$$\sigma * = -1.354 + 1.088(\theta) - 0.271(\theta^2) + 0.0271(\theta^3) - 0.000996(\theta^4) \quad (7.66)$$

The correlation function of a steady time process can be computed now as follows [ ]:

$$K_{\sigma}(\boldsymbol{\theta}) = \frac{1}{10 - \boldsymbol{\theta}} \int_0^{10-\boldsymbol{\theta}} [\boldsymbol{\sigma} * (t) \boldsymbol{\sigma} * (t + \boldsymbol{\theta})] \mathrm{d}t \quad (7.67)$$

where $\boldsymbol{\theta} = 0; 1; 2; \ldots 10$

After substituting (7.66) into (7.67) we have:
The standard deviation $D = K_{\sigma}(0)$ is

$$D = K_{\sigma}(\boldsymbol{\theta} = 0) = \frac{1}{10 - \boldsymbol{\theta}} \int_0^{10-\boldsymbol{\theta}} [\boldsymbol{\sigma} * (t) \boldsymbol{\sigma} * (t + \boldsymbol{\theta})] \mathrm{d}t = \frac{1}{10} \int_0^{10} [\boldsymbol{\sigma} * (t)]^2 \mathrm{d}t$$

$$= 0.1 \int_0^{10} [\mathbf{-1.354} + \mathbf{1.088(\theta)} - \mathbf{0.271(\theta^2)} + \mathbf{0.0271(\theta^3)} - \mathbf{0.000996(\theta^4)}]^2 \mathrm{d}\boldsymbol{\theta} = 0.12$$

$$\boldsymbol{\sigma}_{\sigma} = 0.346 \approx 0.391$$

(7.68)

The integral (7.68) has been calculated using POLYMATH software.

Calculated values of DEQ variables

	Variable	Initial value	Minimal value	Maximal value	Final value
1	t	0	0	10	10
2	y	0	0	1.200701	1.200701

Differential equations

1 $\mathrm{d}(y)/\mathrm{d}(t) = (-1.354 * 1 + 1.088 * (t) - 0.271 * (t\wedge 2) + 0.0271 * (t\wedge 3) - 0.000996 * (t\wedge 4))\wedge 2$

The computations of autocorrelation function based on Eq. (7.67) are presented below (see Table 7.8).

The computations of autocorrelation function values at each discrete dimensionless value $\theta = 0, 1, 2, \ldots, 10$ using again POLYMATH software are presented below.

Table 7.8 Autocorrelation function

θ	0	1	2	3	4	5	6	7	8	9	10
$K_\sigma(\theta)$	0.12	0.047	6.44×10^{-4}	−0.0236	−0.03	−0.0204	−0.024	−0.0224	−0.0082	0.0292	1.8×10^{-4}

Calculated values of DEQ variables

	Variable	Initial value	Minimal value	Maximal value	Final value
1	t	0	0	9	9
2	y	0	0	0.0467313	0.0467313

Differential equations

1 $$d(y)/d(t) = (1/(10-1)) * (-1.354 * 1 + 1.088 * (t) - 0.271 * (t^\wedge 2) + 0.0271 * (t^\wedge 3) - 0.000996 * (t^\wedge 4)) * (-1.354 * 1 + 1.088 * (t+1) - 0.271 * ((t+1)^\wedge 2) + 0.0271 * ((t+1)^\wedge 3) - 0.000996 * ((t+1)^\wedge 4))$$

Calculated values of DEQ variables

	Variable	Initial value	Minimal value	Maximal value	Final value
1	t	0	0	8	8
2	y	0	−0.0086101	0.0010564	0.0006441

Calculated values of DEQ variables

	Variable	Initial value	Minimal value	Maximal value	Final value
1	t	0	0	7	7
2	y	0	−0.0272586	0	−0.0236157

Calculated values of DEQ variables

	Variable	Initial value	Minimal value	Maximal value	Final value
1	t	0	0	6	6
2	y	0	−0.0301313	0	−0.0300032

Calculated values of DEQ variables

	Variable	Initial value	Minimal value	Maximal value	Final value
1	t	0	0	5	5
2	y	0	−0.020383	0	−0.020383

Calculated values of DEQ variables

	Variable	Initial value	Minimal value	Maximal value	Final value
1	t	0	0	4	4
2	y	0	−0.0239568	0	−0.0239568

Calculated values of DEQ variables

	Variable	Initial value	Minimal value	Maximal value	Final value
1	t	0	0	3	3
2	y	0	−0.0225026	0	−0.0224421

Calculated values of DEQ variables

	Variable	Initial value	Minimal value	Maximal value	Final value
1	t	0	0	2	2
2	y	0	−0.0082529	0.0010711	−0.0082529

Calculated values of DEQ variables

	Variable	Initial value	Minimal value	Maximal value	Final value
1	t	0	0	1	1
2	y	0	0	0.0292163	0.0292163

Calculated values of DEQ variables

	Variable	Initial value	Minimal value	Maximal value	Final value
1	t	0	0	0.001	0.001
2	y	0	0	0.0001831	0.0001831

The normalized autocorrelation function is (see Table 7.9)
Let us approximate the data from Table 7.9 by the same type of formulae (7.47):

$$\boldsymbol{\rho}_{\boldsymbol{\sigma}}(\boldsymbol{\theta}) = (\exp(-a*x))*(\cos(b*x)) \tag{7.69}$$

Model: $y = (\exp(-a*t))*(\cos(b*t))$

Variable	Initial guess	Value
a	2	0.3389246
b	1	0.6782014

$$\rho_\sigma(\theta) = (\exp(-0.34 * \theta)) * \cos(0.678 * \theta) \tag{7.70}$$

The ergodic function method is much simpler then the "classical" method of obtaining the correlation functions and it will be used in analysis of other fire severity scenarios ("Fast Fire"; "Medium Fire"; and "Slow" Fire). The corresponding spectral function is [32] (Fig. 7.11):

$$S(\boldsymbol{\omega}) = \frac{1}{2}\frac{\boldsymbol{\sigma}_\sigma^2}{\pi}\left[\frac{a}{(\boldsymbol{\omega}-b)^2+a^2}+\frac{a}{(\boldsymbol{\omega}+b)^2+a^2}\right] \tag{7.71}$$

The graphic presentation of spectral density is (see Fig. 7.12)

Table 7.9 Normalized autocorrelation function

	0	1	2	3	4	5	6	7	8	9	10
$\boldsymbol{\rho_\sigma(\theta)}$	1.0	0.392	0	−0.2	−0.25	−0.17	−0.2	−0.187	−0.068	0.243	0

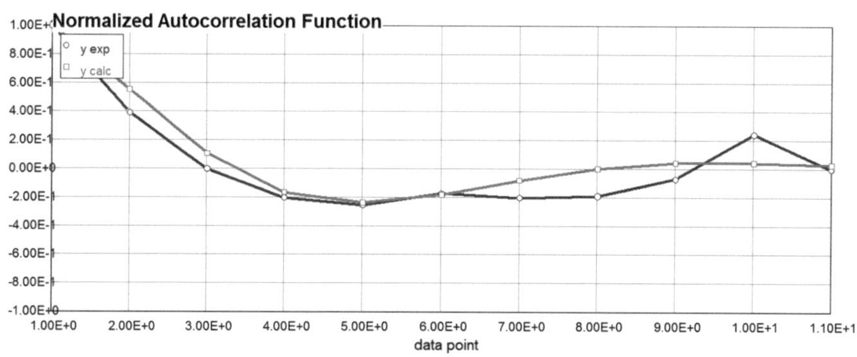

Fig. 7.11 Normalized correlation function $\rho_\sigma(\theta)$

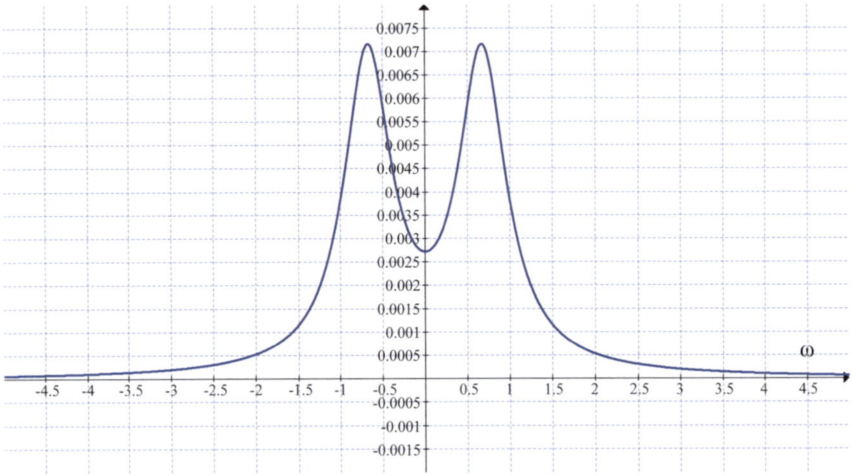

Fig. 7.12 Spectral density function

7.9 The First-Occurrence Time Problem and the Probability Density $P(a_0, \theta)$

The average first-occurrence of stress σ below given level "$\sigma = a_0$" for stationary stress–temperature processes is defined as follows [32]:

$$\overline{\boldsymbol{\sigma}}_{a_0} = \boldsymbol{\theta}_{\max} \int_{a_0}^{\infty} f(x)\mathrm{d}x \tag{7.72}$$

where $f(x)$—probability density of the minimum stress ordinates, which not dependent of time for a stationary processes.

The average number of occurrences below a given level "$\sigma = a_0$" for stationary processes during the same temperature range $0 < \theta < \theta_{\max}$ is defined as follows:

$$\overline{n}_{a_0} = \boldsymbol{\theta}_{\max} \int_{0}^{\infty} vf(a_0, v)\mathrm{d}v \tag{7.73}$$

The average of temperature subintervals for all stress occurrences below given level "$\sigma = a_0$" for stationary processes during the same temperature range $0 < \theta < \theta_{\max}$ is defined as follows [32]:

$$\widetilde{\boldsymbol{\theta}} = \frac{\int_a^{\infty} f(x)\mathrm{d}x}{\int_0^{\infty} vf(a_0, v)\mathrm{d}v} \tag{7.74}$$

The average area between stationary random curve and the horizontal line $\sigma = a_0$ for the average first-occurrence of stress σ below a given level "a_0" is given as [32]:

$$\overline{s\mathbf{v}}_{a_0} = \bar{x} - \int_{-\infty}^{a} xf(x)\mathrm{d}x - a_0\overline{\mathbf{\tau v}}_a \quad \text{where } \bar{\mathbf{v}}_{a_0} = \frac{\bar{n}_{a_0}}{\mathbf{\theta}_{\max}} \tag{7.75}$$

Formulas (7.74); (7.75) and (7.76) are reduced as follows in our case of normal stationary process:

$$\bar{\mathbf{\sigma}}_{a_0} = \mathbf{\theta}_{\max}\int_{a}^{\infty} f(x)\mathrm{d}x = \frac{\mathbf{\theta}_{\max}}{\mathbf{\sigma}_{\mathbf{\sigma}}\sqrt{2\pi}}\int_{a}^{\infty}\exp(-\frac{(x-\mathbf{\mu}_{\mathbf{\sigma}})^2}{2\mathbf{\sigma}_{\mathbf{\sigma}}^2})\mathrm{d}x \tag{7.76}$$

$$\bar{n}_a = \mathbf{\theta}_{\max}\int_{0}^{\infty} vf(a_0, v)\mathrm{d}v = \frac{\mathbf{\theta}_{\max}\mathbf{\sigma}_{v\mathbf{\sigma}}}{2\pi\mathbf{\sigma}_{\mathbf{\sigma}}}\exp\left(-\frac{(a_0-\mathbf{\mu}_{\mathbf{\sigma}})^2}{2\mathbf{\sigma}_{\mathbf{\sigma}}^2}\right) \tag{7.77}$$

where

$$\mathbf{\sigma}_v^2 = -\frac{\mathrm{d}^2}{\mathrm{d}\mathbf{\theta}^2}K_{\mathbf{\sigma}}(\mathbf{\theta})|_{\mathbf{\theta}=0}$$

$$\tilde{\mathbf{\theta}} = \frac{\int_a^\infty f(x)\mathrm{d}x}{\int_0^\infty vf(a,v)\mathrm{d}v} = \pi\frac{\mathbf{\sigma}_{\mathbf{\sigma}}}{\mathbf{\sigma}_{v\mathbf{\sigma}}}\exp\left(+\frac{(a_0-\mathbf{\mu}_{\mathbf{\sigma}})^2}{2\mathbf{\sigma}_{\mathbf{\sigma}}^2}\right)\left[1-\Phi*\left(\frac{a_0-\mathbf{\mu}_{\mathbf{\sigma}}}{\mathbf{\sigma}_{\mathbf{\sigma}}}\right)\right] \tag{7.78}$$

$$\bar{s} = \frac{\mathbf{\sigma}_{\mathbf{\sigma}}^2\sqrt{2\pi}}{\mathbf{\sigma}_{v\mathbf{\sigma}}} + \frac{\pi(\mathbf{\mu}_{\mathbf{\sigma}}-a_0)\mathbf{\sigma}_{\mathbf{\sigma}}}{\mathbf{\sigma}_{v\mathbf{\sigma}}}\left[1-\Phi*\left(\frac{a_0-\mathbf{\mu}_{\mathbf{\sigma}}}{\mathbf{\sigma}_{\mathbf{\sigma}}}\right)\right]\left[\exp(\frac{(a_0-\mathbf{\mu}_{\mathbf{\sigma}})^2}{2\mathbf{\sigma}_{\mathbf{\sigma}}^2}\right] \tag{7.79}$$

For correlation function given by (7.70)

$$\mathbf{\sigma}_{v\mathbf{\sigma}}^2 = K_{\dot{\mathbf{\sigma}}}(0) = -\frac{\mathrm{d}^2}{\mathrm{d}\mathbf{\theta}^2}K_{\dot{\mathbf{\sigma}}}(\mathbf{\theta})|_{\mathbf{\theta}=0} = \mathbf{\sigma}_{\mathbf{\sigma}}^2(b^2-a^2)$$

$$\bar{n}_a = \mathbf{\theta}_{\max}\frac{\sqrt{b^2-a^2}}{2\pi}\exp\left(-\frac{a_0^2}{2\mathbf{\sigma}_{\mathbf{\sigma}}^2}\right) = 10\frac{\sqrt{0.678^2-0.34^2}}{2\pi}\exp\left(-\frac{1.0^2}{2(0.346)^2}\right) = 0.01448$$

$$\bar{\mathbf{v}}_a = \frac{\bar{n}_a}{\mathbf{\theta}_{\max}} = \frac{0.01448}{10} = 0.001448$$

The average temperature interval for the occurrence below given level "a":

$$\tilde{\theta} = \frac{\int_{-\infty}^{a} f(x)\mathrm{d}x}{\int_{0}^{\infty} v f(a,v)\mathrm{d}v} = \pi\frac{\sigma_\sigma}{\sigma_v}\exp\left(+\frac{(a-\bar{x})^2}{2\sigma_\sigma^2}\right)\left[\Phi * \left(\frac{a-\bar{x}}{\sigma_\sigma}\right)\right]$$

$$= \frac{\pi}{\sqrt{0.678^2 - 0.34^2}}\exp\left(\frac{0.37^2}{2(0.346)^2}\right)[\Phi * (-1.07)] = 5.356(1.77)[0.1423] = 1.349 < 10$$

The average first-occurrence temperature below given level "a" for stationary processes is

$$\bar{\theta}_a = \theta_{\max}\int_{-\infty}^{a} f(x)\mathrm{d}x = \frac{10}{\sigma_\sigma\sqrt{2\pi}}\int_{-\infty}^{a}\exp\left(-\frac{(x-\bar{x})^2}{2\sigma_\theta^2}\right)\mathrm{d}x$$

$$= \frac{10}{0.346\sqrt{2\pi}}\left[\Phi * \left(\frac{0-0.37}{0.346}\right)\right] = 11.53(0.1423) = 1.64 < 10$$

The average area between stationary random curve and the horizontal line $\sigma = a$

$$\bar{s} = \frac{\sigma_\sigma^2\sqrt{2\pi}}{\sigma_v} + \frac{\pi(\bar{x}-a)\sigma_\sigma}{\sigma_v}\left[1 - \Phi * \left(\frac{a-\bar{x}}{\sigma_\sigma}\right)\right]\left[\exp\left(\frac{(a-\bar{x})^2}{2\sigma_\sigma^2}\right)\right]$$

$$= \frac{0.346^2}{0.346\sqrt{0.678^2 - 0.34^2}} + \frac{\pi(0-0.37)0.346}{0.346\sqrt{0.678^2-0.34^2}}[1 - \Phi * (1.07)]\left[\exp\left(\frac{(0.37-0)^2}{2(0.346)^2}\right)\right]$$

$$= 0.59 - 1.98(1 - 0.8577)1.77 = 0.091$$

σ_σ—standard deviation for the centered stationary function and "a" is the distance from $\bar{x} = 0$ to level "a".

Now let us compare this area with the area under dimensionless stress–temperature curve.

$$r = \frac{\bar{s}}{\mathrm{AR}} = \frac{0.091}{13.62} = 0.00668 = 0.668\,\%$$

AR---area of stress–temperature curve ($\alpha = 100$)

If this number is suitable for the structural engineer, then one might state that the structural failure did not occur and the allowable stress is acceptable. Now, based on Poisson formulae of the probability to not having the minimum allowable stress ordinates crossing downwards the level "a = 1.0" is: $P_{\mathrm{rel}} = P_o = \exp(-\bar{n}_a) = \exp(-0.01448) = 0.9856$.

This probability characterizes the reliability of the structure (one element or the whole structure).

Example 7.4 Data: see Example 7.2.

Compute the reliability index β.

The limit state equation in this case is as follows:

$$g = \frac{72.76 + 0.041(\mu_{sT})}{50(11.8)} + \frac{13.07 + 0.214(12)(\mu_{sT})}{(50)73} - \mu_{\sigma R} = 0$$

$$\mu_{\sigma S} = 0.166 + 0.000773(\mu_{sT}); \quad \sigma_{\sigma S} = 0.000773(\sigma_{sT}) \quad \mu_{sT} = 1000\ ^{\circ}\mathrm{F}$$

where $\mu_{\sigma R} = 1.838; \quad \sigma_{\sigma R} = 0.346 \rightarrow$ see computations above

$$\beta = \frac{|\mu_{\sigma R} - \mu_{\sigma S}|}{\sqrt{(\sigma_{\sigma S})^2 + (\sigma_{\sigma R})^2}} = \frac{1.838 - 0.939}{\sqrt{0.0773^2 + 0.346^2}} = 2.533$$

$$P_0 = P_{\mathrm{rel}} = 1 - \Phi * (-2.533) = 1 - 0.0057 = 0.9943 \approx 0.9856$$

The reliability index is β = 2.533 and the probability of failure—P_{f} = 0.57 %.

References

1. Kolmogorov AN (1950) Foundations of the theory of probability. The modern measure-theoretic foundation of probability theory; the original German version (Grundbegriffe der Wahrscheinlichkeitrechnung) appeared in 1933
2. Kolmogorov AN (1956) Foundations of the theory of probability. Second English edition (Translation edited by Morrison N). University of Oregon Chelsea Publishing Company, New York
3. Accardi L, von Waldenfels W (eds) Quantum probability and applications V. In: Proceedings of the fourth workshop, held in Heidelberg, FRG. Series: lecture notes in mathematics, vol 1442, 26–30 Sept 1988
4. Mackey GW (2004) Mathematical foundations of quantum mechanics. Courier Dover Publications
5. Shafer G, Vovk V (2012) The origins and legacy of Kolmogorov's Grundbegriffe (PDF). Retrieved 2012-02-12
6. Ang A, Tang WH (2006) Probability concepts in engineering: emphasis on applications to civil and environmental engineering. Wiley, New York. ISBN 0-471-72064-X
7. Papoulis A (2002) Probability, random variables and stochastic process. McGraw-Hill Publishing Co. ISBN 0-07-119981-0
8. Clausing D (1994) Total quality development: a step-by-step guide to world-class concurrent engineering. American Society of Mechanical Engineers. ISBN 0-7918-0035-0
9. Siddall JN (1982) Optimal engineering design. CRC. ISBN 0-8247-1633-7
10. Ash C (1993) The probability tutoring book: an intuitive course for engineers and scientists (and everyone else). Wiley-IEEE Press. ISBN 0-7803-1051-9
11. Houston AI, McNamara JM (1999) Models of adaptive behavior: an approach based on state. Cambridge University Press, Cambridge. ISBN 0-521-65539-0
12. Ruszczyński A; Shapiro A (2003) Stochastic programming. Handbooks in operations research and management science 10. Elsevier, Philadelphia, p 700. ISBN 978-0444508546
13. Razdolsky L (2012) Structural fire load. McGraw-Hill, New York
14. Pugsley AG (1942) A philosophy of aeroplane strength factors. Report and memorandum No. 1906, British Aeronautical Research Committee

15. Freudenthal AM (1945) The safety of structures, Paper No. 2296, American Society of Civil Engineers: Transactions
16. Weibull W (1951) A statistical distribution function of wide applicability. J Appl Mech 18
17. Cornell CA (1967) Bounds on the reliability of structural systems. J Struct Div 93
18. Lind NC (1973) The design of structural design norms. J Struct Mech 1(3)
19. Hasofer AM, Lind NC (1974) Exact and invariant second-moment code format. J Eng Mech Div 100
20. Nakayasu H, Murotsu Y, Mori K, Kase S (1978) Effect of distribution and sample size in material testing on probabilistic design of structure. In: ICQC, TS. D1-20, Tokyo
21. Mahadevan S (1994) Modern structural reliability methods. NASA Marshall Space Flight Center, Oct 1994
22. Ellingwood B et al (1982) Probability-based load criteria: load factors and load combinations. J Struct Div 108(5):978–997
23. Galambos TV et al (1982) Probability-based load criteria: assessment of current design practice. J Struct Div 108(5):959–977
24. Ravindra MK, Galambos TV, Cornell CA (1978) Wind and snow load factors for use in LRFD. J Struct Div 104(9):1443–1457
25. Ellingwood B, MacGregor JG, Galambos TV, Cornell A (1982) Probability based load criteria: load factors and load combinations. J Struct Div 108(5):978–997
26. Turkstra CJ, Madsen HO (1980) Load combinations in codified structural design. J Struct Div 106(12):2527–2543
27. Ellingwood BR (2001) Acceptable risk bases for design of structures. Prog Struct Mat Eng 3:170–179
28. NISTIR 7563 (2009) Best practice guidelines for structural fire resistance design of concrete and steel buildings. U.S. Department of Commerce National Institute of Standards and Technology, Feb 2009
29. Pate-Cornell E (1994) Quantitative safety goals for risk management of industrial facilities. Struct Safety 13(3):145–157
30. McCormac JC (2008) Structural steel design, 5th edn. Prentice Hall, New York
31. Rice JR, Beer FP (1966) First-occurrence time of high-level crossings in a continuous random process. J Acoust Soc Am 39(2)
32. Sveshnokov AA (1978) Problems in probability theory, mathematical statistics and theory of random functions. Dover Publications, New York
33. Razdolsky L, Gandhi PD (2012) Performance-based design and thermal analysis of passive fire protection materials. In: Proceedings of 9th international conference codes and fire safety design methods, Hong Kong, June 2012
34. Razdolsky L (2014) Probability based structural fire load. Cambridge University Press, London
35. SEI/ASCE 7-10 (2010) Minimum design loads for buildings and other structures, American Society of Civil Engineers, Reston, VA

Chapter 8
Probability-Based Engineering Creep and Design Fire Exposure

Notation

k	The thermal conductivity that has the dimensions W/m K or J/m s K
T	Temperature
d	Thickness in the direction if heat flow
c	Specific heat capacity
Q	Activation energy
R	Ideal gas constant
P_l	Losses of heat due to thermal radiation
e	Emissivity factor
σ	Stefan-Boltzmann constant (σ = 5.6703(10^{-8}) W/m^2 K^4)
T_o	Ambient temperature
A_v	Area of openings
c_p	Average specific heat at constant pressure
t	Time
$\vec{v}(u; v; w)$	Velocity vector
D	Diffusion coefficient (m^2/s)
p	Pressure
v	Kinematic viscosity
θ	Dimensionless temperature
τ	Dimensionless time
h	Height of the compartment (m)
a	Thermal diffusivity (m^2/s)
Time	$t = \frac{h^2}{a}\tau$ (s)
Temperature	$T = \frac{RT_*^2}{E}\theta + T_*$ (K), where T_* = 600 K is the base line temperature
Coordinates	$\overline{x} = x/h$ and $\overline{z} = z/h$— "x" and "z"—dimensionless coordinates
Velocities	$\overline{u} = \frac{v}{h}u$ (m/s) and $\overline{w} = \frac{v}{h}w$ (m/s)—horizontal and vertical components velocity accordingly; v—kine-

L. Razdolsky, *Probability Based High Temperature Engineering*,
DOI 10.1007/978-3-319-41909-1_8

	matic viscosity (m^2/s); "u" and "w"—dimensionless velocities
$Pr = \nu/a$	Prandtl number
$Fr = \frac{gh^3}{\nu a}$	Froude number
g	Gravitational acceleration
$Le = a/D = Sc/Pr$	The Lewis number
$Sc = \nu/D$	The Schmidt number
$\overline{\beta} = \frac{RT_*}{E}$	Dimensionless parameter
$\overline{\gamma} = \frac{c_p RT_*^2}{QE}$	Dimensionless parameter
$P = \frac{e\sigma K_\nu (\beta T_*)^3 h}{\lambda}$	Thermal radiation dimensionless coefficient
$\sigma = 5.67(10^{-8})$ (W/m^2 K^4)	Stefan-Boltzman constant
$K_\nu = A_o h/V$	Dimensionless opening factor
A_o	Total area of vertical and horizontal openings
$\delta = \left(\frac{E}{RT_*^2}\right) Qz \left(\exp\left(-\frac{E}{RT_*}\right)\right)$	Frank-Kamenetskii's parameter
$C = [1 - P(t)/P_o]$	Concentration of the burned fuel product in the fire compartment
$\overline{W} = \frac{\nu}{h} W$	Vertical component of gas velocity
$\overline{U} = \frac{\nu}{h} U$	Horizontal component of gas velocity
$b = L/h$	"L" and "h" length (width) and height of fire compartment accordingly
W; U	Dimensionless velocities
R_n	Nominal strength
S_i	Nominal load
φ	Resistance factor
γ	Load factor
R_c	Characteristic value for the resistance
A	Cross-sectional area
I	Moment of inertia
W	Total weight
G_c	Characteristic value for the permanent load
S	Characteristic value for the variable load
ψ_1	Partial safety factor for the permanent load
ψ_2	Partial safety factor for the variable load
β	Reliability index
S	Probability space
A	Set of outcomes (events) to which a probability is assigned
$P(E_2 \vert E_1)$	Conditional probability
$\Phi^*(.)$	Denotes the cumulative distribution function of standard normal distribution $\Phi^*(z) = \frac{1}{\sqrt{2\pi}} \int_{-\infty}^{z} e^{-\frac{z^2}{2}} dz$

$\mu_A, \mu_B, \sigma_A, \sigma_B$	Are mean and standard deviation of A and B, respectively
$J(t, t')$	Compliance function
T_M	Melting point of the metal matrix material
$\varepsilon(t)$	Strain
$\sigma(t)$	Stress
$\overline{\sigma}(t) = E(t)\varepsilon(t)$	Instantaneous stress
ε_e	Instantaneous (elastic) strain
ε_c	Creep strain
ε_T	Thermal expansion due to high-temperature effect
$K(t,t') = \partial J(t,t')/\partial t'$	Retardation function (memory function)
$R(t, t')$	Relaxation function (also called the relaxation modulus)
$M(\theta)$	Bending moment
$V(\theta)$	Shear force
$P(\theta)$	Axial force
$y(\theta)$	Deformation
P_f	Probability of failure
P_{rel}	Reliability
κ	Curvature
ω	Frequency
$S(\omega)$	Spectral density
μ	Poisson coefficient
D	Diffusion rate (Flick's law)
η	*Viscosity* parameter of the material
E	Modulus of elasticity
$n = \eta/E$	Relaxation time
n	Power law exponent
α_i	Material property parameter
L	Span (spring spacing)
k_0	Subgrade modulus

8.1 Probability-Based Stress–Strain Diagram. Fast Fire

8.1.1 Introduction

Even though inelastic deformation in metallic materials is mostly driven by dislocation multiplication and glide, this changes gradually with an increase in temperature. These changes occur mostly above the softening temperature of a material, which usually is between 0.3 and 0.5 of the melting temperature (in degrees Kelvin). Above this temperature, thermal activation becomes important enough to significantly assist dislocations for passing obstacles. Thus work hardening, which hinders and eventually impedes low temperature deformation, does not operate the same way at high temperatures. On the one hand, at elevated temperature recovery and recrystallization anneal rapidly an important proportion of work hardening. On the other hand, when a sample is maintained under load, thermal activation helps dislocations to increase their mobility.

Thermal activation does not completely anneal all work hardening. Especially when the creep rate increases because of a large imposed thermal stress, more dislocations are produced than annihilated. The dislocation density then increases. As already stated, however, this does not translate into a homogenous hardening that will eventually stop any further deformation. In general, the observations show that dislocations, initially arranged in a homogenous network, start to organize themselves. It is particularly dislocation climb, governed by thermal activation, which now limits the rate of deformation. In this state, the creep rate is at its minimum or stationary value (equilibrium between work hardening and recovery). The ongoing excess of stress acting on the subgrain boundaries causes microstructures damage, which weakens the metal. The creep rate finally increases and the sample fractures when damage becomes critical. The description given above is only one mechanism among others. Indeed, from the applied engineering creep points of view the micro-mechanism of creep phenomena is not critically important, because only the end result, stress–temperature–strain diagram, is used in structural engineering analyses and designs, and these relationships must be represented mathematically, preferably in closed analytical form.

8.1.2 Phenomenological Laws and Coefficients

The analytical description of a creep curve as a function of temperature-time or strain, which is based solely on elementary mechanisms, is not easy at all. However in practice, in most cases, measured curves are used. To estimate the effect of stress σ and temperature T, one focuses generally on stage II (constant creep rate). For those materials that do not exhibit a distinct stage II, the minimum creep rate is used. Consideration of the constant creep rate stage is justified by the fact that it often represents a substantial part of the lifespan of a structure during the fire event. In the following, we focus on stage II. The steady state creep rate ε_{ss} is a function of both stress σ (a) and temperature T (b) [1–3] as shown in Fig. 8.1.

The stress dependence can be described as $\varepsilon'_{ss} = B\sigma^n$ and the temperature dependence as $\varepsilon'_{ss} = \exp(-E/RT)$. Combining gives for the minimum creep rate (i.e., in steady state creep, stage II), $\varepsilon'_{ss} = A\sigma^n\exp(-E/RT)$, where A: constant, σ: stress (MPa), n: exponent, E: activation energy (J/mol), $R = 8.314$ J/mol K (ideal gas constant).

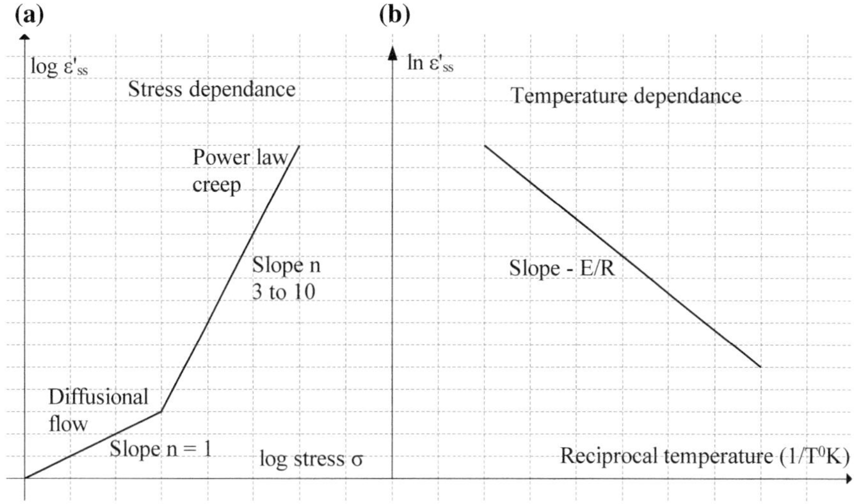

Fig. 8.1 Stress and temperature dependence of the steady state creep rate

When the applied stress is not too high, this law usually gives a satisfactory description of the creep rate as a function of temperature and applied stress. The constant A, the power law exponent n, and the activation energy for creep E are material characteristics and can be determined from a series of creep tests. Several tests have to be carried out since temperature has to be varied at constant applied stress in order to determine the activation energy, and stress has to be varied at constant temperature to determine the power law exponent n. This is consequently always a long-term project. The phenomenological power law is not precise enough to calculate (by extrapolation) the creep rate at low temperatures from measurements taken at high stress near the melting temperature, where things happen very fast.

The power law exponent n can be easily determined through a double-logarithmic representation of the creep rate (at T = constant) and the stress. When the observed results satisfy the law mentioned above, a straight line of slope n should be obtained. Similarly, the logarithm of the creep rate (at constant stress) can be plotted as a function of $1/T$. The slope of the straight line should equal E/R. For pure metals only a few control measures need to be taken, since it is well established that the activation energy for creep corresponds to the activation energy for auto diffusion.

This fact is also confirmed by theory, since the stationary creep rate is limited by rearrangements that require the climb of dislocations in the subgrain boundaries.

In pure metals, the exponent n varies between 4 and 5 (Norton power law). For single-phased alloys (solid solutions), n equals rather 3. In these two metal groups, results obtained at very high temperatures show an exponential dependence of stress such that the creep curve deviates from Norton's law. In order to "save"

Norton's law with reasonable exponents (3–5), one may argue that the applied stress in Norton's law should be included in integral type constitutive creep equation, which corresponds to the main idea of phenomenological creep model (see Chap. 7 for detailed discussions).

8.1.3 High-Temperature Creep and Structural Fire Resistance

Generally, neglecting high-temperature creep effect stiffens the structural response and leads to reduced deflections but larger restraint forces. Therefore, neglecting high-temperature creep in fire resistance analyses of structural systems can lead to the design that might be not conservative. The transient behavior of creep at elevated temperature has a significant influence on the structural fire resistance. However, a review of the literature clearly indicates that the effect of high-temperature creep on structural response did not receive much attention. Much of the reported creep studies were mainly focused at the material level [4–6].

Further, the commonly used creep models are primarily based on Dorn's creep formulation [7], which basically assumes invariant stress level ($\mathrm{d}\sigma/\mathrm{d}t = 0$). Due to the lack of data, creep models used in most previous studies, are derived based on material tests, under constant stress conditions. The effect of stress and temperature variation with time ($\mathrm{d}\sigma/\mathrm{d}t \neq 0$, $\mathrm{d}T/\mathrm{d}t \neq 0$) on creep were considered independently, i.e., either under constant stress but variable temperature, or constant temperature but variable stress. The interaction between variable thermal rate and variable stress rate on high-temperature creep was not investigated. This is because of the severe difficulties associated with experimenting and modeling such complex interaction. However such interaction significantly influences creep effects in structural fire resistance design, where the increase in temperature with time induces higher rates of stress development with time.

To calculate the resistance of a structure exposed to high temperatures, typically tensile tests are required. There are no established theories available that can be used to derive the values of the material property parameters (MPP) that are required for solving the constitutive creep equation. They must be obtained experimentally. This is achieved by fitting mathematical creep function to the appropriate parts of the experimental data and obtaining their values as the curve fitting coefficients. During a creep test, plastic deformation is measured as a function of time, at constant temperature, and at a constant applied stress. The different deformation modes that can be imposed are tension, compression, bending, and torsion. Under fire conditions, creep becomes a dominant factor and influences fire resistance of structural members. The significant forces develop in statically indeterminate structural systems and these forces consequently induce high stresses. The extent of creep deformations is affected by magnitude and rate of development of stress and temperature in structural elements. The rate of development of these stresses depends on many factors including high-temperature material properties, fire scenario, and the structural system composition. These factors that critically affect fire response are mutually time-dependent.

For the purpose of analyzing the behavior of structures, various σ–ε–T relationships are being used throughout the world. The simplest are probably ones based on multi-linear approximations of the stress–strain (σ–ε) behavior. This may be bilinear [8, 9], trilinear [10], or nonlinear [11]. Multi-linear approximations tend to be somewhat coarse in representing the complex shape of the σ–ε curves.

Therefore, in order to more closely represent the steel behavior others use a combination of linear and smooth curves to represent the σ–ε curves. Examples of these are ones proposed by Furamura et al. [12], Lie [13], and Eurocode 3 [14].

For the purpose of general analysis and design, the Lie and Eurocode 3 relationships are probably the most commonly used. The Lie relationship is more widely accepted in North America, whereas, amongst the European countries, the Eurocode 3 relationship tends to be recommended. The features of these two relationships are outlined below

1. The Lie relationship essentially uses a bilinear curve, with a small transition between the linear portions, to represent the σ–ε behavior. The curve comprises two separate equations. The first describes the linear elastic portion, and the second describes the rest of the σ–ε curve. The variation of the σ–ε behavior with temperature is represented using another four separate equations. Therefore, altogether the relationship uses six separate equations and contains 13 independent coefficients.
2. The Eurocode 3 relationship is more complicated than Lie's in that it attempts to fit the various portions of the σ–ε curve. It uses seven linear and parabolic equations to represent the σ–ε curve, including the strain-hardening portion. The variation of the σ–ε behavior with temperature is represented using another 19 separate linear equations. In total, the relationship uses 26 separate equations and contains 42 independent coefficients.

It is important to note that all σ–ε–T relationships are derived based on specific sets of experimental data. Therefore, in situations where the steel properties differ from those of the original data, the relevant coefficients in the relationships must be reevaluated. It is clear that both approaches to construct a stress–strain relationship are very cumbersome and hardly suitable for the daily engineering practice. Therefore in this chapter it is proposed phenomenological model of the creep of the material written in the dimensionless integral form, the solution of which is the stress–temperature–strain function. Obviously the effect of stress and temperature variation for each fire severity scenario with time ($d\sigma/dt \neq 0$, $dT/dt \neq 0$) on creep deformation is considered simultaneously acting in this model. Uniaxial dimensionless model of creep deformation in this case has the form (see Chap. 5 for details)

$$\overline{\sigma}(\theta) = E(\theta)\varepsilon(\theta) = \sigma(\theta) + \sum_{i=1}^{N} \int_{0}^{\theta} a_i[\varphi(\theta)]b_i[\varphi(\tau)]\varphi'(\tau)\sigma^n(\tau)\mathrm{d}\tau$$
$$b_i[\varphi(\tau)] = (\exp(t)/(1+0.1t))[((\exp(\alpha_i m))]m1; a_i[\varphi(\tau)] = [(\exp(-\alpha_i m))]m1$$
$$m = \varphi(\theta) = 0.00245 + 0.0375 * \theta - 0.00934 * \theta\hat{}2 + 0.00104 * \theta\hat{}3 - 0.000041 * \theta\hat{}4$$
$$m1 = \varphi'(\theta) = 0.0375 - 0.01868 * \theta + 0.00312 * \theta\hat{}2 - 0.000164 * \theta\hat{}3 \qquad (8.1)$$

The functions m and $m1$ represent the inverse function $\theta^{-1}(\tau)$ for a given fire severity scenario (Very Fast Fire; Fast Fire; Medium Fire, and Slow Fire) [15–17]. The computations of stress–temperature–strain function in case of Very Fast Fire and corresponding discussions are presented in Chap. 7. All subsequent calculations for each case of fire severity are presented in a step-by-step procedural form in accordance with the main points from Chap. 7.

8.1.4 Temperature–Time Function

Fast Fire; Temperature Increase (882 K < T < 1022 K)
Data $T_* = 600$ K; $0 < \tau < 0.2$; $0.05 < \gamma < 0.175$
Select $\gamma = 0.05$

$$\frac{\partial\theta}{\partial\tau} = \delta(1 - C)^k \exp\left(\frac{\theta}{1+\beta\theta}\right) - P\theta^4 \qquad (8.2)$$

$$\frac{\partial C}{\partial\tau} = \gamma\delta(1 - C)^k \exp\left(\frac{\theta}{1+\beta\theta}\right) \qquad (8.3)$$

Differential equations (8.2) and (8.3) are rewritten as an input for "Polymath" software

$$\mathrm{d}(y0)/\mathrm{d}(t) = 20 * (1 - y2) * \exp(y0/(1 + .1 * y0)) - .157 * y0\hat{}4$$
$$\mathrm{d}(y2)/\mathrm{d}(t) = 3.5 * (1 - y2) * \exp(y0/(1 + .1 * y0))$$
$$\mathrm{d}(y1)/\mathrm{d}(t) = 1.0 * (1 - y1)\hat{}1.0 * \exp(y/(1 + .1 * y))$$
$$\mathrm{d}(y)/\mathrm{d}(t) = (1) * 20 * (1 - y1)\hat{}1.0 * \exp(y/(1 + .1 * y)) - .157 * y\hat{}4$$

where

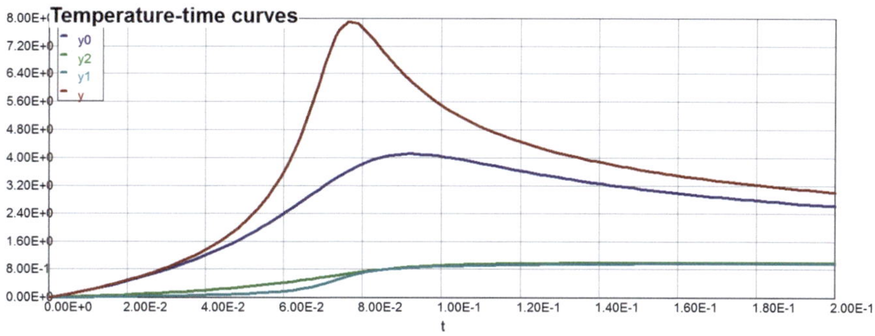

Fig. 8.2 Dimensionless time–temperature curves

"y" is the dimensionless temperature "θ" with the corresponding parameter "$\gamma = 0.05$".
"y0" is the dimensionless temperature "θ" with the corresponding parameter "$\gamma = 0.175$".
"y1" is the concentration of the product of the first-order chemical reaction with "$\gamma = 0.05$".
"y2" is the concentration of the product of the first-order chemical reaction with "$\gamma = 0.175$".

Calculated values of DEQ variables

	Variable	Initial value	Minimal value	Maximal value	Final value
1	*t*	0	0	0.2	0.2
2	*y*	0	0	7.916172	3.026519
3	*y*0	0	0	4.121423	2.629209
4	*y*1	0	0	0.9812653	0.9812653
5	*y*2	0	0	0.9987195	0.9987195

Tabulated solution of the Eqs. (5.3) and (5.4) are presented in [16, 18] and the graphs are shown on Fig. 8.2.

Analytical expression of the inverse function θ^{-1} *and its first derivative*

Based on the tabulated data shown above, the final approximation of the dimensionless temperature–time curve can be presented as follows (Fig 8.3):

Monotonically increasing temperature–time function: $(0 < \theta < 8)$; $(0 < \tau 1 < 0.0763)$

$$\textbf{Model}: \theta = 39.35 * \tau - 688.9 * \tau^{\wedge}2 + 6463.5 * \tau^{\wedge}3 + 197{,}000 * \tau^{\wedge}4 \quad (8.4)$$

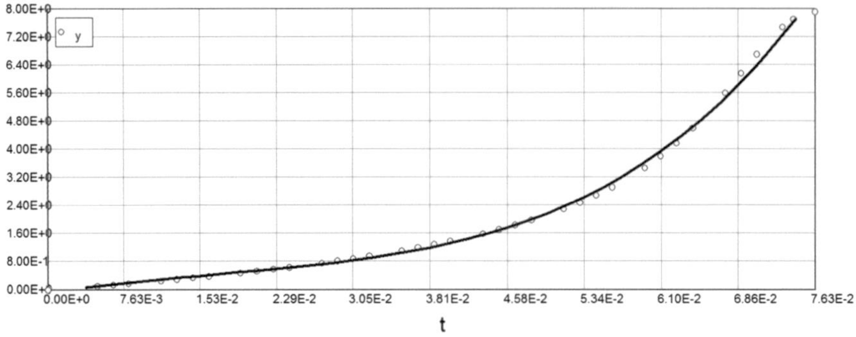

Fig. 8.3 Monotonically increasing temperature–time function

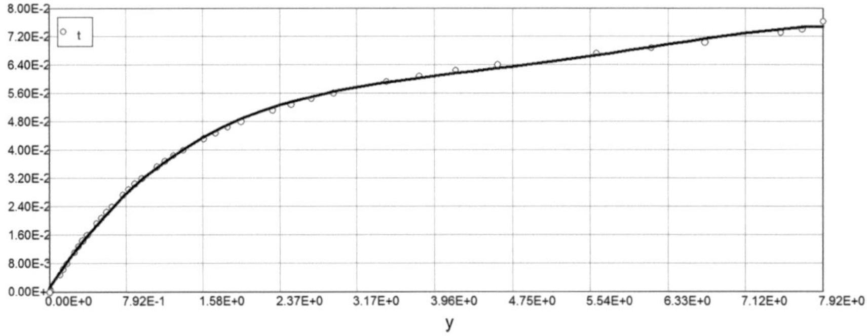

Fig. 8.4 Monotonically increasing inverse function θ^{-1}

Variable	Value
$a0$	0.0
$a1$	39.35
$a2$	−688.9
$a3$	6463.5
$a4$	1.97E+05

The inverse function θ^{-1} and its first derivative (monotonically increasing functions) are (Fig. 8.4)

$$\textbf{Model}: \mathbf{m} = \tau = 0.00154 + 0.0411 * \theta - 0.0113 * \theta^\wedge 2 + 0.00147 * \theta^\wedge 3 - 0.0000697 * \theta^\wedge 4 \quad (8.5)$$

Variable	Value
$a0$	0.00154
$a1$	0.0411
$a2$	−0.0113
$a3$	0.00147
$a4$	−6.97E−05

$$\mathbf{m1} = \tau' = 0.0411 - 0.0226 * \theta + 0.00441 * \theta^\wedge 2 - 0.000279 * \theta^\wedge 3 \quad (8.6)$$

8.1.5 Fast Fire: Statistical Data

Differential equations (DEQs): $\alpha = 0.001$

Calculated values of DEQ variables

	Variable	Initial value	Minimal value	Maximal value	Final value
1	A	0	0	8.983211	8.983211
2	m	0.00154	0.00154	0.0742923	0.0742888
3	$m1$	0.0411	−0.000308	0.0411	−0.000308
4	n	3	3	3	3
5	r	0	0	2.45251	2.409554
6	t	0	0	8	8
7	y	0	0	*1.653418*	0.8486171

Differential equations

$$1 \quad \mathrm{d}(y)/\mathrm{d}(t) = -0.001 * y * m1 + (\exp(-0.15 * t)) * (1 - 0.15 * t) - (\exp(t/(1.0 + 0.1 * t))) * m1 * y^\wedge n + 0.001 * m1 * t * (\exp(-0.15 * t))$$

$$2 \quad \mathrm{d}(A)/\mathrm{d}(t) = 1.432 * t - 0.383 * t^\wedge 2 + 0.034 * t^\wedge 3 - 0.00096 * t^\wedge 4$$

Differential equations (DEQs): $\alpha = 0.01$

Calculated values of DEQ variables

	Variable	Initial value	Minimal value	Maximal value	Final value
1	A	0	0	8.983211	8.983211
2	m	0.00154	0.00154	0.0742923	0.0742888
3	$m1$	0.0411	−0.000308	0.0411	−0.000308

(continued)

(continued)

	Variable	Initial value	Minimal value	Maximal value	Final value
4	n	3	3	3	3
5	r	0	0	2.45251	2.409554
6	t	0	0	8	8
7	y	0	0	*1.653438*	0.8486759

Differential equations

$$1 \quad d(y)/d(t) = -0.01 * y * m1 + (\exp(-0.15 * t)) * (1 - 0.15 * t) - (\exp(t/(1.0 + 0.1 * t))) * m1 * y^\wedge n + 0.01 * m1 * t * (\exp(-0.15 * t))$$

$$2 \quad d(A)/d(t) = 1.432 * t - 0.383 * t^\wedge 2 + 0.034 * t^\wedge 3 - 0.00096 * t^\wedge 4$$

Differential equations (DEQs): $\alpha = 0.1$

Calculated values of DEQ variables

	Variable	Initial value	Minimal value	Maximal value	Final value
1	A	0	0	8.983211	8.983211
2	m	0.00154	0.00154	0.0742925	0.0742888
3	$m1$	0.0411	−0.000308	0.0411	−0.000308
4	n	3	3	3	3
5	r	0	0	2.452516	2.409554
6	t	0	0	8	8
7	y	0	0	*1.653602*	0.8492636

Differential equations

$$1 \quad d(y)/d(t) = -0.1 * y * m1 + (\exp(-0.15 * t)) * (1 - 0.15 * t) - (\exp(t/(1.0 + 0.1 * t))) * m1 * y^\wedge n + 0.1 * m1 * t * (\exp(-0.15 * t))$$

$$2 \quad d(A)/d(t) = 1.432 * t - 0.383 * t^\wedge 2 + 0.034 * t^\wedge 3 - 0.00096 * t^\wedge 4$$

Differential equations (DEQs): $\alpha = 1$

Calculated values of DEQ variables

	Variable	Initial value	Minimal value	Maximal value	Final value
1	*A*	0	0	8.983211	8.983211
2	*m*	0.00154	0.00154	0.0742925	0.0742888
3	*m*1	0.0411	−0.000308	0.0411	−0.000308
4	*n*	3	3	3	3
5	*r*	0	0	2.452516	2.409554
6	*t*	0	0	8	8
7	*y*	0	0	*1.655589*	0.855097

Differential equations

$$1\quad d(y)/d(t) = -1 * y * m1 + (\exp(-0.15 * t)) * (1 - 0.15 * t) - (\exp(t/(1.0 + 0.1 * t))) * m1 * y^\wedge n + 1 * m1 * t * (\exp(-0.15 * t))$$

$$2\quad d(A)/d(t) = 1.432 * t - 0.383 * t^\wedge 2 + 0.034 * t^\wedge 3 - 0.00096 * t^\wedge 4$$

Differential equations (DEQs): $\alpha = 10$

Calculated values of DEQ variables

	Variable	Initial value	Minimal value	Maximal value	Final value
1	*A*	0	0	8.983211	8.983211
2	*m*	0.00154	0.00154	0.0742925	0.0742888
3	*m*1	0.0411	−0.000308	0.0411	−0.000308
4	*n*	3	3	3	3
5	*r*	0	0	2.452516	2.409554
6	*t*	0	0	8	8
7	*y*	0	0	*1.674952*	0.9095045

Differential equations

$$1\quad d(y)/d(t) = -10 * y * m1 + (\exp(-0.15 * t)) * (1 - 0.15 * t) - (\exp(t/(1.0 + 0.1 * t))) * m1 * y^\wedge n + 10 * m1 * t * (\exp(-0.15 * t))$$

$$2\quad d(A)/d(t) = 1.432 * t - 0.383 * t^\wedge 2 + 0.034 * t^\wedge 3 - 0.00096 * t^\wedge 4$$

Differential equations (DEQs): α = 100

Calculated values of DEQ variables

	Variable	Initial value	Minimal value	Maximal value	Final value
1	A	0	0	8.983211	8.983211
2	m	0.00154	0.00154	0.0742917	0.0742888
3	$m1$	0.0411	−0.000308	0.0411	−0.000308
4	n	3	3	3	3
5	r	0	0	2.452498	2.409554
6	t	0	0	8	8
7	y	0	0	*1.805355*	1.238978

Differential equations

1 $\mathrm{d}(y)/\mathrm{d}(t) = -100 * y * m1 + (\exp(-0.15 * t)) * (1 - 0.15 * t) - (\exp(t/(1.0 + 0.1 * t))) * m1 * y\^{}n + 100 * m1 * t * (\exp(-0.15 * t))$

2 $\mathrm{d}(A)/\mathrm{d}(t) = 1.432 * t - 0.383 * t\^{}2 + 0.034 * t\^{}3 - 0.00096 * t\^{}4$

Differential equations (DEQs): α = 1000

Calculated values of DEQ variables

	Variable	Initial value	Minimal value	Maximal value	Final value
1	A	0	0	8.983211	8.983211
2	m	0.00154	0.00154	0.0742922	0.0742888
3	$m1$	0.0411	−0.000308	0.0411	−0.000308
4	n	3	3	3	3
5	r	0	0	2.452529	2.409554
6	t	0	0	8	8
7	y	0	0	*2.12684*	1.895772

Differential equations

1 $\mathrm{d}(y)/\mathrm{d}(t) = -1000 * y * m1 + (\exp(-0.15 * t)) * (1 - 0.15 * t) - (\exp(t/(1.0 + 0.1 * t))) * m1 * y\^{}n + 1000 * m1 * t * (\exp(-0.15 * t))$

2 $\mathrm{d}(A)/\mathrm{d}(t) = 1.432 * t - 0.383 * t\^{}2 + 0.034 * t\^{}3 - 0.00096 * t\^{}4$

Differential equations (DEQs): α = 10,000

Calculated values of DEQ variables

	Variable	Initial value	Minimal value	Maximal value	Final value
1	A	0	0	8.983211	8.983211
2	m	0.00154	0.00154	0.0742925	0.0742888
3	$m1$	0.0411	−0.000308	0.0411	−0.000308
4	n	3	3	3	3
5	r	0	0	2.452526	2.409554
6	t	0	0	8	8
7	y	0	0	2.384656	2.310501

Differential equations

$$1\quad d(y)/d(t) = -10,000 * y * m1 + (\exp(-0.15 * t)) * (1 - 0.15 * t) - (\exp(t/(1.0 + 0.1 * t))) * m1 * y^\wedge n + 10,000 * m1 * t * (\exp(-0.15 * t))$$

$$2\quad d(A)/d(t) = 1.432 * t - 0.383 * t^\wedge 2 + 0.034 * t^\wedge 3 - 0.00096 * t^\wedge 4$$

Differential equations (DEQs): α = 100,000

Calculated values of DEQ variables

	Variable	Initial value	Minimal value	Maximal value	Final value
1	A	0	0	8.983211	8.983211
2	m	0.00154	0.00154	0.0742916	0.0742888
3	$m1$	0.0411	−0.000308	0.0411	−0.000308
4	n	3	3	3	3
5	r	0	0	2.452514	2.409554
6	t	0	0	8	8
7	y	0	0	2.444614	2.398451

Differential equations

$$1\quad d(y)/d(t) = -100,000 * y * m1 + (\exp(-0.15 * t)) * (1 - 0.15 * t) - (\exp(t/(1.0 + 0.1 * t))) * m1 * y^\wedge n + 100,000 * m1 * t * (\exp(-0.15 * t))$$

$$2\quad d(A)/d(t) = 1.432 * t - 0.383 * t^\wedge 2 + 0.034 * t^\wedge 3 - 0.00096 * t^\wedge 4$$

The summary of computations above is presented in Table 8.1.

Table 8.1 Maximum dimensionless stresses (statistical data)

Value	Maximum creep stress	Deviation	Variance	Mean value	Standard deviation	α value
1	*1.6534*	−0.2416	0.0584			0.001
2	*1.6534*	−0.2416	0.0584			0.01
3	*1.6536*	−0.2414	0.0583			0.1
4	*1.6556*	−0.263	0.0692			1.0
5	*1.6750*	−0.2394	0.0573			10
6	*1.8054*	−0.0896	0.00803			100
7	*2.127*	0.232	0.0538			1000
8	*2.3847*	0.4897	0.240			10,000
9	*2.445*	0.55	0.302			100,000
Aver.	*1.895*		Total: 0.906/8 = = 0.1132	*1.895*	*0.336*	

Therefore, the probability that the real fire dimensionless allowable stress "**σ**" will be less than 1.65; 1.2; 1.1 or 1.0 is

$$\begin{aligned}
P(\boldsymbol{\sigma}<1.65) &= \Phi^*\left(\frac{1.65-1.895}{0.336}\right) = \Phi^*(-0.729) = 0.23 = 23\,\% \\
P(\boldsymbol{\sigma}<1.2) &= \Phi^*\left(\frac{1.2-1.895}{0.336}\right) = \Phi^*(-2.068) = 0.021 = 2.1\,\% \\
P(\boldsymbol{\sigma}<1.1) &= \Phi^*\left(\frac{1.1-1.895}{0.336}\right) = \Phi^*(-2.37) = 0.01 = 1.0\,\% \\
P_{\mathrm{f}}(\boldsymbol{\sigma}<1.0) &= \Phi^*\left(\frac{1.0-1.895}{0.336}\right) = \Phi^*(-2.664) = 0.0035 = 0.35\,\%
\end{aligned} \tag{8.7}$$

If probability of failure $P_{\mathrm{f}} = 1$ % is acceptable then $\sigma_{\mathrm{all}} = 1.1(7.02)$ $2.9 = 22.4$ ksi = 154.4 MPa; $\mu_{\sigma} = 1.895$ and $\sigma_{\sigma} = 0.336$, then $\beta_{\mathrm{all}} = 2.366$ (reliability index).

8.1.6 The First-Order Reliability Method (FORM)

For detailed information regarding applicability of this method please see Chap. 7. The computation of the reliability index β and corresponding design temperature (based on Example 7.2) is presented below without any additional comments or explanations in order to save space.

Example 8.1 Data: Fast Fire: $\mu_{sT} = 1000$ °F; $\sigma_{sT} = 100$ °F. Find the reliability index β; probability of failure P_{f} and design temperature T °F.

The deterministic performance function in this case is defined by Eq. H2-1 [35]

$$g = \frac{72.76 + 0.041(\mu_{sT})}{50(11.8)} + \frac{13.07 + 0.214(12)(\mu_{sT})}{(50)73} - \mu_{\sigma R} = 0$$
$$\mu_{\sigma S} = 0.166 + 0.000773(\mu_{sT});\ \sigma_{\sigma S} = 0.000773(\sigma_{sT})\ \mu_{sT} = 1000\,^{\circ}\mathrm{F}$$

where $\mu_{\sigma R} = 1.895;\ \sigma_{\sigma R} = 0.336 \rightarrow$ see computations above

$$\beta = \frac{|\mu_{\sigma R} - \mu_{\sigma S}|}{\sqrt{(\sigma_{\sigma S})^2 + (\sigma_{\sigma R})^2}} = \frac{1.895 - 0.939}{\sqrt{0.0773^2 + 0.336^2}} = 2.773$$

The probability of structural failure and the design temperature in this case are computed as follows:

$$P_{\mathrm{f}} = \Phi^*(-\boldsymbol{\beta}) = \Phi^*(-2.773) = 0.01;$$
$$T_{\mathrm{design}} = 1000 + (2.773)100 = 1277.3\,^{\circ}\mathrm{F}$$

8.1.7 Confidence Interval—Fast Fire

Again, for detailed information regarding applicability of this method please see Chap. 7. The computation of the reliability index β and corresponding design temperature (based on Example 7.3) is presented below

Example 8.2 Data: Fast Fire: $\alpha = 0.95$—given target confidence probability
From Table 8.1: $\mu_{\sigma R} = 1.895;\ \sigma_{\sigma R} = 0.336$

The confidence limit (interval) I_α of mean value (μ_σ) and standard deviation (σ_σ) is as follows:

$$\begin{aligned} \boldsymbol{\sigma}_{\boldsymbol{\mu}_\sigma} &= \sqrt{\frac{\boldsymbol{\sigma}_\sigma^2}{n-1}} = \sqrt{\frac{(0.336)^2}{8}} = 0.119 \\ \boldsymbol{\varepsilon}_\alpha &= \boldsymbol{\sigma}_{\boldsymbol{\mu}_\sigma} \arg\, \Phi^*\left(\frac{1+\boldsymbol{\alpha}}{2}\right) = 0.119\left[\arg\, \Phi^*\left(\frac{1+0.95}{2}\right)\right] = 0.124(1.96) = 0.233 \\ t_\alpha &= \arg\, \Phi^*\left(\frac{1+\boldsymbol{\alpha}}{2}\right) = 1.96 \end{aligned} \tag{8.8}$$

Finally, the confidence limit (interval) I_α for mean value is calculated as follows:

$$I_\alpha = (\boldsymbol{\mu}_\sigma - t_\alpha \boldsymbol{\sigma}_{\boldsymbol{\mu}_\sigma}; \boldsymbol{\mu}_\sigma + t_\alpha \boldsymbol{\sigma}_{\boldsymbol{\mu}_\sigma}) = (1.895 - 0.233; 1.895 + 0.233) = (1.662; 2.128) \tag{8.9}$$

The confidence limit (interval) I_α for standard deviation value is calculated as follows:

$$I_\alpha = \left(\boldsymbol{\sigma}_{\boldsymbol{\sigma}}^2 - t_\alpha \boldsymbol{\sigma}_{\boldsymbol{\sigma}_{\boldsymbol{\sigma}}}; \boldsymbol{\sigma}_{\boldsymbol{\sigma}}^2 + t_\alpha \boldsymbol{\sigma}_{\boldsymbol{\sigma}_{\boldsymbol{\sigma}}}\right) = \left(0.336^2 - (1.96)0.0603; 0.336^2 + (1.96)0.0603\right) = (0; 0.231) \quad (8.10)$$

where $\boldsymbol{\sigma}_{\boldsymbol{\sigma}_{\boldsymbol{\sigma}}} = \sqrt{\frac{2}{n-1}\boldsymbol{\sigma}_{\boldsymbol{\sigma}}^2} = \sqrt{\frac{2}{9-1}(0.336)^2} = 0.0603$ and $I_\alpha^{\boldsymbol{\sigma}_{\boldsymbol{\sigma}_{\boldsymbol{\sigma}}}} = \left(\sqrt{\boldsymbol{\sigma}_{\boldsymbol{\sigma}}^2 - t_\alpha \boldsymbol{\sigma}_{\boldsymbol{\sigma}_{\boldsymbol{\sigma}}}}; \sqrt{\boldsymbol{\sigma}_{\boldsymbol{\sigma}}^2 + t_\alpha \boldsymbol{\sigma}_{\boldsymbol{\sigma}_{\boldsymbol{\sigma}}}}\right) = (0; 0.481)$

The minimum allowable dimensionless stress is

$$\boldsymbol{\sigma}_{\text{min,all}} = 1.662 - 0.481 = 1.181$$

The difference in minimum allowable dimensionless stress is $\bar{\boldsymbol{\sigma}}_{\text{min}} = 1.181/1.1 = 1.07 = 7\ \%$.

8.1.8 Creep Stress–Strain Diagrams (Ergodic Process)

The statistical data (stress-temperature-strain relationship) in this fire severity case are taken from Chap. 5 and presented below.

$$\begin{aligned} &\alpha = 0.001 \\ &\sigma = 0.993 * \theta - 0.16 * \theta^\wedge 2 + 0.00108 * \theta^\wedge 3 - 0.000633 * \theta^\wedge 4 \end{aligned} \quad (8.11)$$

$$\begin{aligned} &\alpha = 0.01 \\ &\sigma = 0.992 * \theta - 0.16 * \theta^\wedge 2 + 0.00108 * \theta^\wedge 3 - 0.000633 * \theta^\wedge 4 \end{aligned} \quad (8.12)$$

$$\begin{aligned} &\alpha = 0.1 \\ &\sigma = 0.992 * \theta - 0.16 * \theta^\wedge 2 + 0.00107 * \theta^\wedge 3 - 0.000633 * \theta^\wedge 4 \end{aligned} \quad (8.13)$$

$$\begin{aligned} &\alpha = 1 \\ &\text{Model: } \sigma = 0.992 * \theta - 0.16 * \theta^\wedge 2 + 0.00104 * \theta^\wedge 3 - 0.000635 * \theta^\wedge 4 \end{aligned} \quad (8.14)$$

$$\begin{aligned} &\alpha = 10 \\ &\text{Model: } \sigma = 0.993 * \theta - 0.16 * \theta^\wedge 2 + 0.000792 * \theta^\wedge 3 - 0.000648 * \theta^\wedge 4 \end{aligned} \quad (8.15)$$

Table 8.2 Stress–temperature data (fast fire)

$\alpha\|\theta$		1	2	3	4	5	6	7	8
0.001	0	0.852	1.379	1.610	1.648	1.513	1.246	0.987	0.849
0.01	0	0.852	1.379	1.610	1.648	1.513	1.246	0.987	0.849
0.1	0	0.852	1.379	1.610	1.648	1.514	1.250	0.988	0.849
1.0	0	0.852	1.379	1.611	1.650	1.516	1.250	0.993	0.855
10	0	0.852	1.385	1.625	1.671	1.543	1.286	1.042	0.910
100	0	0.855	1.418	1.713	1.805	1.723	1.530	1.349	1.239
1000	0	0.859	1.467	1.859	2.068	2.127	2.084	1.991	1.896
10,000	0	0.860	1.480	1.906	2.177	2.327	2.382	2.368	2.31
100,000	0	0.861	1.481	1.912	2.193	2.358	2.433	2.441	2.398
r	0	0.861	1.482	1.913	2.195	2.362	2.439	2.450	2.41

$$\begin{aligned}&\alpha = 100\\ &\text{Model: } \sigma = 0.996 * \theta - 0.148 * \theta^{\wedge}2 + 0.000172 * \theta^{\wedge}3 - 0.000641 * \theta^{\wedge}4\end{aligned} \tag{8.16}$$

$$\begin{aligned}&\alpha = 1000\\ &\text{Model: } \sigma = 0.986 * \theta - 0.135 * \theta^{\wedge}2 + 0.00337 * \theta^{\wedge}3 - 0.000217 * \theta^{\wedge}4\end{aligned} \tag{8.17}$$

$$\begin{aligned}&\alpha = 10,000\\ &\text{Model: } \sigma = 0.991 * \theta - 0.142 * \theta^{\wedge}2 + 0.00831 * \theta^{\wedge}3 - 0.000193 * \theta^{\wedge}4\end{aligned} \tag{8.18}$$

$$\begin{aligned}&\alpha = 100,000\\ &\text{Model: } \sigma = 0.994 * \theta - 0.144 * \theta^{\wedge}2 + 0.0093 * \theta^{\wedge}3 - 0.000261 * \theta^{\wedge}4\end{aligned} \tag{8.19}$$

Finally, the results are summarized and presented in Table 8.2.

From the entire above, one can see that the steady time random function has an ergodic character, therefore only *one* random realization function is sufficient enough in order to obtain the correlation function in this case. The summary of all stress–temperature–strain relationships graphically are presented in Fig. 8.5.

The curve "*r*" (see Fig. 8.1) indicates that the material is creep resistant. It should be noted again that α is inverse value proportional to retardation time. Let us choose the random realization function that corresponds to $\alpha = 100$.

$$\begin{aligned}&\alpha = 100\\ &\sigma = 0.996 * \theta - 0.148 * \theta^{\wedge}2 + 0.000172 * \theta^{\wedge}3 - 0.000641 * \theta^{\wedge}4\end{aligned} \tag{8.20}$$

The mean value can be calculated now as follows:

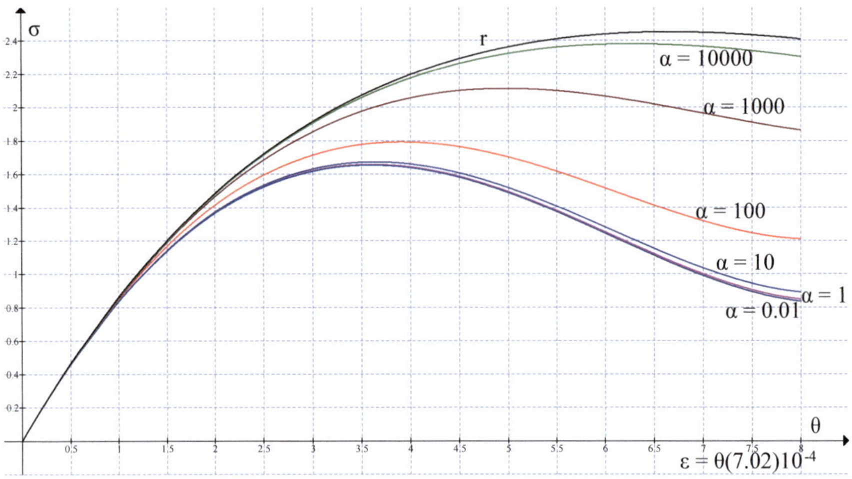

Fig. 8.5 Stress–temperature–strain relationships (fast fire)

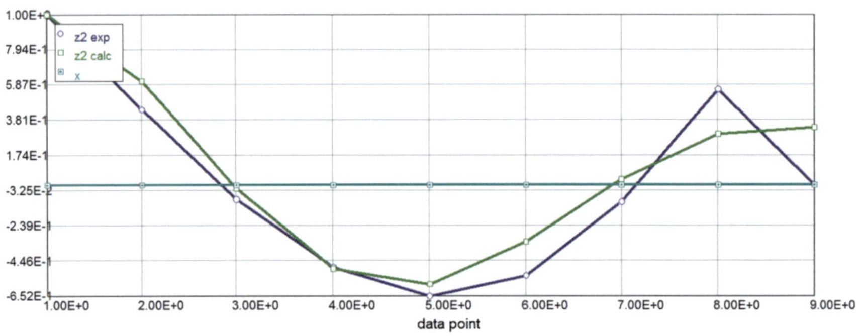

Fig. 8.6 Normalized autocorrelation function

$$\boldsymbol{\mu}_{\boldsymbol{\sigma}} = \frac{1}{8}\int_0^8 \boldsymbol{\sigma}(\theta)\mathrm{d}\boldsymbol{\theta}$$

$$= \frac{1}{8}\int_0^8 (0.996 * \boldsymbol{\theta} - \mathbf{0}.148 * \boldsymbol{\theta}\hat{\ }\mathbf{2} + \mathbf{0}.000172 * \boldsymbol{\theta}\hat{\ }\mathbf{3} + \mathbf{0.000}641 * \boldsymbol{\theta}\hat{\ }\mathbf{4})\mathrm{d}\boldsymbol{\theta} = \frac{10.99}{8} = 1.374 \tag{8.21}$$

In order to calculate the correlation function in this case the chosen function (8.2) has to be centered (see Table 8.3). Again, after using the POLYMATH software we have

Table 8.3 Centered correlation function

$\alpha\|\theta$	0	1	2	3	4	5	6	7	8
100	0	0.855	1.418	1.713	1.805	1.723	1.530	1.349	1.239
σ*	−1.374	−0.519	0.044	0.339	0.431	0.349	0.156	−0.025	−0.135

Model: $z2 = a0 + a1 * x2 + a2 * x2\text{^}2 + a3 * x2\text{^}3 + a4 * x2\text{^}4$

Variable	Value
$a0$	−1.373193
$a1$	1.001478
$a2$	−0.1507506
$a3$	0.0008411
$a4$	0.0005967

$$\sigma^* = -1.373 + 1.0(\theta) - 0.151(\theta^2) + 0.000841(\theta^3) + 0.000597(\theta^4) \tag{8.22}$$

The correlation function of a steady time process can be computed now as follows [19]:

$$K_{\boldsymbol{\sigma}}(\boldsymbol{\theta}) = \frac{1}{8-\boldsymbol{\theta}} \int_0^{8-\boldsymbol{\theta}} [\boldsymbol{\sigma}^*(t)\boldsymbol{\sigma}^*(t+\boldsymbol{\theta})]\mathrm{d}t \tag{8.23}$$

where $\boldsymbol{\theta} = 0; 1; 2; \ldots 8$

After substituting (8.22) into (8.23) we have

The standard deviation $D = K_\theta(0)$ is

$$D = K_{\boldsymbol{\sigma}}(\boldsymbol{\theta} = 0) = \frac{1}{8-\boldsymbol{\theta}} \int_0^{8-\boldsymbol{\theta}} [\boldsymbol{\sigma}^*(t)\boldsymbol{\sigma}^*(t+\boldsymbol{\theta})]\mathrm{d}t = \frac{1}{8}\int_0^8 [\boldsymbol{\sigma}^*(t)]^2\mathrm{d}t$$

$$= 0.125 \int_0^{10} [-\mathbf{1.373} + \mathbf{1.0}(\boldsymbol{\theta}) - \mathbf{0.151}(\boldsymbol{\theta}^2) + \mathbf{0.000841}(\boldsymbol{\theta}^3) + \mathbf{0.000597}(\boldsymbol{\theta}^4)]^2\mathrm{d}\boldsymbol{\theta}$$

$$= 0.125(1.425) = 0.178; \quad \boldsymbol{\sigma}_{\boldsymbol{\sigma}} = 0.422$$

The computations of autocorrelation function based on Eq. (8.23) are presented in Table 8.4).

Table 8.4 Autocorrelation function

θ	0	1	2	3	4	5	6	7	8
$K_\sigma(\theta)$	0.178	0.079	−0.015	−0.0857	−0.116	−0.0947	−0.0178	0.100	0.220

Table 8.5 Normalized autocorrelation function

	0	1	2	3	4	5	6	7	8
$\rho_\sigma(\theta)$	1.0	0.444	−0.084	−0.482	−0.652	−0.532	−0.1	0.562	0

The normalized autocorrelation function is shown in Table 8.5.

Let us approximate the data from Table 8.5 by the type of formulae (Fig. 8.6)

Model: $z2 = (\exp(-a * x2)) * (\cos(b * x2))$

Variable	Initial guess	Value
a	1	0.135
b	1	0.799

$$\rho_\sigma(\theta) = (\exp(-0.135 * x)) * (\cos(0.799 * x)) \tag{8.24}$$

The corresponding spectral function is [20]

$$\begin{aligned} S(\omega) &= \frac{1}{2}\frac{\sigma_\sigma^2}{\pi}\left[\frac{a}{(\omega - b)^2 + a^2} + \frac{a}{(\omega + b)^2 + a^2}\right] \\ &= 3.82(10^{-3})\left[\frac{1}{(\omega - 0.799)^2 + 0.135^2} + \frac{1}{(\omega + 0.799)^2 + 0.135^2}\right] \end{aligned} \tag{8.25}$$

The graphic presentation of spectral density is shown in Fig. 8.7.

8.1.9 The First-Occurrence Time Problem and the Probability Density $P(a_0, \theta)$

The average first-occurrence of stress σ below given level "$\sigma = a_0$" for stationary stress–temperature processes is defined as follows (see Chap. 7) [20]:

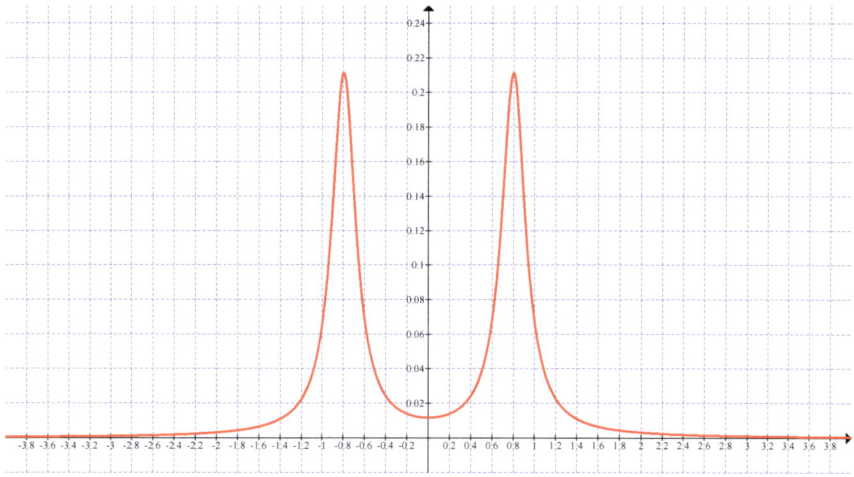

Fig. 8.7 Spectral density function

For correlation function given by (8.24)

$$\sigma_{v\sigma}^2 = K_{\dot{\sigma}}(0) = -\frac{d^2}{d\theta^2}K_{\dot{\sigma}}(\theta)|_{\theta=0} = \sigma_{\sigma}^2(b^2 - a^2)$$
$$\bar{n}_a = \theta_{max}\frac{\sqrt{b^2 - a^2}}{2\pi}\exp\left(-\frac{a_0^2}{2\sigma_{\sigma}^2}\right) = 8\frac{\sqrt{0.799^2 - 0.135^2}}{2\pi}\exp\left(-\frac{1.1^2}{2(0.422)^2}\right) = 0.0335$$
$$\bar{v}_a = \frac{\bar{n}_a}{\theta_{max}} = \frac{0.0335}{8} = 0.00418 \tag{8.26}$$

The average interval for the stress occurrence below given level "*a*":

$$\tilde{\theta} = \frac{\int_a^{\infty} f(x)dx}{\int_0^{\infty} vf(a,v)\mathrm{d}v} = \pi\frac{\sigma_{\sigma}}{\sigma_v}\exp\left(+\frac{(a-\bar{x})^2}{2\sigma_{\sigma}^2}\right)\left[\Phi^*(\frac{a-\bar{x}}{\sigma_{\sigma}})\right]$$
$$= \frac{\pi}{\sqrt{0.799^2 - 0.135^2}}\exp\left(\frac{0.274^2}{2(0.422)^2}\right)[\Phi^*(-0.649)] = 3.989(1.235)[0.2578] = 1.27 < 8 \tag{8.27}$$

The average first-occurrence temperature below given level "*a*" for stationary processes is

$$\overline{\theta}_a = \boldsymbol{\theta}_{\max} \int_a^{\infty} f(x)\mathrm{d}x = \frac{10}{\boldsymbol{\sigma}_{\sigma}\sqrt{2\pi}} \int_a^{\infty} \exp\left(-\frac{(x-\overline{x})^2}{2\boldsymbol{\sigma}_{\theta}^2}\right)\mathrm{d}x$$
$$= \frac{8}{0.422\sqrt{2\pi}}[\Phi^*(-0.649)] = 7.563(0.2578) = 1.95 < 8 \tag{8.28}$$

Now based on Poisson formulae of the probability to not having the minimum allowable stress ordinates crossing downwards the level "a = 1.0" is

$$P_{\mathrm{rel}} = P_o = \exp(-\overline{n}_a) = \exp(-0.0335) = 0.967$$

This probability characterizes the reliability of the structure (one element or the whole structure).

It should be noted that the references to Chap. 7 remain the same for all other fire scenarios below, therefore they will not be repeated.

8.2 Probability-Based Stress–Strain Diagram. Medium Fire

8.2.1 Temperature–Time Function. Medium Fire

Medium Fire; Temperature Increase (822 K < T < 882 K)

Data T_* = 600 K; $0 < \tau < 0.2$; $0.175 < \gamma < 0.275$

Select γ = 0.175

Differential equations (8.2) and (8.3) are rewritten as an input for "Polymath" software

1 $\mathrm{d}(y0)/\mathrm{d}(t) = 20 * (1 - y2) * \exp(y0/(1 + 0.1 * y0)) - 0.157 * y0\hat{\ }4$

2 $\mathrm{d}(y2)/\mathrm{d}(t) = 5.5 * (1 - y2) * \exp(y0/(1 + 0.1 * y0))$

3 $\mathrm{d}(y1)/\mathrm{d}(t) = 3.5 * (1 - y1)\hat{\ }1.0 * \exp(y/(1 + 0.1 * y))$

4 $\mathrm{d}(y)/\mathrm{d}(t) = (1) * 20 * (1 - y1)\hat{\ }1.0 * \exp(y/(1 + 0.1 * y)) - 0.157 * y\hat{\ }4$

where

"y" is the dimensionless temperature "θ" with the corresponding parameter "γ = 0.175".
"y0" is the dimensionless temperature "θ" with the corresponding parameter "γ = 0.275".
"y1" is the concentration of the product of the first-order chemical reaction with "γ = 0.175".
"y2" is the concentration of the product of the first-order chemical reaction with "γ = 0.275".

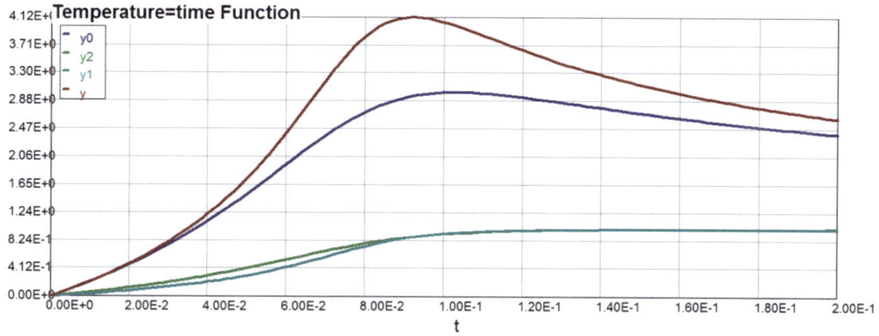

Fig. 8.8 Dimensionless temperature–time curves

Calculated values of DEQ variables

	Variable	Initial value	Minimal value	Maximal value	Final value
1	*t*	0	0	0.2	0.2
2	*y*	0	0	4.121	2.629
3	*y*0	0	0	3.013	2.398
4	*y*1	0	0	0.9987	0.9987
5	*y*2	0	0	0.9993	0.9993

Tabulated solution of the Eqs. (8.2) and (8.3) are presented above and the graphs are shown on Fig. 8.8.

8.2.2 *Analytical Expression of the Inverse Function* θ^{-1} *and Its First Derivative*

First, based on the data shown above, the "best-to-fit" (polynomial regression) approximation of the dimensionless temperature–time curve can be presented as follows:

Model: $y = a0 + a1 * t + a2 * t^\wedge 2 + a3 * t^\wedge 3 + a4 * t^\wedge 4 + a5^* t^\wedge 5$

Variable	Value
*a*0	0
*a*1	−75.87
*a*2	3910
*a*3	−4.542E+04
*a*4	2.06E+05
*a*5	−3.29E+05

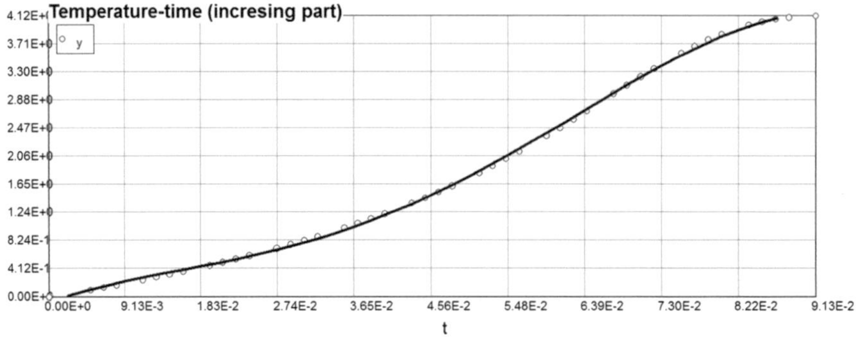

Fig. 8.9 Monotonically increasing temperature–time function

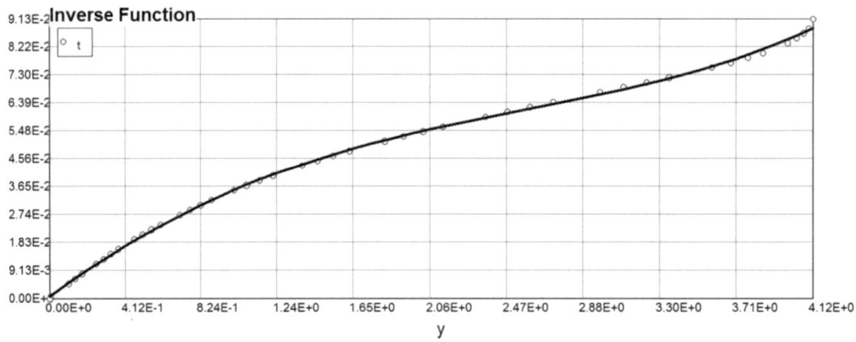

Fig. 8.10 Monotonically increasing inverse function θ^{-1}

$$\textbf{Model}: \theta = -75.87 * \tau + 3910 * \tau^\wedge 2 - 45{,}420 * \tau^\wedge 3 + 206{,}000 * \tau^\wedge 4 - 32{,}900 * \tau^\wedge 5 \tag{8.29}$$

Monotonically increasing temperature–time function: $(0 < \theta < 4)$; $(0 < \tau 1 < 0.0913)$.

As noted above, a function of the temperature–time should be divided into two portions monotonically increasing and monotonically decreasing. Let us start with a monotonically increasing area. The range of temperature values in this case is the interval from 0 to 4 and the time interval from 0 to 0.0913. Therefore using polynomial regression method again we have (Fig. 8.9)

Model: $y = a0 + a1*t + a2*t^\wedge 2 + a3*t^\wedge 3 + a4*t^\wedge 4$

Variable	Value
$a0$	0
$a1$	42.27
$a2$	−1347
$a3$	3.603E+04
$a4$	−2.277E+05

$$\textbf{Model}: \theta = 42.27 * \tau - 1347 * \tau^\wedge 2 + 36{,}030 * \tau^\wedge 3 - 227{,}700 * \tau^\wedge 4 \tag{8.30}$$

The inverse function θ^{-1} and its first derivative (monotonically increasing functions) are

$$\textbf{Model}: t = a0 + a1 * y + a2 * y^\wedge 2 + a3^* y^\wedge 3 + a4 * y^\wedge 4$$

Variable	Value
$a0$	0.0008285
$a1$	0.0438492
$a2$	−0.010139
$a3$	0.0004935
$a4$	0.0001537

The graph of the inverse function θ^{-1} is shown in Fig. 8.10.
Therefore

$$\begin{aligned}\textbf{Model}:\ \mathbf{m} = \tau = {} & 0.000928 + 0.0438 * \theta - 0.0101 * \theta^\wedge 2 \\ & + 0.000494 * \theta^\wedge 3 + 0.000154 * \theta^\wedge 4\end{aligned} \tag{8.31}$$

$$\mathbf{m1} = \tau' = 0.0438 - 0.0202 * \theta + 0.00148 * \theta^\wedge 2 + 0.000616 * \theta^\wedge 3 \tag{8.32}$$

8.2.3 Medium Fire: Statistical Data

Differential equations (DEQs): $\alpha = 0.001$

Calculated values of DEQ variables

	Variable	Initial value	Minimal value	Maximal value	Final value
1	A	0	0	7.723008	7.723008
2	m	0.000928	0.000928	0.206416	0.206416
3	$m1$	0.0438	0.0121152	0.108936	0.108936
4	n	3	3	3	3
5	r	0	0	2.439418	2.439418
6	t	0	0	6	6
7	y	0	0	*1.501322*	0.3841946

Differential equations

$$1 \quad \mathrm{d}(y)/\mathrm{d}(t) = -0.001 * y * m1 + (\exp(-0.15 * t)) * (1 - 0.15 * t) - (\exp(t/(1.0 + 0.1 * t))) * m1 * y^\wedge n + 0.001 * m1 * t * (\exp(-0.15 * t))$$

$$2 \quad \mathrm{d}(A)/\mathrm{d}(t) = 1.432 * t - 0.383 * t^\wedge 2 + 0.034 * t^\wedge 3 - 0.00096 * t^\wedge 4$$

Explicit equations

1 $n = 3$

2 $r = t * (\exp(-0.15 * t))$

3 $m1 = 0.0438 - 0.0202 * t + 0.00148 * t^\wedge 2 + 0.000616 * t^\wedge 3$

4 $m = 0.000928 + 0.0438 * t - 0.0101 * t^\wedge 2 + 0.000494 * t^\wedge 3 + 0.000154 * t^\wedge 4$

Differential equations (DEQs): $\alpha = 0.01$

Calculated values of DEQ variables

	Variable	Initial value	Minimal value	Maximal value	Final value
1	A	0	0	7.723008	7.723008
2	m	0.000928	0.000928	0.206416	0.206416
3	$m1$	0.0438	0.0121152	0.108936	0.108936
4	n	3	3	3	3
5	r	0	0	2.439418	2.439418
6	t	0	0	6	6
7	y	0	0	*1.501342*	0.3848787

Differential equations

1 $d(y)/d(t) = -0.01 * y * m1 + (\exp(-0.15 * t)) * (1 - 0.15 * t) - (\exp(t/(1.0 + 0.1 * t))) * m1 * y^\wedge n + 0.01 * m1 * t * (\exp(-0.15 * t))$

2 $d(A)/d(t) = 1.432 * t - 0.383 * t^\wedge 2 + 0.034 * t^\wedge 3 - 0.00096 * t^\wedge 4$

Explicit equations

1 $n = 3$

2 $r = t * (\exp(-0.15 * t))$

3 $m1 = 0.0438 - 0.0202 * t + 0.00148 * t^\wedge 2 + 0.000616 * t^\wedge 3$

4 $m = 0.000928 + 0.0438 * t - 0.0101 * t^\wedge 2 + 0.000494 * t^\wedge 3 + 0.000154 * t^\wedge 4$

Differential equations (DEQs): $\alpha = 0.1$

Calculated values of DEQ variables

	Variable	Initial value	Minimal value	Maximal value	Final value
1	*A*	0	0	7.723008	7.723008
2	*m*	0.000928	0.000928	0.206416	0.206416
3	*m*1	0.0438	0.0121141	0.108936	0.108936
4	*n*	3	3	3	3
5	*r*	0	0	2.439418	2.439418
6	*t*	0	0	6	6
7	*y*	0	0	*1.50164*	0.3916381

Differential equations

1 $d(y)/d(t) = -0.1 * y * m1 + (\exp(-0.15 * t)) * (1 - 0.15 * t) - (\exp(t/(1.0 + 0.1 * t))) * m1 * y^\wedge n + 0.1 * m1 * t * (\exp(-0.15 * t))$

2 $d(A)/d(t) = 1.432 * t - 0.383 * t^\wedge 2 + 0.034 * t^\wedge 3 - 0.00096 * t^\wedge 4$

Explicit equations

1 $n = 3$

2 $r = t * (\exp(-0.15 * t))$

3 $m1 = 0.0438 - 0.0202 * t + 0.00148 * t^\wedge 2 + 0.000616 * t^\wedge 3$

4 $m = 0.000928 + 0.0438 * t - 0.0101 * t^\wedge 2 + 0.000494 * t^\wedge 3 + 0.000154 * t^\wedge 4$

Differential equations (DEQs): $\alpha = 1$

Calculated values of DEQ variables

	Variable	Initial value	Minimal value	Maximal value	Final value
1	*A*	0	0	7.723008	7.723008
2	*m*	0.000928	0.000928	0.206416	0.206416
3	*m*1	0.0438	0.0121152	0.108936	0.108936
4	*n*	3	3	3	3
5	*r*	0	0	2.439418	2.439418
6	*t*	0	0	6	6
7	*y*	0	0	*1.503757*	0.4520366

Differential equations

$$1 \quad d(y)/d(t) = -1 * y * m1 + (\exp(-0.15 * t)) * (1 - 0.15 * t) - (\exp(t/(1.0 + 0.1 * t))) * m1 * y^\wedge n + 1 * m1 * t * (\exp(-0.15 * t))$$

$$2 \quad d(A)/d(t) = 1.432 * t - 0.383 * t^\wedge 2 + 0.034 * t^\wedge 3 - 0.00096 * t^\wedge 4$$

Explicit equations

1 $n = 3$

2 $r = t * (\exp(-0.15 * t))$

3 $m1 = 0.0438 - 0.0202 * t + 0.00148 * t^\wedge 2 + 0.000616 * t^\wedge 3$

4 $m = 0.000928 + 0.0438 * t - 0.0101 * t^\wedge 2 + 0.000494 * t^\wedge 3 + 0.000154 * t^\wedge 4$

Differential equations (DEQs): $\alpha = 10$

Calculated values of DEQ variables

	Variable	Initial value	Minimal value	Maximal value	Final value
1	*A*	0	0	7.723008	7.723008
2	*m*	0.000928	0.000928	0.206416	0.206416
3	*m*1	0.0438	0.0121141	0.108936	0.108936
4	*n*	3	3	3	3
5	*r*	0	0	2.439418	2.439418
6	*t*	0	0	6	6
7	*y*	0	0	*1.524353*	0.7534053

Differential equations

1 $d(y)/d(t) = -10 * y * m1 + (\exp(-0.15 * t)) * (1 - 0.15 * t) - (\exp(t/(1.0 + 0.1 * t))) * m1 * y^\wedge n + 10 * m1 * t * (\exp(-0.15 * t))$

2 $d(A)/d(t) = 1.432 * t - 0.383 * t^\wedge 2 + 0.034 * t^\wedge 3 - 0.00096 * t^\wedge 4$

Explicit equations

1 $n = 3$

2 $r = t * (\exp(-0.15 * t))$

3 $m1 = 0.0438 - 0.0202 * t + 0.00148 * t^\wedge 2 + 0.000616 * t^\wedge 3$

4 $m = 0.000928 + 0.0438 * t - 0.0101 * t^\wedge 2 + 0.000494 * t^\wedge 3 + 0.000154 * t^\wedge 4$

Differential equations (DEQs): $\alpha = 100$

Calculated values of DEQ variables

	Variable	Initial value	Minimal value	Maximal value	Final value
1	*A*	0	0	7.723008	7.723008
2	*m*	0.000928	0.000928	0.206416	0.206416
3	*m*1	0.0438	0.0121148	0.108936	0.108936
4	*n*	3	3	3	3
5	*r*	0	0	2.439418	2.439418
6	*t*	0	0	6	6
7	*y*	0	0	*1.668804*	1.367226

Differential equations

1 $d(y)/d(t) = -100 * y * m1 + (\exp(-0.15 * t)) * (1 - 0.15 * t) - (\exp(t/(1.0 + 0.1 * t))) * m1 * y^\wedge n + 100 * m1 * t * (\exp(-0.15 * t))$

2 $d(A)/d(t) = 1.432 * t - 0.383 * t^\wedge 2 + 0.034 * t^\wedge 3 - 0.00096 * t^\wedge 4$

Explicit equations

1 $n = 3$

2 $r = t * (\exp(-0.15 * t))$

3 $m1 = 0.0438 - 0.0202 * t + 0.00148 * t^\wedge 2 + 0.000616 * t^\wedge 3$

4 $m = 0.000928 + 0.0438 * t - 0.0101 * t^\wedge 2 + 0.000494 * t^\wedge 3 + 0.000154 * t^\wedge 4$

Differential equations (DEQs): $\alpha = 1000$

Calculated values of DEQ variables

	Variable	Initial value	Minimal value	Maximal value	Final value
1	*A*	0	0	7.723008	7.723008
2	*m*	0.000928	0.000928	0.206416	0.206416
3	*m*1	0.0438	0.0121116	0.108936	0.108936
4	*n*	3	3	3	3
5	*r*	0	0	2.439418	2.439418
6	*t*	0	0	6	6
7	*y*	0	0	*2.103128*	2.065647

Differential equations

$$1 \quad d(y)/d(t) = -1000 * y * m1 + (\exp(-0.15 * t)) * (1 - 0.15 * t) - (\exp(t/(1.0 + 0.1 * t))) * m1 * y\hat{}n + 1000 * m1 * t * (\exp(-0.15 * t))$$

$$2 \quad d(A)/d(t) = 1.432 * t - 0.383 * t\hat{}2 + 0.034 * t\hat{}3 - 0.00096 * t\hat{}4$$

Explicit equations

1 $n = 3$

2 $r = t * (\exp(-0.15 * t))$

3 $m1 = 0.0438 - 0.0202 * t + 0.00148 * t\hat{}2 + 0.000616 * t\hat{}3$

4 $m = 0.000928 + 0.0438 * t - 0.0101 * t\hat{}2 + 0.000494 * t\hat{}3 + 0.000154 * t\hat{}4$

Differential equations (DEQs): $\alpha = 10{,}000$

Calculated values of DEQ variables

	Variable	Initial value	Minimal value	Maximal value	Final value
1	*A*	0	0	7.723008	7.723008
2	*m*	0.000928	0.000928	0.206416	0.206416
3	*m*1	0.0438	0.0121133	0.108936	0.108936
4	*n*	3	3	3	3
5	*r*	0	0	2.439418	2.439418
6	*t*	0	0	6	6
7	*y*	0	0	*2.381973*	*2.381973*

Differential equations

$$
\begin{aligned}
1\quad d(y)/d(t) &= -10,000 * y * m1 + (\exp(-0.15 * t)) * (1 - 0.15 * t) \\
&\quad - (\exp(t/(1.0 + 0.1 * t))) * m1 * y^{\wedge}n \\
&\quad + 10,000 * m1 * t * (\exp(-0.15 * t)) \\
2\quad d(A)/d(t) &= 1.432 * t - 0.383 * t^{\wedge}2 + 0.034 * t^{\wedge}3 - 0.00096 * t^{\wedge}4
\end{aligned}
$$

Explicit equations

1 $n = 3$

2 $r = t * (\exp(-0.15 * t))$

3 $m1 = 0.0438 - 0.0202 * t + 0.00148 * t^{\wedge}2 + 0.000616 * t^{\wedge}3$

4 $m = 0.000928 + 0.0438 * t - 0.0101 * t^{\wedge}2 + 0.000494 * t^{\wedge}3 + 0.000154 * t^{\wedge}4$

Differential equations (DEQs): α = 100,000

Calculated values of DEQ variables

	Variable	Initial value	Minimal value	Maximal value	Final value
1	*A*	0	0	7.723008	7.723008
2	*m*	0.000928	0.000928	0.206416	0.206416
3	*m*1	0.0438	0.0121141	0.108936	0.108936
4	*n*	3	3	3	3
5	*r*	0	0	*2.439418*	*2.439418*
6	*t*	0	0	6	6
7	*y*	0	0	*2.433292*	*2.433292*

Differential equations

$$
\begin{aligned}
1\quad d(y)/d(t) &= -100,000 * y * m1 + (\exp(-0.15 * t)) * (1 - 0.15 * t) \\
&\quad - (\exp(t/(1.0 + 0.1 * t))) * m1 * y^{\wedge}n \\
&\quad + 100,000 * m1 * t * (\exp(-0.15 * t)) \\
2\quad d(A)/d(t) &= 1.432 * t - 0.383 * t^{\wedge}2 + 0.034 * t^{\wedge}3 - 0.00096 * t^{\wedge}4
\end{aligned}
$$

Explicit equations

1 $n = 3$

2 $r = t * (\exp(-0.15 * t))$

3 $m1 = 0.0438 - 0.0202 * t + 0.00148 * t^{\wedge}2 + 0.000616 * t^{\wedge}3$

4 $m = 0.000928 + 0.0438 * t - 0.0101 * t^{\wedge}2 + 0.000494 * t^{\wedge}3 + 0.000154 * t^{\wedge}4$

Table 8.6 Maximum dimensionless stresses (statistical data)

Value	Maximum Creep Stress	Deviation	Variance	Mean Value	Standard Deviation	α value
1	*1.501*	−0.29	0.0841			0.001
2	*1.501*	−0.29	0.0841			0.01
3	*1.502*	−0.289	0.0835			0.1
4	*1.504*	−0.287	0.0713			1.0
5	*1.524*	−0.267	0.0573			10
6	*1.669*	−0.122	0.0149			100
7	*2.103*	0.312	0.0973			1000
8	*2.382*	0.591	0.349			10,000
9	*2.433*	0.642	0.412			100,000
Aver.	*1.791*		Total: 1.2535/8 = = 0.1567	*1.791*	*0.396*	

The summary of computations above is presented in Table 8.6.

Therefore, the probability that the real fire dimensionless allowable stress "**σ**" will be less than 1.5; 1.2; 1.1, or 1.0 is

$$
\begin{aligned}
P(\boldsymbol{\sigma}<1.65) &= \Phi^*\left(\frac{1.5-1.791}{0.396}\right) = \Phi^*(-0.735) = 0.226 = 22.6\,\% \\
P(\boldsymbol{\sigma}<1.2) &= \Phi^*\left(\frac{1.2-1.791}{0.396}\right) = \Phi^*(-1.5) = 0.067 = 6.7\,\% \\
P(\boldsymbol{\sigma}<1.1) &= \Phi^*\left(\frac{1.1-1.791}{0.396}\right) = \Phi^*(-1.75) = 0.04 = 4.0\,\% \\
P_{\mathrm{f}}(\boldsymbol{\sigma}<1.0) &= \Phi^*\left(\frac{1.0-1.791}{0.396}\right) = \Phi^*(-2.0) = 0.023 = 2.3\,\%
\end{aligned}
\tag{8.33}
$$

If probability of failure $P_{\mathrm{f}} = 4\ \%$ is acceptable, then $\sigma_{\mathrm{all}} = 1.1(7.02)$ 2.9 = 22.4 ksi = 154.4 MPa; $\mu_{\sigma} = 1.791$ and $\sigma_{\sigma} = 0.396$, then $\beta_{\mathrm{all}} = 1.745$ (reliability index).

8.2.4 The First-Order Reliability Method (FORM)

For detailed information regarding applicability of this method please see Chap. 7. The computation of the reliability index β and corresponding design temperature (based on Example 7.2) is presented below without any additional comments or explanations in order to save space.

Example 8.3 Data: Fast Fire: $\mu_{sT} = 1000$ °F; $\sigma_{sT} = 100$ °F. Find the reliability index β; probability of failure P_{f} and design temperature T °F.

The deterministic performance function in this case is defined by Eq. H2-1 [15]

$$g = \frac{72.76 + 0.041(\mu_{sT})}{50(11.8)} + \frac{13.07 + 0.214(12)(\mu_{sT})}{(50)73} - \mu_R = 0$$
$$\mu_{\sigma S} = 0.166 + 0.000773(\mu_{sT});\ \sigma_{\sigma S} = 0.000773(\sigma_{sT})\ \mu_{sT} = 1000\,^{\circ}\mathrm{F}$$

where $\mu_{\sigma R} = 1.791;\ \sigma_{\sigma R} = 0.396 \rightarrow$ see computations above

$$\beta = \frac{|\mu_{\sigma R} - \mu_{\sigma S}|}{\sqrt{(\sigma_{\sigma S})^2 + (\sigma_{\sigma R})^2}} = \frac{1.791 - 0.939}{\sqrt{0.0773^2 + 0.396^2}} = 2.11$$

The probability of structural failure and the design temperature in this case are computed as follows:

$$P_{\mathrm{f}} = \Phi^*(-\boldsymbol{\beta}) = \Phi^*(-2.11) = 0.0179;$$
$$T_{\mathrm{design}} = 1000 + (2.11)100 = 1211\,^{\circ}\mathrm{F}$$

8.2.5 Confidence Interval—Fast Fire

Again, for detailed information regarding applicability of this method please see Chap. 7. The computation of the reliability index β and corresponding design temperature (based on Example 7.3) is presented below

Example 8.4 Data: Fast Fire: $\alpha = 0.95$—given target confidence probability
From Table 8.1: $\mu_{\sigma R} = 1.791;\ \sigma_{\sigma R} = 0.396$

The confidence limit (interval) I_α of mean value (μ_σ) and standard deviation (σ_σ) is as follows

$$\boldsymbol{\sigma}_{\boldsymbol{\mu}_\sigma} = \sqrt{\frac{\boldsymbol{\sigma}_\sigma^2}{n-1}} = \sqrt{\frac{(0.396)^2}{8}} = 0.14$$
$$\boldsymbol{\varepsilon}_\alpha = \boldsymbol{\sigma}_{\boldsymbol{\mu}_\sigma} \arg\ \Phi^*\left(\frac{1+\boldsymbol{\alpha}}{2}\right) = 0.14\left[\arg\ \Phi^*\left(\frac{1+0.95}{2}\right)\right] = 0.14(1.96) = 0.274$$
$$t_\alpha = \arg\ \Phi^*\left(\frac{1+\boldsymbol{\alpha}}{2}\right) = 1.96$$

Finally, the confidence limit (interval) I_α for mean value is calculated as follows:

$$I_\alpha = (\boldsymbol{\mu}_\sigma - t_\alpha \boldsymbol{\sigma}_{\boldsymbol{\mu}_\sigma};\ \boldsymbol{\mu}_\sigma + t_\alpha \boldsymbol{\sigma}_{\boldsymbol{\mu}_\sigma}) = (1.791 - 0.274;\ 1.791 + 0.274) = (1.517;\ 2.065)$$

The confidence limit (interval) I_α for standard deviation value is calculated as follows:

$$I_\alpha = (\boldsymbol{\sigma}_{\boldsymbol{\sigma}}^2 - t_\alpha \boldsymbol{\sigma}_{\boldsymbol{\sigma}_\sigma}; \boldsymbol{\sigma}_{\boldsymbol{\sigma}}^2 + t_\alpha \boldsymbol{\sigma}_{\boldsymbol{\sigma}_\sigma}) = (0.396^2 - (1.96)0.0564;\ 0.396^2 + (1.96)0.0564)$$
$$= (0.046;\ 0.267)$$

where $\boldsymbol{\sigma}_{\boldsymbol{\sigma}_\sigma} = \sqrt{\frac{2}{n-1}\sigma_{\boldsymbol{\sigma}}^2} = \sqrt{\frac{2}{9-1}(0.336)^2} = 0.0564$ and $I_\alpha^{\boldsymbol{\sigma}_{\boldsymbol{\sigma}_\sigma}} = \left(\sqrt{\boldsymbol{\sigma}_{\boldsymbol{\sigma}}^2 - t_\alpha \boldsymbol{\sigma}_{\boldsymbol{\sigma}_\sigma}};\ \sqrt{\boldsymbol{\sigma}_{\boldsymbol{\sigma}}^2 + t_\alpha \boldsymbol{\sigma}_{\boldsymbol{\sigma}_\sigma}}\right) = (0.214;\ 0.517)$.

The minimum allowable dimensionless stress is

$$\boldsymbol{\sigma}_{\text{min,all}} = 1.517 - 0.517 = 1.0$$

8.2.6 Creep Stress–Strain Diagrams (Ergodic Process)

The statistical data (stress–temperature–strain relationship) in this fire severity case are taken from Chap. 5 and presented below.

$$\begin{aligned} &\alpha = 0.001 \\ &\text{Model: } \sigma = 0.86 * \theta + 0.0189 * \theta\hat{}2 - 0.0713 * \theta\hat{}3 + 0.00767 * \theta\hat{}4 \end{aligned} \tag{8.34}$$

$$\begin{aligned} &\alpha = 0.01 \\ &\text{Model: } \sigma = 0.861 * \theta + 0.019 * \theta\hat{}2 - 0.0713 * \theta\hat{}3 + 0.00767 * \theta\hat{}4 \end{aligned} \tag{8.35}$$

$$\begin{aligned} &\alpha = 0.1 \\ &\text{Model: } \sigma = 0.861 * \theta + 0.0193 * \theta\hat{}2 - 0.0714 * \theta\hat{}3 + 0.00769 * \theta\hat{}4 \end{aligned} \tag{8.36}$$

$$\begin{aligned} &\alpha = 1 \\ &\text{Model: } \sigma = 0.86 * \theta + 0.0208 * \theta\hat{}2 - 0.0721 * \theta\hat{}3 + 0.00782 * \theta\hat{}4 \end{aligned} \tag{8.37}$$

$$\begin{aligned} &\alpha = 10 \\ &\text{Model: } \sigma = 0.905 * \theta - 0.0169 * \theta\hat{}2 - 0.0611 * \theta\hat{}3 + 0.00707 * \theta\hat{}4 \end{aligned} \tag{8.38}$$

$$\begin{aligned} &\alpha = 100 \\ &\text{Model: } \sigma = 1.0 * \theta - 0.12 * \theta\hat{}2 - 0.0184 * \theta\hat{}3 + 0.00286 * \theta\hat{}4 \end{aligned} \tag{8.39}$$

$$\begin{aligned} &\alpha = 1000 \\ &\text{Model: } \sigma = 0.987 * \theta - 0.132 * \theta\hat{}2 + 0.00172 * \theta\hat{}3 + 0.000408 * \theta\hat{}4 \end{aligned} \tag{8.40}$$

Table 8.7 Stress–temperature data (medium fire)

$\alpha\|\theta$	0	1	2	3	4	5	6
0.001	0	0.850	1.362	1.494	1.166	0.687	0.384
0.01	0	0.850	1.362	1.494	1.166	0.688	0.385
0.1	0	0.850	1.362	1.494	1.167	0.690	0.392
1.0	0	0.850	1.363	1.497	1.175	0.715	0.452
10	0	0.851	1.370	1.520	1.246	0.904	0.735
100	0	0.854	1.412	1.659	1.606	1.485	1.367
1000	0	0.860	1.467	1.854	2.049	2.103	2.066
10,000	0	0.860	1.480	1.906	2.177	2.327	2.382
100,000	0	0.861	1.481	1.912	2.193	2.358	2.433
r	0	0.861	1.482	1.913	2.195	2.362	2.439

$$\begin{aligned}&\alpha = 10,000\\&\text{Model: } \sigma = 0.996 * \theta - 0.146 * \theta^\wedge 2 + 0.0092 * \theta^\wedge 3 - 0.000265 * \theta^\wedge 4\end{aligned} \tag{8.41}$$

$$\begin{aligned}&\alpha = 100,000\\&\text{Model: } \sigma = 0.998 * \theta - 0.147 * \theta^\wedge 2 + 0.01 * \theta^\wedge 3 - 0.000317 * \theta^\wedge 4\end{aligned} \tag{8.42}$$

Finally, the results are summarized and presented in Table 8.7.

The summary of all stress–temperature–strain relationships are presented graphically in Fig. 8.11.

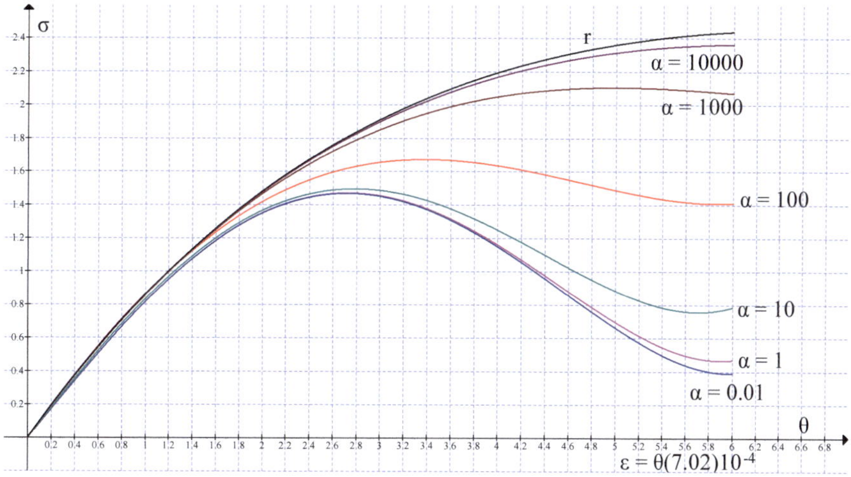

Fig. 8.11 Stress–temperature–strain relationships (fast fire)

Table 8.8 Centered correlation function

$\alpha\|\theta$	0	1	2	3	4	5	6
100	0	0.854	1.412	1.659	1.606	1.485	1.367
σ^*	−1.388	−0.534	0.024	0.271	0.218	0.097	−0.021

The curve "*r*" (see Fig. 8.11) indicates that the material is creep resistant. It should note again that α is inverse value proportional to retardation time. Let us choose the random realization function that corresponds to $\alpha = 100$.

$$\begin{aligned} &\alpha = 100 \\ &\sigma = 1.0 * \theta - 0.12 * \theta^\wedge 2 - 0.0184 * \theta^\wedge 3 - 0.00286 * \theta^\wedge 4 \end{aligned} \tag{8.43}$$

The mean value can be calculated now as follows [19]:

$$\begin{aligned} \boldsymbol{\mu}_{\boldsymbol{\sigma}} &= \frac{1}{8}\int_0^8 \boldsymbol{\sigma}(\boldsymbol{\theta})\mathrm{d}\boldsymbol{\theta} \\ &= \frac{1}{6}\int_0^6 (\mathbf{0.996} * \boldsymbol{\theta} - \mathbf{0.148} * \boldsymbol{\theta}^\wedge\mathbf{2} + \mathbf{0.000172} * \boldsymbol{\theta}^\wedge\mathbf{3} + \mathbf{0.000641} * \boldsymbol{\theta}^\wedge\mathbf{4})\mathrm{d}\boldsymbol{\theta} \\ &= \frac{8.325}{6} = 1.388 \end{aligned} \tag{8.44}$$

In order to calculate the correlation function in this case the chosen function (8.43) has to be centered (see Table 8.8). Again, after using the POLYMATH software we have

Model: $z1 = a0 + a1 * x1 + a2 * x1^\wedge 2 + a3 * x1^\wedge 3 + a4 * x1^\wedge 4$

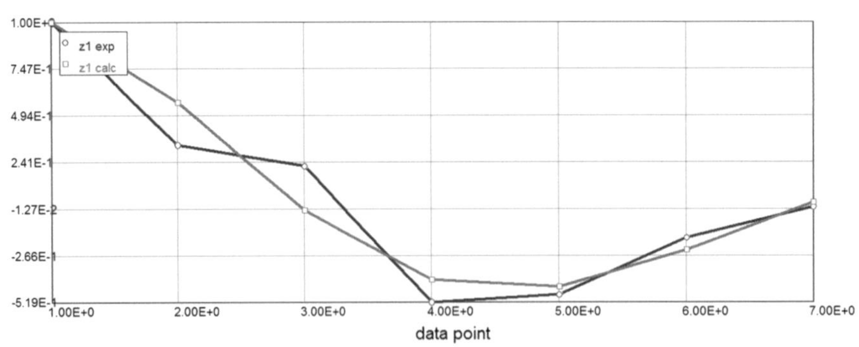

Fig. 8.12 Normalized autocorrelation function

Variable	Value
$a0$	−1.389602
$a1$	0.9979845
$a2$	−0.1231136
$a3$	−0.0167626
$a4$	0.0026515

$$\sigma^* = -1.39 + 1.0(\theta) - 0.123(\theta^2) - 0.0168(\theta^3) + 0.00256(\theta^4) \tag{8.45}$$

The correlation function of a steady time process can be computed now as follows [20]:

$$K_{\boldsymbol{\sigma}}(\boldsymbol{\theta}) = \frac{1}{6-\boldsymbol{\theta}} \int_0^{6-\boldsymbol{\theta}} [\boldsymbol{\sigma}^*(t)\boldsymbol{\sigma}^*(t+\boldsymbol{\theta})]\mathrm{d}t \tag{8.46}$$

where $\boldsymbol{\theta} = 0; 1; 2; \ldots 6.$

After substituting (8.45) into (8.46) we have

The standard deviation $D = K_\theta(0)$ is

$$D = K_{\boldsymbol{\sigma}}(\boldsymbol{\theta} = 0) = \frac{1}{8-\boldsymbol{\theta}} \int_0^{8-\boldsymbol{\theta}} [\boldsymbol{\sigma}^*(t)\boldsymbol{\sigma}^*(t+\boldsymbol{\theta})]\mathrm{d}t = \frac{1}{8}\int_0^8 [\boldsymbol{\sigma}^*(t)]^2\mathrm{d}t$$

$$= \frac{1}{6}\int_0^6 [\mathbf{-1.39 + 1.0(\boldsymbol{\theta}) - 0.123(\boldsymbol{\theta}^2) - 0.0168(\boldsymbol{\theta}^3) + 0.00265(\boldsymbol{\theta}^4)}]^2\mathrm{d}\boldsymbol{\theta}$$

$$= 0.192; \quad \boldsymbol{\sigma}_{\boldsymbol{\sigma}} = 0.438$$

Calculated values of DEQ variables

	Variable	Initial value	Minimal value	Maximal value	Final value
1	A	0	0	*0.0642533*	0.0641997
2	m	0.000928	0.000928	0.125428	0.125428
3	$m1$	0.0438	0.0121195	0.0568	0.0568
4	n	3	3	3	3
5	r	0	0	2.361833	2.361833
6	t	0	0	5	5
7	y	0	0	1.668809	1.485117

Table 8.9 Autocorrelation function

θ	0	1	2	3	4	5	6
$K_\sigma(\theta)$	0.192	0.0642	−0.0425	−0.0997	−0.0918	−0.032	0

Table 8.10 Normalized autocorrelation function

	0	1	2	3	4	5	6
$\rho_\sigma(\theta)$	1.0	0.334	0.221	−0.519	−0.478	−0.167	0

Differential equations

$$1\quad d(y)/d(t) = -100*y*m1 + (\exp(-0.15*t))*(1-0.15*t), \\ -(\exp(t/(1.0+0.1*t)))*m1*y\hat{}n + 100*m1*t*(\exp(-0.15*t))$$

$$2\quad d(A)/d(t) = (1/(6-1))*(-1.39+1.0*(t)-0.123*(t\hat{}2)-0.0168*(t\hat{}3) \\ +0.00265*(t\hat{}4))*(-1.39+1.0*(t+1)-0.123*((t+1)\hat{}2) \\ -0.0168*((t+1)\hat{}3)+0.00265*((t+1)\hat{}4))$$

The computations of autocorrelation function based on Eq. (8.46) are presented in Table 8.9.

The normalized autocorrelation function is shown in Table 8.10.

Let us approximate the data from Table 8.10 by the type of formulae (Fig. 8.12)

Model: $z1 = (\cos(b*x1))*\exp(-a*x1)$

Variable	Initial guess	Value
b	1	0.7989306
a	1	0.207364

$$\rho_\sigma(\theta) = (\exp(-0.207{*}x)){*}(\cos(0.8{*}x)) \tag{8.47}$$

The corresponding spectral function is [20]

$$\begin{aligned} S(\omega) &= \frac{1}{2}\frac{\sigma_\sigma^2}{\pi}\left[\frac{a}{(\omega-b)^2+a^2}+\frac{a}{(\omega+b)^2+a^2}\right] \\ &= 6.32(10^{-3})\left[\frac{1}{(\omega-0.8)^2+0.207^2}+\frac{1}{(\omega+0.8)^2+0.207^2}\right] \end{aligned} \tag{8.48}$$

The graphic presentation of spectral density is shown in Fig. 8.13.

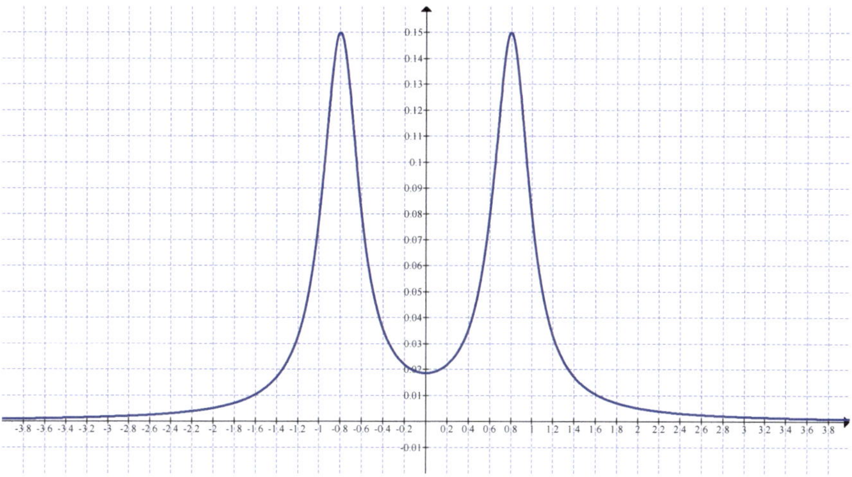

Fig. 8.13 Spectral density function

8.2.7 The First-Occurrence Time Problem and the Probability Density P(a$_0$, θ)

The average first-occurrence of stress σ below given level "$\sigma = a_0$" for stationary stress–temperature processes is defined as follows (see Chap. 7) [20]:

For correlation function given by (8.47)

$$\boldsymbol{\sigma}_{v\boldsymbol{\sigma}}^2 = K_{\dot{\sigma}}(0) = -\frac{d^2}{d\boldsymbol{\theta}^2}K_{\dot{\sigma}}(\boldsymbol{\theta})|_{\boldsymbol{\theta}=0} = \boldsymbol{\sigma}_{\boldsymbol{\sigma}}^2(b^2 - a^2)$$

$$\overline{n}_a = \boldsymbol{\theta}_{\max}\frac{\sqrt{b^2 - a^2}}{2\pi}\exp\left(-\frac{a_0^2}{2\boldsymbol{\sigma}_{\boldsymbol{\sigma}}^2}\right) = 6\frac{\sqrt{0.8^2 - 0.207^2}}{2\pi}\exp\left(-\frac{1.0^2}{2(0.438)^2}\right) = 0.0545$$

$$\overline{\mathbf{v}}_a = \frac{\overline{n}_a}{\boldsymbol{\theta}_{\max}} = \frac{0.0545}{6} = 0.00908$$

The average interval for the stress occurrence below given level "a"

$$\widetilde{\boldsymbol{\theta}} = \frac{\int_a^{\infty} f(x)\mathrm{d}x}{\int_0^{\infty} vf(a,v)\mathrm{d}v} = \boldsymbol{\pi}\frac{\boldsymbol{\sigma}_{\boldsymbol{\sigma}}}{\boldsymbol{\sigma}_v}\exp\left(+\frac{(a-\overline{x})^2}{2\boldsymbol{\sigma}_{\boldsymbol{\sigma}}^2}\right)\left[\Phi^*\left(\frac{a-\overline{x}}{\boldsymbol{\sigma}_{\boldsymbol{\sigma}}}\right)\right]$$

$$= \frac{\boldsymbol{\pi}}{\sqrt{0.8^2 - 0.207^2}}\exp\left(\frac{0.388^2}{2(0.438)^2}\right)[\Phi^*(-0.886)] = 4.065(1.48)[0.1867] = 1.123 < 6$$

The average first-occurrence temperature below given level "a" for stationary processes is

$$\widetilde{\theta} = \frac{\int_a^{\infty} f(x)\mathrm{d}x}{\int_0^{\infty} vf(a,v)\mathrm{d}v} = \pi\frac{\sigma_\sigma}{\sigma_v}\exp\left(+\frac{(a-\bar{x})^2}{2\sigma_\sigma^2}\right)\left[\Phi^*\left(\frac{a-\bar{x}}{\sigma_\sigma}\right)\right]$$

$$= \frac{\pi}{\sqrt{0.8^2 - 0.207^2}}\exp\left(\frac{0.388^2}{2(0.438)^2}\right)[\Phi^*(-0.886)] = 4.065(1.48)[0.1867] = 1.123 < 6$$

Now based on Poisson formulae of the probability to not having the minimum allowable stress ordinates crossing downwards the level "a = 1.0" is

$$P_{\mathrm{rel}} = P_o = \exp(-\bar{n}_a) = \exp(-0.0545) = 0.947$$

This probability characterizes the reliability of the structure (one element or the whole structure).

It should be noted that the references to Chap. 7 remain the same for all other fire scenarios below, therefore they will not be repeated.

8.3 Probability-Based Stress–Strain Diagram. Slow Fire

8.3.1 Temperature–Time Function. Slow Fire

Slow Fire; Temperature Increase (715 K < T < 822 K)

Data $T_* = 600$ K; $0 < \tau < 0.2$; $0.275 < \gamma < 1$

Select $\gamma = 0.275$

Differential equations (8.2) and (8.3) are rewritten as an input for "Polymath" software

```
d(y0)/d(t) = 20 * (1 − y2) * exp(y0/(1 + 0.1 * y0)) − 2.53 * 0 − 0.157 * y0^4
d(y2)/d(t) = 20 * (1 − y2) * exp(y0/(1 + 0.1 * y0))
d(y1)/d(t) = 5.5 * (1 − y1)^1.0 * exp(y/(1 + 0.1 * y))
d(y)/d(t) = 20 * (1 − y1)^1.0 * exp(y/(1 + 0.1 * y)) − 2.53 * 0 − 0.157 * y^4
```

where

"y" is the dimensionless temperature "θ" with the corresponding parameter "γ = 0.275".

"y0" is the dimensionless temperature "θ" with the corresponding parameter "γ = 1.0".

"y1" is the concentration of the product of the first-order chemical reaction with "γ = 0.275".

"y2" is the concentration of the product of the first-order chemical reaction with "γ = 1.0".

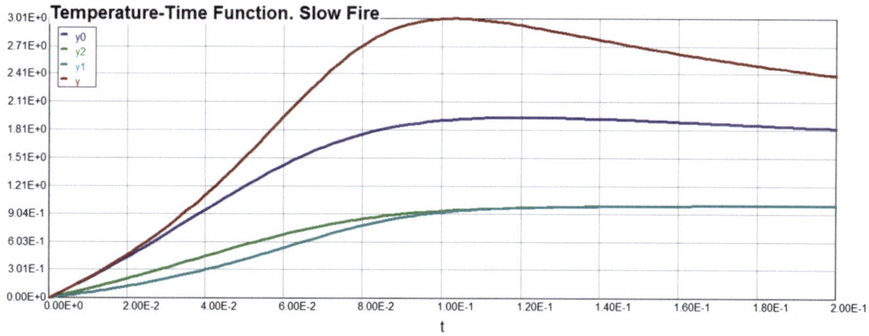

Fig. 8.14 Dimensionless temperature–time curve

Calculated values of DEQ variables

	Variable	Initial value	Minimal value	Maximal value	Final value
1	t	0	0	0.2	0.2
2	y	0	0	*3.01*	*2.4*
3	$y0$	0	0	0.9848585	0.9788831
4	$y1$	0	0	0.9993049	0.9993049
5	$y2$	0	0	0.9998429	0.9998429

Differential equations

1 $d(y0)/d(t) = 20 * (1 - y2) * \exp(y0/(1 + 0.1 * y0)) - 0.157 * y0\^{}4$

2 $d(y2)/d(t) = 20 * (1 - y2) * \exp(y0/(1 + 0.1 * y0))$

3 $d(y1)/d(t) = 5.5 * (1 - y1)\^{}1.0 * \exp(y/(1 + 0.1 * y))$

4 $d(y)/d(t) = 20 * (1 - y1)\^{}1.0 * \exp(y/(1 + 0.1 * y)) - 0.157 * y\^{}4$

Tabulated solution of the Eqs. (8.2) and (8.3) is presented above and the graphs are shown on Fig. 8.14.

Now based on the data shown above, the "best-to-fit" (polynomial regression) approximation of the dimensionless temperature–time curve can be presented as follows:

Model: $y = a0 + a1 * t + a2 * t\^{}2 + a3 * t\^{}3 + a4 * t\^{}4$

Variable	Value
$a0$	0.0
$a1$	9.924
$a2$	779.8

(continued)

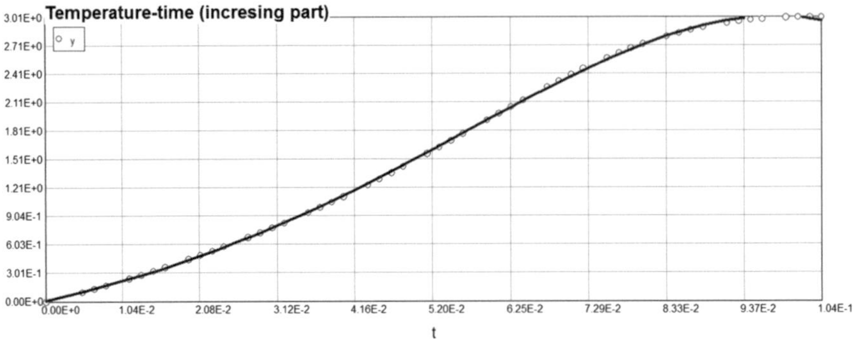

Fig. 8.15 Monotonically increasing temperature–time function

(continued)

Variable	Value
*a*3	−7836
*a*4	2.006E+04

$$\theta(\tau) = 9.924\tau + 779.8\tau^2 - 7836\tau^3 + 20{,}060\tau^4 \tag{8.49}$$

8.3.2 *Analytical Expression of the Inverse Function* θ^{-1} *and Its First Derivative*

Monotonically increasing temperature–time function: $(0 < \theta < 3)$; $(0 < \tau 1 < 0.104)$.

As noted above, a function of the temperature–time should be divided into two portions monotonically increasing and monotonically decreasing. Let us start with a monotonically increasing area. The range of temperature values in this case is the interval from 0 to 3 and the time interval from 0 to 0.104. Therefore using polynomial regression method again we have (Fig. 8.15)

Model: $y = a0 + a1 * t + a2 * t^\wedge 2 + a3 * t^\wedge 3 + a4 * t^\wedge 4$

Variable	Value
*a*0	0
*a*1	18.08
*a*2	137.7
*a*3	4376.9
*a*4	−4.56E+04

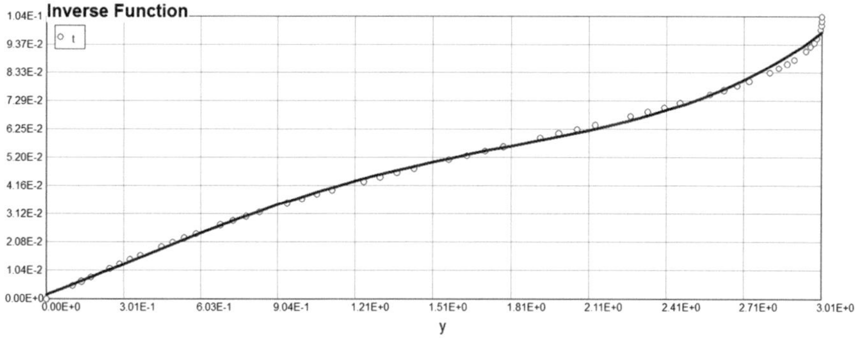

Fig. 8.16 Monotonically increasing inverse function θ^{-1}

$$\textbf{Model}: \theta = 18.08 * \tau + 137.7 * \tau\hat{}2 + 4376.9 * \tau\hat{}3 - 45{,}600 * \tau\hat{}4 \quad (8.50)$$

The inverse function θ^{-1} and its first derivative (monotonically increasing functions) are

$$\textbf{Model}: t = a0 + a1 * y + a2 * y\hat{}2 + a3 * y\hat{}3 + a4 * y\hat{}4$$

Variable	Value
$a0$	0
$a1$	0.0351
$a2$	0.0104
$a3$	−0.0124
$a4$	0.00286

The graph of the inverse function θ^{-1} is shown in Fig. 8.16.

Therefore the analytical expression of the inverse function and its first derivative is as follows:

$$\textbf{Model}: \mathbf{m} = \tau = 0.351 * \theta + 0.0104 * \theta\hat{}2 - 0.0124 * \theta\hat{}3 + 0.00286 * \theta\hat{}4 \quad (8.51)$$

$$\mathbf{m1} = \tau' = 0.351 + 0.0208 * \theta\hat{}1 + 0.0372 * \theta\hat{}2 + 0.001144 * \theta\hat{}3 \quad (8.52)$$

8.3.3 Fast Fire: Statistical Data

Differential equations (DEQs): $\alpha = 0.001$

Calculated values of DEQ variables

	Variable	Initial value	Minimal value	Maximal value	Final value
1	A	0	0	0	0
2	m	0	0	1.50896	1.50896
3	$m1$	0.351	0.351	1.76156	1.76156
4	n	3	3	3	3
5	r	0	0	2.195247	2.195247
6	t	0	0	4	4
7	y	0	0	*0.7828498*	0.2219964

Differential equations

1 $d(y)/d(t) = -0.001 * y * m1 + (\exp(-0.15 * t)) * (1 - 0.15 * t)$
$- (\exp(t/(1.0 + 0.1 * t))) * m1 * y^\wedge n + 0.001 * m1 * t * (\exp(-0.15 * t))$

2 $d(A)/d(t) = (1/(6 - 5.5)) * (-1.39 + 1.0 * (t) - 0.123 * (t^\wedge 2) - 0.0168 * (t^\wedge 3)$
$+ 0.00265 * (t^\wedge 4)) * (-1.39 + 1.0 * (t + 5.5) - 0.123 * ((t + 5.5)^\wedge 2)$
$- 0.0168 * ((t + 5.5)^\wedge 3) + 0.00265 * ((t + 5.5)^\wedge 4)) * 0$

Explicit equations

1 $n = 3$
2 $r = t * (\exp(-0.15 * t))$
3 $m1 = 0.351 + 0.0208 * t^\wedge 1 + 0.0372 * t^\wedge 2 + 0.01144 * t^\wedge 3$
4 $m = 0.351 * t + 0.0104 * t^\wedge 2 - 0.0124 * t^\wedge 3 + 0.00286 * t^\wedge 4$

Differential equations (DEQs): $\alpha = 0.01$

Calculated values of DEQ variables

	Variable	Initial value	Minimal value	Maximal value	Final value
1	A	0	0	0	0
2	m	0	0	1.50896	1.50896
3	$m1$	0.351	0.351	1.76156	1.76156
4	n	3	3	3	3
5	r	0	0	2.195247	2.195247
6	t	0	0	4	4
7	y	0	0	*0.7830289*	0.2278537

Differential equations

1 $d(y)/d(t) = -0.01 * y * m1 + (\exp(-0.15 * t)) * (1 - 0.15 * t) - (\exp(t/(1.0 + 0.1 * t))) * m1 * y^\wedge n + 0.01 * m1 * t * (\exp(-0.15 * t))$

2 $d(A)/d(t) = (1/(6 - 5.5)) * (-1.39 + 1.0 * (t) - 0.123 * (t^\wedge 2) - 0.0168 * (t^\wedge 3) + 0.00265 * (t^\wedge 4)) * (-1.39 + 1.0 * (t + 5.5) - 0.123 * ((t + 5.5)^\wedge 2) - 0.0168 * ((t + 5.5)^\wedge 3) + 0.00265 * ((t + 5.5)^\wedge 4)) * 0$

Explicit equations

1 $n = 3$

2 $r = t * (\exp(-0.15 * t))$

3 $m1 = 0.351 + 0.0208 * t^\wedge 1 + 0.0372 * t^\wedge 2 + 0.01144 * t^\wedge 3$

4 $m = 0.351 * t + 0.0104 * t^\wedge 2 - 0.0124 * t^\wedge 3 + 0.00286 * t^\wedge 4$

Differential equations (DEQs): $\alpha = 0.1$

Calculated values of DEQ variables

	Variable	Initial value	Minimal value	Maximal value	Final value
1	A	0	0	0	0
2	m	0	0	1.50896	1.50896
3	$m1$	0.351	0.351	1.76156	1.76156
4	n	3	3	3	3
5	r	0	0	2.195247	2.195247
6	t	0	0	4	4
7	y	0	0	*0.7848064*	0.2758059

Differential equations

1 $d(y)/d(t) = -0.1 * y * m1 + (\exp(-0.15 * t)) * (1 - 0.15 * t) - (\exp(t/(1.0 + 0.1 * t))) * m1 * y^\wedge n + 0.1 * m1 * t * (\exp(-0.15 * t))$

2 $d(A)/d(t) = (1/(6 - 5.5)) * (-1.39 + 1.0 * (t) - 0.123 * (t^\wedge 2) - 0.0168 * (t^\wedge 3) + 0.00265 * (t^\wedge 4)) * (-1.39 + 1.0 * (t + 5.5) - 0.123 * ((t + 5.5)^\wedge 2) - 0.0168 * ((t + 5.5)^\wedge 3) + 0.00265 * ((t + 5.5)^\wedge 4)) * 0$

Explicit equations

1 $n = 3$

2 $r = t * (\exp(-0.15 * t))$

3 $m1 = 0.351 + 0.0208 * t^\wedge 1 + 0.0372 * t^\wedge 2 + 0.01144 * t^\wedge 3$

4 $m = 0.351 * t + 0.0104 * t^\wedge 2 - 0.0124 * t^\wedge 3 + 0.00286 * t^\wedge 4$

Differential equations (DEQs): $\alpha = 1$

Calculated values of DEQ variables

	Variable	Initial value	Minimal value	Maximal value	Final value
1	A	0	0	0	0
2	m	0	0	1.50896	1.50896
3	$m1$	0.351	0.351	1.76156	1.76156
4	n	3	3	3	3
5	r	0	0	2.195247	2.195247
6	t	0	0	4	4
7	y	0	0	*0.8038112*	0.4764483

Differential equations

1 $$d(y)/d(t) = -1000 * y * m1 + (\exp(-0.15 * t)) * (1 - 0.15 * t) - (\exp(t/(1.0 + 0.1 * t))) * m1 * y^\wedge n + 1000 * m1 * t * (\exp(-0.15 * t))$$

2 $$d(A)/d(t) = (1/(6 - 5.5)) * (-1.39 + 1.0 * (t) - 0.123 * (t^\wedge 2) - 0.0168 * (t^\wedge 3) + 0.00265 * (t^\wedge 4)) * (-1.39 + 1.0 * (t + 5.5) - 0.123 * ((t + 5.5)^\wedge 2) - 0.0168 * ((t + 5.5)^\wedge 3) + 0.00265 * ((t + 5.5)^\wedge 4)) * 0$$

Explicit equations

1 $n = 3$

2 $r = t * (\exp(-0.15 * t))$

3 $m1 = 0.351 + 0.0208 * t^\wedge 1 + 0.0372 * t^\wedge 2 + 0.01144 * t^\wedge 3$

4 $m = 0.351 * t + 0.0104 * t^\wedge 2 - 0.0124 * t^\wedge 3 + 0.00286 * t^\wedge 4$

Differential equations (DEQs): $\alpha = 10$

Calculated values of DEQ variables

	Variable	Initial value	Minimal value	Maximal value	Final value
1	A	0	0	0	0
2	m	0	0	1.50896	1.50896
3	$m1$	0.351	0.351	1.76156	1.76156
4	n	3	3	3	3
5	r	0	0	2.195247	2.195247
6	t	0	0	4	4
7	y	0	0	*1.015137*	0.9082207

Differential equations

1 $\mathrm{d}(y)/\mathrm{d}(t) = -10,000 * y * m1 + (\exp(-0.15 * t)) * (1 - 0.15 * t)$
$- (\exp(t/(1.0 + 0.1 * t))) * m1 * y\hat{}n + 10,000 * m1 * t * (\exp(-0.15 * t))$

2 $\mathrm{d}(A)/\mathrm{d}(t) = (1/(6 - 5.5)) * (-1.39 + 1.0 * (t) - 0.123 * (t\hat{}2) - 0.0168 * (t\hat{}3)$
$+ 0.00265 * (t\hat{}4)) * (-1.39 + 1.0 * (t + 5.5) - 0.123 * ((t + 5.5)\hat{}2)$
$- 0.0168 * ((t + 5.5)\hat{}3) + 0.00265 * ((t + 5.5)\hat{}4)) * 0$

Explicit equations

1 $n = 3$

2 $r = t * (\exp(-0.15 * t))$

3 $m1 = 0.351 + 0.0208 * t\hat{}1 + 0.0372 * t\hat{}2 + 0.01144 * t\hat{}3$

4 $m = 0.351 * t + 0.0104 * t\hat{}2 - 0.0124 * t\hat{}3 + 0.00286 * t\hat{}4$

Differential equations (DEQs): α = 100

Calculated values of DEQ variables

	Variable	Initial value	Minimal value	Maximal value	Final value
1	A	0	0	0	0
2	m	0	0	1.50896	1.50896
3	$m1$	0.351	0.351	1.76156	1.76156
4	n	3	3	3	3
5	r	0	0	2.195247	2.195247
6	t	0	0	4	4
7	y	0	0	*1.562581*	1.549283

Differential equations

1 $d(y)/d(t) = -100,000 * y * m1 + (\exp(-0.15 * t)) * (1 - 0.15 * t) - (\exp(t/(1.0 + 0.1 * t))) * m1 * y^\wedge n + 100,000 * m1 * t * (\exp(-0.15 * t))$

2 $d(A)/d(t) = (1/(6 - 5.5)) * (-1.39 + 1.0 * (t) - 0.123 * (t^\wedge 2) - 0.0168 * (t^\wedge 3) + 0.00265 * (t^\wedge 4)) * (-1.39 + 1.0 * (t + 5.5) - 0.123 * ((t + 5.5)^\wedge 2) - 0.0168 * ((t + 5.5)^\wedge 3) + 0.00265 * ((t + 5.5)^\wedge 4)) * 0$

Explicit equations

1 $n = 3$

2 $r = t * (\exp(-0.15 * t))$

3 $m1 = 0.351 + 0.0208 * t^\wedge 1 + 0.0372 * t^\wedge 2 + 0.01144 * t^\wedge 3$

4 $m = 0.351 * t + 0.0104 * t^\wedge 2 - 0.0124 * t^\wedge 3 + 0.00286 * t^\wedge 4$

Differential equations (DEQs): $\alpha = 1000$

Calculated values of DEQ variables

	Variable	Initial value	Minimal value	Maximal value	Final value
1	A	0	0	0	0
2	m	0	0	1.50896	1.50896
3	$m1$	0.351	0.351	1.76156	1.76156
4	n	3	3	3	3
5	r	0	0	2.195247	2.195247
6	t	0	0	4	4
7	y	0	0	*2.046145*	2.046145

Differential equations

1 $d(y)/d(t) = -1000 * y * m1 + (\exp(-0.15 * t)) * (1 - 0.15 * t) - (\exp(t/(1.0 + 0.1 * t))) * m1 * y^\wedge n + 1000 * m1 * t * (\exp(-0.15 * t))$

2 $d(A)/d(t) = (1/(6 - 5.5)) * (-1.39 + 1.0 * (t) - 0.123 * (t^\wedge 2) - 0.0168 * (t^\wedge 3) + 0.00265 * (t^\wedge 4)) * (-1.39 + 1.0 * (t + 5.5) - 0.123 * ((t + 5.5)^\wedge 2) - 0.0168 * ((t + 5.5)^\wedge 3) + 0.00265 * ((t + 5.5)^\wedge 4)) * 0$

Explicit equations

1 $n = 3$

2 $r = t * (\exp(-0.15 * t))$

3 $m1 = 0.351 + 0.0208 * t^\wedge 1 + 0.0372 * t^\wedge 2 + 0.01144 * t^\wedge 3$

4 $m = 0.351 * t + 0.0104 * t^\wedge 2 - 0.0124 * t^\wedge 3 + 0.00286 * t^\wedge 4$

Differential equations (DEQs): α = 10,000

Calculated values of DEQ variables

	Variable	Initial value	Minimal value	Maximal value	Final value
1	A	0	0	0	0
2	m	0	0	1.50896	1.50896
3	$m1$	0.351	0.351	1.76156	1.76156
4	n	3	3	3	3
5	r	0	0	2.195247	2.195247
6	t	0	0	4	4
7	y	0	0	2.177276	2.177276

Differential equations

1 $d(y)/d(t) = -1000 * y * m1 + (\exp(-0.15 * t)) * (1 - 0.15 * t) - (\exp(t/(1.0 + 0.1 * t))) * m1 * y^\wedge n + 1000 * m1 * t * (\exp(-0.15 * t))$

2 $d(A)/d(t) = (1/(6 - 5.5)) * (-1.39 + 1.0 * (t) - 0.123 * (t^\wedge 2) - 0.0168 * (t^\wedge 3) + 0.00265 * (t^\wedge 4)) * (-1.39 + 1.0 * (t + 5.5) - 0.123 * ((t + 5.5)^\wedge 2) - 0.0168 * ((t + 5.5)^\wedge 3) + 0.00265 * ((t + 5.5)^\wedge 4)) * 0$

Explicit equations

1 $n = 3$

2 $r = t * (\exp(-0.15 * t))$

3 $m1 = 0.351 + 0.0208 * t^\wedge 1 + 0.0372 * t^\wedge 2 + 0.01144 * t^\wedge 3$

4 $m = 0.351 * t + 0.0104 * t^\wedge 2 - 0.0124 * t^\wedge 3 + 0.00286 * t^\wedge 4$

Differential equations (DEQs): α = 100,000

Calculated values of DEQ variables

	Variable	Initial value	Minimal value	Maximal value	Final value
1	*A*	0	0	0	0
2	*m*	0	0	1.50896	1.50896
3	*m*1	0.351	0.351	1.76156	1.76156
4	*n*	3	3	3	3
5	*r*	0	0	*2.195247*	2.195247
6	*t*	0	0	4	4
7	*y*	0	0	*2.193409*	2.193409

Differential equations

$$1 \quad \mathrm{d}(y)/\mathrm{d}(t) = -100{,}000 * y * m1 + (\exp(-0.15 * t)) * (1 - 0.15 * t) - (\exp(t/(1.0 + 0.1 * t))) * m1 * y^\wedge n + 100{,}000 * m1 * t * (\exp(-0.15 * t))$$

$$2 \quad \mathrm{d}(A)/\mathrm{d}(t) = (1/(6 - 5.5)) * (-1.39 + 1.0 * (t) - 0.123 * (t^\wedge 2) - 0.0168 * (t^\wedge 3) + 0.00265 * (t^\wedge 4)) * (-1.39 + 1.0 * (t + 5.5) - 0.123 * ((t + 5.5)^\wedge 2) - 0.0168 * ((t + 5.5)^\wedge 3) + 0.00265 * ((t + 5.5)^\wedge 4)) * 0$$

Explicit equations

$$1 \quad n = 3$$
$$2 \quad r = t * (\exp(-0.15 * t))$$
$$3 \quad m1 = 0.351 + 0.0208 * t^\wedge 1 + 0.0372 * t^\wedge 2 + 0.01144 * t^\wedge 3$$
$$4 \quad m = 0.351 * t + 0.0104 * t^\wedge 2 - 0.0124 * t^\wedge 3 + 0.00286 * t^\wedge 4$$

The summary of computations above is presented in Table 8.11.

Therefore, the probability that the real fire dimensionless allowable stress "$\boldsymbol{\sigma}$" will be less than 0.78; 1.2; 1.1 or 1.0 is

$$\begin{aligned}
P(\boldsymbol{\sigma} < 0.78) &= \Phi^*\left(\frac{0.78 - 1.35}{0.642}\right) = \Phi^*(-0.888) = 0.187 = 18.7\,\% \\
P(\boldsymbol{\sigma} < 0/35) &= \Phi^*\left(\frac{0.35 - 1.35}{0.642}\right) = \Phi^*(-1.558) = 0.0594 = 5.94\,\% \\
P(\boldsymbol{\sigma} < 0.25) &= \Phi^*\left(\frac{0.25 - 1.35}{0.642}\right) = \Phi^*(-1.713) = 0.044 = 4.4\,\% \\
P_{\mathrm{f}}(\boldsymbol{\sigma} < 0.15) &= \Phi^*\left(\frac{0.15 - 1.35}{0.642}\right) = \Phi^*(-1.869) = 0.0307 = 3.07\,\%
\end{aligned} \tag{8.53}$$

Table 8.11 Maximum dimensionless stresses (statistical data)

Value	Maximum creep stress	Deviation	Variance	Mean value	Standard deviation	α value
1	*0.783*	−0.567	0.322			0.001
2	*0.783*	−0.567	0.322			0.01
3	*0.785*	−0.565	0.319			0.1
4	*0.804*	−0.546	0.298			1.0
5	*1.015*	−0.335	0.112			10
6	*1.563*	0.213	0.0458			100
7	*2.046*	0.696	0.484			1000
8	*2.177*	0.827	0.684			10,000
9	*2.193*	0.843	0.711			100,000
Aver.	*1.35*		Total: 3.3/8 = = 0.412	*1.791*	*0.642*	

If probability of failure $P_{\mathrm{f}} = 4.4\ \%$ is acceptable then $\sigma_{\mathrm{all}} = 0.25(7.02)$ 2.9 = 5.1 ksi = 35.2 MPa; $\mu_{\sigma} = 1.35$ and $\sigma_{\sigma} = 0.642$, then $\beta_{\mathrm{all}} = 1.713$ (reliability index).

8.3.4 The First-Order Reliability Method (FORM)

For detailed information regarding applicability of this method please see Chap. 7. The computation of the reliability index β and corresponding design temperature (based on Example 7.3) is presented below without any additional comments or explanations in order to save space.

Example 8.5 Data: Fast Fire: $\mu_{sT} = 1000\ °\mathrm{F}$; $\sigma_{sT} = 100\ °\mathrm{F}$. Find the reliability index β; probability of failure P_{f} and design temperature T °F.

The deterministic performance function in this case is defined by Eq. H2-1 [15]

$$g = \frac{72.76 + 0.041(\mu_{sT})}{50(11.8)} + \frac{13.07 + 0.214(12)(\mu_{sT})}{(50)73} - \mu_{\sigma R} = 0$$

$$\mu_{\sigma S} = 0.166 + 0.000773(\mu_{sT});\ \sigma_{\sigma S} = 0.000773(\sigma_{sT}) \quad \mu_{sT} = 1000\,°\mathrm{F}$$

where $\mu_{\sigma R} = 1.35; \quad \sigma_{\sigma R} = 0.642 \rightarrow$ see computations above

$$\beta = \frac{|\mu_{\sigma R} - \mu_{\sigma S}|}{\sqrt{(\sigma_{\sigma S})^2 + (\sigma_{\sigma R})^2}} = \frac{1.35 - 0.939}{\sqrt{0.0773^2 + 0.642^2}} = 0.636$$

The probability of structural failure and the design temperature in this case are computed as follows:

$$P_{\mathrm{f}} = \Phi^*(-\boldsymbol{\beta}) = \Phi^*(-0.636) = 0.26;$$
$$T_{\mathrm{design}} = 1000 + (0.636)100 = 1063.6\,^{\circ}\mathrm{F}$$

8.3.5 Confidence Interval—Fast Fire

Again, for detailed information regarding applicability of this method please see Chap. 7. The computation of the reliability index β and corresponding design temperature (based on Example 7.3) is presented below

Example 8.6 Data: Fast Fire: $\alpha = 0.95$—given target confidence probability
From Table 8.1: $\mu_{\sigma R} = 1.35$; $\mu_{\sigma R} = 0.642$

The confidence limit (interval) I_{α} of mean value (μ_{σ}) and standard deviation (σ_{σ}) is as follows:

$$\boldsymbol{\sigma}_{\boldsymbol{\mu}_\sigma} = \sqrt{\frac{\boldsymbol{\sigma}_{\boldsymbol{\sigma}}^2}{n}} = \sqrt{\frac{(0.642)^2}{8}} = 0.227$$
$$\boldsymbol{\varepsilon}_{\boldsymbol{\alpha}} = \boldsymbol{\sigma}_{\boldsymbol{\mu}_\sigma} \arg\,\Phi^*\left(\frac{1+\boldsymbol{\alpha}}{2}\right) = 0.119\left[\arg\,\Phi^*\left(\frac{1+0.95}{2}\right)\right] = 0.227(1.96) = 0.445$$
$$t_{\boldsymbol{\alpha}} = \arg\,\Phi^*\left(\frac{1+\boldsymbol{\alpha}}{2}\right) = 1.96$$

Finally, the confidence limit (interval) I_{α} for mean value is calculated as follows:

$$I_{\boldsymbol{\alpha}} = (\boldsymbol{\mu}_{\boldsymbol{\sigma}} - t_{\boldsymbol{\alpha}}\boldsymbol{\sigma}_{\boldsymbol{\mu}_\sigma}; \boldsymbol{\mu}_{\boldsymbol{\sigma}} + t_{\boldsymbol{\alpha}}\boldsymbol{\sigma}_{\boldsymbol{\mu}_\sigma}) = (1.35 - 0.445; 1.35 + 0.445) = (0.905; 1.795)$$

The confidence limit (interval) I_{α} for standard deviation value is calculated as follows:

$$I_{\boldsymbol{\alpha}} = (\boldsymbol{\sigma}_{\boldsymbol{\sigma}}^2 - t_{\boldsymbol{\alpha}}\boldsymbol{\sigma}_{\boldsymbol{\sigma}_\sigma}; \boldsymbol{\sigma}_{\boldsymbol{\sigma}}^2 + t_{\boldsymbol{\alpha}}\boldsymbol{\sigma}_{\boldsymbol{\sigma}_\sigma}) = (0.642^2 - (1.96)0.206; 0.642^2 + (1.96)0.0603) = (0; 0.816)$$

where $\boldsymbol{\sigma}_{\boldsymbol{\sigma}_\sigma} = \sqrt{\frac{2}{n-1}\boldsymbol{\sigma}_{\boldsymbol{\sigma}}^2} = \sqrt{\frac{2}{9-1}(0.642)^2} = 0.206$ and $I_{\boldsymbol{\alpha}}^{\sigma_{\sigma\sigma}} = \left(\sqrt{\boldsymbol{\sigma}_{\boldsymbol{\sigma}}^2 - t_{\boldsymbol{\alpha}}\boldsymbol{\sigma}_{\boldsymbol{\sigma}_\sigma}}; \sqrt{\boldsymbol{\sigma}_{\boldsymbol{\sigma}}^2 + t_{\boldsymbol{\alpha}}\boldsymbol{\sigma}_{\boldsymbol{\sigma}_\sigma}}\right); 0; \sqrt{0.816}) = (0; 0.903)$

The minimum allowable dimensionless stress is

$$\boldsymbol{\sigma}_{\text{min,all}} = 1.35 - 0.903 = 0.447$$

8.3.6 *Creep Stress–Strain Diagrams (Ergodic Process)*

The statistical data (stress-temperature-strain relationship) in this fire severity case are taken from Chap. 5 and presented below.

$$\begin{aligned}&\alpha = 0.001\\&\sigma = 1.585 * \theta - 0.974 * \theta^\wedge 2 + 0.212 * \theta^\wedge 3 - 0.0157 * \theta^\wedge 4\end{aligned} \tag{8.54}$$

$$\begin{aligned}&\alpha = 0.01\\&\sigma = 1.585 * \theta - 0.974 * \theta^\wedge 2 + 0.212 * \theta^\wedge 3 - 0.0157 * \theta^\wedge 4\end{aligned} \tag{8.55}$$

$$\begin{aligned}&\alpha = 0.1\\&\sigma = 1.583 * \theta - 0.973 * \theta^\wedge 2 + 0.214 * \theta^\wedge 3 - 0.0159 * \theta^\wedge 4\end{aligned} \tag{8.56}$$

$$\begin{aligned}&\alpha = 1\\&\sigma = 1.538 * \theta - 0.924 * \theta^\wedge 2 + 0.211 * \theta^\wedge 3 - 0.0168 * \theta^\wedge 4\end{aligned} \tag{8.57}$$

$$\begin{aligned}&\alpha = 10\\&\sigma = 1.249 * \theta - 0.505 * \theta^\wedge 2 + 0.0801 * \theta^\wedge 3 - 0.00427 * \theta^\wedge 4\end{aligned} \tag{8.58}$$

$$\begin{aligned}&\alpha = 100\\&\sigma = 1.0 * \theta - 0.142 * \theta^\wedge 2 + 0.0193 * \theta^\wedge 3 - 0.00406 * \theta^\wedge 4\end{aligned} \tag{8.59}$$

$$\begin{aligned}&\alpha = 1000\\&\sigma = 0.995 * \theta - 0.141 * \theta^\wedge 2 + 0.00463 * \theta^\wedge 3 + 0.000095 * \theta^\wedge 4\end{aligned} \tag{8.60}$$

$$\begin{aligned}&\alpha = 10{,}000\\&\sigma = 0.999 * \theta - 0.148 * \theta^\wedge 2 + 0.0102 * \theta^\wedge 3 - 0.000378 * \theta^\wedge 4\end{aligned} \tag{8.61}$$

Finally, the results are summarized and presented in Table 8.12.

From the entire above, one can see that the steady time random function has an ergodic character, therefore only *one* random realization function is sufficient enough in order to obtain the correlation function in this case. The summary of all stress–temperature–strain relationships graphically are presented in Fig. 8.17.

Table 8.12 Stress–temperature data (slow fire)

$\alpha\|\theta$	0	1	1.5	2	2.5	3	3.5	4
0.001	0	0.746	0.822	0.624	0.711	0.372	0.349	0.222
0.01	0	0.746	0.822	0.625	0.711	0.375	0.349	0.228
0.1	0	0.747	0.841	0.634	0.584	0.40	0.367	0.276
1.0	0	0.753	0.855	0.709	0.574	0.562	0.589	0.476
10	0	0.793	0.986	1.011	1.051	0.987	0.979	0.908
100	0	0.847	1.136	1.353	1.469	1.545	1.542	1.549
1000	0	0.859	1.191	1.465	1.682	1.849	1.968	2.046
10,000	0	0.860	1.198	1.480	1.717	1.906	2.064	2.177
r	0	0.861	1.198	1.482	1.717	1.913	2.064	2.195

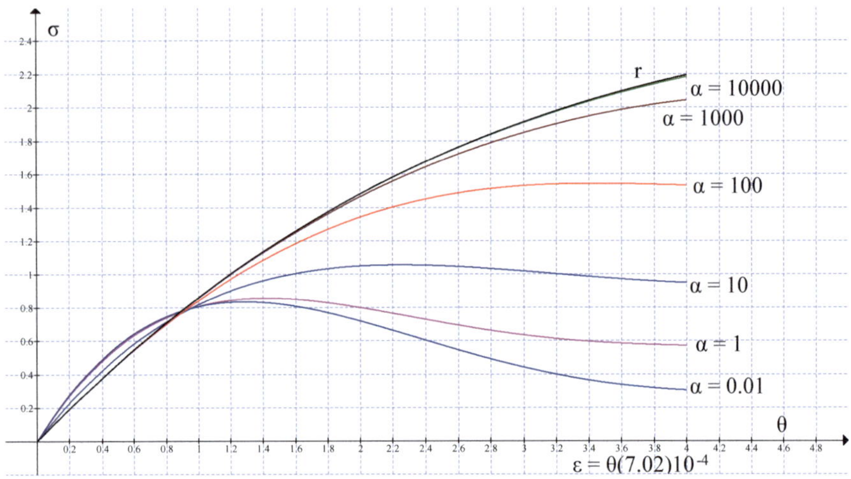

Fig. 8.17 Stress–temperature–strain relationships (slow fire)

The curve "*r*" (see Fig. 8.5) indicates that the material is creep resistant. It should be noted again that α is inverse value proportional to retardation time. Let us choose the random realization function that corresponds to α = 100.

$$\begin{aligned} &\alpha = 100 \\ &\sigma = 1.0 * \theta - 0.142 * \theta^{\wedge}2 - 0.0193 * \theta^{\wedge}3 + 0.00406 * \theta^{\wedge}4 \end{aligned} \tag{8.62}$$

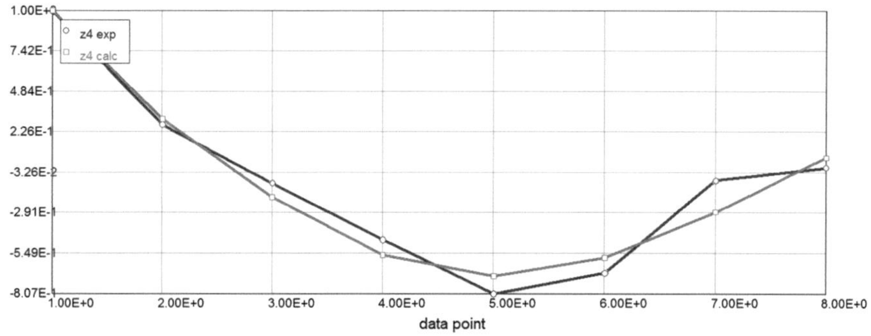

Fig. 8.18 Normalized autocorrelation function

The mean value can be calculated now as follows:

$$\begin{aligned} \boldsymbol{\mu}_{\boldsymbol{\sigma}} &= \frac{1}{4}\int_0^4 \boldsymbol{\sigma}(\boldsymbol{\theta})d\boldsymbol{\theta} \\ &= \frac{1}{4}\int_0^4 (\mathbf{0.996} * \boldsymbol{\theta} - \mathbf{0.148} * \boldsymbol{\theta}\hat{\ }\mathbf{2} + \mathbf{0.000172} * \boldsymbol{\theta}\hat{\ }\mathbf{3} + \mathbf{0.000641} * \boldsymbol{\theta}\hat{\ }\mathbf{4})d\boldsymbol{\theta} \\ &= \frac{4.953}{4} = 1.238 \end{aligned} \tag{8.63}$$

In order to calculate the correlation function in this case, the chosen function (8.62) has to be centered (see Table 8.13). Again, after using the POLYMATH software we have

Model: $z3 = a0 + a1 * x2 + a2 * x2\hat{\ }2 + a3 * x2\hat{\ }3 + a4 * x2\hat{\ }4$

Variable	Value
$a0$	−1.237638
$a1$	0.9848999
$a2$	−0.1164608
$a3$	−0.0304494
$a4$	0.0055433

$$\sigma^* = -1.238 + 0.985(\theta) - 0.116(\theta^2) - 0.0304(\theta^3) + 0.00554(\theta^4) \tag{8.64}$$

Table 8.13 Centered correlation function

$\alpha\vert\theta$	0	1	1.5	2	2.5	3	3.5	4
100	0	0.847	1.136	1.353	1.469	1.545	1.542	1.549
σ^*	−1.238	−0.391	−0.102	0.115	0.231	0.307	0.301	0.311

The correlation function of a steady time process can be computed now as follows [ ]:

$$K_{\boldsymbol{\sigma}}(\boldsymbol{\theta}) = \frac{1}{8-\boldsymbol{\theta}} \int_0^{8-\boldsymbol{\theta}} [\boldsymbol{\sigma}^*(t)\boldsymbol{\sigma}^*(t+\boldsymbol{\theta})]\mathrm{d}t \tag{8.65}$$

where$\boldsymbol{\theta} = 0; 1; 2; \ldots 8$.

After substituting (7.62) into (7.63) we have

The standard deviation $D = K_\theta(0)$ is

$$D = K_{\boldsymbol{\sigma}}(\boldsymbol{\theta} = 0) = \frac{1}{4-\boldsymbol{\theta}} \int_0^{4-\boldsymbol{\theta}} [\boldsymbol{\sigma}^*(t)\boldsymbol{\sigma}^*(t+\boldsymbol{\theta})]\mathrm{d}t = \frac{1}{8}\int_0^8 [\boldsymbol{\sigma}^*(t)]^2\mathrm{d}t$$

$$= 0.25 \int_0^4 [\mathbf{-1.238 + 0.985(\boldsymbol{\theta}) - 0.116(\boldsymbol{\theta}^2) - 0.0304(\boldsymbol{\theta}^3) + 0.00554(\boldsymbol{\theta}^4)}]^2 \mathrm{d}\boldsymbol{\theta}$$

$$= 0.25(0.873) = 0.218;\ \boldsymbol{\sigma}_{\boldsymbol{\sigma}} = 0.467$$

The computations of autocorrelation function based on Eq. (8.65) are presented in Table 8.14.

The normalized autocorrelation function is shown in Table 8.15.

Let us approximate the data from Table 8.15 by the type of formulae (Fig. 8.18)

Model: $z4 = (\cos(b * x2)) * \exp(-a * x2)$

Variable	Initial guess	Value
b	1	1.206
a	1	0.143

$$\boldsymbol{\rho}_{\boldsymbol{\sigma}}(\boldsymbol{\theta}) = (\exp(-0.143 * x)) * (\cos(1.206 * x)) \tag{8.66}$$

The corresponding spectral function is [20]

Table 8.14 Autocorrelation function

θ	0	1	1.5	2	2.5	3	3.5	4
$K_\sigma(\theta)$	0.218	0.0595	−0.0221	−0.101	−0.176	−0.147	−0.018	0

Table 8.15 Normalized autocorrelation function

θ	0	1	1.5	2	2.5	3	3.5	4
$\boldsymbol{\rho}_{\boldsymbol{\sigma}}(\boldsymbol{\theta})$	1.0	0.273	−0.101	−0.463	−0.807	−0.674	−0.082	0

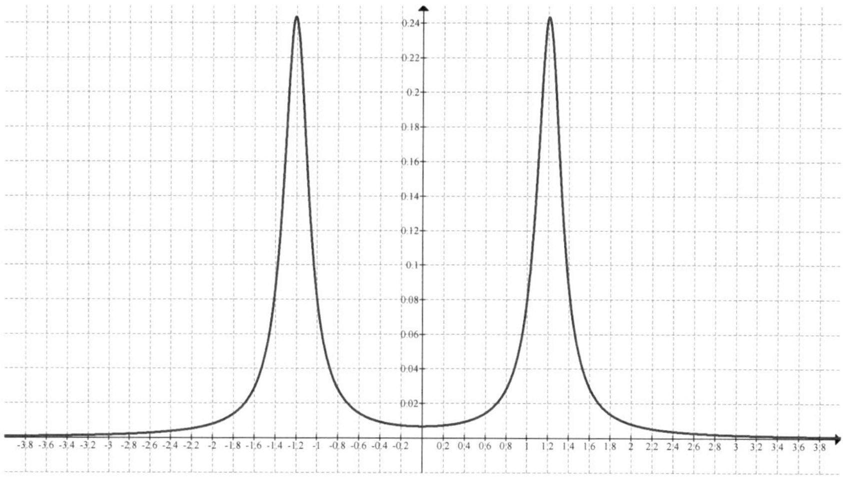

Fig. 8.19 Spectral density function

$$
\begin{aligned}
S(\omega) &= \frac{1}{2}\frac{\sigma_\sigma^2}{\pi}\left[\frac{a}{(\omega-b)^2+a^2}+\frac{a}{(\omega+b)^2+a^2}\right] \\
&= 4.96\left(10^{-3}\right)\left[\frac{1}{(\omega-1.206)^2+0.143^2}+\frac{1}{(\omega+1.206)^2+0.143^2}\right]
\end{aligned}
\tag{8.67}
$$

The graphic presentation of spectral density is shown in Fig. 8.19.

8.3.7 The First-Occurrence Time Problem and the Probability Density $\mathbf{P}(\mathbf{a}_0, \theta)$

The average first-occurrence of stress σ below given level "$\sigma = a_0$" for stationary stress–temperature processes is defined as follows (see Chap. 7) [19]:

For correlation function given by (8.66):

$$
\sigma_{v\sigma}^2 = K_{\dot\sigma}(0) = -\frac{d^2}{d\theta^2}K_{\dot\sigma}(\theta)|_{\theta=0} = \sigma_\sigma^2(b^2-a^2)
$$

$$
\bar n_a = \theta_{\max}\frac{\sqrt{b^2-a^2}}{2\pi}\exp\left(-\frac{a_0^2}{2\sigma_\sigma^2}\right) = 4\frac{\sqrt{1.206^2-0.143^2}}{2\pi}\exp\left(-\frac{0.447^2}{2(0.467)^2}\right) = 0.482
$$

$$
\bar v_a = \frac{\bar n_a}{\theta_{\max}} = \frac{0.482}{4} = 0.1205
$$

The average interval for the stress occurrence below given level "a"

$$\tilde{\theta} = \frac{\int_a^{\infty} f(x)\mathrm{d}x}{\int_0^{\infty} vf(a,v)\mathrm{d}v} = \pi\frac{\boldsymbol{\sigma}_{\boldsymbol{\sigma}}}{\boldsymbol{\sigma}_v}\exp\left(+\frac{(a-\bar{x})^2}{2\boldsymbol{\sigma}_{\boldsymbol{\sigma}}^2}\right)\left[\Phi^*\left(\frac{a-\bar{x}}{\boldsymbol{\sigma}_{\boldsymbol{\sigma}}}\right)\right]$$

$$= \frac{\pi}{\sqrt{1.206^2 - 0.143^2}}\exp\left(\frac{(0.447-1.206)}{2(0.467)^2}\right)[\Phi^*(-1.625)] = 2.623(3.746)[0.052] = 0.511 < 4$$

The average first-occurrence temperature below given level "a" for stationary processes is

$$\bar{\boldsymbol{\theta}}_a = \boldsymbol{\theta}_{\max}\int_{-\infty}^{a} f(x)\mathrm{d}x = \frac{10}{\boldsymbol{\sigma}_{\boldsymbol{\sigma}}\sqrt{2\pi}}\int_{-\infty}^{a}\exp\left(-\frac{(x-\bar{x})^2}{2\boldsymbol{\sigma}_{\boldsymbol{\theta}}^2}\right)\mathrm{d}x$$

$$= \frac{4}{0.467\sqrt{2\pi}}[\Phi^*(-1.625)] = 3.417(0.052) = 0.178 < 4$$

Now based on Poisson formulae of the probability to not having the minimum allowable stress ordinates crossing downwards the level "a = 0.447" is

$$P_{\text{rel}} = P_o = \exp(-\bar{n}_a) = \exp(-0.482) = 0.618$$

This probability characterizes the reliability of the structure (one element or the whole structure).

It should be noted that the reliability of the structure computed above using FORM method is as follows: $P_{\text{rel}} = 1 - 0.26 = 0.74 \approx 0.618$. This is a very low reliability of the structure (in most practical engineering design cases) comparable to other fire severity scenarios (for example, Fast Fire—$P_{\text{rel}} = 0.967$—for the same structural system and temperature loading condition). In addition, it is worth mentioning that the allowable stress—$\sigma_{\text{all}} = 0.447 = 9.1$ ksi [62.7 MPa] in Slow Fire scenario—is much smaller than $\sigma_{\text{all}} = 1.1 = 22.4$ ksi [154.4 MPa] in Fast Fire scenario. Thus, the quantitative comparison findings above confirm the well-known phenomenological statement, consisting in the fact that the creep of the material manifests itself more when the heat flux is applied with less intensity.

It can be seen from the results that the influence of creep is inversely proportional to the heating rate. Lower heating rates increase the influence of high-temperature creep. This is because under slow heating rate, steel will experience high temperatures under high stress for longer time durations. While under faster heating, steel experiences high temperature for short time durations. Therefore the effect of high-temperature creep might be neglected for fast heating rates, such as the case in unprotected steel. While for insulated members where heating rate is slow, the effect of high-temperature creep becomes dominant, and should be accounted for in the structural fire resistance analysis.

References

1. Ashby M, Shercliff H, Cebon D (2007) Materials. engineering, science, processing and design. Butterworth-Heinemann, Oxford
2. Cadek J (1988) Creep in metallic materials. Elsevier, Amsterdam
3. Callister WD (1997) Materials science and engineering. An introduction. Wiley, New York
4. Harmathy T (1967) A comprehensive creep model. J Basic Eng 89(D-3):496–502
5. Williams-Leir G (1983) Creep of structural steel in fire: analytical expressions. Fire Mater 7 (2):73–80
6. Anderberg Y (1988) Modeling steel behavior. Fire Saf J 13(1):17–26
7. Dorn JE (1954) Some fundamental experiments on high temperature creep. J Mech Phys Solids 3:85
8. Jeanes DC (1985) Application of the computer in modeling fire endurance of structural steel floor systems. Fire Saf J 9:119–135
9. Contro R, Poggi C, Cazzani A (1988) Numerical analysis of fire effects on beam structures. Eng Comp 5:53–55
10. Corradi L, Poggi C, Setti P (1990) Interaction domains for steel beam-columns in fire conditions. J Constr Steel Res 17:217–235
11. Ianizzi R, Schleich JB (1991) Mechanical properties of structural steel at elevated temperatures —comparisons by numerical simulations. RPS Rep. No. 05/90, C.E.C. Res. 7210-SA/112, ARBED Recherches, Esch-sur-Alzette, Luxembourg
12. Furamura F, Ave, T, Kim WT (1985) An experimental formula of stress–strain relations of structural steels at fire temperature range. Summaries of Technical Papers of Annual Meeting, Architectural Institute of Japan, Tokyo, pp 621–622
13. Lie TT (ed) (1992) Structural fire protection. ASCE, New York
14. Commission of the European Communities (CEC) (1995) Eurocode 3: design of steel structures. Part 1-2: general rules. Structural fire design. EC3-1.2, Brussels, Belgium
15. Razdolsky L (2014) Probability based structural fire load. Cambridge University Press, London, UK
16. Razdolsky L (2012) Structural fire load. Theory and principals. McGraw-Hill
17. Razdolsky L (2009) Mathematical modeling of fire dynamics. In: Proceedings of the world congress on engineering and computer science 2009. WCE, London
18. Frank-Kamenestkii DA (1969) Diffusion and heat transfer in chemical kinetics. Plenum Press, New York
19. Wentzel E, Ovcharov L (1986) Applied problems in probability theory. Mir Publishers, Moscow
20. Sveshnokov AA (1978) Problems in probability theory, mathematical statistics and theory of random functions. Dover Publications, New York

Chapter 9
Fire Severity and Structural Creep Analysis/Design

Notation

k	The thermal conductivity that has the dimensions W/m * K or J/m * s * K
T	Temperature
d	Thickness in the direction of heat flow
c	Specific heat capacity
Q	Activation energy
R	Ideal gas constant
P_1	Losses of heat due to thermal radiation
e	Emissivity factor
σ	Stefan–Boltzmann constant (σ = 5.6703(10^{-8}) W/m^2 K^4)
T_o	Ambient temperature
A_v	Area of openings
c_p	Average specific heat at constant pressure
t	Time
$\vec{v}(u; v; w)$	Velocity vector
D	Diffusion coefficient (m^2/sec)
p	Pressure
v	Kinematic viscosity
θ	Dimensionless temperature
τ	Dimensionless time
h	Height of the compartment (m)
a	Thermal diffusivity (m^2/sec)
Time	$t = \frac{h^2}{a}\tau$ (sec)
Temperature	$T = \frac{RT_*^2}{E}\theta + T_*$ (K), where T_* = 600 K is the base line temperature
Coordinates	$\bar{x} = x/h$ and $\bar{z} = z/h$ - "x" and "z"—dimensionless coordinates
Velocities	$\bar{u} = \frac{v}{h}u$ (m/sec) and $\bar{w} = \frac{v}{h}w$ (m/sec) horizontal and vertical components velocity accordingly
v	Kinematic viscosity (m^2/sec)

L. Razdolsky, *Probability Based High Temperature Engineering*,
DOI 10.1007/978-3-319-41909-1_9

u and w	Dimensionless velocities
$Pr = v/a$	Prandtl number
$Fr = \frac{gh^3}{va}$	Froude number
g	Gravitational acceleration
$Le = a/D = Sc/Pr$	The Lewis number
$Sc = v/D$	The Schmidt number
$\bar{\beta} = \frac{RT_*}{E}$	Dimensionless parameter
$\bar{\gamma} = \frac{c_p RT_*^2}{QE}$	Dimensionless parameter
$P = \frac{e\sigma K_v(\beta T_*)^3 h}{\lambda}$	Thermal radiation dimensionless coefficient
$\sigma = 5.67(10^{-8})$ (W/m^{2}K^4)	Stefan–Boltzman constant
$K_v = A_o\ h/V$	Dimensionless opening factor
A_o	Total area of vertical and horizontal openings
$\delta = \left(\frac{E}{RT_*^2}\right)Qz\left(\exp\left(-\frac{E}{RT_*}\right)\right)$	Frank-Kamenetskii's parameter
$C = [1 - P(t)/P_o]$	Concentration of the burned fuel product in the fire compartment
$\bar{W} = \frac{v}{h}W$	Vertical component of gas velocity
$\bar{U} = \frac{v}{h}U$	Horizontal component of gas velocity
$b = L/h$, "L" and "h"	Length (width) and height of fire compartment accordingly
W; U	Dimensionless velocities
R_n	Nominal strength
S_i	Nominal load
Φ	Resistance factor
γ	Load factor
R_c	Characteristic value for the resistance
A	Cross-sectional area
I	Moment of inertia
W	Total weight
G_c	Characteristic value for the permanent load
S	Characteristic value for the variable load
ψ_1	Partial safety factor for the permanent load
ψ_2	Partial safety factor for the variable load
β	Reliability index
S	Probability space
A	Set of outcomes (events) to which a probability is assigned
$P(E_2 \mid E_1)$	Conditional probability
$\Phi^*(.)$	Denotes the cumulative distribution function of standard normal distribution $\Phi^*(z) = \frac{1}{\sqrt{2\pi}}\int_{-\infty}^{z} e^{-\frac{z^2}{2}}\mathrm{d}z$.
μ_A, μ_B, σ_A, σ_B	Mean and standard deviation of A and B, respectively

$J(t, t')$	Compliance function
T_M	Melting point of the metal matrix material
$\varepsilon(t)$	Strain
$\sigma(t)$	Stress
$\bar{\sigma}(t) = E(t)\varepsilon(t)$	Instantaneous stress
ε_e	Instantaneous (elastic) strain
ε_c	Creep strain
ε_T	Thermal expansion due to high temperature effect
$K(t, t') = - \partial J(t, t')/\partial t'$	Retardation function (memory function)
$R(t, t')$	Relaxation function (also called the relaxation modulus)
$M(\theta)$	Bending moment
$V(\theta)$	Shear force
$P(\theta)$	Axial force
$y(\theta)$	Deformation
P_f	Probability of failure
P_{rel}	Reliability
Θ	Dimensionless temperature
κ	Curvature
τ	Dimensionless time
ω	Frequency
$S(\omega)$	Spectral density
μ	Poisson coefficient
D	Diffusion rate (Flick's law)
η	*Viscosity* parameter of the material
E	Modulus of elasticity
$n = \eta/E$	Relaxation time
n	Power law exponent
α_i	Material property parameter
L	Span (spring spacing)
k_0	Subgrade modulus

9.1 Introduction

This chapter introduces the fundamental relationships of computing the probability of failure from interfering and compatible stress-strength probability functions as the most promising approach to the reliability analysis of structural elements.

Stress analysis for creep has a long history in engineering mechanics driven by the needs of design for high temperature in many industries but primarily power generation and aerospace. In the absence of the computing power required for detailed finite element analysis of time-dependent nonlinear creep with complex loading histories, robust simplified methods of analysis were developed [1–3] for simple constitutive models. These models, for example the time- and strain-hardening constitutive equations, were based on adaptations for time-varying stress of equally simple models for the secondary creep stage from constant load/stress uniaxial tests where minimum creep rate is constant. The most common secondary creep constitutive model has been the Norton-Bailey Law which gives a power-law relationship between minimum creep rate and (constant) stress. The unique mathematical properties of the power law allowed the development of robust simplified methods, many of which can be found in high temperature design codes. Now that detailed finite element analysis for creep is readily accomplished on the desktop it is perhaps surprising that the simple time- or strain-hardening constitutive models based on power law creep remain the most widely available in common commercial finite element software, such as ANSYS or ABAQUS, even though more comprehensive time-dependent nonlinear constitutive models are available (and can be included as user-defined materials). The most common reason for persisting with the more simple constitutive models is the ease with which material constants can be derived from experiments, the ability to check detailed solutions with simplified (robust) methods and an underlying understanding of the expected behavior of simple (but fundamental) structures subject to power law creep. Nevertheless, it has long been known that creep over a range of stress does not follow one simple power law relationship, typically (approximately) following one power law at low stress and another at high stress—a phenomenon known as 'power-law breakdown'. A common observation is a shift from a power law (usually dislocation) mechanism at 'moderate' stress to a diffusion mechanism at 'low' stress, characterized by a linear viscous relationship between creep rate and stress [4, 5] with a more significant power-law breakdown at 'high' stress. Such a stress range dependent constitutive model, with a transition from linear to power law behavior, has recently been studied by [6–8]. This chapter presents a fast numerical algorithm for the implementation of material models in structural engineering design practice. Details of the numerical algorithm are discussed. The chapter includes several numerical examples which illustrate the speed, robustness and accuracy of the proposed procedure. The stresses produced by differences in creep among various parts of the structure, or due to a change of the structural system during fire event, can cause deleterious cracking, accompanied by degradation of structural stiffness. This may further facilitate increase the deformations of structural system and ultimately result in loss of serviceability of the structure.

By altering the temperature-time stress state, creep indirectly causes a change in the stress maxima for superimposed live loads. Due to the continuous degradation of the materials creep may exert in this manner a significant influence on the failures mode of structures. Thus, creep and may alter the safety margin against the collapse of a structure under short-time overloads. In slender or thin structures, creep also causes a slow longtime growth of buckling deflections. Consequently, the critical loads for longtime instability may be much less than the elastic critical loads.

9.1.1 Basic Assumptions and Code Recommendations

Although many strides have recently been made towards nonlinear creep analysis of structures, nearly all practical applications still rely on the linearity assumption, which is sanctioned by the current design code recommendations.

For time-variable stresses, the linearity assumption implies the principle of superposition, which may be expressed by the memory integral in creep constitutive equation.

As shown in Chap. 2, the integral-type stress–strain relation can be reduced to a rate-type stress–strain relation if the compliance function (or, more generally, the kernel of the hereditary integral) is approximated by a degenerate kernel.

9.1.2 Non-linearity Due to Cracking or Strain-Softening

These effects are normally taken into account in step-by-step time integration by adjusting the incremental stiffness according to the stress and strain values in the previous iteration of the same step or, for the first iteration, in the last iteration of the previous step. There are, nevertheless, serious problems with convergence of the iterations, as well as with convergence when the structural discretization is refined. These problems are due to strain-localization instabilities and border on fracture mechanics. A vast amount of research is being done in this field (for a review see [9]), most of it not motivated by concrete creep and shrinkage but still applicable to it.

A particular difficulty arises with strain-softening when the time steps are decreased substantially beyond the shortest relaxation time, as is necessary when long-time creep is analyzed. It was found [10] that an arbitrary increase of the time step is made possible by an exponential algorithm which is analogous to the exponential algorithm for the Maxwell chain model and is based on an exact solution of the incremental stress–strain relation for a strain-softening element under the assumption that the coefficients of the associated differential equation are constant.

9.1.3 Non-linearity at Unloading and Adaptation

In metal matrix materials significant deviations from linearity (i.e., from the principle of superposition) are observed at decreasing strain, especially at large sudden unloading. Generally, the unloading deformation behavior is stiffer than that predicted according to linearity. The deviations from linearity are not caused by the decrease of stress, since stress relaxation follows the predictions of linear analysis, but by a decrease of strain. There are some stress–strain relations, who are calibrated according to the behavior at unloading, but then they sacrifice the possibility of good representation of test data at constant stress for a wide range of loading conditions and for temperature load durations. In these approaches, one cannot of course speak of non-linearity at unloading.

The non-linearity which is responsible for the reduced recoverability of strain at unloading has been called the adaptation. Described by an adaptation parameter in the kernel of the iterative integral, this non-linearity also causes an increased stiffness for load increments after a longer period under sustained compressive load. It appears that sustained compressive stress gradually stiffens material. On the other hand, tensile stress should cause a decrease of stiffness, although experimental data to verify this are still lacking.

For the integral-type constitutive law with a kernel that includes a stress dependent adaptation parameter, the structural creep analysis can be again carried out in a step-by-step manner, with iterations in each time step in which the value of the adaptation parameter is updated from one iteration to the next until a certain tolerance is met. The integral type formulation would of course be inefficient for large structures, and for that purpose it would be preferable to develop a rate-type form of the constitutive equation with a stress-dependent adaptation parameter.

9.1.4 Composite and Inhomogeneous Cross-Sections of Beams

When a beam structure consists of parts that have significantly different modulus of elasticity, stress redistributions occur due to creep. Other types of structural in homogeneity, which can be analyzed linearly as a good approximation, are differences in the cross-section size. Overall, the stresses in an inhomogeneous structure are transferred due to creep from the parts which creep more into the parts which creep less (composite materials).

The exact analysis of a composite or inhomogeneous cross-section according to time-dependant linear viscoelasticity leads to a system of two Volterra integral equations if the cross-section resultants (axial force and bending moment) are known. When the structure is redundant, the cross-section resultants are normally not known. The exact analysis leads to the system of Volterra integral equations coupled with a differential equation of the beam (see examples below). This can also be reduced to a system of Volterra integral equations which contain no unknown functions, only the redundant forces. By solving this system, one obtains the cross-section resultants, and then the above-mentioned method of cross-section analysis should be applied. This method is complicated, although the exact solutions for redundant composite structures can be obtained for the rate-of-creep method.

Experimental data obtained from uniaxial tests allow to establish basic features of the creep behavior and to find relations between strain rate, stress, temperature, time, etc. Most structural members are, however, subjected to multi-axial stress and strain conditions. In order to analyze the influence of the stress state on the time dependent material behavior, multi-axial creep tests are required.

Numerical simulation plays an indispensable role in the manufacturing process, speeding product design time while improving quality and performance. Recently, analysts and designers have begun to use numerical simulation alone as an acceptable means of validation. In many disciplines, virtual prototyping—employing numerical simulation tools based on finite element methods—has replaced traditional build-and-break prototyping.

Can one reliably simulate the collapse of a shell, interaction of multiple parts, behavior of a rubber seal, post-yield strength of metals, manufacturing process and so on using linear approximation? The answer is not really. With the trend toward ever-improving simulation accuracy, approximations of linear behavior have become less acceptable; even so, costs associated with a nonlinear analysis prohibited its wider use in the past. Today, rapid increases in computing power and concurrent advances in analysis methods have made it possible to perform nonlinear analysis and design more often while minimizing approximations. Analysts and designers now expect nonlinear analysis capabilities in general-purpose programs such as ANSYS Mechanical.

9.2 Axial Compression. Linear Creep Deformations

9.2.1 The Standard Linear Model

The equations for this model are [11]:

$$\begin{aligned} &\varepsilon = \varepsilon_1 + \varepsilon_2; \quad \sigma = \sigma_1 + \sigma_2; \quad \sigma = E_1 \varepsilon_1; \\ &\sigma_1 = E_2 \varepsilon_2; \quad \sigma_2 = \eta \dot{\varepsilon}_2 \end{aligned} \tag{9.1}$$

We can eliminate the four unknowns from these five equations and after simplifications we have:

$$\sigma + \frac{\eta}{E_1 + E_2} \dot{\sigma} = \frac{E_1 E_2}{E_1 + E_2} \varepsilon + \frac{E_1 \eta}{E_1 + E_2} \dot{\varepsilon} \tag{9.2}$$

The creep compliance function in this case is:

$$\begin{aligned} &J(t) = \frac{1}{E_1} \mathrm{e}^{-(E_2/\eta)t} + \frac{E_1 + E_2}{E_1 E_2} \left(1 - \mathrm{e}^{-(E_2/\eta)t}\right) \\ &\text{where: } E = E_1; \;\; H = \frac{E_1 E_2}{E_1 + E_2}; \;\; n = \frac{\eta}{E_1 + E_2} \end{aligned} \tag{9.3}$$

The Eq. (9.2) will have a standard form now:

$$\begin{aligned} n\dot{\sigma}+\sigma &= En\dot{\varepsilon}+H\varepsilon \\ \varepsilon &= \kappa z \end{aligned} \tag{9.4}$$

Stiffness parameter E in Eq. (9.4) represents the instantaneous modulus of elasticity, and H—long-term elastic modulus (prolonged modulus of elasticity) and κ is the curvature.

Substituting ε into first Eq. (9.4) and integrating with respect to cross-sectional area we have

$$\begin{aligned} &H\boldsymbol{\kappa}\int z^2 \mathrm{d}F+nE\dot{\boldsymbol{\kappa}}\int z^2 \mathrm{d}F=\int \boldsymbol{\sigma} z \mathrm{d}F+n\int \dot{\boldsymbol{\sigma}} z \mathrm{d}F \\ &H\boldsymbol{\kappa} I+nEI\dot{\boldsymbol{\kappa}}=M+n\dot{M} \end{aligned} \tag{9.5}$$

Replacing the curvature of κ through the second derivative of the deflection along the length of the beam, taken with the opposite sign, we obtain.

$$HIy''+nEI\dot{y}''+M+n\dot{M}=0 \tag{9.6}$$

If the beam is loaded by only longitudinal axial force P, the bending moment (in the presence of deflection y) becomes equal to the Py and the Eq. (9.6) will have the form

$$HIy''+nEI\dot{y}''+Py+nP\dot{y}+n\dot{P}y=0 \tag{9.7}$$

We seek a solution of Eq. (9.7) in the form

$$y(x,\,t)=T(t)Y(x)$$

$Y(x)$ is the function that satisfies the boundary conditions of the rod. Thus, from Eq. (9.7) we obtain.

$$\begin{aligned} &HIY''T+nEIY''\dot{T}+PTY+nP\dot{T}Y+n\dot{P}TY=0\Rightarrow :TY \\ &\frac{HIY''}{Y}+\frac{nEIY''}{Y}\frac{\dot{T}}{T}+P+\frac{nP\dot{T}}{T}+n\dot{P}=0 \\ &\frac{HIY''}{Y}=C\Rightarrow Y''+\frac{C}{HJ}Y=0\Rightarrow Y=\sin\frac{\pi x}{L}\Rightarrow C=\frac{\pi^2 HI}{L^2}=P_d \\ &\frac{Y''}{Y}=\frac{C}{HJ}=\frac{\pi^2}{L^2}\Rightarrow\frac{nEI\pi^2}{L^2}\frac{\dot{T}}{T}+P+\frac{nP\dot{T}}{T}+n\dot{P}=\frac{\pi^2 HI}{L^2}\Rightarrow \\ &\left[\frac{nEI\pi^2}{L^2}+nP\right]\dot{T}+n\dot{P}T+PT=0\Rightarrow n[P^*-P]\dot{T}+[P_d-n\dot{P}-P]T=0 \end{aligned} \tag{9.8}$$

From Eq. (9.8) we have

$$n[P^* - P]\dot{T} + [P_d - n\dot{P} - P]T = 0$$

$$\frac{\mathrm{d}T}{\mathrm{d}t} = -\frac{[P_d - n\dot{P} - P]}{n[P^* - P]}T$$

$$T(0) = y_{st}$$

$$P(\boldsymbol{\theta}) \approx \frac{qL}{\sqrt{24\boldsymbol{\varepsilon}}} = \frac{W}{\sqrt{24(7.02)10^{-4}\boldsymbol{\theta}}} = 7.7W\boldsymbol{\theta}^{-0.5} = 1.16(10^3)\boldsymbol{\theta}^{-0.5} \tag{9.9}$$

$$P^*(\boldsymbol{\theta}) = \frac{\pi^2 IE_0}{L^2}\boldsymbol{\rho}(\boldsymbol{\theta}) = \frac{\pi^2 IE_0}{L^2}\mathrm{e}^{-0.15\boldsymbol{\theta}} = 48.5(10^3)\mathrm{e}^{-0.15\boldsymbol{\theta}}$$

$$P_d = \frac{\pi^2 HI}{L^2} = 24(10^3); \quad n = 0.01$$

Now, we make the change of variables [11] in the Eq. (9.9)

$$\begin{aligned} \frac{\mathrm{d}T}{\mathrm{d}\boldsymbol{\theta}} &= -\frac{[P_d - n\dot{P} - P]}{n[P^* - P]}T[m1]; \quad \dot{P}(\boldsymbol{\theta}) = -0.58(10^3)\boldsymbol{\theta}^{-1.5} \\ m1 &= 0.351 + 0.0208 * t^\wedge 1 - 0.0372 * t^\wedge 2 + 0.01144 * t^\wedge 3 \\ m &= 0.351 * t + 0.0104 * t^\wedge 2 - 0.0124 * t^\wedge 3 + 0.00286 * t^\wedge 4 \end{aligned} \tag{9.10}$$

Where for instance in case of Slow Fire scenario we have the inverse temperature—time relationship as follows:

$$t'(\theta) = m1 = 0.351 + 0.0208 * \theta^\wedge 1 - 0.0372 * \theta^\wedge 2 + 0.01144 * \theta^\wedge 3 \tag{9.11}$$

$$t(\theta) = m = 0.351 * \theta + 0.0104 * \theta^\wedge 2 - 0.0124 * \theta^\wedge 3 + 0.00286 * \theta^\wedge 4 \tag{9.12}$$

The change of deflections rate is now determined by the Eq. (9.10). It should be noted that according to (9.10) the deformation is always positive. Deflections rate is determined by the sign of the numerator in formula (9.10) and $P^* > P$. If the numerator in formula (9.10) is positive then the deformation will decrease monotonically. If the numerator is negative the deformation will increase monotonically. Finally, if the numerator is zero, the deformation will be permanent, that is independent of temperature (time). Here are some examples illustrating the stability of axially loaded structural elements.

Example 9.1 Data: Steel column design

$P_{\mathrm{D.L.}}$ = 100 k
$P_{\mathrm{L.L.}}$ = 100 k
Span L = 38′-0″
F_y = 50 ksi

W14x90 (ASTM A992)
A = 26.5 in^2
F_u = 65 ksi
S_x = 143 in^3
I_x = 999 in^4
Z_x = 157 in^3
P_{all} = 261 k
$T_{\max}$ = 548 °C
E = 2.9(10^4)exp(−0.15θ) ksi
ε = 7.02(10^{-4})θ

Slow Fire Severity Case
Original Structural Design (ASD method):

1. Allowable Design Load:

$$P = (100) + (100) = 200\,\text{k} < 261\,\text{k O.K.}$$

   Column is restrained at both ends, span L = 38′.
2. Total elongation of the column: $\Delta L = \alpha_o T_m L = 7.02(4)(10^{-4})456 = 1.28''$.
3. Since the deformed length of the beam is known ($L_{\text{tot.}} = L + \Delta L = L(1 + \alpha_o T_{\max})$, the maximum deflection of the beam can be approximated as follows (large deformations):

$$\Delta_b = \frac{L}{2}\sqrt{2(7.02)(10^{-4})\theta_{\max}} = 228(10^{-2})\sqrt{2(7.02)4} = 17.1'' = 1.42'.$$

4. Equivalent load q = 17.1(384)999(29,000)/(5)456^4 = 0.88 klf W = 0.45 (38) = 33.4 k.
5. The maximum trust force in this case can be approximated as follows:

$$P = W\sqrt{\frac{1}{24(\alpha_o T_{\max})}} = 33.4\sqrt{\frac{1}{24(7.02)10^{-4}\theta}} = 0.258(10^3)[\theta^{-0.5}]$$

$$P^*(\theta) = \frac{\pi^2 I E_0}{L^2}\rho(\theta) = \frac{\pi^2 I E_0}{L^2}\mathrm{e}^{-0.15\theta} = 1.375(10^3)\mathrm{e}^{-0.15\theta}$$

$$P_d = \frac{\pi^2 HJ}{L^2} = 0.69(10^3); \quad H = E_0/2; \quad n = 0.01.$$

6. Differential Eq. (9.17) is:

$$\frac{\mathrm{d}T}{\mathrm{d}\boldsymbol{\theta}} = -\frac{[P_d - n\dot{P} - P]}{n[P^* - P]}T[m1]; \quad n\dot{P}(\boldsymbol{\theta}) = -0.00066(10^3)\boldsymbol{\theta}^{-1.5}$$

$$m1 = 0.351 + 0.0208 * t\hat{}1 - 0.0372 * t\hat{}2 + 0.01144 * t\hat{}3$$

$$m = 0.351 * t + 0.0104 * t\hat{}2 - 0.0124 * t\hat{}3 + 0.00286 * t\hat{}4.$$

7. Solution of (9.17) is (assuming $H = E_0/2$ and using POLYMATH software):

Calculated values of DEQ variables

	Variable	Initial value	Minimal value	Maximal value	Final value
1	t	0.1	0.1	4	4
2	T	1	0.0001107	1.006672	0.0001107

Differential equations

1 $d(T)/d(t) = -(10\wedge 1) * m1 * T * (0.69 * 1 * (\exp(-0.15 * t * 0)) + 0.00129 * (t\wedge -1.5) - 0.258 * (t\wedge -0.5))/((1.375 * (\exp(-0.15 * t))) - 0.258 * (t\wedge -0.5))$

We can see from Fig. 9.1 that the structure is stable.

Example 9.2 Data: Steel column design (see Example 9.1 but the long-term modulus of elasticity

$H \to 0$. The solution of (9.17) is (assuming $H = 0$ and using POLYMATH software):

Calculated values of DEQ variables

	Variable	Initial value	Minimal value	Maximal value	Final value
1	t	0.1	0.1	4	4
2	T	1	1	53.02455	53.02455

Differential equations

1 $d(T)/d(t) = -(10\wedge 1) * m1 * T * (0.69 * 0 * (\exp(-0.15 * t * 0)) + 0.00129 * (t\wedge -1.5) - 0.258 * (t\wedge -0.5))/((1.375 * (\exp(-0.15 * t))) - 0.258 * (t\wedge -0.5))$

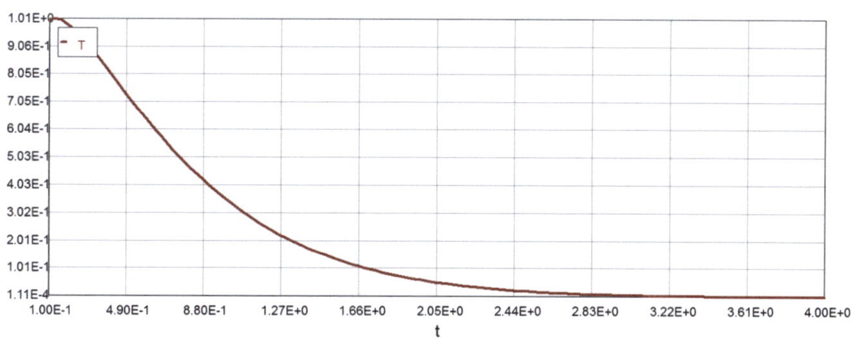

Fig. 9.1 Deflection—temperature—time function ($H > 0$)

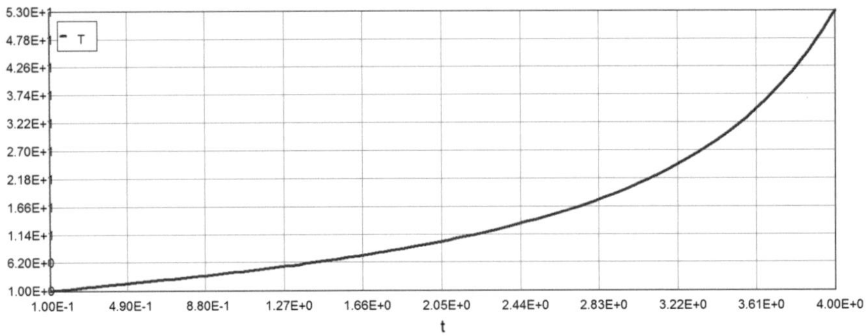

Fig. 9.2 Deflection—temperature—time function ($H = 0$)

We can see now from Fig. 9.2 that the deflections are increasing with the temperature (time) rise and the structure is unstable.

9.2.2 Axial Compression. Linear Creep Constitutive Equation

The general integro-differential equation in this case is as follows:

$$E(t)I\frac{\partial^4 y(x,t)}{\partial x^4} + P(t)\frac{\partial^2 y(x,t)}{\partial x^2} + \int_0^t P(\tau)\frac{\partial^2 y(x,\tau)}{\partial x^2}K(t,\tau)\mathrm{d}\tau = q(x,t) + \int_0^t q(\tau)K(t,\tau)\mathrm{d}\tau \tag{9.13}$$

If $q \equiv 0$ and $y(x, t) = Y(x)T(t)$ then

$$Y^{\mathrm{IV}}(x)E(t)IT(t) + P(t)Y^{\mathrm{II}}(x)T(t) + Y^{\mathrm{II}}(x)\int_0^t P(\tau)T(\tau)K(t,\tau)\mathrm{d}\tau = 0 \tag{9.14}$$

$$-\frac{Y^{\mathrm{IV}}(x)}{Y^{\mathrm{II}}(x)} = \frac{1}{E(t)I}\left\{P(t) + \frac{1}{T(t)}\int_0^t P(\tau)T(\tau)K(t,\tau)\mathrm{d}\tau\right\} = C = \mathrm{const.} \tag{9.15}$$

First equation:

$$Y^{\mathrm{IV}}(x) + CY^{\mathrm{II}}(x) = 0$$

This equation indicates forms of stability: $Y(x) = A[\sin(\pi x/L)]$, therefore $C = \pi^2/L^2$

Second equation:

$$\frac{1}{E(t)I}\left\{P(t) + \frac{1}{T(t)}\int_0^t P(\tau)T(\tau)K(t,\tau)\mathrm{d}\tau\right\} = \pi^2/L^2; \quad P^* = \frac{\pi^2 E(t)I}{L^2}$$

$$\text{or}: \quad [P^* - P(t)]T(t) - \int_0^t P(\tau)T(\tau)K(t,\tau)\mathrm{d}\tau = 0 \tag{9.16}$$

In case of fire we have: modulus of elasticity E is a function of temperature $E(\theta)$ and time $t = m = \varphi(\theta) = \theta^{-1}$ for each fire severity scenario [10], therefore Eq. (9.16) has a form:

$$[P^*(\theta) - P(\theta)]T(\theta) - \int_0^{\theta} P(\tau)T(\tau)K(\theta,\tau)m1\mathrm{d}\tau = 0$$

$$P(\theta) \approx \frac{qL}{\sqrt{24\varepsilon}} = \frac{W}{\sqrt{24(7.02)10^{-4}\theta}} = 7.7W\theta^{-0.5}$$

$$P^*(\theta) = \frac{\pi^2 IE_0}{L^2}\rho(\theta) = \frac{\pi^2 IE_0}{L^2}\mathrm{e}^{-0.15\theta} \tag{9.17}$$

$$T(\theta) = \frac{b}{[a\mathrm{e}^{-0.15\theta} - b(\theta^{-0.5})]}\int_0^{\theta}(\tau^{-0.5})T(\tau)K(\theta,\tau)m1\mathrm{d}\tau$$

$$a = \frac{\pi^2 IE_0}{L^2}; \quad b = 7.7W; \quad \theta \approx b^2/a^2 \approx 0.0005 \quad \therefore \quad P*(\theta) \approx \frac{\pi^2 IE_0}{L^2}$$

If the structural system is stiff (moment of inertia I is large) and the span L is not large (regular framing system) than $\theta \approx (b/a)^2 \approx 0$ and P^* is constant. On the other hand, if I is small (e.g., steel bar joist) and span L is large, than it is possible that the structural system becomes unstable for the whole range of dimensionless temperatures, since the Euler's force P^* must be greater than P (see Fig. 9.3).

Consider now two cases: $P^* > P$ and $P^* < P$

Case 1 $P^* > P$ for all $0 < \theta < 10$

The solution of Eq. (9.17) is as follows (for different material property parameters α) (Fig. 9.4):

$\alpha = 0.1$

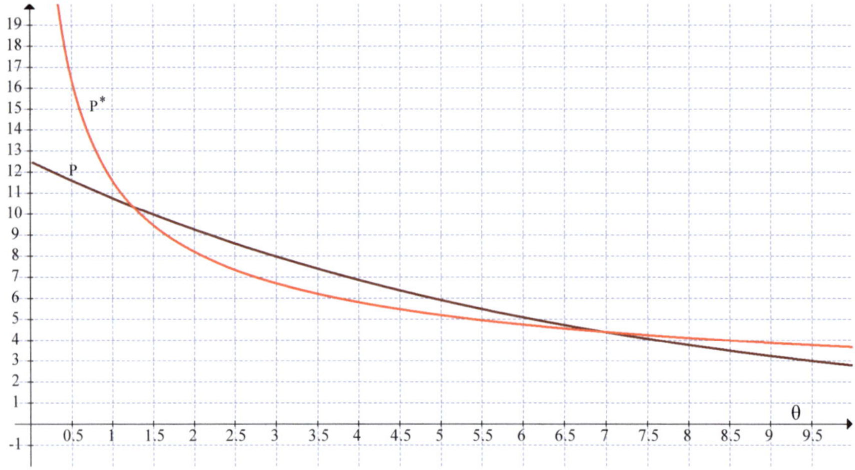

Fig. 9.3 Euler's force P^* versus dimensionless temperature

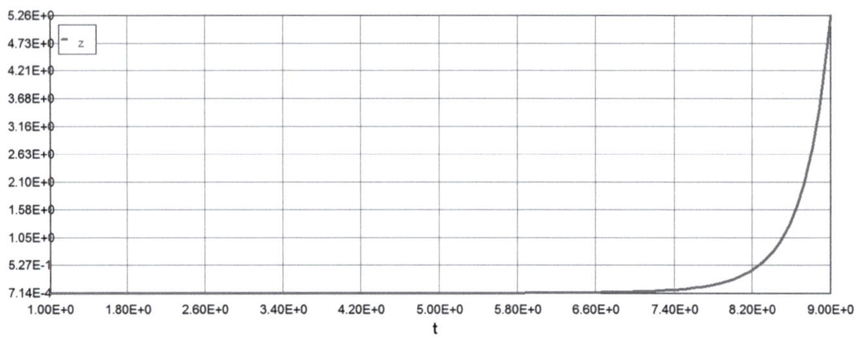

Fig. 9.4 Deflection—temperature relationship ($\alpha = 0.1$)

Calculated values of DEQ variables

	Variable	Initial value	Minimal value	Maximal value	Final value
1	m	0.35186	0.35186	13.72626	13.72626
2	$m1$	0.34604	0.3345935	5.86476	5.86476
3	t	1	1	9	9
4	Y	0.03	0.03	115.2147	115.2147
5	z	0.0007141	0.0007141	*5.25928*	*5.25928*

Differential equations

1 $d(Y)/d(t) = 1.16 * t^\wedge(-0.5) * z * (\exp(0.1 * m1)) * (\exp(t/(1 + 0.1 * t)))$

Explicit equations

1 $m1 = 0.351 + 0.0208 * t^\wedge 1 - 0.0372 * t^\wedge 2 + 0.01144 * t^\wedge 3$

2 $m = 0.351 * t + 0.0104 * t^\wedge 2 - 0.0124 * t^\wedge 3 + 0.00286 * t^\wedge 4$

3 $z = 1/(48.5 * (\exp(-0.15 * t)) - 1.16 * (t^\wedge(-0.5))) * Y * (\exp(-0.1 * m1))$

Deflection increases rapidly at $\theta = 8.5$ and structural system becomes unstable (Fig. 9.5).

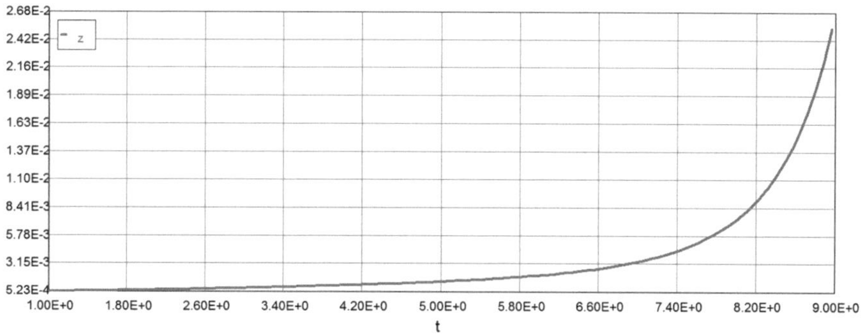

Fig. 9.5 Deflection—temperature relationship ($\alpha = 1$)

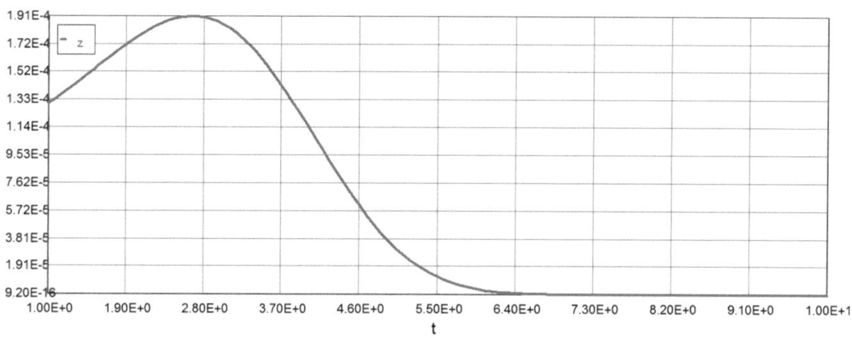

Fig. 9.6 Deflection—temperature relationship ($\alpha = 5$)

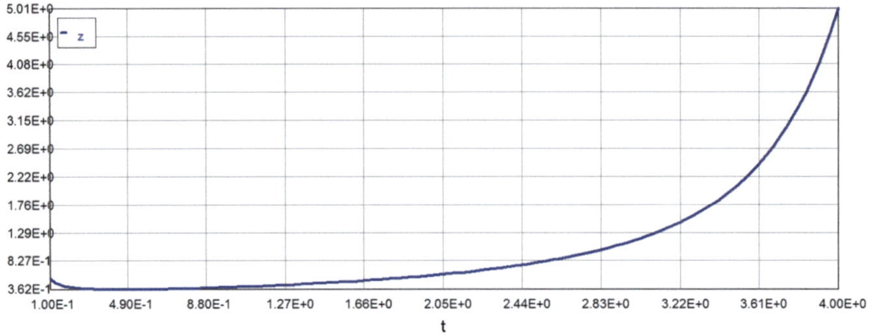

Fig. 9.7 Deformation—temperature (time) curve (α = 0.01)

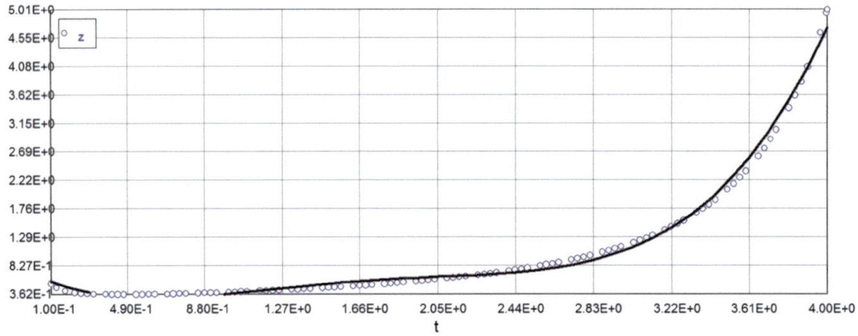

Fig. 9.8 Deformation—temperature (time) curve (α = 0.01—approximation)

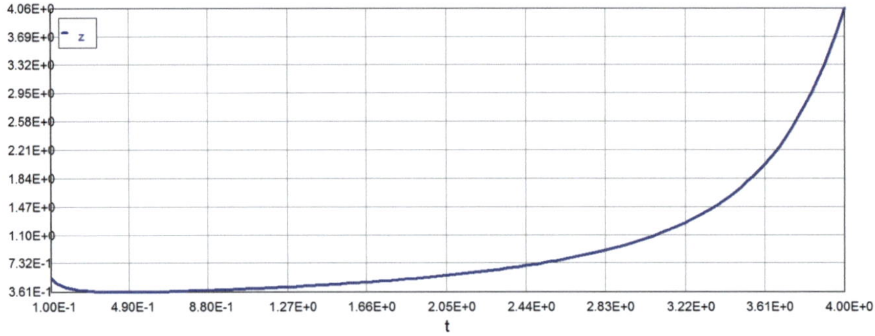

Fig. 9.9 Deformation—temperature (time) curve (α = 0.1)

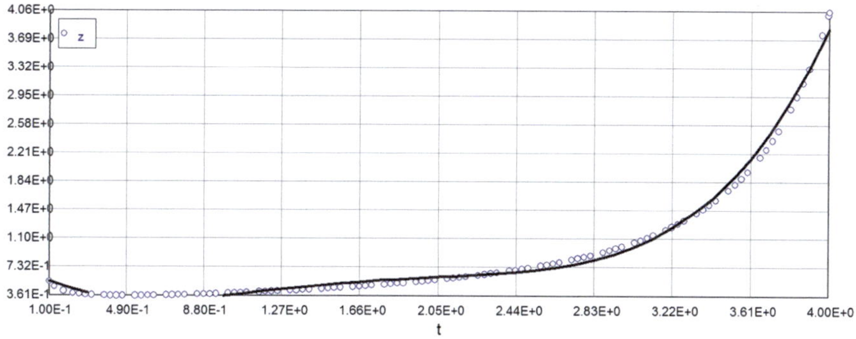

Fig. 9.10 Deformation—temperature (time) curve ($\alpha = 0.1$—approximation)

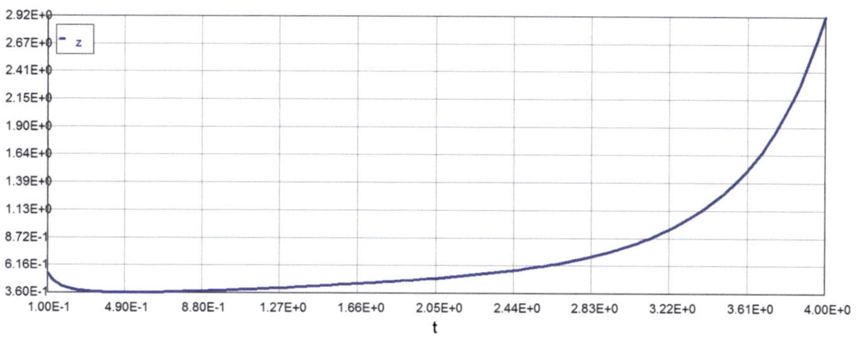

Fig. 9.11 Deformation—temperature (time) curve ($\alpha = 0.33$)

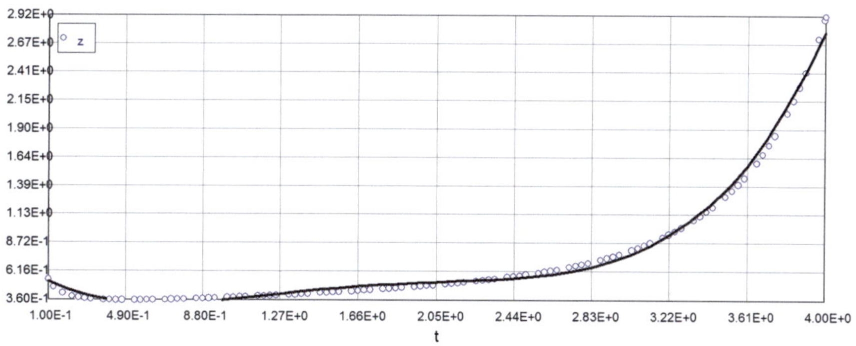

Fig. 9.12 Deformation—temperature (time) curve ($\alpha = 0.33$—approximation)

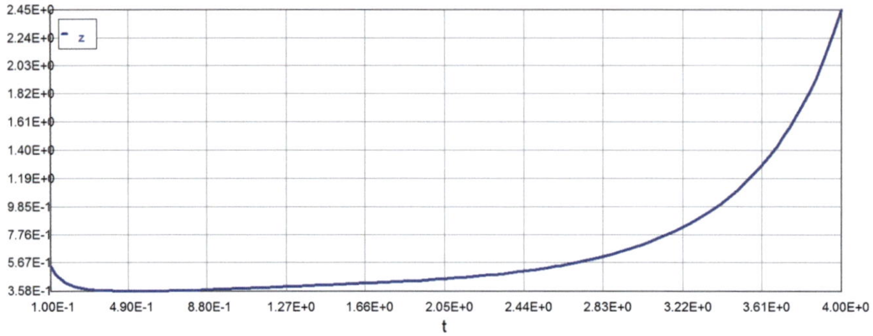

Fig. 9.13 Deformation—temperature (time) curve ($\alpha = 0.5$)

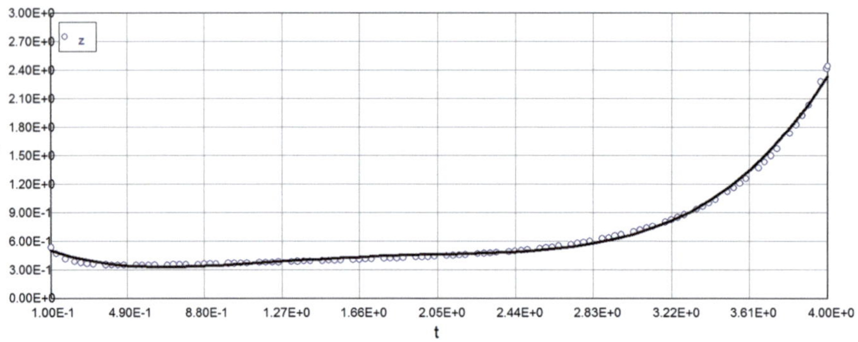

Fig. 9.14 Deformation—temperature (time) curve ($\alpha = 0.5$—approximation)

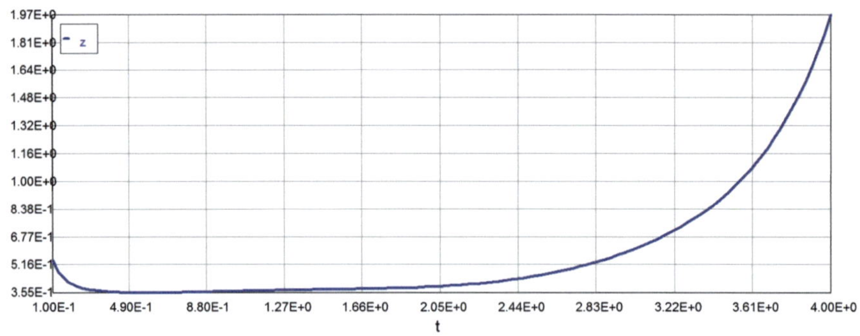

Fig. 9.15 Deformation—temperature (time) curve ($\alpha = 0.75$)

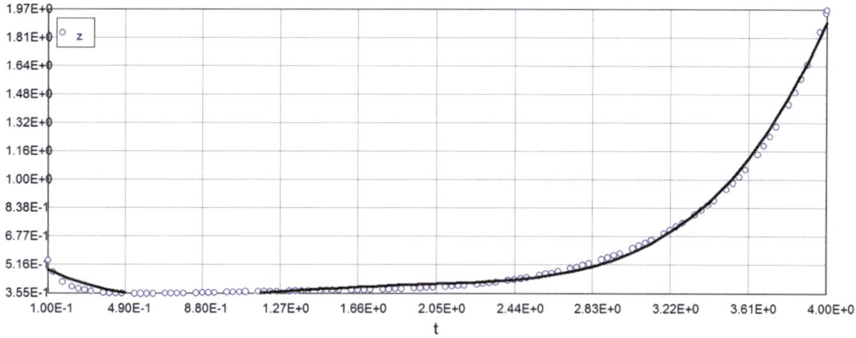

Fig. 9.16 Deformation—temperature (time) curve (α = 0.75—approximation)

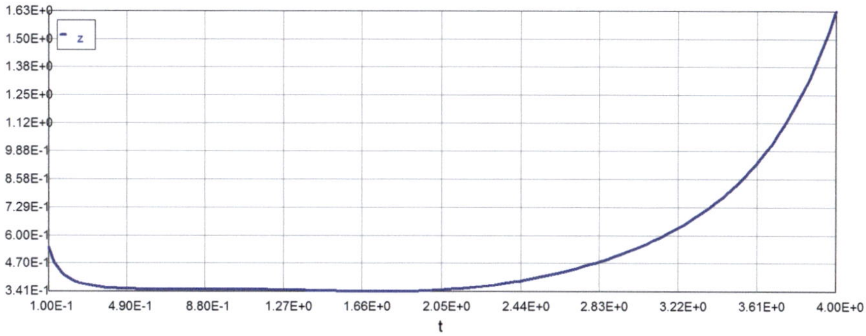

Fig. 9.17 Deformation—temperature (time) curve (α = 1)

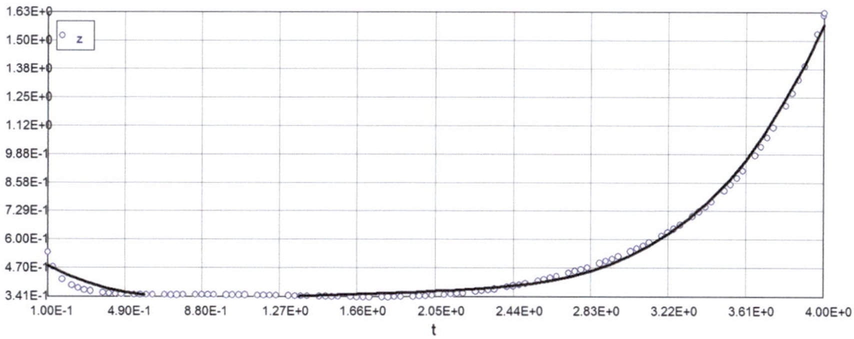

Fig. 9.18 Deformation—temperature (time) curve (α = 1—approximation)

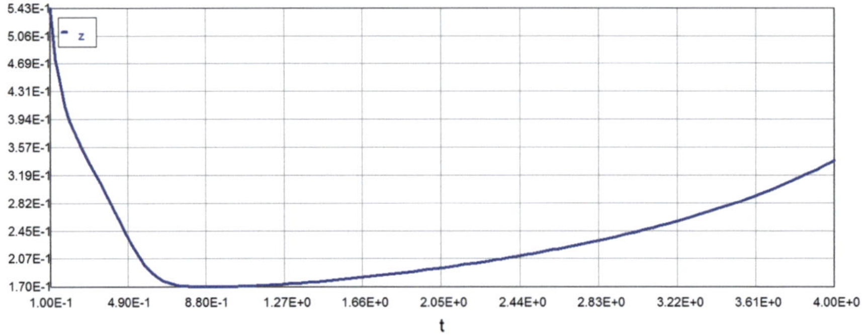

Fig. 9.19 Deformation—temperature (time) curve ($\alpha = 10$)

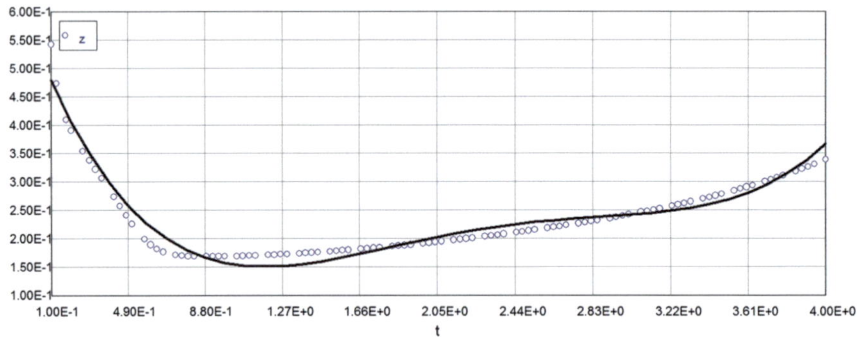

Fig. 9.20 Deformation—temperature (time) curve ($\alpha = 10$—approximation)

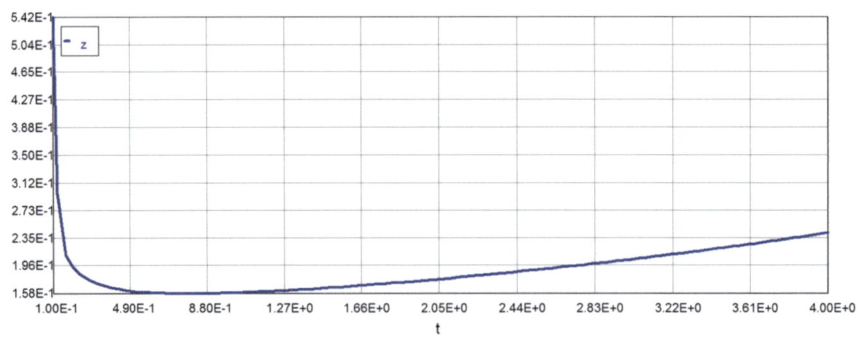

Fig. 9.21 Deformation—temperature (time) curve ($\alpha = 100$)

Model: $z = a0 + a1 * t + a2 * t^\wedge 2 + a3 * t^\wedge 3 + a4 * t^\wedge 4 + a5 * t^\wedge 5$

Variable	Value
$a0$	−3.154449
$a1$	5.628834
$a2$	−3.459841
$a3$	0.9402373
$a4$	−0.11571
$a5$	0.005261

$$y_{\max} = -3.15 + 5.62\theta - 3.46\theta^2 + 0.94\theta^3 - 0.116\theta^4 + 0.0053\theta^5 \quad (9.18)$$

$\alpha = 1$

Calculated values of DEQ variables

	Variable	Initial value	Minimal value	Maximal value	Final value
1	A	0	−0.0991422	1499.549	1499.549
2	m	0.35186	0.35186	13.72626	13.72626
3	$m1$	0.34604	0.3345935	5.86476	5.86476
4	n	3	3	3	3
5	r	0.860708	0.860708	2.452518	2.333162
6	t	1	1	9	9
7	y	0	0	0.7146508	0.2623553
8	Y	0.03	0.03	115.2147	115.2147
9	z	0.000523	0.000523	*0.0268285*	*0.0268285*

Differential equations

1 $d(y)/d(t) = -1 * y * m1 + (\exp(-0.15 * t)) * (1 - 0.15 * t) - (\exp(t/(1.0 + 0.1 * t))) * m1 * y^\wedge n + 1 * m1 * t * (\exp(-0.15 * t))$

2 $d(A)/d(t) = (1/(4 - 3.5)) * (-1.238 + 0.985 * (t) - 0.116 * (t^\wedge 2) - 0.0304 * (t^\wedge 3) + 0.00554 * (t^\wedge 4))^\wedge 1 * (-1.238 + 0.985 * (t + 3.5) - 0.116 * ((t + 3.5)^\wedge 2) - 0.0304 * ((t + 3.5)^\wedge 3) + 0.00554 * ((t + 3.5)^\wedge 4))^\wedge 1$

3 $d(Y)/d(t) = 1.16 * t^\wedge(-0.5) * z * (\exp(1 * m1)) * (\exp(t/(1 + 0.1 * t)))$

Explicit equations

1 $n = 3$

2 $r = t * (\exp(-0.15 * t))$

3 $m1 = 0.351 + 0.0208 * t^\wedge 1 - 0.0372 * t^\wedge 2 + 0.01144 * t^\wedge 3$

4 $m = 0.351 * t + 0.0104 * t^\wedge 2 - 0.0124 * t^\wedge 3 + 0.00286 * t^\wedge 4$

5 $z = 1/(48.5 * (\exp(-0.15 * t)) - 1.16 * (t^\wedge(-0.5))) * Y * (\exp(-1 * m1))$

Deflection increases at $\theta = 7.4$ but structural system is stable (Fig. 9.6).
$\alpha = 5$

Calculated values of DEQ variables

	Variable	Initial value	Minimal value	Maximal value	Final value
1	A	0	0	18	18
2	m	0.35186	0.35186	20.75	20.75
3	$m1$	0.34604	0.3346245	8.279	8.279
4	n	3	3	3	3
5	r	0.860708	0.860708	2.452502	2.231302
6	t	1	1	10	10
7	y	0	−0.0153191	0.49855	−0.0134001
8	Y	0.03	0.03	9133.616	9133.616
9	z	0.000131	9.198E−16	0.0001906	9.198E−16

Differential equations

1 $\mathrm{d}(y)/\mathrm{d}(t) = -1 * y * m1 + (\exp(-0.15 * t)) * (1 - 0.15 * t) - (\exp(t/(1.0 + 0.1 * t))) * m1 * y\hat{}n + 1 * m1 * t * (\exp(-0.15 * t)) * 0$

2 $\mathrm{d}(A)/\mathrm{d}(t) = (1/(4 - 3.5)) * (-1.238 + 0.985 * (t) - 0.116 * (t\hat{}2) - 0.0304 * (t\hat{}3) + 0.00554 * (t\hat{}4))\hat{}0 * (-1.238 + 0.985 * (t + 3.5) - 0.116 * ((t + 3.5)\hat{}2) - 0.0304 * ((t + 3.5)\hat{}3) + 0.00554 * ((t + 3.5)\hat{}4))\hat{}0$

3 $\mathrm{d}(Y)/\mathrm{d}(t) = 1.16 * t\hat{}(-0.5) * z * (\exp(5 * m1)) * (\exp(t/(1 + 0.1 * t)))$

Explicit equations

1 $n = 3$

2 $r = t * (\exp(-0.15 * t))$

3 $m1 = 0.351 + 0.0208 * t\hat{}1 - 0.0372 * t\hat{}2 + 0.01144 * t\hat{}3$

4 $m = 0.351 * t + 0.0104 * t\hat{}2 - 0.0124 * t\hat{}3 + 0.00286 * t\hat{}4$

5 $z = 1/(48.5 * (\exp(-0.15 * t)) - 1.16 * (t\hat{}(-0.5))) * Y * (\exp(-5 * m1))$

Deflection increases up to $\theta = 2.8$; after that—decreases and structural system is stable.

Model: $z = a0 + a1 * t + a2 * t\hat{}2 + a3 * t\hat{}3 + a4 * t\hat{}4 + a5 * t\hat{}5$

Variable	Value
$a0$	0
$a1$	0.00033
$a2$	−0.000115
$a3$	1.371E−05

(continued)

(continued)

Variable	Value
*a*4	−5.168E−07
*a*5	−2.623E−09

$$y_{\max} = 0.00033\theta - 0.000115\theta^2 + 1.37\theta^3 - 5.168E{-}07\theta^4 + 0 * \theta^5 \tag{9.19}$$

9.3 Combined Flexure and Axial Load Resistance with High Temperature Creep Effect

Applying method of change of variable to the Eq. (9.13) we have:

$$\begin{aligned}
&E(\boldsymbol{\theta})I\frac{\partial^4 y(x,\boldsymbol{\theta})}{\partial x^4} + P(\boldsymbol{\theta})\frac{\partial^2 y(x,\boldsymbol{\theta})}{\partial x^2} + \int_0^{\boldsymbol{\theta}} P(\tau)\frac{\partial^2 y(x,\tau)}{\partial x^2}K(\boldsymbol{\theta},\tau)m1\mathrm{d}\tau \\
&= q(x,\boldsymbol{\theta}) + \int_0^{\boldsymbol{\theta}} q(\tau)K(\boldsymbol{\theta},\tau)m1\mathrm{d}\tau \\
&K(\boldsymbol{\theta},\tau,\boldsymbol{\alpha}_i) = \mathrm{e}^{-\boldsymbol{\alpha}_i\boldsymbol{\theta}}[\exp(\boldsymbol{\theta}/(1+0.1\boldsymbol{\theta})]\mathrm{e}^{\boldsymbol{\alpha}_i\tau}m1 \\
&E(\boldsymbol{\theta}) = E_0\boldsymbol{\rho}(\boldsymbol{\theta}) = E_0[\exp(-0.15\boldsymbol{\theta})]; \\
&P(\boldsymbol{\theta}) \approx \frac{qL}{\sqrt{24\varepsilon}} = \frac{W}{\sqrt{24(7.02)10^{-4}\boldsymbol{\theta}}} = 7.7W\boldsymbol{\theta}^{-0.5}
\end{aligned} \tag{9.20}$$

Consider $y(x,\ \theta) = \varphi(\theta)[\sin(\pi x/L)]$ and $q(x,\ \theta) = q_0(\theta)[\sin(\pi x/L)]$. Substituting this into (9.20) we have:

$$\begin{aligned}
&E(\boldsymbol{\theta})I\frac{\pi^4}{L^4}\varphi(\boldsymbol{\theta}) - P(\boldsymbol{\theta})\frac{\pi^2}{L^2}\varphi(\boldsymbol{\theta}) - \frac{\pi^2}{L^2}\int_0^{\boldsymbol{\theta}} P(\tau)\varphi(\tau)K(\boldsymbol{\theta},\tau)m1\mathrm{d}\tau \\
&= q_0(\boldsymbol{\theta}) + \int_0^{\boldsymbol{\theta}} q_0(\tau)K(\boldsymbol{\theta},\tau)m1\mathrm{d}\tau \\
&\varphi(\boldsymbol{\theta})\left[P(\boldsymbol{\theta}) - E(\boldsymbol{\theta})I\frac{\pi^2}{L^2}\right] + \int_0^{\boldsymbol{\theta}} P(\tau)\varphi(\tau)K(\boldsymbol{\theta},\tau)m1\mathrm{d}\tau \\
&+ \frac{L^2}{\pi^2}\left[q_0(\boldsymbol{\theta}) + \int_0^{\boldsymbol{\theta}} q_0(\tau)K(\boldsymbol{\theta},\tau)m1\mathrm{d}\tau\right] = 0
\end{aligned} \tag{9.21}$$

Denote $P^* = E(\theta)I\frac{\pi^2}{L^2}$ and $q_0(\theta) = q_0 = \text{const.}\ f(\theta) = \frac{q_0L^2}{\pi^2}\left[1 + \int_0^\theta K(\theta,\tau)m1\mathrm{d}\tau\right]$ than Eq. (9.21) has a form:

$$\varphi(\theta)[P^*(\theta)-P(\theta)] - \int_0^\theta P(\tau)\varphi(\tau)K(\theta,\tau)m1\mathrm{d}\tau = f(\theta) \tag{9.22}$$

We can see from Eq. (9.22) that $\varphi(\theta)$ is the dimensionless function; both P^* and P have dimension of force and $f(\theta)$ also has dimension of force. The creep function $K(\theta, \tau, \alpha_i)$ is a function of temperature and material property parameter (MPP) α_i that can substantially affect the deformation—temperature (time) curve (for the detailed discussion regarding the effect of α_i on creep deformations in general see Chap. 8). The effect of α_i on deformation—temperature (time) curve is presented below via examples (Figs. 9.7, 9.8, 9.9, 9.10, 9.11, 9.12, 9.13, 9.14, 9.15, 9.16, 9.17, 9.18, 9.19, 9.20, 9.21 and 9.22).

Example 9.3 $\alpha = 0.01$

Calculated values of DEQ variables

	Variable	Initial value	Minimal value	Maximal value	Final value
1	*B*	0.9996481	0.9996481	2864.854	2864.854
2	*C*	292	269.7057	292	269.7057
3	*D*	538.6613	538.6613	931.0558	625.616
4	*m*	0.0351919	0.0351919	1.50896	1.50896
5	*m*1	0.3527194	0.3345935	0.57116	0.57116
6	*t*	0.1	0.1	4	4
7	*T*1	1	1	2908.411	2908.411
8	*T*2	0	0	10.98113	10.98113
9	*y*	0	−0.0738527	11.47169	11.47169
10	*z*	0.5439404	0.362198	5.010357	5.010357

Differential equations

1 $\mathrm{d}(T1)/\mathrm{d}(t) = 258 * m1 * z * (\exp(0.01 * m)) * (\exp(t/(1+0.1 * t))) * ((t\hat{}-0.5))$

2 $\mathrm{d}(T2)/\mathrm{d}(t) = 1 * m1 * (\exp(0.01 * m)) * (\exp(t/(1+0.1 * t)))$

Explicit equations

1 $m1 = 0.351 + 0.0208 * t^\wedge 1 - 0.0372 * t^\wedge 2 + 0.01144 * t^\wedge 3$

2 $m = 0.351 * t + 0.0104 * t^\wedge 2 - 0.0124 * t^\wedge 3 + 0.00286 * t^\wedge 4$

3 $D = (1375 * (\exp(-0.15 * t)) - 258 * (t^\wedge -0.5))$

4 $C = 146 * (1 + (\exp(-0.01 * m * T2)))$

5 $B = (\exp(-0.01 * m)) * T1$

6 $z = (C + B)/D$

Model: $z = a0 + a1 * t + a2 * t^\wedge 2 + a3 * t^\wedge 3 + a4 * t^\wedge 4$

Variable	Value
$a0$	0.7231618
$a1$	−1.655858
$a2$	2.083999
$a3$	−0.9214914
$a4$	0.1415819

Model: $\varphi(\theta) = 0.723 - 1.656 * \theta + 2.084 * \theta^\wedge 2 - 0.921 * \theta^\wedge 3 + 0.142 * \theta^\wedge 4$ (9.23)

Example 9.4 $\alpha = 0.1$

Calculated values of DEQ variables

	Variable	Initial value	Minimal value	Maximal value	Final value
1	B	0.996487	0.996487	2372.617	2372.617
2	C	292	169.5857	292	169.5857
3	D	538.6613	538.6613	931.0562	625.616
4	m	0.0351919	0.0351919	1.50896	1.50896
5	$m1$	0.3527194	0.3345935	0.57116	0.57116
6	t	0.1	0.1	4	4
7	$T1$	1	1	2759.058	2759.058
8	$T2$	0	0	12.08096	12.08096
9	y	0	−0.0738526	11.47169	11.47169
10	z	0.5439346	0.3614855	4.063518	4.063518

Differential equations

1 $d(T1)/d(t) = 258 * m1 * z * (\exp(0.1 * m)) * (\exp(t/(1 + 0.1 * t))) * ((t^\wedge -0.5))$

2 $d(T2)/d(t) = 1 * m1 * (\exp(0.1 * m)) * (\exp(t/(1 + 0.1 * t)))$

Explicit equations

1 $m1 = 0.351 + 0.0208 * t^\wedge 1 - 0.0372 * t^\wedge 2 + 0.01144 * t^\wedge 3$

2 $m = 0.351 * t + 0.0104 * t^\wedge 2 - 0.0124 * t^\wedge 3 + 0.00286 * t^\wedge 4$

3 $D = (1375 * (\exp(-0.15 * t)) - 258 * (t^\wedge -0.5))$

4 $\mathrm{C} = 146 * (1 + (\exp(-0.1 * \mathrm{m} * \mathrm{T2})))$

5 $\mathrm{B} = (\exp(-0.1 * \mathrm{m})) * \mathrm{T1}$

6 $\mathrm{z} = (\mathrm{C} + \mathrm{B})/\mathrm{D}$

Model: $z = a0 + a1 * t + a2 * t^\wedge 2 + a3 * t^\wedge 3 + a4 * t^\wedge 4$

Variable	Value
$a0$	0.6763257
$a1$	−1.394036
$a2$	1.726777
$a3$	−0.7531454
$a4$	0.1144738

$$\textbf{Model}: \varphi(\theta) = 0.676 - 1.394 * \theta + 1.727 * \theta^\wedge 2 - 0.753 * \theta^\wedge 3 + 0.114 * \theta^\wedge 4 \tag{9.24}$$

Example 9.5 $\alpha = 0.33$

Calculated values of DEQ variables

	Variable	Initial value	Minimal value	Maximal value	Final value
1	B	0.9884539	0.9884539	1683.239	1683.239
2	C	292	146.0654	292	146.0654
3	D	538.6613	538.6613	931.0569	625.616
4	m	0.0351919	0.0351919	1.50896	1.50896
5	$m1$	0.3527194	0.3345935	0.57116	0.57116
6	t	0.1	0.1	4	4
7	$T1$	1	1	2769.527	2769.527
8	$T2$	0	0	15.48483	15.48483
9	y	0	−0.0738525	11.47169	11.47169
10	z	0.5439196	0.3596013	2.924005	2.924005

Differential equations

1 $d(T1)/d(t) = 258 * m1 * z * (\exp(0.33 * m)) * (\exp(t/(1 + 0.1 * t))) * ((t\wedge{-0.5}))$

2 $d(T2)/d(t) = 1 * m1 * (\exp(0.33 * m)) * (\exp(t/(1 + 0.1 * t)))$

3 $d(y)/d(t) = 1.035 - 5.74 * t + 8.015 * t\wedge 2 - 3.8 * t\wedge 3 + 0.613 * t\wedge 4$

Explicit equations

1 $m1 = 0.351 + 0.0208 * t\wedge 1 - 0.0372 * t\wedge 2 + 0.01144 * t\wedge 3$

2 $m = 0.351 * t + 0.0104 * t\wedge 2 - 0.0124 * t\wedge 3 + 0.00286 * t\wedge 4$

3 $D = (1375 * (\exp(-0.15 * t)) - 258 * (t\wedge{-0.5}))$

4 $C = 146 * (1 + (\exp(-0.33 * m * T2)))$

5 $B = (\exp(-0.33 * m)) * T1$

6 $z = (C + B)/D$

Model: $z = a0 + a1 * t + a2 * t\wedge 2 + a3 * t\wedge 3 + a4 * t\wedge 4$

Variable	Value
$a0$	0.6186426
$a1$	−1.076411
$a2$	1.300275
$a3$	−0.5631797
$a4$	0.0847898

Model: $\varphi(\theta) = 0.619 - 1.076 * \theta + 1.3 * \theta\wedge 2 - 0.563 * \theta\wedge 3 + 0.085 * \theta\wedge 4$ (9.25)

Example 9.6 $\alpha = 0.5$

Calculated values of DEQ variables

	Variable	Initial value	Minimal value	Maximal value	Final value
1	B	0.982558	0.982558	1385.418	1385.418
2	C	292	146.0001	292	146.0001
3	D	538.6613	538.6613	931.057	625.616
4	m	0.0351919	0.0351919	1.50896	1.50896
5	$m1$	0.3527194	0.3345935	0.57116	0.57116
6	t	0.1	0.1	4	4
7	$T1$	1	1	2946.1	2946.1

(continued)

(continued)

	Variable	Initial value	Minimal value	Maximal value	Final value
8	*T2*	0	0	18.67366	18.67366
9	*y*	0	−0.0738525	11.47169	11.47169
10	*z*	0.5439087	0.3577559	2.447857	2.447857

Differential equations

1 $\mathrm{d}(T1)/\mathrm{d}(t) = 258 * m1 * z * (\exp(0.5 * m)) * (\exp(t/(1 + 0.1 * t))) * ((t^\wedge -0.5))$

2 $\mathrm{d}(T2)/\mathrm{d}(t) = 1 * m1 * (\exp(0.5 * m)) * (\exp(t/(1 + 0.1 * t)))$

3 $\mathrm{d}(y)/\mathrm{d}(t) = 1.035 - 5.74 * t + 8.015 * t^\wedge 2 - 3.8 * t^\wedge 3 + 0.613 * t^\wedge 4$

Explicit equations

1 $m1 = 0.351 + 0.0208 * t^\wedge 1 - 0.0372 * t^\wedge 2 + 0.01144 * t^\wedge 3$

2 $m = 0.351 * t + 0.0104 * t^\wedge 2 - 0.0124 * t^\wedge 3 + 0.00286 * t^\wedge 4$

3 $D = (1375 * (\exp(-0.15 * t)) - 258 * (t^\wedge -0.5))$

4 $C = 146 * (1 + (\exp(-0.5 * m * T2)))$

5 $B = (\exp(-0.5 * m)) * T1$

6 $z = (C + B)/D$

Model: $z = a0 + a1 * t + a2 * t^\wedge 2 + a3 * t^\wedge 3 + a4 * t^\wedge 4$

Variable	Value
a0	0.5851962
a1	−0.8899544
a2	1.04724
a3	−0.4532186
a4	0.0686218

$$\varphi(\theta) = 0.585 - 0.89 * \theta + 1.047 * \theta^\wedge 2 - 0.453 * \theta^\wedge 3 + 0.0686 * \theta^\wedge 4 \quad (9.26)$$

Example 9.7 $\alpha = 0.75$

Calculated values of DEQ variables

	Variable	Initial value	Minimal value	Maximal value	Final value
1	*B*	0.9739514	0.9739514	1084.564	1084.564
2	*C*	292	146	292	146

(continued)

(continued)

	Variable	Initial value	Minimal value	Maximal value	Final value
3	D	538.6613	538.6613	931.0579	625.616
4	m	0.0351919	0.0351919	1.50896	1.50896
5	$m1$	0.3527194	0.3345936	0.57116	0.57116
6	t	0.1	0.1	4	4
7	$T1$	1	1	3363.217	3363.217
8	$T2$	0	0	24.72791	24.72791
9	y	0	−0.0738522	11.47169	11.47169
10	z	0.5438927	0.3548104	1.966963	1.966963

Differential equations

1 $\mathrm{d}(T1)/\mathrm{d}(t) = 258 * m1 * z * (\exp(0.75 * m)) * (\exp(t/(1 + 0.1 * t))) * ((t^\wedge -0.5))$

2 $\mathrm{d}(T2)/\mathrm{d}(t) = 1 * m1 * (\exp(0.75 * m)) * (\exp(t/(1 + 0.1 * t)))$

3 $\mathrm{d}(y)/\mathrm{d}(t) = 1.035 - 5.74 * t + 8.015 * t^\wedge 2 - 3.8 * t^\wedge 3 + 0.613 * t^\wedge 4$

Explicit equations

1 $m1 = 0.351 + 0.0208 * t^\wedge 1 - 0.0372 * t^\wedge 2 + 0.01144 * t^\wedge 3$

2 $m = 0.351 * t + 0.0104 * t^\wedge 2 - 0.0124 * t^\wedge 3 + 0.00286 * t^\wedge 4$

3 $D = (1375 * (\exp(-0.15 * t)) - 258 * (t^\wedge -0.5))$

4 $C = 146 * (1 + (\exp(-0.75 * m * T2)))$

5 $B = (\exp(-0.75 * m)) * T1$

6 $z = (C + B)/D$

Model: $z = a0 + a1 * t + a2 * t^\wedge 2 + a3 * t^\wedge 3 + a4 * t^\wedge 4$

Variable	Value
$a0$	0.5519859
$a1$	−0.6967452
$a2$	0.7697312
$a3$	−0.329472
$a4$	0.0503852

$$\varphi(\theta) = 0.552 - 0.697 * \theta + 0.77 * \theta^\wedge 2 - 0.33 * \theta^\wedge 3 + 0.05 * \theta^\wedge 4 \quad (9.27)$$

Example 9.8 $\alpha = 1$

Calculated values of DEQ variables

	Variable	Initial value	Minimal value	Maximal value	Final value
1	B	0.9654201	0.9654201	876.3342	876.3342
2	C	292	146	292	146
3	D	538.6613	538.6613	931.0585	625.616
4	m	0.0351919	0.0351919	1.50896	1.50896
5	$m1$	0.3527194	0.3345936	0.57116	0.57116
6	t	0.1	0.1	4	4
7	$T1$	1	1	3962.806	3962.806
8	$T2$	0	0	32.94283	32.94283
9	y	0	−0.073852	11.47169	11.47169
10	z	0.5438769	0.3411298	1.634124	1.634124

Differential equations

1 $\mathrm{d}(T1)/\mathrm{d}(t) = 258 * m1 * z * (\exp(1 * m)) * (\exp(t/(1 + 0.1 * t))) * ((t^\wedge{-0.5}))$

2 $\mathrm{d}(T2)/\mathrm{d}(t) = 1 * m1 * (\exp(1 * m)) * (\exp(t/(1 + 0.1 * t)))$

3 $\mathrm{d}(y)/\mathrm{d}(t) = 1.035 - 5.74 * t + 8.015 * t^\wedge 2 - 3.8 * t^\wedge 3 + 0.613 * t^\wedge 4$

Explicit equations

1 $m1 = 0.351 + 0.0208 * t^\wedge 1 - 0.0372 * t^\wedge 2 + 0.01144 * t^\wedge 3$

2 $m = 0.351 * t + 0.0104 * t^\wedge 2 - 0.0124 * t^\wedge 3 + 0.00286 * t^\wedge 4$

3 $D = (1375 * (\exp(-0.15 * t)) - 258 * (t^\wedge{-0.5}))$

4 $C = 146 * (1 + (\exp(-1 * m * T2)))$

5 $B = (\exp(-1 * m)) * T1$

6 $z = (C + B)/D$

Model: $z = a0 + a1 * t + a2 * t^\wedge 2 + a3 * t^\wedge 3 + a4 * t^\wedge 4$

Variable	Value
$a0$	0.5342948
$a1$	−0.5860701
$a2$	0.5931241
$a3$	−0.2470101
$a4$	0.0379162

Model: $\varphi(\theta) = 0.534 - 0.586 * \theta + 0.593 * \theta^\wedge 2 - 0.247 * \theta^\wedge 3 + 0.038 * \theta^\wedge 4$ (9.28)

Example 9.9 $\alpha = 10$

Calculated values of DEQ variables

	Variable	Initial value	Minimal value	Maximal value	Final value
1	B	0.7033372	0.7033372	66.51355	66.51355
2	C	292	146	292	146
3	D	538.6613	538.6613	931.0732	625.616
4	m	0.0351919	0.0351919	1.50896	1.50896
5	$m1$	0.3527194	0.3345969	0.57116	0.57116
6	t	0.1	0.1	4	4
7	$T1$	1	1	2.378E+08	2.378E+08
8	$T2$	0	0	5.663E+06	5.663E+06
9	y	0	−0.073806	11.47169	11.47169
10	z	0.5433903	0.1698848	0.5433903	0.3396869

Differential equations

1 $d(T1)/d(t) = 258 * m1 * z * (\exp(10 * m)) * (\exp(t/(1 + 0.1 * t))) * ((t^\wedge - 0.5))$

2 $d(T2)/d(t) = 1 * m1 * (\exp(10 * m)) * (\exp(t/(1 + 0.1 * t)))$

3 $d(y)/d(t) = 1.035 - 5.74 * t + 8.015 * t^\wedge 2 - 3.8 * t^\wedge 3 + 0.613 * t^\wedge 4$

Explicit equations

1 $m1 = 0.351 + 0.0208 * t^\wedge 1 - 0.0372 * t^\wedge 2 + 0.01144 * t^\wedge 3$

2 $m = 0.351 * t + 0.0104 * t^\wedge 2 - 0.0124 * t^\wedge 3 + 0.00286 * t^\wedge 4$

3 $D = (1375 * (\exp(-0.15 * t)) - 258 * (t^\wedge - 0.5))$

4 $C = 146 * (1 + (\exp(-10 * m * T2)))$

5 $B = (\exp(-10 * m)) * T1$

6 $z = (C + B)/D$

Model: $z = a0 + a1 * t + a2 * t^\wedge 2 + a3 * t^\wedge 3 + a4 * t^\wedge 4$

Variable	Value
$a0$	0.5638095
$a1$	−0.9147177
$a2$	0.6940812
$a3$	−0.2083114
$a4$	0.0222236

Model: $\varphi(\theta) = 0.564 - 0.915 * \theta + 0.694 * \theta^\wedge 2 - 0.208 * \theta^\wedge 3 + 0.022 * \theta^\wedge 4$ (9.29)

Example 9.10 $\alpha = 100$

Calculated values of DEQ variables

	Variable	Initial value	Minimal value	Maximal value	Final value
1	B	0.0296235	0.0296235	5.381669	5.381669
2	C	292	146	292	146
3	D	538.6613	538.6613	931.0694	625.616
4	m	0.0351919	0.0351919	1.50896	1.50896
5	$m1$	0.3527194	0.3345931	0.57116	0.57116
6	t	0.1	0.1	4	4
7	$T1$	1	1	1.837E+66	1.837E+66
8	$T2$	0	0	5.892E+64	5.892E+64
9	y	0	−0.0738286	11.47169	11.47169
10	z	0.5421396	0.1578258	0.5421396	0.2419722

Differential equations

1 $d(T1)/d(t) = 258 * m1 * z * (\exp(100 * m)) * (\exp(t/(1 + 0.1 * t))) * ((t^\wedge -0.5))$

2 $d(T2)/d(t) = 1 * m1 * (\exp(100 * m)) * (\exp(t/(1 + 0.1 * t)))$

3 $d(y)/d(t) = 1.035 - 5.74 * t + 8.015 * t^\wedge 2 - 3.8 * t^\wedge 3 + 0.613 * t^\wedge 4$

Explicit equations

1 $m1 = 0.351 + 0.0208 * t^\wedge 1 - 0.0372 * t^\wedge 2 + 0.01144 * t^\wedge 3$

2 $m = 0.351 * t + 0.0104 * t^\wedge 2 - 0.0124 * t^\wedge 3 + 0.00286 * t^\wedge 4$

3 $D = (1375 * (\exp(-0.15 * t)) - 258 * (t^\wedge -0.5))$

4 $C = 146 * (1 + (\exp(-100 * m * T2)))$

5 $B = (\exp(-100 * m)) * T1$

6 $z = (C + B)/D$

Model: $z = a0 + a1 * t + a2 * t^\wedge 2 + a3 * t^\wedge 3 + a4 * t^\wedge 4$

Variable	Value
$a0$	0.383539
$a1$	−0.6015205
$a2$	0.4934402
$a3$	−0.1545575
$a4$	0.0167409

Model: $\varphi(\theta) = 0.384 - 0.602 * \theta + 0.493 * \theta^\wedge 2 - 0.155 * \theta^\wedge 3 + 0.0167 * \theta^\wedge 4$ (9.30)

The summary of all computations above graphically presented in Fig. 9.23. The deformation process in this case is considered again (similar to stress–temperature–strain relationships in Chap. 8) as an ergodic random process, therefore one realization is sufficient enough to compute the mean value; standard deviation and autocorrelation function [12, 13].

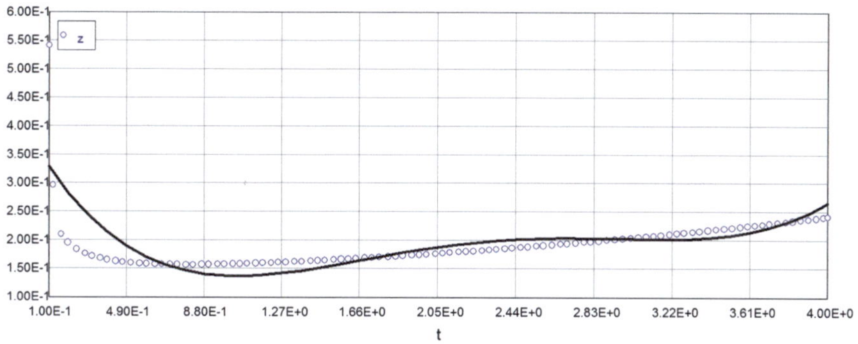

Fig. 9.22 Deformation—temperature (time) curve (α = 100—approximation)

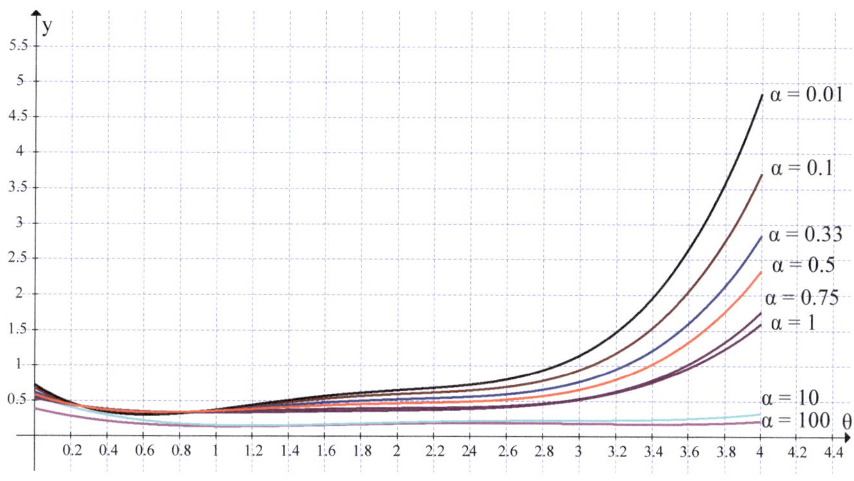

Fig. 9.23 Deformation—temperature (time) curve (summary)

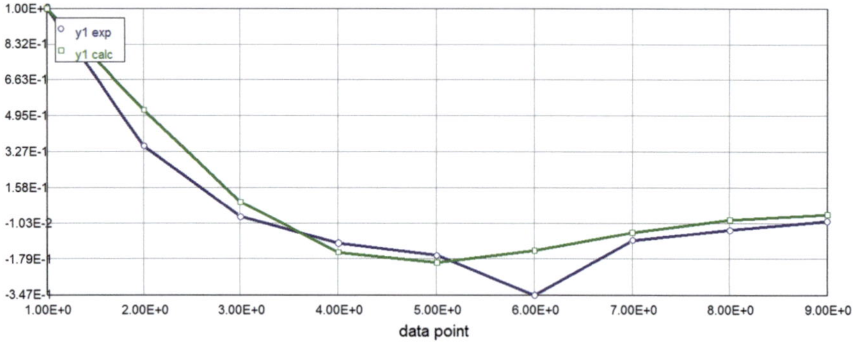

Fig. 9.24 Normalized autocorrelation function

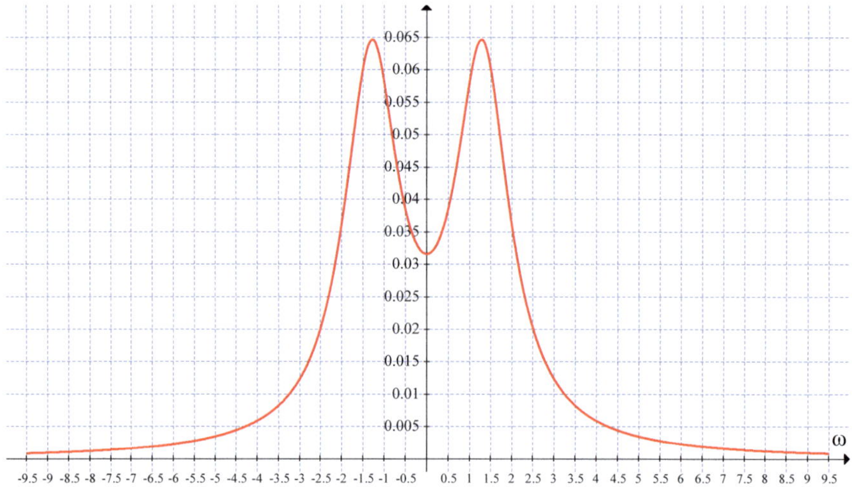

Fig. 9.25 Spectral density function

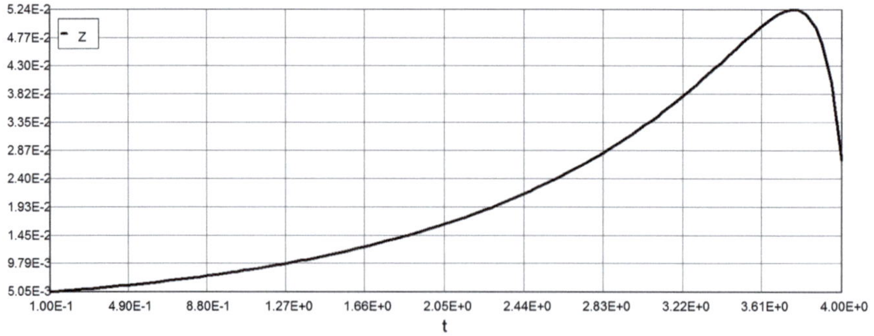

Fig. 9.26 Beam deflections (maximum) versus temperature. Case 3

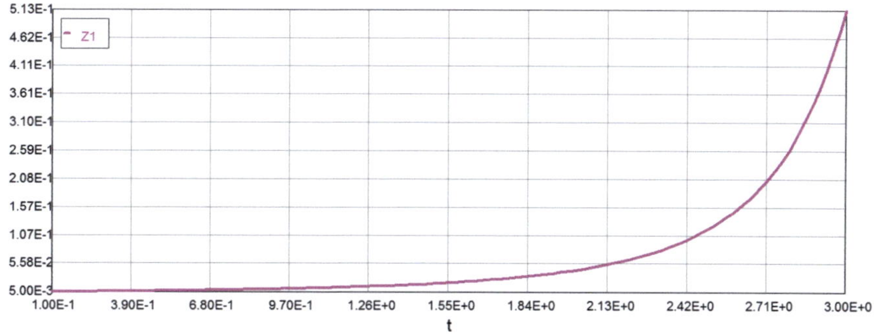

Fig. 9.27 Soil reactions versus temperature. Case 3

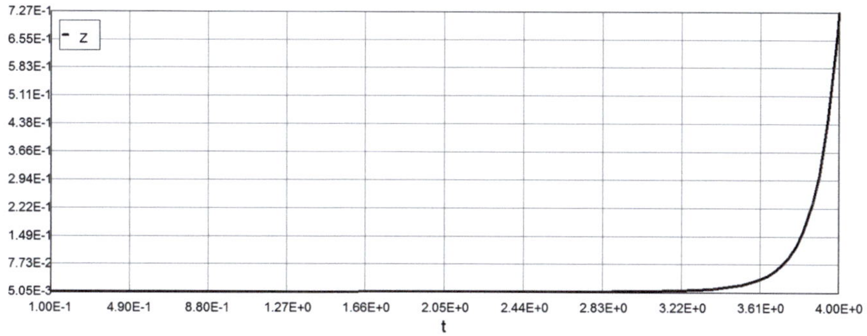

Fig. 9.28 Beam deflections (maximum) versus temperature. Case 2

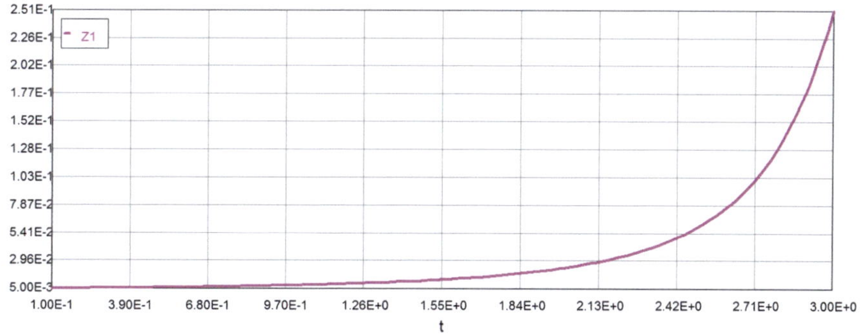

Fig. 9.29 Soil reactions versus temperature. Case 2

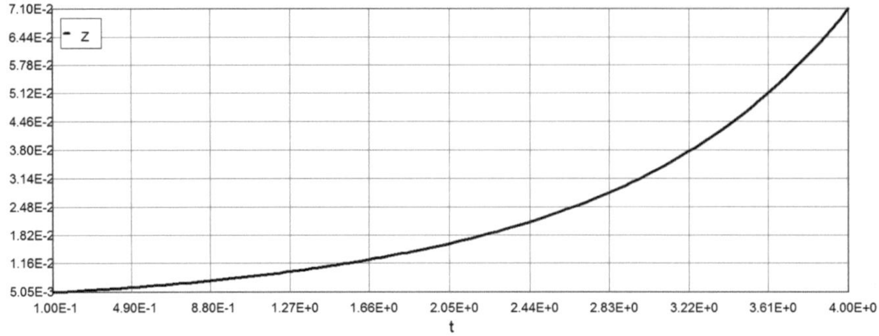

Fig. 9.30 Beam deflections (maximum) versus temperature. Case 1

9.4 Design for Stochastic Behavior of Structures

9.4.1 Introduction

Eventually, further progress will require the design engineers to recognize that all the effects in structures discussed above are in reality stochastic. The statistical variability of creep is enormous, because the fire is the most unpredictable phenomena. It is desirable to develop design methods which are based on certain extreme deformation values that are exceeded only with a small specified probability such as 5 %.

The deterministic approach again is the base for obtaining the statistical information. The ergodic random process is assumed in this chapter by inspecting the graphical presentation on Fig. 9.23.

Another practical method based on response surface fitting has been developed by Wiurn and Buyukozturk [10]. They applied the method to calculate the variability of the long-term deformation in a multiple-span bridge. The results indicate a large variability in the response due to randomness of the factors that influence creep. The major source of uncertainty is found to be the variability of the material properties parameters (MPP).

Various observations of creep effects in structures have been made in the past. However, comparisons of the results of structural fire resistance analysis with measurements on structures have so far been of limited usefulness since typically only some of the intervening effects were analyzed while others have been either ignored or taken into account in an oversimplified manner. Successful verification of probability-based approach in structural design in general still is the very important issue and the fundamental question arises as to the meaning of such verification since we are not looking anymore for any given test result, but we are looking rather for an unknown value that is covered by the so-called confidence limit (interval) I_α of mean value (μ_σ) and standard deviation (σ_σ), where confidence probability P_α is given and cannot be computed by any means of applied probability

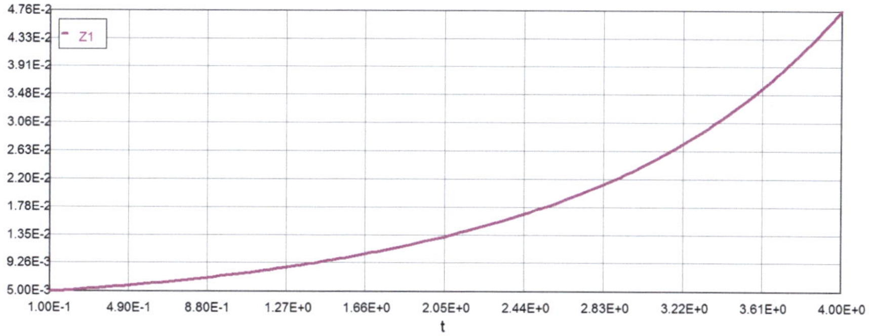

Fig. 9.31 Soil reactions versus temperature. Case 1

theory. Consider now the deflection—temperature function corresponding to parameter $\alpha = 0.5$.

$$y = \varphi(\theta) = 1.035 - 5.74 * \theta + 8.015 * \theta\wedge 2 - 3.8 * \theta\wedge 3 + 0.613 * \theta\wedge 4 \quad (9.31)$$

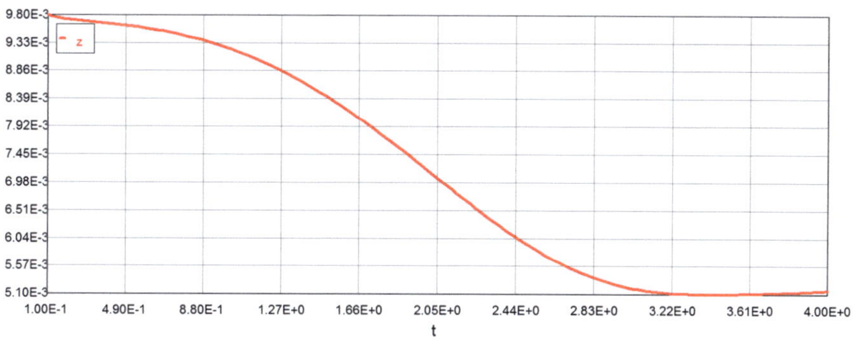

Fig. 9.32 Deformation—temperature (time) curve. Ergodic process

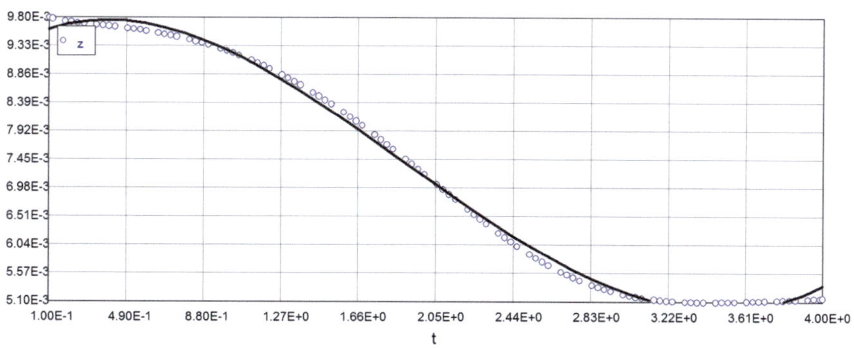

Fig. 9.33 Deformation—temperature (time) curve. Ergodic process approximation

9.4.2 Stochastic Behavior of Structures

The mean value can be calculated now as follows:

$$\begin{aligned}\boldsymbol{\mu}_y &= \frac{1}{4}\int_0^4 y(\boldsymbol{\theta})d\boldsymbol{\theta}\\ &= \frac{1}{4}\int_0^4 \left(\mathbf{1.035} - \mathbf{5.74} * \boldsymbol{\theta} + \mathbf{8.015} * \boldsymbol{\theta}\hat{}\mathbf{2} - \mathbf{3.8} * \boldsymbol{\theta}\hat{}\mathbf{3} + \mathbf{0.613} * \boldsymbol{\theta}\hat{}\mathbf{4}\right)d\boldsymbol{\theta} \quad (9.32)\\ &= \frac{2.56}{4} = 0.64; \quad \boldsymbol{\mu}_y = q_0(s)0.64 = 0.1(10)(0.64) = 0.64' = 7.68''\end{aligned}$$

$s = 10'$ – beam spacing

In order to calculate the correlation function in this case the chosen function (9.31) has to be centered (see Table 9.1). Again, after using the POLYMATH software we have:

Model: $y = a0 + a1 * x + a2 * x\hat{}2 + a3 * x\hat{}3 + a4 * x\hat{}4$

Variable	Value
$a0$	−0.608
$a1$	0.529
$a2$	−0.053
$a3$	−0.125
$a4$	0.0356

$$y^* = -0.608 + 0.529(\theta) - 0.053(\theta^2) - 0.125(\theta^3) + 0.0356(\theta^4) \quad (9.33)$$

The correlation function of a steady time process can be computed now as follows [14]:

$$\mathrm{K_y}(\boldsymbol{\theta}) = \frac{1}{4-\boldsymbol{\theta}}\int_0^{4-\boldsymbol{\theta}} [\mathrm{y} * (\mathrm{t})\mathrm{y} * (t+\boldsymbol{\theta})]dt \quad (9.34)$$

where $\boldsymbol{\theta} = 0; 0.5; 1; \ldots 4$

Table 9.1 Centered correlation function

$\alpha\|\theta$	0	0.5	1	1.5	2	2.5	3	3.5	4
0.5	0	0.358	0.376	0.407	0.446	0.519	0.693	1.129	2.447
y^*	−0.64	−0.282	−0.264	−0.233	−0.194	−0.121	0.053	0.489	1.807

Table 9.2 Autocorrelation function

θ	0	0.5	1	1.5	2	2.5	3	3.5	4
$K_y(\theta)$	0.228	0.081	0.0047	−0.0234	−0.0364	−0.0792	−0.0205	−0.0436	−0.0835

Table 9.3 Normalized autocorrelation function

	0	0.5	1	1.5	2	2.5	3	3.5	4
$\rho_y(\tau)$	1.0	0.355	0.021	−0.103	−0.16	−0.347	−0.09	−0.191	−0.366

After substituting (9.33) into (9.34) we have:
The standard deviation $D = K_y(0)$ is

$$D = K_y(\theta = 0) = \frac{1}{4-\theta}\int_0^{4-\theta}[y*(t)y*(t+\theta)]\mathrm{d}t = \frac{1}{4}\int_0^4 [y*(t)]^2\mathrm{d}t$$

$$= 0.25\int_0^4 \left[\mathbf{-0.608 + 0.529(\theta) - 0.053(\theta^2) - 0.125(\theta^3) + 0.0356(\theta^4)}\right]^2\mathrm{d}\theta$$

$$= 0.25(0.91) = 0.228; \quad \sigma_y = 0.477$$

The computations of autocorrelation function based on Eq. (9.34) are presented below (see Table 9.2).

The normalized autocorrelation function is (see Table 9.3)

Let us approximate the data from Table 9.3 by the type of formulae (Fig. 9.24):

Model: $y1 = (\exp(-a*x))*(\cos(b*x))$

Variable	Initial guess	Value
a	1	0.78
b	1	1.372

$$\rho_y(\tau) = (\cos(1.372*\tau))*(\exp(-0.78*\tau)) \tag{9.35}$$

The corresponding spectral function is [13]:

$$\begin{aligned} S(\omega) &= \frac{1}{2}\frac{\sigma_y^2}{\pi}\left[\frac{a}{(\omega-b)^2+a^2} + \frac{a}{(\omega+b)^2+a^2}\right] \\ &= 3.63(10^{-2})\left[\frac{1}{(\omega-1.372)^2+0.78^2} + \frac{1}{(\omega+1.372)^2+0.78^2}\right] \end{aligned} \tag{9.36}$$

The graphic presentation of spectral density is (see Fig. 9.25).

9.4.3 *The First-Occurrence Time Problem and the Probability Density* $\mathbf{P(a_0, \theta)}$

The average first-occurrence of deflection y above given level "y = a_0" for stationary deflection—temperature processes is defined as follows (see Chap. 7) [13]:

For correlation function given by (9.35):

$$\begin{aligned}
\sigma_{vy}^2 &= K_{\dot{y}}(0) = -\frac{d^2}{d\theta^2}K_y(\theta)|_{\theta=0} = \sigma_\sigma^2(b^2 - a^2) \\
\bar{n}_a &= \theta_{\max}\frac{\sqrt{b^2 - a^2}}{2\pi}\exp\left(-\frac{a_0^2}{2\sigma_\sigma^2}\right) = 4\frac{\sqrt{1.372^2 - 0.78^2}}{2\pi}\exp\left(-\frac{1.2^2}{2(0.477)^2}\right) = 0.03 \\
\bar{\mathbf{v}}_a &= \frac{\bar{n}_a}{\theta_{\max}} = \frac{0.03}{4} = 0.0075
\end{aligned} \tag{9.37}$$

Now based on Poisson formulae of the probability to not having the maximum beam deflection ordinates crossing upwards the level a = 1.2 is: $P_{\text{rel}} = P_o = \exp(-\bar{n}_a) = \exp(-0.03) = 0.97$

This probability characterizes the reliability of the structure (one element or the whole structure).

9.5 Beams on Elastic Foundation

Equation (9.13) can be rewritten for a beam on elastic (Winkler type) foundation as follows:

Case 1 If the beam with the properties of the linear creep is supported on an elastic foundation Winkler type, then Eq. (9.13) can be written as

$$\begin{aligned}
&E(\theta)I\frac{\partial^4 y(x,\theta)}{\partial x^4} + k(\theta)y(x,\theta) + \int_0^\theta k(\tau)y(x,\tau)K(\theta,\tau)m1\,d\tau \\
&= q(x,\theta) + \int_0^\theta q(x,\tau)K(\theta,\tau)m1\,d\tau
\end{aligned} \tag{9.38}$$

$$K(\theta,\tau,\alpha_i) = e^{-\alpha_i\theta}[\exp(\theta/(1+0.1\theta)]e^{\alpha_i\tau}m1$$

$$E(\theta) = E_0\rho(\theta) = E_0[\exp(-0.15\theta)];$$

$q(x,\theta) = X(x)T(\theta)$—distributed load is a given function of x—coordinate and temperature $k(\theta) = k_0[\exp(-0.1\theta)]$—Winkler's coefficient is a given function of temperature only.

$$y(x,\boldsymbol{\theta}) = y_0 T(\boldsymbol{\theta}) \sum_{n=1}^{\infty} \sin\frac{n\pi x}{L}$$
$$q(x,\boldsymbol{\theta}) = q_0 Q(\boldsymbol{\theta}) \sum_{s=1}^{\infty} \sin\frac{s\pi x}{L} \tag{9.39}$$

$$\left\{E(\boldsymbol{\theta})\mathbf{I}\frac{y_0 n^4 \boldsymbol{\pi}^4}{L^4} + y_0 k_0[\exp(-0.1\boldsymbol{\theta})]\right\}T(\boldsymbol{\theta}) + y_0 \int_0^{\boldsymbol{\theta}} k_0[\exp(-0.1\boldsymbol{\tau})]T(\boldsymbol{\tau})K(\boldsymbol{\theta},\boldsymbol{\tau})m1\mathrm{d}\boldsymbol{\tau}$$
$$= q_0\left[Q(\boldsymbol{\theta}) + \int_0^{\boldsymbol{\theta}} Q(\boldsymbol{\tau})K(\boldsymbol{\theta},\boldsymbol{\tau})m1\mathrm{d}\boldsymbol{\tau}\right]$$
$$\{1+A\}T(\boldsymbol{\theta}) + B\int_0^{\boldsymbol{\theta}} [\exp(-0.1\boldsymbol{\tau})]T(\boldsymbol{\tau})K(\boldsymbol{\theta},\boldsymbol{\tau})m1\mathrm{d}\boldsymbol{\tau}$$
$$= C\left[Q(\boldsymbol{\theta}) + \int_0^{\boldsymbol{\theta}} Q(\boldsymbol{\tau})K(\boldsymbol{\theta},\boldsymbol{\tau})m1\mathrm{d}\boldsymbol{\tau}\right];$$
$$A = \frac{L^4 k_0[\exp(-0.1\boldsymbol{\theta})]}{n^4\boldsymbol{\pi}^4 I E_0 \boldsymbol{\rho}(\boldsymbol{\theta})}; \quad B = \frac{A}{[\exp(-0.1\boldsymbol{\theta})]}; \quad C = \frac{L^4 q_0}{n^4\boldsymbol{\pi}^4 y_0 E(\boldsymbol{\theta}) I}$$
$$K(\boldsymbol{\theta},\boldsymbol{\tau},\boldsymbol{\alpha}_i) = \mathrm{e}^{-\boldsymbol{\alpha}_i\boldsymbol{\theta}}[\exp(\boldsymbol{\theta}/(1+0.1\boldsymbol{\theta})]\mathrm{e}^{\boldsymbol{\alpha}_i\boldsymbol{\tau}}m1$$
$$E(\boldsymbol{\theta}) = E_0\boldsymbol{\rho}(\boldsymbol{\theta}) = E_0[\exp(-0.15\boldsymbol{\theta})]; \quad n = 1,2,\ldots,N \tag{9.40}$$

Case 2 If the elastic beam is supported on an elastic foundation with the properties of the linear creep, then Eq. (9.13) can be written as

$$E(\boldsymbol{\theta})\mathbf{I}\frac{\partial^4 y(x,\boldsymbol{\theta})}{\partial x^4} = q(x,\boldsymbol{\theta}) - p(x,\boldsymbol{\theta}) \tag{9.41}$$

where

$$p(x,\theta) = k(\theta)y(x,\theta) - \int_0^{\theta} p(x,\tau)K_0(\theta,\tau)m1\mathrm{d}\tau$$
$$K_0(\theta,\tau,\beta_i) = \mathrm{e}^{-\beta_i\theta}[\exp(\theta/(1+0.1\theta)]e^{\beta_i\tau}m1$$

Expressing $p(x, \theta)$ from the second Eq. (9.41) and substituting in the first equation, we obtain the following integral-differential equation ($k(\theta)$—instantaneous subgrade modulus of elasticity).

$$IE(\theta)\frac{\partial^4 y(x,\theta)}{\partial x^4}+k(\theta)y(x,\theta)+\int_0^{\theta} IE(\tau)\frac{\partial^4 y(x,\tau)}{\partial x^4}K_0(\theta,\tau)m1\mathrm{d}\tau$$
$$=q(x,\theta)+\int_0^{\theta} q(x,\tau)K_0(\theta,\tau)m1\mathrm{d}\tau \tag{9.42}$$

We use again the load and deflections expansion in the series (9.39). Then for nth term of the expansion after canceling out $\sin(n\pi x/L)$, we obtain:

$$\left\{E(\theta)I\frac{y_0 n^4\pi^4}{L^4}+y_0k_0[\exp(-0.1\theta)]\right\}T(\theta)-\frac{IE_0n^2\pi^2}{y_0L^2}\int_0^{\theta}[\exp(-0.15\tau)]T(\tau)K(\theta,\tau)m1\mathrm{d}\tau$$
$$=q_0\left[Q(\theta)+\int_0^{\theta}Q(\tau)K(\theta,\tau)m1\mathrm{d}\tau\right]$$
$$\{1+A\}T(\theta)-B\int_0^{\theta}[\exp(-0.15\tau)]T(\tau)K(\theta,\tau)m1\mathrm{d}\tau$$
$$=C\left[Q(\theta)+\int_0^{\theta}Q(\tau)K(\theta,\tau)m1\mathrm{d}\tau\right]$$
$$A=\frac{L^4k_0[\exp(-0.1\theta)]}{n^4\pi^4IE_0\rho(\theta)};\quad B=A\frac{IE_0n^2\pi^2}{y_0L^2[\exp(-0.15\theta)]}=\frac{L^2k_0[\exp(-0.1\theta)]}{y_0n^2\pi^2\rho(\theta)};$$
$$C=\frac{L^4q_0}{n^4\pi^4y_0E(\theta)I};\quad n=1,2,\ldots,N$$
$$K_0(\theta,\tau,\beta_i)=\mathrm{e}^{-\beta_i\theta}[\exp(\theta/(1+0.1\theta)]\mathrm{e}^{\beta_i\tau}m1 \tag{9.43}$$

Case 3 (*K* $\neq$ *0; K*$_0$ $\neq$ *0)* If both the elastic beam and the foundation have the properties of the linear creep, then Eq. (9.12) can be written as:

$$IE(\theta)\frac{\partial^4 y(x,\theta)}{\partial x^4}+p(x,\theta)+\int_0^{\theta}p(x,\tau)K(\theta,\tau)m1\mathrm{d}\tau$$
$$=q(x,\theta)+\int_0^{\theta}q(x,\tau)K(\theta,\tau)m1\mathrm{d}\tau \tag{9.44}$$

$$k(\theta)y(x,\theta)=p(x,\theta)+\int_0^{\theta}p(x,\tau)K_0(\theta,\tau)m1\mathrm{d}\tau \tag{9.45}$$

We use again the load, soil reaction and deflections expansion in the series (9.39). Then for nth term of the expansion after canceling out $\sin(n\pi x/L)$, we obtain:

$$\frac{IE(\theta)n^4\pi^4}{L^4}T(\theta)+p_0[P(\theta)]+p_0\int_0^{\theta}P(\tau)K(\theta,\tau)m1\mathrm{d}\tau$$
$$=q_0\left[Q(\theta)+\int_0^{\theta}Q(\tau)K(\theta,\tau)m1\mathrm{d}\tau\right]$$
$$\frac{y_0k_0}{p_0}[k(\theta)T(\theta)]=P(\theta)+\int_0^{\theta}P(\tau)K_0(\theta,\tau)m1\mathrm{d}\tau \tag{9.46}$$
$$K(\theta,\tau,\alpha_i)=\mathrm{e}^{-\alpha_i\theta}[\exp(\theta/(1+0.1\theta)]\mathrm{e}^{\alpha_i\tau}m1$$
$$K_0(\theta,\tau,\beta_i)=\mathrm{e}^{-\beta_i\theta}[\exp(\theta/(1+0.1\theta)]\mathrm{e}^{\beta_i\tau}m1$$
$$n=1,2,\ldots,N$$

Here are some design examples illustrating the effect of linear creep on beam design supported on a Winkler type elastic foundation.

Example 9.11 Data:

For beam with width b, we use $p = k_0y = k_0by = 200(1.0)y = 200y$ (k/ft^2)
Beam W21x50; $L = 30'$; $q_0 = 1$ k/ft; $I = 984$ in^4; $E_0 = 2.9(10^4)$ ksi; $b = 1.0'$

$$E(\theta)=E_0[\exp(-0.15\theta)];\quad k(\theta)=k_0[\exp(-0.1\theta)]$$

Computer Code and the solution of Eq. (9.46) in this case (using POLYMATH software) is as follows:

Case 3 **Calculated values of DEQ variables**

	Variable	Initial value	Minimal value	Maximal value	Final value
1	m	0.0351919	0.0351919	1.50896	1.50896
2	$m1$	0.3527194	0.3345937	0.57116	0.57116
3	Q	0	0	18.67366	18.67366
4	t	0.1	0.1	4	4
5	Z	0.0050461	0.0050461	0.052439	0.027211
6	$Z1$	0.0049959	0.0049959	194.4128	194.4128
7	$Z2$	0	0	426.4185	426.4185
8	$Z3$	0	0	879.0572	879.0572

Differential equations

1 $d(Z2)/d(t) = 1 * m1 * Z1 * (\exp(0.5 * m)) * (\exp(t/(1 + 0.1 * t)))$
2 $d(Z3)/d(t) = 1 * m1 * Z1 * (\exp(1 * m)) * (\exp(t/(1 + 0.1 * t)))$
3 $d(Q)/d(t) = 1 * m1 * 1 * (\exp(0.5 * m)) * (\exp(t/(1 + 0.1 * t)))$

Explicit equations

1 $m1 = 0.351 + 0.0208 * t\wedge 1 - 0.0372 * t\wedge 2 + 0.01144 * t\wedge 3$
2 $m = 0.351 * t + 0.0104 * t\wedge 2 - 0.0124 * t\wedge 3 + 0.00286 * t\wedge 4$
3 $Z1 = (\exp(-1 * m)) * Z3 * 1 + 1 * (\exp(-0.10 * t)) * (1 + (\exp(-0.5 * m)) * Q + (\exp(-1 * m)) * Z3 * 1 - (\exp(-0.5 * m)) * Z2) * (0.166 * (\exp(-0.15 * t)) + 200 * (\exp(-0.10 * t)))\wedge -1$
4 $Z = (1 + (\exp(-0.5 * m)) * Q + (\exp(-1 * m)) * Z3 * 1 - (\exp(-0.5 * m)) * Z2) * (0.166 * (\exp(-0.15 * t)) + 200 * (\exp(-0.10 * t)))\wedge -1$

Beam deflections (maximum) are (Fig. 9.26):

$$\textbf{Model}: T(\theta) = Z = 0 + 0.0283 * \theta - 0.0318 * \theta\wedge 2 + 0.0145 * \theta\wedge 3 - 0.0019 * \theta\wedge 4 \quad (9.47)$$

Variable	Value
$a0$	0
$a1$	0.0283
$a2$	−0.0318
$a3$	0.0145
$a4$	−0.0019

Soil reactions versus temperature are (Fig. 9.27):

$$\textbf{Model}: P(\theta) = Z1 = 0.0346 - 0.212 * \theta + 0.387 * \theta\wedge 2 - 0.243 * \theta\wedge 3 + 0.051 * \theta\wedge 4 \quad (9.48)$$

Variable	Value
$a0$	0.0346
$a1$	−0.212
$a2$	0.387

(continued)

(continued)

Variable	Value
$a3$	−0.243
$a4$	0.0510

Case 2 (K = 0)

Calculated values of DEQ variables

	Variable	Initial value	Minimal value	Maximal value	Final value
1	m	0.0351919	0.0351919	1.50896	1.50896
2	$m1$	0.3527194	0.3345937	0.57116	0.57116
3	Q	0	0	0	0
4	t	0.1	0.1	4	4
5	Z	0.0050461	0.0050461	0.7271618	0.7271618
6	$Z1$	0.0049959	0.0049959	97.03991	97.03991
7	$Z2$	0	0	0	0
8	$Z3$	0	0	436.6128	436.6128

Differential equations

1 $\mathrm{d}(Z2)/\mathrm{d}(t) = 1 * m1 * Z1 * (\exp(0.5 * m)) * (\exp(t/(1 + 0.1 * t)))$

2 $\mathrm{d}(Z3)/\mathrm{d}(t) = 0 * m1 * Z1 * (\exp(1 * m)) * (\exp(t/(1 + 0.1 * t)))$

3 $\mathrm{d}(Q)/\mathrm{d}(t) = 1 * m1 * 1 * (\exp(0.5 * m)) * (\exp(t/(1 + 0.1 * t)))$

Explicit equations

1 $m1 = 0.351 + 0.0208 * t\hat{}1 - 0.0372 * t\hat{}2 + 0.01144 * t\hat{}3$

2 $m = 0.351 * t + 0.0104 * t\hat{}2 - 0.0124 * t\hat{}3 + 0.00286 * t\hat{}4$

3 $Z1 = (\exp(-1 * m)) * Z3 * 1 + 1 * (\exp(-0.10 * t)) * (1 + (\exp(-0.5 * m)) * Q + (\exp(-1 * m)) * Z3 * 1 - (\exp(-0.5 * m)) * Z2) * (0.166 * (\exp(-0.15 * t)) + 200 * (\exp(-0.10 * t)))\hat{}-1$

4 $Z = (1 + (\exp(-0.5 * m)) * Q + (\exp(-1 * m)) * Z3 * 1 - (\exp(-0.5 * m)) * Z2) * (0.166 * (\exp(-0.15 * t)) + 200 * (\exp(-0.10 * t)))\hat{}-1$

Beam deflections (maximum) are (Fig. 9.28):

Model: $T(\theta) = Z = 0.0893 - 0.452 * \theta + 0.58 * \theta\hat{}2 - 0.258 * \theta\hat{}3 + 0.0367 * \theta\hat{}4$ (9.49)

Variable	Value
$a0$	0.0893
$a1$	−0.452
$a2$	0.580
$a3$	−0.258
$a4$	0.0367

Soil reactions versus temperature are (Fig. 9.29):

$$\textbf{Model}: P(\theta) = Z1 = 11.36 - 60.99 * \theta + 78.14 * \theta^\wedge 2 - 34.76 * \theta^\wedge 3 + 4.952 * \theta^\wedge 4 \tag{9.50}$$

Variable	Value
$a0$	11.36
$a1$	−60.99
$a2$	78.14
$a3$	−34.76
$a4$	4.952

Case 1 ($K_0 = 0$)

Calculated values of DEQ variables

	Variable	Initial value	Minimal value	Maximal value	Final value
1	m	0.0351919	0.0351919	1.50896	1.50896
2	$m1$	0.3527194	0.3345937	0.57116	0.57116
3	Q	0	0	18.67366	18.67366
4	t	0.1	0.1	4	4
5	Z	0.0050461	0.0050461	0.0710063	0.0710063
6	$Z1$	0.0049959	0.0049959	0.047597	0.047597
7	$Z2$	0	0	0.5433591	0.5433591
8	$Z3$	0	0	0	0

Differential equations

1 $\mathrm{d}(Z2)/\mathrm{d}(t) = 1 * m1 * Z1 * (\exp(0.5 * m)) * (\exp(t/(1 + 0.1 * t)))$

2 $\mathrm{d}(Z3)/\mathrm{d}(t) = 0 * m1 * Z1 * (\exp(1 * m)) * (\exp(t/(1 + 0.1 * t)))$

3 $\mathrm{d}(Q)/\mathrm{d}(t) = 1 * m1 * 1 * (\exp(0.5 * m)) * (\exp(t/(1 + 0.1 * t)))$

Explicit equations

1 $m1 = 0.351 + 0.0208 * t^\wedge 1 - 0.0372 * t^\wedge 2 + 0.01144 * t^\wedge 3$

2 $m = 0.351 * t + 0.0104 * t^\wedge 2 - 0.0124 * t^\wedge 3 + 0.00286 * t^\wedge 4$

3 $Z1 = (\exp(-1 * m)) * Z3 * 1 + 1 * (\exp(-0.10 * t)) * (1 + (\exp(-0.5 * m)) * Q + (\exp(-1 * m)) * Z3 * 1 - (\exp(-0.5 * m)) * Z2) * (0.166 * (\exp(-0.15 * t)) + 200 * (\exp(-0.10 * t)))^\wedge -1$

4 $Z = (1 + (\exp(-0.5 * m)) * Q + (\exp(-1 * m)) * Z3 * 1 - (\exp(-0.5 * m)) * Z2) * (0.166 * (\exp(-0.15 * t)) + 200 * (\exp(-0.10 * t)))^\wedge -1$

Beam deflections (maximum) are (Fig. 9.30):

$$\textbf{Model}: T(\theta) = Z = 0.005 - 0.000184 * \theta + 0.00462 * \theta^\wedge 2 - 0.00175 * \theta^\wedge 3 + 0.000408 * \theta^\wedge 4 \quad (9.51)$$

Variable	Value
$a0$	0.005
$a1$	−0.000184
$a2$	0.00462
$a3$	−0.00175
$a4$	0.000408

Soil reactions versus temperature are (Fig. 9.31):

$$\textbf{Model}: P(\theta) = Z1 = 0.005 - 0.000513 * \theta + 0.00269 * \theta^\wedge 2 - 0.00092 * \theta^\wedge 3 + 0.00022 * \theta^\wedge 4 \quad (9.52)$$

Variable	Value
$a0$	0.005
$a1$	0.000513
$a2$	0.00269
$a3$	−0.00092
$a4$	0.00022

9.6 Beams on Elastic Foundation Subjected to High Temperature Load

Equation (9.44) differs from the formula (9.13) only by the presence of the additional term K_w. Solutions of the Eq. (9.44) also represent a realization of a random process, depending on parameters α_i. Also as before, we assume that the process is ergodic, so we can limit our analysis to only one realization ($\alpha_i = 0.5$) for the subsequent calculation of the mean value, standard deviation and correlation function parameters.

$$E(\theta)I\frac{\partial^4 y(x,\theta)}{\partial x^4} + P(\theta)\frac{\partial^2 y(x,\theta)}{\partial x^2} + yk_w + \int_0^\theta P(\tau)\frac{\partial^2 y(x,\tau)}{\partial x^2}K(\theta,\tau)m1\mathrm{d}\tau$$

$$= q(x,\theta) + \int_0^\theta q(\tau)K(\theta,\tau)m1\mathrm{d}\tau \tag{9.53}$$

$$K(\theta,\tau,\alpha_i) = \mathrm{e}^{-\alpha_i\theta}[\exp(\theta/(1+0.1\theta)]\mathrm{e}^{\alpha_i\tau}m1$$

$$E(\theta) = E_0\rho(\theta) = E_0[\exp(-0.15\theta)];$$

$$P(\theta) \approx \frac{qL}{\sqrt{24\varepsilon}} = \frac{W}{\sqrt{24(7.02)10^{-4}\theta}} = 7.7W\theta^{-0.5}$$

k_w – Winkler foundation modulus of elasticity

$$E(\theta)I\frac{\pi^4}{L^4}y(\theta) - P(\theta)\frac{\pi^2}{L^2}y(\theta) + k_w y(\theta) - \frac{\pi^2}{L^2}\int_0^\theta P(\tau)y(\tau)K(\theta,\tau)m1\mathrm{d}\tau$$

$$= q(\theta) + \int_0^\theta q(\tau)K(\theta,\tau)m1\mathrm{d}\tau$$

$$y(\theta)\left[P(\theta) - E(\theta)I\frac{\pi^2}{L^2} - k_w\frac{L^2}{\pi^2}\right] + \int_0^\theta P(\tau)y(\tau)K(\theta,\tau)m1\mathrm{d}\tau$$

$$+ \frac{q_0L^2}{\pi^2}\left[q(\theta) + \int_0^\theta q(\tau)K(\theta,\tau)m1\mathrm{d}\tau\right] = 0$$

Denote

$$P^* = \frac{\pi^2 E_0 I}{L^2}\boldsymbol{\rho}(\boldsymbol{\theta}); \quad \boldsymbol{\rho}(\boldsymbol{\theta}) = \exp(-0.15\boldsymbol{\theta}); K_w = k_w \frac{L^2}{\pi^2} \text{ and } q_0(\boldsymbol{\theta}) = q_0 = \text{const.}$$

$$f(\boldsymbol{\theta}) = \frac{q_0 L^2}{\pi^2}\left[1 + \int_0^{\boldsymbol{\theta}} K(\boldsymbol{\theta}, \tau) m1 \mathrm{d}\tau\right]$$

Than Eq. (9.53) has a form:

$$y(\boldsymbol{\theta})[P^*(\boldsymbol{\theta}) - P(\boldsymbol{\theta}) + K_w] - \int_0^{\boldsymbol{\theta}} P(\tau) y(\tau) K(\boldsymbol{\theta}, \tau) m1 \mathrm{d}\tau = f(\boldsymbol{\theta}) \tag{9.54}$$

The effect of α_i on deformation—temperature (time) curve is presented below again via example.

Example 9.12 Data: see Example 9.1 and K_z = 200 ksf

The Computer Code and solution of Eq. (9.10) using POLYMATH software is as follows (Figs. 9.32 and 9.33):

Calculated values of DEQ variables

	Variable	Initial value	Minimal value	Maximal value	Final value
1	*B*	0	0	8.218488	8.218488
2	*C*	292	146.0001	292	146.0001
3	*D*	2.98E+04	2.98E+04	3.019E+04	2.989E+04
4	*m*	0.0351919	0.0351919	1.50896	1.50896
5	*m*1	0.3527194	0.3345936	0.57116	0.57116
6	*Q*	0	0	18.67366	18.67366
7	*t*	0.1	0.1	4	4
8	*T*1	0	0	17.47666	17.47666
9	*T*2	0	0	18.67366	18.67366
10	*y*	0	−6.494103	35.20191	35.20191
11	*z*	0.0097991	0.0050963	0.0097991	0.0051603

Differential equations

1 $d(T1)/d(t) = 258 * m1 * z * (\exp(0.5 * m)) * (\exp(t/(1+0.1 * t))) * ((t\hat{}-0.5))$
2 $d(T2)/d(t) = 1 * m1 * (\exp(0.5 * m)) * (\exp(t/(1+0.1 * t)))$
3 $d(y)/d(t) = (-0.608 + 0.529 * (t) - 0.053 * (t\hat{}2) - 0.125 * (t\hat{}3) + 0.0356 * (t\hat{}4)) * (-0.608 + 0.529 * (t+3.8) - 0.053 * ((t+3.8)\hat{}2) - 0.125 * ((t+3.8)\hat{}3) + 0.0356 * ((t+3.8)\hat{}4))$
4 $d(Z2)/d(t) = 1 * m1 * Z1 * (\exp(0.5 * m)) * (\exp(t/(1+0.1 * t)))$
5 $d(Z3)/d(t) = 0 * m1 * Z1 * (\exp(1 * m)) * (\exp(t/(1+0.1 * t)))$
6 $d(Q)/d(t) = 1 * m1 * 1 * (\exp(0.5 * m)) * (\exp(t/(1+0.1 * t)))$

Explicit equations

1 $m1 = 0.351 + 0.0208 * t\hat{}1 - 0.0372 * t\hat{}2 + 0.01144 * t\hat{}3$
2 $m = 0.351 * t + 0.0104 * t\hat{}2 - 0.0124 * t\hat{}3 + 0.00286 * t\hat{}4$
3 $D = (1375 * (\exp(-0.15 * t)) - 258 * (t\hat{}-0.5) + 200 * 146.3 * (\exp(-0.10 * t * 0)))$
4 $C = 146 * (1 + (\exp(-0.5 * m * T2)))$
5 $B = (\exp(-0.5 * m)) * T1$
6 $z = (C+B)/D$

Model: $y(\theta) = z = 0.01 + 0.0015 * \theta - 0.0021 * \theta\hat{}2 + 0.00042 * \theta\hat{}3 - 0.00012 * \theta\hat{}4$ (9.55)

Variable	Value
$a0$	0.01
$a1$	0.0015
$a2$	−0.0021
$a3$	0.00042
$a4$	−1.158E−05

9.7 Design for Stochastic Behavior of Beams on Elastic Foundation

9.7.1 *Probability-Based Data*

The mean value can be calculated now as follows:

$$\begin{aligned}\mu_y &= \frac{1}{4}\int_0^4 y(\theta)d\theta\\ &= \frac{1}{4}\int_0^4 (\mathbf{0.01+0.0015*\theta-0.0021*\theta^\wedge 2+0.00042*\theta^\wedge 3-0.000012*\theta^\wedge 4})d\theta\\ &= \frac{0.0306}{4}=0.00765;\quad \mu_y=q_0(s)0.64=0.1(10)(0.00765)=0.00765'=0.092''\\ s&=10'-\text{beam spacing}\end{aligned} \tag{9.56}$$

In order to calculate the correlation function in this case the chosen function (9.55) has to be centered (see Table 9.4). Again, after using the POLYMATH software we have:

Model: $y2=a0+a1*x+a2*x^\wedge 2+a3*x^\wedge 3+a4*x^\wedge 4$

Variable	Value
*a*0	0.0023559
*a*1	0.0014916
*a*2	−0.0020853
*a*3	0.0004117
*a*4	−1.058E−05

$$y^*=0.00236+0.00149(\theta)-0.0021(\theta)^2+0.000412(\theta)^3-0.000011(\theta)^4 \tag{9.57}$$

The correlation function of a steady time process can be computed now as follows [14]:

Table 9.4 Centered correlation function. Beams on elastic foundation

$\alpha\vert\theta$	0	0.5	1	1.5	2	2.5	3	3.5	4
0.5	0.01	0.0103	0.0098	0.00888	0.00777	0.00672	0.00597	0.00573	0.00620
y^*	0.00235	0.00265	0.00215	0.00123	0.00012	−0.0009	−0.0017	−0.0019	−0.0014

Table 9.5 Autocorrelation function. Beams on elastic foundation

θ	0	0.5	1	1.5	2	2.5	3	3.5	4
$K_y(\theta)$	3.2E −06	2.58E −06	1.215E −06	−0.568E −06	−2.4E −06	−2.882E −06	−2.611E −06	−2.048E −06	0

Table 9.6 Normalized Autocorrelation Function. Beams on elastic foundation

	0	0.5	1	1.5	2	2.5	3	3.5	4
$\boldsymbol{\rho}_y(\boldsymbol{\tau})$	1.0E −06	0.806E −06	0.38E −06	−0.178E −06	−0.75E −06	−0.9E −06	−0.81E −06	−0.64E −06	0

$$K_y(\boldsymbol{\theta}) = \frac{1}{4-\boldsymbol{\theta}} \int_0^{4-\boldsymbol{\theta}} [y^*(t)y^*(t+\boldsymbol{\theta})]\mathrm{d}t \tag{9.58}$$

where $\boldsymbol{\theta} = 0; 0.5; 1; \ldots 4$

After substituting (9.57) into (9.58) we have:

The standard deviation $D = K_y(0)$ is

$$D = K_y(\boldsymbol{\theta} = 0) = \frac{1}{4-\boldsymbol{\theta}} \int_0^{4-\boldsymbol{\theta}} [y*(t)y*(t+\boldsymbol{\theta})]\mathrm{d}t = \frac{1}{4}\int_0^4 [\sigma*(t)]^2 \mathrm{d}t$$

$$= 0.25 \int_0^4 \left[\mathbf{0.00236} + \mathbf{0.00149}(\boldsymbol{\theta}) - \mathbf{0.0021}(\boldsymbol{\theta}^2) + \mathbf{0.000412}(\boldsymbol{\theta}^3) - \mathbf{0.000011}(\boldsymbol{\theta}^4)\right]^2 \mathrm{d}\boldsymbol{\theta}$$

$$= 0.25(1.282\text{E}{-}05) = 0.000003205; \quad \boldsymbol{\sigma}_y = 0.00179$$

The computations of autocorrelation function based on Eq. (8.23) are presented Table 9.5.

The normalized autocorrelation function is shown in Table 9.6

Let us approximate the data from Table 9.6 by the (9.35) type of formulae (Fig. 9.34):

Model: $z = \left(10^{-6}\right)(\exp(-a*x))*(\cos(b*x))$

Variable	Initial guess	Value
a	1	0.0176
b	1	1.171

$$\boldsymbol{\rho}_{\boldsymbol{\theta}}(\boldsymbol{\tau}) = (\cos(0.0176*\boldsymbol{\tau}))*(\exp(-1.171*\boldsymbol{\tau})) \tag{9.59}$$

The corresponding spectral function is [14]:

$$S(\omega) = \frac{1}{2}\frac{\sigma_y^2}{\pi}\left[\frac{a}{(\omega - b)^2 + a^2} + \frac{a}{(\omega + b)^2 + a^2}\right]$$
$$= 0.00896\left(10^{-6}\right)\left[\frac{1}{(\omega - 1.171)^2 + 0.0176^2} + \frac{1}{(\omega + 1.171)^2 + 0.0176^2}\right] \tag{9.60}$$

The graphic presentation of spectral density is Fig. 9.35

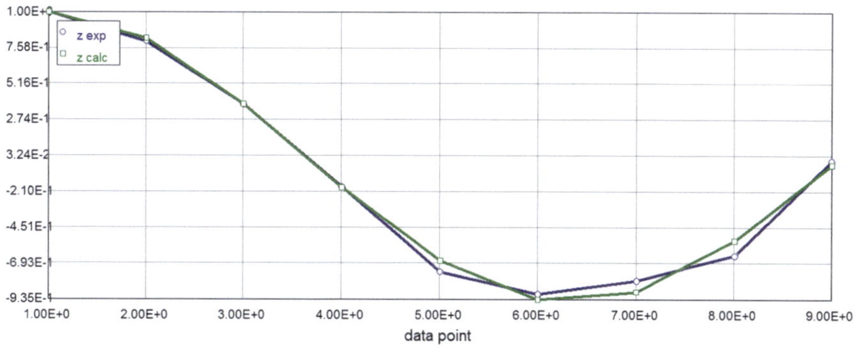

Fig. 9.34 Normalized autocorrelation function. Beams on elastic foundation

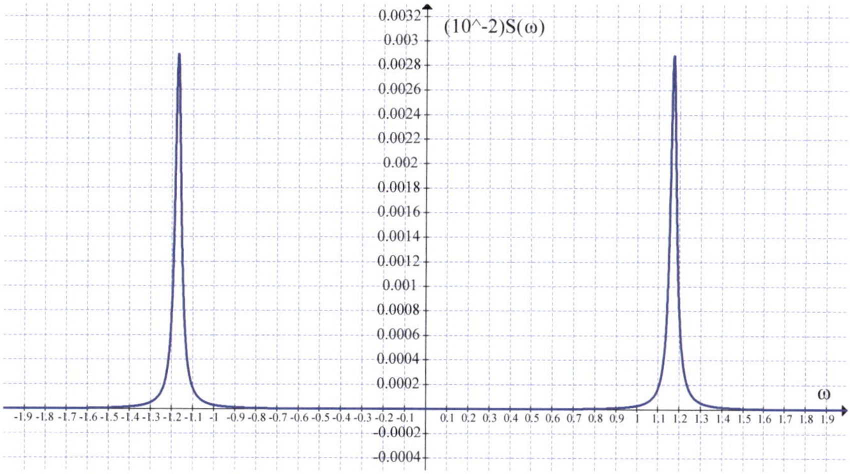

Fig. 9.35 Spectral density function. Beams on elastic foundation

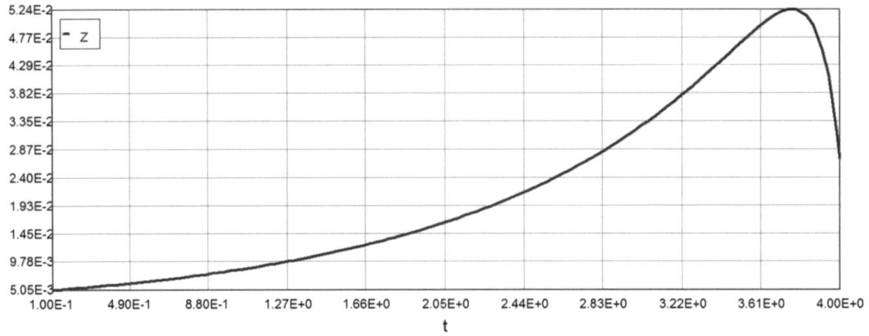

Fig. 9.36 Core deformation—temperature curve ($n = 1$)

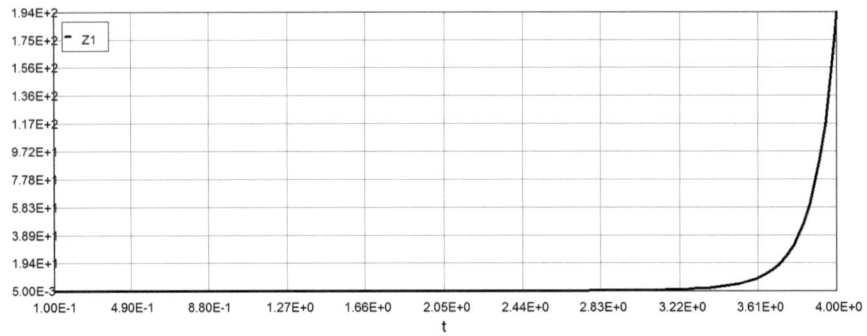

Fig. 9.37 Soil pressure—temperature (time) curve ($n = 1$)

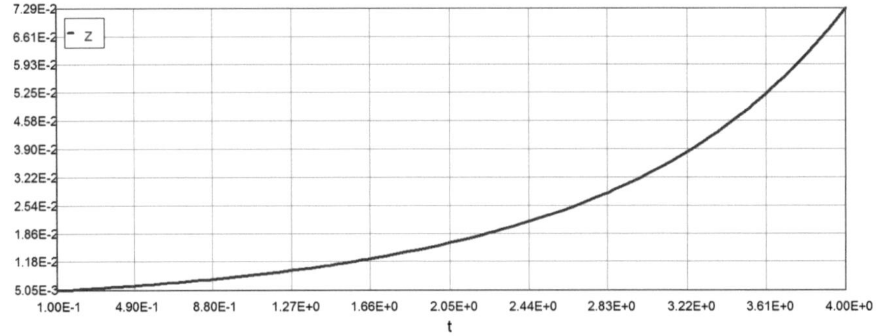

Fig. 9.38 Core deformation—temperature curve ($n = 2$)

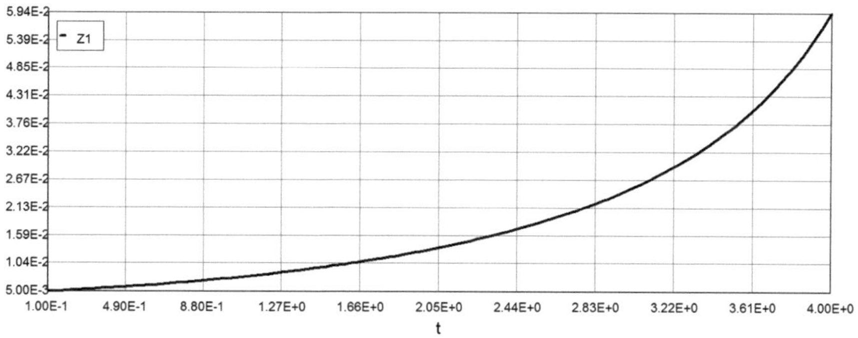

Fig. 9.39 Soil pressure—temperature (time) curve (n = 2)

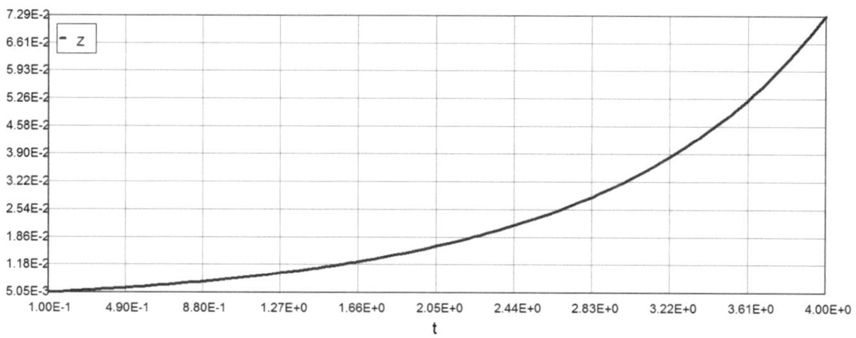

Fig. 9.40 Core deformation—temperature curve (n = 3)

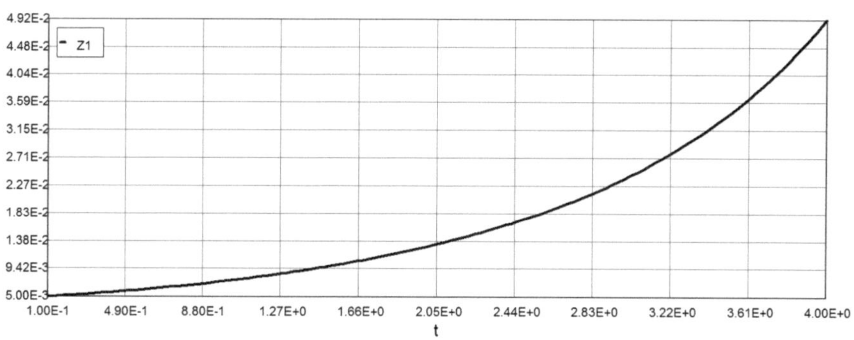

Fig. 9.41 Soil pressure—temperature (time) curve (n = 3)

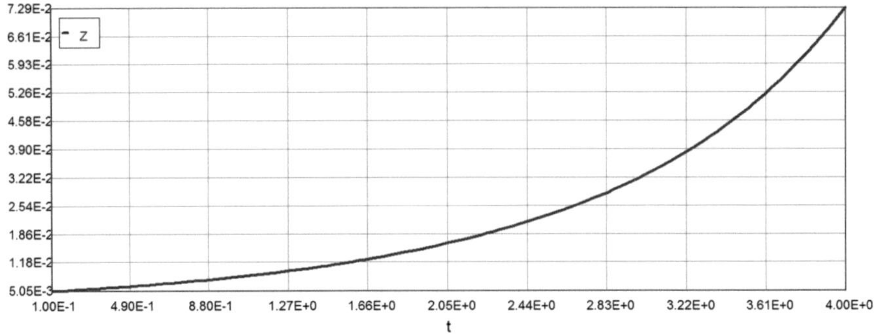

Fig. 9.42 Core deformation—temperature curve (n = 4)

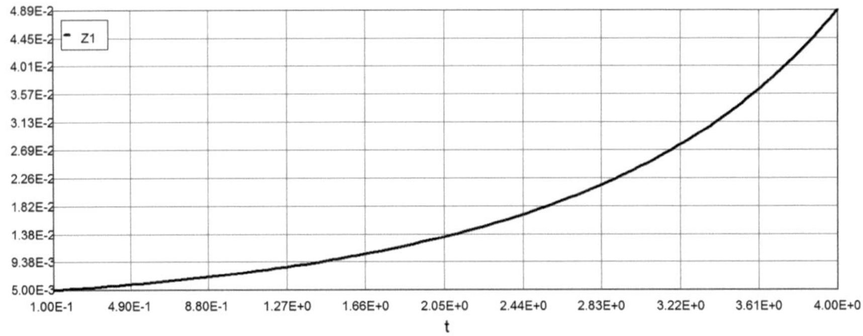

Fig. 9.43 Soil pressure—temperature (time) curve (n = 4)

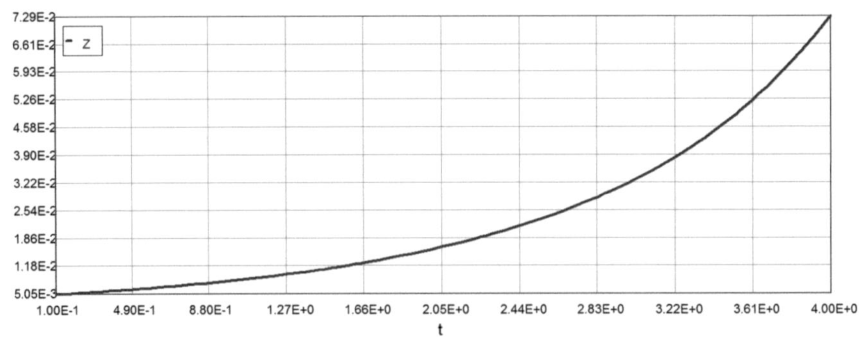

Fig. 9.44 Core deformation—temperature curve (n = 6)

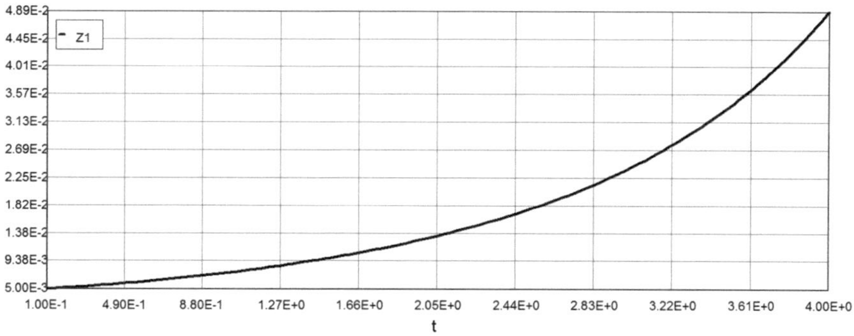

Fig. 9.45 Soil pressure—temperature (time) curve (n = 6)

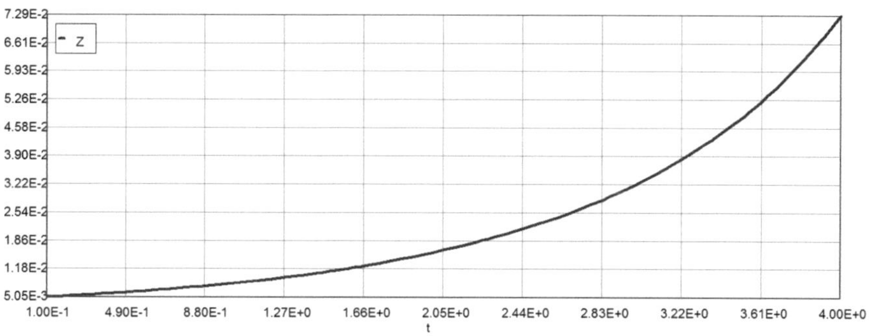

Fig. 9.46 Core deformation—temperature curve (n = 8)

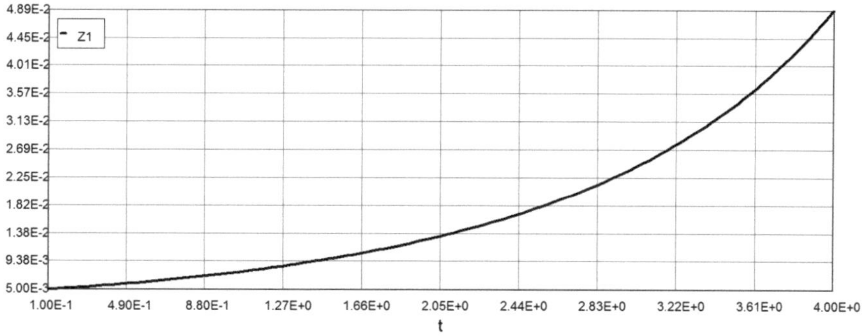

Fig. 9.47 Soil pressure—temperature (time) curve (n = 8)

9.7.2 The First-Occurrence Time Problem and the Probability Density $\mathbf{P(a_0, \theta)}$

The average first-occurrence of deflection y above given level "y = a_0" for stationary deflection—temperature processes is defined as follows (see Chap. 7) [15]:

For correlation function given by (9.59):

$$\sigma_{vy}^2 = K_{\dot{y}}(0) = -\frac{d^2}{d\theta^2}K_y(\theta)|_{\theta=0} = \sigma_\sigma^2(b^2 - a^2)$$
$$\bar{n}_a = \theta_{max}\frac{\sqrt{b^2 - a^2}}{2\pi}\exp\left(-\frac{a_0^2}{2\sigma_\sigma^2}\right) = 4\frac{\sqrt{1.171^2 - 0.0176^2}}{2\pi}\exp\left(-\frac{7.65^2(10^{-6})}{2(3.2)(10^{-6})}\right) = 0.0574$$
$$\bar{v}_a = \frac{\bar{n}_a}{\theta_{max}} = \frac{0.0574}{4} = 0.0144 \tag{9.61}$$

Now, based on Poisson formulae of the probability to not having the maximum beam deflection ordinates crossing upwards the level $a = 0.00765$ is: $P_{rel} = P_o = \exp(-\bar{n}_a) = \exp(-0.0574) = 0.944$.

This probability characterizes the reliability of the structure (one element or the whole structure).

9.8 Composite Materials and Structures

9.8.1 Introduction

Research efforts are continuously looking for new, better and efficient construction materials. The main goal of these researches is to improve the structural efficiency, performance and durability. New materials typically bring new challenges to designer who utilizes these new materials. In the past decades, various sandwich panels have been implemented in aerospace, marine, architectural, and transportation industry. Light-weight, excellent corrosion characteristics and rapid installation capabilities has created tremendous opportunities for these sandwich panels in industry. Work on the theoretical description of sandwich structure behavior began after World War II. Reference [16] is the first book published about sandwich structures, followed by [17] and more recently [18]. Reference [19] developed a method to design for minimum weight, and reported the failure mode map of sandwich construction, without considering the post yield state of the sandwich structure.

Reference [20] reported that the load carried by sandwich structures continue to increase after core yielding. Assuming that the core could not carry additional load after yield, this increasing load carrying capacity of post yield sandwich structure initiates the postulation that the additional shear load was transferred to the face sheets. To account for the above-mentioned phenomenon, [20] developed a higher order theory by including a bilinear stress—strain relationship for the core material. This theory yields a fairly accurate prediction on the deflection of a foam cored sandwich structure [21].

However, this theory does not take into account the core compression under localized load, or any geometric nonlinearity. The classical sandwich beam theory also assumes that in-plane displacements of the core through its depth are linear. In other words, it was assumed that the core thickness remains constant and cross-sections perpendicular to the neutral axis remain plane after deformation. This assumption is generally true for traditional core material such as metallic honeycomb. However, this assumption is not suitable for soft, foam-based cores, especially when the sandwich structure is subjected to a concentrated load [22]. With a much lower rigidity compared to metallic honeycomb, foam-based cored sandwich structures are susceptible to localized failure. Insufficient support to the face sheets due to core compression near the application points of concentrated loads can lead to failures such as face sheet/core delaminating, face sheet buckling, and face sheet yielding. This localized non-linearity is reported by many researchers such as [22–26]. The shear distribution at localized failure points has not been well defined. Reference [26] investigated the effect of localized strengthening inserts on the overall stiffness of a sandwich structure. This localized strengthening increases the rigidity of the sandwich structure, but the addition of high stiffness inserts complicates the manufacturing process of sandwich structure. To design an efficient sandwich structure, it is vital to understand the behavior of each layer in the structure. Classical sandwich theory, higher order theory by [27] could predict the sandwich panel behavior fairly accurate in the linear range.

However, these theories could not give an accurate prediction of the sandwich structure behavior after core yielding.

The following assumptions are made to simplify the model without losing the physics of the problem

- Face sheets and core are perfectly bonded: no delaminating occurs between layers.
- Face sheets remain elastic all the time: Due to the significantly higher yield strength and modulus of elasticity of the face sheets compared to the core, face sheets are assumed to remain elastic throughout the loading.
- Geometric non-linearity is considered to have significant effect on the load distribution on each layer of the sandwich structure.
- The panel is simply supported on all sides.
- Out of plane structural load is applied and varied in quasi-static manner.

As the core starts to yield, its maximum-stress increment rate starts to decrease while the maximum-stress incremental-rate of the bottom—face—sheet starts to increase. This means that the load is being transferred to the face sheet-metal. This is the main advantage of increasing the load beyond the yield limit of the core material.

It can be seen that the softer material is the more load is transferred from core material to the sheet metal as the core starts to yield. It is obvious that the load carrying capacity of the panel increases as its core material stiffness increases.

The increment of shear stress with respect to strain decreases as the load increases. The nonlinear deformation rate of the core material is higher than that of the elastic range for the same load increment. This deformation works as a mechanism of transferring the excess load to the face sheets.

The elastic deflection of a sandwich beam is the sum of the bending deflection of the beam, which depends on the overall flexural rigidity of the beam, and of its shear deflection, which depends on the shear rigidity of the core. We propose first to analyze the creep of a sandwich beam in which the foam core creeps but faces do not (e.g., a beam made with a polyurethane foam core with aluminum faces).

The ASTM defines a sandwich structure as follows:

A structural sandwich is a special form of a laminated composite comprising of a combination of different materials that are bonded to each other so as to utilize the properties of each separate component to the structural advantage of the whole assembly.

A sandwich consists of three main parts: two thin, stiff and strong faces are separated by a thick, light and weaker core. The faces are adhesively bonded to the core to obtain a load transfer between the components. The modus operandi of a sandwich is much the same as that of an I-beam, which is an efficient structural shape because as much as possible of the material is placed in the flanges situated farthest from the center of bending or neutral axis.

Only enough material is left in the connecting web to make the flanges work together and to resist shear and buckling. In a sandwich, the faces take the place of the flanges and the core takes the place of the web.

In all cases the primary loading, both in-plane and bending, are carried by the faces, while the core resists transverse shear loads (analogous to the web of an I-beam), and keeps the faces in place. In most foam-core and honeycomb-core sandwiches, one can assume that all of the in-plane and bending loads are carried by the faces only. However, in web-core and truss-core construction, a portion of the in-plane and bending loads are also carried by the core elements.

A simple assumption regarding the core as a medium supporting the sandwich face is the *Winkler foundation model*. This assumes that the support material consists of an array of continuously distributed linear springs perpendicular to the faces (E_{cz}) but with no out-of-plane shear stiffness, i.e., $G_{czx} = 0$ (springs are not cross-linked). Core may be modeled by elastic foundation with subgrade modulus k_z. Approximately, $k_z \approx 2E_{cz}/t_c$ (t_c—thickness of the core) and correspondingly the soil reaction is $K_z = k_z y$. It will be also assumed that the face elements behavior due to high temperature load corresponds to linear creep constitutive law while the core material will follow the nonlinear integral—type creep constitutive law presented in Chap. 8 with the power law type exponent parameter n.

9.8.2 *Nonlinear Creep of Core Material*

The solution of the composite cross-section is analogous to its elastic analysis for combined action of axial force and bending, Eqs. (9.45) and (9.46). As an extension of the creep solutions according to *Winkler foundation model* the above equations have a form.

$$IE(\theta)\frac{\partial^4 y(x,\theta)}{\partial x^4}+p(x,\theta)+\int_0^{\theta} p^n(x,\tau)K(\theta,\tau)m1\,d\tau = q(x,\theta)+\int_0^{\theta} q(x,\tau)K(\theta,\tau)m1\,d\tau$$

$$k(\theta)y(x,\theta) = p(x,\theta)+\int_0^{\theta} p^n(x,\tau)K_0(\theta,\tau)m1\,d\tau \tag{9.62}$$

$k(\theta)$—instantaneous subgrade modulus; n—power law exponential.

In a cross-section in which the creep properties of the material are not the same at all the points, creep causes stress redistributions. Further, temperature-dependent stress variations are caused when the cross-section is built out of two different materials and the temperature load continuously changing with the time.

The exact analysis of a composite or inhomogeneous cross-section according to nonlinear viscoelasticity leads to a system of two Volterra integral Eq. (9.62)

We assume here that the necessary data regarding the Winkler subgrade modulus and exponential parameter n obtained experimentally. Let us analyze the effect of nonlinear creep on core deformation and subsequently on sandwich panel as a whole. The Computer Code and the solution of Eq. (9.62) are similar to Example 9.11 (Figs. 9.36, 9.37, 9.38, 9.39, 9.40, 9.41, 9.42, 9.43, 9.44, 9.45, 9.46 and 9.47):

Example 9.13 $n = 1$

Calculated values of DEQ variables

	Variable	Initial value	Minimal value	Maximal value	Final value
1	m	0.0351919	0.0351919	1.50896	1.50896
2	$m1$	0.3527194	0.3345936	0.57116	0.57116
3	n	1	1	1	1
4	Q	0	0	18.67366	18.67366
5	t	0.1	0.1	4	4
6	Z	0.0050461	0.0050461	0.0524201	0.027211
7	$Z1$	0.0049959	0.0049959	194.4128	194.4128
8	$Z2$	0	0	426.4185	426.4185
9	$Z3$	0	0	879.0572	879.0572

Differential equations

1 $d(Z2)/d(t) = 1 * m1 * Z1 * (\exp(0.5 * m)) * (\exp(t/(1 + 0.1 * t)))$
2 $d(Z3)/d(t) = 1 * m1 * Z1^{\wedge}n * (\exp(1 * m)) * (\exp(t/(1 + 0.1 * t)))$
3 $d(Q)/d(t) = 1 * m1 * 1 * (\exp(0.5 * m)) * (\exp(t/(1 + 0.1 * t)))$

Explicit equations

1 $m1 = 0.351 + 0.0208 * t^{\wedge}1 - 0.0372 * t^{\wedge}2 + 0.01144 * t^{\wedge}3$
2 $m = 0.351 * t + 0.0104 * t^{\wedge}2 - 0.0124 * t^{\wedge}3 + 0.00286 * t^{\wedge}4$
3 $Z1 = (\exp(-1 * m)) * Z3 * 1 + 1 * (\exp(-0.10 * t)) * (1 + (\exp(-0.5 * m)) * Q + (\exp(-1 * m)) * Z3 * 1 - (\exp(-0.5 * m)) * Z2) * (0.166 * (\exp(-0.15 * t)) + 200 * (\exp(-0.10 * t)))^{\wedge}-1$
4 $Z = (1 + (\exp(-0.5 * m)) * Q + (\exp(-1 * m)) * Z3 * 1 - (\exp(-0.5 * m)) * Z2) * (0.166 * (\exp(-0.15 * t)) + 200 * (\exp(-0.10 * t)))^{\wedge}-1$
5 $n = 1$

Model: $Z = a0 + a1 * t + a2 * t^{\wedge}2 + a3 * t^{\wedge}3 + a4 * t^{\wedge}4$

Variable	Value
*a*0	2.098E−05
*a*1	0.0278015
*a*2	−0.0312983
*a*3	0.0142662
*a*4	−0.0018758

$$T(\theta) = Z = 0.00003 + 0.0278 * \theta - 0.0313 * \theta^{\wedge}2 + 0.0413 * \theta^{\wedge}3 - 0.00188 * \theta^{\wedge}4 \tag{9.63}$$

Model: $Z1 = a0 + a1 * t + a2 * t^{\wedge}2 + a3 * t^{\wedge}3 + a4 * t^{\wedge}4$

Variable	Value
*a*0	22.49356
*a*1	−121.0303
*a*2	155.2813
*a*3	−69.16873
*a*4	9.864638

$$P(\theta) = Z1 = 22.5 - 121 * \theta + 155.3 * \theta^\wedge 2 - 69.2 * \theta^\wedge 3 + 9.86 * \theta^\wedge 4 \quad (9.64)$$

Example 9.14 n = 2

Calculated values of DEQ variables

	Variable	Initial value	Minimal value	Maximal value	Final value
1	m	0.0351919	0.0351919	1.50896	1.50896
2	$m1$	0.3527194	0.3345936	0.57116	0.57116
3	n	2	2	2	2
4	Q	0	0	18.67366	18.67366
5	t	0.1	0.1	4	4
6	Z	0.0050461	0.0050461	0.0729039	0.0729039
7	$Z1$	0.0049959	0.0049959	0.059367	0.059367
8	$Z2$	0	0	0.0243483	0.0243483
9	$Z3$	0	0	0.0474724	0.0474724

Differential equations

1 $d(Z2)/d(t) = 1 * m1 * Z1^\wedge n * (\exp(0.5 * m)) * (\exp(t/(1 + 0.1 * t)))$

2 $d(Z3)/d(t) = 1 * m1 * Z1^\wedge n * (\exp(1 * m)) * (\exp(t/(1 + 0.1 * t)))$

3 $d(Q)/d(t) = 1 * m1 * 1 * (\exp(0.5 * m)) * (\exp(t/(1 + 0.1 * t)))$

Explicit equations

1 $m1 = 0.351 + 0.0208 * t^\wedge 1 - 0.0372 * t^\wedge 2 + 0.01144 * t^\wedge 3$

2 $m = 0.351 * t + 0.0104 * t^\wedge 2 - 0.0124 * t^\wedge 3 + 0.00286 * t^\wedge 4$

3 $Z1 = (\exp(-1 * m)) * Z3 * 1 + 1 * (\exp(-0.10 * t)) * (1 + (\exp(-0.5 * m)) * Q + (\exp(-1 * m)) * Z3 * 1 - (\exp(-0.5 * m)) * Z2) * (0.166 * (\exp(-0.15 * t)) + 200 * (\exp(-0.10 * t)))^\wedge -1$

4 $Z = (1 + (\exp(-0.5 * m)) * Q + (\exp(-1 * m)) * Z3 * 1 - (\exp(-0.5 * m)) * Z2) * (0.166 * (\exp(-0.15 * t)) + 200 * (\exp(-0.10 * t)))^\wedge -1$

5 $n = 2$

Model: $Z = a0 + a1 * t + a2 * t^\wedge 2 + a3 * t^\wedge 3 + a4 * t^\wedge 4$

Variable	Value
$a0$	0.0052647
$a1$	−0.0004653

(continued)

(continued)

Variable	Value
$a2$	0.0050394
$a3$	−0.0019582
$a4$	0.0004448

$$T(\theta) = Z \\ = 0.005 - 0.000465 * \theta + 0.00504 * \theta^\wedge 2 - 0.00196 * \theta^\wedge 3 + 0.00445 * \theta^\wedge 4 \quad (9.65)$$

Model: $Z1 = a0 + a1 * t + a2 * t^\wedge 2 + a3 * t^\wedge 3 + a4 * t^\wedge 4$

Variable	Value
$a0$	0.0056773
$a1$	−0.0031991
$a2$	0.0077856
$a3$	−0.0033587
$a4$	0.0006088

$$P(\theta) = Z1 \\ = 0.0057 - 0.0032 * \theta + 0.0078 * \theta^\wedge 2 - 0.00336 * \theta^\wedge 3 + 0.00061 * \theta^\wedge 4 \quad (9.66)$$

Example 9.15 $n = 3$

Calculated values of DEQ variables

	Variable	Initial value	Minimal value	Maximal value	Final value
1	m	0.0351919	0.0351919	1.50896	1.50896
2	$m1$	0.3527194	0.3345936	0.57116	0.57116
3	n	3	3	3	3
4	Q	0	0	18.67366	18.67366
5	t	0.1	0.1	4	4
6	Z	0.0050461	0.0050461	0.0729108	0.0729108
7	$Z1$	0.0049959	0.0049959	0.0491941	0.0491941
8	$Z2$	0	0	0.0007346	0.0007346
9	$Z3$	0	0	0.0014495	0.0014495

Differential equations

1 $d(Z2)/d(t) = 1 * m1 * Z1^\wedge n * (\exp(0.5 * m)) * (\exp(t/(1 + 0.1 * t)))$
2 $d(Z3)/d(t) = 1 * m1 * Z1^\wedge n * (\exp(1 * m)) * (\exp(t/(1 + 0.1 * t)))$
3 $d(Q)/d(t) = 1 * m1 * 1 * (\exp(0.5 * m)) * (\exp(t/(1 + 0.1 * t)))$

Explicit equations

1 $m1 = 0.351 + 0.0208 * t^\wedge 1 - 0.0372 * t^\wedge 2 + 0.01144 * t^\wedge 3$
2 $m = 0.351 * t + 0.0104 * t^\wedge 2 - 0.0124 * t^\wedge 3 + 0.00286 * t^\wedge 4$
3 $Z1 = (\exp(-1 * m)) * Z3 * 1 + 1 * (\exp(-0.10 * t)) * (1 + (\exp(-0.5 * m)) * Q + (\exp(-1 * m)) * Z3 * 1 - (\exp(-0.5 * m)) * Z2) * (0.166 * (\exp(-0.15 * t)) + 200 * (\exp(-0.10 * t)))^\wedge -1$
4 $Z = (1 + (\exp(-0.5 * m)) * Q + (\exp(-1 * m)) * Z3 * 1 - (\exp(-0.5 * m)) * Z2) * (0.166 * (\exp(-0.15 * t)) + 200 * (\exp(-0.10 * t)))^\wedge -1$
5 $n = 3$

Model: $Z = a0 + a1 * t + a2 * t^\wedge 2 + a3 * t^\wedge 3 + a4 * t^\wedge 4$

Variable	Value
$a0$	0.0052652
$a1$	−0.0004679
$a2$	0.0050429
$a3$	−0.0019598
$a4$	0.0004451

$$T(\theta) = Z = 0.005 - 0.00047 * \theta + 0.005 * \theta^\wedge 2 - 0.002 * \theta^\wedge 3 + 0.000445 * \theta^\wedge 4 \tag{9.67}$$

Model: $Z1 = a0 + a1 * t + a2 * t^\wedge 2 + a3 * t^\wedge 3 + a4 * t^\wedge 4$

Variable	Value
$a0$	0.0050707
$a1$	0.0002129
$a2$	0.0031266
$a3$	−0.0011308
$a4$	0.0002556

$$P(\theta) = Z1 = 0.005 - 0.000513 * \theta + 0.00269 * \theta\hat{}2 - 0.00092 * \theta\hat{}3 + 0.00022 * \theta\hat{}4 \tag{9.52}$$

Example 9.16 $n = 4$

Calculated values of DEQ variables

	Variable	Initial value	Minimal value	Maximal value	Final value
1	m	0.0351919	0.0351919	1.50896	1.50896
2	$m1$	0.3527194	0.3345936	0.57116	0.57116
3	n	4	4	4	4
4	Q	0	0	18.67366	18.67366
5	t	0.1	0.1	4	4
6	Z	0.0050461	0.0050461	0.072911	0.072911
7	$Z1$	0.0049959	0.0049959	0.0488865	0.0488865
8	$Z2$	0	0	2.898E−05	2.898E−05
9	$Z3$	0	0	5.79E−05	5.79E−05

Differential equations

1 $d(Z2)/d(t) = 1 * m1 * Z1\hat{}n * (\exp(0.5 * m)) * (\exp(t/(1 + 0.1 * t)))$

2 $d(Z3)/d(t) = 1 * m1 * Z1\hat{}n * (\exp(1 * m)) * (\exp(t/(1 + 0.1 * t)))$

3 $d(Q)/d(t) = 1 * m1 * 1 * (\exp(0.5 * m)) * (\exp(t/(1 + 0.1 * t)))$

Explicit equations

1 $m1 = 0.351 + 0.0208 * t\hat{}1 - 0.0372 * t\hat{}2 + 0.01144 * t\hat{}3$

2 $m = 0.351 * t + 0.0104 * t\hat{}2 - 0.0124 * t\hat{}3 + 0.00286 * t\hat{}4$

3 $Z1 = (\exp(-1 * m)) * Z3 * 1 + 1 * (\exp(-0.10 * t)) * (1 + (\exp(-0.5 * m)) * Q + (\exp(-1 * m)) * Z3 * 1 - (\exp(-0.5 * m)) * Z2) * (0.166 * (\exp(-0.15 * t)) + 200 * (\exp(-0.10 * t)))\hat{}-1$

4 $Z = (1 + (\exp(-0.5 * m)) * Q + (\exp(-1 * m)) * Z3 * 1 - (\exp(-0.5 * m)) * Z2) * (0.166 * (\exp(-0.15 * t)) + 200 * (\exp(-0.10 * t)))\hat{}-1$

9 $n = 4$

Model: $Z = a0 + a1 * t + a2 * t\wedge 2 + a3 * t\wedge 3 + a4 * t\wedge 4$

Variable	Value
$a0$	0.0052652
$a1$	−0.000468
$a2$	0.005043
$a3$	−0.0019599
$a4$	0.0004451

$$T(\theta) = Z = 0.005 - 0.000468 * \theta + 0.005 * \theta\wedge 2 - 0.002 * \theta\wedge 3 + 0.000445 * \theta\wedge 4 \quad (9.68)$$

Model: $Z1 = a0 + a1 * t + a2 * t\wedge 2 + a3 * t\wedge 3 + a4 * t\wedge 4$

Variable	Value
$a0$	0.0050482
$a1$	0.0003391
$a2$	0.002956
$a3$	−0.0010499
$a4$	0.0002431

$$P(\theta) = Z1 = 0.005 - 0.00034 * \theta + 0.00296 * \theta\wedge 2 - 0.001 * \theta\wedge 3 + 0.00024 * \theta\wedge 4 \quad (9.69)$$

Example 9.17 $n = 6$

Calculated values of DEQ variables

	Variable	Initial value	Minimal value	Maximal value	Final value
1	m	0.0351919	0.0351919	1.50896	1.50896
2	$m1$	0.3527194	0.3345936	0.57116	0.57116
3	n	6	6	6	6
4	Q	0	0	18.67366	18.67366
5	t	0.1	0.1	4	4
6	Z	0.0050461	0.0050461	0.072911	0.072911
7	$Z1$	0.0049959	0.0049959	0.0488737	0.0488737
8	$Z2$	0	0	5.061E−08	5.061E−08
9	$Z3$	0	0	1.027E−07	1.027E−07

Differential equations

1 $d(Z2)/d(t) = 1 * m1 * Z1^n * (\exp(0.5 * m)) * (\exp(t/(1 + 0.1 * t)))$
2 $d(Z3)/d(t) = 1 * m1 * Z1^n * (\exp(1 * m)) * (\exp(t/(1 + 0.1 * t)))$
3 $d(Q)/d(t) = 1 * m1 * 1 * (\exp(0.5 * m)) * (\exp(t/(1 + 0.1 * t)))$

Explicit equations

1 $m1 = 0.351 + 0.0208 * t^1 - 0.0372 * t^2 + 0.01144 * t^3$
2 $m = 0.351 * t + 0.0104 * t^2 - 0.0124 * t^3 + 0.00286 * t^4$
3 $Z1 = (\exp(-1 * m)) * Z3 * 1 + 1 * (\exp(-0.10 * t)) * (1 + (\exp(-0.5 * m)) * Q + (\exp(-1 * m)) * Z3 * 1 - (\exp(-0.5 * m)) * Z2) * (0.166 * (\exp(-0.15 * t)) + 200 * (\exp(-0.10 * t)))^{-1}$
4 $Z = (1 + (\exp(-0.5 * m)) * Q + (\exp(-1 * m)) * Z3 * 1 - (\exp(-0.5 * m)) * Z2) * (0.166 * (\exp(-0.15 * t)) + 200 * (\exp(-0.10 * t)))^{-1}$
5 $n = 6$

Model: $Z = a0 + a1 * t + a2 * t^2 + a3 * t^3 + a4 * t^4$

Variable	Value
$a0$	0.0052652
$a1$	−0.000468
$a2$	0.005043
$a3$	−0.0019599
$a4$	0.0004451

$$T(\theta) = Z = 0.005 - 0.0005 * \theta + 0.005 * \theta^2 - 0.002 * \theta^3 + 0.000445 * \theta^4 \tag{9.70}$$

Model: $Z1 = a0 + a1 * t + a2 * t^2 + a3 * t^3 + a4 * t^4$

Variable	Value
$a0$	0.005047
$a1$	0.0003455
$a2$	0.0029474
$a3$	−0.0010459
$a4$	0.0002425

$$P(\theta) = Z1 = 0.005 - 0.00035 * \theta + 0.00295 * \theta\hat{}2 - 0.001 * \theta\hat{}3 + 0.00024 * \theta\hat{}4 \tag{9.71}$$

Example 9.18 $n = 8$

Calculated values of DEQ variables

	Variable	Initial value	Minimal value	Maximal value	Final value
1	m	0.0351919	0.0351919	1.50896	1.50896
2	$m1$	0.3527194	0.3345936	0.57116	0.57116
3	n	8	8	8	8
4	Q	0	0	18.67366	18.67366
5	t	0.1	0.1	4	4
6	Z	0.0050461	0.0050461	0.072911	0.072911
7	$Z1$	0.0049959	0.0049959	0.0488737	0.0488737
8	$Z2$	0	0	9.534E−11	9.534E−11
9	$Z3$	0	0	1.953E−10	1.953E−10

Differential equations

1 $\mathrm{d}(Z2)/\mathrm{d}(t) = 1 * m1 * Z1\hat{}n * (\exp(0.5 * m)) * (\exp(t/(1 + 0.1 * t)))$

2 $\mathrm{d}(Z3)/\mathrm{d}(t) = 1 * m1 * Z1\hat{}n * (\exp(1 * m)) * (\exp(t/(1 + 0.1 * t)))$

3 $\mathrm{d}(Q)/\mathrm{d}(t) = 1 * m1 * 1 * (\exp(0.5 * m)) * (\exp(t/(1 + 0.1 * t)))$

Explicit equations

1 $m1 = 0.351 + 0.0208 * t\hat{}1 - 0.0372 * t\hat{}2 + 0.01144 * t\hat{}3$

2 $m = 0.351 * t + 0.0104 * t\hat{}2 - 0.0124 * t\hat{}3 + 0.00286 * t\hat{}4$

3 $Z1 = (\exp(-1 * m)) * Z3 * 1 + 1 * (\exp(-0.10 * t)) * (1 + (\exp(-0.5 * m)) * Q + (\exp(-1 * m)) * Z3 * 1 - (\exp(-0.5 * m)) * Z2) * (0.166 * (\exp(-0.15 * t)) + 200 * (\exp(-0.10 * t)))\hat{}-1$

4 $Z = (1 + (\exp(-0.5 * m)) * Q + (\exp(-1 * m)) * Z3 * 1 - (\exp(-0.5 * m)) * Z2) * (0.166 * (\exp(-0.15 * t)) + 200 * (\exp(-0.10 * t)))\hat{}-1$

5 $n = 8$

Model: $Z = a0 + a1 * t + a2 * t^{\wedge}2 + a3 * t^{\wedge}3 + a4 * t^{\wedge}4$

Variable	Value
$a0$	0.0052652
$a1$	−0.000468
$a2$	0.005043
$a3$	−0.0019599
$a4$	0.0004451

$$T(\theta) = Z = 0.005 - 0.0005 * \theta + 0.005 * \theta^{\wedge}2 - 0.002 * \theta^{\wedge}3 + 0.000445 * \theta^{\wedge}4 \tag{9.72}$$

Model: $Z1 = a0 + a1 * t + a2 * t^{\wedge}2 + a3 * t^{\wedge}3 + a4 * t^{\wedge}4$

Variable	Value
$a0$	0.005047
$a1$	0.0003455
$a2$	0.0029474
$a3$	−0.0010459
$a4$	0.0002425

$$P(\theta) = Z1 = 0.005 + 0.00035 * \theta + 0.00295 * \theta^{\wedge}2 - 0.001 * \theta^{\wedge}3 + 0.00024 * \theta^{\wedge}4 \tag{9.73}$$

Example 9.19 An infinite beam rest on equally spaced linear coil springs located every 1.1 m along the beam. A concentrated load of 18 kN is applied to the beam, over one of the springs. *EI* of the beam is 441 × 10^9 Nmm^2, K = 275 N/mm for each spring. Compute the largest spring force and largest bending moment in the beam.

(1) To "smear" the springs into a Winkler foundation: force applied to the beam by a spring with deflection w is Ky, so if the spring spacing is L, the associated force in each span L is Ky, then the hypothetical distributed force is therefore K/L. The "equivalent" Winkler foundation modulus is $k = K/L$ and the $\beta = [k/4EI]^{1/4} = 6.136 \times 10^{-4}$/mm.
(2) According to [28] for infinite beam with concentrated load P, we have 22.1 mm, the maximum spring force is $F_{max} = Kw_{max}$ = 6075 N.

If the beam length is finite with several springs, then the problem can be solved as static indeterminate beam. Finally, it should be emphasized here that all the creep effects in structures are in reality stochastic. The properties of materials exhibit a large variability due to randomness of the factors that influence creep. These variations introduce large scatter in the response of structures.

References

1. Boyle JT, Spence J (1983) Stress analysis for creep. Butterworth's, London
2. Penny RK, Marriott DL (1995) Design for Creep. Chapman & Hall, London
3. Kraus H (1980) Creep analysis. Wiley, New York
4. Wilshire B (2002) Observations, theories and predictions of high temperature creep behavior. Mettal Mater Trans 33A:241–248
5. Frost HJ, Ashby MF (1982) Deformation-mechanism maps. Pergamon, Oxford
6. Naumenko K, Altenbach H, Gorash Y (2009) Creep analysis with a stress range dependent constitutive model. Arch Appl Mech 79:619–630
7. Altenbach H, Gorash Y, Naumenko K (2008) Steady-state creep of a pressurized thick cylinder in both the linear and power law ranges. Acta Mech 196:263–274
8. Altenbach H, Naumenko K, Gorash Y (2008) creep analysis for a wide stress range based on stress relaxation experiments. Int J Mod Phys 22B:5413–5418
9. Kfistek V, Bazant ZP (1987) Shear lag effect and uncertainty in concrete beam girder creep. J Struct Eng ASCE 113(3):557–574
10. Wium DJW, Buyukozturk O (1985) Variability in long-term concrete deformations. J Eng Mech Div ASCE 111(8):1792–1809
11. Razdolsky L (2015) ASCE/SEI Congress
12. Razdolsky L (2014) Probability based structural fire load. Cambridge University Press, UK
13. Sveshnokov AA (1978) Problems in probability theory, mathematical statistics and theory of random functions. Dover Publications, New York
14. Wentzel E, Ovcharov L (1986) Applied problems in probability theory. Mir Publishers, Moscow
15. Rice JR, Beer FP (1966) First-occurrence time of high-level crossings in a continuous random process. J Acoust Soc Am 39:323–335
16. Plantema FJ (1966) Sandwich construction. Wiley, New York
17. Allen HG (1969) Analysis and design of structural sandwich panels. Pergamon Press, Oxford
18. Zenkert D (1995) An introduction to sandwich construction, (EMAS). The Chameleon Press Ltd., London
19. Triantafillou TC, Gibson LJ (1987) Minimum weight of foam core sandwich panels for a given strength. Mater Sci Eng 95:55–62
20. Mercado LL, Sikarskie DL, Miskioglu I (2000) Higher order theory for sandwich beams with yielded core. In: Proceedings of ICSS-5 conference, Zurich; September 2000, pp 141–153
21. Mercado LL, Sikarskie DL (1999) On response of a sandwich panel with a bilinear core. Mech Compos Mater Struct 6:57–67
22. Thomsen OT (1995) Theoretical and experimental investigation of local bending effects in sandwich plates. Compos Struct 30(1):85–101
23. Thomsen OT (1997) Localized bending effects in sandwich panel, photoelastic investigation versus high order sandwich theory results. Compos Struct 37(1):97–108
24. Caprino G, Langelan A (2000) Study of 3 pt. bending specimen for shear characterization of sandwich cores. J Compos Mater 34(9):791–814
25. Gdoutos EE, Daniel IM, Wang KA, Abot JL (2001) Non-linear behavior of composite sandwich beams in three-point bending. Exp Mech 41(2):182–189

26. Miers SA (2001) Analysis and design of edge inserts in sandwich beams. Master of Science Thesis, Michigan Technological University
27. Frostig Y, Baruch M, Vilnai O, Sheinman I (1992) Higher-order theory for sandwich beam behavior with transversely flexible core. J Eng Mech 118(5):1026–1043
28. Warren C (1989) Young, Roark's formulas for stress and strain, 6th edn. McGraw-Hill Company, NY

Erratum to: Probability Based High Temperature Engineering

Erratum to:
L. Razdolsky, *Probability Based High Temperature Engineering,* https://doi.org/10.1007/978-3-319-41909-1_10

In the original version of the book, the following list of references have been included in the erratum chapter as a belated correction from author:

Chapter 1

1. Pg. 26 and Fig. 1.5
Razdolsky, Leo. "Probability Based Rheological Models of High Temperature Structural Creep", AIAA SPACE 2014 Conference and Exposition, 2015

Chapter 2

1. Pgs. 58–59
Atkinson KE (1976) A survey of numerical methods for the solution of Fredholm integral equations of the second kind, SIAM

and

Baker CTH (1977) The numerical treatment of integral equations, Oxford University Press, Chap. 4
2. Pg. 67
V.I. Smirnov, "A course of higher mathematics", 4, Addison-Wesley (1964) (Translated from Russian)

An erratum to this publication is available online at https://doi.org/10.1007/978-3-319-41909-1

L. Razdolsky, *Probability Based High Temperature Engineering*,
https://doi.org/10.1007/978-3-319-41909-1_10

Chapter 3

1. Pgs. 102–106
www.me.umn.edu/labs/composites/Projects/
2. Pg. 106, Paragraph 3.2
academic.csuohio.edu/duffy_s/Linear_Visco.pdf
3. Pg. 107
Razdolsky, L., "High Temperature Creep and Structural Fire Resistance", Structures Congress 2015, 2015
4. Pg. 109
academic.csuohio.edu/duffy_s/Linear_Visco.pdf David Roylance, ENGINEERING VISCOELASTICITY, 2001
5. Pg. 112
academic.csuohio.edu/duffy
6. Pg. 112
academic.csuohio.edu/duffy
7. Pgs. 112–113
www2.esc.auckland.ac.nz
8. Pg. 117
www.mate.tue.nl/mate/pdfs/4944.pdf Klompen, Edwin T.J. Mechanical properties of solid polymers
9. Pg. 118
http://ksuweb.kennesaw.edu/~plaval/math2203/jacobian.pdf

Chapter 4

1. Pgs. 162 & 174
http://courses.washington.edu/mengr354/jenkins/notes/chap8.pdf
2. Pg. 163
Holm Altenbach, Serge Kruch: Advanced Materials Modelling for Structures, Heidelberg, Springer-Verlag. 2013
3. Pg. 163
Konstantin Naumenko and Holm Altenbach: Modeling of Creep for Structural Analysis, Heidelberg, Springer-Verlag. 2007
4. Pg. 163
courses.washington.edu/me354a/chap8.pdf TIME DEPENDENT BEHAVIOUR: CREEP
5. Pg. 164
www.hep.caltech.edu/~fcp/math/integralEquations/integralEquations.pdf
6. Pg. 168
L. Razdolsky "Probability Based Rheological Models…" AIAA Space 2014
7. Pg. 176
extras.springer.com/2007/978-1-4020-6238-4/data/full_papers/220_Oh.pdf
HIGH TEMPERATURE BEHAVIOR OF AL THIN FILM

8. Pg. 179
Sarkar A., and J.K. Chakravartty "prediction of Floww Stress..", Journal of Materials… 2013
9. Pg. 231
L. Razdolsky "High Temperature Creep and Structural Fire Resistance", Structures Congress 2015, 2015

Chapter 6

1. Pg. 391
aeweb.tamu.edu/haisler/engr214/Word_Lecture_Notes_by_Chapter/chapter9.doc
2. Pg. 394
https://www.coursehero.com › Rutgers › ENG MECH › ENG MECH 407

Chapter 8

1. Pg. 520
Ana Fernandez, H.-U. Künzi, A. Athanasiou Ioannou, L. Michelet and A. Rossoll: Lab course on Deformation and Fracture–Creep Test 2014. http://lmm.epfl.ch/files/content/sites/lmm/files/shared/TPDefRupt2014/TPDefFract14_Creep.pdf
2. Pg. 523
K. W. Poh STRESS-STRAIN-TEMPERATURE RELATIONSHIP FOR STRUCTURAL STEEL. http://www.egr.msu.edu/firestruct/Fire%20Research%20PhD/Stress%20Strain%20Temperature%20Relationship%20for%20Steel%20by%20Poh.pdf

Chapter 9

1. Pg. 581
Boyle, J.T.: The creep behavior of simple structures with a stress range dependent constitutive model. https://pure.strath.ac.uk/portal/files/4531224/Boyle_JT_Pure_The_creep_behaviour_of_simple_structures…_constitutive_model_Jun_2011.pdf
2. Pgs. 636 & 637
Salih N. Akour and Hussein Z. Maaitah "Effect of Core Material Stiffness on Sandwich Panel Behavior Beyond the Yield Limit." Proceedings of the World Congress on Engineering 2010 Vol II WCE 2010, June 30–July 2, 2010, London, U.K
3. Pg. 638
American Society of Testing and Materials. ASTM Standard Terminology of Structural Sandwich Constructions (C274-99). West Conshohocken, PA: ASTM International 1999
4. Pg. 638
Fagerberg, Linus. (2003). Wrinkling in sandwich panels for marine applications. KTH Royal Institute of Technology. https://www.researchgate.net/publication/228891097_Wrinkling_in_sandwich_panels_for_marine_applications

Index

L. Razdolsky, *Probability Based High Temperature Engineering*,
DOI 10.1007/978-3-319-41909-1

MIX
Papier aus verantwortungsvollen Quellen
Paper from responsible sources
FSC® C105338

If you have any concerns about our products,
you can contact us on
ProductSafety@springernature.com

In case Publisher is established outside the EU,
the EU authorized representative is:
Springer Nature Customer Service Center GmbH
Europaplatz 3, 69115 Heidelberg, Germany

Printed by Libri Plureos GmbH
in Hamburg, Germany